Tables of Dominant Weight Multiplicities for Representations of Simple Lie Algebras

PURE AND APPLIED MATHEMATICS

A Program of Monographs, Textbooks, and Lecture Notes

MONOGRAPHS AND TEXTBOOKS IN PURE AND APPLIED MATHEMATICS

1. *K. Yano,* Integral Formulas in Riemannian Geometry (1970) *(out of print)*
2. *S. Kobayashi,* Hyperbolic Manifolds and Holomorphic Mappings (1970) *(out of print)*
3. *V. S. Vladimirov,* Equations of Mathematical Physics (A. Jeffrey, editor; A. Littlewood, translator) (1970) *(out of print)*
4. *B. N. Pshenichnyi,* Necessary Conditions for an Extremum (L. Neustadt, translation editor; K. Makowski, translator) (1971)
5. *L. Narici, E. Beckenstein, and G. Bachman,* Functional Analysis and Valuation Theory (1971)
6. *D. S. Passman,* Infinite Group Rings (1971)
7. *L. Dornhoff,* Group Representation Theory (in two parts). Part A: Ordinary Representation Theory. Part B: Modular Representation Theory (1971, 1972)
8. *W. Boothby and G. L. Weiss (eds.),* Symmetric Spaces: Short Courses Presented at Washington University (1972)
9. *Y. Matsushima,* Differentiable Manifolds (E. T. Kobayashi, translator) (1972)
10. *L. E. Ward, Jr.,* Topology: An Outline for a First Course (1972) *(out of print)*
11. *A. Babakhanian,* Cohomological Methods in Group Theory (1972)
12. *R. Gilmer,* Multiplicative Ideal Theory (1972)
13. *J. Yeh,* Stochastic Processes and the Wiener Integral (1973) *(out of print)*
14. *J. Barros-Neto,* Introduction to the Theory of Distributions (1973) *(out of print)*
15. *R. Larsen,* Functional Analysis: An Introduction (1973) *(out of print)*
16. *K. Yano and S. Ishihara,* Tangent and Cotangent Bundles: Differential Geometry (1973) *(out of print)*
17. *C. Procesi,* Rings with Polynomial Identities (1973)
18. *R. Hermann,* Geometry, Physics, and Systems (1973)
19. *N. R. Wallach,* Harmonic Analysis on Homogeneous Spaces (1973) *(out of print)*
20. *J. Dieudonné,* Introduction to the Theory of Formal Groups (1973)
21. *I. Vaisman,* Cohomology and Differential Forms (1973)
22. *B.-Y. Chen,* Geometry of Submanifolds (1973)
23. *M. Marcus,* Finite Dimensional Multilinear Algebra (in two parts) (1973, 1975)
24. *R. Larsen,* Banach Algebras: An Introduction (1973)
25. *R. O. Kujala and A. L. Vitter (eds.),* Value Distribution Theory: Part A; Part B: Deficit and Bezout Estimates by Wilhelm Stoll (1973)
26. *K. B. Stolarsky,* Algebraic Numbers and Diophantine Approximation (1974)
27. *A. R. Magid,* The Separable Galois Theory of Commutative Rings (1974)
28. *B. R. McDonald,* Finite Rings with Identity (1974)
29. *J. Satake,* Linear Algebra (S. Koh, T. A. Akiba, and S. Ihara, translators) (1975)

30. *J. S. Golan,* Localization of Noncommutative Rings (1975)
31. *G. Klambauer,* Mathematical Analysis (1975)
32. *M. K. Agoston,* Algebraic Topology: A First Course (1976)
33. *K. R. Goodearl,* Ring Theory: Nonsingular Rings and Modules (1976)
34. *L. E. Mansfield,* Linear Algebra with Geometric Applications: Selected Topics (1976)
35. *N. J. Pullman,* Matrix Theory and Its Applications (1976)
36. *B. R. McDonald,* Geometric Algebra Over Local Rings (1976)
37. *C. W. Groetsch,* Generalized Inverses of Linear Operators: Representation and Approximation (1977)
38. *J. E. Kuczkowski and J. L. Gersting,* Abstract Algebra: A First Look (1977)
39. *C. O. Christenson and W. L. Voxman,* Aspects of Topology (1977)
40. *M. Nagata,* Field Theory (1977)
41. *R. L. Long,* Algebraic Number Theory (1977)
42. *W. F. Pfeffer,* Integrals and Measures (1977)
43. *R. L. Wheeden and A. Zygmund,* Measure and Integral: An Introduction to Real Analysis (1977)
44. *J. H. Curtiss,* Introduction to Functions of a Complex Variable (1978)
45. *K. Hrbacek and T. Jech,* Introduction to Set Theory (1978)
46. *W. S. Massey,* Homology and Cohomology Theory (1978)
47. *M. Marcus,* Introduction to Modern Algebra (1978)
48. *E. C. Young,* Vector and Tensor Analysis (1978)
49. *S. B. Nadler, Jr.,* Hyperspaces of Sets (1978)
50. *S. K. Segal,* Topics in Group Rings (1978)
51. *A. C. M. van Rooij,* Non-Archimedean Functional Analysis (1978)
54. *L. Corwin and R. Szczarba,* Calculus in Vector Spaces (1979)
53. *C. Sadosky,* Interpolation of Operators and Singular Integrals: An Introduction to Harmonic Analysis (1979)
54. *J. Cronin,* Differential Equations: Introduction and Quantitative Theory (1980)
55. *C. W. Groetsch,* Elements of Applicable Functional Analysis (1980)
56. *I. Vaisman,* Foundations of Three-Dimensional Euclidean Geometry (1980)
57. *H. I. Freedman,* Deterministic Mathematical Models in Population Ecology (1980)
58. *S. B. Chae,* Lebesgue Integration (1980)
59. *C. S. Rees, S. M. Shah, and C. V. Stanojević,* Theory and Applications of Fourier Analysis (1981)
60. *L. Nachbin,* Introduction to Functional Analysis: Banach Spaces and Differential Calculus (R. M. Aron, translator) (1981)
61. *G. Orzech and M. Orzech,* Plane Algebraic Curves: An Introduction Via Valuations (1981)
62. *R. Johnsonbaugh and W. E. Pfaffenberger,* Foundations of Mathematical Analysis (1981)
63. *W. L. Voxman and R. H. Goetschel,* Advanced Calculus: An Introduction to Modern Analysis (1981)
64. *L. J. Corwin and R. H. Szcarba,* Multivariable Calculus (1982)
65. *V. I. Istrătescu,* Introduction to Linear Operator Theory (1981)
66. *R. D. Järvinen,* Finite and Infinite Dimensional Linear Spaces: A Comparative Study in Algebraic and Analytic Settings (1981)

67. *J. K. Beem and P. E. Ehrlich,* Global Lorentzian Geometry (1981)
68. *D. L. Armacost,* The Structure of Locally Compact Abelian Groups (1981)
69. *J. W. Brewer and M. K. Smith, eds.,* Emmy Noether: A Tribute to Her Life and Work (1981)
70. *K. H. Kim,* Boolean Matrix Theory and Applications (1982)
71. *T. W. Wieting,* The Mathematical Theory of Chromatic Plane Ornaments (1982)
72. *D. B. Gauld,* Differential Topology: An Introduction (1982)
73. *R. L. Faber,* Foundations of Euclidean and Non-Euclidean Geometry (1983)
74. *M. Carmeli,* Statistical Theory and Random Matrices (1983)
75. *J. H. Carruth, J. A. Hildebrant, and R. J. Koch,* The Theory of Topological Semigroups (1983)
76. *R. L. Faber,* Differential Geometry and Relativity Theory: An Introduction (1983)
77. *S. Barnett,* Polynomials and Linear Control Systems (1983)
78. *G. Karpilovsky,* Commutative Group Algebras (1983)
79. *F. Van Oystaeyen and A. Verschoren,* Relative Invariants of Rings: The Commutative Theory (1983)
80. *I. Vaisman,* A First Course in Differential Geometry (1984)
81. *G. W. Swan,* Applications of Optimal Control Theory in Biomedicine (1984)
82. *T. Petrie and J. D. Randall,* Transformation Groups on Manifolds (1984)
83. *K. Goebel and S. Reich,* Uniform Convexity, Hyperbolic Geometry, and Nonexpansive Mappings (1984)
84. *T. Albu and C. Năstăsescu,* Relative Finiteness in Module Theory (1984)
85. *K. Hrbacek and T. Jech,* Introduction to Set Theory, Second Edition, Revised and Expanded (1984)
86. *F. Van Oystaeyen and A. Verschoren,* Relative Invariants of Rings: The Noncommutative Theory (1984)
87. *B. R. McDonald,* Linear Algebra Over Commutative Rings (1984)
88. *M. Namba,* Geometry of Projective Algebraic Curves (1984)
89. *G. F. Webb,* Theory of Nonlinear Age-Dependent Population Dynamics (1985)
90. *M. R. Bremner, R. V. Moody, and J. Patera,* Tables of Dominant Weight Multiplicities for Representations of Simple Lie Algebras (1985)
91. *A. E. Fekete,* Real Linear Algebra (1985)

Other Volumes in Preparation

Tables of Dominant Weight Multiplicities for Representations of Simple Lie Algebras

M. R. Bremner

New Haven, Connecticut

R. V. Moody

Department of Mathematics
University of Saskatchewan
Saskatoon, Saskatchewan, Canada

J. Patera

Centre de Recherche de Mathématiques Appliquées
Université de Montréal
Montréal, Québec, Canada

MARCEL DEKKER, INC.   New York and Basel

Library of Congress Cataloging in Publication Data

Bremner, M. R., [date]
Tables of dominant weight multiplicities for representations of simple Lie algebras.

(Monographs and textbooks in pure and applied mathematics ; v. 90)
Bibliography: p.
1. Lie algebras–Tables. 2. Representations of algebras–Tables. 3. Lie groups–Tables. 4. Representations of groups–Tables. I. Moody, R. V., [date]. II. Patera, Jiri. III. Title. IV. Series.
QA252.3.B74 1985 512'.55 84-19984
ISBN 0-8247-7270-9

MARCEL DEKKER, INC.
270 Madison Avenue, New York, New York 10016

Current printing (last digit):
10 9 8 7 6 5 4 3 2 1

PRINTED IN THE UNITED STATES OF AMERICA

Preface

The number of applications of the theory of simple and semisimple Lie groups both to mathematics and physics continues to grow with no end in sight. Although in recent years many of the new developments have been in the direction of infinite dimensional representations, the finite dimensional representations continue to be the basic tool in the majority of applied situations. Here there are a number of standard, but difficult, computational problems, among which we might mention the finding of suitable bases for representation spaces, the splitting of various symmetry classes of tensors, the computation of characters, character decomposition of class functions, and the determination of subgroup reductions.

The first question to arise in almost any finite dimensional problem (in particular in those mentioned above) is the determination of the weights and the weight multiplicities of various representations of the group at hand. There are several documented [1-4] (and no doubt many undocumented) computer implementations of weight algorithms which attest to the fundamental nature of this problem. This book will, we hope, obviate the need for many of these programs, the main part of it being precisely tables of weights and weight multiplicities for the simple Lie groups of rank n for $2 \leqslant n \leqslant 12$.

This work is an early part of a more ambitious program of developing efficient and solidly based methods of general applicability for computing in compact Lie groups [5-9]. The algorithms need to be fast enough to compute useful information about the higher rank groups which are presently of interest to physicists and mathematicians, and general enough that they do not have to rely on special features of particular groups or series of groups. As a consequence there are two basic premises in this particular work:

(1) all the simple Lie groups are placed on equal footing;
(2) inherent symmetry is exploited as much as possible.

For anyone simply using the tables, (1) is not especially relevant. On the other hand, the effect of the second premise is immediately evident in the tables, where one sees that only dominant weights appear. Every other weight is linked to one of these by a sequence of Weyl reflections. Each weight on a given Weyl orbit has the same multiplicity and each orbit has a unique dominant representative. Restricting to dominant weights not only makes it possible to succinctly write down the relevant information but also represents an enormous increase in computational efficiency. In this respect the algorithm which we use (see Chapter 2) is a substantial improvement over any other in the literature. Basically it is a modified version of Freudenthal's classic formula but preceded by a direct computation of all the dominant weights of a given representation. Our implementation is in Pascal.

Time was not an important factor in the computation (the table of 104 representations of E_8 in this volume took about 18 minutes of computing time on a Cyber 835) and there is no difficulty either in extending the rank or the number of representations beyond what we have listed here.

The volume contains three tables. The first two offer useful information about each simple Lie group and algebra. These are the Biographical and Positive Roots Tables. In one form or another this information can be found elsewhere in the literature. The main content of the volume is the original Dominant Weight Multiplicity Table. It is set out with one subtable for each congruence class (see Chapter 1) of each simple Lie group of rank n with $2 \leqslant n \leqslant 12$. Within each subtable there are the "first" 52 representations of the given congruence class (note that the ordering is not by dimension!). In each of the cases G_2, F_4, and E_8, since there is only one congruence class, we have given the first 104 representations. Thus every group has at least 104 representations listed.

Chapter 1 is a compendium of relevant facts about representations related to the tables of the volume and their use. No proofs and only a few references are given. The aim of the section is to systematize the facts for readers who may have experience in working with the representations but whose principal interests are elsewhere. A thorough explanation can be found, for instance, in Ref. 10.

In Chapter 2 we describe the weight space and weight multiplicity algorithms.

Chapter 3, the Biographical Table and the Positive Roots Table, contains basic biographical information about each type of Lie group (A, B, etc.). Included are facts about the roots, exponents, Cartan matrices, etc. To paraphrase a remark in the Introduction of Freudenthal and deVries [11], it probably contains every possible fact except the one that is actually needed at the moment.

The main part of the book is the Multiplicity Table of Chapter 4, preceded by brief instructions on how to use it.

The development of the program constituted the major part of the Master's thesis of one of us (M.B.) at the Department of Computer Science, Concordia University, Montreal, where the computation also took place (Cyber 835). We would like to thank Cliff Grossner, Lily Lam, and Larry Thiel for their assistance with file transfers and the final printing of the tables, and Randy Funk for his help in early stages of the program development. It is also a pleasure to thank Wendy McKay, who assisted in formatting the Dominant Weight Multiplicity Tables and the Biographical Table.

Two of us (R.M. and J.P.) would like to thank the University of Montreal, the California Institute of Technology, the University of Saskatchewan, and the Aspen Center for Physics for their hospitality during the work on the book. We also acknowledge the Special Collaborative Project Grant from the Natural Sciences and Engineering Research Council of Canada which was essential for successful completion of the project.

M. R. Bremner
R. V. Moody
J. Patera

Contents

1

Finite Dimensional Representations of Simple Lie Groups

Let K be an arbitrary simply connected simple compact Lie group of rank n with Lie algebra $\mathbf{k}$. Set $\mathbf{S} = \{1, \cdots, n\}$. The structure theory of K and its unitary representations proceeds through the choice of a maximal torus T, i.e., maximal connected abelian subgroup of K. Let $\mathbf{g}$ be the complexification $\mathbf{k}_{\mathbb{C}}$ of $\mathbf{k}$ and let $\mathbf{h} \subset \mathbf{g}$ be the Cartan subalgebra of $\mathbf{g}$ determined by T. Thus $\mathbf{h} = \mathbf{t}_{\mathbb{C}}$, where $\mathbf{t}$ is the subalgebra of $\mathbf{k}$ corresponding to T, and $\mathbf{h}_{\mathbb{R}} := \sqrt{-1}\,\mathbf{t}$ is a real Euclidean space (under the Killing form) of dimension n. The exponential mapping determines an exact sequence

$$O \to \hat{Q} \to \mathbf{h}_{\mathbb{R}} \xrightarrow{\exp 2\pi i(\cdot)} T \to 1. \tag{1.1}$$

The subgroup $\hat{Q}$ is the *co-root lattice* and is a discrete subgroup of $\mathbf{h}_{\mathbb{R}}$ of rank n.

Let $\pi : K \to SU(V)$ be any (unitary) representation of K on a finite dimensional complex vector space V and let $d\pi : \mathbf{k} \to End(V)$ be the corresponding representation of $\mathbf{k}$. Complexification determines the representation $(d\pi)_{\mathbb{C}}: \mathbf{g} \to End(V)$ of $\mathbf{g}$. Then V decomposes into weight subspaces relative to T:

$$V = \bigoplus_{\lambda \in \Omega(V)} V(\lambda), \tag{1.2}$$

where $\Omega(V) \subset \mathbf{h}^{\circ}_{\mathbb{R}}$ (real dual space of $\mathbf{h}_{\mathbb{R}}$) is the *weight system* of V and the $V(\lambda)$ are the corresponding weight subspaces of V. Thus

$$V(\lambda) = \{v \in V \mid \text{for all } x \in \mathbf{h}_{\mathbb{R}},\ \pi(\exp 2\pi i x)v = e^{2\pi i\lambda(x)}v\}. \tag{1.3}$$

We are primarily concerned with determining $\Omega(V)$ and the corresponding *weight multiplicities* $\dim\ V(\lambda)$, $\lambda \in \Omega(V)$.

The adjoint representation Ad: $K \to Aut\,(\mathbf{g})$ leads to the root space decomposition

$$\mathbf{g} = \mathbf{h} \oplus \bigoplus_{\alpha \in \Delta} \mathbf{g}(\alpha), \tag{1.4}$$

where Δ is the *root system* of $\mathbf{g}$ relative to T. We choose a basis

$$\Pi = \{\alpha_1, \cdots, \alpha_n\} \tag{1.5}$$

of simple roots of Δ once and for all. Let Δ^+ denote the corresponding system of positive roots.

Let

$$\langle \cdot, \cdot \rangle : \mathbf{h}^{\circ}_{\mathbb{R}} \times \mathbf{h}_{\mathbb{R}} \to \mathbb{R} \tag{1.6}$$

be the standard pairing and denote by

$$(\cdot,\cdot):\ \mathbf{h}^{\circ}_{\mathbb{R}} \times \mathbf{h}^{\circ}_{\mathbb{R}} \to \mathbb{R}, \tag{1.7}$$

the transpose of the Killing form. It is standard to assume that this has been so scaled that $(\alpha,\alpha) = 2$ for the long roots of Δ. (Warning: In the multiplicity tables of Chapter 4 a different convention has been adopted.) The Cartan matrix A of K is defined by

$$A_{ij} = 2(\alpha_i, \alpha_j)/(\alpha_j, \alpha_j). \tag{1.8}$$

The matrix A is of course an invariant of K. For the purpose of setting the notation we have listed the Cartan matrices in Table 3 (Chapter 3).

The $\mathbb{Z}$-dual of $\hat{Q}$ relative to $<\cdot,\ \cdot>$ is the *weight lattice P.* All weight systems $\Omega(V)$ lie in P and P is generated as a subgroup of $\mathbf{h}^{\circ}_{\mathbb{R}}$ by them. In particular, the $\mathbb{Z}$-span Q of the roots lies in P. The index $[P\!:\!Q]$ of Q in P, the *index of connection,* is the finite number

$$[P\!:\!Q] = \det(A). \tag{1.9}$$

The weight system $\Omega(V)$ of any given irreducible representation (π, V) of K lies entirely in one coset of P/Q. This results in the partitioning of the irreducible representations of K into $[P\!:\!Q]$ classes called *congruence classes.* In the construction of weight multiplicity tables it is always possible, and is always most efficient, to work within a single congruence class at any one time.

An element $\lambda \in P$ is called *dominant* if $(\lambda, \alpha_i) \geqslant 0$ for $i \in \mathbf{S}$, and the set of dominant weights is denoted by P^{++}. Then

$$P^{++} = \mathbb{N}\,\omega_1 + \cdots + \mathbb{N}\,\omega_n, \tag{1.10}$$

where $\omega_1, \cdots, \omega_n$, the *fundamental weights,* are defined by

$$2(\omega_i, \alpha_j)/(\alpha_j, \alpha_j) = \delta_{ij}. \tag{1.11}$$

Here $\mathbb{N}$ is the set of nonnegative integers and δ_{ij} is the Kronecker function. Weights λ of P are labeled by their coordinates $\lambda_1, \cdots, \lambda_n$ in terms of the basis $\omega_1, \cdots, \omega_n$. In every irreducible representation (π, V) there is a unique weight subspace $V(\lambda)$ such that

$$\pi(\mathbf{g}(\alpha))\, V(\lambda) = 0 \qquad \text{for all } \alpha \in \Delta^+. \tag{1.12}$$

This *highest weight,* $\lambda = \lambda_V$, is dominant and its multiplicity is 1. The correspondence $V \mapsto \lambda_V$ is a bijective correspondence between irreducible representations (up to isomorphism) and elements of P^{++}. This is Weyl's classification of irreducible unitary representations of K. We denote the irreducible representation with highest weight $\lambda \in P^{++}$ by V^λ. The weight system $\Omega(V^\lambda)$ of V^λ lies in the set

$$\{\lambda - \Sigma\, c_i\alpha_i \mid c_i \in \mathbb{N},\ i \in \mathbf{S}\}. \tag{1.13}$$

In the case of the adjoint representation the highest weight is the *highest root* (the unique root $\Sigma\, c_i\alpha_i$ for which Σc_i is maximized).

The lattice P is partially ordered by the *level function L.* Define $\hat{\rho} \in \mathbf{h}_{\mathbb{R}}$ by $(\alpha_i, \hat{\rho}) = 1$ for all $i \in \mathbf{S}$. Then $L: P \to \mathbb{Q}$ is defined by

$$L(\lambda) = 2(\lambda, \hat{\rho}) + 1. \tag{1.14}$$

We write $\lambda > \zeta$ if $L(\lambda) > L(\zeta)$. Specific expressions for the *level vector* $2\hat{\rho} = \boldsymbol{l} = (l_1, \cdots, l_n)^T$ are given in Table 1. Putting $\lambda = (\lambda_1, \lambda_2, \cdots, \lambda_n)$, one rewrites (1.14) as

$$L(\lambda) = 1 + \sum_{i=1}^{n} l_i \lambda_i. \tag{1.15}$$

If we imagine that the weight system is partitioned into *levels* by the values of $\sum_i c_i$ in (1.13), then in fact the *number* of levels is precisely $L(\lambda)$.

For each $\alpha \in \Delta$ we may choose $e_{\pm\alpha} \in \mathfrak{g}(\alpha)$ and $h_\alpha \in \mathbf{h}_{\mathbb{R}}$ so that

$$[h_\alpha, e_{\pm\alpha}] = \pm 2e_{\pm\alpha}, \qquad [e_\alpha, e_{-\alpha}] = h_\alpha. \tag{1.16}$$

The h_α are uniquely determined by (1.16) and satisfy

$$\lambda(h_\alpha) = 2(\lambda, \alpha)/(\alpha,\alpha) \qquad \text{for all } \lambda \in \mathbf{h}^\circ_{\mathbb{R}}. \tag{1.17}$$

The choice of $e_{\pm\alpha}$ has one degree of freedom. The Lie subalgebra $\mathbb{C}e_{-\alpha} + \mathbb{C}e_\alpha + \mathbb{C}h_\alpha$ of $\mathfrak{g}$ is isomorphic to $sl(2,\mathbb{C})$. It is possible to choose the $e_{\pm\alpha}$ uniquely to within sign (Chevalley basis) so that under the action of $sl(2,\mathbb{C})$ on V in any representation (π, V) of $\mathfrak{g}$,

$$R_\alpha := \exp e_{-\alpha} \exp(-e_\alpha) \exp e_{-\alpha} \tag{1.18}$$

acts as an element of order 4 and

$$\pi(R_\alpha) V(\lambda) = V(\lambda - \lambda(h_\alpha)\alpha) \qquad \text{for all } \lambda \in P. \tag{1.19}$$

There is a resulting action on the weights

$$\lambda \mapsto \lambda - \lambda(h_\alpha)\alpha. \tag{1.20}$$

Define for each $\alpha \in \Delta$

$$r_\alpha \colon \mathbf{h}^\circ_{\mathbb{R}} \to \mathbf{h}^\circ_{\mathbb{R}},$$

by

$$r_\alpha(\lambda) = \lambda - \lambda(h_\alpha)\alpha \qquad \text{for all } \lambda \in \mathbf{h}^\circ_{\mathbb{R}}. \tag{1.21}$$

The operators r_α are genuine reflections (the *Weyl reflections*) of $\mathbf{h}^\circ_{\mathbb{R}}$ relative to the Euclidean geometry $(\cdot, \cdot)$ of the transposed Killing form. They generate a finite group W, the *Weyl group* of K (relative to T). Alternatively W arises as the quotient group N/T, where N is the normalizer of T in K. According to (1.19)

$$\pi(R_\alpha) V(\lambda) = V(r_\alpha \lambda). \tag{1.22}$$

Thus

(1) $W\Omega(V) = \Omega(V)$

(2) for all $w \in W$ and for all $\lambda \in \Omega(V)$

$$\dim V(\lambda) = \dim V(w\lambda). \tag{1.23}$$

This internal symmetry of the weight systems is an important ingredient in the theory of simple Lie groups. It is an *indispensable* ingredient in our approach to computational methods and theory [5, 7, 8, 9]. This will become apparent as we go on.

The *dominant chamber* $\mathbf{C}$ in $\mathbf{h}^{\circ}_{\mathbb{R}}$ is defined by

$$\mathbf{C} := \left\{\varphi \in \mathbf{h}^{\circ}_{\mathbb{R}} \mid (\varphi, \alpha_i) \geqslant 0, i \in \mathbf{S}\right\}. \tag{1.24}$$

$\mathbf{C}$ is a closed convex cone the boundaries of which lie in the reflecting hyperplanes $H_i := \left\{\varphi \in \mathbf{h}^{\circ}_{\mathbb{R}} \mid (\varphi, \alpha_i) = 0\right\}$ of the $r_i := r_{\alpha_i}$, $i \in \mathbf{S}$. The group W is generated by $r_1, \cdots, r_n$. Each W-orbit $W\lambda$, $\lambda \in \mathbf{h}^{\circ}_{\mathbb{R}}$, has exactly one point in $\mathbf{C}$. In particular, each W-orbit of a weight system $\Omega(V)$ has precisely one dominant weight. The entire orbit lies in $\Omega(V)$ and the multiplicities are constant along the orbit. Of course, the highest weight heads an orbit of multiplicity one.

The determination of the dominant weight ζ in an orbit $W\lambda$, $\lambda \in P$, is a purely mechanical procedure. One writes the coordinates $(\lambda_1, \cdots, \lambda_n)$ of λ relative to the basis of fundamental weights, i.e., $\lambda = \Sigma\lambda_i\omega_i$. Then λ is dominant if and only if $\lambda_i \geqslant 0$ for each i. If some $\lambda_i < 0$, then perform the reflection r_i on λ. The level $L(r_i\lambda)$ of $r_i\lambda$ is greater than that of λ. At most $|\Delta^+|$ reflections are required to bring λ into P^{++}. Conversely, given $\zeta \in P^{++}$ the entire orbit $W\zeta$ can be constructed by the following procedure [7]:

We define a graph $\Sigma(\zeta)$ with nodes corresponding to distinct elements of $W\zeta$ and edges "colored" by the index set $\mathbf{S}$. Distinct λ, λ' of $W\zeta$ are joined by an i-edge ($i \in \mathbf{S}$) if and only if $\lambda' = r_i\lambda$. Then $\Sigma(\zeta)$ is connected. Define the *depth* of a node λ to be the minimal number of edges in a path from ζ to λ. Now $\Sigma(\zeta)$ is created by induction on depth starting from the head node ζ. For each node λ of depth d, consider the coordinates $(\lambda_1, \cdots, \lambda_n)$ of λ. For each i for which $\lambda_i > 0$ there is an i-edge down from λ to $r_i\lambda$ and the depth of $r_i\lambda$ is $d + 1$.

From the point of view of efficiency it is clear that one should restrict attention to dominant weights and their multiplicities, only going out of the dominant chamber when it is unavoidable. For example, consider the case of the representation of D_8 whose highest weight is $\overset{2}{0000000}$. Its dimension is 6435. There are 5 orbits; corresponding multiplicities and orbit sizes are as follows:

Dominant weight	Multiplicity	Orbit size
$\overset{2}{0000000}$	1	128
$\overset{0}{0000010}$	1	1792
$\overset{0}{0001000}$	3	1120
$\overset{0}{0100000}$	10	112
$\overset{0}{0000000}$	35	1

This little table summarizes the essential information about the representation. The main body of the volume provides precisely this information about each representation. The reader can easily see how the multiplicity tables (Chapter 4) are used by looking under representation #10 in the D_8 class 0 table and also looking at the orbit size row, labeled O.S.

Determination of orbit sizes is a straightforward procedure. If λ is the unique dominant weight of an orbit $W\lambda$ in P then define $\mathbf{J} = \mathbf{J}(\lambda) \subset \mathbf{S}$ in terms of the labels $(\lambda_1, \cdots, \lambda_n)$ of λ by

$$\mathbf{J} = \{j \in \mathbf{S} \mid \lambda_j = 0\}. \tag{1.25}$$

Define $W_{\mathbf{J}}$ to be the group generated by $\{r_j \mid j \in \mathbf{J}\}$. Then $W_{\mathbf{J}}$ is the stabilizer of λ in W and $W\lambda$ has precisely

$$[W : W_{\mathbf{J}}] = \frac{|W|}{|W_{\mathbf{J}}|} \tag{1.26}$$

elements. Now the subgroups $W_{\mathbf{J}}$ of W (so-called *parabolic subgroups*) are again Weyl groups. In fact $W_{\mathbf{J}}$ is the Weyl group of the semisimple Lie algebra whose Dynkin diagram $\Gamma_{\mathbf{J}}$ is obtained from the diagram Γ of K by deleting all nodes and incident edges which are not indexed from the subset $\mathbf{J}$ of $\mathbf{S}$. The group $W_{\mathbf{J}}$ is the direct product of the Weyl groups of the connected components of $\Gamma_{\mathbf{J}}$. For example, in D_8 for $\lambda = \overset{0}{0001000}$, $\Gamma_{\mathbf{J}}$ is

 , (1.27)

and we have the decomposition $W_{\mathbf{J}} \cong W(A_3) \times W(D_4)$ into the Weyl groups of A_3 and D_4 with order $|W_{\mathbf{J}}| = 2^3\,(4!)^2$. A list of the orders of the Weyl groups for the simple Lie algebras appears in Table 1. In the multiplicity tables (Chapter 4) the orbit sizes (1.26) are shown in the row labeled O.S.

2

Weight Algorithms

In this chapter we describe the algorithm by which the dominant weight multiplicities were computed. The mathematical development of this is found in [5], where one may also find a worked-out example.

Given an irreducible representation V^λ with highest weight $\lambda = (\lambda_1, \cdots, \lambda_n)$, the determination of its dominant weight multiplicities proceeds in two steps:

(1) determination of the dominant weights;
(2) determination of the multiplicities.

(1a). Inductively we construct subsets $L_0, L_1, L_2, \cdots$ of P^{++} as follows:

$$L_0 = \{\lambda\}$$
$$L_k = \{\gamma \epsilon P^{++} - \bigcup_{j<k} L_j \mid \gamma = \zeta - \beta, \zeta \in L_{k-1}, \beta \in \Delta^+\}. \tag{2.1}$$

At a certain point these subsets become empty. The set of dominant weights of V^λ is precisely the union $\bigcup_k L_k$. Given a list of the positive roots written in terms of the fundamental weights, it is an easy matter to compute the L_k. In the table of the positive roots (Table 2) one finds under each algebra a list of the positive roots in the basis of fundamental weights (ω-basis), in the basis of simple roots (α-basis), and in the orthogonal basis (e-basis).

Example. Consider the D_8 -representation as above, where $\lambda = (0\,0\,0\,0\,0\,0\,0\,2)$. Using the root list for D_8 in the fundamental weights basis (Table 2), one finds

$$L_i = \{\Lambda_i\}, \quad i = 0, 1, \cdots, 4, \quad \text{and} \quad L_5 = \emptyset.$$

where

$$\begin{aligned}
\Lambda_0 &= (0\,0\,0\,0\,0\,0\,0\,2)\\
\Lambda_1 &= (0\,0\,0\,0\,0\,1\,0\,0) = \Lambda_0 - \alpha_8\\
\Lambda_2 &= (0\,0\,0\,1\,0\,0\,0\,0) = \Lambda_1 - \alpha_5 - 2\alpha_6 - \alpha_7 - \alpha_8\\
\Lambda_3 &= (0\,1\,0\,0\,0\,0\,0\,0) = \Lambda_2 - \alpha_3 - 2\alpha_4 - 2\alpha_5 - 2\alpha_6 - \alpha_7 - \alpha_8\\
\Lambda_4 &= (0\,0\,0\,0\,0\,0\,0\,0) = \Lambda_3 - \alpha_1 - 2\alpha_2 - 2\alpha_3 - 2\alpha_4 - 2\alpha_5 - 2\alpha_6 - \alpha_7 - \alpha_8.
\end{aligned}$$

In this case each L_k contains only one element.

(1b). Next we compute the level $L(\zeta)$ of each dominant weight and reorder (if necessary) the weights in decreasing level. For the example above we use the D_8 level vector l of Table 2:

$$l = (14, 26, 36, 44, 50, 54, 28, 28). \tag{2.2}$$

Then the levels of $\Lambda_0, \ldots, \Lambda_4$, according to (1.15), are 57, 55, 45, 27, 1 already in descending order.

We assume, after step (1) is completed, that the dominant weights $\lambda = \Lambda_1, \Lambda_2, \ldots, \Lambda_r$ of V^λ are known and arranged in descending order according to levels.

(2) The multiplicities $m_0, m_1, \ldots, m_r$ of $\Lambda_0, \Lambda_1, \ldots, \Lambda_r$ are computed inductively starting from the known fact that $m_0 = 1$. The algorithm is a form of the well-known Freudenthal formula [11] modified to exploit Weyl group symmetry as much as possible [5].

To write down the formula we need to introduce the group

$$\hat{W} = \langle W, -1_P \rangle \tag{2.3}$$

generated by W and -1_P. Here -1_P is the map $-1_P(x) = -x$ for all $x \in P$. For the algebras $A_1, B_l, C_l, D_{2k}, E_7, E_8, F_4, G_2$ the Weyl group W already contains -1_P, i.e., $\hat{W} = W$. Similarly we define $\hat{W}_\mathbf{J} = \langle W_\mathbf{J}, -1_P \rangle$ for every subset $\mathbf{J}$ of $\mathbf{S}$. Each group $\hat{W}_\mathbf{J}$ decomposes the root system Δ into orbits $O_1, O_2, \ldots, O_q$ and each orbit has a unique *subdominant* representative $\zeta_i = \sum_{j=0}^{n} n_{ij}\omega_j \in \Delta^+$, with $n_{ij} \geq 0$ for all $j \in \mathbf{J}$.

Suppose now that ζ is one of the weights Λ_k, $k = 1, \ldots, r$, and define $\mathbf{J} = \mathbf{J}_k = \{i \mid (\Lambda_k, \alpha_i) = 0\}$. Let $\zeta_1, \cdots, \zeta_q$ be defined accordingly. Then the fast recursion formula for the multiplicities m_ζ is written as

$$\sum_{i=1}^{q} |O_i| \sum_{p=1}^{\infty} (\zeta + p\zeta_i, \zeta_i) m_{\zeta + p\zeta_i} = (c_\lambda - c_\zeta) m_\zeta. \tag{2.4}$$

where for all ζ,

$$c_\zeta := (\zeta + \rho, \zeta + \rho) - (\rho, \rho) \quad \text{and} \quad \rho := \frac{1}{2} \sum_{\alpha \in \Delta^+} \alpha, \tag{2.5}$$

or equivalently

$$c_\zeta = (\zeta, \zeta + 2\rho) \quad \text{and} \quad \rho = (1, 1, \ldots, 1). \tag{2.5'}$$

For each i, the members p for which $\zeta + p\zeta_i$ is a weight form a string with $p = 0, 1, \ldots, t$. The weights $\zeta + p\zeta_i$, $p > 0$, are all of level higher than ζ but are not necessarily dominant. Using the method outlined in the preceding chapter, each $\zeta + p\zeta_i$ is transformed into a dominant weight. This only increases the level and hence $m_{\zeta + p\zeta_i} = m_{\Lambda_k + p\zeta_i} = m_{\Lambda_j}$, where $j < k$ and m_{Λ_j} can be assumed already known.

3

Biographical Table of Simple Lie Groups and the Positive Roots Table

Table 1 gives basic biographical details for each of the great series of simple Lie groups/algebras A, B, C, D and for each of the five exceptional ones, E_6, E_7, E_8, F_4, G_2. Both Dynkin diagrams Γ and extended Dynkin diagrams $\widetilde{\Gamma}$ appear. The extended diagrams play no part in the rest of the book but they have many uses elsewhere. The extension node of $\widetilde{\Gamma}$ can be interpreted as representing $\alpha_0 = -\gamma$, where $\gamma = \Sigma_{i=1}^{n} c_i \alpha_i$ is the highest root. The *marks* are 1 on the 0th node and the coefficient c_i on the ith node. The convention is that blackened nodes stand for *short* roots.

There are two commonly used indexings (numberings) of the nodes of the diagram Γ. The first, due to E. B. Dynkin, is the one adopted as standard throughout the book. The second, due to N. Bourbaki [12], is shown only when it differs from Dynkin's.

There are many reasons why the B series should begin at $n = 3$ and the C series at $n = 2$ (for example see the extended Dynkin diagrams). We have adopted this convention here.

The level vector $l = (l_1, \ldots, l_n)^T$ is given in such a way that the number of levels of the irreducible representation of K with highest weight λ is given by

$$\sum_{i=1}^{n} \lambda_i l_i + 1 \tag{3.1}$$

where

$$\lambda = \sum_{i=1}^{n} \lambda_i \omega_i \tag{3.2}$$

is written in the basis of fundamental weights ω_i. The sum of positive roots is given in terms of its coordinates in the α_i-basis.

The congruence vector $\varphi = (\varphi_1, \ldots, \varphi_n)$ is used to determine to which congruence class an irreducible module with the highest weight λ belongs (zero values of φ_i are not shown) [13]. With λ given as in (3.2), the scalar product $\lambda \cdot \varphi = \Sigma_{i=1}^{n} \lambda_i \varphi_i$ evaluated mod(k) with an appropriate k specifies the congruence class In all but the D_{2n} cases, k is the index of connection (1.9). For the D-series we use two congruence vectors φ and φ' and interpret $(\lambda \cdot \varphi, \lambda \cdot \varphi')$ in $\mathbb{Z}/2\mathbb{Z} \times \mathbb{Z}/4\mathbb{Z}$. For D_n with n odd the congruence class is labeled by the second component alone. In the case of D_n with n even the second component is 0 or 2 (mod 4) and the congruence class can be interpreted in $\mathbb{Z}/2\mathbb{Z} \times \mathbb{Z}/2\mathbb{Z}$. The congruence classes for D_n are labeled in the multiplicity tables in such a way that classes 0 and 1 correspond respectively to $(0, 0)$ and $(0, *)$, $* \neq 0$.

The Cartan matrix is equal to $[2(\alpha_i, \alpha_j)/(\alpha_j, \alpha_j)]$. The quadratic form matrix is the matrix $[(\omega_i, \omega_j)]$, $1 \leqslant i, j \leqslant n$, with the assumption that $(\alpha, \alpha) = 2$ for *long* roots.

A brief description of the Weyl group is included in each table. We use W^+ to denote the subgroup of even elements of W and $Z(W)$ to denote its center. Some of the groups have been given in terms of the classical groups (see for instance [12; 14, Chapter 9; 15]).

The positive roots for ranks $n \leqslant 12$ are given in the Positive Roots Table (Table 2) by their coordinates relative to three often used bases:

the fundamental weight basis (ω_i);
the simple root basis (α_i);
an orthonormal basis (e_i) in $\mathbb{R}^m$, $m \geqslant n$.

In the last case the roots are given as a subset of a quadratic lattice inside $(1/2)(\mathbb{Z}e_1 + \cdots + \mathbb{Z}e_m)$. The action of the Weyl group is, in most cases, easily described in terms of permutations and sign changes of these e_i.

The *tips* of an extended diagram $\tilde{\Gamma}$ are the nodes which are on the same orbit as the extension node under the action of the automorphism group $Aut(\tilde{\Gamma})$ of $\tilde{\Gamma}$ equivalently their marks equal 1. There is an explicit way [7] of identifying the center Z of the Lie group with a subgroup of $Aut(\tilde{\Gamma})$ so that $Aut(\tilde{\Gamma}) \cong Z \cdot Aut(\Gamma)$. The action of Z on $\tilde{\Gamma}$ is described by giving the action of a generator (or two generators for D_{2n}) as a permutation of the tips.

Table 1. Biographical Table

A_n $(n \geq 2)$

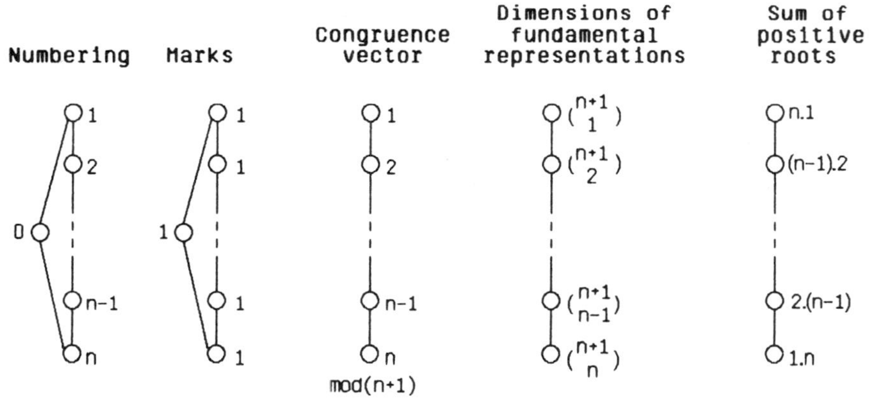

Cartan matrix

$$\left\|\begin{matrix} 2 & -1 & 0 & 0 & \ldots & 0 & 0 \\ -1 & 2 & -1 & 0 & \ldots & 0 & 0 \\ 0 & -1 & 2 & -1 & \ldots & 0 & 0 \\ . & . & . & . & \ldots & . & . \\ 0 & 0 & 0 & 0 & \ldots & 2 & -1 \\ 0 & 0 & 0 & 0 & \ldots & -1 & 2 \end{matrix}\right\|$$

Quadratic form matrix

$$\frac{1}{n+1}\left\|\begin{matrix} 1.n & 1.(n-1) & 1.(n-2) & \ldots & 1.2 & 1.1 \\ 1.(n-1) & 2.(n-1) & 2.(n-2) & \ldots & 2.2 & 2.1 \\ 1.(n-2) & 2.(n-2) & 3.(n-2) & \ldots & 3.2 & 3.1 \\ . & . & . & \ldots & . & . \\ 1.2 & 2.2 & (n-1).2 & \ldots & (n-1).2 & (n-1).1 \\ 1.1 & 2.1 & (n-1).1 & \cdots & (n-1).1 & n.1 \end{matrix}\right\|$$

Index of connection: $n+1$

Roots: $e_i - e_j$ $(1 \leq i, j \leq n+1, i \neq j)$

Positive roots: $e_i - e_j$ $(1 \leq i, j \leq n+1)$

Level vector: use sum of positive roots

Exponents: $1, 2, \ldots, n$

Action of centre: $(0\ 1\ \ldots\ n)$

Coxeter number: $n+1$

Weyl group order: $(n+1)!$

Weyl group structure: S_{n+1} operating permutations on the e_i

Table 1. *(continued)*

B_n $(n \geq 3)$

Numbering	Marks	Congruence vector	Dimensions of fundamental representations	Level vector	Sum of positive roots

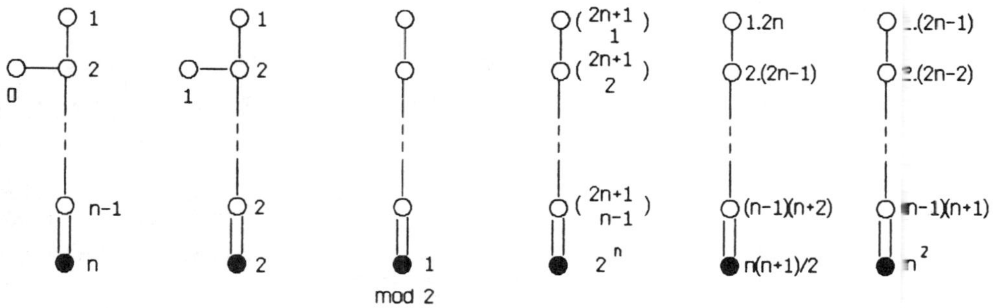

Cartan matrix

$$\begin{Vmatrix} 2 & -1 & 0 & \cdots & 0 & 0 \\ -1 & 2 & -1 & \cdots & 0 & 0 \\ 0 & -1 & 2 & \cdots & 0 & 0 \\ \cdot & \cdot & \cdot & \cdots & \cdot & \cdot \\ 0 & 0 & 0 & \cdots & 2 & -2 \\ 0 & 0 & 0 & \cdots & -1 & 2 \end{Vmatrix}$$

Quadratic form matrix

$$\frac{1}{2}\begin{Vmatrix} 2 & 2 & 2 & \cdots & 2 & 1 \\ 2 & 4 & 4 & \cdots & 4 & 2 \\ 2 & 4 & 6 & \cdots & 6 & 3 \\ \cdot & \cdot & \cdot & \cdots & \cdot & \cdot \\ 2 & 4 & 6 & \cdots & 2(n-1) & n-1 \\ 1 & 2 & 3 & \cdots & n-1 & n/2 \end{Vmatrix}$$

Index of connection: 2

Roots: $\pm e_i$ $(1 \leq i \leq n)$; $\pm e_i \pm e_j$ $(1 \leq i < j \leq n)$

Positive roots: e_i $(1 \leq i \leq n)$; $e_i \pm e_j$ $(1 \leq i < j \leq n)$

Exponents: 1, 3, 5, ..., 2n-1

Action of centre: (0 1)

Coxeter number: 2n

Weyl group order: $n!\,.\,2^n$

Weyl group structure: semidirect product of S_n operating as permutations on the e_i and $(Z_2)^n$ operating as $e_i \rightarrow (\pm 1)e_i$

Table 1. *(continued)*

C_n $(n \geq 2)$

Numbering	Marks	Congruence vector	Dimensions of fundamental representations	Level vector	Sum of positive roots

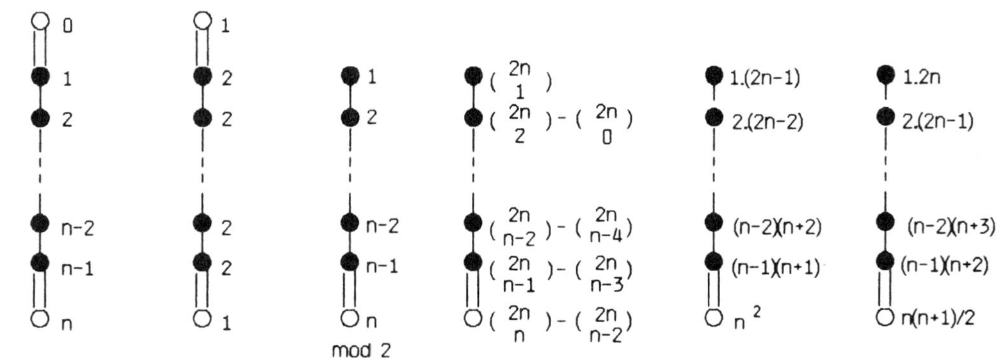

Cartan matrix

$$\begin{Vmatrix} 2 & -1 & 0 & \dots & 0 & 0 \\ -1 & 2 & -1 & \dots & 0 & 0 \\ 0 & -1 & 2 & \dots & 0 & 0 \\ . & . & . & \dots & . & . \\ 0 & 0 & 0 & \dots & 2 & -1 \\ 0 & 0 & 0 & \dots & -2 & 2 \end{Vmatrix}$$

Quadratic form matrix

$$\frac{1}{2}\begin{Vmatrix} 1 & 1 & 1 & \dots & 1 & 1 \\ 1 & 2 & 2 & \dots & 2 & 2 \\ 1 & 2 & 3 & \dots & 3 & 3 \\ . & . & . & \dots & . & . \\ 1 & 2 & 3 & \dots & n-1 & n-1 \\ 1 & 2 & 3 & \dots & n-1 & n \end{Vmatrix}$$

Index of connection: 2

Roots: $\pm 2e_i$ $(1 \leq i \leq n)$; $\pm e_i \pm e_j$ $(1 \leq i < j \leq n)$

Positive roots: $2e_i$ $(1 \leq i \leq n)$; $e_i \pm e_j$ $(1 \leq i < j \leq n)$

Exponents: 1, 3, 5, ... , 2n-1

Action of centre: (0 n)

Coxeter number: 2n

Weyl group order: $n!.2^n$

Weyl group structure: semidirect product of S_n operating as permutations on the e_i and $(Z_2)^n$ operating as $e_i \longrightarrow (\pm 1)e_i$

Table 1. *(continued)*

D_n $(n \geq 4)$

Numbering | Marks | Congruence vectors | Dimensions of fundamental representations | Sum of positive roots

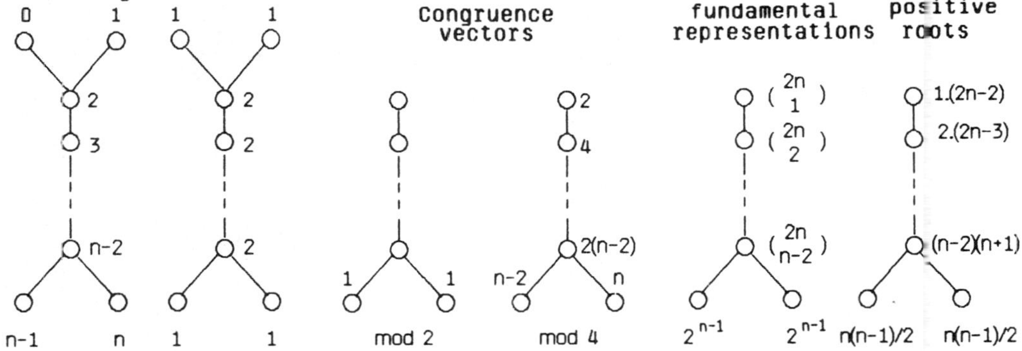

Cartan matrix

$$\begin{Vmatrix} 2 & -1 & 0 & 0 & \cdots & 0 & 0 & 0 \\ -1 & 2 & -1 & 0 & \cdots & 0 & 0 & 0 \\ 0 & -1 & 2 & -1 & \cdots & 0 & 0 & 0 \\ \cdot & \cdot & \cdot & \cdot & \cdots & \cdot & \cdot & \cdot \\ 0 & 0 & 0 & 0 & \cdots & 2 & -1 & -1 \\ 0 & 0 & 0 & 0 & \cdots & -1 & 2 & 0 \\ 0 & 0 & 0 & 0 & \cdots & -1 & 0 & 2 \end{Vmatrix}$$

Quadratic form matrix

$$\frac{1}{2}\begin{Vmatrix} 2 & 2 & 2 & \cdots & 2 & 1 & 1 \\ 2 & 4 & 4 & \cdots & 4 & 2 & 2 \\ 2 & 4 & 6 & \cdots & 6 & 3 & 3 \\ \cdot & \cdot & \cdot & \cdots & \cdot & \cdot & \\ 2 & 4 & 6 & \cdots & 2(n-2) & n-2 & n-2 \\ 1 & 2 & 3 & \cdots & n-2 & n/2 & (n-2)/2 \\ 1 & 2 & 3 & \cdots & n-2 & (n-2)/2 & n/2 \end{Vmatrix}$$

Index of connection: 4

Roots: $\pm e_i \pm e_j$ $(1 \leq i < j \leq n)$

Positive roots: $e_i \pm e_j$ $(1 \leq i < j \leq n)$

Level vector: use sum of positive roots

Exponents: 1, 3, 5, ..., 2n-5, 2n-3, n-1

Action of centre: (0 n 1 n-1) for n odd;

(0 1)(n-1 n) and (0 n)(1 n-1) for n even

Coxeter number: 2n-2

Weyl group order: $n!.2^{n-1}$

Weyl group structure: semidirect product of S_n operating by permutations on the e_i and $(Z_2)^{n-1}$ operating as $e_i \rightarrow (\pm 1)e_i$ with $\prod(\pm 1)_i = 1$;
$Z(W) = 1$ if n odd; $Z(W) \approx Z_2$ if n even

Table 1. *(continued)*

$\mathbf{E}_6$

Numbering (Dynkin)	Numbering (Bourbaki)	Marks	Congruence vector	Dimensions of fundamental representations	Sum of positive roots

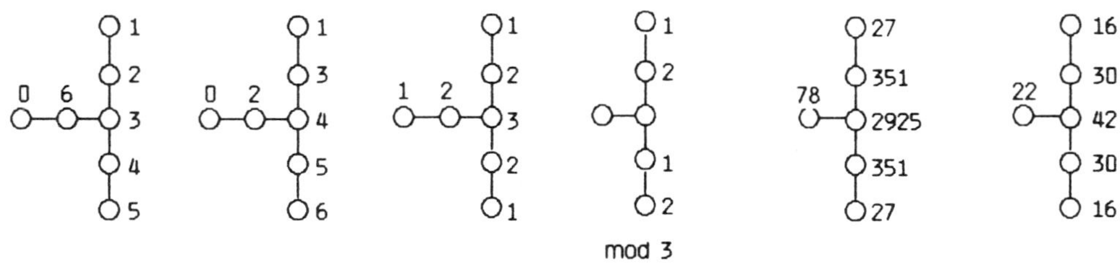

Cartan matrix

$$\begin{Vmatrix} 2 & -1 & 0 & 0 & 0 & 0 \\ -1 & 2 & -1 & 0 & 0 & 0 \\ 0 & -1 & 2 & -1 & 0 & -1 \\ 0 & 0 & -1 & 2 & -1 & 0 \\ 0 & 0 & 0 & -1 & 2 & 0 \\ 0 & 0 & -1 & 0 & 0 & 2 \end{Vmatrix}$$

Quadratic form matrix

$$\frac{1}{3}\begin{Vmatrix} 4 & 5 & 6 & 4 & 2 & 3 \\ 5 & 10 & 12 & 8 & 4 & 6 \\ 6 & 12 & 18 & 12 & 6 & 9 \\ 4 & 8 & 12 & 10 & 5 & 6 \\ 2 & 4 & 6 & 5 & 4 & 3 \\ 3 & 6 & 9 & 6 & 3 & 6 \end{Vmatrix}$$

Index of connection: 3

Roots: $\pm e_i \pm e_j \ (1 \le i < j \le 5)$; $\pm(e_8 - e_7 - e_6 + \sum_{k=1}^{5}(-1)^{a(k)}e_k)/2$, $(\sum_{k=1}^{5} a(k) = 0 \pmod 2))$

Positive roots: $\pm e_i + e_j \ (1 \le i < j \le 5)$; $(e_8 - e_7 - e_6 + \sum_{k=1}^{5}(-1)^{a(k)}e_k)/2$

Level vector: use sum of positive roots

Exponents: 1, 4, 5, 7, 8, 11

Action of centre: (0 1 5)

Coxeter number: 12

Weyl group order: $2^7.3^4.5 = 51840$

Weyl group structure: $Z(W)=1$; $W \approx O(Q/2Q, q)$ where q is the quadratic form induced on $Q/2Q$ by $(\cdot,\cdot)/2$

Table 1. *(continued)*

E_7

Numbering (Dynkin)	Numbering (Bourbaki)	Marks	Congruence vector	Dimensions of fundamental representations	Sum of positive roots

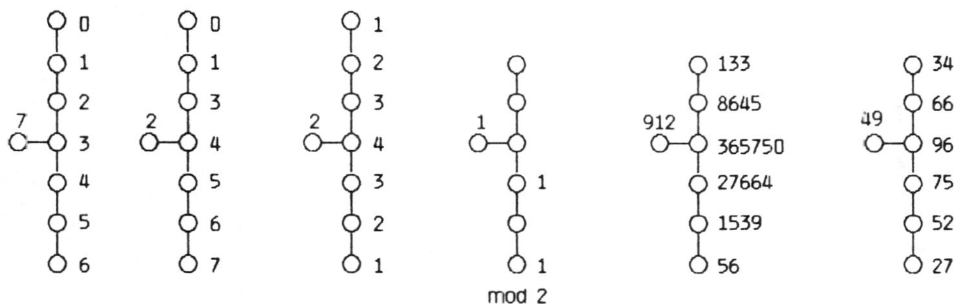

Cartan matrix

2	-1	0	0	0	0	0
-1	2	-1	0	0	0	0
0	-1	2	-1	0	0	-1
0	0	-1	2	-1	0	0
0	0	0	-1	2	-1	0
0	0	0	0	-1	2	0
0	0	-1	0	0	0	2

Quadratic form matrix ($\frac{1}{2}$ ×)

4	6	8	6	4	2	4
6	12	16	12	8	4	8
8	16	24	18	12	6	12
6	12	18	15	10	5	9
4	8	12	10	8	4	6
2	4	6	5	4	3	3
4	8	12	9	6	3	7

Index of connection: 2

Roots: $\pm e_i \pm e_j$ $(1 \le i < j \le 6)$; $\pm(e_8 - e_7)$; $\pm(e_8 - e_7 + \sum_{k=1}^{6} (-1)^{a(k)} e_k)/2$ $(\sum_{k=1} a(k) = 1 \pmod 2)$

Positive roots: $\pm e_i + e_j$ $(1 \le i < j \le 6)$; $(e_8 - e_7)$; $(e_8 - e_7 + \sum_{k=1}^{6} (-1)^{a(k)} e_k)/2$

Level vector: use sum of positive roots

Exponents: 1, 5, 7, 9, 11, 13, 17

Action of centre: (0 6)

Coxeter number: 18

Weyl group order: $2^{10}.3^4.5.7 = 2903040$

Weyl group structure: $Z(W) \approx \{1, -1\}$; $W \approx W^+ \times Z(W)$;
$1 \longrightarrow \{1, -1\} \longrightarrow W \longrightarrow Sp_6(F_2) \longrightarrow 1$ (exact)
through the natural mapping of W onto $Sp(Q/2P, a)$ where
a is the alternating form on $Q/2P$ induced by $(\cdot, \cdot)$

Table 1. *(continued)*

E_8

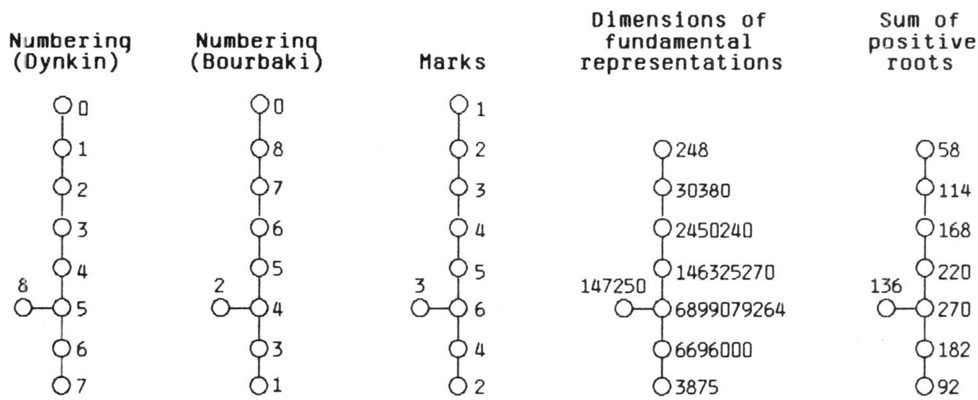

Cartan matrix

2	-1	0	0	0	0	0	0
-1	2	-1	0	0	0	0	0
0	-1	2	-1	0	0	0	0
0	0	-1	2	-1	0	0	0
0	0	0	-1	2	-1	0	-1
0	0	0	0	-1	2	-1	0
0	0	0	0	0	-1	2	0
0	0	0	0	-1	0	0	2

Quadratic form matrix

2	3	4	5	6	4	2	3
3	6	8	10	12	8	4	6
4	8	12	15	18	12	6	9
5	10	15	20	24	16	8	12
6	12	18	24	30	20	10	15
4	8	12	16	20	14	7	10
2	4	6	8	10	7	4	5
3	6	9	12	15	10	5	8

Index of connection: 1

Roots: $\pm e_i \pm e_j \ (1 \le i < j \le 8)$; $\pm(\sum_{k=1}^{8}(-1)^{a(k)} e_k)/2$, $(\sum_{k=1}^{8} a(k) = 0 \pmod 2)$

Positive roots: $\pm e_i + e_j \ (1 \le i < j \le 8)$; $(e_8 + \sum_{k=1}^{7}(-1)^{a(k)} e_k)/2$, $(\sum_{k=1}^{7} a(k) = 0 \pmod 2)$

Level vector: use sum of positive roots

Exponents: 1, 7, 11, 13, 17, 19, 23, 29

Coxeter number: 30

Weyl group order: $2^{14}.3^5.5^2.7 = 696729600$

Weyl group structure: $Z(W) \approx \{1, -1\}$;

$1 \longrightarrow \{1, -1\} \longrightarrow W \longrightarrow O(q) \longrightarrow 1$ (exact)

through the natural action of W on $Q/2Q$, where

q is the quadratic form induced by $(\cdot,\cdot)/2$

Table 1. *(continued)*

F_4

Numbering	Marks	Dimensions of fundamental representations	Level vector	Sum of positive roots

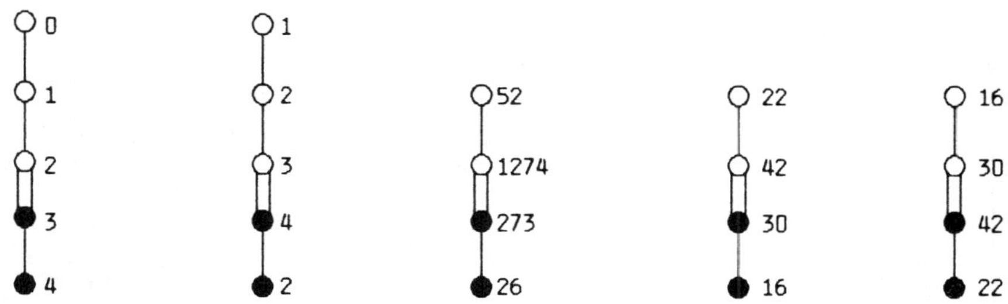

Cartan matrix

$$\begin{Vmatrix} 2 & -1 & 0 & 0 \\ -1 & 2 & -2 & 0 \\ 0 & -1 & 2 & -1 \\ 0 & 0 & -1 & 2 \end{Vmatrix}$$

Quadratic form matrix

$$\begin{Vmatrix} 2 & 3 & 2 & 1 \\ 3 & 6 & 4 & 2 \\ 2 & 4 & 3 & 3/2 \\ 1 & 2 & 3/2 & 1 \end{Vmatrix}$$

Index of connection: 1

Roots: $\pm e_i \pm e_j$ $(1 \le i < j \le 4)$; $\pm e_i (1 \le i \le 4)$; $(\pm e_1 \pm e_2 \pm e_3 \pm e_4)/2$

Positive roots: $e_i \pm e_j$ $(1 \le i < j \le 4)$; $e_i (1 \le i \le 4)$; $(e_1 \pm e_2 \pm e_3 \pm e_4)/2$

Exponents: 1, 5, 7, 11

Coxeter number: 12

Weyl group order: $2^7 . 3^2 = 1152$

Weyl group structure: $S_3 . S_4 . (Z_2)^3$ - semidirect product of $S_4 . (Z_2)^3 \approx W(D_4)$ acting as the Weyl group on the long roots (resp. short roots of F_4) and S_3 acting as outer automorphisms on the D_4 root systems;

$\{1, -1\} \approx Z(W) < W^+$

Table 1. *(continued)*

G_2

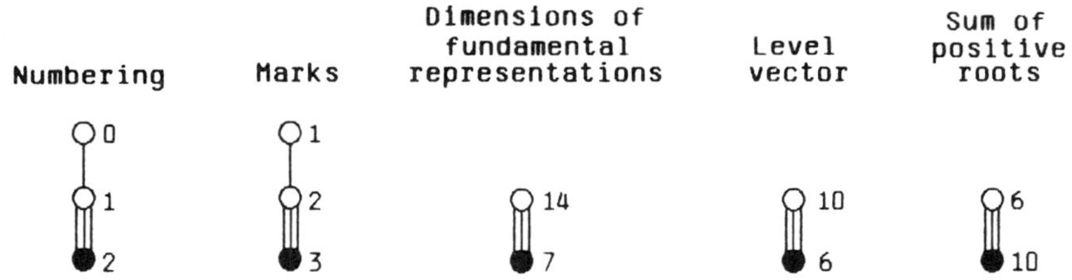

Cartan matrix	Quadratic form matrix
$\begin{Vmatrix} 2 & -3 \\ -1 & 2 \end{Vmatrix}$	$\frac{1}{3}\begin{Vmatrix} 6 & 3 \\ 3 & 2 \end{Vmatrix}$

Index of connection: 1

Roots: $\pm(e_i - e_j)$ $(1 \le i < j \le 3)$; $\pm(2e_i - e_j - e_k)$ $(i \ne j \ne k)$;

Positive roots: e_1-e_2; $-2e_1+e_2+e_3$; $-e_1+e_3$; $-e_2+e_3$; $e_1-2e_2+e_3$; $-e_1-e_2+2e_3$;

Exponents: 1, 5

Coxeter number: 6

Weyl group order: 12

Weyl group structure: dihedral group $D_6 \approx S_3 \times \{1,-1\}$ with S_3 operating as permutations on the e_i and -1 acting by $-1(e_i) = -e_i$; $Z(W) \approx Z_2$

Table 2. Positive Roots Table

ALGEBRA A2

#	OMEGA	ALPHA	EPSILON
# 1	1 1	1 1	1 0 -1
# 2	-1 2	0 1	0 1 -1
# 3	2 -1	1 0	1 -1 0

ALGEBRA A3

#	OMEGA	ALPHA	EPSILON
# 1	1 0 1	1 1 1	1 0 0 -1
# 2	-1 1 1	0 1 1	0 1 0 -1
# 3	1 1 -1	1 1 0	1 0 -1 0
# 4	0 -1 2	0 0 1	0 0 1 -1
# 5	-1 2 -1	0 1 0	0 1 -1 0
# 6	2 -1 0	1 0 0	1 -1 0 0

ALGEBRA A4

#	OMEGA	ALPHA	EPSILON
# 1	1 0 0 1	1 1 1 1	1 0 0 0 -1
# 2	-1 1 0 1	0 1 1 1	0 1 0 0 -1
# 3	1 0 1 -1	1 1 1 0	1 0 0 -1 0
# 4	0 -1 1 1	0 0 1 1	0 0 1 0 -1
# 5	-1 1 1 -1	0 1 1 0	0 1 0 -1 0
# 6	1 1 -1 0	1 1 0 0	1 0 -1 0 0
# 7	0 0 -1 2	0 0 0 1	0 0 0 1 -1
# 8	0 -1 2 -1	0 0 1 0	0 0 1 -1 0
# 9	-1 2 -1 0	0 1 0 0	0 1 -1 0 0
# 10	2 -1 0 0	1 0 0 0	1 -1 0 0 0

ALGEBRA A5

#	OMEGA	ALPHA	EPSILON
# 1	1 0 0 0 1	1 1 1 1 1	1 0 0 0 0 -1
# 2	-1 1 0 0 1	0 1 1 1 1	0 1 0 0 0 -1
# 3	1 0 0 1 -1	1 1 1 1 0	1 0 0 0 -1 0
# 4	0 -1 1 0 1	0 0 1 1 1	0 0 1 0 0 -1
# 5	-1 1 0 1 -1	0 1 1 1 0	0 1 0 0 -1 0
# 6	1 0 1 -1 0	1 1 1 0 0	1 0 0 -1 0 0
# 7	0 0 -1 1 1	0 0 0 1 1	0 0 0 1 0 -1
# 8	0 -1 1 1 -1	0 0 1 1 0	0 0 1 0 -1 0
# 9	-1 1 1 -1 0	0 1 1 0 0	0 1 0 -1 0 0
# 10	1 1 -1 0 0	1 1 0 0 0	1 0 -1 0 0 0
# 11	0 0 0 -1 2	0 0 0 0 1	0 0 0 0 1 -1
# 12	0 0 -1 2 -1	0 0 0 1 0	0 0 0 1 -1 0
# 13	0 -1 2 -1 0	0 0 1 0 0	0 0 1 -1 0 0
# 14	-1 2 -1 0 0	0 1 0 0 0	0 1 -1 0 0 0
# 15	2 -1 0 0 0	1 0 0 0 0	1 -1 0 0 0 0

Table 2. *(continued)*

ALGEBRA A6

#	OMEGA	ALPHA	EPSILON
# 1	1 0 0 0 0 1	1 1 1 1 1 1	1 0 0 0 0 0 -1
# 2	-1 1 0 0 0 1	0 1 1 1 1 1	0 1 0 0 0 0 -1
# 3	1 0 0 0 1 -1	1 1 1 1 1 0	1 0 0 0 0 -1 0
# 4	0 -1 1 0 0 1	0 0 1 1 1 1	0 0 1 0 0 0 -1
# 5	-1 1 0 0 1 -1	0 1 1 1 1 0	0 1 0 0 0 -1 0
# 6	1 0 0 1 -1 0	1 1 1 1 0 0	1 0 0 0 -1 0 0
# 7	0 0 -1 1 0 1	0 0 0 1 1 1	0 0 0 1 0 0 -1
# 8	0 -1 1 0 1 -1	0 0 1 1 1 0	0 0 1 0 0 -1 0
# 9	-1 1 0 1 -1 0	0 1 1 1 0 0	0 1 0 0 -1 0 0
# 10	1 0 1 -1 0 0	1 1 1 0 0 0	1 0 0 -1 0 0 0
# 11	0 0 0 -1 1 1	0 0 0 0 1 1	0 0 0 0 1 0 -1
# 12	0 0 -1 1 1 -1	0 0 0 1 1 0	0 0 0 1 0 -1 0
# 13	0 -1 1 1 -1 0	0 0 1 1 0 0	0 0 1 0 -1 0 0
# 14	-1 1 1 -1 0 0	0 1 1 0 0 0	0 1 0 -1 0 0 0
# 15	1 1 -1 0 0 0	1 1 0 0 0 0	1 0 -1 0 0 0 0
# 16	0 0 0 0 -1 2	0 0 0 0 0 1	0 0 0 0 0 1 -1
# 17	0 0 0 -1 2 -1	0 0 0 0 1 0	0 0 0 0 1 -1 0
# 18	0 0 -1 2 -1 0	0 0 0 1 0 0	0 0 0 1 -1 0 0
# 19	0 -1 2 -1 0 0	0 0 1 0 0 0	0 0 1 -1 0 0 0
# 20	-1 2 -1 0 0 0	0 1 0 0 0 0	0 1 -1 0 0 0 0
# 21	2 -1 0 0 0 0	1 0 0 0 0 0	1 -1 0 0 0 0 0

ALGEBRA A7

#	OMEGA	ALPHA	EPSILON
# 1	1 0 0 0 0 0 1	1 1 1 1 1 1 1	1 0 0 0 0 0 0 -1
# 2	-1 1 0 0 0 0 1	0 1 1 1 1 1 1	0 1 0 0 0 0 0 -1
# 3	1 0 0 0 0 1 -1	1 1 1 1 1 1 0	1 0 0 0 0 0 -1 0
# 4	0 -1 1 0 0 0 1	0 0 1 1 1 1 1	0 0 1 0 0 0 0 -1
# 5	-1 1 0 0 0 1 -1	0 1 1 1 1 1 0	0 1 0 0 0 0 -1 0
# 6	1 0 0 0 1 -1 0	1 1 1 1 1 0 0	1 0 0 0 0 -1 0 0
# 7	0 0 -1 1 0 0 1	0 0 0 1 1 1 1	0 0 0 1 0 0 0 -1
# 8	0 -1 1 0 0 1 -1	0 0 1 1 1 1 0	0 0 1 0 0 0 -1 0
# 9	-1 1 0 0 1 -1 0	0 1 1 1 1 0 0	0 1 0 0 0 -1 0 0
# 10	1 0 0 1 -1 0 0	1 1 1 1 0 0 0	1 0 0 0 -1 0 0 0
# 11	0 0 0 -1 1 0 1	0 0 0 0 1 1 1	0 0 0 0 1 0 0 -1
# 12	0 0 -1 1 0 1 -1	0 0 0 1 1 1 0	0 0 0 1 0 0 -1 0
# 13	0 -1 1 0 1 -1 0	0 0 1 1 1 0 0	0 0 1 0 0 -1 0 0
# 14	-1 1 0 1 -1 0 0	0 1 1 1 0 0 0	0 1 0 0 -1 0 0 0
# 15	1 0 1 -1 0 0 0	1 1 1 0 0 0 0	1 0 0 -1 0 0 0 0
# 16	0 0 0 0 -1 1 1	0 0 0 0 0 1 1	0 0 0 0 0 1 0 -1
# 17	0 0 0 -1 1 1 -1	0 0 0 0 1 1 0	0 0 0 0 1 0 -1 0
# 18	0 0 -1 1 1 -1 0	0 0 0 1 1 0 0	0 0 0 1 0 -1 0 0
# 19	0 -1 1 1 -1 0 0	0 0 1 1 0 0 0	0 0 1 0 -1 0 0 0
# 20	-1 1 1 -1 0 0 0	0 1 1 0 0 0 0	0 1 0 -1 0 0 0 0
# 21	1 1 -1 0 0 0 0	1 1 0 0 0 0 0	1 0 -1 0 0 0 0 0
# 22	0 0 0 0 0 -1 2	0 0 0 0 0 0 1	0 0 0 0 0 0 1 -1
# 23	0 0 0 0 -1 2 -1	0 0 0 0 0 1 0	0 0 0 0 0 1 -1 0
# 24	0 0 0 -1 2 -1 0	0 0 0 0 1 0 0	0 0 0 0 1 -1 0 0
# 25	0 0 -1 2 -1 0 0	0 0 0 1 0 0 0	0 0 0 1 -1 0 0 0
# 26	0 -1 2 -1 0 0 0	0 0 1 0 0 0 0	0 0 1 -1 0 0 0 0
# 27	-1 2 -1 0 0 0 0	0 1 0 0 0 0 0	0 1 -1 0 0 0 0 0
# 28	2 -1 0 0 0 0 0	1 0 0 0 0 0 0	1 -1 0 0 0 0 0 0

ALGEBRA AB

#	OMEGA								ALPHA								EPSILON								
# 1	1	0	0	0	0	0	0	1	1	1	1	1	1	1	1	1	1	0	0	0	0	0	0	0	-1
# 2	-1	1	0	0	0	0	0	1	0	1	1	1	1	1	1	1	0	1	0	0	0	0	0	0	-1
# 3	1	0	0	0	0	0	1	-1	1	1	1	1	1	1	1	0	1	0	0	0	0	0	0	-1	0
# 4	0	-1	1	0	0	0	0	1	0	0	1	1	1	1	1	1	0	0	1	0	0	0	0	0	-1
# 5	-1	1	0	0	0	0	1	-1	0	1	1	1	1	1	1	0	0	1	0	0	0	0	0	-1	0
# 6	1	0	0	0	0	1	-1	0	1	1	1	1	1	1	0	0	1	0	0	0	0	0	-1	0	0
# 7	0	0	-1	1	0	0	0	1	0	0	0	1	1	1	1	1	0	0	0	1	0	0	0	0	-1
# 8	0	-1	1	0	0	0	1	-1	0	0	1	1	1	1	1	0	0	0	1	0	0	0	0	-1	0
# 9	-1	1	0	0	0	1	-1	0	0	1	1	1	1	1	0	0	0	1	0	0	0	0	-1	0	0
# 10	1	0	0	0	1	-1	0	0	1	1	1	1	1	0	0	0	1	0	0	0	0	-1	0	0	0
# 11	0	0	0	-1	1	0	0	1	0	0	0	0	1	1	1	1	0	0	0	0	1	0	0	0	-1
# 12	0	0	-1	1	0	0	1	-1	0	0	0	1	1	1	1	0	0	0	0	1	0	0	0	-1	0
# 13	0	-1	1	0	0	1	-1	0	0	0	1	1	1	1	0	0	0	0	1	0	0	0	-1	0	0
# 14	-1	1	0	0	1	-1	0	0	0	1	1	1	1	0	0	0	0	1	0	0	0	-1	0	0	0
# 15	1	0	0	1	-1	0	0	0	1	1	1	1	0	0	0	0	1	0	0	0	-1	0	0	0	0
# 16	0	0	0	0	-1	1	0	1	0	0	0	0	0	1	1	1	0	0	0	0	0	1	0	0	-1
# 17	0	0	0	-1	1	0	1	-1	0	0	0	0	1	1	1	0	0	0	0	0	1	0	0	-1	0
# 18	0	0	-1	1	0	1	-1	0	0	0	0	1	1	1	0	0	0	0	0	1	0	0	-1	0	0
# 19	0	-1	1	0	1	-1	0	0	0	0	1	1	1	0	0	0	0	0	1	0	0	-1	0	0	0
# 20	-1	1	0	1	-1	0	0	0	0	1	1	1	0	0	0	0	0	1	0	0	-1	0	0	0	0
# 21	1	0	1	-1	0	0	0	0	1	1	1	0	0	0	0	0	1	0	0	-1	0	0	0	0	0
# 22	0	0	0	0	0	-1	1	1	0	0	0	0	0	0	1	1	0	0	0	0	0	0	1	0	-1
# 23	0	0	0	0	-1	1	1	-1	0	0	0	0	0	1	1	0	0	0	0	0	0	1	0	-1	0
# 24	0	0	0	-1	1	1	-1	0	0	0	0	0	1	1	0	0	0	0	0	0	1	0	-1	0	0
# 25	0	0	-1	1	1	-1	0	0	0	0	0	1	1	0	0	0	0	0	0	1	0	-1	0	0	0
# 26	0	-1	1	1	-1	0	0	0	0	0	1	1	0	0	0	0	0	0	1	0	-1	0	0	0	0
# 27	-1	1	1	-1	0	0	0	0	0	1	1	0	0	0	0	0	0	1	0	-1	0	0	0	0	0
# 28	1	1	-1	0	0	0	0	0	1	1	0	0	0	0	0	0	1	0	-1	0	0	0	0	0	0
# 29	0	0	0	0	0	0	-1	2	0	0	0	0	0	0	0	1	0	0	0	0	0	0	0	1	-1
# 30	0	0	0	0	0	-1	2	-1	0	0	0	0	0	0	1	0	0	0	0	0	0	0	1	-1	0
# 31	0	0	0	0	-1	2	-1	0	0	0	0	0	0	1	0	0	0	0	0	0	0	1	-1	0	0
# 32	0	0	0	-1	2	-1	0	0	0	0	0	0	1	0	0	0	0	0	0	0	1	-1	0	0	0
# 33	0	0	-1	2	-1	0	0	0	0	0	0	1	0	0	0	0	0	0	0	1	-1	0	0	0	0
# 34	0	-1	2	-1	0	0	0	0	0	0	1	0	0	0	0	0	0	0	1	-1	0	0	0	0	0
# 35	-1	2	-1	0	0	0	0	0	0	1	0	0	0	0	0	0	0	1	-1	0	0	0	0	0	0
# 36	2	-1	0	0	0	0	0	0	1	0	0	0	0	0	0	0	1	-1	0	0	0	0	0	0	0

Table 2. *(continued)*

ALGEBRA A9

#	OMEGA									ALPHA									EPSILON									
# 1	1	0	0	0	0	0	0	0	1	1	1	1	1	1	1	1	1	1	1	0	0	0	0	0	0	0	0	-1
# 2	-1	1	0	0	0	0	0	0	1	0	1	1	1	1	1	1	1	1	0	1	0	0	0	0	0	0	0	-1
# 3	1	0	0	0	0	0	0	1	-1	1	1	1	1	1	1	1	1	0	1	0	0	0	0	0	0	0	-1	0
# 4	0	-1	1	0	0	0	0	0	1	0	0	1	1	1	1	1	1	1	0	0	1	0	0	0	0	0	0	-1
# 5	-1	1	0	0	0	0	0	1	-1	0	1	1	1	1	1	1	1	0	0	1	0	0	0	0	0	0	-1	0
# 6	1	0	0	0	0	0	1	-1	0	1	1	1	1	1	1	1	0	0	1	0	0	0	0	0	0	-1	0	0
# 7	0	0	-1	1	0	0	0	0	1	0	0	0	1	1	1	1	1	1	0	0	0	1	0	0	0	0	0	-1
# 8	0	-1	1	0	0	0	0	1	-1	0	0	1	1	1	1	1	1	0	0	0	1	0	0	0	0	0	-1	0
# 9	-1	1	0	0	0	0	1	-1	0	0	1	1	1	1	1	1	0	0	0	1	0	0	0	0	0	-1	0	0
# 10	1	0	0	0	0	1	-1	0	0	1	1	1	1	1	1	0	0	0	1	0	0	0	0	0	-1	0	0	0
# 11	0	0	0	-1	1	0	0	0	1	0	0	0	0	1	1	1	1	1	0	0	0	0	1	0	0	0	0	-1
# 12	0	0	-1	1	0	0	0	1	-1	0	0	0	1	1	1	1	1	0	0	0	0	1	0	0	0	0	-1	0
# 13	0	-1	1	0	0	0	1	-1	0	0	0	1	1	1	1	1	0	0	0	0	1	0	0	0	0	-1	0	0
# 14	-1	1	0	0	0	1	-1	0	0	0	1	1	1	1	1	0	0	0	0	1	0	0	0	0	-1	0	0	0
# 15	1	0	0	0	1	-1	0	0	0	1	1	1	1	1	0	0	0	0	1	0	0	0	0	-1	0	0	0	0
# 16	0	0	0	0	-1	1	0	0	1	0	0	0	0	0	1	1	1	1	0	0	0	0	0	1	0	0	0	-1
# 17	0	0	0	-1	1	0	0	1	-1	0	0	0	0	1	1	1	1	0	0	0	0	0	1	0	0	0	-1	0
# 18	0	0	-1	1	0	0	1	-1	0	0	0	0	1	1	1	1	0	0	0	0	0	1	0	0	0	-1	0	0
# 19	0	-1	1	0	0	1	-1	0	0	0	0	1	1	1	1	0	0	0	0	0	1	0	0	0	-1	0	0	0
# 20	-1	1	0	0	1	-1	0	0	0	0	1	1	1	1	0	0	0	0	0	1	0	0	0	-1	0	0	0	0
# 21	1	0	0	1	-1	0	0	0	0	1	1	1	1	0	0	0	0	0	1	0	0	0	-1	0	0	0	0	0
# 22	0	0	0	0	0	-1	1	0	1	0	0	0	0	0	0	1	1	1	0	0	0	0	0	0	1	0	0	-1
# 23	0	0	0	0	-1	1	0	1	-1	0	0	0	0	0	1	1	1	0	0	0	0	0	0	1	0	0	-1	0
# 24	0	0	0	-1	1	0	1	-1	0	0	0	0	0	1	1	1	0	0	0	0	0	0	1	0	0	-1	0	0
# 25	0	0	-1	1	0	1	-1	0	0	0	0	0	1	1	1	0	0	0	0	0	0	1	0	0	-1	0	0	0
# 26	0	-1	1	0	1	-1	0	0	0	0	0	1	1	1	0	0	0	0	0	0	1	0	0	-1	0	0	0	0
# 27	-1	1	0	1	-1	0	0	0	0	0	1	1	1	0	0	0	0	0	0	1	0	0	-1	0	0	0	0	0
# 28	1	0	1	-1	0	0	0	0	0	1	1	1	0	0	0	0	0	0	1	0	0	-1	0	0	0	0	0	0
# 29	0	0	0	0	0	0	-1	1	1	0	0	0	0	0	0	0	1	1	0	0	0	0	0	0	0	1	0	-1
# 30	0	0	0	0	0	-1	1	1	-1	0	0	0	0	0	0	1	1	0	0	0	0	0	0	0	1	0	-1	0
# 31	0	0	0	0	-1	1	1	-1	0	0	0	0	0	0	1	1	0	0	0	0	0	0	0	1	0	-1	0	0
# 32	0	0	0	-1	1	1	-1	0	0	0	0	0	0	1	1	0	0	0	0	0	0	0	1	0	-1	0	0	0
# 33	0	0	-1	1	1	-1	0	0	0	0	0	0	1	1	0	0	0	0	0	0	0	1	0	-1	0	0	0	0
# 34	0	-1	1	1	-1	0	0	0	0	0	0	1	1	0	0	0	0	0	0	0	1	0	-1	0	0	0	0	0
# 35	-1	1	1	-1	0	0	0	0	0	0	1	1	0	0	0	0	0	0	0	1	0	-1	0	0	0	0	0	0
# 36	1	1	-1	0	0	0	0	0	0	1	1	0	0	0	0	0	0	0	1	0	-1	0	0	0	0	0	0	0
# 37	0	0	0	0	0	0	0	-1	2	0	0	0	0	0	0	0	0	1	0	0	0	0	0	0	0	0	1	-1
# 38	0	0	0	0	0	0	-1	2	-1	0	0	0	0	0	0	0	1	0	0	0	0	0	0	0	0	1	-1	0
# 39	0	0	0	0	0	-1	2	-1	0	0	0	0	0	0	0	1	0	0	0	0	0	0	0	0	1	-1	0	0
# 40	0	0	0	0	-1	2	-1	0	0	0	0	0	0	0	1	0	0	0	0	0	0	0	0	1	-1	0	0	0
# 41	0	0	0	-1	2	-1	0	0	0	0	0	0	0	1	0	0	0	0	0	0	0	0	1	-1	0	0	0	0
# 42	0	0	-1	2	-1	0	0	0	0	0	0	0	1	0	0	0	0	0	0	0	0	1	-1	0	0	0	0	0
# 43	0	-1	2	-1	0	0	0	0	0	0	0	1	0	0	0	0	0	0	0	0	1	-1	0	0	0	0	0	0
# 44	-1	2	-1	0	0	0	0	0	0	0	1	0	0	0	0	0	0	0	0	1	-1	0	0	0	0	0	0	0
# 45	2	-1	0	0	0	0	0	0	0	1	0	0	0	0	0	0	0	0	1	-1	0	0	0	0	0	0	0	0

ALGEBRA A10

#	OMEGA									
# 1	1	0	0	0	0	0	0	0	0	1
# 2	-1	1	0	0	0	0	0	0	0	1
# 3	1	0	0	0	0	0	0	0	1	-1
# 4	0	-1	1	0	0	0	0	0	0	1
# 5	-1	1	0	0	0	0	0	0	1	-1
# 6	1	0	0	0	0	0	0	1	-1	0
# 7	0	0	-1	1	0	0	0	0	0	1
# 8	0	-1	1	0	0	0	0	0	1	-1
# 9	-1	1	0	0	0	0	0	1	-1	0
# 10	1	0	0	0	0	0	1	-1	0	0
# 11	0	0	0	-1	1	0	0	0	0	1
# 12	0	0	-1	1	0	0	0	0	1	-1
# 13	0	-1	1	0	0	0	0	1	-1	0
# 14	-1	1	0	0	0	0	1	-1	0	0
# 15	1	0	0	0	0	1	-1	0	0	0
# 16	0	0	0	0	-1	1	0	0	0	1
# 17	0	0	0	-1	1	0	0	0	1	-1
# 18	0	0	-1	1	0	0	0	1	-1	0
# 19	0	-1	1	0	0	0	1	-1	0	0
# 20	-1	1	0	0	0	1	-1	0	0	0
# 21	1	0	0	0	1	-1	0	0	0	0
# 22	0	0	0	0	0	-1	1	0	0	1
# 23	0	0	0	0	-1	1	0	0	1	-1
# 24	0	0	0	-1	1	0	0	1	-1	0
# 25	0	0	-1	1	0	0	1	-1	0	0
# 26	0	-1	1	0	0	1	-1	0	0	0
# 27	-1	1	0	0	1	-1	0	0	0	0
# 28	1	0	0	1	-1	0	0	0	0	0
# 29	0	0	0	0	0	0	-1	1	0	1
# 30	0	0	0	0	0	-1	1	0	1	-1
# 31	0	0	0	0	-1	1	0	1	-1	0
# 32	0	0	0	-1	1	0	1	-1	0	0
# 33	0	0	-1	1	0	1	-1	0	0	0
# 34	0	-1	1	0	1	-1	0	0	0	0
# 35	-1	1	0	1	-1	0	0	0	0	0
# 36	1	0	1	-1	0	0	0	0	0	0
# 37	0	0	0	0	0	0	0	-1	1	1
# 38	0	0	0	0	0	0	-1	1	1	-1
# 39	0	0	0	0	0	-1	1	1	-1	0
# 40	0	0	0	0	-1	1	1	-1	0	0
# 41	0	0	0	-1	1	1	-1	0	0	0
# 42	0	0	-1	1	1	-1	0	0	0	0
# 43	0	-1	1	1	-1	0	0	0	0	0
# 44	-1	1	1	-1	0	0	0	0	0	0
# 45	1	1	-1	0	0	0	0	0	0	0
# 46	0	0	0	0	0	0	0	0	-1	2
# 47	0	0	0	0	0	0	0	-1	2	-1
# 48	0	0	0	0	0	0	-1	2	-1	0
# 49	0	0	0	0	0	-1	2	-1	0	0
# 50	0	0	0	0	-1	2	-1	0	0	0
# 51	0	0	0	-1	2	-1	0	0	0	0
# 52	0	0	-1	2	-1	0	0	0	0	0
# 53	0	-1	2	-1	0	0	0	0	0	0
# 54	-1	2	-1	0	0	0	0	0	0	0
# 55	2	-1	0	0	0	0	0	0	0	0

#	ALPHA									
# 1	1	1	1	1	1	1	1	1	1	1
# 2	0	1	1	1	1	1	1	1	1	1
# 3	1	1	1	1	1	1	1	1	1	0
# 4	0	0	1	1	1	1	1	1	1	1
# 5	0	1	1	1	1	1	1	1	1	0
# 6	1	1	1	1	1	1	1	1	0	0
# 7	0	0	0	1	1	1	1	1	1	1
# 8	0	0	1	1	1	1	1	1	1	0
# 9	0	1	1	1	1	1	1	1	0	0
# 10	1	1	1	1	1	1	1	0	0	0
# 11	0	0	0	0	1	1	1	1	1	1
# 12	0	0	0	1	1	1	1	1	1	0
# 13	0	0	1	1	1	1	1	1	0	0
# 14	0	1	1	1	1	1	1	0	0	0
# 15	1	1	1	1	1	1	0	0	0	0
# 16	0	0	0	0	0	1	1	1	1	1
# 17	0	0	0	0	1	1	1	1	1	0
# 18	0	0	0	1	1	1	1	1	0	0
# 19	0	0	1	1	1	1	1	0	0	0
# 20	0	1	1	1	1	1	0	0	0	0
# 21	1	1	1	1	1	0	0	0	0	0
# 22	0	0	0	0	0	0	1	1	1	1
# 23	0	0	0	0	0	1	1	1	1	0
# 24	0	0	0	0	1	1	1	1	0	0
# 25	0	0	0	1	1	1	1	0	0	0
# 26	0	0	1	1	1	1	0	0	0	0
# 27	0	1	1	1	1	0	0	0	0	0
# 28	1	1	1	1	0	0	0	0	0	0
# 29	0	0	0	0	0	0	0	1	1	1
# 30	0	0	0	0	0	0	1	1	1	0
# 31	0	0	0	0	0	1	1	1	0	0
# 32	0	0	0	0	1	1	1	0	0	0
# 33	0	0	0	1	1	1	0	0	0	0
# 34	0	0	1	1	1	0	0	0	0	0
# 35	0	1	1	1	0	0	0	0	0	0
# 36	1	1	1	0	0	0	0	0	0	0
# 37	0	0	0	0	0	0	0	0	1	1
# 38	0	0	0	0	0	0	0	1	1	0
# 39	0	0	0	0	0	0	1	1	0	0
# 40	0	0	0	0	0	1	1	0	0	0
# 41	0	0	0	0	1	1	0	0	0	0
# 42	0	0	0	1	1	0	0	0	0	0
# 43	0	0	1	1	0	0	0	0	0	0
# 44	0	1	1	0	0	0	0	0	0	0
# 45	1	1	0	0	0	0	0	0	0	0
# 46	0	0	0	0	0	0	0	0	0	1
# 47	0	0	0	0	0	0	0	0	1	0
# 48	0	0	0	0	0	0	0	1	0	0
# 49	0	0	0	0	0	0	1	0	0	0
# 50	0	0	0	0	0	1	0	0	0	0
# 51	0	0	0	0	1	0	0	0	0	0
# 52	0	0	0	1	0	0	0	0	0	0
# 53	0	0	1	0	0	0	0	0	0	0
# 54	0	1	0	0	0	0	0	0	0	0
# 55	1	0	0	0	0	0	0	0	0	0

#	EPSILON										
# 1	1	0	0	0	0	0	0	0	0	0	-1
# 2	0	1	0	0	0	0	0	0	0	0	-1
# 3	1	0	0	0	0	0	0	0	0	-1	0
# 4	0	0	1	0	0	0	0	0	0	0	-1
# 5	0	1	0	0	0	0	0	0	0	-1	0
# 6	1	0	0	0	0	0	0	0	-1	0	0
# 7	0	0	0	1	0	0	0	0	0	0	-1
# 8	0	0	1	0	0	0	0	0	0	-1	0
# 9	0	1	0	0	0	0	0	0	-1	0	0
# 10	1	0	0	0	0	0	0	-1	0	0	0
# 11	0	0	0	0	1	0	0	0	0	0	-1
# 12	0	0	0	1	0	0	0	0	0	-1	0
# 13	0	0	1	0	0	0	0	0	-1	0	0
# 14	0	1	0	0	0	0	0	-1	0	0	0
# 15	1	0	0	0	0	0	-1	0	0	0	0
# 16	0	0	0	0	0	1	0	0	0	0	-1
# 17	0	0	0	0	1	0	0	0	0	-1	0
# 18	0	0	0	1	0	0	0	0	-1	0	0
# 19	0	0	1	0	0	0	0	-1	0	0	0
# 20	0	1	0	0	0	0	-1	0	0	0	0
# 21	1	0	0	0	0	-1	0	0	0	0	0
# 22	0	0	0	0	0	0	1	0	0	0	-1
# 23	0	0	0	0	0	1	0	0	0	-1	0
# 24	0	0	0	0	1	0	0	0	-1	0	0
# 25	0	0	0	1	0	0	0	-1	0	0	0
# 26	0	0	1	0	0	0	-1	0	0	0	0
# 27	0	1	0	0	0	-1	0	0	0	0	0
# 28	1	0	0	0	-1	0	0	0	0	0	0
# 29	0	0	0	0	0	0	0	1	0	0	-1
# 30	0	0	0	0	0	0	1	0	0	-1	0
# 31	0	0	0	0	0	1	0	0	-1	0	0
# 32	0	0	0	0	1	0	0	-1	0	0	0
# 33	0	0	0	1	0	0	-1	0	0	0	0
# 34	0	0	1	0	0	-1	0	0	0	0	0
# 35	0	1	0	0	-1	0	0	0	0	0	0
# 36	1	0	0	-1	0	0	0	0	0	0	0
# 37	0	0	0	0	0	0	0	0	1	0	-1
# 38	0	0	0	0	0	0	0	1	0	-1	0
# 39	0	0	0	0	0	0	1	0	-1	0	0
# 40	0	0	0	0	0	1	0	-1	0	0	0
# 41	0	0	0	0	1	0	-1	0	0	0	0
# 42	0	0	0	1	0	-1	0	0	0	0	0
# 43	0	0	1	0	-1	0	0	0	0	0	0
# 44	0	1	0	-1	0	0	0	0	0	0	0
# 45	1	0	-1	0	0	0	0	0	0	0	0
# 46	0	0	0	0	0	0	0	0	0	1	-1
# 47	0	0	0	0	0	0	0	0	1	-1	0
# 48	0	0	0	0	0	0	0	1	-1	0	0
# 49	0	0	0	0	0	0	1	-1	0	0	0
# 50	0	0	0	0	0	1	-1	0	0	0	0
# 51	0	0	0	0	1	-1	0	0	0	0	0
# 52	0	0	0	1	-1	0	0	0	0	0	0
# 53	0	0	1	-1	0	0	0	0	0	0	0
# 54	0	1	-1	0	0	0	0	0	0	0	0
# 55	1	-1	0	0	0	0	0	0	0	0	0

Table 2. *(continued)*

ALGEBRA A11

#	OMEGA										
# 1	1	0	0	0	0	0	0	0	0	0	1
# 2	-1	1	0	0	0	0	0	0	0	0	1
# 3	1	0	0	0	0	0	0	0	0	1	-1
# 4	0	-1	1	0	0	0	0	0	0	0	1
# 5	-1	1	0	0	0	0	0	0	0	1	-1
# 6	1	0	0	0	0	0	0	0	1	-1	0
# 7	0	0	-1	1	0	0	0	0	0	0	1
# 8	0	-1	1	0	0	0	0	0	0	1	-1
# 9	-1	1	0	0	0	0	0	0	1	-1	0
# 10	1	0	0	0	0	0	0	1	-1	0	0
# 11	0	0	0	-1	1	0	0	0	0	0	1
# 12	0	0	-1	1	0	0	0	0	0	1	-1
# 13	0	-1	1	0	0	0	0	0	1	-1	0
# 14	-1	1	0	0	0	0	0	1	-1	0	0
# 15	1	0	0	0	0	0	1	-1	0	0	0
# 16	0	0	0	0	-1	1	0	0	0	0	1
# 17	0	0	0	-1	1	0	0	0	0	1	-1
# 18	0	0	-1	1	0	0	0	0	1	-1	0
# 19	0	-1	1	0	0	0	0	1	-1	0	0
# 20	-1	1	0	0	0	0	1	-1	0	0	0
# 21	1	0	0	0	0	1	-1	0	0	0	0
# 22	0	0	0	0	0	-1	1	0	0	0	1
# 23	0	0	0	0	-1	1	0	0	0	1	-1
# 24	0	0	0	-1	1	0	0	0	1	-1	0
# 25	0	0	-1	1	0	0	0	1	-1	0	0
# 26	0	-1	1	0	0	0	1	-1	0	0	0
# 27	-1	1	0	0	0	1	-1	0	0	0	0
# 28	1	0	0	0	1	-1	0	0	0	0	0
# 29	0	0	0	0	0	0	-1	1	0	0	1
# 30	0	0	0	0	0	-1	1	0	0	1	-1
# 31	0	0	0	0	-1	1	0	0	1	-1	0
# 32	0	0	0	-1	1	0	0	1	-1	0	0
# 33	0	0	-1	1	0	0	1	-1	0	0	0
# 34	0	-1	1	0	0	1	-1	0	0	0	0
# 35	-1	1	0	0	1	-1	0	0	0	0	0
# 36	1	0	0	1	-1	0	0	0	0	0	0
# 37	0	0	0	0	0	0	0	-1	1	0	1
# 38	0	0	0	0	0	0	-1	1	0	1	-1
# 39	0	0	0	0	0	-1	1	0	1	-1	0
# 40	0	0	0	0	-1	1	0	1	-1	0	0
# 41	0	0	0	-1	1	0	1	-1	0	0	0
# 42	0	0	-1	1	0	1	-1	0	0	0	0
# 43	0	-1	1	0	1	-1	0	0	0	0	0
# 44	-1	1	0	1	-1	0	0	0	0	0	0
# 45	1	0	1	-1	0	0	0	0	0	0	0
# 46	0	0	0	0	0	0	0	0	-1	1	1
# 47	0	0	0	0	0	0	0	-1	1	1	-1
# 48	0	0	0	0	0	0	-1	1	1	-1	0
# 49	0	0	0	0	0	-1	1	1	-1	0	0
# 50	0	0	0	0	-1	1	1	-1	0	0	0
# 51	0	0	0	-1	1	1	-1	0	0	0	0
# 52	0	0	-1	1	1	-1	0	0	0	0	0
# 53	0	-1	1	1	-1	0	0	0	0	0	0
# 54	-1	1	1	-1	0	0	0	0	0	0	0
# 55	1	1	-1	0	0	0	0	0	0	0	0
# 56	0	0	0	0	0	0	0	0	0	-1	2
# 57	0	0	0	0	0	0	0	0	-1	2	-1
# 58	0	0	0	0	0	0	0	-1	2	-1	0
# 59	0	0	0	0	0	0	-1	2	-1	0	0
# 60	0	0	0	0	0	-1	2	-1	0	0	0
# 61	0	0	0	0	-1	2	-1	0	0	0	0
# 62	0	0	0	-1	2	-1	0	0	0	0	0
# 63	0	0	-1	2	-1	0	0	0	0	0	0
# 64	0	-1	2	-1	0	0	0	0	0	0	0
# 65	-1	2	-1	0	0	0	0	0	0	0	0
# 66	2	-1	0	0	0	0	0	0	0	0	0

#	ALPHA										
# 1	1	1	1	1	1	1	1	1	1	1	1
# 2	0	1	1	1	1	1	1	1	1	1	1
# 3	1	1	1	1	1	1	1	1	1	1	0
# 4	0	0	1	1	1	1	1	1	1	1	1
# 5	0	1	1	1	1	1	1	1	1	1	0
# 6	1	1	1	1	1	1	1	1	1	0	0
# 7	0	0	0	1	1	1	1	1	1	1	1
# 8	0	0	1	1	1	1	1	1	1	1	0
# 9	0	1	1	1	1	1	1	1	1	0	0
# 10	1	1	1	1	1	1	1	1	0	0	0
# 11	0	0	0	0	1	1	1	1	1	1	1
# 12	0	0	0	1	1	1	1	1	1	1	0
# 13	0	0	1	1	1	1	1	1	1	0	0
# 14	0	1	1	1	1	1	1	1	0	0	0
# 15	1	1	1	1	1	1	1	0	0	0	0
# 16	0	0	0	0	0	1	1	1	1	1	1
# 17	0	0	0	0	1	1	1	1	1	1	0
# 18	0	0	0	1	1	1	1	1	1	0	0
# 19	0	0	1	1	1	1	1	1	0	0	0
# 20	0	1	1	1	1	1	1	0	0	0	0
# 21	1	1	1	1	1	1	0	0	0	0	0
# 22	0	0	0	0	0	0	1	1	1	1	1
# 23	0	0	0	0	0	1	1	1	1	1	0
# 24	0	0	0	0	1	1	1	1	1	0	0
# 25	0	0	0	1	1	1	1	1	0	0	0
# 26	0	0	1	1	1	1	1	0	0	0	0
# 27	0	1	1	1	1	1	0	0	0	0	0
# 28	1	1	1	1	1	0	0	0	0	0	0
# 29	0	0	0	0	0	0	0	1	1	1	1
# 30	0	0	0	0	0	0	1	1	1	1	0
# 31	0	0	0	0	0	1	1	1	1	0	0
# 32	0	0	0	0	1	1	1	1	0	0	0
# 33	0	0	0	1	1	1	1	0	0	0	0
# 34	0	0	1	1	1	1	0	0	0	0	0
# 35	0	1	1	1	1	0	0	0	0	0	0
# 36	1	1	1	1	0	0	0	0	0	0	0
# 37	0	0	0	0	0	0	0	0	1	1	1
# 38	0	0	0	0	0	0	0	1	1	1	0
# 39	0	0	0	0	0	0	1	1	1	0	0
# 40	0	0	0	0	0	1	1	1	0	0	0
# 41	0	0	0	0	1	1	1	0	0	0	0
# 42	0	0	0	1	1	1	0	0	0	0	0
# 43	0	0	1	1	1	0	0	0	0	0	0
# 44	0	1	1	1	0	0	0	0	0	0	0
# 45	1	1	1	0	0	0	0	0	0	0	0
# 46	0	0	0	0	0	0	0	0	0	1	1
# 47	0	0	0	0	0	0	0	0	1	1	0
# 48	0	0	0	0	0	0	0	1	1	0	0
# 49	0	0	0	0	0	0	1	1	0	0	0
# 50	0	0	0	0	0	1	1	0	0	0	0
# 51	0	0	0	0	1	1	0	0	0	0	0
# 52	0	0	0	1	1	0	0	0	0	0	0
# 53	0	0	1	1	0	0	0	0	0	0	0
# 54	0	1	1	0	0	0	0	0	0	0	0
# 55	1	1	0	0	0	0	0	0	0	0	0
# 56	0	0	0	0	0	0	0	0	0	0	1
# 57	0	0	0	0	0	0	0	0	0	1	0
# 58	0	0	0	0	0	0	0	0	1	0	0
# 59	0	0	0	0	0	0	0	1	0	0	0
# 60	0	0	0	0	0	0	1	0	0	0	0
# 61	0	0	0	0	0	1	0	0	0	0	0
# 62	0	0	0	0	1	0	0	0	0	0	0
# 63	0	0	0	1	0	0	0	0	0	0	0
# 64	0	0	1	0	0	0	0	0	0	0	0
# 65	0	1	0	0	0	0	0	0	0	0	0
# 66	1	0	0	0	0	0	0	0	0	0	0

#	EPSILON											
# 1	1	0	0	0	0	0	0	0	0	0	0	-1
# 2	0	1	0	0	0	0	0	0	0	0	0	-1
# 3	1	0	0	0	0	0	0	0	0	0	-1	0
# 4	0	0	1	0	0	0	0	0	0	0	0	-1
# 5	0	1	0	0	0	0	0	0	0	0	-1	0
# 6	1	0	0	0	0	0	0	0	0	-1	0	0
# 7	0	0	0	1	0	0	0	0	0	0	0	-1
# 8	0	0	1	0	0	0	0	0	0	0	-1	0
# 9	0	1	0	0	0	0	0	0	0	-1	0	0
# 10	1	0	0	0	0	0	0	0	-1	0	0	0
# 11	0	0	0	0	1	0	0	0	0	0	0	-1
# 12	0	0	0	1	0	0	0	0	0	0	-1	0
# 13	0	0	1	0	0	0	0	0	0	-1	0	0
# 14	0	1	0	0	0	0	0	0	-1	0	0	0
# 15	1	0	0	0	0	0	0	-1	0	0	0	0
# 16	0	0	0	0	0	1	0	0	0	0	0	-1
# 17	0	0	0	0	1	0	0	0	0	0	-1	0
# 18	0	0	0	1	0	0	0	0	0	-1	0	0
# 19	0	0	1	0	0	0	0	0	-1	0	0	0
# 20	0	1	0	0	0	0	0	-1	0	0	0	0
# 21	1	0	0	0	0	0	-1	0	0	0	0	0
# 22	0	0	0	0	0	0	1	0	0	0	0	-1
# 23	0	0	0	0	0	1	0	0	0	0	-1	0
# 24	0	0	0	0	1	0	0	0	0	-1	0	0
# 25	0	0	0	1	0	0	0	0	-1	0	0	0
# 26	0	0	1	0	0	0	0	-1	0	0	0	0
# 27	0	1	0	0	0	0	-1	0	0	0	0	0
# 28	1	0	0	0	0	-1	0	0	0	0	0	0
# 29	0	0	0	0	0	0	0	1	0	0	0	-1
# 30	0	0	0	0	0	0	1	0	0	0	-1	0
# 31	0	0	0	0	0	1	0	0	0	-1	0	0
# 32	0	0	0	0	1	0	0	0	-1	0	0	0
# 33	0	0	0	1	0	0	0	-1	0	0	0	0
# 34	0	0	1	0	0	0	-1	0	0	0	0	0
# 35	0	1	0	0	0	-1	0	0	0	0	0	0
# 36	1	0	0	0	-1	0	0	0	0	0	0	0
# 37	0	0	0	0	0	0	0	0	1	0	0	-1
# 38	0	0	0	0	0	0	0	1	0	0	-1	0
# 39	0	0	0	0	0	0	1	0	0	-1	0	0
# 40	0	0	0	0	0	1	0	0	-1	0	0	0
# 41	0	0	0	0	1	0	0	-1	0	0	0	0
# 42	0	0	0	1	0	0	-1	0	0	0	0	0
# 43	0	0	1	0	0	-1	0	0	0	0	0	0
# 44	0	1	0	0	-1	0	0	0	0	0	0	0
# 45	1	0	0	-1	0	0	0	0	0	0	0	0
# 46	0	0	0	0	0	0	0	0	0	1	0	-1
# 47	0	0	0	0	0	0	0	0	1	0	-1	0
# 48	0	0	0	0	0	0	0	1	0	-1	0	0
# 49	0	0	0	0	0	0	1	0	-1	0	0	0
# 50	0	0	0	0	0	1	0	-1	0	0	0	0
# 51	0	0	0	0	1	0	-1	0	0	0	0	0
# 52	0	0	0	1	0	-1	0	0	0	0	0	0
# 53	0	0	1	0	-1	0	0	0	0	0	0	0
# 54	0	1	0	-1	0	0	0	0	0	0	0	0
# 55	1	0	-1	0	0	0	0	0	0	0	0	0
# 56	0	0	0	0	0	0	0	0	0	0	1	-1
# 57	0	0	0	0	0	0	0	0	0	1	-1	0
# 58	0	0	0	0	0	0	0	0	1	-1	0	0
# 59	0	0	0	0	0	0	0	1	-1	0	0	0
# 60	0	0	0	0	0	0	1	-1	0	0	0	0
# 61	0	0	0	0	0	1	-1	0	0	0	0	0
# 62	0	0	0	0	1	-1	0	0	0	0	0	0
# 63	0	0	0	1	-1	0	0	0	0	0	0	0
# 64	0	0	1	-1	0	0	0	0	0	0	0	0
# 65	0	1	-1	0	0	0	0	0	0	0	0	0
# 66	1	-1	0	0	0	0	0	0	0	0	0	0

ALGEBRA A12

	OMEGA												ALPHA												EPSILON												
# 1	1	0	0	0	0	0	0	0	0	0	0	1	1	1	1	1	1	1	1	1	1	1	1	1	1	0	0	0	0	0	0	0	0	0	0	0	-1
# 2	-1	1	0	0	0	0	0	0	0	0	0	1	0	1	1	1	1	1	1	1	1	1	1	1	0	1	0	0	0	0	0	0	0	0	0	0	-1
# 3	1	0	0	0	0	0	0	0	0	0	1	-1	1	1	1	1	1	1	1	1	1	1	1	0	1	0	0	0	0	0	0	0	0	0	0	-1	0
# 4	0	-1	1	0	0	0	0	0	0	0	0	1	0	0	1	1	1	1	1	1	1	1	1	1	0	0	1	0	0	0	0	0	0	0	0	0	-1
# 5	-1	1	0	0	0	0	0	0	0	0	1	-1	0	1	1	1	1	1	1	1	1	1	1	0	0	1	0	0	0	0	0	0	0	0	0	-1	0
# 6	1	0	0	0	0	0	0	0	0	1	-1	0	1	1	1	1	1	1	1	1	1	1	0	0	1	0	0	0	0	0	0	0	0	0	-1	0	0
# 7	0	0	-1	1	0	0	0	0	0	0	0	1	0	0	0	1	1	1	1	1	1	1	1	1	0	0	0	1	0	0	0	0	0	0	0	0	-1
# 8	0	-1	1	0	0	0	0	0	0	0	1	-1	0	0	1	1	1	1	1	1	1	1	1	0	0	0	1	0	0	0	0	0	0	0	0	-1	0
# 9	-1	1	0	0	0	0	0	0	0	1	-1	0	0	1	1	1	1	1	1	1	1	1	0	0	0	1	0	0	0	0	0	0	0	0	-1	0	0
# 10	1	0	0	0	0	0	0	0	1	-1	0	0	1	1	1	1	1	1	1	1	1	0	0	0	1	0	0	0	0	0	0	0	0	-1	0	0	0
# 11	0	0	0	-1	1	0	0	0	0	0	0	1	0	0	0	0	1	1	1	1	1	1	1	1	0	0	0	0	1	0	0	0	0	0	0	0	-1
# 12	0	0	-1	1	0	0	0	0	0	0	1	-1	0	0	0	1	1	1	1	1	1	1	1	0	0	0	0	1	0	0	0	0	0	0	0	-1	0
# 13	0	-1	1	0	0	0	0	0	0	1	-1	0	0	0	1	1	1	1	1	1	1	1	0	0	0	0	1	0	0	0	0	0	0	0	-1	0	0
# 14	-1	1	0	0	0	0	0	0	1	-1	0	0	0	1	1	1	1	1	1	1	1	0	0	0	0	1	0	0	0	0	0	0	0	-1	0	0	0
# 15	1	0	0	0	0	0	0	1	-1	0	0	0	1	1	1	1	1	1	1	1	0	0	0	0	1	0	0	0	0	0	0	0	-1	0	0	0	0
# 16	0	0	0	0	-1	1	0	0	0	0	0	1	0	0	0	0	0	1	1	1	1	1	1	1	0	0	0	0	0	1	0	0	0	0	0	0	-1
# 17	0	0	0	-1	1	0	0	0	0	0	1	-1	0	0	0	0	1	1	1	1	1	1	1	0	0	0	0	0	1	0	0	0	0	0	0	-1	0
# 18	0	0	-1	1	0	0	0	0	0	1	-1	0	0	0	0	1	1	1	1	1	1	1	0	0	0	0	0	1	0	0	0	0	0	0	-1	0	0
# 19	0	-1	1	0	0	0	0	0	1	-1	0	0	0	0	1	1	1	1	1	1	1	0	0	0	0	0	1	0	0	0	0	0	0	-1	0	0	0
# 20	-1	1	0	0	0	0	0	1	-1	0	0	0	0	1	1	1	1	1	1	1	0	0	0	0	0	1	0	0	0	0	0	0	-1	0	0	0	0
# 21	1	0	0	0	0	0	1	-1	0	0	0	0	1	1	1	1	1	1	1	0	0	0	0	0	1	0	0	0	0	0	0	-1	0	0	0	0	0
# 22	0	0	0	0	0	-1	1	0	0	0	0	1	0	0	0	0	0	0	1	1	1	1	1	1	0	0	0	0	0	0	1	0	0	0	0	0	-1
# 23	0	0	0	0	-1	1	0	0	0	0	1	-1	0	0	0	0	0	1	1	1	1	1	1	0	0	0	0	0	0	1	0	0	0	0	0	-1	0
# 24	0	0	0	-1	1	0	0	0	0	1	-1	0	0	0	0	0	1	1	1	1	1	1	0	0	0	0	0	0	1	0	0	0	0	0	-1	0	0
# 25	0	0	-1	1	0	0	0	0	1	-1	0	0	0	0	0	1	1	1	1	1	1	0	0	0	0	0	0	1	0	0	0	0	0	-1	0	0	0
# 26	0	-1	1	0	0	0	0	1	-1	0	0	0	0	0	1	1	1	1	1	1	0	0	0	0	0	0	1	0	0	0	0	0	-1	0	0	0	0
# 27	-1	1	0	0	0	0	1	-1	0	0	0	0	0	1	1	1	1	1	1	0	0	0	0	0	0	1	0	0	0	0	0	-1	0	0	0	0	0
# 28	1	0	0	0	0	1	-1	0	0	0	0	0	1	1	1	1	1	1	0	0	0	0	0	0	1	0	0	0	0	0	-1	0	0	0	0	0	0
# 29	0	0	0	0	0	0	-1	1	0	0	0	1	0	0	0	0	0	0	0	1	1	1	1	1	0	0	0	0	0	0	0	1	0	0	0	0	-1
# 30	0	0	0	0	0	-1	1	0	0	0	1	-1	0	0	0	0	0	0	1	1	1	1	1	0	0	0	0	0	0	0	1	0	0	0	0	-1	0
# 31	0	0	0	0	-1	1	0	0	0	1	-1	0	0	0	0	0	0	1	1	1	1	1	0	0	0	0	0	0	0	1	0	0	0	0	-1	0	0
# 32	0	0	0	-1	1	0	0	0	1	-1	0	0	0	0	0	0	1	1	1	1	1	0	0	0	0	0	0	0	1	0	0	0	0	-1	0	0	0
# 33	0	0	-1	1	0	0	0	1	-1	0	0	0	0	0	0	1	1	1	1	1	0	0	0	0	0	0	0	1	0	0	0	0	-1	0	0	0	0
# 34	0	-1	1	0	0	0	1	-1	0	0	0	0	0	0	1	1	1	1	1	0	0	0	0	0	0	0	1	0	0	0	0	-1	0	0	0	0	0
# 35	-1	1	0	0	0	1	-1	0	0	0	0	0	0	1	1	1	1	1	0	0	0	0	0	0	0	1	0	0	0	0	-1	0	0	0	0	0	0
# 36	1	0	0	0	1	-1	0	0	0	0	0	0	1	1	1	1	1	0	0	0	0	0	0	0	1	0	0	0	0	-1	0	0	0	0	0	0	0
# 37	0	0	0	0	0	0	0	-1	1	0	0	1	0	0	0	0	0	0	0	0	1	1	1	1	0	0	0	0	0	0	0	0	1	0	0	0	-1
# 38	0	0	0	0	0	0	-1	1	0	0	1	-1	0	0	0	0	0	0	0	1	1	1	1	0	0	0	0	0	0	0	0	1	0	0	0	-1	0
# 39	0	0	0	0	0	-1	1	0	0	1	-1	0	0	0	0	0	0	0	1	1	1	1	0	0	0	0	0	0	0	0	1	0	0	0	-1	0	0
# 40	0	0	0	0	-1	1	0	0	1	-1	0	0	0	0	0	0	0	1	1	1	1	0	0	0	0	0	0	0	0	1	0	0	0	-1	0	0	0
# 41	0	0	0	-1	1	0	0	1	-1	0	0	0	0	0	0	0	1	1	1	1	0	0	0	0	0	0	0	0	1	0	0	0	-1	0	0	0	0
# 42	0	0	-1	1	0	0	1	-1	0	0	0	0	0	0	0	1	1	1	1	0	0	0	0	0	0	0	0	1	0	0	0	-1	0	0	0	0	0
# 43	0	-1	1	0	0	1	-1	0	0	0	0	0	0	0	1	1	1	1	0	0	0	0	0	0	0	0	1	0	0	0	-1	0	0	0	0	0	0
# 44	-1	1	0	0	1	-1	0	0	0	0	0	0	0	1	1	1	1	0	0	0	0	0	0	0	0	1	0	0	0	-1	0	0	0	0	0	0	0
# 45	1	0	0	1	-1	0	0	0	0	0	0	0	1	1	1	1	0	0	0	0	0	0	0	0	1	0	0	0	-1	0	0	0	0	0	0	0	0
# 46	0	0	0	0	0	0	0	0	-1	1	0	1	0	0	0	0	0	0	0	0	0	1	1	1	0	0	0	0	0	0	0	0	0	1	0	0	-1
# 47	0	0	0	0	0	0	0	-1	1	0	1	-1	0	0	0	0	0	0	0	0	1	1	1	0	0	0	0	0	0	0	0	0	1	0	0	-1	0
# 48	0	0	0	0	0	0	-1	1	0	1	-1	0	0	0	0	0	0	0	0	1	1	1	0	0	0	0	0	0	0	0	0	1	0	0	-1	0	0
# 49	0	0	0	0	0	-1	1	0	1	-1	0	0	0	0	0	0	0	0	1	1	1	0	0	0	0	0	0	0	0	0	1	0	0	-1	0	0	0
# 50	0	0	0	0	-1	1	0	1	-1	0	0	0	0	0	0	0	0	1	1	1	0	0	0	0	0	0	0	0	0	1	0	0	-1	0	0	0	0
# 51	0	0	0	-1	1	0	1	-1	0	0	0	0	0	0	0	0	1	1	1	0	0	0	0	0	0	0	0	0	1	0	0	-1	0	0	0	0	0
# 52	0	0	-1	1	0	1	-1	0	0	0	0	0	0	0	0	1	1	1	0	0	0	0	0	0	0	0	0	1	0	0	-1	0	0	0	0	0	0
# 53	0	-1	1	0	1	-1	0	0	0	0	0	0	0	0	1	1	1	0	0	0	0	0	0	0	0	0	1	0	0	-1	0	0	0	0	0	0	0
# 54	-1	1	0	1	-1	0	0	0	0	0	0	0	0	1	1	1	0	0	0	0	0	0	0	0	0	1	0	0	-1	0	0	0	0	0	0	0	0
# 55	1	0	1	-1	0	0	0	0	0	0	0	0	1	1	1	0	0	0	0	0	0	0	0	0	1	0	0	-1	0	0	0	0	0	0	0	0	0
# 56	0	0	0	0	0	0	0	0	0	-1	1	1	0	0	0	0	0	0	0	0	0	0	1	1	0	0	0	0	0	0	0	0	0	0	1	0	-1
# 57	0	0	0	0	0	0	0	0	-1	1	1	-1	0	0	0	0	0	0	0	0	0	1	1	0	0	0	0	0	0	0	0	0	0	1	0	-1	0
# 58	0	0	0	0	0	0	0	-1	1	1	-1	0	0	0	0	0	0	0	0	0	1	1	0	0	0	0	0	0	0	0	0	0	1	0	-1	0	0
# 59	0	0	0	0	0	0	-1	1	1	-1	0	0	0	0	0	0	0	0	0	1	1	0	0	0	0	0	0	0	0	0	0	1	0	-1	0	0	0
# 60	0	0	0	0	0	-1	1	1	-1	0	0	0	0	0	0	0	0	0	1	1	0	0	0	0	0	0	0	0	0	0	1	0	-1	0	0	0	0
# 61	0	0	0	0	-1	1	1	-1	0	0	0	0	0	0	0	0	0	1	1	0	0	0	0	0	0	0	0	0	0	1	0	-1	0	0	0	0	0
# 62	0	0	0	-1	1	1	-1	0	0	0	0	0	0	0	0	0	1	1	0	0	0	0	0	0	0	0	0	0	1	0	-1	0	0	0	0	0	0
# 63	0	0	-1	1	1	-1	0	0	0	0	0	0	0	0	0	1	1	0	0	0	0	0	0	0	0	0	0	1	0	-1	0	0	0	0	0	0	0
# 64	0	-1	1	1	-1	0	0	0	0	0	0	0	0	0	1	1	0	0	0	0	0	0	0	0	0	0	1	0	-1	0	0	0	0	0	0	0	0
# 65	-1	1	1	-1	0	0	0	0	0	0	0	0	0	1	1	0	0	0	0	0	0	0	0	0	0	1	0	-1	0	0	0	0	0	0	0	0	0
# 66	1	1	-1	0	0	0	0	0	0	0	0	0	1	1	0	0	0	0	0	0	0	0	0	0	1	0	-1	0	0	0	0	0	0	0	0	0	0
# 67	0	0	0	0	0	0	0	0	0	0	-1	2	0	0	0	0	0	0	0	0	0	0	0	1	0	0	0	0	0	0	0	0	0	0	0	1	-1
# 68	0	0	0	0	0	0	0	0	0	-1	2	-1	0	0	0	0	0	0	0	0	0	0	1	0	0	0	0	0	0	0	0	0	0	0	1	-1	0
# 69	0	0	0	0	0	0	0	0	-1	2	-1	0	0	0	0	0	0	0	0	0	0	1	0	0	0	0	0	0	0	0	0	0	0	1	-1	0	0
# 70	0	0	0	0	0	0	0	-1	2	-1	0	0	0	0	0	0	0	0	0	0	1	0	0	0	0	0	0	0	0	0	0	0	1	-1	0	0	0
# 71	0	0	0	0	0	0	-1	2	-1	0	0	0	0	0	0	0	0	0	0	1	0	0	0	0	0	0	0	0	0	0	0	1	-1	0	0	0	0
# 72	0	0	0	0	0	-1	2	-1	0	0	0	0	0	0	0	0	0	0	1	0	0	0	0	0	0	0	0	0	0	0	1	-1	0	0	0	0	0
# 73	0	0	0	0	-1	2	-1	0	0	0	0	0	0	0	0	0	0	1	0	0	0	0	0	0	0	0	0	0	0	1	-1	0	0	0	0	0	0
# 74	0	0	0	-1	2	-1	0	0	0	0	0	0	0	0	0	0	1	0	0	0	0	0	0	0	0	0	0	0	1	-1	0	0	0	0	0	0	0
# 75	0	0	-1	2	-1	0	0	0	0	0	0	0	0	0	0	1	0	0	0	0	0	0	0	0	0	0	0	1	-1	0	0	0	0	0	0	0	0
# 76	0	-1	2	-1	0	0	0	0	0	0	0	0	0	0	1	0	0	0	0	0	0	0	0	0	0	0	1	-1	0	0	0	0	0	0	0	0	0
# 77	-1	2	-1	0	0	0	0	0	0	0	0	0	0	1	0	0	0	0	0	0	0	0	0	0	0	1	-1	0	0	0	0	0	0	0	0	0	0
# 78	2	-1	0	0	0	0	0	0	0	0	0	0	1	0	0	0	0	0	0	0	0	0	0	0	1	-1	0	0	0	0	0	0	0	0	0	0	0

Table 2. *(continued)*

ALGEBRA B2

#	OMEGA		ALPHA		EPSILON	
# 1	0	2	1	2	1	1
# 2	1	0	1	1	1	0
# 3	2	-2	1	0	1	-1
# 4	-1	2	0	1	0	1

ALGEBRA B3

#	OMEGA			ALPHA			EPSILON		
# 1	0	1	0	1	2	2	1	1	0
# 2	1	-1	2	1	1	2	1	0	1
# 3	-1	0	2	0	1	2	0	1	1
# 4	1	0	0	1	1	1	1	0	0
# 5	1	1	-2	1	1	0	1	0	-1
# 6	-1	1	0	0	1	1	0	1	0
# 7	-1	2	-2	0	1	0	0	1	-1
# 8	2	-1	0	1	0	0	1	-1	0
# 9	0	-1	2	0	0	1	0	0	1

ALGEBRA B4

#	OMEGA				ALPHA				EPSILON			
# 1	0	1	0	0	1	2	2	2	1	1	0	0
# 2	1	-1	1	0	1	1	2	2	1	0	1	0
# 3	-1	0	1	0	0	1	2	2	0	1	1	0
# 4	1	0	-1	2	1	1	1	2	1	0	0	1
# 5	-1	1	-1	2	0	1	1	2	0	1	0	1
# 6	1	0	0	0	1	1	1	1	1	0	0	0
# 7	1	0	1	-2	1	1	1	0	1	0	0	-1
# 8	0	-1	0	2	0	0	1	2	0	0	1	1
# 9	-1	1	0	0	0	1	1	1	0	1	0	0
# 10	-1	1	1	-2	0	1	1	0	0	1	0	-1
# 11	1	1	-1	0	1	1	0	0	1	0	-1	0
# 12	0	-1	1	0	0	0	1	1	0	0	1	0
# 13	0	-1	2	-2	0	0	1	0	0	0	1	-1
# 14	-1	2	-1	0	0	1	0	0	0	1	-1	0
# 15	2	-1	0	0	1	0	0	0	1	-1	0	0
# 16	0	0	-1	2	0	0	0	1	0	0	0	1

ALGEBRA B5

#	OMEGA					ALPHA					EPSILON				
# 1	0	1	0	0	0	1	2	2	2	2	1	1	0	0	0
# 2	1	-1	1	0	0	1	1	2	2	2	1	0	1	0	0
# 3	-1	0	1	0	0	0	1	2	2	2	0	1	1	0	0
# 4	1	0	-1	1	0	1	1	1	2	2	1	0	0	1	0
# 5	-1	1	-1	1	0	0	1	1	2	2	0	1	0	1	0
# 6	1	0	0	-1	2	1	1	1	1	2	1	0	0	0	1
# 7	0	-1	0	1	0	0	0	1	2	2	0	0	1	1	0
# 8	-1	1	0	-1	2	0	1	1	1	2	0	1	0	0	1
# 9	1	0	0	0	0	1	1	1	1	1	1	0	0	0	0
# 10	1	0	0	1	-2	1	1	1	1	0	1	0	0	0	-1
# 11	0	-1	1	-1	2	0	0	1	1	2	0	0	1	0	1
# 12	-1	1	0	0	0	0	1	1	1	1	0	1	0	0	0
# 13	-1	1	0	1	-2	0	1	1	1	0	0	1	0	0	-1
# 14	1	0	1	-1	0	1	1	1	0	0	1	0	0	-1	0
# 15	0	0	-1	0	2	0	0	0	1	2	0	0	0	1	1
# 16	0	-1	1	0	0	0	0	1	1	1	0	0	1	0	0
# 17	0	-1	1	1	-2	0	0	1	1	0	0	0	1	0	-1
# 18	-1	1	1	-1	0	0	1	1	0	0	0	1	0	-1	0
# 19	1	1	-1	0	0	1	1	0	0	0	1	0	-1	0	0
# 20	0	0	-1	1	0	0	0	0	1	1	0	0	0	1	0
# 21	0	0	-1	2	-2	0	0	0	1	0	0	0	0	1	-1
# 22	0	-1	2	-1	0	0	0	1	0	0	0	0	1	-1	0
# 23	-1	2	-1	0	0	0	1	0	0	0	0	1	-1	0	0
# 24	2	-1	0	0	0	1	0	0	0	0	1	-1	0	0	0
# 25	0	0	0	-1	2	0	0	0	0	1	0	0	0	0	1

ALGEBRA B6

#	OMEGA						ALPHA						EPSILON					
# 1	0	1	0	0	0	0	1	2	2	2	2	2	1	1	0	0	0	0
# 2	1	-1	1	0	0	0	1	1	2	2	2	2	1	0	1	0	0	0
# 3	-1	0	1	0	0	0	0	1	2	2	2	2	0	1	1	0	0	0
# 4	1	0	-1	1	0	0	1	1	1	2	2	2	1	0	0	1	0	0
# 5	-1	1	-1	1	0	0	0	1	1	2	2	2	0	1	0	1	0	0
# 6	1	0	0	-1	1	0	1	1	1	1	2	2	1	0	0	0	1	0
# 7	0	-1	0	1	0	0	0	0	1	2	2	2	0	0	1	1	0	0
# 8	-1	1	0	-1	1	0	0	1	1	1	2	2	0	1	0	0	1	0
# 9	1	0	0	0	-1	2	1	1	1	1	1	2	1	0	0	0	0	1
# 10	0	-1	1	-1	1	0	0	0	1	1	2	2	0	0	1	0	1	0
# 11	-1	1	0	0	-1	2	0	1	1	1	1	2	0	1	0	0	0	1
# 12	1	0	0	0	0	0	1	1	1	1	1	1	1	0	0	0	0	0
# 13	1	0	0	0	1	-2	1	1	1	1	1	0	1	0	0	0	0	-1
# 14	0	0	-1	0	1	0	0	0	0	1	2	2	0	0	0	1	1	0
# 15	0	-1	1	0	-1	2	0	0	1	1	1	2	0	0	1	0	0	1
# 16	-1	1	0	0	0	0	0	1	1	1	1	1	0	1	0	0	0	0
# 17	-1	1	0	0	1	-2	0	1	1	1	1	0	0	1	0	0	0	-1
# 18	1	0	0	1	-1	0	1	1	1	1	0	0	1	0	0	0	-1	0
# 19	0	0	-1	1	-1	2	0	0	0	1	1	2	0	0	0	1	0	1
# 20	0	-1	1	0	0	0	0	0	1	1	1	1	0	0	1	0	0	0
# 21	0	-1	1	0	1	-2	0	0	1	1	1	0	0	0	1	0	0	-1
# 22	-1	1	0	1	-1	0	0	1	1	1	0	0	0	1	0	0	-1	0
# 23	1	0	1	-1	0	0	1	1	1	0	0	0	1	0	0	-1	0	0
# 24	0	0	0	-1	0	2	0	0	0	0	1	2	0	0	0	0	1	1
# 25	0	0	-1	1	0	0	0	0	0	1	1	1	0	0	0	1	0	0
# 26	0	0	-1	1	1	-2	0	0	0	1	1	0	0	0	0	1	0	-1
# 27	0	-1	1	1	-1	0	0	0	1	1	0	0	0	0	1	0	-1	0
# 28	-1	1	1	-1	0	0	0	1	1	0	0	0	0	1	0	-1	0	0
# 29	1	1	-1	0	0	0	1	1	0	0	0	0	1	0	-1	0	0	0
# 30	0	0	0	-1	1	0	0	0	0	0	1	1	0	0	0	0	1	0
# 31	0	0	0	-1	2	-2	0	0	0	0	1	0	0	0	0	0	1	-1
# 32	0	0	-1	2	-1	0	0	0	0	1	0	0	0	0	0	1	-1	0
# 33	0	-1	2	-1	0	0	0	0	1	0	0	0	0	0	1	-1	0	0
# 34	-1	2	-1	0	0	0	0	1	0	0	0	0	0	1	-1	0	0	0
# 35	2	-1	0	0	0	0	1	0	0	0	0	0	1	-1	0	0	0	0
# 36	0	0	0	0	-1	2	0	0	0	0	0	1	0	0	0	0	0	1

Table 2. *(continued)*

ALGEBRA B7

	OMEGA							ALPHA							EPSILON						
# 1	0	1	0	0	0	0	0	1	2	2	2	2	2	2	1	1	0	0	0	0	0
# 2	1	-1	1	0	0	0	0	1	1	2	2	2	2	2	1	0	1	0	0	0	0
# 3	-1	0	1	0	0	0	0	0	1	2	2	2	2	2	0	1	1	0	0	0	0
# 4	1	0	-1	1	0	0	0	1	1	1	2	2	2	2	1	0	0	1	0	0	0
# 5	-1	1	-1	1	0	0	0	0	1	1	2	2	2	2	0	1	0	1	0	0	0
# 6	1	0	0	-1	1	0	0	1	1	1	1	2	2	2	1	0	0	0	1	0	0
# 7	0	-1	0	1	0	0	0	0	0	1	2	2	2	2	0	0	1	1	0	0	0
# 8	-1	1	0	-1	1	0	0	0	1	1	1	2	2	2	0	1	0	0	1	0	0
# 9	1	0	0	0	-1	1	0	1	1	1	1	1	2	2	1	0	0	0	0	1	0
# 10	0	-1	1	-1	1	0	0	0	0	1	1	2	2	2	0	0	1	0	1	0	0
# 11	-1	1	0	0	-1	1	0	0	1	1	1	1	2	2	0	1	0	0	0	1	0
# 12	1	0	0	0	0	-1	2	1	1	1	1	1	1	2	1	0	0	0	0	0	1
# 13	0	0	-1	0	1	0	0	0	0	0	1	2	2	2	0	0	0	1	1	0	0
# 14	0	-1	1	0	-1	1	0	0	0	1	1	1	2	2	0	0	1	0	0	1	0
# 15	-1	1	0	0	0	-1	2	0	1	1	1	1	1	2	0	1	0	0	0	0	1
# 16	1	0	0	0	0	0	0	1	1	1	1	1	1	1	1	0	0	0	0	0	0
# 17	1	0	0	0	0	1	-2	1	1	1	1	1	1	0	1	0	0	0	0	0	-1
# 18	0	0	-1	1	-1	1	0	0	0	0	1	1	2	2	0	0	0	1	0	1	0
# 19	0	-1	1	0	0	-1	2	0	0	1	1	1	1	2	0	0	1	0	0	0	1
# 20	-1	1	0	0	0	0	0	0	1	1	1	1	1	1	0	1	0	0	0	0	0
# 21	-1	1	0	0	0	1	-2	0	1	1	1	1	1	0	0	1	0	0	0	0	-1
# 22	1	0	0	0	1	-1	0	1	1	1	1	1	0	0	1	0	0	0	0	-1	0
# 23	0	0	0	-1	0	1	0	0	0	0	0	1	2	2	0	0	0	0	1	1	0
# 24	0	0	-1	1	0	-1	2	0	0	0	1	1	1	2	0	0	0	1	0	0	1
# 25	0	-1	1	0	0	0	0	0	0	1	1	1	1	1	0	0	1	0	0	0	0
# 26	0	-1	1	0	0	1	-2	0	0	1	1	1	1	0	0	0	1	0	0	0	-1
# 27	-1	1	0	0	1	-1	0	0	1	1	1	1	0	0	0	1	0	0	0	-1	0
# 28	1	0	0	1	-1	0	0	1	1	1	1	0	0	0	1	0	0	0	-1	0	0
# 29	0	0	0	-1	1	-1	2	0	0	0	0	1	1	2	0	0	0	0	1	0	1
# 30	0	0	-1	1	0	0	0	0	0	0	1	1	1	1	0	0	0	1	0	0	0
# 31	0	0	-1	1	0	1	-2	0	0	0	1	1	1	0	0	0	0	1	0	0	-1
# 32	0	-1	1	0	1	-1	0	0	0	1	1	1	0	0	0	0	1	0	0	-1	0
# 33	-1	1	0	1	-1	0	0	0	1	1	1	0	0	0	0	1	0	0	-1	0	0
# 34	1	0	1	-1	0	0	0	1	1	1	0	0	0	0	1	0	0	-1	0	0	0
# 35	0	0	0	0	-1	0	2	0	0	0	0	0	1	2	0	0	0	0	0	1	1
# 36	0	0	0	-1	1	0	0	0	0	0	0	1	1	1	0	0	0	0	1	0	0
# 37	0	0	0	-1	1	1	-2	0	0	0	0	1	1	0	0	0	0	0	1	0	-1
# 38	0	0	-1	1	1	-1	0	0	0	0	1	1	0	0	0	0	0	1	0	-1	0
# 39	0	-1	1	1	-1	0	0	0	0	1	1	0	0	0	0	0	1	0	-1	0	0
# 40	-1	1	1	-1	0	0	0	0	1	1	0	0	0	0	0	1	0	-1	0	0	0
# 41	1	1	-1	0	0	0	0	1	1	0	0	0	0	0	1	0	-1	0	0	0	0
# 42	0	0	0	0	-1	1	0	0	0	0	0	0	1	1	0	0	0	0	0	1	0
# 43	0	0	0	0	-1	2	-2	0	0	0	0	0	1	0	0	0	0	0	0	1	-1
# 44	0	0	0	-1	2	-1	0	0	0	0	0	1	0	0	0	0	0	0	1	-1	0
# 45	0	0	-1	2	-1	0	0	0	0	0	1	0	0	0	0	0	0	1	-1	0	0
# 46	0	-1	2	-1	0	0	0	0	0	1	0	0	0	0	0	0	1	-1	0	0	0
# 47	-1	2	-1	0	0	0	0	0	1	0	0	0	0	0	0	1	-1	0	0	0	0
# 48	2	-1	0	0	0	0	0	1	0	0	0	0	0	0	1	-1	0	0	0	0	0
# 49	0	0	0	0	0	-1	2	0	0	0	0	0	0	1	0	0	0	0	0	0	1

ALGEBRA B8

#	OMEGA								ALPHA								EPSILON							
# 1	0	1	0	0	0	0	0	0	1	2	2	2	2	2	2	2	1	1	0	0	0	0	0	0
# 2	1	-1	1	0	0	0	0	0	1	1	2	2	2	2	2	2	1	0	1	0	0	0	0	0
# 3	-1	0	1	0	0	0	0	0	0	1	2	2	2	2	2	2	0	1	1	0	0	0	0	0
# 4	1	0	-1	1	0	0	0	0	1	1	1	2	2	2	2	2	1	0	0	1	0	0	0	0
# 5	-1	1	-1	1	0	0	0	0	0	1	1	2	2	2	2	2	0	1	0	1	0	0	0	0
# 6	1	0	0	-1	1	0	0	0	1	1	1	1	2	2	2	2	1	0	0	0	1	0	0	0
# 7	0	-1	0	1	0	0	0	0	0	0	1	2	2	2	2	2	0	0	1	1	0	0	0	0
# 8	-1	1	0	-1	1	0	0	0	0	1	1	1	2	2	2	2	0	1	0	0	1	0	0	0
# 9	1	0	0	0	-1	1	0	0	1	1	1	1	1	2	2	2	1	0	0	0	0	1	0	0
# 10	0	-1	1	-1	1	0	0	0	0	0	1	1	2	2	2	2	0	0	1	0	1	0	0	0
# 11	-1	1	0	0	-1	1	0	0	0	1	1	1	1	2	2	2	0	1	0	0	0	1	0	0
# 12	1	0	0	0	0	-1	1	0	1	1	1	1	1	1	2	2	1	0	0	0	0	0	1	0
# 13	0	0	-1	0	1	0	0	0	0	0	0	1	2	2	2	2	0	0	0	1	1	0	0	0
# 14	0	-1	1	0	-1	1	0	0	0	0	1	1	1	2	2	2	0	0	1	0	0	1	0	0
# 15	-1	1	0	0	0	-1	1	0	0	1	1	1	1	1	2	2	0	1	0	0	0	0	1	0
# 16	1	0	0	0	0	0	-1	2	1	1	1	1	1	1	1	2	1	0	0	0	0	0	0	1
# 17	0	0	-1	1	-1	1	0	0	0	0	0	1	1	2	2	2	0	0	0	1	0	1	0	0
# 18	0	-1	1	0	0	-1	1	0	0	0	1	1	1	1	2	2	0	0	1	0	0	0	1	0
# 19	-1	1	0	0	0	0	-1	2	0	1	1	1	1	1	1	2	0	1	0	0	0	0	0	1
# 20	1	0	0	0	0	0	0	0	1	1	1	1	1	1	1	1	1	0	0	0	0	0	0	0
# 21	1	0	0	0	0	0	1	-2	1	1	1	1	1	1	1	0	1	0	0	0	0	0	0	-1
# 22	0	0	0	-1	0	1	0	0	0	0	0	0	1	2	2	2	0	0	0	0	1	1	0	0
# 23	0	0	-1	1	0	-1	1	0	0	0	0	1	1	1	2	2	0	0	0	1	0	0	1	0
# 24	0	-1	1	0	0	0	-1	2	0	0	1	1	1	1	1	2	0	0	1	0	0	0	0	1
# 25	-1	1	0	0	0	0	0	0	0	1	1	1	1	1	1	1	0	1	0	0	0	0	0	0
# 26	-1	1	0	0	0	0	1	-2	0	1	1	1	1	1	1	0	0	1	0	0	0	0	0	-1
# 27	1	0	0	0	0	1	-1	0	1	1	1	1	1	1	0	0	1	0	0	0	0	0	-1	0
# 28	0	0	0	-1	1	-1	1	0	0	0	0	0	1	1	2	2	0	0	0	0	1	0	1	0
# 29	0	0	-1	1	0	0	-1	2	0	0	0	1	1	1	1	2	0	0	0	1	0	0	0	1
# 30	0	-1	1	0	0	0	0	0	0	0	1	1	1	1	1	1	0	0	1	0	0	0	0	0
# 31	0	-1	1	0	0	0	1	-2	0	0	1	1	1	1	1	0	0	0	1	0	0	0	0	-1
# 32	-1	1	0	0	0	1	-1	0	0	1	1	1	1	1	0	0	0	1	0	0	0	0	-1	0
# 33	1	0	0	0	1	-1	0	0	1	1	1	1	1	0	0	0	1	0	0	0	0	-1	0	0
# 34	0	0	0	0	-1	0	1	0	0	0	0	0	0	1	2	2	0	0	0	0	0	1	1	0
# 35	0	0	0	-1	1	0	-1	2	0	0	0	0	1	1	1	2	0	0	0	0	1	0	0	1
# 36	0	0	-1	1	0	0	0	0	0	0	0	1	1	1	1	1	0	0	0	1	0	0	0	0
# 37	0	0	-1	1	0	0	1	-2	0	0	0	1	1	1	1	0	0	0	0	1	0	0	0	-1
# 38	0	-1	1	0	0	1	-1	0	0	0	1	1	1	1	0	0	0	0	1	0	0	0	-1	0
# 39	-1	1	0	0	1	-1	0	0	0	1	1	1	1	0	0	0	0	1	0	0	0	-1	0	0
# 40	1	0	0	1	-1	0	0	0	1	1	1	1	0	0	0	0	1	0	0	0	-1	0	0	0
# 41	0	0	0	0	-1	1	-1	2	0	0	0	0	0	1	1	2	0	0	0	0	0	1	0	1
# 42	0	0	0	-1	1	0	0	0	0	0	0	0	1	1	1	1	0	0	0	0	1	0	0	0
# 43	0	0	0	-1	1	0	1	-2	0	0	0	0	1	1	1	0	0	0	0	0	1	0	0	-1
# 44	0	0	-1	1	0	1	-1	0	0	0	0	1	1	1	0	0	0	0	0	1	0	0	-1	0
# 45	0	-1	1	0	1	-1	0	0	0	0	1	1	1	0	0	0	0	0	1	0	0	-1	0	0
# 46	-1	1	0	1	-1	0	0	0	0	1	1	1	0	0	0	0	0	1	0	0	-1	0	0	0
# 47	1	0	1	-1	0	0	0	0	1	1	1	0	0	0	0	0	1	0	0	-1	0	0	0	0
# 48	0	0	0	0	0	-1	0	2	0	0	0	0	0	0	1	2	0	0	0	0	0	0	1	1
# 49	0	0	0	0	-1	1	0	0	0	0	0	0	0	1	1	1	0	0	0	0	0	1	0	0
# 50	0	0	0	0	-1	1	1	-2	0	0	0	0	0	1	1	0	0	0	0	0	0	1	0	-1
# 51	0	0	0	-1	1	1	-1	0	0	0	0	0	1	1	0	0	0	0	0	0	1	0	-1	0
# 52	0	0	-1	1	1	-1	0	0	0	0	0	1	1	0	0	0	0	0	0	1	0	-1	0	0
# 53	0	-1	1	1	-1	0	0	0	0	0	1	1	0	0	0	0	0	0	1	0	-1	0	0	0
# 54	-1	1	1	-1	0	0	0	0	0	1	1	0	0	0	0	0	0	1	0	-1	0	0	0	0
# 55	1	1	-1	0	0	0	0	0	1	1	0	0	0	0	0	0	1	0	-1	0	0	0	0	0
# 56	0	0	0	0	0	-1	1	0	0	0	0	0	0	0	1	1	0	0	0	0	0	0	1	0
# 57	0	0	0	0	0	-1	2	-2	0	0	0	0	0	0	1	0	0	0	0	0	0	0	1	-1
# 58	0	0	0	0	-1	2	-1	0	0	0	0	0	0	1	0	0	0	0	0	0	0	1	-1	0
# 59	0	0	0	-1	2	-1	0	0	0	0	0	0	1	0	0	0	0	0	0	0	1	-1	0	0
# 60	0	0	-1	2	-1	0	0	0	0	0	0	1	0	0	0	0	0	0	0	1	-1	0	0	0
# 61	0	-1	2	-1	0	0	0	0	0	0	1	0	0	0	0	0	0	0	1	-1	0	0	0	0
# 62	-1	2	-1	0	0	0	0	0	0	1	0	0	0	0	0	0	0	1	-1	0	0	0	0	0
# 63	2	-1	0	0	0	0	0	0	1	0	0	0	0	0	0	0	1	-1	0	0	0	0	0	0
# 64	0	0	0	0	0	0	-1	2	0	0	0	0	0	0	0	1	0	0	0	0	0	0	0	1

Table 2. *(continued)*

ALGEBRA B9

	OMEGA									ALPHA									EPSILON								
# 1	0	1	0	0	0	0	0	0	0	1	2	2	2	2	2	2	2	2	1	1	0	0	0	0	0	0	0
# 2	1	-1	1	0	0	0	0	0	0	1	1	2	2	2	2	2	2	2	1	0	1	0	0	0	0	0	0
# 3	-1	0	1	0	0	0	0	0	0	0	1	2	2	2	2	2	2	2	0	1	1	0	0	0	0	0	0
# 4	1	0	-1	1	0	0	0	0	0	1	1	1	2	2	2	2	2	2	1	0	0	1	0	0	0	0	0
# 5	-1	1	-1	1	0	0	0	0	0	0	1	1	2	2	2	2	2	2	0	1	0	1	0	0	0	0	0
# 6	1	0	0	-1	1	0	0	0	0	1	1	1	1	2	2	2	2	2	1	0	0	0	1	0	0	0	0
# 7	0	-1	0	1	0	0	0	0	0	0	0	1	2	2	2	2	2	2	0	0	1	1	0	0	0	0	0
# 8	-1	1	0	-1	1	0	0	0	0	0	1	1	1	2	2	2	2	2	0	1	0	0	1	0	0	0	0
# 9	1	0	0	0	-1	1	0	0	0	1	1	1	1	1	2	2	2	2	1	0	0	0	0	1	0	0	0
# 10	0	-1	1	-1	1	0	0	0	0	0	0	1	1	2	2	2	2	2	0	0	1	0	1	0	0	0	0
# 11	-1	1	0	0	-1	1	0	0	0	0	1	1	1	1	2	2	2	2	0	1	0	0	0	1	0	0	0
# 12	1	0	0	0	0	-1	1	0	0	1	1	1	1	1	1	2	2	2	1	0	0	0	0	0	1	0	0
# 13	0	0	-1	0	1	0	0	0	0	0	0	0	1	2	2	2	2	2	0	0	0	1	1	0	0	0	0
# 14	0	-1	1	0	-1	1	0	0	0	0	0	1	1	1	2	2	2	2	0	0	1	0	0	1	0	0	0
# 15	-1	1	0	0	0	-1	1	0	0	0	1	1	1	1	1	2	2	2	0	1	0	0	0	0	1	0	0
# 16	1	0	0	0	0	0	-1	1	0	1	1	1	1	1	1	1	2	2	1	0	0	0	0	0	0	1	0
# 17	0	0	-1	1	-1	1	0	0	0	0	0	0	1	1	2	2	2	2	0	0	0	1	0	1	0	0	0
# 18	0	-1	1	0	0	-1	1	0	0	0	0	1	1	1	1	2	2	2	0	0	1	0	0	0	1	0	0
# 19	-1	1	0	0	0	0	-1	1	0	0	1	1	1	1	1	1	2	2	0	1	0	0	0	0	0	1	0
# 20	1	0	0	0	0	0	0	-1	2	1	1	1	1	1	1	1	1	2	1	0	0	0	0	0	0	0	1
# 21	0	0	0	-1	0	1	0	0	0	0	0	0	0	1	2	2	2	2	0	0	0	0	1	1	0	0	0
# 22	0	0	-1	1	0	-1	1	0	0	0	0	0	1	1	1	2	2	2	0	0	0	1	0	0	1	0	0
# 23	0	-1	1	0	0	0	-1	1	0	0	0	1	1	1	1	1	2	2	0	0	1	0	0	0	0	1	0
# 24	-1	1	0	0	0	0	0	-1	2	0	1	1	1	1	1	1	1	2	0	1	0	0	0	0	0	0	1
# 25	1	0	0	0	0	0	0	0	0	1	1	1	1	1	1	1	1	1	1	0	0	0	0	0	0	0	0
# 26	1	0	0	0	0	0	0	1	-2	1	1	1	1	1	1	1	1	0	1	0	0	0	0	0	0	0	-1
# 27	0	0	0	-1	1	-1	1	0	0	0	0	0	0	1	1	2	2	2	0	0	0	0	1	0	1	0	0
# 28	0	0	-1	1	0	0	-1	1	0	0	0	0	1	1	1	1	2	2	0	0	0	1	0	0	0	1	0
# 29	0	-1	1	0	0	0	0	-1	2	0	0	1	1	1	1	1	1	2	0	0	1	0	0	0	0	0	1
# 30	-1	1	0	0	0	0	0	0	0	0	1	1	1	1	1	1	1	1	0	1	0	0	0	0	0	0	0
# 31	-1	1	0	0	0	0	0	1	-2	0	1	1	1	1	1	1	1	0	0	1	0	0	0	0	0	0	-1
# 32	1	0	0	0	0	0	1	-1	0	1	1	1	1	1	1	1	0	0	1	0	0	0	0	0	0	-1	0
# 33	0	0	0	0	-1	0	1	0	0	0	0	0	0	0	1	2	2	2	0	0	0	0	0	1	1	0	0
# 34	0	0	0	-1	1	0	-1	1	0	0	0	0	0	1	1	1	2	2	0	0	0	0	1	0	0	1	0
# 35	0	0	-1	1	0	0	0	-1	2	0	0	0	1	1	1	1	1	2	0	0	0	1	0	0	0	0	1
# 36	0	-1	1	0	0	0	0	0	0	0	0	1	1	1	1	1	1	1	0	0	1	0	0	0	0	0	0
# 37	0	-1	1	0	0	0	0	1	-2	0	0	1	1	1	1	1	1	0	0	0	1	0	0	0	0	0	-1

# 38	-1	1	0	0	0	0	1	-1	0	0	1	1	1	1	1	1	0	0	0	1	0	0	0	0	0	-1	0
# 39	1	0	0	0	0	1	-1	0	0	1	1	1	1	1	1	0	0	0	1	0	0	0	0	0	-1	0	0
# 40	0	0	0	0	-1	1	-1	1	0	0	0	0	0	0	1	1	2	2	0	0	0	0	0	1	0	1	0
# 41	0	0	0	-1	1	0	0	-1	2	0	0	0	0	1	1	1	1	2	0	0	0	0	1	0	0	0	1
# 42	0	0	-1	1	0	0	0	0	0	0	0	0	1	1	1	1	1	1	0	0	0	1	0	0	0	0	0
# 43	0	0	-1	1	0	0	0	1	-2	0	0	0	1	1	1	1	1	0	0	0	0	1	0	0	0	0	-1
# 44	0	-1	1	0	0	0	1	-1	0	0	0	1	1	1	1	1	0	0	0	0	1	0	0	0	0	-1	0
# 45	-1	1	0	0	0	1	-1	0	0	0	1	1	1	1	1	0	0	0	0	1	0	0	0	0	-1	0	0
# 46	1	0	0	0	1	-1	0	0	0	1	1	1	1	1	0	0	0	0	1	0	0	0	0	-1	0	0	0
# 47	0	0	0	0	0	-1	0	1	0	0	0	0	0	0	0	1	2	2	0	0	0	0	0	0	1	1	0
# 48	0	0	0	0	-1	1	0	-1	2	0	0	0	0	0	1	1	1	2	0	0	0	0	0	1	0	0	1
# 49	0	0	0	-1	1	0	0	0	0	0	0	0	0	1	1	1	1	1	0	0	0	0	1	0	0	0	0
# 50	0	0	0	-1	1	0	0	1	-2	0	0	0	0	1	1	1	1	0	0	0	0	0	1	0	0	0	-1
# 51	0	0	-1	1	0	0	1	-1	0	0	0	0	1	1	1	1	0	0	0	0	0	1	0	0	0	-1	0
# 52	0	-1	1	0	0	1	-1	0	0	0	0	1	1	1	1	0	0	0	0	0	1	0	0	0	-1	0	0
# 53	-1	1	0	0	1	-1	0	0	0	0	1	1	1	1	0	0	0	0	0	1	0	0	0	-1	0	0	0
# 54	1	0	0	1	-1	0	0	0	0	1	1	1	1	0	0	0	0	0	1	0	0	0	-1	0	0	0	0
# 55	0	0	0	0	0	-1	1	-1	2	0	0	0	0	0	0	1	1	2	0	0	0	0	0	0	1	0	1
# 56	0	0	0	0	-1	1	0	0	0	0	0	0	0	0	1	1	1	1	0	0	0	0	0	1	0	0	0
# 57	0	0	0	0	-1	1	0	1	-2	0	0	0	0	0	1	1	1	0	0	0	0	0	0	1	0	0	-1
# 58	0	0	0	-1	1	0	1	-1	0	0	0	0	0	1	1	1	0	0	0	0	0	0	1	0	0	-1	0
# 59	0	0	-1	1	0	1	-1	0	0	0	0	0	1	1	1	0	0	0	0	0	0	1	0	0	-1	0	0
# 60	0	-1	1	0	1	-1	0	0	0	0	0	1	1	1	0	0	0	0	0	0	1	0	0	-1	0	0	0
# 61	-1	1	0	1	-1	0	0	0	0	0	1	1	1	0	0	0	0	0	0	1	0	0	-1	0	0	0	0
# 62	1	0	1	-1	0	0	0	0	0	1	1	1	0	0	0	0	0	0	1	0	0	-1	0	0	0	0	0
# 63	0	0	0	0	0	0	-1	0	2	0	0	0	0	0	0	0	1	2	0	0	0	0	0	0	0	1	1
# 64	0	0	0	0	0	-1	1	0	0	0	0	0	0	0	0	1	1	1	0	0	0	0	0	0	1	0	0
# 65	0	0	0	0	0	-1	1	1	-2	0	0	0	0	0	0	1	1	0	0	0	0	0	0	0	1	0	-1
# 66	0	0	0	0	-1	1	1	-1	0	0	0	0	0	0	1	1	0	0	0	0	0	0	0	1	0	-1	0
# 67	0	0	0	-1	1	1	-1	0	0	0	0	0	0	1	1	0	0	0	0	0	0	0	1	0	-1	0	0
# 68	0	0	-1	1	1	-1	0	0	0	0	0	0	1	1	0	0	0	0	0	0	0	1	0	-1	0	0	0
# 69	0	-1	1	1	-1	0	0	0	0	0	0	1	1	0	0	0	0	0	0	0	1	0	-1	0	0	0	0
# 70	-1	1	1	-1	0	0	0	0	0	0	1	1	0	0	0	0	0	0	0	1	0	-1	0	0	0	0	0
# 71	1	1	-1	0	0	0	0	0	0	1	1	0	0	0	0	0	0	0	1	0	-1	0	0	0	0	0	0
# 72	0	0	0	0	0	0	-1	1	0	0	0	0	0	0	0	0	1	1	0	0	0	0	0	0	0	1	0
# 73	0	0	0	0	0	0	-1	2	-2	0	0	0	0	0	0	0	1	0	0	0	0	0	0	0	0	1	-1
# 74	0	0	0	0	0	-1	2	-1	0	0	0	0	0	0	0	1	0	0	0	0	0	0	0	0	1	-1	0
# 75	0	0	0	0	-1	2	-1	0	0	0	0	0	0	0	1	0	0	0	0	0	0	0	0	1	-1	0	0
# 76	0	0	0	-1	2	-1	0	0	0	0	0	0	0	1	0	0	0	0	0	0	0	0	1	-1	0	0	0
# 77	0	0	-1	2	-1	0	0	0	0	0	0	0	1	0	0	0	0	0	0	0	0	1	-1	0	0	0	0
# 78	0	-1	2	-1	0	0	0	0	0	0	0	1	0	0	0	0	0	0	0	0	1	-1	0	0	0	0	0
# 79	-1	2	-1	0	0	0	0	0	0	0	1	0	0	0	0	0	0	0	0	1	-1	0	0	0	0	0	0
# 80	2	-1	0	0	0	0	0	0	0	1	0	0	0	0	0	0	0	0	1	-1	0	0	0	0	0	0	0
# 81	0	0	0	0	0	0	0	-1	2	0	0	0	0	0	0	0	0	1	0	0	0	0	0	0	0	0	1

Table 2. *(continued)*

ALGEBRA B10

	OMEGA										ALPHA										EPSILON									
# 1	0	1	0	0	0	0	0	0	0	0	1	2	2	2	2	2	2	2	2	2	1	1	0	0	0	0	0	0	0	0
# 2	1	-1	1	0	0	0	0	0	0	0	1	1	2	2	2	2	2	2	2	2	1	0	1	0	0	0	0	0	0	0
# 3	-1	0	1	0	0	0	0	0	0	0	0	1	2	2	2	2	2	2	2	2	0	1	1	0	0	0	0	0	0	0
# 4	1	0	-1	1	0	0	0	0	0	0	1	1	1	2	2	2	2	2	2	2	1	0	0	1	0	0	0	0	0	0
# 5	-1	1	-1	1	0	0	0	0	0	0	0	1	1	2	2	2	2	2	2	2	0	1	0	1	0	0	0	0	0	0
# 6	1	0	0	-1	1	0	0	0	0	0	1	1	1	1	2	2	2	2	2	2	1	0	0	0	1	0	0	0	0	0
# 7	0	-1	0	1	0	0	0	0	0	0	0	0	1	2	2	2	2	2	2	2	0	0	1	1	0	0	0	0	0	0
# 8	-1	1	0	-1	1	0	0	0	0	0	0	1	1	1	2	2	2	2	2	2	0	1	0	0	1	0	0	0	0	0
# 9	1	0	0	0	-1	1	0	0	0	0	1	1	1	1	1	2	2	2	2	2	1	0	0	0	0	1	0	0	0	0
# 10	0	-1	1	-1	1	0	0	0	0	0	0	0	1	1	2	2	2	2	2	2	0	0	1	0	1	0	0	0	0	0
# 11	-1	1	0	0	-1	1	0	0	0	0	0	1	1	1	1	2	2	2	2	2	0	1	0	0	0	1	0	0	0	0
# 12	1	0	0	0	0	-1	1	0	0	0	1	1	1	1	1	1	2	2	2	2	1	0	0	0	0	0	1	0	0	0
# 13	0	0	-1	0	1	0	0	0	0	0	0	0	0	1	2	2	2	2	2	2	0	0	0	1	1	0	0	0	0	0
# 14	0	-1	1	0	-1	1	0	0	0	0	0	0	1	1	1	2	2	2	2	2	0	0	1	0	0	1	0	0	0	0
# 15	-1	1	0	0	0	-1	1	0	0	0	0	1	1	1	1	1	2	2	2	2	0	1	0	0	0	0	1	0	0	0
# 16	1	0	0	0	0	0	-1	1	0	0	1	1	1	1	1	1	1	2	2	2	1	0	0	0	0	0	0	1	0	0
# 17	0	0	-1	1	-1	1	0	0	0	0	0	0	0	1	1	2	2	2	2	2	0	0	0	1	0	1	0	0	0	0
# 18	0	-1	1	0	0	-1	1	0	0	0	0	0	1	1	1	1	2	2	2	2	0	0	1	0	0	0	1	0	0	0
# 19	-1	1	0	0	0	0	-1	1	0	0	0	1	1	1	1	1	1	2	2	2	0	1	0	0	0	0	0	1	0	0
# 20	1	0	0	0	0	0	0	-1	1	0	1	1	1	1	1	1	1	1	2	2	1	0	0	0	0	0	0	0	1	0
# 21	0	0	0	-1	0	1	0	0	0	0	0	0	0	0	1	2	2	2	2	2	0	0	0	0	1	1	0	0	0	0
# 22	0	0	-1	1	0	-1	1	0	0	0	0	0	0	1	1	1	2	2	2	2	0	0	0	1	0	0	1	0	0	0
# 23	0	-1	1	0	0	0	-1	1	0	0	0	0	1	1	1	1	1	2	2	2	0	0	1	0	0	0	0	1	0	0
# 24	-1	1	0	0	0	0	0	-1	1	0	0	1	1	1	1	1	1	1	2	2	0	1	0	0	0	0	0	0	1	0
# 25	1	0	0	0	0	0	0	0	-1	2	1	1	1	1	1	1	1	1	1	2	1	0	0	0	0	0	0	0	0	1
# 26	0	0	0	-1	1	-1	1	0	0	0	0	0	0	0	1	1	2	2	2	2	0	0	0	0	1	0	1	0	0	0
# 27	0	0	-1	1	0	0	-1	1	0	0	0	0	0	1	1	1	1	2	2	2	0	0	0	1	0	0	0	1	0	0
# 28	0	-1	1	0	0	0	0	-1	1	0	0	0	1	1	1	1	1	1	2	2	0	0	1	0	0	0	0	0	1	0
# 29	-1	1	0	0	0	0	0	0	-1	2	0	1	1	1	1	1	1	1	1	2	0	1	0	0	0	0	0	0	0	1
# 30	1	0	0	0	0	0	0	0	0	0	1	1	1	1	1	1	1	1	1	1	1	0	0	0	0	0	0	0	0	0
# 31	1	0	0	0	0	0	0	0	1	-2	1	1	1	1	1	1	1	1	1	0	1	0	0	0	0	0	0	0	0	-1
# 32	0	0	0	0	-1	0	1	0	0	0	0	0	0	0	0	1	2	2	2	2	0	0	0	0	0	1	1	0	0	0
# 33	0	0	0	-1	1	0	-1	1	0	0	0	0	0	0	1	1	1	2	2	2	0	0	0	0	1	0	0	1	0	0
# 34	0	0	-1	1	0	0	0	-1	1	0	0	0	0	1	1	1	1	1	2	2	0	0	0	1	0	0	0	0	1	0
# 35	0	-1	1	0	0	0	0	0	-1	2	0	0	1	1	1	1	1	1	1	2	0	0	1	0	0	0	0	0	0	1
# 36	-1	1	0	0	0	0	0	0	0	0	0	1	1	1	1	1	1	1	1	1	0	1	0	0	0	0	0	0	0	0
# 37	-1	1	0	0	0	0	0	0	1	-2	0	1	1	1	1	1	1	1	1	0	0	1	0	0	0	0	0	0	0	-1
# 38	1	0	0	0	0	0	0	1	-1	0	1	1	1	1	1	1	1	1	0	0	1	0	0	0	0	0	0	0	-1	0
# 39	0	0	0	0	-1	1	-1	1	0	0	0	0	0	0	0	1	1	2	2	2	0	0	0	0	0	1	0	1	0	0
# 40	0	0	0	-1	1	0	0	-1	1	0	0	0	0	0	1	1	1	1	2	2	0	0	0	0	1	0	0	0	1	0
# 41	0	0	-1	1	0	0	0	0	-1	2	0	0	0	1	1	1	1	1	1	2	0	0	0	1	0	0	0	0	0	1
# 42	0	-1	1	0	0	0	0	0	0	0	0	0	1	1	1	1	1	1	1	1	0	0	1	0	0	0	0	0	0	0
# 43	0	-1	1	0	0	0	0	0	1	-2	0	0	1	1	1	1	1	1	1	0	0	0	1	0	0	0	0	0	0	-1
# 44	-1	1	0	0	0	0	0	1	-1	0	0	1	1	1	1	1	1	1	0	0	0	1	0	0	0	0	0	0	-1	0
# 45	1	0	0	0	0	0	1	-1	0	0	1	1	1	1	1	1	1	0	0	0	1	0	0	0	0	0	0	-1	0	0
# 46	0	0	0	0	0	-1	0	1	0	0	0	0	0	0	0	0	1	2	2	2	0	0	0	0	0	0	1	1	0	0
# 47	0	0	0	0	-1	1	0	-1	1	0	0	0	0	0	0	1	1	1	2	2	0	0	0	0	0	1	0	0	1	0
# 48	0	0	0	-1	1	0	0	0	-1	2	0	0	0	0	1	1	1	1	1	2	0	0	0	0	1	0	0	0	0	1
# 49	0	0	-1	1	0	0	0	0	0	0	0	0	0	1	1	1	1	1	1	1	0	0	0	1	0	0	0	0	0	0

# 50	0	0	-1	1	0	0	0	0	1	-2	0	0	0	1	1	1	1	1	1	0	0	0	0	1	0	0	0	0	0	-1
# 51	0	-1	1	0	0	0	0	1	-1	0	0	0	1	1	1	1	1	1	0	0	0	0	1	0	0	0	0	0	-1	0
# 52	-1	1	0	0	0	0	1	-1	0	0	0	1	1	1	1	1	1	0	0	0	0	1	0	0	0	0	0	-1	0	0
# 53	1	0	0	0	0	1	-1	0	0	0	1	1	1	1	1	1	0	0	0	0	1	0	0	0	0	0	-1	0	0	0
# 54	0	0	0	0	0	-1	1	-1	1	0	0	0	0	0	0	0	1	1	2	2	0	0	0	0	0	0	1	0	1	0
# 55	0	0	0	0	-1	1	0	0	-1	2	0	0	0	0	0	1	1	1	1	2	0	0	0	0	0	1	0	0	0	1
# 56	0	0	0	-1	1	0	0	0	0	0	0	0	0	0	1	1	1	1	1	1	0	0	0	0	1	0	0	0	0	0
# 57	0	0	0	-1	1	0	0	0	1	-2	0	0	0	0	1	1	1	1	1	0	0	0	0	0	1	0	0	0	0	-1
# 58	0	0	-1	1	0	0	0	1	-1	0	0	0	0	1	1	1	1	1	0	0	0	0	0	1	0	0	0	0	-1	0
# 59	0	-1	1	0	0	0	1	-1	0	0	0	0	1	1	1	1	1	0	0	0	0	0	1	0	0	0	0	-1	0	0
# 60	-1	1	0	0	0	1	-1	0	0	0	0	1	1	1	1	1	0	0	0	0	0	1	0	0	0	0	-1	0	0	0
# 61	1	0	0	0	1	-1	0	0	0	0	1	1	1	1	1	0	0	0	0	0	1	0	0	0	0	-1	0	0	0	0
# 62	0	0	0	0	0	0	-1	0	1	0	0	0	0	0	0	0	0	1	2	2	0	0	0	0	0	0	0	1	1	0
# 63	0	0	0	0	0	-1	1	0	-1	2	0	0	0	0	0	0	1	1	1	2	0	0	0	0	0	0	1	0	0	1
# 64	0	0	0	0	-1	1	0	0	0	0	0	0	0	0	0	1	1	1	1	1	0	0	0	0	0	1	0	0	0	0
# 65	0	0	0	0	-1	1	0	0	1	-2	0	0	0	0	0	1	1	1	1	0	0	0	0	0	0	1	0	0	0	-1
# 66	0	0	0	-1	1	0	0	1	-1	0	0	0	0	0	1	1	1	1	0	0	0	0	0	0	1	0	0	0	-1	0
# 67	0	0	-1	1	0	0	1	-1	0	0	0	0	0	1	1	1	1	0	0	0	0	0	0	1	0	0	0	-1	0	0
# 68	0	-1	1	0	0	1	-1	0	0	0	0	0	1	1	1	1	0	0	0	0	0	0	1	0	0	0	-1	0	0	0
# 69	-1	1	0	0	1	-1	0	0	0	0	0	1	1	1	1	0	0	0	0	0	0	1	0	0	0	-1	0	0	0	0
# 70	1	0	0	1	-1	0	0	0	0	0	1	1	1	1	0	0	0	0	0	0	1	0	0	0	-1	0	0	0	0	0
# 71	0	0	0	0	0	0	-1	1	-1	2	0	0	0	0	0	0	0	1	1	2	0	0	0	0	0	0	0	1	0	1
# 72	0	0	0	0	0	-1	1	0	0	0	0	0	0	0	0	0	1	1	1	1	0	0	0	0	0	0	1	0	0	0
# 73	0	0	0	0	0	-1	1	0	1	-2	0	0	0	0	0	0	1	1	1	0	0	0	0	0	0	0	1	0	0	-1
# 74	0	0	0	0	-1	1	0	1	-1	0	0	0	0	0	0	1	1	1	0	0	0	0	0	0	0	1	0	0	-1	0
# 75	0	0	0	-1	1	0	1	-1	0	0	0	0	0	0	1	1	1	0	0	0	0	0	0	0	1	0	0	-1	0	0
# 76	0	0	-1	1	0	1	-1	0	0	0	0	0	0	1	1	1	0	0	0	0	0	0	0	1	0	0	-1	0	0	0
# 77	0	-1	1	0	1	-1	0	0	0	0	0	0	1	1	1	0	0	0	0	0	0	0	1	0	0	-1	0	0	0	0
# 78	-1	1	0	1	-1	0	0	0	0	0	0	1	1	1	0	0	0	0	0	0	0	1	0	0	-1	0	0	0	0	0
# 79	1	0	1	-1	0	0	0	0	0	0	1	1	1	0	0	0	0	0	0	0	1	0	0	-1	0	0	0	0	0	0
# 80	0	0	0	0	0	0	0	-1	0	2	0	0	0	0	0	0	0	0	1	2	0	0	0	0	0	0	0	0	1	1
# 81	0	0	0	0	0	0	-1	1	0	0	0	0	0	0	0	0	0	1	1	1	0	0	0	0	0	0	0	1	0	0
# 82	0	0	0	0	0	0	-1	1	1	-2	0	0	0	0	0	0	0	1	1	0	0	0	0	0	0	0	0	1	0	-1
# 83	0	0	0	0	0	-1	1	1	-1	0	0	0	0	0	0	0	1	1	0	0	0	0	0	0	0	0	1	0	-1	0
# 84	0	0	0	0	-1	1	1	-1	0	0	0	0	0	0	0	1	1	0	0	0	0	0	0	0	0	1	0	-1	0	0
# 85	0	0	0	-1	1	1	-1	0	0	0	0	0	0	0	1	1	0	0	0	0	0	0	0	0	1	0	-1	0	0	0
# 86	0	0	-1	1	1	-1	0	0	0	0	0	0	0	1	1	0	0	0	0	0	0	0	0	1	0	-1	0	0	0	0
# 87	0	-1	1	1	-1	0	0	0	0	0	0	0	1	1	0	0	0	0	0	0	0	0	1	0	-1	0	0	0	0	0
# 88	-1	1	1	-1	0	0	0	0	0	0	0	1	1	0	0	0	0	0	0	0	0	1	0	-1	0	0	0	0	0	0
# 89	1	1	-1	0	0	0	0	0	0	0	1	1	0	0	0	0	0	0	0	0	1	0	-1	0	0	0	0	0	0	0
# 90	0	0	0	0	0	0	0	-1	1	0	0	0	0	0	0	0	0	0	1	1	0	0	0	0	0	0	0	0	1	0
# 91	0	0	0	0	0	0	0	-1	2	-2	0	0	0	0	0	0	0	0	1	0	0	0	0	0	0	0	0	0	1	-1
# 92	0	0	0	0	0	0	-1	2	-1	0	0	0	0	0	0	0	0	1	0	0	0	0	0	0	0	0	0	1	-1	0
# 93	0	0	0	0	0	-1	2	-1	0	0	0	0	0	0	0	0	1	0	0	0	0	0	0	0	0	0	1	-1	0	0
# 94	0	0	0	0	-1	2	-1	0	0	0	0	0	0	0	0	1	0	0	0	0	0	0	0	0	0	1	-1	0	0	0
# 95	0	0	0	-1	2	-1	0	0	0	0	0	0	0	0	1	0	0	0	0	0	0	0	0	0	1	-1	0	0	0	0
# 96	0	0	-1	2	-1	0	0	0	0	0	0	0	0	1	0	0	0	0	0	0	0	0	0	1	-1	0	0	0	0	0
# 97	0	-1	2	-1	0	0	0	0	0	0	0	0	1	0	0	0	0	0	0	0	0	0	1	-1	0	0	0	0	0	0
# 98	-1	2	-1	0	0	0	0	0	0	0	0	1	0	0	0	0	0	0	0	0	0	1	-1	0	0	0	0	0	0	0
# 99	2	-1	0	0	0	0	0	0	0	0	1	0	0	0	0	0	0	0	0	0	1	-1	0	0	0	0	0	0	0	0
#100	0	0	0	0	0	0	0	0	-1	2	0	0	0	0	0	0	0	0	0	1	0	0	0	0	0	0	0	0	0	1

Table 2. *(continued)*

ALGEBRA B11

#	OMEGA											ALPHA											EPSILON										
# 1	0	1	0	0	0	0	0	0	0	0	0	1	2	2	2	2	2	2	2	2	2	2	1	1	0	0	0	0	0	0	0	0	0
# 2	1	-1	1	0	0	0	0	0	0	0	0	1	1	2	2	2	2	2	2	2	2	2	1	0	1	0	0	0	0	0	0	0	0
# 3	-1	0	1	0	0	0	0	0	0	0	0	0	1	2	2	2	2	2	2	2	2	2	0	1	1	0	0	0	0	0	0	0	0
# 4	1	0	-1	1	0	0	0	0	0	0	0	1	1	1	2	2	2	2	2	2	2	2	1	0	0	1	0	0	0	0	0	0	0
# 5	-1	1	-1	1	0	0	0	0	0	0	0	0	1	1	2	2	2	2	2	2	2	2	0	1	0	1	0	0	0	0	0	0	0
# 6	1	0	0	-1	1	0	0	0	0	0	0	1	1	1	1	2	2	2	2	2	2	2	1	0	0	0	1	0	0	0	0	0	0
# 7	0	-1	0	1	0	0	0	0	0	0	0	0	0	1	2	2	2	2	2	2	2	2	0	0	1	1	0	0	0	0	0	0	0
# 8	-1	1	0	-1	1	0	0	0	0	0	0	0	1	1	1	2	2	2	2	2	2	2	0	1	0	0	1	0	0	0	0	0	0
# 9	1	0	0	0	-1	1	0	0	0	0	0	1	1	1	1	1	2	2	2	2	2	2	1	0	0	0	0	1	0	0	0	0	0
# 10	0	-1	1	-1	1	0	0	0	0	0	0	0	0	1	1	2	2	2	2	2	2	2	0	0	1	0	1	0	0	0	0	0	0
# 11	-1	1	0	0	-1	1	0	0	0	0	0	0	1	1	1	1	2	2	2	2	2	2	0	1	0	0	0	1	0	0	0	0	0
# 12	1	0	0	0	0	-1	1	0	0	0	0	1	1	1	1	1	1	2	2	2	2	2	1	0	0	0	0	0	1	0	0	0	0
# 13	0	0	-1	0	1	0	0	0	0	0	0	0	0	0	1	2	2	2	2	2	2	2	0	0	0	1	1	0	0	0	0	0	0
# 14	0	-1	1	0	-1	1	0	0	0	0	0	0	0	1	1	1	2	2	2	2	2	2	0	0	1	0	0	1	0	0	0	0	0
# 15	-1	1	0	0	0	-1	1	0	0	0	0	0	1	1	1	1	1	2	2	2	2	2	0	1	0	0	0	0	1	0	0	0	0
# 16	1	0	0	0	0	0	-1	1	0	0	0	1	1	1	1	1	1	1	2	2	2	2	1	0	0	0	0	0	0	1	0	0	0
# 17	0	0	-1	1	-1	1	0	0	0	0	0	0	0	0	1	1	2	2	2	2	2	2	0	0	0	1	0	1	0	0	0	0	0
# 18	0	-1	1	0	0	-1	1	0	0	0	0	0	0	1	1	1	1	2	2	2	2	2	0	0	1	0	0	0	1	0	0	0	0
# 19	-1	1	0	0	0	0	-1	1	0	0	0	0	1	1	1	1	1	1	2	2	2	2	0	1	0	0	0	0	0	1	0	0	0
# 20	1	0	0	0	0	0	0	-1	1	0	0	1	1	1	1	1	1	1	1	2	2	2	1	0	0	0	0	0	0	0	1	0	0
# 21	0	0	0	-1	0	1	0	0	0	0	0	0	0	0	0	1	2	2	2	2	2	2	0	0	0	0	1	1	0	0	0	0	0
# 22	0	0	-1	1	0	-1	1	0	0	0	0	0	0	0	1	1	1	2	2	2	2	2	0	0	0	1	0	0	1	0	0	0	0
# 23	0	-1	1	0	0	0	-1	1	0	0	0	0	0	1	1	1	1	1	2	2	2	2	0	0	1	0	0	0	0	1	0	0	0
# 24	-1	1	0	0	0	0	0	-1	1	0	0	0	1	1	1	1	1	1	1	2	2	2	0	1	0	0	0	0	0	0	1	0	0
# 25	1	0	0	0	0	0	0	0	-1	1	0	1	1	1	1	1	1	1	1	1	2	2	1	0	0	0	0	0	0	0	0	1	0
# 26	0	0	0	-1	1	-1	1	0	0	0	0	0	0	0	0	1	1	2	2	2	2	2	0	0	0	0	1	0	1	0	0	0	0
# 27	0	0	-1	1	0	0	-1	1	0	0	0	0	0	0	1	1	1	1	2	2	2	2	0	0	0	1	0	0	0	1	0	0	0
# 28	0	-1	1	0	0	0	0	-1	1	0	0	0	0	1	1	1	1	1	1	2	2	2	0	0	1	0	0	0	0	0	1	0	0
# 29	-1	1	0	0	0	0	0	0	-1	1	0	0	1	1	1	1	1	1	1	1	2	2	0	1	0	0	0	0	0	0	0	1	0
# 30	1	0	0	0	0	0	0	0	0	-1	2	1	1	1	1	1	1	1	1	1	1	2	1	0	0	0	0	0	0	0	0	0	1
# 31	0	0	0	0	-1	0	1	0	0	0	0	0	0	0	0	0	1	2	2	2	2	2	0	0	0	0	0	1	1	0	0	0	0
# 32	0	0	0	-1	1	0	-1	1	0	0	0	0	0	0	0	1	1	1	2	2	2	2	0	0	0	0	1	0	0	1	0	0	0
# 33	0	0	-1	1	0	0	0	-1	1	0	0	0	0	0	1	1	1	1	1	2	2	2	0	0	0	1	0	0	0	0	1	0	0
# 34	0	-1	1	0	0	0	0	0	-1	1	0	0	0	1	1	1	1	1	1	1	2	2	0	0	1	0	0	0	0	0	0	1	0
# 35	-1	1	0	0	0	0	0	0	0	-1	2	0	1	1	1	1	1	1	1	1	1	2	0	1	0	0	0	0	0	0	0	0	1
# 36	1	0	0	0	0	0	0	0	0	0	0	1	1	1	1	1	1	1	1	1	1	1	1	0	0	0	0	0	0	0	0	0	0
# 37	1	0	0	0	0	0	0	0	0	1	-2	1	1	1	1	1	1	1	1	1	1	0	1	0	0	0	0	0	0	0	0	0	-1
# 38	0	0	0	0	-1	1	-1	1	0	0	0	0	0	0	0	0	1	1	2	2	2	2	0	0	0	0	0	1	0	1	0	0	0
# 39	0	0	0	-1	1	0	0	-1	1	0	0	0	0	0	0	1	1	1	1	2	2	2	0	0	0	0	1	0	0	0	1	0	0
# 40	0	0	-1	1	0	0	0	0	-1	1	0	0	0	0	1	1	1	1	1	1	2	2	0	0	0	1	0	0	0	0	0	1	0
# 41	0	-1	1	0	0	0	0	0	0	-1	2	0	0	1	1	1	1	1	1	1	1	2	0	0	1	0	0	0	0	0	0	0	1
# 42	-1	1	0	0	0	0	0	0	0	0	0	0	1	1	1	1	1	1	1	1	1	1	0	1	0	0	0	0	0	0	0	0	0
# 43	-1	1	0	0	0	0	0	0	0	1	-2	0	1	1	1	1	1	1	1	1	1	0	0	1	0	0	0	0	0	0	0	0	-1
# 44	1	0	0	0	0	0	0	0	1	-1	0	1	1	1	1	1	1	1	1	1	0	0	1	0	0	0	0	0	0	0	0	-1	0
# 45	0	0	0	0	0	-1	0	1	0	0	0	0	0	0	0	0	0	1	2	2	2	2	0	0	0	0	0	0	1	1	0	0	0
# 46	0	0	0	0	-1	1	0	-1	1	0	0	0	0	0	0	0	1	1	1	2	2	2	0	0	0	0	0	1	0	0	1	0	0
# 47	0	0	0	-1	1	0	0	0	-1	1	0	0	0	0	0	1	1	1	1	1	2	2	0	0	0	0	1	0	0	0	0	1	0
# 48	0	0	-1	1	0	0	0	0	0	-1	2	0	0	0	1	1	1	1	1	1	1	2	0	0	0	1	0	0	0	0	0	0	1
# 49	0	-1	1	0	0	0	0	0	0	0	0	0	0	1	1	1	1	1	1	1	1	1	0	0	1	0	0	0	0	0	0	0	0
# 50	0	-1	1	0	0	0	0	0	0	1	-2	0	0	1	1	1	1	1	1	1	1	0	0	0	1	0	0	0	0	0	0	0	-1
# 51	-1	1	0	0	0	0	0	0	1	-1	0	0	1	1	1	1	1	1	1	1	0	0	0	1	0	0	0	0	0	0	0	-1	0
# 52	1	0	0	0	0	0	0	1	-1	0	0	1	1	1	1	1	1	1	1	0	0	0	1	0	0	0	0	0	0	0	-1	0	0
# 53	0	0	0	0	0	-1	1	-1	1	0	0	0	0	0	0	0	0	1	1	2	2	2	0	0	0	0	0	0	1	0	1	0	0
# 54	0	0	0	0	-1	1	0	0	-1	1	0	0	0	0	0	0	1	1	1	1	2	2	0	0	0	0	0	1	0	0	0	1	0
# 55	0	0	0	-1	1	0	0	0	0	-1	2	0	0	0	0	1	1	1	1	1	1	2	0	0	0	0	1	0	0	0	0	0	1
# 56	0	0	-1	1	0	0	0	0	0	0	0	0	0	0	1	1	1	1	1	1	1	1	0	0	0	1	0	0	0	0	0	0	0
# 57	0	0	-1	1	0	0	0	0	0	1	-2	0	0	0	1	1	1	1	1	1	1	0	0	0	0	1	0	0	0	0	0	0	-1
# 58	0	-1	1	0	0	0	0	0	1	-1	0	0	0	1	1	1	1	1	1	1	0	0	0	0	1	0	0	0	0	0	0	-1	0

# 59	-1	1	0	0	0	0	0	1	-1	0	0	0	1	1	1	1	1	1	1	0	0	0	0	1	0	0	0	0	0	0	-1	0	0
# 60	1	0	0	0	0	0	1	1	0	0	0	1	1	1	1	1	1	1	0	0	0	0	1	0	0	0	0	0	0	-1	0	0	0
# 61	0	0	0	0	0	0	-1	0	1	0	0	0	0	0	0	0	0	0	1	2	2	2	0	0	0	0	0	0	0	1	1	0	0
# 62	0	0	0	0	0	-1	1	0	-1	1	0	0	0	0	0	0	0	1	1	1	2	2	0	0	0	0	0	0	1	0	0	1	0
# 63	0	0	0	0	-1	1	0	0	0	-1	2	0	0	0	0	0	1	1	1	1	1	2	0	0	0	0	0	1	0	0	0	0	1
# 64	0	0	0	-1	1	0	0	0	0	0	0	0	0	0	0	1	1	1	1	1	1	1	0	0	0	0	1	0	0	0	0	0	0
# 65	0	0	0	-1	1	0	0	0	0	1	-2	0	0	0	0	1	1	1	1	1	1	0	0	0	0	0	1	0	0	0	0	0	-1
# 66	0	0	-1	1	0	0	0	0	1	-1	0	0	0	0	1	1	1	1	1	1	0	0	0	0	0	1	0	0	0	0	0	-1	0
# 67	0	-1	1	0	0	0	0	1	-1	0	0	0	0	1	1	1	1	1	1	0	0	0	0	0	1	0	0	0	0	0	-1	0	0
# 68	-1	1	0	0	0	0	1	-1	0	0	0	0	1	1	1	1	1	1	0	0	0	0	0	1	0	0	0	0	0	-1	0	0	0
# 69	1	0	0	0	0	1	-1	0	0	0	0	1	1	1	1	1	1	0	0	0	0	0	1	0	0	0	0	0	-1	0	0	0	0
# 70	0	0	0	0	0	0	-1	1	-1	1	0	0	0	0	0	0	0	0	1	1	2	2	0	0	0	0	0	0	0	1	0	1	0
# 71	0	0	0	0	0	-1	1	0	0	-1	2	0	0	0	0	0	0	1	1	1	1	2	0	0	0	0	0	0	1	0	0	0	1
# 72	0	0	0	0	-1	1	0	0	0	0	0	0	0	0	0	0	1	1	1	1	1	1	0	0	0	0	0	1	0	0	0	0	0
# 73	0	0	0	0	-1	1	0	0	0	1	-2	0	0	0	0	0	1	1	1	1	1	0	0	0	0	0	0	1	0	0	0	0	-1
# 74	0	0	0	-1	1	0	0	0	1	-1	0	0	0	0	0	1	1	1	1	1	0	0	0	0	0	0	1	0	0	0	0	-1	0
# 75	0	0	-1	1	0	0	0	1	-1	0	0	0	0	0	1	1	1	1	1	0	0	0	0	0	0	1	0	0	0	0	-1	0	0
# 76	0	-1	1	0	0	0	1	-1	0	0	0	0	0	1	1	1	1	1	0	0	0	0	0	0	1	0	0	0	0	-1	0	0	0
# 77	-1	1	0	0	0	1	-1	0	0	0	0	0	1	1	1	1	1	0	0	0	0	0	0	1	0	0	0	0	-1	0	0	0	0
# 78	1	0	0	0	1	-1	0	0	0	0	0	1	1	1	1	1	0	0	0	0	0	0	1	0	0	0	0	-1	0	0	0	0	0
# 79	0	0	0	0	0	0	0	-1	0	1	0	0	0	0	0	0	0	0	0	1	2	2	0	0	0	0	0	0	0	0	1	1	0
# 80	0	0	0	0	0	0	-1	1	0	-1	2	0	0	0	0	0	0	0	1	1	1	2	0	0	0	0	0	0	0	1	0	0	1
# 81	0	0	0	0	0	-1	1	0	0	0	0	0	0	0	0	0	0	1	1	1	1	1	0	0	0	0	0	0	1	0	0	0	0
# 82	0	0	0	0	0	-1	1	0	0	1	-2	0	0	0	0	0	0	1	1	1	1	0	0	0	0	0	0	0	1	0	0	0	-1
# 83	0	0	0	0	-1	1	0	0	1	-1	0	0	0	0	0	0	1	1	1	1	0	0	0	0	0	0	0	1	0	0	0	-1	0
# 84	0	0	0	-1	1	0	0	1	-1	0	0	0	0	0	0	1	1	1	1	0	0	0	0	0	0	0	1	0	0	0	-1	0	0
# 85	0	0	-1	1	0	0	1	-1	0	0	0	0	0	0	1	1	1	1	0	0	0	0	0	0	0	1	0	0	0	-1	0	0	0
# 86	0	-1	1	0	0	1	-1	0	0	0	0	0	0	1	1	1	1	0	0	0	0	0	0	0	1	0	0	0	-1	0	0	0	0
# 87	-1	1	0	0	1	-1	0	0	0	0	0	0	1	1	1	1	0	0	0	0	0	0	0	1	0	0	0	-1	0	0	0	0	0
# 88	1	0	0	1	-1	0	0	0	0	0	0	1	1	1	1	0	0	0	0	0	0	0	1	0	0	0	-1	0	0	0	0	0	0
# 89	0	0	0	0	0	0	0	-1	1	-1	2	0	0	0	0	0	0	0	0	1	1	2	0	0	0	0	0	0	0	0	1	0	1
# 90	0	0	0	0	0	0	-1	1	0	0	0	0	0	0	0	0	0	0	1	1	1	1	0	0	0	0	0	0	0	1	0	0	0
# 91	0	0	0	0	0	0	-1	1	0	1	-2	0	0	0	0	0	0	0	1	1	1	0	0	0	0	0	0	0	0	1	0	0	-1
# 92	0	0	0	0	0	-1	1	0	1	-1	0	0	0	0	0	0	0	1	1	1	0	0	0	0	0	0	0	0	1	0	0	-1	0
# 93	0	0	0	0	-1	1	0	1	-1	0	0	0	0	0	0	0	1	1	1	0	0	0	0	0	0	0	0	1	0	0	-1	0	0
# 94	0	0	0	-1	1	0	1	-1	0	0	0	0	0	0	0	1	1	1	0	0	0	0	0	0	0	0	1	0	0	-1	0	0	0
# 95	0	0	-1	1	0	1	-1	0	0	0	0	0	0	0	1	1	1	0	0	0	0	0	0	0	0	1	0	0	-1	0	0	0	0
# 96	0	-1	1	0	1	-1	0	0	0	0	0	0	0	1	1	1	0	0	0	0	0	0	0	0	1	0	0	-1	0	0	0	0	0
# 97	-1	1	0	1	-1	0	0	0	0	0	0	0	1	1	1	0	0	0	0	0	0	0	0	1	0	0	-1	0	0	0	0	0	0
# 98	1	0	1	-1	0	0	0	0	0	0	0	1	1	1	0	0	0	0	0	0	0	0	1	0	0	-1	0	0	0	0	0	0	0
# 99	0	0	0	0	0	0	0	0	-1	0	2	0	0	0	0	0	0	0	0	0	1	2	0	0	0	0	0	0	0	0	0	1	1
#100	0	0	0	0	0	0	0	-1	1	0	0	0	0	0	0	0	0	0	0	1	1	1	0	0	0	0	0	0	0	0	1	0	0
#101	0	0	0	0	0	0	0	-1	1	1	-2	0	0	0	0	0	0	0	0	1	1	0	0	0	0	0	0	0	0	0	1	0	-1
#102	0	0	0	0	0	0	-1	1	1	-1	0	0	0	0	0	0	0	0	1	1	0	0	0	0	0	0	0	0	0	1	0	-1	0
#103	0	0	0	0	0	-1	1	1	-1	0	0	0	0	0	0	0	0	1	1	0	0	0	0	0	0	0	0	0	1	0	-1	0	0
#104	0	0	0	0	-1	1	1	-1	0	0	0	0	0	0	0	0	1	1	0	0	0	0	0	0	0	0	0	1	0	-1	0	0	0
#105	0	0	0	-1	1	1	-1	0	0	0	0	0	0	0	0	1	1	0	0	0	0	0	0	0	0	0	1	0	-1	0	0	0	0
#106	0	0	-1	1	1	-1	0	0	0	0	0	0	0	0	1	1	0	0	0	0	0	0	0	0	0	1	0	-1	0	0	0	0	0
#107	0	-1	1	1	-1	0	0	0	0	0	0	0	0	1	1	0	0	0	0	0	0	0	0	0	1	0	-1	0	0	0	0	0	0
#108	-1	1	1	-1	0	0	0	0	0	0	0	0	1	1	0	0	0	0	0	0	0	0	0	1	0	-1	0	0	0	0	0	0	0
#109	1	1	-1	0	0	0	0	0	0	0	0	1	1	0	0	0	0	0	0	0	0	0	1	0	-1	0	0	0	0	0	0	0	0
#110	0	0	0	0	0	0	0	0	-1	1	0	0	0	0	0	0	0	0	0	0	1	1	0	0	0	0	0	0	0	0	0	1	0
#111	0	0	0	0	0	0	0	0	-1	2	-2	0	0	0	0	0	0	0	0	0	1	0	0	0	0	0	0	0	0	0	0	1	-1
#112	0	0	0	0	0	0	0	-1	2	-1	0	0	0	0	0	0	0	0	0	1	0	0	0	0	0	0	0	0	0	0	1	-1	0
#113	0	0	0	0	0	0	-1	2	-1	0	0	0	0	0	0	0	0	0	1	0	0	0	0	0	0	0	0	0	0	1	-1	0	0
#114	0	0	0	0	0	-1	2	-1	0	0	0	0	0	0	0	0	0	1	0	0	0	0	0	0	0	0	0	0	1	-1	0	0	0
#115	0	0	0	0	-1	2	-1	0	0	0	0	0	0	0	0	0	1	0	0	0	0	0	0	0	0	0	0	1	-1	0	0	0	0
#116	0	0	0	-1	2	-1	0	0	0	0	0	0	0	0	0	1	0	0	0	0	0	0	0	0	0	0	1	-1	0	0	0	0	0
#117	0	0	-1	2	-1	0	0	0	0	0	0	0	0	0	1	0	0	0	0	0	0	0	0	0	0	1	-1	0	0	0	0	0	0
#118	0	-1	2	-1	0	0	0	0	0	0	0	0	0	1	0	0	0	0	0	0	0	0	0	0	1	-1	0	0	0	0	0	0	0
#119	-1	2	-1	0	0	0	0	0	0	0	0	0	1	0	0	0	0	0	0	0	0	0	0	1	-1	0	0	0	0	0	0	0	0
#120	2	-1	0	0	0	0	0	0	0	0	0	1	0	0	0	0	0	0	0	0	0	0	1	-1	0	0	0	0	0	0	0	0	0
#121	0	0	0	0	0	0	0	0	0	-1	2	0	0	0	0	0	0	0	0	0	0	1	0	0	0	0	0	0	0	0	0	0	1

Table 2. *(continued)*

ALGEBRA B12

	OMEGA												ALPHA												EPSILON											
# 1	0	1	0	0	0	0	0	0	0	0	0	0	1	2	2	2	2	2	2	2	2	2	2	2	1	1	0	0	0	0	0	0	0	0	0	0
# 2	1	-1	1	0	0	0	0	0	0	0	0	0	1	1	2	2	2	2	2	2	2	2	2	2	1	0	1	0	0	0	0	0	0	0	0	0
# 3	-1	0	1	0	0	0	0	0	0	0	0	0	0	1	2	2	2	2	2	2	2	2	2	2	0	1	1	0	0	0	0	0	0	0	0	0
# 4	1	0	-1	1	0	0	0	0	0	0	0	0	1	1	1	2	2	2	2	2	2	2	2	2	1	0	0	1	0	0	0	0	0	0	0	0
# 5	-1	1	-1	1	0	0	0	0	0	0	0	0	0	1	1	2	2	2	2	2	2	2	2	2	0	1	0	1	0	0	0	0	0	0	0	0
# 6	1	0	0	-1	1	0	0	0	0	0	0	0	1	1	1	1	2	2	2	2	2	2	2	2	1	0	0	0	1	0	0	0	0	0	0	0
# 7	0	-1	0	1	0	0	0	0	0	0	0	0	0	0	1	2	2	2	2	2	2	2	2	2	0	0	1	1	0	0	0	0	0	0	0	0
# 8	-1	1	0	-1	1	0	0	0	0	0	0	0	0	1	1	1	2	2	2	2	2	2	2	2	0	1	0	0	1	0	0	0	0	0	0	0
# 9	1	0	0	0	-1	1	0	0	0	0	0	0	1	1	1	1	1	2	2	2	2	2	2	2	1	0	0	0	0	1	0	0	0	0	0	0
# 10	0	-1	1	-1	1	0	0	0	0	0	0	0	0	0	1	1	2	2	2	2	2	2	2	2	0	0	1	0	1	0	0	0	0	0	0	0
# 11	-1	1	0	0	-1	1	0	0	0	0	0	0	0	1	1	1	1	2	2	2	2	2	2	2	0	1	0	0	0	1	0	0	0	0	0	0
# 12	1	0	0	0	0	-1	1	0	0	0	0	0	1	1	1	1	1	1	2	2	2	2	2	2	1	0	0	0	0	0	1	0	0	0	0	0
# 13	0	0	-1	0	1	0	0	0	0	0	0	0	0	0	0	1	2	2	2	2	2	2	2	2	0	0	0	1	1	0	0	0	0	0	0	0
# 14	0	-1	1	0	-1	1	0	0	0	0	0	0	0	0	1	1	1	2	2	2	2	2	2	2	0	0	1	0	0	1	0	0	0	0	0	0
# 15	-1	1	0	0	0	-1	1	0	0	0	0	0	0	1	1	1	1	1	2	2	2	2	2	2	0	1	0	0	0	0	1	0	0	0	0	0
# 16	1	0	0	0	0	0	-1	1	0	0	0	0	1	1	1	1	1	1	1	2	2	2	2	2	1	0	0	0	0	0	0	1	0	0	0	0
# 17	0	0	-1	1	-1	1	0	0	0	0	0	0	0	0	0	1	1	2	2	2	2	2	2	2	0	0	0	1	0	1	0	0	0	0	0	0
# 18	0	-1	1	0	0	-1	1	0	0	0	0	0	0	0	1	1	1	1	2	2	2	2	2	2	0	0	1	0	0	0	1	0	0	0	0	0
# 19	-1	1	0	0	0	0	-1	1	0	0	0	0	0	1	1	1	1	1	1	2	2	2	2	2	0	1	0	0	0	0	0	1	0	0	0	0
# 20	1	0	0	0	0	0	0	-1	1	0	0	0	1	1	1	1	1	1	1	1	2	2	2	2	1	0	0	0	0	0	0	0	1	0	0	0
# 21	0	0	0	-1	0	1	0	0	0	0	0	0	0	0	0	0	1	2	2	2	2	2	2	2	0	0	0	0	1	1	0	0	0	0	0	0
# 22	0	0	-1	1	0	-1	1	0	0	0	0	0	0	0	0	1	1	1	2	2	2	2	2	2	0	0	0	1	0	0	1	0	0	0	0	0
# 23	0	-1	1	0	0	0	-1	1	0	0	0	0	0	0	1	1	1	1	1	2	2	2	2	2	0	0	1	0	0	0	0	1	0	0	0	0
# 24	-1	1	0	0	0	0	0	-1	1	0	0	0	0	1	1	1	1	1	1	1	2	2	2	2	0	1	0	0	0	0	0	0	1	0	0	0
# 25	1	0	0	0	0	0	0	0	-1	1	0	0	1	1	1	1	1	1	1	1	1	2	2	2	1	0	0	0	0	0	0	0	0	1	0	0
# 26	0	0	0	-1	1	-1	1	0	0	0	0	0	0	0	0	0	1	1	2	2	2	2	2	2	0	0	0	0	1	0	1	0	0	0	0	0
# 27	0	0	-1	1	0	0	-1	1	0	0	0	0	0	0	0	1	1	1	1	2	2	2	2	2	0	0	0	1	0	0	0	1	0	0	0	0
# 28	0	-1	1	0	0	0	0	-1	1	0	0	0	0	0	1	1	1	1	1	1	2	2	2	2	0	0	1	0	0	0	0	0	1	0	0	0
# 29	-1	1	0	0	0	0	0	0	-1	1	0	0	0	1	1	1	1	1	1	1	1	2	2	2	0	1	0	0	0	0	0	0	0	1	0	0
# 30	1	0	0	0	0	0	0	0	0	-1	1	0	1	1	1	1	1	1	1	1	1	1	2	2	1	0	0	0	0	0	0	0	0	0	1	0
# 31	0	0	0	0	-1	0	1	0	0	0	0	0	0	0	0	0	0	1	2	2	2	2	2	2	0	0	0	0	0	1	1	0	0	0	0	0
# 32	0	0	0	-1	1	0	-1	1	0	0	0	0	0	0	0	0	1	1	1	2	2	2	2	2	0	0	0	0	1	0	0	1	0	0	0	0
# 33	0	0	-1	1	0	0	0	-1	1	0	0	0	0	0	0	1	1	1	1	1	2	2	2	2	0	0	0	1	0	0	0	0	1	0	0	0
# 34	0	-1	1	0	0	0	0	0	-1	1	0	0	0	0	1	1	1	1	1	1	1	2	2	2	0	0	1	0	0	0	0	0	0	1	0	0
# 35	-1	1	0	0	0	0	0	0	0	-1	1	0	0	1	1	1	1	1	1	1	1	1	2	2	0	1	0	0	0	0	0	0	0	0	1	0
# 36	1	0	0	0	0	0	0	0	0	0	-1	2	1	1	1	1	1	1	1	1	1	1	1	2	1	0	0	0	0	0	0	0	0	0	0	1
# 37	0	0	0	0	-1	1	-1	1	0	0	0	0	0	0	0	0	0	1	1	2	2	2	2	2	0	0	0	0	0	1	0	1	0	0	0	0
# 38	0	0	0	-1	1	0	0	-1	1	0	0	0	0	0	0	0	1	1	1	1	2	2	2	2	0	0	0	0	1	0	0	0	1	0	0	0
# 39	0	0	-1	1	0	0	0	0	-1	1	0	0	0	0	0	1	1	1	1	1	1	2	2	2	0	0	0	1	0	0	0	0	0	1	0	0
# 40	0	-1	1	0	0	0	0	0	0	-1	1	0	0	0	1	1	1	1	1	1	1	1	2	2	0	0	1	0	0	0	0	0	0	0	1	0
# 41	-1	1	0	0	0	0	0	0	0	0	-1	2	0	1	1	1	1	1	1	1	1	1	1	2	0	1	0	0	0	0	0	0	0	0	0	1
# 42	1	0	0	0	0	0	0	0	0	0	0	0	1	1	1	1	1	1	1	1	1	1	1	1	1	0	0	0	0	0	0	0	0	0	0	0
# 43	1	0	0	0	0	0	0	0	0	0	1	-2	1	1	1	1	1	1	1	1	1	1	1	0	1	0	0	0	0	0	0	0	0	0	0	-1
# 44	0	0	0	0	0	-1	0	1	0	0	0	0	0	0	0	0	0	0	1	2	2	2	2	2	0	0	0	0	0	0	1	1	0	0	0	0
# 45	0	0	0	0	-1	1	0	-1	1	0	0	0	0	0	0	0	0	1	1	1	2	2	2	2	0	0	0	0	0	1	0	0	1	0	0	0
# 46	0	0	0	-1	1	0	0	0	-1	1	0	0	0	0	0	0	1	1	1	1	1	2	2	2	0	0	0	0	1	0	0	0	0	1	0	0
# 47	0	0	-1	1	0	0	0	0	0	-1	1	0	0	0	0	1	1	1	1	1	1	1	2	2	0	0	0	1	0	0	0	0	0	0	1	0
# 48	0	-1	1	0	0	0	0	0	0	0	-1	2	0	0	1	1	1	1	1	1	1	1	1	2	0	0	1	0	0	0	0	0	0	0	0	1
# 49	-1	1	0	0	0	0	0	0	0	0	0	0	0	1	1	1	1	1	1	1	1	1	1	1	0	1	0	0	0	0	0	0	0	0	0	0
# 50	-1	1	0	0	0	0	0	0	0	0	1	-2	0	1	1	1	1	1	1	1	1	1	1	0	0	1	0	0	0	0	0	0	0	0	0	-1
# 51	1	0	0	0	0	0	0	0	0	1	-1	0	1	1	1	1	1	1	1	1	1	1	0	0	1	0	0	0	0	0	0	0	0	0	-1	0
# 52	0	0	0	0	0	-1	1	-1	1	0	0	0	0	0	0	0	0	0	1	1	2	2	2	2	0	0	0	0	0	0	1	0	1	0	0	0
# 53	0	0	0	0	-1	1	0	0	-1	1	0	0	0	0	0	0	0	1	1	1	1	2	2	2	0	0	0	0	0	1	0	0	0	1	0	0
# 54	0	0	0	-1	1	0	0	0	0	-1	1	0	0	0	0	0	1	1	1	1	1	1	2	2	0	0	0	0	1	0	0	0	0	0	1	0
# 55	0	0	-1	1	0	0	0	0	0	0	-1	2	0	0	0	1	1	1	1	1	1	1	1	2	0	0	0	1	0	0	0	0	0	0	0	1
# 56	0	-1	1	0	0	0	0	0	0	0	0	0	0	0	1	1	1	1	1	1	1	1	1	1	0	0	1	0	0	0	0	0	0	0	0	0
# 57	0	-1	1	0	0	0	0	0	0	0	1	-2	0	0	1	1	1	1	1	1	1	1	1	0	0	0	1	0	0	0	0	0	0	0	0	-1
# 58	-1	1	0	0	0	0	0	0	0	1	-1	0	0	1	1	1	1	1	1	1	1	1	0	0	0	1	0	0	0	0	0	0	0	0	-1	0
# 59	1	0	0	0	0	0	0	0	1	-1	0	0	1	1	1	1	1	1	1	1	1	0	0	0	1	0	0	0	0	0	0	0	0	-1	0	0
# 60	0	0	0	0	0	0	-1	0	1	0	0	0	0	0	0	0	0	0	0	1	2	2	2	2	0	0	0	0	0	0	0	1	1	0	0	0
# 61	0	0	0	0	0	-1	1	0	-1	1	0	0	0	0	0	0	0	0	1	1	1	2	2	2	0	0	0	0	0	0	1	0	0	1	0	0
# 62	0	0	0	0	-1	1	0	0	0	-1	1	0	0	0	0	0	0	1	1	1	1	1	2	2	0	0	0	0	0	1	0	0	0	0	1	0
# 63	0	0	0	-1	1	0	0	0	0	0	-1	2	0	0	0	0	1	1	1	1	1	1	1	2	0	0	0	0	1	0	0	0	0	0	0	1
# 64	0	0	-1	1	0	0	0	0	0	0	0	0	0	0	0	1	1	1	1	1	1	1	1	1	0	0	0	1	0	0	0	0	0	0	0	0
# 65	0	0	-1	1	0	0	0	0	0	0	1	-2	0	0	0	1	1	1	1	1	1	1	1	0	0	0	0	1	0	0	0	0	0	0	0	-1
# 66	0	-1	1	0	0	0	0	0	0	1	-1	0	0	0	1	1	1	1	1	1	1	1	0	0	0	0	1	0	0	0	0	0	0	0	-1	0
# 67	-1	1	0	0	0	0	0	0	1	-1	0	0	0	1	1	1	1	1	1	1	1	0	0	0	0	1	0	0	0	0	0	0	0	-1	0	0
# 68	1	0	0	0	0	0	0	1	-1	0	0	0	1	1	1	1	1	1	1	1	0	0	0	0	1	0	0	0	0	0	0	0	-1	0	0	0
# 69	0	0	0	0	0	0	-1	1	-1	1	0	0	0	0	0	0	0	0	0	1	1	2	2	2	0	0	0	0	0	0	0	1	0	1	0	0

Root																																				
# 70	0	0	0	0	0	-1	1	0	0	-1	1	0	0	0	0	0	0	0	1	1	1	1	2	2	0	0	0	0	0	0	1	0	0	0	1	0
# 71	0	0	0	0	-1	1	0	0	0	0	-1	2	0	0	0	0	0	1	1	1	1	1	1	2	0	0	0	0	0	1	0	0	0	0	0	1
# 72	0	0	0	-1	1	0	0	0	0	0	0	0	0	0	0	0	1	1	1	1	1	1	1	1	0	0	0	0	1	0	0	0	0	0	0	0
# 73	0	0	0	-1	1	0	0	0	0	0	1	-2	0	0	0	0	1	1	1	1	1	1	1	0	0	0	0	0	1	0	0	0	0	0	0	-1
# 74	0	0	-1	1	0	0	0	0	0	1	-1	0	0	0	0	1	1	1	1	1	1	1	0	0	0	0	0	1	0	0	0	0	0	0	-1	0
# 75	0	-1	1	0	0	0	0	0	1	-1	0	0	0	0	1	1	1	1	1	1	1	0	0	0	0	0	1	0	0	0	0	0	0	-1	0	0
# 76	-1	1	0	0	0	0	0	1	-1	0	0	0	0	1	1	1	1	1	1	1	0	0	0	0	0	1	0	0	0	0	0	0	-1	0	0	0
# 77	1	0	0	0	0	0	1	-1	0	0	0	0	1	1	1	1	1	1	1	0	0	0	0	0	1	0	0	0	0	0	0	-1	0	0	0	0
# 78	0	0	0	0	0	0	0	-1	0	1	0	0	0	0	0	0	0	0	0	0	1	2	2	2	0	0	0	0	0	0	0	0	1	1	0	0
# 79	0	0	0	0	0	0	-1	1	0	-1	1	0	0	0	0	0	0	0	0	1	1	1	2	2	0	0	0	0	0	0	0	1	0	0	1	0
# 80	0	0	0	0	0	-1	1	0	0	0	-1	2	0	0	0	0	0	0	1	1	1	1	1	2	0	0	0	0	0	0	1	0	0	0	0	1
# 81	0	0	0	0	-1	1	0	0	0	0	0	0	0	0	0	0	0	1	1	1	1	1	1	1	0	0	0	0	0	1	0	0	0	0	0	0
# 82	0	0	0	0	-1	1	0	0	0	0	1	-2	0	0	0	0	0	1	1	1	1	1	1	0	0	0	0	0	0	1	0	0	0	0	0	-1
# 83	0	0	0	-1	1	0	0	0	0	1	-1	0	0	0	0	0	1	1	1	1	1	1	0	0	0	0	0	0	1	0	0	0	0	0	-1	0
# 84	0	0	-1	1	0	0	0	0	1	-1	0	0	0	0	0	1	1	1	1	1	1	0	0	0	0	0	0	1	0	0	0	0	0	-1	0	0
# 85	0	-1	1	0	0	0	0	1	-1	0	0	0	0	0	1	1	1	1	1	1	0	0	0	0	0	0	1	0	0	0	0	0	-1	0	0	0
# 86	-1	1	0	0	0	0	1	-1	0	0	0	0	0	1	1	1	1	1	1	0	0	0	0	0	0	1	0	0	0	0	0	-1	0	0	0	0
# 87	1	0	0	0	0	1	-1	0	0	0	0	0	1	1	1	1	1	1	0	0	0	0	0	0	1	0	0	0	0	0	-1	0	0	0	0	0
# 88	0	0	0	0	0	0	0	-1	1	-1	1	0	0	0	0	0	0	0	0	1	1	2	2	0	0	0	0	0	0	0	0	0	1	0	1	0
# 89	0	0	0	0	0	0	-1	1	0	0	-1	2	0	0	0	0	0	0	0	1	1	1	1	2	0	0	0	0	0	0	0	1	0	0	0	1
# 90	0	0	0	0	0	-1	1	0	0	0	0	0	0	0	0	0	0	0	1	1	1	1	1	1	0	0	0	0	0	0	1	0	0	0	0	0
# 91	0	0	0	0	0	-1	1	0	0	0	1	-2	0	0	0	0	0	0	1	1	1	1	1	0	0	0	0	0	0	0	1	0	0	0	0	-1
# 92	0	0	0	0	-1	1	0	0	0	1	-1	0	0	0	0	0	0	1	1	1	1	1	0	0	0	0	0	0	0	1	0	0	0	0	-1	0
# 93	0	0	0	-1	1	0	0	0	1	-1	0	0	0	0	0	0	1	1	1	1	1	0	0	0	0	0	0	0	1	0	0	0	0	-1	0	0
# 94	0	0	-1	1	0	0	0	1	-1	0	0	0	0	0	0	1	1	1	1	1	0	0	0	0	0	0	0	1	0	0	0	0	-1	0	0	0
# 95	0	-1	1	0	0	0	1	-1	0	0	0	0	0	0	1	1	1	1	1	0	0	0	0	0	0	0	1	0	0	0	0	-1	0	0	0	0
# 96	-1	1	0	0	0	1	-1	0	0	0	0	0	0	1	1	1	1	1	0	0	0	0	0	0	0	1	0	0	0	0	-1	0	0	0	0	0
# 97	1	0	0	0	1	-1	0	0	0	0	0	0	1	1	1	1	1	0	0	0	0	0	0	0	1	0	0	0	0	-1	0	0	0	0	0	0
# 98	0	0	0	0	0	0	0	0	-1	0	1	0	0	0	0	0	0	0	0	0	0	1	2	2	0	0	0	0	0	0	0	0	0	1	1	0
# 99	0	0	0	0	0	0	0	-1	1	0	-1	2	0	0	0	0	0	0	0	0	1	1	1	2	0	0	0	0	0	0	0	0	1	0	0	1
#100	0	0	0	0	0	0	-1	1	0	0	0	0	0	0	0	0	0	0	0	1	1	1	1	1	0	0	0	0	0	0	0	1	0	0	0	0
#101	0	0	0	0	0	0	-1	1	0	0	1	-2	0	0	0	0	0	0	0	1	1	1	1	0	0	0	0	0	0	0	0	1	0	0	0	-1
#102	0	0	0	0	0	-1	1	0	0	1	-1	0	0	0	0	0	0	0	1	1	1	1	0	0	0	0	0	0	0	0	1	0	0	0	-1	0
#103	0	0	0	0	-1	1	0	0	1	-1	0	0	0	0	0	0	0	1	1	1	1	0	0	0	0	0	0	0	0	1	0	0	0	-1	0	0
#104	0	0	0	-1	1	0	0	1	-1	0	0	0	0	0	0	0	1	1	1	1	0	0	0	0	0	0	0	0	1	0	0	0	-1	0	0	0
#105	0	0	-1	1	0	0	1	-1	0	0	0	0	0	0	0	1	1	1	1	0	0	0	0	0	0	0	0	1	0	0	0	-1	0	0	0	0
#106	0	-1	1	0	0	1	-1	0	0	0	0	0	0	0	1	1	1	1	0	0	0	0	0	0	0	0	1	0	0	0	-1	0	0	0	0	0
#107	-1	1	0	0	1	-1	0	0	0	0	0	0	0	1	1	1	1	0	0	0	0	0	0	0	0	1	0	0	0	-1	0	0	0	0	0	0
#108	1	0	0	1	-1	0	0	0	0	0	0	0	1	1	1	1	0	0	0	0	0	0	0	0	1	0	0	0	-1	0	0	0	0	0	0	0
#109	0	0	0	0	0	0	0	0	-1	1	-1	2	0	0	0	0	0	0	0	0	0	1	1	2	0	0	0	0	0	0	0	0	0	1	0	1
#110	0	0	0	0	0	0	0	-1	1	0	0	0	0	0	0	0	0	0	0	0	1	1	1	1	0	0	0	0	0	0	0	0	1	0	0	0
#111	0	0	0	0	0	0	0	-1	1	0	1	-2	0	0	0	0	0	0	0	0	1	1	1	0	0	0	0	0	0	0	0	0	1	0	0	-1
#112	0	0	0	0	0	0	-1	1	0	1	-1	0	0	0	0	0	0	0	0	1	1	1	0	0	0	0	0	0	0	0	0	1	0	0	-1	0
#113	0	0	0	0	0	-1	1	0	1	-1	0	0	0	0	0	0	0	0	1	1	1	0	0	0	0	0	0	0	0	0	1	0	0	-1	0	0
#114	0	0	0	0	-1	1	0	1	-1	0	0	0	0	0	0	0	0	1	1	1	0	0	0	0	0	0	0	0	0	1	0	0	-1	0	0	0
#115	0	0	0	-1	1	0	1	-1	0	0	0	0	0	0	0	0	1	1	1	0	0	0	0	0	0	0	0	0	1	0	0	-1	0	0	0	0
#116	0	0	-1	1	0	1	-1	0	0	0	0	0	0	0	0	1	1	1	0	0	0	0	0	0	0	0	0	1	0	0	-1	0	0	0	0	0
#117	0	-1	1	0	1	-1	0	0	0	0	0	0	0	0	1	1	1	0	0	0	0	0	0	0	0	0	1	0	0	-1	0	0	0	0	0	0
#118	-1	1	0	1	-1	0	0	0	0	0	0	0	0	1	1	1	0	0	0	0	0	0	0	0	0	1	0	0	-1	0	0	0	0	0	0	0
#119	1	0	1	-1	0	0	0	0	0	0	0	0	1	1	1	0	0	0	0	0	0	0	0	0	1	0	0	-1	0	0	0	0	0	0	0	0
#120	0	0	0	0	0	0	0	0	0	-1	0	2	0	0	0	0	0	0	0	0	0	0	1	2	0	0	0	0	0	0	0	0	0	0	1	1
#121	0	0	0	0	0	0	0	0	-1	1	0	0	0	0	0	0	0	0	0	0	0	1	1	1	0	0	0	0	0	0	0	0	0	1	0	0
#122	0	0	0	0	0	0	0	0	-1	1	1	-2	0	0	0	0	0	0	0	0	0	1	1	0	0	0	0	0	0	0	0	0	0	1	0	-1
#123	0	0	0	0	0	0	0	-1	1	1	-1	0	0	0	0	0	0	0	0	0	1	1	0	0	0	0	0	0	0	0	0	0	1	0	-1	0
#124	0	0	0	0	0	0	-1	1	1	-1	0	0	0	0	0	0	0	0	0	1	1	0	0	0	0	0	0	0	0	0	0	1	0	-1	0	0
#125	0	0	0	0	0	-1	1	1	-1	0	0	0	0	0	0	0	0	0	1	1	0	0	0	0	0	0	0	0	0	0	1	0	-1	0	0	0
#126	0	0	0	0	-1	1	1	-1	0	0	0	0	0	0	0	0	0	1	1	0	0	0	0	0	0	0	0	0	0	1	0	-1	0	0	0	0
#127	0	0	0	-1	1	1	-1	0	0	0	0	0	0	0	0	0	1	1	0	0	0	0	0	0	0	0	0	0	1	0	-1	0	0	0	0	0
#128	0	0	-1	1	1	-1	0	0	0	0	0	0	0	0	0	1	1	0	0	0	0	0	0	0	0	0	0	1	0	-1	0	0	0	0	0	0
#129	0	-1	1	1	-1	0	0	0	0	0	0	0	0	0	1	1	0	0	0	0	0	0	0	0	0	0	1	0	-1	0	0	0	0	0	0	0
#130	-1	1	1	-1	0	0	0	0	0	0	0	0	0	1	1	0	0	0	0	0	0	0	0	0	0	1	0	-1	0	0	0	0	0	0	0	0
#131	1	1	-1	0	0	0	0	0	0	0	0	0	1	1	0	0	0	0	0	0	0	0	0	0	1	0	-1	0	0	0	0	0	0	0	0	0
#132	0	0	0	0	0	0	0	0	0	-1	1	0	0	0	0	0	0	0	0	0	0	0	1	1	0	0	0	0	0	0	0	0	0	0	1	0
#133	0	0	0	0	0	0	0	0	0	-1	2	-2	0	0	0	0	0	0	0	0	0	0	1	0	0	0	0	0	0	0	0	0	0	0	1	-1
#134	0	0	0	0	0	0	0	0	-1	2	-1	0	0	0	0	0	0	0	0	0	0	1	0	0	0	0	0	0	0	0	0	0	0	1	-1	0
#135	0	0	0	0	0	0	0	-1	2	-1	0	0	0	0	0	0	0	0	0	0	1	0	0	0	0	0	0	0	0	0	0	0	1	-1	0	0
#136	0	0	0	0	0	0	-1	2	-1	0	0	0	0	0	0	0	0	0	0	1	0	0	0	0	0	0	0	0	0	0	0	1	-1	0	0	0
#137	0	0	0	0	0	-1	2	-1	0	0	0	0	0	0	0	0	0	0	1	0	0	0	0	0	0	0	0	0	0	0	1	-1	0	0	0	0
#138	0	0	0	0	-1	2	-1	0	0	0	0	0	0	0	0	0	0	1	0	0	0	0	0	0	0	0	0	0	0	1	-1	0	0	0	0	0
#139	0	0	0	-1	2	-1	0	0	0	0	0	0	0	0	0	0	1	0	0	0	0	0	0	0	0	0	0	0	1	-1	0	0	0	0	0	0
#140	0	0	-1	2	-1	0	0	0	0	0	0	0	0	0	0	1	0	0	0	0	0	0	0	0	0	0	0	1	-1	0	0	0	0	0	0	0
#141	0	-1	2	-1	0	0	0	0	0	0	0	0	0	0	1	0	0	0	0	0	0	0	0	0	0	0	1	-1	0	0	0	0	0	0	0	0
#142	-1	2	-1	0	0	0	0	0	0	0	0	0	0	1	0	0	0	0	0	0	0	0	0	0	0	1	-1	0	0	0	0	0	0	0	0	0
#143	2	-1	0	0	0	0	0	0	0	0	0	0	1	0	0	0	0	0	0	0	0	0	0	0	1	-1	0	0	0	0	0	0	0	0	0	0
#144	0	0	0	0	0	0	0	0	0	0	-1	2	0	0	0	0	0	0	0	0	0	0	0	1	0	0	0	0	0	0	0	0	0	0	0	1

Table 2. *(continued)*

ALGEBRA C2

	OMEGA		ALPHA		EPSILON	
# 1	2	0	2	1	2	0
# 2	0	1	1	1	1	1
# 3	-2	2	0	1	0	2
# 4	2	-1	1	0	1	-1

ALGEBRA C3

	OMEGA			ALPHA			EPSILON		
# 1	2	0	0	2	2	1	2	0	0
# 2	0	1	0	1	2	1	1	1	0
# 3	-2	2	0	0	2	1	0	2	0
# 4	1	-1	1	1	1	1	1	0	1
# 5	-1	0	1	0	1	1	0	1	1
# 6	1	1	-1	1	1	0	1	0	-1
# 7	0	-2	2	0	0	1	0	0	2
# 8	-1	2	-1	0	1	0	0	1	-1
# 9	2	-1	0	1	0	0	1	-1	0

ALGEBRA C4

	OMEGA				ALPHA				EPSILON			
# 1	2	0	0	0	2	2	2	1	2	0	0	0
# 2	0	1	0	0	1	2	2	1	1	1	0	0
# 3	-2	2	0	0	0	2	2	1	0	2	0	0
# 4	1	-1	1	0	1	1	2	1	1	0	1	0
# 5	-1	0	1	0	0	1	2	1	0	1	1	0
# 6	1	0	-1	1	1	1	1	1	1	0	0	1
# 7	0	-2	2	0	0	0	2	1	0	0	2	0
# 8	-1	1	-1	1	0	1	1	1	0	1	0	1
# 9	1	0	1	-1	1	1	1	0	1	0	0	-1
# 10	0	-1	0	1	0	0	1	1	0	0	1	1
# 11	-1	1	1	-1	0	1	1	0	0	1	0	-1
# 12	1	1	-1	0	1	1	0	0	1	0	-1	0
# 13	0	0	-2	2	0	0	0	1	0	0	0	2
# 14	0	-1	2	-1	0	0	1	0	0	0	1	-1
# 15	-1	2	-1	0	0	1	0	0	0	1	-1	0
# 16	2	-1	0	0	1	0	0	0	1	-1	0	0

ALGEBRA C5

#	OMEGA					ALPHA					EPSILON				
# 1	2	0	0	0	0	2	2	2	2	1	2	0	0	0	0
# 2	0	1	0	0	0	1	2	2	2	1	1	1	0	0	0
# 3	-2	2	0	0	0	0	2	2	2	1	0	2	0	0	0
# 4	1	-1	1	0	0	1	1	2	2	1	1	0	1	0	0
# 5	-1	0	1	0	0	0	1	2	2	1	0	1	1	0	0
# 6	1	0	-1	1	0	1	1	1	2	1	1	0	0	1	0
# 7	0	-2	2	0	0	0	0	2	2	1	0	0	2	0	0
# 8	-1	1	-1	1	0	0	1	1	2	1	0	1	0	1	0
# 9	1	0	0	-1	1	1	1	1	1	1	1	0	0	0	1
# 10	0	-1	0	1	0	0	0	1	2	1	0	0	1	1	0
# 11	-1	1	0	-1	1	0	1	1	1	1	0	1	0	0	1
# 12	1	0	0	1	-1	1	1	1	1	0	1	0	0	0	-1
# 13	0	0	-2	2	0	0	0	0	2	1	0	0	0	2	0
# 14	0	-1	1	-1	1	0	0	1	1	1	0	0	1	0	1
# 15	-1	1	0	1	-1	0	1	1	1	0	0	1	0	0	-1
# 16	1	0	1	-1	0	1	1	1	0	0	1	0	0	-1	0
# 17	0	0	-1	0	1	0	0	0	1	1	0	0	0	1	1
# 18	0	-1	1	1	-1	0	0	1	1	0	0	0	1	0	-1
# 19	-1	1	1	-1	0	0	1	1	0	0	0	1	0	-1	0
# 20	1	1	-1	0	0	1	1	0	0	0	1	0	-1	0	0
# 21	0	0	0	-2	2	0	0	0	0	1	0	0	0	0	2
# 22	0	0	-1	2	-1	0	0	0	1	0	0	0	0	1	-1
# 23	0	-1	2	-1	0	0	0	1	0	0	0	0	1	-1	0
# 24	-1	2	-1	0	0	0	1	0	0	0	0	1	-1	0	0
# 25	2	-1	0	0	0	1	0	0	0	0	1	-1	0	0	0

ALGEBRA C6

#	OMEGA						ALPHA						EPSILON					
# 1	2	0	0	0	0	0	2	2	2	2	2	1	2	0	0	0	0	0
# 2	0	1	0	0	0	0	1	2	2	2	2	1	1	1	0	0	0	0
# 3	-2	2	0	0	0	0	0	2	2	2	2	1	0	2	0	0	0	0
# 4	1	-1	1	0	0	0	1	1	2	2	2	1	1	0	1	0	0	0
# 5	-1	0	1	0	0	0	0	1	2	2	2	1	0	1	1	0	0	0
# 6	1	0	-1	1	0	0	1	1	1	2	2	1	1	0	0	1	0	0
# 7	0	-2	2	0	0	0	0	0	2	2	2	1	0	0	2	0	0	0
# 8	-1	1	-1	1	0	0	0	1	1	2	2	1	0	1	0	1	0	0
# 9	1	0	0	-1	1	0	1	1	1	1	2	1	1	0	0	0	1	0
# 10	0	-1	0	1	0	0	0	0	1	2	2	1	0	0	1	1	0	0
# 11	-1	1	0	-1	1	0	0	1	1	1	2	1	0	1	0	0	1	0
# 12	1	0	0	0	-1	1	1	1	1	1	1	1	1	0	0	0	0	1
# 13	0	0	-2	2	0	0	0	0	0	2	2	1	0	0	0	2	0	0
# 14	0	-1	1	-1	1	0	0	0	1	1	2	1	0	0	1	0	1	0
# 15	-1	1	0	0	-1	1	0	1	1	1	1	1	0	1	0	0	0	1
# 16	1	0	0	0	1	-1	1	1	1	1	1	0	1	0	0	0	0	-1
# 17	0	0	-1	0	1	0	0	0	0	1	2	1	0	0	0	1	1	0
# 18	0	-1	1	0	-1	1	0	0	1	1	1	1	0	0	1	0	0	1
# 19	-1	1	0	0	1	-1	0	1	1	1	1	0	0	1	0	0	0	-1
# 20	1	0	0	1	-1	0	1	1	1	1	0	0	1	0	0	0	-1	0
# 21	0	0	0	-2	2	0	0	0	0	0	2	1	0	0	0	0	2	0
# 22	0	0	-1	1	-1	1	0	0	0	1	1	1	0	0	0	1	0	1
# 23	0	-1	1	0	1	-1	0	0	1	1	1	0	0	0	1	0	0	-1
# 24	-1	1	0	1	-1	0	0	1	1	1	0	0	0	1	0	0	-1	0
# 25	1	0	1	-1	0	0	1	1	1	0	0	0	1	0	0	-1	0	0
# 26	0	0	0	-1	0	1	0	0	0	0	1	1	0	0	0	0	1	1
# 27	0	0	-1	1	1	-1	0	0	0	1	1	0	0	0	0	1	0	-1
# 28	0	-1	1	1	-1	0	0	0	1	1	0	0	0	0	1	0	-1	0
# 29	-1	1	1	-1	0	0	0	1	1	0	0	0	0	1	0	-1	0	0
# 30	1	1	-1	0	0	0	1	1	0	0	0	0	1	0	-1	0	0	0
# 31	0	0	0	0	-2	2	0	0	0	0	0	1	0	0	0	0	0	2
# 32	0	0	0	-1	2	-1	0	0	0	0	1	0	0	0	0	0	1	-1
# 33	0	0	-1	2	-1	0	0	0	0	1	0	0	0	0	0	1	-1	0
# 34	0	-1	2	-1	0	0	0	0	1	0	0	0	0	0	1	-1	0	0
# 35	-1	2	-1	0	0	0	0	1	0	0	0	0	0	1	-1	0	0	0
# 36	2	-1	0	0	0	0	1	0	0	0	0	0	1	-1	0	0	0	0

Table 2. *(continued)*

ALGEBRA C7

#	OMEGA							ALPHA							EPSILON						
# 1	2	0	0	0	0	0	0	2	2	2	2	2	2	1	2	0	0	0	0	0	0
# 2	0	1	0	0	0	0	0	1	2	2	2	2	2	1	1	1	0	0	0	0	0
# 3	-2	2	0	0	0	0	0	0	2	2	2	2	2	1	0	2	0	0	0	0	0
# 4	1	-1	1	0	0	0	0	1	1	2	2	2	2	1	1	0	1	0	0	0	0
# 5	-1	0	1	0	0	0	0	0	1	2	2	2	2	1	0	1	1	0	0	0	0
# 6	1	0	-1	1	0	0	0	1	1	1	2	2	2	1	1	0	0	1	0	0	0
# 7	0	-2	2	0	0	0	0	0	0	2	2	2	2	1	0	0	2	0	0	0	0
# 8	-1	1	-1	1	0	0	0	0	1	1	2	2	2	1	0	1	0	1	0	0	0
# 9	1	0	0	-1	1	0	0	1	1	1	1	2	2	1	1	0	0	0	1	0	0
# 10	0	-1	0	1	0	0	0	0	0	1	2	2	2	1	0	0	1	1	0	0	0
# 11	-1	1	0	-1	1	0	0	0	1	1	1	2	2	1	0	1	0	0	1	0	0
# 12	1	0	0	0	-1	1	0	1	1	1	1	1	2	1	1	0	0	0	0	1	0
# 13	0	0	-2	2	0	0	0	0	0	0	2	2	2	1	0	0	0	2	0	0	0
# 14	0	-1	1	-1	1	0	0	0	0	1	1	2	2	1	0	0	1	0	1	0	0
# 15	-1	1	0	0	-1	1	0	0	1	1	1	1	2	1	0	1	0	0	0	1	0
# 16	1	0	0	0	0	-1	1	1	1	1	1	1	1	1	1	0	0	0	0	0	1
# 17	0	0	-1	0	1	0	0	0	0	0	1	2	2	1	0	0	0	1	1	0	0
# 18	0	-1	1	0	-1	1	0	0	0	1	1	1	2	1	0	0	1	0	0	1	0
# 19	-1	1	0	0	0	-1	1	0	1	1	1	1	1	1	0	1	0	0	0	0	1
# 20	1	0	0	0	0	1	-1	1	1	1	1	1	1	0	1	0	0	0	0	0	-1
# 21	0	0	0	-2	2	0	0	0	0	0	0	2	2	1	0	0	0	0	2	0	0
# 22	0	0	-1	1	-1	1	0	0	0	0	1	1	2	1	0	0	0	1	0	1	0
# 23	0	-1	1	0	0	-1	1	0	0	1	1	1	1	1	0	0	1	0	0	0	1
# 24	-1	1	0	0	0	1	-1	0	1	1	1	1	1	0	0	1	0	0	0	0	-1
# 25	1	0	0	0	1	-1	0	1	1	1	1	1	0	0	1	0	0	0	0	-1	0
# 26	0	0	0	-1	0	1	0	0	0	0	0	1	2	1	0	0	0	0	1	1	0
# 27	0	0	-1	1	0	-1	1	0	0	0	1	1	1	1	0	0	0	1	0	0	1
# 28	0	-1	1	0	0	1	-1	0	0	1	1	1	1	0	0	0	1	0	0	0	-1
# 29	-1	1	0	0	1	-1	0	0	1	1	1	1	0	0	0	1	0	0	0	-1	0
# 30	1	0	0	1	-1	0	0	1	1	1	1	0	0	0	1	0	0	0	-1	0	0
# 31	0	0	0	0	-2	2	0	0	0	0	0	0	2	1	0	0	0	0	0	2	0
# 32	0	0	0	-1	1	-1	1	0	0	0	0	1	1	1	0	0	0	0	1	0	1
# 33	0	0	-1	1	0	1	-1	0	0	0	1	1	1	0	0	0	0	1	0	0	-1
# 34	0	-1	1	0	1	-1	0	0	0	1	1	1	0	0	0	0	1	0	0	-1	0
# 35	-1	1	0	1	-1	0	0	0	1	1	1	0	0	0	0	1	0	0	-1	0	0
# 36	1	0	1	-1	0	0	0	1	1	1	0	0	0	0	1	0	0	-1	0	0	0
# 37	0	0	0	0	-1	0	1	0	0	0	0	0	1	1	0	0	0	0	0	1	1
# 38	0	0	0	-1	1	1	-1	0	0	0	0	1	1	0	0	0	0	0	1	0	-1
# 39	0	0	-1	1	1	-1	0	0	0	0	1	1	0	0	0	0	0	1	0	-1	0
# 40	0	-1	1	1	-1	0	0	0	0	1	1	0	0	0	0	0	1	0	-1	0	0
# 41	-1	1	1	-1	0	0	0	0	1	1	0	0	0	0	0	1	0	-1	0	0	0
# 42	1	1	-1	0	0	0	0	1	1	0	0	0	0	0	1	0	-1	0	0	0	0
# 43	0	0	0	0	0	-2	2	0	0	0	0	0	0	1	0	0	0	0	0	0	2
# 44	0	0	0	0	-1	2	-1	0	0	0	0	0	1	0	0	0	0	0	0	1	-1
# 45	0	0	0	-1	2	-1	0	0	0	0	0	1	0	0	0	0	0	0	1	-1	0
# 46	0	0	-1	2	-1	0	0	0	0	0	1	0	0	0	0	0	0	1	-1	0	0
# 47	0	-1	2	-1	0	0	0	0	0	1	0	0	0	0	0	0	1	-1	0	0	0
# 48	-1	2	-1	0	0	0	0	0	1	0	0	0	0	0	0	1	-1	0	0	0	0
# 49	2	-1	0	0	0	0	0	1	0	0	0	0	0	0	1	-1	0	0	0	0	0

ALGEBRA C8

#	OMEGA								ALPHA								EPSILON							
# 1	2	0	0	0	0	0	0	0	2	2	2	2	2	2	2	1	2	0	0	0	0	0	0	0
# 2	0	1	0	0	0	0	0	0	1	2	2	2	2	2	2	1	1	1	0	0	0	0	0	0
# 3	-2	2	0	0	0	0	0	0	0	2	2	2	2	2	2	1	0	2	0	0	0	0	0	0
# 4	1	-1	1	0	0	0	0	0	1	1	2	2	2	2	2	1	1	0	1	0	0	0	0	0
# 5	-1	0	1	0	0	0	0	0	0	1	2	2	2	2	2	1	0	1	1	0	0	0	0	0
# 6	1	0	-1	1	0	0	0	0	1	1	1	2	2	2	2	1	1	0	0	1	0	0	0	0
# 7	0	-2	2	0	0	0	0	0	0	0	2	2	2	2	2	1	0	0	2	0	0	0	0	0
# 8	-1	1	-1	1	0	0	0	0	0	1	1	2	2	2	2	1	0	1	0	1	0	0	0	0
# 9	1	0	0	-1	1	0	0	0	1	1	1	1	2	2	2	1	1	0	0	0	1	0	0	0
# 10	0	-1	0	1	0	0	0	0	0	0	1	2	2	2	2	1	0	0	1	1	0	0	0	0
# 11	-1	1	0	-1	1	0	0	0	0	1	1	1	2	2	2	1	0	1	0	0	1	0	0	0
# 12	1	0	0	0	-1	1	0	0	1	1	1	1	1	2	2	1	1	0	0	0	0	1	0	0
# 13	0	0	-2	2	0	0	0	0	0	0	0	2	2	2	2	1	0	0	0	2	0	0	0	0
# 14	0	-1	1	-1	1	0	0	0	0	0	1	1	2	2	2	1	0	0	1	0	1	0	0	0
# 15	-1	1	0	0	-1	1	0	0	0	1	1	1	1	2	2	1	0	1	0	0	0	1	0	0
# 16	1	0	0	0	0	-1	1	0	1	1	1	1	1	1	2	1	1	0	0	0	0	0	1	0
# 17	0	0	-1	0	1	0	0	0	0	0	0	1	2	2	2	1	0	0	0	1	1	0	0	0
# 18	0	-1	1	0	-1	1	0	0	0	0	1	1	1	2	2	1	0	0	1	0	0	1	0	0
# 19	-1	1	0	0	0	-1	1	0	0	1	1	1	1	1	2	1	0	1	0	0	0	0	1	0
# 20	1	0	0	0	0	0	-1	1	1	1	1	1	1	1	1	1	1	0	0	0	0	0	0	1
# 21	0	0	0	-2	2	0	0	0	0	0	0	0	2	2	2	1	0	0	0	0	2	0	0	0
# 22	0	0	-1	1	-1	1	0	0	0	0	0	1	1	2	2	1	0	0	0	1	0	1	0	0
# 23	0	-1	1	0	0	-1	1	0	0	0	1	1	1	1	2	1	0	0	1	0	0	0	1	0
# 24	-1	1	0	0	0	0	-1	1	0	1	1	1	1	1	1	1	0	1	0	0	0	0	0	1
# 25	1	0	0	0	0	0	1	-1	1	1	1	1	1	1	1	0	1	0	0	0	0	0	0	-1
# 26	0	0	0	-1	0	1	0	0	0	0	0	0	1	2	2	1	0	0	0	0	1	1	0	0
# 27	0	0	-1	1	0	-1	1	0	0	0	0	1	1	1	2	1	0	0	0	1	0	0	1	0
# 28	0	-1	1	0	0	0	-1	1	0	0	1	1	1	1	1	1	0	0	1	0	0	0	0	1
# 29	-1	1	0	0	0	0	1	-1	0	1	1	1	1	1	1	0	0	1	0	0	0	0	0	-1
# 30	1	0	0	0	0	1	-1	0	1	1	1	1	1	1	0	0	1	0	0	0	0	0	-1	0
# 31	0	0	0	0	-2	2	0	0	0	0	0	0	0	2	2	1	0	0	0	0	0	2	0	0
# 32	0	0	0	-1	1	-1	1	0	0	0	0	0	1	1	2	1	0	0	0	0	1	0	1	0
# 33	0	0	-1	1	0	0	-1	1	0	0	0	1	1	1	1	1	0	0	0	1	0	0	0	1
# 34	0	-1	1	0	0	0	1	-1	0	0	1	1	1	1	1	0	0	0	1	0	0	0	0	-1
# 35	-1	1	0	0	0	1	-1	0	0	1	1	1	1	1	0	0	0	1	0	0	0	0	-1	0
# 36	1	0	0	0	1	-1	0	0	1	1	1	1	1	0	0	0	1	0	0	0	0	-1	0	0
# 37	0	0	0	0	-1	0	1	0	0	0	0	0	0	1	2	1	0	0	0	0	0	1	1	0
# 38	0	0	0	-1	1	0	-1	1	0	0	0	0	1	1	1	1	0	0	0	0	1	0	0	1
# 39	0	0	-1	1	0	0	1	-1	0	0	0	1	1	1	1	0	0	0	0	1	0	0	0	-1
# 40	0	-1	1	0	0	1	-1	0	0	0	1	1	1	1	0	0	0	0	1	0	0	0	-1	0
# 41	-1	1	0	0	1	-1	0	0	0	1	1	1	1	0	0	0	0	1	0	0	0	-1	0	0
# 42	1	0	0	1	-1	0	0	0	1	1	1	1	0	0	0	0	1	0	0	0	-1	0	0	0
# 43	0	0	0	0	0	-2	2	0	0	0	0	0	0	0	2	1	0	0	0	0	0	0	2	0
# 44	0	0	0	0	-1	1	-1	1	0	0	0	0	0	1	1	1	0	0	0	0	0	1	0	1
# 45	0	0	0	-1	1	0	1	-1	0	0	0	0	1	1	1	0	0	0	0	0	1	0	0	-1
# 46	0	0	-1	1	0	1	-1	0	0	0	0	1	1	1	0	0	0	0	0	1	0	0	-1	0
# 47	0	-1	1	0	1	-1	0	0	0	0	1	1	1	0	0	0	0	0	1	0	0	-1	0	0
# 48	-1	1	0	1	-1	0	0	0	0	1	1	1	0	0	0	0	0	1	0	0	-1	0	0	0
# 49	1	0	1	-1	0	0	0	0	1	1	1	0	0	0	0	0	1	0	0	-1	0	0	0	0
# 50	0	0	0	0	0	-1	0	1	0	0	0	0	0	0	1	1	0	0	0	0	0	0	1	1
# 51	0	0	0	0	-1	1	1	-1	0	0	0	0	0	1	1	0	0	0	0	0	0	1	0	-1
# 52	0	0	0	-1	1	1	-1	0	0	0	0	0	1	1	0	0	0	0	0	0	1	0	-1	0
# 53	0	0	-1	1	1	-1	0	0	0	0	0	1	1	0	0	0	0	0	0	1	0	-1	0	0
# 54	0	-1	1	1	-1	0	0	0	0	0	1	1	0	0	0	0	0	0	1	0	-1	0	0	0
# 55	-1	1	1	-1	0	0	0	0	0	1	1	0	0	0	0	0	0	1	0	-1	0	0	0	0
# 56	1	1	-1	0	0	0	0	0	1	1	0	0	0	0	0	0	1	0	-1	0	0	0	0	0
# 57	0	0	0	0	0	0	-2	2	0	0	0	0	0	0	0	1	0	0	0	0	0	0	0	2
# 58	0	0	0	0	0	-1	2	-1	0	0	0	0	0	0	1	0	0	0	0	0	0	0	1	-1
# 59	0	0	0	0	-1	2	-1	0	0	0	0	0	0	1	0	0	0	0	0	0	0	1	-1	0
# 60	0	0	0	-1	2	-1	0	0	0	0	0	0	1	0	0	0	0	0	0	0	1	-1	0	0
# 61	0	0	-1	2	-1	0	0	0	0	0	0	1	0	0	0	0	0	0	0	1	-1	0	0	0
# 62	0	-1	2	-1	0	0	0	0	0	0	1	0	0	0	0	0	0	0	1	-1	0	0	0	0
# 63	-1	2	-1	0	0	0	0	0	0	1	0	0	0	0	0	0	0	1	-1	0	0	0	0	0
# 64	2	-1	0	0	0	0	0	0	1	0	0	0	0	0	0	0	1	-1	0	0	0	0	0	0

Table 2. *(continued)*

ALGEBRA C9

	OMEGA									ALPHA									EPSILON								
# 1	2	0	0	0	0	0	0	0	0	2	2	2	2	2	2	2	2	1	2	0	0	0	0	0	0	0	0
# 2	0	1	0	0	0	0	0	0	0	1	2	2	2	2	2	2	2	1	1	1	0	0	0	0	0	0	0
# 3	-2	2	0	0	0	0	0	0	0	0	2	2	2	2	2	2	2	1	0	2	0	0	0	0	0	0	0
# 4	1	-1	1	0	0	0	0	0	0	1	1	2	2	2	2	2	2	1	1	0	1	0	0	0	0	0	0
# 5	-1	0	1	0	0	0	0	0	0	0	1	2	2	2	2	2	2	1	0	1	1	0	0	0	0	0	0
# 6	1	0	-1	1	0	0	0	0	0	1	1	1	2	2	2	2	2	1	1	0	0	1	0	0	0	0	0
# 7	0	-2	2	0	0	0	0	0	0	0	0	2	2	2	2	2	2	1	0	0	2	0	0	0	0	0	0
# 8	-1	1	-1	1	0	0	0	0	0	0	1	1	2	2	2	2	2	1	0	1	0	1	0	0	0	0	0
# 9	1	0	0	-1	1	0	0	0	0	1	1	1	1	2	2	2	2	1	1	0	0	0	1	0	0	0	0
# 10	0	-1	0	1	0	0	0	0	0	0	0	1	2	2	2	2	2	1	0	0	1	1	0	0	0	0	0
# 11	-1	1	0	-1	1	0	0	0	0	0	1	1	1	2	2	2	2	1	0	1	0	0	1	0	0	0	0
# 12	1	0	0	0	-1	1	0	0	0	1	1	1	1	1	2	2	2	1	1	0	0	0	0	1	0	0	0
# 13	0	0	-2	2	0	0	0	0	0	0	0	0	2	2	2	2	2	1	0	0	0	2	0	0	0	0	0
# 14	0	-1	1	-1	1	0	0	0	0	0	0	1	1	2	2	2	2	1	0	0	1	0	1	0	0	0	0
# 15	-1	1	0	0	-1	1	0	0	0	0	1	1	1	1	2	2	2	1	0	1	0	0	0	1	0	0	0
# 16	1	0	0	0	0	-1	1	0	0	1	1	1	1	1	1	2	2	1	1	0	0	0	0	0	1	0	0
# 17	0	0	-1	0	1	0	0	0	0	0	0	0	1	2	2	2	2	1	0	0	0	1	1	0	0	0	0
# 18	0	-1	1	0	-1	1	0	0	0	0	0	1	1	1	2	2	2	1	0	0	1	0	0	1	0	0	0
# 19	-1	1	0	0	0	-1	1	0	0	0	1	1	1	1	1	2	2	1	0	1	0	0	0	0	1	0	0
# 20	1	0	0	0	0	0	-1	1	0	1	1	1	1	1	1	1	2	1	1	0	0	0	0	0	0	1	0
# 21	0	0	0	-2	2	0	0	0	0	0	0	0	0	2	2	2	2	1	0	0	0	0	2	0	0	0	0
# 22	0	0	-1	1	-1	1	0	0	0	0	0	0	1	1	2	2	2	1	0	0	0	1	0	1	0	0	0
# 23	0	-1	1	0	0	-1	1	0	0	0	0	1	1	1	1	2	2	1	0	0	1	0	0	0	1	0	0
# 24	-1	1	0	0	0	0	-1	1	0	0	1	1	1	1	1	1	2	1	0	1	0	0	0	0	0	1	0
# 25	1	0	0	0	0	0	0	-1	1	1	1	1	1	1	1	1	1	1	1	0	0	0	0	0	0	0	1
# 26	0	0	0	-1	0	1	0	0	0	0	0	0	0	1	2	2	2	1	0	0	0	0	1	1	0	0	0
# 27	0	0	-1	1	0	-1	1	0	0	0	0	0	1	1	1	2	2	1	0	0	0	1	0	0	1	0	0
# 28	0	-1	1	0	0	0	-1	1	0	0	0	1	1	1	1	1	2	1	0	0	1	0	0	0	0	1	0
# 29	-1	1	0	0	0	0	0	-1	1	0	1	1	1	1	1	1	1	1	0	1	0	0	0	0	0	0	1
# 30	1	0	0	0	0	0	0	1	-1	1	1	1	1	1	1	1	1	0	1	0	0	0	0	0	0	0	-1
# 31	0	0	0	0	-2	2	0	0	0	0	0	0	0	0	2	2	2	1	0	0	0	0	0	2	0	0	0
# 32	0	0	0	-1	1	-1	1	0	0	0	0	0	0	1	1	2	2	1	0	0	0	0	1	0	1	0	0
# 33	0	0	-1	1	0	0	-1	1	0	0	0	0	1	1	1	1	2	1	0	0	0	1	0	0	0	1	0
# 34	0	-1	1	0	0	0	0	-1	1	0	0	1	1	1	1	1	1	1	0	0	1	0	0	0	0	0	1
# 35	-1	1	0	0	0	0	0	1	-1	0	1	1	1	1	1	1	1	0	0	1	0	0	0	0	0	0	-1
# 36	1	0	0	0	0	0	1	-1	0	1	1	1	1	1	1	1	0	0	1	0	0	0	0	0	0	-1	0
# 37	0	0	0	0	-1	0	1	0	0	0	0	0	0	0	1	2	2	1	0	0	0	0	0	1	1	0	0

# 38	0	0	0	-1	1	0	-1	1	0	0	0	0	0	1	1	1	2	1	0	0	0	0	1	0	0	1	0
# 39	0	0	-1	1	0	0	0	-1	1	0	0	0	1	1	1	1	1	1	0	0	0	1	0	0	0	0	1
# 40	0	-1	1	0	0	0	0	1	-1	0	0	1	1	1	1	1	1	0	0	0	1	0	0	0	0	0	-1
# 41	-1	1	0	0	0	0	1	-1	0	0	1	1	1	1	1	1	0	0	0	1	0	0	0	0	0	-1	0
# 42	1	0	0	0	0	1	-1	0	0	1	1	1	1	1	1	0	0	0	1	0	0	0	0	0	-1	0	0
# 43	0	0	0	0	0	-2	2	0	0	0	0	0	0	0	0	2	2	1	0	0	0	0	0	0	2	0	0
# 44	0	0	0	0	-1	1	-1	1	0	0	0	0	0	0	1	1	2	1	0	0	0	0	0	1	0	1	0
# 45	0	0	0	-1	1	0	0	-1	1	0	0	0	0	1	1	1	1	1	0	0	0	0	1	0	0	0	1
# 46	0	0	-1	1	0	0	0	1	-1	0	0	0	1	1	1	1	1	0	0	0	0	1	0	0	0	0	-1
# 47	0	-1	1	0	0	0	1	-1	0	0	0	1	1	1	1	1	0	0	0	0	1	0	0	0	0	-1	0
# 48	-1	1	0	0	0	1	-1	0	0	0	1	1	1	1	1	0	0	0	0	1	0	0	0	0	-1	0	0
# 49	1	0	0	0	1	-1	0	0	0	1	1	1	1	1	0	0	0	0	1	0	0	0	0	-1	0	0	0
# 50	0	0	0	0	0	-1	0	1	0	0	0	0	0	0	0	1	2	1	0	0	0	0	0	0	1	1	0
# 51	0	0	0	0	-1	1	0	-1	1	0	0	0	0	0	1	1	1	1	0	0	0	0	0	1	0	0	1
# 52	0	0	0	-1	1	0	0	1	-1	0	0	0	0	1	1	1	1	0	0	0	0	0	1	0	0	0	-1
# 53	0	0	-1	1	0	0	1	-1	0	0	0	0	1	1	1	1	0	0	0	0	0	1	0	0	0	-1	0
# 54	0	-1	1	0	0	1	-1	0	0	0	0	1	1	1	1	0	0	0	0	0	1	0	0	0	-1	0	0
# 55	-1	1	0	0	1	-1	0	0	0	0	1	1	1	1	0	0	0	0	0	1	0	0	0	-1	0	0	0
# 56	1	0	0	1	-1	0	0	0	0	1	1	1	1	0	0	0	0	0	1	0	0	0	-1	0	0	0	0
# 57	0	0	0	0	0	0	-2	2	0	0	0	0	0	0	0	0	2	1	0	0	0	0	0	0	0	2	0
# 58	0	0	0	0	0	-1	1	-1	1	0	0	0	0	0	0	1	1	1	0	0	0	0	0	0	1	0	1
# 59	0	0	0	0	-1	1	0	1	-1	0	0	0	0	0	1	1	1	0	0	0	0	0	0	1	0	0	-1
# 60	0	0	0	-1	1	0	1	-1	0	0	0	0	0	1	1	1	0	0	0	0	0	0	1	0	0	-1	0
# 61	0	0	-1	1	0	1	-1	0	0	0	0	0	1	1	1	0	0	0	0	0	0	1	0	0	-1	0	0
# 62	0	-1	1	0	1	-1	0	0	0	0	0	1	1	1	0	0	0	0	0	0	1	0	0	-1	0	0	0
# 63	-1	1	0	1	-1	0	0	0	0	0	1	1	1	0	0	0	0	0	0	1	0	0	-1	0	0	0	0
# 64	1	0	1	-1	0	0	0	0	0	1	1	1	0	0	0	0	0	0	1	0	0	-1	0	0	0	0	0
# 65	0	0	0	0	0	0	-1	0	1	0	0	0	0	0	0	0	1	1	0	0	0	0	0	0	0	1	1
# 66	0	0	0	0	0	-1	1	1	-1	0	0	0	0	0	0	1	1	0	0	0	0	0	0	0	1	0	-1
# 67	0	0	0	0	-1	1	1	-1	0	0	0	0	0	0	1	1	0	0	0	0	0	0	0	1	0	-1	0
# 68	0	0	0	-1	1	1	-1	0	0	0	0	0	0	1	1	0	0	0	0	0	0	0	1	0	-1	0	0
# 69	0	0	-1	1	1	-1	0	0	0	0	0	0	1	1	0	0	0	0	0	0	0	1	0	-1	0	0	0
# 70	0	-1	1	1	-1	0	0	0	0	0	0	1	1	0	0	0	0	0	0	0	1	0	-1	0	0	0	0
# 71	-1	1	1	-1	0	0	0	0	0	0	1	1	0	0	0	0	0	0	0	1	0	-1	0	0	0	0	0
# 72	1	1	-1	0	0	0	0	0	0	1	1	0	0	0	0	0	0	0	1	0	-1	0	0	0	0	0	0
# 73	0	0	0	0	0	0	0	-2	2	0	0	0	0	0	0	0	0	1	0	0	0	0	0	0	0	0	2
# 74	0	0	0	0	0	0	-1	2	-1	0	0	0	0	0	0	0	1	0	0	0	0	0	0	0	0	1	-1
# 75	0	0	0	0	0	-1	2	-1	0	0	0	0	0	0	0	1	0	0	0	0	0	0	0	0	1	-1	0
# 76	0	0	0	0	-1	2	-1	0	0	0	0	0	0	0	1	0	0	0	0	0	0	0	0	1	-1	0	0
# 77	0	0	0	-1	2	-1	0	0	0	0	0	0	0	1	0	0	0	0	0	0	0	0	1	-1	0	0	0
# 78	0	0	-1	2	-1	0	0	0	0	0	0	0	1	0	0	0	0	0	0	0	0	1	-1	0	0	0	0
# 79	0	-1	2	-1	0	0	0	0	0	0	0	1	0	0	0	0	0	0	0	0	1	-1	0	0	0	0	0
# 80	-1	2	-1	0	0	0	0	0	0	0	1	0	0	0	0	0	0	0	0	1	-1	0	0	0	0	0	0
# 81	2	-1	0	0	0	0	0	0	0	1	0	0	0	0	0	0	0	0	1	-1	0	0	0	0	0	0	0

Table 2. *(continued)*

ALGEBRA C10

OMEGA

#										
# 1	2	0	0	0	0	0	0	0	0	0
# 2	0	1	0	0	0	0	0	0	0	0
# 3	-2	2	0	0	0	0	0	0	0	0
# 4	1	-1	1	0	0	0	0	0	0	0
# 5	-1	0	1	0	0	0	0	0	0	0
# 6	1	0	-1	1	0	0	0	0	0	0
# 7	0	-2	2	0	0	0	0	0	0	0
# 8	-1	1	-1	1	0	0	0	0	0	0
# 9	1	0	0	-1	1	0	0	0	0	0
# 10	0	-1	0	1	0	0	0	0	0	0
# 11	-1	1	0	-1	1	0	0	0	0	0
# 12	1	0	0	0	-1	1	0	0	0	0
# 13	0	0	-2	2	0	0	0	0	0	0
# 14	0	-1	1	-1	1	0	0	0	0	0
# 15	-1	1	0	0	-1	1	0	0	0	0
# 16	1	0	0	0	0	-1	1	0	0	0
# 17	0	0	-1	0	1	0	0	0	0	0
# 18	0	-1	1	0	-1	1	0	0	0	0
# 19	-1	1	0	0	0	-1	1	0	0	0
# 20	1	0	0	0	0	0	-1	1	0	0
# 21	0	0	0	-2	2	0	0	0	0	0
# 22	0	0	-1	1	-1	1	0	0	0	0
# 23	0	-1	1	0	0	-1	1	0	0	0
# 24	-1	1	0	0	0	0	-1	1	0	0
# 25	1	0	0	0	0	0	0	-1	1	0
# 26	0	0	0	-1	0	1	0	0	0	0
# 27	0	0	-1	1	0	-1	1	0	0	0
# 28	0	-1	1	0	0	0	-1	1	0	0
# 29	-1	1	0	0	0	0	0	-1	1	0
# 30	1	0	0	0	0	0	0	0	-1	1
# 31	0	0	0	0	-2	2	0	0	0	0
# 32	0	0	0	-1	1	-1	1	0	0	0
# 33	0	0	-1	1	0	0	-1	1	0	0
# 34	0	-1	1	0	0	0	0	-1	1	0
# 35	-1	1	0	0	0	0	0	0	-1	1
# 36	1	0	0	0	0	0	0	0	1	-1
# 37	0	0	0	0	-1	0	1	0	0	0
# 38	0	0	0	-1	1	0	-1	1	0	0
# 39	0	0	-1	1	0	0	0	-1	1	0
# 40	0	-1	1	0	0	0	0	0	-1	1
# 41	-1	1	0	0	0	0	0	0	1	-1
# 42	1	0	0	0	0	0	0	1	-1	0
# 43	0	0	0	0	0	-2	2	0	0	0
# 44	0	0	0	0	-1	1	-1	1	0	0
# 45	0	0	0	-1	1	0	0	-1	1	0
# 46	0	0	-1	1	0	0	0	0	-1	1

ALPHA

#										
# 1	2	2	2	2	2	2	2	2	2	1
# 2	1	2	2	2	2	2	2	2	2	1
# 3	0	2	2	2	2	2	2	2	2	1
# 4	1	1	2	2	2	2	2	2	2	1
# 5	0	1	2	2	2	2	2	2	2	1
# 6	1	1	1	2	2	2	2	2	2	1
# 7	0	0	2	2	2	2	2	2	2	1
# 8	0	1	1	2	2	2	2	2	2	1
# 9	1	1	1	1	2	2	2	2	2	1
# 10	0	0	1	2	2	2	2	2	2	1
# 11	0	1	1	1	2	2	2	2	2	1
# 12	1	1	1	1	1	2	2	2	2	1
# 13	0	0	0	2	2	2	2	2	2	1
# 14	0	0	1	1	2	2	2	2	2	1
# 15	0	1	1	1	1	2	2	2	2	1
# 16	1	1	1	1	1	1	2	2	2	1
# 17	0	0	0	1	2	2	2	2	2	1
# 18	0	0	1	1	1	2	2	2	2	1
# 19	0	1	1	1	1	1	2	2	2	1
# 20	1	1	1	1	1	1	1	2	2	1
# 21	0	0	0	0	2	2	2	2	2	1
# 22	0	0	0	1	1	2	2	2	2	1
# 23	0	0	1	1	1	1	2	2	2	1
# 24	0	1	1	1	1	1	1	2	2	1
# 25	1	1	1	1	1	1	1	1	2	1
# 26	0	0	0	0	1	2	2	2	2	1
# 27	0	0	0	1	1	1	2	2	2	1
# 28	0	0	1	1	1	1	1	2	2	1
# 29	0	1	1	1	1	1	1	1	2	1
# 30	1	1	1	1	1	1	1	1	1	1
# 31	0	0	0	0	0	2	2	2	2	1
# 32	0	0	0	0	1	1	2	2	2	1
# 33	0	0	0	1	1	1	1	2	2	1
# 34	0	0	1	1	1	1	1	1	2	1
# 35	0	1	1	1	1	1	1	1	1	1
# 36	1	1	1	1	1	1	1	1	1	0
# 37	0	0	0	0	0	1	2	2	2	1
# 38	0	0	0	0	1	1	1	2	2	1
# 39	0	0	0	1	1	1	1	1	2	1
# 40	0	0	1	1	1	1	1	1	1	1
# 41	0	1	1	1	1	1	1	1	1	0
# 42	1	1	1	1	1	1	1	1	0	0
# 43	0	0	0	0	0	0	2	2	2	1
# 44	0	0	0	0	0	1	1	2	2	1
# 45	0	0	0	0	1	1	1	1	2	1
# 46	0	0	0	1	1	1	1	1	1	1

EPSILON

#										
# 1	2	0	0	0	0	0	0	0	0	0
# 2	1	1	0	0	0	0	0	0	0	0
# 3	0	2	0	0	0	0	0	0	0	0
# 4	1	0	1	0	0	0	0	0	0	0
# 5	0	1	1	0	0	0	0	0	0	0
# 6	1	0	0	1	0	0	0	0	0	0
# 7	0	0	2	0	0	0	0	0	0	0
# 8	0	1	0	1	0	0	0	0	0	0
# 9	1	0	0	0	1	0	0	0	0	0
# 10	0	0	1	1	0	0	0	0	0	0
# 11	0	1	0	0	1	0	0	0	0	0
# 12	1	0	0	0	0	1	0	0	0	0
# 13	0	0	0	2	0	0	0	0	0	0
# 14	0	0	1	0	1	0	0	0	0	0
# 15	0	1	0	0	0	1	0	0	0	0
# 16	1	0	0	0	0	0	1	0	0	0
# 17	0	0	0	1	1	0	0	0	0	0
# 18	0	0	1	0	0	1	0	0	0	0
# 19	0	1	0	0	0	0	1	0	0	0
# 20	1	0	0	0	0	0	0	1	0	0
# 21	0	0	0	0	2	0	0	0	0	0
# 22	0	0	0	1	0	1	0	0	0	0
# 23	0	0	1	0	0	0	1	0	0	0
# 24	0	1	0	0	0	0	0	1	0	0
# 25	1	0	0	0	0	0	0	0	1	0
# 26	0	0	0	0	1	1	0	0	0	0
# 27	0	0	0	1	0	0	1	0	0	0
# 28	0	0	1	0	0	0	0	1	0	0
# 29	0	1	0	0	0	0	0	0	1	0
# 30	1	0	0	0	0	0	0	0	0	1
# 31	0	0	0	0	0	2	0	0	0	0
# 32	0	0	0	0	1	0	1	0	0	0
# 33	0	0	0	1	0	0	0	1	0	0
# 34	0	0	1	0	0	0	0	0	1	0
# 35	0	1	0	0	0	0	0	0	0	1
# 36	1	0	0	0	0	0	0	0	0	-1
# 37	0	0	0	0	0	1	1	0	0	0
# 38	0	0	0	0	1	0	0	1	0	0
# 39	0	0	0	1	0	0	0	0	1	0
# 40	0	0	1	0	0	0	0	0	0	1
# 41	0	1	0	0	0	0	0	0	0	-1
# 42	1	0	0	0	0	0	0	0	-1	0
# 43	0	0	0	0	0	0	2	0	0	0
# 44	0	0	0	0	0	1	0	1	0	0
# 45	0	0	0	0	1	0	0	0	1	0
# 46	0	0	0	1	0	0	0	0	0	1

# 47	0	-1	1	0	0	0	0	0	1	-1	0	0	1	1	1	1	1	1	1	0	0	0	1	0	0	0	0	0	0	-1
# 48	-1	1	0	0	0	0	0	1	-1	0	0	1	1	1	1	1	1	1	0	0	0	1	0	0	0	0	0	0	-1	0
# 49	1	0	0	0	0	0	1	-1	0	0	1	1	1	1	1	1	1	0	0	0	1	0	0	0	0	0	0	-1	0	0
# 50	0	0	0	0	0	-1	0	1	0	0	0	0	0	0	0	0	1	2	2	1	0	0	0	0	0	0	1	1	0	0
# 51	0	0	0	0	-1	1	0	-1	1	0	0	0	0	0	0	1	1	1	2	1	0	0	0	0	0	1	0	0	1	0
# 52	0	0	0	-1	1	0	0	0	-1	1	0	0	0	0	1	1	1	1	1	1	0	0	0	0	1	0	0	0	0	1
# 53	0	0	-1	1	0	0	0	0	1	-1	0	0	0	1	1	1	1	1	1	0	0	0	0	1	0	0	0	0	0	-1
# 54	0	-1	1	0	0	0	0	1	-1	0	0	0	1	1	1	1	1	1	0	0	0	0	1	0	0	0	0	0	-1	0
# 55	-1	1	0	0	0	0	1	-1	0	0	0	1	1	1	1	1	1	0	0	0	0	1	0	0	0	0	0	-1	0	0
# 56	1	0	0	0	0	1	-1	0	0	0	1	1	1	1	1	1	0	0	0	0	1	0	0	0	0	0	-1	0	0	0
# 57	0	0	0	0	0	0	-2	2	0	0	0	0	0	0	0	0	0	2	2	1	0	0	0	0	0	0	0	2	0	0
# 58	0	0	0	0	0	-1	1	-1	1	0	0	0	0	0	0	0	1	1	2	1	0	0	0	0	0	0	1	0	1	0
# 59	0	0	0	0	-1	1	0	0	-1	1	0	0	0	0	0	1	1	1	1	1	0	0	0	0	0	1	0	0	0	1
# 60	0	0	0	-1	1	0	0	0	1	-1	0	0	0	0	1	1	1	1	1	0	0	0	0	0	1	0	0	0	0	-1
# 61	0	0	-1	1	0	0	0	1	-1	0	0	0	0	1	1	1	1	1	0	0	0	0	0	1	0	0	0	0	-1	0
# 62	0	-1	1	0	0	0	1	-1	0	0	0	0	1	1	1	1	1	0	0	0	0	0	1	0	0	0	0	-1	0	0
# 63	-1	1	0	0	0	1	-1	0	0	0	0	1	1	1	1	1	0	0	0	0	0	1	0	0	0	0	-1	0	0	0
# 64	1	0	0	0	1	-1	0	0	0	0	1	1	1	1	1	0	0	0	0	0	1	0	0	0	0	-1	0	0	0	0
# 65	0	0	0	0	0	0	-1	0	1	0	0	0	0	0	0	0	0	1	2	1	0	0	0	0	0	0	0	1	1	0
# 66	0	0	0	0	0	-1	1	0	-1	1	0	0	0	0	0	0	1	1	1	1	0	0	0	0	0	0	1	0	0	1
# 67	0	0	0	0	-1	1	0	0	1	-1	0	0	0	0	0	1	1	1	1	0	0	0	0	0	0	1	0	0	0	-1
# 68	0	0	0	-1	1	0	0	1	-1	0	0	0	0	0	1	1	1	1	0	0	0	0	0	0	1	0	0	0	-1	0
# 69	0	0	-1	1	0	0	1	-1	0	0	0	0	0	1	1	1	1	0	0	0	0	0	0	1	0	0	0	-1	0	0
# 70	0	-1	1	0	0	1	-1	0	0	0	0	0	1	1	1	1	0	0	0	0	0	0	1	0	0	0	-1	0	0	0
# 71	-1	1	0	0	1	-1	0	0	0	0	0	1	1	1	1	0	0	0	0	0	0	1	0	0	0	-1	0	0	0	0
# 72	1	0	0	1	-1	0	0	0	0	0	1	1	1	1	0	0	0	0	0	0	1	0	0	0	-1	0	0	0	0	0
# 73	0	0	0	0	0	0	0	-2	2	0	0	0	0	0	0	0	0	0	2	1	0	0	0	0	0	0	0	0	2	0
# 74	0	0	0	0	0	0	-1	1	-1	1	0	0	0	0	0	0	0	1	1	1	0	0	0	0	0	0	0	1	0	1
# 75	0	0	0	0	0	-1	1	0	1	-1	0	0	0	0	0	0	1	1	1	0	0	0	0	0	0	0	1	0	0	-1
# 76	0	0	0	0	-1	1	0	1	-1	0	0	0	0	0	0	1	1	1	0	0	0	0	0	0	0	1	0	0	-1	0
# 77	0	0	0	-1	1	0	1	-1	0	0	0	0	0	0	1	1	1	0	0	0	0	0	0	0	1	0	0	-1	0	0
# 78	0	0	-1	1	0	1	-1	0	0	0	0	0	0	1	1	1	0	0	0	0	0	0	0	1	0	0	-1	0	0	0
# 79	0	-1	1	0	1	-1	0	0	0	0	0	0	1	1	1	0	0	0	0	0	0	0	1	0	0	-1	0	0	0	0
# 80	-1	1	0	1	-1	0	0	0	0	0	0	1	1	1	0	0	0	0	0	0	0	1	0	0	-1	0	0	0	0	0
# 81	1	0	1	-1	0	0	0	0	0	0	1	1	1	0	0	0	0	0	0	0	1	0	0	-1	0	0	0	0	0	0
# 82	0	0	0	0	0	0	0	-1	0	1	0	0	0	0	0	0	0	0	1	1	0	0	0	0	0	0	0	0	1	1
# 83	0	0	0	0	0	0	-1	1	1	-1	0	0	0	0	0	0	0	1	1	0	0	0	0	0	0	0	0	1	0	-1
# 84	0	0	0	0	0	-1	1	1	-1	0	0	0	0	0	0	0	1	1	0	0	0	0	0	0	0	0	1	0	-1	0
# 85	0	0	0	0	-1	1	1	-1	0	0	0	0	0	0	0	1	1	0	0	0	0	0	0	0	0	1	0	-1	0	0
# 86	0	0	0	-1	1	1	-1	0	0	0	0	0	0	0	1	1	0	0	0	0	0	0	0	0	1	0	-1	0	0	0
# 87	0	0	-1	1	1	-1	0	0	0	0	0	0	0	1	1	0	0	0	0	0	0	0	0	1	0	-1	0	0	0	0
# 88	0	-1	1	1	-1	0	0	0	0	0	0	0	1	1	0	0	0	0	0	0	0	0	1	0	-1	0	0	0	0	0
# 89	-1	1	1	-1	0	0	0	0	0	0	0	1	1	0	0	0	0	0	0	0	0	1	0	-1	0	0	0	0	0	0
# 90	1	1	-1	0	0	0	0	0	0	0	1	1	0	0	0	0	0	0	0	0	1	0	-1	0	0	0	0	0	0	0
# 91	0	0	0	0	0	0	0	0	-2	2	0	0	0	0	0	0	0	0	0	1	0	0	0	0	0	0	0	0	0	2
# 92	0	0	0	0	0	0	0	-1	2	-1	0	0	0	0	0	0	0	0	1	0	0	0	0	0	0	0	0	0	1	-1
# 93	0	0	0	0	0	0	-1	2	-1	0	0	0	0	0	0	0	0	1	0	0	0	0	0	0	0	0	0	1	-1	0
# 94	0	0	0	0	0	-1	2	-1	0	0	0	0	0	0	0	0	1	0	0	0	0	0	0	0	0	0	1	-1	0	0
# 95	0	0	0	0	-1	2	-1	0	0	0	0	0	0	0	0	1	0	0	0	0	0	0	0	0	0	1	-1	0	0	0
# 96	0	0	0	-1	2	-1	0	0	0	0	0	0	0	0	1	0	0	0	0	0	0	0	0	0	1	-1	0	0	0	0
# 97	0	0	-1	2	-1	0	0	0	0	0	0	0	0	1	0	0	0	0	0	0	0	0	0	1	-1	0	0	0	0	0
# 98	0	-1	2	-1	0	0	0	0	0	0	0	0	1	0	0	0	0	0	0	0	0	0	1	-1	0	0	0	0	0	0
# 99	-1	2	-1	0	0	0	0	0	0	0	0	1	0	0	0	0	0	0	0	0	0	1	-1	0	0	0	0	0	0	0
#100	2	-1	0	0	0	0	0	0	0	0	1	0	0	0	0	0	0	0	0	0	1	-1	0	0	0	0	0	0	0	0

Table 2. *(continued)*

ALGEBRA C11

	OMEGA											ALPHA											EPSILON										
# 1	2	0	0	0	0	0	0	0	0	0	0	2	2	2	2	2	2	2	2	2	2	1	2	0	0	0	0	0	0	0	0	0	0
# 2	0	1	0	0	0	0	0	0	0	0	0	1	2	2	2	2	2	2	2	2	2	1	1	1	0	0	0	0	0	0	0	0	0
# 3	-2	2	0	0	0	0	0	0	0	0	0	0	2	2	2	2	2	2	2	2	2	1	0	2	0	0	0	0	0	0	0	0	0
# 4	1	-1	1	0	0	0	0	0	0	0	0	1	1	2	2	2	2	2	2	2	2	1	1	0	1	0	0	0	0	0	0	0	0
# 5	-1	0	1	0	0	0	0	0	0	0	0	0	1	2	2	2	2	2	2	2	2	1	0	1	1	0	0	0	0	0	0	0	0
# 6	1	0	-1	1	0	0	0	0	0	0	0	1	1	1	2	2	2	2	2	2	2	1	1	0	0	1	0	0	0	0	0	0	0
# 7	0	-2	2	0	0	0	0	0	0	0	0	0	0	2	2	2	2	2	2	2	2	1	0	0	2	0	0	0	0	0	0	0	0
# 8	-1	1	-1	1	0	0	0	0	0	0	0	0	1	1	2	2	2	2	2	2	2	1	0	1	0	1	0	0	0	0	0	0	0
# 9	1	0	0	-1	1	0	0	0	0	0	0	1	1	1	1	2	2	2	2	2	2	1	1	0	0	0	1	0	0	0	0	0	0
# 10	0	-1	0	1	0	0	0	0	0	0	0	0	0	1	2	2	2	2	2	2	2	1	0	0	1	1	0	0	0	0	0	0	0
# 11	-1	1	0	-1	1	0	0	0	0	0	0	0	1	1	1	2	2	2	2	2	2	1	0	1	0	0	1	0	0	0	0	0	0
# 12	1	0	0	0	-1	1	0	0	0	0	0	1	1	1	1	1	2	2	2	2	2	1	1	0	0	0	0	1	0	0	0	0	0
# 13	0	0	-2	2	0	0	0	0	0	0	0	0	0	0	2	2	2	2	2	2	2	1	0	0	0	2	0	0	0	0	0	0	0
# 14	0	-1	1	-1	1	0	0	0	0	0	0	0	0	1	1	2	2	2	2	2	2	1	0	0	1	0	1	0	0	0	0	0	0
# 15	-1	1	0	0	-1	1	0	0	0	0	0	0	1	1	1	1	2	2	2	2	2	1	0	1	0	0	0	1	0	0	0	0	0
# 16	1	0	0	0	0	-1	1	0	0	0	0	1	1	1	1	1	1	2	2	2	2	1	1	0	0	0	0	0	1	0	0	0	0
# 17	0	0	-1	0	1	0	0	0	0	0	0	0	0	0	1	2	2	2	2	2	2	1	0	0	0	1	1	0	0	0	0	0	0
# 18	0	-1	1	0	-1	1	0	0	0	0	0	0	0	1	1	1	2	2	2	2	2	1	0	0	1	0	0	1	0	0	0	0	0
# 19	-1	1	0	0	0	-1	1	0	0	0	0	0	1	1	1	1	1	2	2	2	2	1	0	1	0	0	0	0	1	0	0	0	0
# 20	1	0	0	0	0	0	-1	1	0	0	0	1	1	1	1	1	1	1	2	2	2	1	1	0	0	0	0	0	0	1	0	0	0
# 21	0	0	0	-2	2	0	0	0	0	0	0	0	0	0	0	2	2	2	2	2	2	1	0	0	0	0	2	0	0	0	0	0	0
# 22	0	0	-1	1	-1	1	0	0	0	0	0	0	0	0	1	1	2	2	2	2	2	1	0	0	0	1	0	1	0	0	0	0	0
# 23	0	-1	1	0	0	-1	1	0	0	0	0	0	0	1	1	1	1	2	2	2	2	1	0	0	1	0	0	0	1	0	0	0	0
# 24	-1	1	0	0	0	0	-1	1	0	0	0	0	1	1	1	1	1	1	2	2	2	1	0	1	0	0	0	0	0	1	0	0	0
# 25	1	0	0	0	0	0	0	-1	1	0	0	1	1	1	1	1	1	1	1	2	2	1	1	0	0	0	0	0	0	0	1	0	0
# 26	0	0	0	-1	0	1	0	0	0	0	0	0	0	0	0	1	2	2	2	2	2	1	0	0	0	0	1	1	0	0	0	0	0
# 27	0	0	-1	1	0	-1	1	0	0	0	0	0	0	0	1	1	1	2	2	2	2	1	0	0	0	1	0	0	1	0	0	0	0
# 28	0	-1	1	0	0	0	-1	1	0	0	0	0	0	1	1	1	1	1	2	2	2	1	0	0	1	0	0	0	0	1	0	0	0
# 29	-1	1	0	0	0	0	0	-1	1	0	0	0	1	1	1	1	1	1	1	2	2	1	0	1	0	0	0	0	0	0	1	0	0
# 30	1	0	0	0	0	0	0	0	-1	1	0	1	1	1	1	1	1	1	1	1	2	1	1	0	0	0	0	0	0	0	0	1	0
# 31	0	0	0	0	-2	2	0	0	0	0	0	0	0	0	0	0	2	2	2	2	2	1	0	0	0	0	0	2	0	0	0	0	0
# 32	0	0	0	-1	1	-1	1	0	0	0	0	0	0	0	0	1	1	2	2	2	2	1	0	0	0	0	1	0	1	0	0	0	0
# 33	0	0	-1	1	0	0	-1	1	0	0	0	0	0	0	1	1	1	1	2	2	2	1	0	0	0	1	0	0	0	1	0	0	0
# 34	0	-1	1	0	0	0	0	-1	1	0	0	0	0	1	1	1	1	1	1	2	2	1	0	0	1	0	0	0	0	0	1	0	0
# 35	-1	1	0	0	0	0	0	0	-1	1	0	0	1	1	1	1	1	1	1	1	2	1	0	1	0	0	0	0	0	0	0	1	0
# 36	1	0	0	0	0	0	0	0	0	-1	1	1	1	1	1	1	1	1	1	1	1	1	1	0	0	0	0	0	0	0	0	0	1
# 37	0	0	0	0	-1	0	1	0	0	0	0	0	0	0	0	0	1	2	2	2	2	1	0	0	0	0	0	1	1	0	0	0	0
# 38	0	0	0	-1	1	0	-1	1	0	0	0	0	0	0	0	1	1	1	2	2	2	1	0	0	0	0	1	0	0	1	0	0	0
# 39	0	0	-1	1	0	0	0	-1	1	0	0	0	0	0	1	1	1	1	1	2	2	1	0	0	0	1	0	0	0	0	1	0	0
# 40	0	-1	1	0	0	0	0	0	-1	1	0	0	0	1	1	1	1	1	1	1	2	1	0	0	1	0	0	0	0	0	0	1	0
# 41	-1	1	0	0	0	0	0	0	0	-1	1	0	1	1	1	1	1	1	1	1	1	1	0	1	0	0	0	0	0	0	0	0	1
# 42	1	0	0	0	0	0	0	0	0	1	-1	1	1	1	1	1	1	1	1	1	1	0	1	0	0	0	0	0	0	0	0	0	-1
# 43	0	0	0	0	0	-2	2	0	0	0	0	0	0	0	0	0	0	2	2	2	2	1	0	0	0	0	0	0	2	0	0	0	0
# 44	0	0	0	0	-1	1	-1	1	0	0	0	0	0	0	0	0	1	1	2	2	2	1	0	0	0	0	0	1	0	1	0	0	0
# 45	0	0	0	-1	1	0	0	-1	1	0	0	0	0	0	0	1	1	1	1	2	2	1	0	0	0	0	1	0	0	0	1	0	0
# 46	0	0	-1	1	0	0	0	0	-1	1	0	0	0	0	1	1	1	1	1	1	2	1	0	0	0	1	0	0	0	0	0	1	0
# 47	0	-1	1	0	0	0	0	0	0	-1	1	0	0	1	1	1	1	1	1	1	1	1	0	0	1	0	0	0	0	0	0	0	1
# 48	-1	1	0	0	0	0	0	0	0	1	-1	0	1	1	1	1	1	1	1	1	1	0	0	1	0	0	0	0	0	0	0	0	-1
# 49	1	0	0	0	0	0	0	0	1	-1	0	1	1	1	1	1	1	1	1	1	0	0	1	0	0	0	0	0	0	0	0	-1	0
# 50	0	0	0	0	0	-1	0	1	0	0	0	0	0	0	0	0	0	1	2	2	2	1	0	0	0	0	0	0	1	1	0	0	0
# 51	0	0	0	0	-1	1	0	-1	1	0	0	0	0	0	0	0	1	1	1	2	2	1	0	0	0	0	0	1	0	0	1	0	0
# 52	0	0	0	-1	1	0	0	0	-1	1	0	0	0	0	0	1	1	1	1	1	2	1	0	0	0	0	1	0	0	0	0	1	0
# 53	0	0	-1	1	0	0	0	0	0	-1	1	0	0	0	1	1	1	1	1	1	1	1	0	0	0	1	0	0	0	0	0	0	1
# 54	0	-1	1	0	0	0	0	0	0	1	-1	0	0	1	1	1	1	1	1	1	1	0	0	0	1	0	0	0	0	0	0	0	-1
# 55	-1	1	0	0	0	0	0	0	1	-1	0	0	1	1	1	1	1	1	1	1	0	0	0	1	0	0	0	0	0	0	0	-1	0
# 56	1	0	0	0	0	0	0	1	-1	0	0	1	1	1	1	1	1	1	1	0	0	0	1	0	0	0	0	0	0	0	-1	0	0
# 57	0	0	0	0	0	0	-2	2	0	0	0	0	0	0	0	0	0	0	2	2	2	1	0	0	0	0	0	0	0	2	0	0	0

# 58	0	0	0	0	0	-1	1	-1	1	0	0	0	0	0	0	0	0	1	1	2	2	1	0	0	0	0	0	0	1	0	1	0	0
# 59	0	0	0	0	-1	1	0	0	-1	1	0	0	0	0	0	0	1	1	1	1	2	1	0	0	0	0	0	1	0	0	0	1	0
# 60	0	0	0	-1	1	0	0	0	0	-1	1	0	0	0	0	1	1	1	1	1	1	1	0	0	0	0	1	0	0	0	0	0	1
# 61	0	0	-1	1	0	0	0	0	0	1	-1	0	0	0	1	1	1	1	1	1	1	0	0	0	0	1	0	0	0	0	0	0	-1
# 62	0	-1	1	0	0	0	0	0	1	-1	0	0	0	1	1	1	1	1	1	1	0	0	0	0	1	0	0	0	0	0	0	-1	0
# 63	-1	1	0	0	0	0	0	1	-1	0	0	0	1	1	1	1	1	1	1	0	0	0	0	1	0	0	0	0	0	0	-1	0	0
# 64	1	0	0	0	0	0	1	-1	0	0	0	1	1	1	1	1	1	1	0	0	0	0	1	0	0	0	0	0	0	-1	0	0	0
# 65	0	0	0	0	0	0	-1	0	1	0	0	0	0	0	0	0	0	0	1	2	2	1	0	0	0	0	0	0	0	1	1	0	0
# 66	0	0	0	0	0	-1	1	0	-1	1	0	0	0	0	0	0	0	1	1	1	2	1	0	0	0	0	0	0	1	0	0	1	0
# 67	0	0	0	0	-1	1	0	0	0	-1	1	0	0	0	0	0	1	1	1	1	1	1	0	0	0	0	0	1	0	0	0	0	1
# 68	0	0	0	-1	1	0	0	0	0	1	-1	0	0	0	0	1	1	1	1	1	1	0	0	0	0	0	1	0	0	0	0	0	-1
# 69	0	0	-1	1	0	0	0	0	1	-1	0	0	0	0	1	1	1	1	1	1	0	0	0	0	0	1	0	0	0	0	0	-1	0
# 70	0	-1	1	0	0	0	0	1	-1	0	0	0	0	1	1	1	1	1	1	0	0	0	0	0	1	0	0	0	0	0	-1	0	0
# 71	-1	1	0	0	0	0	1	-1	0	0	0	0	1	1	1	1	1	1	0	0	0	0	0	1	0	0	0	0	0	-1	0	0	0
# 72	1	0	0	0	0	1	-1	0	0	0	0	1	1	1	1	1	1	0	0	0	0	0	1	0	0	0	0	0	-1	0	0	0	0
# 73	0	0	0	0	0	0	0	-2	2	0	0	0	0	0	0	0	0	0	0	2	2	1	0	0	0	0	0	0	0	0	2	0	0
# 74	0	0	0	0	0	0	-1	1	-1	1	0	0	0	0	0	0	0	0	1	1	2	1	0	0	0	0	0	0	0	1	0	1	0
# 75	0	0	0	0	0	-1	1	0	0	-1	1	0	0	0	0	0	0	1	1	1	1	1	0	0	0	0	0	0	1	0	0	0	1
# 76	0	0	0	0	-1	1	0	0	0	1	-1	0	0	0	0	0	1	1	1	1	1	0	0	0	0	0	0	1	0	0	0	0	-1
# 77	0	0	0	-1	1	0	0	0	1	-1	0	0	0	0	0	1	1	1	1	1	0	0	0	0	0	0	1	0	0	0	0	-1	0
# 78	0	0	-1	1	0	0	0	1	-1	0	0	0	0	0	1	1	1	1	1	0	0	0	0	0	0	1	0	0	0	0	-1	0	0
# 79	0	-1	1	0	0	0	1	-1	0	0	0	0	0	1	1	1	1	1	0	0	0	0	0	0	1	0	0	0	0	-1	0	0	0
# 80	-1	1	0	0	0	1	-1	0	0	0	0	0	1	1	1	1	1	0	0	0	0	0	0	1	0	0	0	0	-1	0	0	0	0
# 81	1	0	0	0	1	-1	0	0	0	0	0	1	1	1	1	1	0	0	0	0	0	0	1	0	0	0	0	-1	0	0	0	0	0
# 82	0	0	0	0	0	0	0	-1	0	1	0	0	0	0	0	0	0	0	0	1	2	1	0	0	0	0	0	0	0	0	1	1	0
# 83	0	0	0	0	0	0	-1	1	0	-1	1	0	0	0	0	0	0	0	1	1	1	1	0	0	0	0	0	0	0	1	0	0	1
# 84	0	0	0	0	0	-1	1	0	0	1	-1	0	0	0	0	0	0	1	1	1	1	0	0	0	0	0	0	0	1	0	0	0	-1
# 85	0	0	0	0	-1	1	0	0	1	-1	0	0	0	0	0	0	1	1	1	1	0	0	0	0	0	0	0	1	0	0	0	-1	0
# 86	0	0	0	-1	1	0	0	1	-1	0	0	0	0	0	0	1	1	1	1	0	0	0	0	0	0	0	1	0	0	0	-1	0	0
# 87	0	0	-1	1	0	0	1	-1	0	0	0	0	0	0	1	1	1	1	0	0	0	0	0	0	0	1	0	0	0	-1	0	0	0
# 88	0	-1	1	0	0	1	-1	0	0	0	0	0	0	1	1	1	1	0	0	0	0	0	0	0	1	0	0	0	-1	0	0	0	0
# 89	-1	1	0	0	1	-1	0	0	0	0	0	0	1	1	1	1	0	0	0	0	0	0	0	1	0	0	0	-1	0	0	0	0	0
# 90	1	0	0	1	-1	0	0	0	0	0	0	1	1	1	1	0	0	0	0	0	0	0	1	0	0	0	-1	0	0	0	0	0	0
# 91	0	0	0	0	0	0	0	0	-2	2	0	0	0	0	0	0	0	0	0	0	2	1	0	0	0	0	0	0	0	0	0	2	0
# 92	0	0	0	0	0	0	0	-1	1	-1	1	0	0	0	0	0	0	0	0	1	1	1	0	0	0	0	0	0	0	0	1	0	1
# 93	0	0	0	0	0	0	-1	1	0	1	-1	0	0	0	0	0	0	0	1	1	1	0	0	0	0	0	0	0	0	1	0	0	-1
# 94	0	0	0	0	0	-1	1	0	1	-1	0	0	0	0	0	0	0	1	1	1	0	0	0	0	0	0	0	0	1	0	0	-1	0
# 95	0	0	0	0	-1	1	0	1	-1	0	0	0	0	0	0	0	1	1	1	0	0	0	0	0	0	0	0	1	0	0	-1	0	0
# 96	0	0	0	-1	1	0	1	-1	0	0	0	0	0	0	0	1	1	1	0	0	0	0	0	0	0	0	1	0	0	-1	0	0	0
# 97	0	0	-1	1	0	1	-1	0	0	0	0	0	0	0	1	1	1	0	0	0	0	0	0	0	0	1	0	0	-1	0	0	0	0
# 98	0	-1	1	0	1	-1	0	0	0	0	0	0	0	1	1	1	0	0	0	0	0	0	0	0	1	0	0	-1	0	0	0	0	0
# 99	-1	1	0	1	-1	0	0	0	0	0	0	0	1	1	1	0	0	0	0	0	0	0	0	1	0	0	-1	0	0	0	0	0	0
#100	1	0	1	-1	0	0	0	0	0	0	0	1	1	1	0	0	0	0	0	0	0	0	1	0	0	-1	0	0	0	0	0	0	0
#101	0	0	0	0	0	0	0	0	-1	0	1	0	0	0	0	0	0	0	0	0	1	1	0	0	0	0	0	0	0	0	0	1	1
#102	0	0	0	0	0	0	0	-1	1	1	-1	0	0	0	0	0	0	0	0	1	1	0	0	0	0	0	0	0	0	0	1	0	-1
#103	0	0	0	0	0	0	-1	1	1	-1	0	0	0	0	0	0	0	0	1	1	0	0	0	0	0	0	0	0	0	1	0	-1	0
#104	0	0	0	0	0	-1	1	1	-1	0	0	0	0	0	0	0	0	1	1	0	0	0	0	0	0	0	0	0	1	0	-1	0	0
#105	0	0	0	0	-1	1	1	-1	0	0	0	0	0	0	0	0	1	1	0	0	0	0	0	0	0	0	0	1	0	-1	0	0	0
#106	0	0	0	-1	1	1	-1	0	0	0	0	0	0	0	0	1	1	0	0	0	0	0	0	0	0	0	1	0	-1	0	0	0	0
#107	0	0	-1	1	1	-1	0	0	0	0	0	0	0	0	1	1	0	0	0	0	0	0	0	0	0	1	0	-1	0	0	0	0	0
#108	0	-1	1	1	-1	0	0	0	0	0	0	0	0	1	1	0	0	0	0	0	0	0	0	0	1	0	-1	0	0	0	0	0	0
#109	-1	1	1	-1	0	0	0	0	0	0	0	0	1	1	0	0	0	0	0	0	0	0	0	1	0	-1	0	0	0	0	0	0	0
#110	1	1	-1	0	0	0	0	0	0	0	0	1	1	0	0	0	0	0	0	0	0	0	1	0	-1	0	0	0	0	0	0	0	0
#111	0	0	0	0	0	0	0	0	0	-2	2	0	0	0	0	0	0	0	0	0	0	1	0	0	0	0	0	0	0	0	0	0	2
#112	0	0	0	0	0	0	0	0	-1	2	-1	0	0	0	0	0	0	0	0	0	1	0	0	0	0	0	0	0	0	0	0	1	-1
#113	0	0	0	0	0	0	0	-1	2	-1	0	0	0	0	0	0	0	0	0	1	0	0	0	0	0	0	0	0	0	0	1	-1	0
#114	0	0	0	0	0	0	-1	2	-1	0	0	0	0	0	0	0	0	0	1	0	0	0	0	0	0	0	0	0	0	1	-1	0	0
#115	0	0	0	0	0	-1	2	-1	0	0	0	0	0	0	0	0	0	1	0	0	0	0	0	0	0	0	0	0	1	-1	0	0	0
#116	0	0	0	0	-1	2	-1	0	0	0	0	0	0	0	0	0	1	0	0	0	0	0	0	0	0	0	0	1	-1	0	0	0	0
#117	0	0	0	-1	2	-1	0	0	0	0	0	0	0	0	0	1	0	0	0	0	0	0	0	0	0	0	1	-1	0	0	0	0	0
#118	0	0	-1	2	-1	0	0	0	0	0	0	0	0	0	1	0	0	0	0	0	0	0	0	0	0	1	-1	0	0	0	0	0	0
#119	0	-1	2	-1	0	0	0	0	0	0	0	0	0	1	0	0	0	0	0	0	0	0	0	0	1	-1	0	0	0	0	0	0	0
#120	-1	2	-1	0	0	0	0	0	0	0	0	0	1	0	0	0	0	0	0	0	0	0	0	1	-1	0	0	0	0	0	0	0	0
#121	2	-1	0	0	0	0	0	0	0	0	0	1	0	0	0	0	0	0	0	0	0	0	1	-1	0	0	0	0	0	0	0	0	0

Table 2. *(continued)*

ALGEBRA C12

	OMEGA											
# 1	2	0	0	0	0	0	0	0	0	0	0	0
# 2	0	1	0	0	0	0	0	0	0	0	0	0
# 3	-2	2	0	0	0	0	0	0	0	0	0	0
# 4	1	-1	1	0	0	0	0	0	0	0	0	0
# 5	-1	0	1	0	0	0	0	0	0	0	0	0
# 6	1	0	-1	1	0	0	0	0	0	0	0	0
# 7	0	-2	2	0	0	0	0	0	0	0	0	0
# 8	-1	1	-1	1	0	0	0	0	0	0	0	0
# 9	1	0	0	-1	1	0	0	0	0	0	0	0
# 10	0	-1	0	1	0	0	0	0	0	0	0	0
# 11	-1	1	0	-1	1	0	0	0	0	0	0	0
# 12	1	0	0	0	-1	1	0	0	0	0	0	0
# 13	0	0	-2	2	0	0	0	0	0	0	0	0
# 14	0	-1	1	-1	1	0	0	0	0	0	0	0
# 15	-1	1	0	0	-1	1	0	0	0	0	0	0
# 16	1	0	0	0	0	-1	1	0	0	0	0	0
# 17	0	0	-1	0	1	0	0	0	0	0	0	0
# 18	0	-1	1	0	-1	1	0	0	0	0	0	0
# 19	-1	1	0	0	0	-1	1	0	0	0	0	0
# 20	1	0	0	0	0	0	-1	1	0	0	0	0
# 21	0	0	0	-2	2	0	0	0	0	0	0	0
# 22	0	0	-1	1	-1	1	0	0	0	0	0	0
# 23	0	-1	1	0	0	-1	1	0	0	0	0	0
# 24	-1	1	0	0	0	0	-1	1	0	0	0	0
# 25	1	0	0	0	0	0	0	-1	1	0	0	0
# 26	0	0	0	-1	0	1	0	0	0	0	0	0
# 27	0	0	-1	1	0	-1	1	0	0	0	0	0
# 28	0	-1	1	0	0	0	-1	1	0	0	0	0
# 29	-1	1	0	0	0	0	0	-1	1	0	0	0
# 30	1	0	0	0	0	0	0	0	-1	1	0	0
# 31	0	0	0	0	-2	2	0	0	0	0	0	0
# 32	0	0	0	-1	1	-1	1	0	0	0	0	0
# 33	0	0	-1	1	0	0	-1	1	0	0	0	0
# 34	0	-1	1	0	0	0	0	-1	1	0	0	0
# 35	-1	1	0	0	0	0	0	0	-1	1	0	0
# 36	1	0	0	0	0	0	0	0	0	-1	1	0
# 37	0	0	0	0	-1	0	1	0	0	0	0	0
# 38	0	0	0	-1	1	0	-1	1	0	0	0	0
# 39	0	0	-1	1	0	0	0	-1	1	0	0	0
# 40	0	-1	1	0	0	0	0	0	-1	1	0	0
# 41	-1	1	0	0	0	0	0	0	0	-1	1	0
# 42	1	0	0	0	0	0	0	0	0	0	-1	1
# 43	0	0	0	0	0	-2	2	0	0	0	0	0
# 44	0	0	0	0	-1	1	-1	1	0	0	0	0
# 45	0	0	0	-1	1	0	0	-1	1	0	0	0
# 46	0	0	-1	1	0	0	0	0	-1	1	0	0
# 47	0	-1	1	0	0	0	0	0	0	-1	1	0
# 48	-1	1	0	0	0	0	0	0	0	0	-1	1
# 49	1	0	0	0	0	0	0	0	0	0	1	-1
# 50	0	0	0	0	0	-1	0	1	0	0	0	0
# 51	0	0	0	0	-1	1	0	-1	1	0	0	0
# 52	0	0	0	-1	1	0	0	0	-1	1	0	0
# 53	0	0	-1	1	0	0	0	0	0	-1	1	0
# 54	0	-1	1	0	0	0	0	0	0	0	-1	1
# 55	-1	1	0	0	0	0	0	0	0	0	1	-1
# 56	1	0	0	0	0	0	0	0	0	1	-1	0
# 57	0	0	0	0	0	0	-2	2	0	0	0	0
# 58	0	0	0	0	0	-1	1	-1	1	0	0	0
# 59	0	0	0	0	-1	1	0	0	-1	1	0	0
# 60	0	0	0	-1	1	0	0	0	0	-1	1	0
# 61	0	0	-1	1	0	0	0	0	0	0	-1	1
# 62	0	-1	1	0	0	0	0	0	0	0	1	-1
# 63	-1	1	0	0	0	0	0	0	0	1	-1	0
# 64	1	0	0	0	0	0	0	0	1	-1	0	0
# 65	0	0	0	0	0	0	-1	0	1	0	0	0
# 66	0	0	0	0	0	-1	1	0	-1	1	0	0
# 67	0	0	0	0	-1	1	0	0	0	-1	1	0
# 68	0	0	0	-1	1	0	0	0	0	0	-1	1

	ALPHA											
# 1	2	2	2	2	2	2	2	2	2	2	2	1
# 2	1	2	2	2	2	2	2	2	2	2	2	1
# 3	0	2	2	2	2	2	2	2	2	2	2	1
# 4	1	1	2	2	2	2	2	2	2	2	2	1
# 5	0	1	2	2	2	2	2	2	2	2	2	1
# 6	1	1	1	2	2	2	2	2	2	2	2	1
# 7	0	0	2	2	2	2	2	2	2	2	2	1
# 8	0	1	1	2	2	2	2	2	2	2	2	1
# 9	1	1	1	1	2	2	2	2	2	2	2	1
# 10	0	0	1	2	2	2	2	2	2	2	2	1
# 11	0	1	1	1	2	2	2	2	2	2	2	1
# 12	1	1	1	1	1	2	2	2	2	2	2	1
# 13	0	0	0	2	2	2	2	2	2	2	2	1
# 14	0	0	1	1	2	2	2	2	2	2	2	1
# 15	0	1	1	1	1	2	2	2	2	2	2	1
# 16	1	1	1	1	1	1	2	2	2	2	2	1
# 17	0	0	0	1	2	2	2	2	2	2	2	1
# 18	0	0	1	1	1	2	2	2	2	2	2	1
# 19	0	1	1	1	1	1	2	2	2	2	2	1
# 20	1	1	1	1	1	1	1	2	2	2	2	1
# 21	0	0	0	0	2	2	2	2	2	2	2	1
# 22	0	0	0	1	1	2	2	2	2	2	2	1
# 23	0	0	1	1	1	1	2	2	2	2	2	1
# 24	0	1	1	1	1	1	1	2	2	2	2	1
# 25	1	1	1	1	1	1	1	1	2	2	2	1
# 26	0	0	0	0	1	2	2	2	2	2	2	1
# 27	0	0	0	1	1	1	2	2	2	2	2	1
# 28	0	0	1	1	1	1	1	2	2	2	2	1
# 29	0	1	1	1	1	1	1	1	2	2	2	1
# 30	1	1	1	1	1	1	1	1	1	2	2	1
# 31	0	0	0	0	0	2	2	2	2	2	2	1
# 32	0	0	0	0	1	1	2	2	2	2	2	1
# 33	0	0	0	1	1	1	1	2	2	2	2	1
# 34	0	0	1	1	1	1	1	1	2	2	2	1
# 35	0	1	1	1	1	1	1	1	1	2	2	1
# 36	1	1	1	1	1	1	1	1	1	1	2	1
# 37	0	0	0	0	0	1	2	2	2	2	2	1
# 38	0	0	0	0	1	1	1	2	2	2	2	1
# 39	0	0	0	1	1	1	1	1	2	2	2	1
# 40	0	0	1	1	1	1	1	1	1	2	2	1
# 41	0	1	1	1	1	1	1	1	1	1	2	1
# 42	1	1	1	1	1	1	1	1	1	1	1	1
# 43	0	0	0	0	0	0	2	2	2	2	2	1
# 44	0	0	0	0	0	1	1	2	2	2	2	1
# 45	0	0	0	0	1	1	1	1	2	2	2	1
# 46	0	0	0	1	1	1	1	1	1	2	2	1
# 47	0	0	1	1	1	1	1	1	1	1	2	1
# 48	0	1	1	1	1	1	1	1	1	1	1	1
# 49	1	1	1	1	1	1	1	1	1	1	1	0
# 50	0	0	0	0	0	0	1	2	2	2	2	1
# 51	0	0	0	0	0	1	1	1	2	2	2	1
# 52	0	0	0	0	1	1	1	1	1	2	2	1
# 53	0	0	0	1	1	1	1	1	1	1	2	1
# 54	0	0	1	1	1	1	1	1	1	1	1	1
# 55	0	1	1	1	1	1	1	1	1	1	1	0
# 56	1	1	1	1	1	1	1	1	1	1	0	0
# 57	0	0	0	0	0	0	0	2	2	2	2	1
# 58	0	0	0	0	0	0	1	1	2	2	2	1
# 59	0	0	0	0	0	1	1	1	1	2	2	1
# 60	0	0	0	0	1	1	1	1	1	1	2	1
# 61	0	0	0	1	1	1	1	1	1	1	1	1
# 62	0	0	1	1	1	1	1	1	1	1	1	0
# 63	0	1	1	1	1	1	1	1	1	1	0	0
# 64	1	1	1	1	1	1	1	1	1	0	0	0
# 65	0	0	0	0	0	0	0	1	2	2	2	1
# 66	0	0	0	0	0	0	1	1	1	2	2	1
# 67	0	0	0	0	0	1	1	1	1	1	2	1
# 68	0	0	0	0	1	1	1	1	1	1	1	1

	EPSILON											
# 1	2	0	0	0	0	0	0	0	0	0	0	0
# 2	1	1	0	0	0	0	0	0	0	0	0	0
# 3	0	2	0	0	0	0	0	0	0	0	0	0
# 4	1	0	1	0	0	0	0	0	0	0	0	0
# 5	0	1	1	0	0	0	0	0	0	0	0	0
# 6	1	0	0	1	0	0	0	0	0	0	0	0
# 7	0	0	2	0	0	0	0	0	0	0	0	0
# 8	0	1	0	1	0	0	0	0	0	0	0	0
# 9	1	0	0	0	1	0	0	0	0	0	0	0
# 10	0	0	1	1	0	0	0	0	0	0	0	0
# 11	0	1	0	0	1	0	0	0	0	0	0	0
# 12	1	0	0	0	0	1	0	0	0	0	0	0
# 13	0	0	0	2	0	0	0	0	0	0	0	0
# 14	0	0	1	0	1	0	0	0	0	0	0	0
# 15	0	1	0	0	0	1	0	0	0	0	0	0
# 16	1	0	0	0	0	0	1	0	0	0	0	0
# 17	0	0	0	1	1	0	0	0	0	0	0	0
# 18	0	0	1	0	0	1	0	0	0	0	0	0
# 19	0	1	0	0	0	0	1	0	0	0	0	0
# 20	1	0	0	0	0	0	0	1	0	0	0	0
# 21	0	0	0	0	2	0	0	0	0	0	0	0
# 22	0	0	0	1	0	1	0	0	0	0	0	0
# 23	0	0	1	0	0	0	1	0	0	0	0	0
# 24	0	1	0	0	0	0	0	1	0	0	0	0
# 25	1	0	0	0	0	0	0	0	1	0	0	0
# 26	0	0	0	0	1	1	0	0	0	0	0	0
# 27	0	0	0	1	0	0	1	0	0	0	0	0
# 28	0	0	1	0	0	0	0	1	0	0	0	0
# 29	0	1	0	0	0	0	0	0	1	0	0	0
# 30	1	0	0	0	0	0	0	0	0	1	0	0
# 31	0	0	0	0	0	2	0	0	0	0	0	0
# 32	0	0	0	0	1	0	1	0	0	0	0	0
# 33	0	0	0	1	0	0	0	1	0	0	0	0
# 34	0	0	1	0	0	0	0	0	1	0	0	0
# 35	0	1	0	0	0	0	0	0	0	1	0	0
# 36	1	0	0	0	0	0	0	0	0	0	1	0
# 37	0	0	0	0	0	1	1	0	0	0	0	0
# 38	0	0	0	0	1	0	0	1	0	0	0	0
# 39	0	0	0	1	0	0	0	0	1	0	0	0
# 40	0	0	1	0	0	0	0	0	0	1	0	0
# 41	0	1	0	0	0	0	0	0	0	0	1	0
# 42	1	0	0	0	0	0	0	0	0	0	0	1
# 43	0	0	0	0	0	0	2	0	0	0	0	0
# 44	0	0	0	0	0	1	0	1	0	0	0	0
# 45	0	0	0	0	1	0	0	0	1	0	0	0
# 46	0	0	0	1	0	0	0	0	0	1	0	0
# 47	0	0	1	0	0	0	0	0	0	0	1	0
# 48	0	1	0	0	0	0	0	0	0	0	0	1
# 49	1	0	0	0	0	0	0	0	0	0	0	-1
# 50	0	0	0	0	0	0	1	1	0	0	0	0
# 51	0	0	0	0	0	1	0	0	1	0	0	0
# 52	0	0	0	0	1	0	0	0	0	1	0	0
# 53	0	0	0	1	0	0	0	0	0	0	1	0
# 54	0	0	1	0	0	0	0	0	0	0	0	1
# 55	0	1	0	0	0	0	0	0	0	0	0	-1
# 56	1	0	0	0	0	0	0	0	0	0	-1	0
# 57	0	0	0	0	0	0	0	2	0	0	0	0
# 58	0	0	0	0	0	0	1	0	1	0	0	0
# 59	0	0	0	0	0	1	0	0	0	1	0	0
# 60	0	0	0	0	1	0	0	0	0	0	1	0
# 61	0	0	0	1	0	0	0	0	0	0	0	1
# 62	0	0	1	0	0	0	0	0	0	0	0	-1
# 63	0	1	0	0	0	0	0	0	0	0	-1	0
# 64	1	0	0	0	0	0	0	0	0	-1	0	0
# 65	0	0	0	0	0	0	0	1	1	0	0	0
# 66	0	0	0	0	0	0	1	0	0	1	0	0
# 67	0	0	0	0	0	1	0	0	0	0	1	0
# 68	0	0	0	0	1	0	0	0	0	0	0	1

# 69	0	0	-1	1	0	0	0	0	0	0	1	-1	0	0	0	1	1	1	1	1	1	1	1	0	0	0	0	1	0	0	0	0	0	0	0	-1
# 70	0	-1	1	0	0	0	0	0	0	1	-1	0	0	0	1	1	1	1	1	1	1	1	0	0	0	0	1	0	0	0	0	0	0	0	-1	0
# 71	-1	1	0	0	0	0	0	0	1	-1	0	0	0	1	1	1	1	1	1	1	1	0	0	0	0	1	0	0	0	0	0	0	0	-1	0	0
# 72	1	0	0	0	0	0	0	1	-1	0	0	0	1	1	1	1	1	1	1	1	0	0	0	0	1	0	0	0	0	0	0	0	-1	0	0	0
# 73	0	0	0	0	0	0	0	-2	2	0	0	0	0	0	0	0	0	0	0	0	2	2	2	1	0	0	0	0	0	0	0	0	2	0	0	0
# 74	0	0	0	0	0	0	-1	1	-1	1	0	0	0	0	0	0	0	0	0	1	1	2	2	1	0	0	0	0	0	0	0	1	0	1	0	0
# 75	0	0	0	0	0	-1	1	0	0	-1	1	0	0	0	0	0	0	0	1	1	1	1	2	1	0	0	0	0	0	0	1	0	0	0	1	0
# 76	0	0	0	0	-1	1	0	0	0	0	-1	1	0	0	0	0	0	1	1	1	1	1	1	1	0	0	0	0	0	1	0	0	0	0	0	1
# 77	0	0	0	-1	1	0	0	0	0	0	1	-1	0	0	0	0	1	1	1	1	1	1	1	0	0	0	0	0	1	0	0	0	0	0	0	-1
# 78	0	0	-1	1	0	0	0	0	0	1	-1	0	0	0	0	1	1	1	1	1	1	1	0	0	0	0	0	1	0	0	0	0	0	0	-1	0
# 79	0	-1	1	0	0	0	0	0	1	-1	0	0	0	0	1	1	1	1	1	1	1	0	0	0	0	0	1	0	0	0	0	0	0	-1	0	0
# 80	-1	1	0	0	0	0	0	1	-1	0	0	0	0	1	1	1	1	1	1	1	0	0	0	0	0	1	0	0	0	0	0	0	-1	0	0	0
# 81	1	0	0	0	0	0	1	-1	0	0	0	0	1	1	1	1	1	1	1	0	0	0	0	0	1	0	0	0	0	0	0	-1	0	0	0	0
# 82	0	0	0	0	0	0	0	-1	0	1	0	0	0	0	0	0	0	0	0	0	1	2	2	1	0	0	0	0	0	0	0	0	1	1	0	0
# 83	0	0	0	0	0	0	-1	1	0	-1	1	0	0	0	0	0	0	0	0	1	1	1	2	1	0	0	0	0	0	0	0	1	0	0	1	0
# 84	0	0	0	0	0	-1	1	0	0	0	-1	1	0	0	0	0	0	0	1	1	1	1	1	1	0	0	0	0	0	0	1	0	0	0	0	1
# 85	0	0	0	0	-1	1	0	0	0	0	1	-1	0	0	0	0	0	1	1	1	1	1	1	0	0	0	0	0	0	1	0	0	0	0	0	-1
# 86	0	0	0	-1	1	0	0	0	0	1	-1	0	0	0	0	0	1	1	1	1	1	1	0	0	0	0	0	0	1	0	0	0	0	0	-1	0
# 87	0	0	-1	1	0	0	0	0	1	-1	0	0	0	0	0	1	1	1	1	1	1	0	0	0	0	0	0	1	0	0	0	0	0	-1	0	0
# 88	0	-1	1	0	0	0	0	1	-1	0	0	0	0	0	1	1	1	1	1	1	0	0	0	0	0	0	1	0	0	0	0	0	-1	0	0	0
# 89	-1	1	0	0	0	0	1	-1	0	0	0	0	0	1	1	1	1	1	1	0	0	0	0	0	0	1	0	0	0	0	0	-1	0	0	0	0
# 90	1	0	0	0	0	1	-1	0	0	0	0	0	1	1	1	1	1	1	0	0	0	0	0	0	1	0	0	0	0	0	-1	0	0	0	0	0
# 91	0	0	0	0	0	0	0	0	-2	2	0	0	0	0	0	0	0	0	0	0	0	2	2	1	0	0	0	0	0	0	0	0	0	2	0	0
# 92	0	0	0	0	0	0	0	-1	1	-1	1	0	0	0	0	0	0	0	0	0	1	1	2	1	0	0	0	0	0	0	0	0	1	0	1	0
# 93	0	0	0	0	0	0	-1	1	0	0	-1	1	0	0	0	0	0	0	0	1	1	1	1	1	0	0	0	0	0	0	0	1	0	0	0	1
# 94	0	0	0	0	0	-1	1	0	0	0	1	-1	0	0	0	0	0	0	1	1	1	1	1	0	0	0	0	0	0	0	1	0	0	0	0	-1
# 95	0	0	0	0	-1	1	0	0	0	1	-1	0	0	0	0	0	0	1	1	1	1	1	0	0	0	0	0	0	0	1	0	0	0	0	-1	0
# 96	0	0	0	-1	1	0	0	0	1	-1	0	0	0	0	0	0	1	1	1	1	1	0	0	0	0	0	0	0	1	0	0	0	0	-1	0	0
# 97	0	0	-1	1	0	0	0	1	-1	0	0	0	0	0	0	1	1	1	1	1	0	0	0	0	0	0	0	1	0	0	0	0	-1	0	0	0
# 98	0	-1	1	0	0	0	1	-1	0	0	0	0	0	0	1	1	1	1	1	0	0	0	0	0	0	0	1	0	0	0	0	-1	0	0	0	0
# 99	-1	1	0	0	0	1	-1	0	0	0	0	0	0	1	1	1	1	1	0	0	0	0	0	0	0	1	0	0	0	0	-1	0	0	0	0	0
#100	1	0	0	0	1	-1	0	0	0	0	0	0	1	1	1	1	1	0	0	0	0	0	0	0	1	0	0	0	0	-1	0	0	0	0	0	0
#101	0	0	0	0	0	0	0	0	-1	0	1	0	0	0	0	0	0	0	0	0	0	1	2	1	0	0	0	0	0	0	0	0	0	1	1	0
#102	0	0	0	0	0	0	0	-1	1	0	-1	1	0	0	0	0	0	0	0	0	1	1	1	1	0	0	0	0	0	0	0	0	1	0	0	1
#103	0	0	0	0	0	0	-1	1	0	0	1	-1	0	0	0	0	0	0	0	1	1	1	1	0	0	0	0	0	0	0	0	1	0	0	0	-1
#104	0	0	0	0	0	-1	1	0	0	1	-1	0	0	0	0	0	0	0	1	1	1	1	0	0	0	0	0	0	0	0	1	0	0	0	-1	0
#105	0	0	0	0	-1	1	0	0	1	-1	0	0	0	0	0	0	0	1	1	1	1	0	0	0	0	0	0	0	0	1	0	0	0	-1	0	0
#106	0	0	0	-1	1	0	0	1	-1	0	0	0	0	0	0	0	1	1	1	1	0	0	0	0	0	0	0	0	1	0	0	0	-1	0	0	0
#107	0	0	-1	1	0	0	1	-1	0	0	0	0	0	0	0	1	1	1	1	0	0	0	0	0	0	0	0	1	0	0	0	-1	0	0	0	0
#108	0	-1	1	0	0	1	-1	0	0	0	0	0	0	0	1	1	1	1	0	0	0	0	0	0	0	0	1	0	0	0	-1	0	0	0	0	0
#109	-1	1	0	0	1	-1	0	0	0	0	0	0	0	1	1	1	1	0	0	0	0	0	0	0	0	1	0	0	0	-1	0	0	0	0	0	0
#110	1	0	0	1	-1	0	0	0	0	0	0	0	1	1	1	1	0	0	0	0	0	0	0	0	1	0	0	0	-1	0	0	0	0	0	0	0
#111	0	0	0	0	0	0	0	0	0	-2	2	0	0	0	0	0	0	0	0	0	0	0	2	1	0	0	0	0	0	0	0	0	0	0	2	0
#112	0	0	0	0	0	0	0	0	-1	1	-1	1	0	0	0	0	0	0	0	0	0	1	1	1	0	0	0	0	0	0	0	0	0	1	0	1
#113	0	0	0	0	0	0	0	-1	1	0	1	-1	0	0	0	0	0	0	0	0	1	1	1	0	0	0	0	0	0	0	0	0	1	0	0	-1
#114	0	0	0	0	0	0	-1	1	0	1	-1	0	0	0	0	0	0	0	0	1	1	1	0	0	0	0	0	0	0	0	0	1	0	0	-1	0
#115	0	0	0	0	0	-1	1	0	1	-1	0	0	0	0	0	0	0	0	1	1	1	0	0	0	0	0	0	0	0	0	1	0	0	-1	0	0
#116	0	0	0	0	-1	1	0	1	-1	0	0	0	0	0	0	0	0	1	1	1	0	0	0	0	0	0	0	0	0	1	0	0	-1	0	0	0
#117	0	0	0	-1	1	0	1	-1	0	0	0	0	0	0	0	0	1	1	1	0	0	0	0	0	0	0	0	0	1	0	0	-1	0	0	0	0
#118	0	0	-1	1	0	1	-1	0	0	0	0	0	0	0	0	1	1	1	0	0	0	0	0	0	0	0	0	1	0	0	-1	0	0	0	0	0
#119	0	-1	1	0	1	-1	0	0	0	0	0	0	0	0	1	1	1	0	0	0	0	0	0	0	0	0	1	0	0	-1	0	0	0	0	0	0
#120	-1	1	0	1	-1	0	0	0	0	0	0	0	0	1	1	1	0	0	0	0	0	0	0	0	0	1	0	0	-1	0	0	0	0	0	0	0
#121	1	0	1	-1	0	0	0	0	0	0	0	0	1	1	1	0	0	0	0	0	0	0	0	0	1	0	0	-1	0	0	0	0	0	0	0	0
#122	0	0	0	0	0	0	0	0	0	-1	0	1	0	0	0	0	0	0	0	0	0	0	1	1	0	0	0	0	0	0	0	0	0	0	1	1
#123	0	0	0	0	0	0	0	0	-1	1	1	-1	0	0	0	0	0	0	0	0	0	1	1	0	0	0	0	0	0	0	0	0	0	1	0	-1
#124	0	0	0	0	0	0	0	-1	1	1	-1	0	0	0	0	0	0	0	0	0	1	1	0	0	0	0	0	0	0	0	0	0	1	0	-1	0
#125	0	0	0	0	0	0	-1	1	1	-1	0	0	0	0	0	0	0	0	0	1	1	0	0	0	0	0	0	0	0	0	0	1	0	-1	0	0
#126	0	0	0	0	0	-1	1	1	-1	0	0	0	0	0	0	0	0	0	1	1	0	0	0	0	0	0	0	0	0	0	1	0	-1	0	0	0
#127	0	0	0	0	-1	1	1	-1	0	0	0	0	0	0	0	0	0	1	1	0	0	0	0	0	0	0	0	0	0	1	0	-1	0	0	0	0
#128	0	0	0	-1	1	1	-1	0	0	0	0	0	0	0	0	0	1	1	0	0	0	0	0	0	0	0	0	0	1	0	-1	0	0	0	0	0
#129	0	0	-1	1	1	-1	0	0	0	0	0	0	0	0	0	1	1	0	0	0	0	0	0	0	0	0	0	1	0	-1	0	0	0	0	0	0
#130	0	-1	1	1	-1	0	0	0	0	0	0	0	0	0	1	1	0	0	0	0	0	0	0	0	0	0	1	0	-1	0	0	0	0	0	0	0
#131	-1	1	1	-1	0	0	0	0	0	0	0	0	0	1	1	0	0	0	0	0	0	0	0	0	0	1	0	-1	0	0	0	0	0	0	0	0
#132	1	1	-1	0	0	0	0	0	0	0	0	0	1	1	0	0	0	0	0	0	0	0	0	0	1	0	-1	0	0	0	0	0	0	0	0	0
#133	0	0	0	0	0	0	0	0	0	0	-2	2	0	0	0	0	0	0	0	0	0	0	0	1	0	0	0	0	0	0	0	0	0	0	0	2
#134	0	0	0	0	0	0	0	0	0	-1	2	-1	0	0	0	0	0	0	0	0	0	0	1	0	0	0	0	0	0	0	0	0	0	0	1	-1
#135	0	0	0	0	0	0	0	0	-1	2	-1	0	0	0	0	0	0	0	0	0	0	1	0	0	0	0	0	0	0	0	0	0	0	1	-1	0
#136	0	0	0	0	0	0	0	-1	2	-1	0	0	0	0	0	0	0	0	0	0	1	0	0	0	0	0	0	0	0	0	0	0	1	-1	0	0
#137	0	0	0	0	0	0	-1	2	-1	0	0	0	0	0	0	0	0	0	0	1	0	0	0	0	0	0	0	0	0	0	0	1	-1	0	0	0
#138	0	0	0	0	0	-1	2	-1	0	0	0	0	0	0	0	0	0	0	1	0	0	0	0	0	0	0	0	0	0	0	1	-1	0	0	0	0
#139	0	0	0	0	-1	2	-1	0	0	0	0	0	0	0	0	0	0	1	0	0	0	0	0	0	0	0	0	0	0	1	-1	0	0	0	0	0
#140	0	0	0	-1	2	-1	0	0	0	0	0	0	0	0	0	0	1	0	0	0	0	0	0	0	0	0	0	0	1	-1	0	0	0	0	0	0
#141	0	0	-1	2	-1	0	0	0	0	0	0	0	0	0	0	1	0	0	0	0	0	0	0	0	0	0	0	1	-1	0	0	0	0	0	0	0
#142	0	-1	2	-1	0	0	0	0	0	0	0	0	0	0	1	0	0	0	0	0	0	0	0	0	0	0	1	-1	0	0	0	0	0	0	0	0
#143	-1	2	-1	0	0	0	0	0	0	0	0	0	0	1	0	0	0	0	0	0	0	0	0	0	0	1	-1	0	0	0	0	0	0	0	0	0
#144	2	-1	0	0	0	0	0	0	0	0	0	0	1	0	0	0	0	0	0	0	0	0	0	0	1	-1	0	0	0	0	0	0	0	0	0	0

Table 2. *(continued)*

ALGEBRA D4

#	OMEGA				ALPHA				EPSILON			
# 1	0	1	0	0	1	2	1	1	1	1	0	0
# 2	1	-1	1	1	1	1	1	1	1	0	1	0
# 3	-1	0	1	1	0	1	1	1	0	1	1	0
# 4	1	0	-1	1	1	1	0	1	1	0	0	1
# 5	1	0	1	-1	1	1	1	0	1	0	0	-1
# 6	-1	1	-1	1	0	1	0	1	0	1	0	1
# 7	-1	1	1	-1	0	1	1	0	0	1	0	-1
# 8	1	1	-1	-1	1	1	0	0	1	0	-1	0
# 9	0	-1	0	2	0	0	0	1	0	0	1	1
# 10	-1	2	-1	-1	0	1	0	0	0	1	-1	0
# 11	0	-1	2	0	0	0	1	0	0	0	1	-1
# 12	2	-1	0	0	1	0	0	0	1	-1	0	0

ALGEBRA D5

#	OMEGA					ALPHA					EPSILON				
# 1	0	1	0	0	0	1	2	2	1	1	1	1	0	0	0
# 2	1	-1	1	0	0	1	1	2	1	1	1	0	1	0	0
# 3	-1	0	1	0	0	0	1	2	1	1	0	1	1	0	0
# 4	1	0	-1	1	1	1	1	1	1	1	1	0	0	1	0
# 5	-1	1	-1	1	1	0	1	1	1	1	0	1	0	1	0
# 6	1	0	0	-1	1	1	1	1	0	1	1	0	0	0	1
# 7	1	0	0	1	-1	1	1	1	1	0	1	0	0	0	-1
# 8	0	-1	0	1	1	0	0	1	1	1	0	0	1	1	0
# 9	-1	1	0	-1	1	0	1	1	0	1	0	1	0	0	1
# 10	-1	1	0	1	-1	0	1	1	1	0	0	1	0	0	-1
# 11	1	0	1	-1	-1	1	1	1	0	0	1	0	0	-1	0
# 12	0	-1	1	-1	1	0	0	1	0	1	0	0	1	0	1
# 13	0	-1	1	1	-1	0	0	1	1	0	0	0	1	0	-1
# 14	-1	1	1	-1	-1	0	1	1	0	0	0	1	0	-1	0
# 15	1	1	-1	0	0	1	1	0	0	0	1	0	-1	0	0
# 16	0	0	-1	0	2	0	0	0	0	1	0	0	0	1	1
# 17	0	-1	2	-1	-1	0	0	1	0	0	0	0	1	-1	0
# 18	0	0	-1	2	0	0	0	0	1	0	0	0	0	1	-1
# 19	-1	2	-1	0	0	0	1	0	0	0	0	1	-1	0	0
# 20	2	-1	0	0	0	1	0	0	0	0	1	-1	0	0	0

ALGEBRA D6

#	OMEGA						ALPHA						EPSILON					
# 1	0	1	0	0	0	0	1	2	2	2	1	1	1	1	0	0	0	0
# 2	1	-1	1	0	0	0	1	1	2	2	1	1	1	0	1	0	0	0
# 3	-1	0	1	0	0	0	0	1	2	2	1	1	0	1	1	0	0	0
# 4	1	0	-1	1	0	0	1	1	1	2	1	1	1	0	0	1	0	0
# 5	-1	1	-1	1	0	0	0	1	1	2	1	1	0	1	0	1	0	0
# 6	1	0	0	-1	1	1	1	1	1	1	1	1	1	0	0	0	1	0
# 7	0	-1	0	1	0	0	0	0	1	2	1	1	0	0	1	1	0	0
# 8	-1	1	0	-1	1	1	0	1	1	1	1	1	0	1	0	0	1	0
# 9	1	0	0	0	-1	1	1	1	1	1	0	1	1	0	0	0	0	1
# 10	1	0	0	0	1	-1	1	1	1	1	1	0	1	0	0	0	0	-1
# 11	0	-1	1	-1	1	1	0	0	1	1	1	1	0	0	1	0	1	0
# 12	-1	1	0	0	-1	1	0	1	1	1	0	1	0	1	0	0	0	1
# 13	-1	1	0	0	1	-1	0	1	1	1	1	0	0	1	0	0	0	-1
# 14	1	0	0	1	-1	-1	1	1	1	1	0	0	1	0	0	0	-1	0

#																		
# 15	0	0	-1	0	1	1	0	0	0	1	1	1	0	0	0	1	1	0
# 16	0	-1	1	0	-1	1	0	0	1	1	0	1	0	0	1	0	0	1
# 17	0	-1	1	0	1	-1	0	0	1	1	1	0	0	0	1	0	0	-1
# 18	-1	1	0	1	-1	-1	0	1	1	1	0	0	0	1	0	0	-1	0
# 19	1	0	1	-1	0	0	1	1	1	0	0	0	1	0	0	-1	0	0
# 20	0	0	-1	1	-1	1	0	0	0	1	0	1	0	0	0	1	0	1
# 21	0	0	-1	1	1	-1	0	0	0	1	1	0	0	0	0	1	0	-1
# 22	0	-1	1	1	-1	-1	0	0	1	1	0	0	0	0	1	0	-1	0
# 23	-1	1	1	-1	0	0	0	1	1	0	0	0	0	1	0	-1	0	0
# 24	1	1	-1	0	0	0	1	1	0	0	0	0	1	0	-1	0	0	0
# 25	0	0	0	-1	0	2	0	0	0	0	0	1	0	0	0	0	1	1
# 26	0	0	-1	2	-1	-1	0	0	0	1	0	0	0	0	0	1	-1	0
# 27	0	0	0	-1	2	0	0	0	0	0	1	0	0	0	0	0	1	-1
# 28	0	-1	2	-1	0	0	0	0	1	0	0	0	0	0	1	-1	0	0
# 29	-1	2	-1	0	0	0	0	1	0	0	0	0	0	1	-1	0	0	0
# 30	2	-1	0	0	0	0	1	0	0	0	0	0	1	-1	0	0	0	0

ALGEBRA D7

#	OMEGA							ALPHA							EPSILON						
# 1	0	1	0	0	0	0	0	1	2	2	2	2	1	1	1	1	0	0	0	0	0
# 2	1	-1	1	0	0	0	0	1	1	2	2	2	1	1	1	0	1	0	0	0	0
# 3	-1	0	1	0	0	0	0	0	1	2	2	2	1	1	0	1	1	0	0	0	0
# 4	1	0	-1	1	0	0	0	1	1	1	2	2	1	1	1	0	0	1	0	0	0
# 5	-1	1	-1	1	0	0	0	0	1	1	2	2	1	1	0	1	0	1	0	0	0
# 6	1	0	0	-1	1	0	0	1	1	1	1	2	1	1	1	0	0	0	1	0	0
# 7	0	-1	0	1	0	0	0	0	0	1	2	2	1	1	0	0	1	1	0	0	0
# 8	-1	1	0	-1	1	0	0	0	1	1	1	2	1	1	0	1	0	0	1	0	0
# 9	1	0	0	0	-1	1	1	1	1	1	1	1	1	1	1	0	0	0	0	1	0
# 10	0	-1	1	-1	1	0	0	0	0	1	1	2	1	1	0	0	1	0	1	0	0
# 11	-1	1	0	0	-1	1	1	0	1	1	1	1	1	1	0	1	0	0	0	1	0
# 12	1	0	0	0	0	-1	1	1	1	1	1	1	0	1	1	0	0	0	0	0	1
# 13	1	0	0	0	0	1	-1	1	1	1	1	1	1	0	1	0	0	0	0	0	-1
# 14	0	0	-1	0	1	0	0	0	0	0	1	2	1	1	0	0	0	1	1	0	0
# 15	0	-1	1	0	-1	1	1	0	0	1	1	1	1	1	0	0	1	0	0	1	0
# 16	-1	1	0	0	0	-1	1	0	1	1	1	1	0	1	0	1	0	0	0	0	1
# 17	-1	1	0	0	0	1	-1	0	1	1	1	1	1	0	0	1	0	0	0	0	-1
# 18	1	0	0	0	1	-1	-1	1	1	1	1	1	0	0	1	0	0	0	0	-1	0
# 19	0	0	-1	1	-1	1	1	0	0	0	1	1	1	1	0	0	0	1	0	1	0
# 20	0	-1	1	0	0	-1	1	0	0	1	1	1	0	1	0	0	1	0	0	0	1
# 21	0	-1	1	0	0	1	-1	0	0	1	1	1	1	0	0	0	1	0	0	0	-1
# 22	-1	1	0	0	1	-1	-1	0	1	1	1	1	0	0	0	1	0	0	0	-1	0
# 23	1	0	0	1	-1	0	0	1	1	1	1	0	0	0	1	0	0	0	-1	0	0
# 24	0	0	0	-1	0	1	1	0	0	0	0	1	1	1	0	0	0	0	1	1	0
# 25	0	0	-1	1	0	-1	1	0	0	0	1	1	0	1	0	0	0	1	0	0	1
# 26	0	0	-1	1	0	1	-1	0	0	0	1	1	1	0	0	0	0	1	0	0	-1
# 27	0	-1	1	0	1	-1	-1	0	0	1	1	1	0	0	0	0	1	0	0	-1	0
# 28	-1	1	0	1	-1	0	0	0	1	1	1	0	0	0	0	1	0	0	-1	0	0
# 29	1	0	1	-1	0	0	0	1	1	1	0	0	0	0	1	0	0	-1	0	0	0
# 30	0	0	0	-1	1	-1	1	0	0	0	0	1	0	1	0	0	0	0	1	0	1
# 31	0	0	0	-1	1	1	-1	0	0	0	0	1	1	0	0	0	0	0	1	0	-1
# 32	0	0	-1	1	1	-1	-1	0	0	0	1	1	0	0	0	0	0	1	0	-1	0
# 33	0	-1	1	1	-1	0	0	0	0	1	1	0	0	0	0	0	1	0	-1	0	0
# 34	-1	1	1	-1	0	0	0	0	1	1	0	0	0	0	0	1	0	-1	0	0	0
# 35	1	1	-1	0	0	0	0	1	1	0	0	0	0	0	1	0	-1	0	0	0	0
# 36	0	0	0	0	-1	0	2	0	0	0	0	0	0	1	0	0	0	0	0	1	1
# 37	0	0	0	-1	2	-1	-1	0	0	0	0	1	0	0	0	0	0	0	1	-1	0
# 38	0	0	0	0	-1	2	0	0	0	0	0	0	1	0	0	0	0	0	0	1	-1
# 39	0	0	-1	2	-1	0	0	0	0	0	1	0	0	0	0	0	0	1	-1	0	0
# 40	0	-1	2	-1	0	0	0	0	0	1	0	0	0	0	0	0	1	-1	0	0	0
# 41	-1	2	-1	0	0	0	0	0	1	0	0	0	0	0	0	1	-1	0	0	0	0
# 42	2	-1	0	0	0	0	0	1	0	0	0	0	0	0	1	-1	0	0	0	0	0

Table 2. *(continued)*

ALGEBRA DB

	OMEGA								ALPHA								EPSILON							
# 1	0	1	0	0	0	0	0	0	1	2	2	2	2	2	1	1	1	1	0	0	0	0	0	0
# 2	1	-1	1	0	0	0	0	0	1	1	2	2	2	2	1	1	1	0	1	0	0	0	0	0
# 3	-1	0	1	0	0	0	0	0	0	1	2	2	2	2	1	1	0	1	1	0	0	0	0	0
# 4	1	0	-1	1	0	0	0	0	1	1	1	2	2	2	1	1	1	0	0	1	0	0	0	0
# 5	-1	1	-1	1	0	0	0	0	0	1	1	2	2	2	1	1	0	1	0	1	0	0	0	0
# 6	1	0	0	-1	1	0	0	0	1	1	1	1	2	2	1	1	1	0	0	0	1	0	0	0
# 7	0	-1	0	1	0	0	0	0	0	0	1	2	2	2	1	1	0	0	1	1	0	0	0	0
# 8	-1	1	0	-1	1	0	0	0	0	1	1	1	2	2	1	1	0	1	0	0	1	0	0	0
# 9	1	0	0	0	-1	1	0	0	1	1	1	1	1	2	1	1	1	0	0	0	0	1	0	0
# 10	0	-1	1	-1	1	0	0	0	0	0	1	1	2	2	1	1	0	0	1	0	1	0	0	0
# 11	-1	1	0	0	-1	1	0	0	0	1	1	1	1	2	1	1	0	1	0	0	0	1	0	0
# 12	1	0	0	0	0	-1	1	1	1	1	1	1	1	1	1	1	1	0	0	0	0	0	1	0
# 13	0	0	-1	0	1	0	0	0	0	0	0	1	2	2	1	1	0	0	0	1	1	0	0	0
# 14	0	-1	1	0	-1	1	0	0	0	0	1	1	1	2	1	1	0	0	1	0	0	1	0	0
# 15	-1	1	0	0	0	-1	1	1	0	1	1	1	1	1	1	1	0	1	0	0	0	0	1	0
# 16	1	0	0	0	0	0	-1	1	1	1	1	1	1	1	0	1	1	0	0	0	0	0	0	1
# 17	1	0	0	0	0	0	1	-1	1	1	1	1	1	1	1	0	1	0	0	0	0	0	0	-1
# 18	0	0	-1	1	-1	1	0	0	0	0	0	1	1	2	1	1	0	0	0	1	0	1	0	0
# 19	0	-1	1	0	0	-1	1	1	0	0	1	1	1	1	1	1	0	0	1	0	0	0	1	0
# 20	-1	1	0	0	0	0	-1	1	0	1	1	1	1	1	0	1	0	1	0	0	0	0	0	1
# 21	-1	1	0	0	0	0	1	-1	0	1	1	1	1	1	1	0	0	1	0	0	0	0	0	-1
# 22	1	0	0	0	0	1	-1	-1	1	1	1	1	1	1	0	0	1	0	0	0	0	0	-1	0
# 23	0	0	0	-1	0	1	0	0	0	0	0	0	1	2	1	1	0	0	0	0	1	1	0	0
# 24	0	0	-1	1	0	-1	1	1	0	0	0	1	1	1	1	1	0	0	0	1	0	0	1	0
# 25	0	-1	1	0	0	0	-1	1	0	0	1	1	1	1	0	1	0	0	1	0	0	0	0	1
# 26	0	-1	1	0	0	0	1	-1	0	0	1	1	1	1	1	0	0	0	1	0	0	0	0	-1
# 27	-1	1	0	0	0	1	-1	-1	0	1	1	1	1	1	0	0	0	1	0	0	0	0	-1	0
# 28	1	0	0	0	1	-1	0	0	1	1	1	1	1	0	0	0	1	0	0	0	0	-1	0	0
# 29	0	0	0	-1	1	-1	1	1	0	0	0	0	1	1	1	1	0	0	0	0	1	0	1	0
# 30	0	0	-1	1	0	0	-1	1	0	0	0	1	1	1	0	1	0	0	0	1	0	0	0	1
# 31	0	0	-1	1	0	0	1	-1	0	0	0	1	1	1	1	0	0	0	0	1	0	0	0	-1
# 32	0	-1	1	0	0	1	-1	-1	0	0	1	1	1	1	0	0	0	0	1	0	0	0	-1	0
# 33	-1	1	0	0	1	-1	0	0	0	1	1	1	1	0	0	0	0	1	0	0	0	-1	0	0
# 34	1	0	0	1	-1	0	0	0	1	1	1	1	0	0	0	0	1	0	0	0	-1	0	0	0
# 35	0	0	0	0	-1	0	1	1	0	0	0	0	0	1	1	1	0	0	0	0	0	1	1	0
# 36	0	0	0	-1	1	0	-1	1	0	0	0	0	1	1	0	1	0	0	0	0	1	0	0	1
# 37	0	0	0	-1	1	0	1	-1	0	0	0	0	1	1	1	0	0	0	0	0	1	0	0	-1
# 38	0	0	-1	1	0	1	-1	-1	0	0	0	1	1	1	0	0	0	0	0	1	0	0	-1	0
# 39	0	-1	1	0	1	-1	0	0	0	0	1	1	1	0	0	0	0	0	1	0	0	-1	0	0
# 40	-1	1	0	1	-1	0	0	0	0	1	1	1	0	0	0	0	0	1	0	0	-1	0	0	0
# 41	1	0	1	-1	0	0	0	0	1	1	1	0	0	0	0	0	1	0	0	-1	0	0	0	0
# 42	0	0	0	0	-1	1	-1	1	0	0	0	0	0	1	0	1	0	0	0	0	0	1	0	1
# 43	0	0	0	0	-1	1	1	-1	0	0	0	0	0	1	1	0	0	0	0	0	0	1	0	-1
# 44	0	0	0	-1	1	1	-1	-1	0	0	0	0	1	1	0	0	0	0	0	0	1	0	-1	0
# 45	0	0	-1	1	1	-1	0	0	0	0	0	1	1	0	0	0	0	0	0	1	0	-1	0	0
# 46	0	-1	1	1	-1	0	0	0	0	0	1	1	0	0	0	0	0	0	1	0	-1	0	0	0
# 47	-1	1	1	-1	0	0	0	0	0	1	1	0	0	0	0	0	0	1	0	-1	0	0	0	0
# 48	1	1	-1	0	0	0	0	0	1	1	0	0	0	0	0	0	1	0	-1	0	0	0	0	0
# 49	0	0	0	0	0	-1	0	2	0	0	0	0	0	0	0	1	0	0	0	0	0	0	1	1
# 50	0	0	0	0	-1	2	-1	-1	0	0	0	0	0	1	0	0	0	0	0	0	0	1	-1	0
# 51	0	0	0	0	0	-1	2	0	0	0	0	0	0	0	1	0	0	0	0	0	0	0	1	-1
# 52	0	0	0	-1	2	-1	0	0	0	0	0	0	1	0	0	0	0	0	0	0	1	-1	0	0
# 53	0	0	-1	2	-1	0	0	0	0	0	0	1	0	0	0	0	0	0	0	1	-1	0	0	0
# 54	0	-1	2	-1	0	0	0	0	0	0	1	0	0	0	0	0	0	0	1	-1	0	0	0	0
# 55	-1	2	-1	0	0	0	0	0	0	1	0	0	0	0	0	0	0	1	-1	0	0	0	0	0
# 56	2	-1	0	0	0	0	0	0	1	0	0	0	0	0	0	0	1	-1	0	0	0	0	0	0

ALGEBRA D9

#	OMEGA									ALPHA									EPSILON								
# 1	0	1	0	0	0	0	0	0	0	1	2	2	2	2	2	2	1	1	1	1	0	0	0	0	0	0	0
# 2	1	-1	1	0	0	0	0	0	0	1	1	2	2	2	2	2	1	1	1	0	1	0	0	0	0	0	0
# 3	-1	0	1	0	0	0	0	0	0	0	1	2	2	2	2	2	1	1	0	1	1	0	0	0	0	0	0
# 4	1	0	-1	1	0	0	0	0	0	1	1	1	2	2	2	2	1	1	1	0	0	1	0	0	0	0	0
# 5	-1	1	-1	1	0	0	0	0	0	0	1	1	2	2	2	2	1	1	0	1	0	1	0	0	0	0	0
# 6	1	0	0	-1	1	0	0	0	0	1	1	1	1	2	2	2	1	1	1	0	0	0	1	0	0	0	0
# 7	0	-1	0	1	0	0	0	0	0	0	0	1	2	2	2	2	1	1	0	0	1	1	0	0	0	0	0
# 8	-1	1	0	-1	1	0	0	0	0	0	1	1	1	2	2	2	1	1	0	1	0	0	1	0	0	0	0
# 9	1	0	0	0	-1	1	0	0	0	1	1	1	1	1	2	2	1	1	1	0	0	0	0	1	0	0	0
# 10	0	-1	1	-1	1	0	0	0	0	0	0	1	1	2	2	2	1	1	0	0	1	0	1	0	0	0	0
# 11	-1	1	0	0	-1	1	0	0	0	0	1	1	1	1	2	2	1	1	0	1	0	0	0	1	0	0	0
# 12	1	0	0	0	0	-1	1	0	0	1	1	1	1	1	1	2	1	1	1	0	0	0	0	0	1	0	0
# 13	0	0	-1	0	1	0	0	0	0	0	0	0	1	2	2	2	1	1	0	0	0	1	1	0	0	0	0
# 14	0	-1	1	0	-1	1	0	0	0	0	0	1	1	1	2	2	1	1	0	0	1	0	0	1	0	0	0
# 15	-1	1	0	0	0	-1	1	0	0	0	1	1	1	1	1	2	1	1	0	1	0	0	0	0	1	0	0
# 16	1	0	0	0	0	0	-1	1	1	1	1	1	1	1	1	1	1	1	1	0	0	0	0	0	0	1	0
# 17	0	0	-1	1	-1	1	0	0	0	0	0	0	1	1	2	2	1	1	0	0	0	1	0	1	0	0	0
# 18	0	-1	1	0	0	-1	1	0	0	0	0	1	1	1	1	2	1	1	0	0	1	0	0	0	1	0	0
# 19	-1	1	0	0	0	0	-1	1	1	0	1	1	1	1	1	1	1	1	0	1	0	0	0	0	0	1	0
# 20	1	0	0	0	0	0	0	-1	1	1	1	1	1	1	1	1	0	1	1	0	0	0	0	0	0	0	1
# 21	1	0	0	0	0	0	0	1	-1	1	1	1	1	1	1	1	1	0	1	0	0	0	0	0	0	0	-1
# 22	0	0	0	-1	0	1	0	0	0	0	0	0	0	1	2	2	1	1	0	0	0	0	1	1	0	0	0
# 23	0	0	-1	1	0	-1	1	0	0	0	0	0	1	1	1	2	1	1	0	0	0	1	0	0	1	0	0
# 24	0	-1	1	0	0	0	-1	1	1	0	0	1	1	1	1	1	1	1	0	0	1	0	0	0	0	1	0
# 25	-1	1	0	0	0	0	0	-1	1	0	1	1	1	1	1	1	0	1	0	1	0	0	0	0	0	0	1
# 26	-1	1	0	0	0	0	0	1	-1	0	1	1	1	1	1	1	1	0	0	1	0	0	0	0	0	0	-1
# 27	1	0	0	0	0	0	1	-1	-1	1	1	1	1	1	1	1	0	0	1	0	0	0	0	0	0	-1	0
# 28	0	0	0	-1	1	-1	1	0	0	0	0	0	0	1	1	2	1	1	0	0	0	0	1	0	1	0	0
# 29	0	0	-1	1	0	0	-1	1	1	0	0	0	1	1	1	1	1	1	0	0	0	1	0	0	0	1	0
# 30	0	-1	1	0	0	0	0	-1	1	0	0	1	1	1	1	1	0	1	0	0	1	0	0	0	0	0	1
# 31	0	-1	1	0	0	0	0	1	-1	0	0	1	1	1	1	1	1	0	0	0	1	0	0	0	0	0	-1
# 32	-1	1	0	0	0	0	1	-1	-1	0	1	1	1	1	1	1	0	0	0	1	0	0	0	0	0	-1	0
# 33	1	0	0	0	0	1	-1	0	0	1	1	1	1	1	1	0	0	0	1	0	0	0	0	0	-1	0	0
# 34	0	0	0	0	-1	0	1	0	0	0	0	0	0	0	1	2	1	1	0	0	0	0	0	1	1	0	0
# 35	0	0	0	-1	1	0	-1	1	1	0	0	0	0	1	1	1	1	1	0	0	0	0	1	0	0	1	0
# 36	0	0	-1	1	0	0	0	-1	1	0	0	0	1	1	1	1	0	1	0	0	0	1	0	0	0	0	1
# 37	0	0	-1	1	0	0	0	1	-1	0	0	0	1	1	1	1	1	0	0	0	0	1	0	0	0	0	-1
# 38	0	-1	1	0	0	0	1	-1	-1	0	0	1	1	1	1	1	0	0	0	0	1	0	0	0	0	-1	0
# 39	-1	1	0	0	0	1	-1	0	0	0	1	1	1	1	1	0	0	0	0	1	0	0	0	0	-1	0	0
# 40	1	0	0	0	1	-1	0	0	0	1	1	1	1	1	0	0	0	0	1	0	0	0	0	-1	0	0	0
# 41	0	0	0	0	-1	1	-1	1	1	0	0	0	0	0	1	1	1	1	0	0	0	0	0	1	0	1	0
# 42	0	0	0	-1	1	0	0	-1	1	0	0	0	0	1	1	1	0	1	0	0	0	0	1	0	0	0	1
# 43	0	0	0	-1	1	0	0	1	-1	0	0	0	0	1	1	1	1	0	0	0	0	0	1	0	0	0	-1
# 44	0	0	-1	1	0	0	1	-1	-1	0	0	0	1	1	1	1	0	0	0	0	0	1	0	0	0	-1	0
# 45	0	-1	1	0	0	1	-1	0	0	0	0	1	1	1	1	0	0	0	0	0	1	0	0	0	-1	0	0
# 46	-1	1	0	0	1	-1	0	0	0	0	1	1	1	1	0	0	0	0	0	1	0	0	0	-1	0	0	0
# 47	1	0	0	1	-1	0	0	0	0	1	1	1	1	0	0	0	0	0	1	0	0	0	-1	0	0	0	0
# 48	0	0	0	0	0	-1	0	1	1	0	0	0	0	0	0	1	1	1	0	0	0	0	0	0	1	1	0
# 49	0	0	0	0	-1	1	0	-1	1	0	0	0	0	0	1	1	0	1	0	0	0	0	0	1	0	0	1
# 50	0	0	0	0	-1	1	0	1	-1	0	0	0	0	0	1	1	1	0	0	0	0	0	0	1	0	0	-1
# 51	0	0	0	-1	1	0	1	-1	-1	0	0	0	0	1	1	1	0	0	0	0	0	0	1	0	0	-1	0
# 52	0	0	-1	1	0	1	-1	0	0	0	0	0	1	1	1	0	0	0	0	0	0	1	0	0	-1	0	0
# 53	0	-1	1	0	1	-1	0	0	0	0	0	1	1	1	0	0	0	0	0	0	1	0	0	-1	0	0	0
# 54	-1	1	0	1	-1	0	0	0	0	0	1	1	1	0	0	0	0	0	0	1	0	0	-1	0	0	0	0
# 55	1	0	1	-1	0	0	0	0	0	1	1	1	0	0	0	0	0	0	1	0	0	-1	0	0	0	0	0
# 56	0	0	0	0	0	-1	1	-1	1	0	0	0	0	0	0	1	0	1	0	0	0	0	0	0	1	0	1
# 57	0	0	0	0	0	-1	1	1	-1	0	0	0	0	0	0	1	1	0	0	0	0	0	0	0	1	0	-1
# 58	0	0	0	0	-1	1	1	-1	-1	0	0	0	0	0	1	1	0	0	0	0	0	0	0	1	0	-1	0
# 59	0	0	0	-1	1	1	-1	0	0	0	0	0	0	1	1	0	0	0	0	0	0	0	1	0	-1	0	0
# 60	0	0	-1	1	1	-1	0	0	0	0	0	0	1	1	0	0	0	0	0	0	0	1	0	-1	0	0	0
# 61	0	-1	1	1	-1	0	0	0	0	0	0	1	1	0	0	0	0	0	0	0	1	0	-1	0	0	0	0
# 62	-1	1	1	-1	0	0	0	0	0	0	1	1	0	0	0	0	0	0	0	1	0	-1	0	0	0	0	0
# 63	1	1	-1	0	0	0	0	0	0	1	1	0	0	0	0	0	0	0	1	0	-1	0	0	0	0	0	0
# 64	0	0	0	0	0	0	-1	0	2	0	0	0	0	0	0	0	0	1	0	0	0	0	0	0	0	1	1
# 65	0	0	0	0	0	-1	2	-1	-1	0	0	0	0	0	0	1	0	0	0	0	0	0	0	0	1	-1	0
# 66	0	0	0	0	0	0	-1	2	0	0	0	0	0	0	0	0	1	0	0	0	0	0	0	0	0	1	-1
# 67	0	0	0	0	-1	2	-1	0	0	0	0	0	0	0	1	0	0	0	0	0	0	0	0	1	-1	0	0
# 68	0	0	0	-1	2	-1	0	0	0	0	0	0	0	1	0	0	0	0	0	0	0	0	1	-1	0	0	0
# 69	0	0	-1	2	-1	0	0	0	0	0	0	0	1	0	0	0	0	0	0	0	0	1	-1	0	0	0	0
# 70	0	-1	2	-1	0	0	0	0	0	0	0	1	0	0	0	0	0	0	0	0	1	-1	0	0	0	0	0
# 71	-1	2	-1	0	0	0	0	0	0	0	1	0	0	0	0	0	0	0	0	1	-1	0	0	0	0	0	0
# 72	2	-1	0	0	0	0	0	0	0	1	0	0	0	0	0	0	0	0	1	-1	0	0	0	0	0	0	0

Table 2. *(continued)*

ALGEBRA D10

	OMEGA									
# 1	0	1	0	0	0	0	0	0	0	0
# 2	1	-1	1	0	0	0	0	0	0	0
# 3	-1	0	1	0	0	0	0	0	0	0
# 4	1	0	-1	1	0	0	0	0	0	0
# 5	-1	1	-1	1	0	0	0	0	0	0
# 6	1	0	0	-1	1	0	0	0	0	0
# 7	0	-1	0	1	0	0	0	0	0	0
# 8	-1	1	0	-1	1	0	0	0	0	0
# 9	1	0	0	0	-1	1	0	0	0	0
# 10	0	-1	1	-1	1	0	0	0	0	0
# 11	-1	1	0	0	-1	1	0	0	0	0
# 12	1	0	0	0	0	-1	1	0	0	0
# 13	0	0	-1	0	1	0	0	0	0	0
# 14	0	-1	1	0	-1	1	0	0	0	0
# 15	-1	1	0	0	0	-1	1	0	0	0
# 16	1	0	0	0	0	0	-1	1	0	0
# 17	0	0	-1	1	-1	1	0	0	0	0
# 18	0	-1	1	0	0	-1	1	0	0	0
# 19	-1	1	0	0	0	0	-1	1	0	0
# 20	1	0	0	0	0	0	0	-1	1	1
# 21	0	0	0	-1	0	1	0	0	0	0
# 22	0	0	-1	1	0	-1	1	0	0	0
# 23	0	-1	1	0	0	0	-1	1	0	0
# 24	-1	1	0	0	0	0	0	-1	1	1
# 25	1	0	0	0	0	0	0	0	-1	1
# 26	1	0	0	0	0	0	0	0	1	-1
# 27	0	0	0	-1	1	-1	1	0	0	0
# 28	0	0	-1	1	0	0	-1	1	0	0
# 29	0	-1	1	0	0	0	0	-1	1	1
# 30	-1	1	0	0	0	0	0	0	-1	1
# 31	-1	1	0	0	0	0	0	0	1	-1
# 32	1	0	0	0	0	0	0	1	-1	-1
# 33	0	0	0	0	-1	0	1	0	0	0
# 34	0	0	0	-1	1	0	-1	1	0	0
# 35	0	0	-1	1	0	0	0	-1	1	1
# 36	0	-1	1	0	0	0	0	0	-1	1
# 37	0	-1	1	0	0	0	0	0	1	-1
# 38	-1	1	0	0	0	0	0	1	-1	-1
# 39	1	0	0	0	0	0	1	-1	0	0
# 40	0	0	0	0	-1	1	-1	1	0	0
# 41	0	0	0	-1	1	0	0	-1	1	1

	ALPHA									
# 1	1	2	2	2	2	2	2	2	1	1
# 2	1	1	2	2	2	2	2	2	1	1
# 3	0	1	2	2	2	2	2	2	1	1
# 4	1	1	1	2	2	2	2	2	1	1
# 5	0	1	1	2	2	2	2	2	1	1
# 6	1	1	1	1	2	2	2	2	1	1
# 7	0	0	1	2	2	2	2	2	1	1
# 8	0	1	1	1	2	2	2	2	1	1
# 9	1	1	1	1	1	2	2	2	1	1
# 10	0	0	1	1	2	2	2	2	1	1
# 11	0	1	1	1	1	2	2	2	1	1
# 12	1	1	1	1	1	1	2	2	1	1
# 13	0	0	0	1	2	2	2	2	1	1
# 14	0	0	1	1	1	2	2	2	1	1
# 15	0	1	1	1	1	1	2	2	1	1
# 16	1	1	1	1	1	1	1	2	1	1
# 17	0	0	0	1	1	2	2	2	1	1
# 18	0	0	1	1	1	1	2	2	1	1
# 19	0	1	1	1	1	1	1	2	1	1
# 20	1	1	1	1	1	1	1	1	1	1
# 21	0	0	0	0	1	2	2	2	1	1
# 22	0	0	0	1	1	1	2	2	1	1
# 23	0	0	1	1	1	1	1	2	1	1
# 24	0	1	1	1	1	1	1	1	1	1
# 25	1	1	1	1	1	1	1	1	0	1
# 26	1	1	1	1	1	1	1	1	1	0
# 27	0	0	0	0	1	1	2	2	1	1
# 28	0	0	0	1	1	1	1	2	1	1
# 29	0	0	1	1	1	1	1	1	1	1
# 30	0	1	1	1	1	1	1	1	0	1
# 31	0	1	1	1	1	1	1	1	1	0
# 32	1	1	1	1	1	1	1	1	0	0
# 33	0	0	0	0	0	1	2	2	1	1
# 34	0	0	0	0	1	1	1	2	1	1
# 35	0	0	0	1	1	1	1	1	1	1
# 36	0	0	1	1	1	1	1	1	0	1
# 37	0	0	1	1	1	1	1	1	1	0
# 38	0	1	1	1	1	1	1	1	0	0
# 39	1	1	1	1	1	1	1	0	0	0
# 40	0	0	0	0	0	1	1	2	1	1
# 41	0	0	0	0	1	1	1	1	1	1

	EPSILON									
# 1	1	1	0	0	0	0	0	0	0	0
# 2	1	0	1	0	0	0	0	0	0	0
# 3	0	1	1	0	0	0	0	0	0	0
# 4	1	0	0	1	0	0	0	0	0	0
# 5	0	1	0	1	0	0	0	0	0	0
# 6	1	0	0	0	1	0	0	0	0	0
# 7	0	0	1	1	0	0	0	0	0	0
# 8	0	1	0	0	1	0	0	0	0	0
# 9	1	0	0	0	0	1	0	0	0	0
# 10	0	0	1	0	1	0	0	0	0	0
# 11	0	1	0	0	0	1	0	0	0	0
# 12	1	0	0	0	0	0	1	0	0	0
# 13	0	0	0	1	1	0	0	0	0	0
# 14	0	0	1	0	0	1	0	0	0	0
# 15	0	1	0	0	0	0	1	0	0	0
# 16	1	0	0	0	0	0	0	1	0	0
# 17	0	0	0	1	0	1	0	0	0	0
# 18	0	0	1	0	0	0	1	0	0	0
# 19	0	1	0	0	0	0	0	1	0	0
# 20	1	0	0	0	0	0	0	0	1	0
# 21	0	0	0	0	1	1	0	0	0	0
# 22	0	0	0	1	0	0	1	0	0	0
# 23	0	0	1	0	0	0	0	1	0	0
# 24	0	1	0	0	0	0	0	0	1	0
# 25	1	0	0	0	0	0	0	0	0	1
# 26	1	0	0	0	0	0	0	0	0	-1
# 27	0	0	0	0	1	0	1	0	0	0
# 28	0	0	0	1	0	0	0	1	0	0
# 29	0	0	1	0	0	0	0	0	1	0
# 30	0	1	0	0	0	0	0	0	0	1
# 31	0	1	0	0	0	0	0	0	0	-1
# 32	1	0	0	0	0	0	0	0	-1	0
# 33	0	0	0	0	0	1	1	0	0	0
# 34	0	0	0	0	1	0	0	1	0	0
# 35	0	0	0	1	0	0	0	0	1	0
# 36	0	0	1	0	0	0	0	0	0	1
# 37	0	0	1	0	0	0	0	0	0	-1
# 38	0	1	0	0	0	0	0	0	-1	0
# 39	1	0	0	0	0	0	0	-1	0	0
# 40	0	0	0	0	0	1	0	1	0	0
# 41	0	0	0	0	1	0	0	0	1	0

#			
# 42	0 0 -1 1 0 0 0 0 -1 1	0 0 0 1 1 1 1 1 0 1	0 0 0 1 0 0 0 0 0 1
# 43	0 0 -1 1 0 0 0 0 1 -1	0 0 0 1 1 1 1 1 1 0	0 0 0 1 0 0 0 0 0 -1
# 44	0 -1 1 0 0 0 0 1 -1 -1	0 0 1 1 1 1 1 1 0 0	0 0 1 0 0 0 0 0 -1 0
# 45	-1 1 0 0 0 0 1 -1 0 0	0 1 1 1 1 1 1 0 0 0	0 1 0 0 0 0 0 -1 0 0
# 46	1 0 0 0 0 1 -1 0 0 0	1 1 1 1 1 1 0 0 0 0	1 0 0 0 0 0 -1 0 0 0
# 47	0 0 0 0 0 -1 0 1 0 0	0 0 0 0 0 0 1 2 1 1	0 0 0 0 0 0 1 1 0 0
# 48	0 0 0 0 -1 1 0 -1 1 1	0 0 0 0 0 1 1 1 1 1	0 0 0 0 0 1 0 0 1 0
# 49	0 0 0 -1 1 0 0 0 -1 1	0 0 0 0 1 1 1 1 0 1	0 0 0 0 1 0 0 0 0 1
# 50	0 0 0 -1 1 0 0 0 1 -1	0 0 0 0 1 1 1 1 1 0	0 0 0 0 1 0 0 0 0 -1
# 51	0 0 -1 1 0 0 0 1 -1 -1	0 0 0 1 1 1 1 1 0 0	0 0 0 1 0 0 0 0 -1 0
# 52	0 -1 1 0 0 0 1 -1 0 0	0 0 1 1 1 1 1 0 0 0	0 0 1 0 0 0 0 -1 0 0
# 53	-1 1 0 0 0 1 -1 0 0 0	0 1 1 1 1 1 0 0 0 0	0 1 0 0 0 0 -1 0 0 0
# 54	1 0 0 0 1 -1 0 0 0 0	1 1 1 1 1 0 0 0 0 0	1 0 0 0 0 -1 0 0 0 0
# 55	0 0 0 0 0 -1 1 -1 1 1	0 0 0 0 0 0 1 1 1 1	0 0 0 0 0 0 1 0 1 0
# 56	0 0 0 0 -1 1 0 0 -1 1	0 0 0 0 0 1 1 1 0 1	0 0 0 0 0 1 0 0 0 1
# 57	0 0 0 0 -1 1 0 0 1 -1	0 0 0 0 0 1 1 1 1 0	0 0 0 0 0 1 0 0 0 -1
# 58	0 0 0 -1 1 0 0 1 -1 -1	0 0 0 0 1 1 1 1 0 0	0 0 0 0 1 0 0 0 -1 0
# 59	0 0 -1 1 0 0 1 -1 0 0	0 0 0 1 1 1 1 0 0 0	0 0 0 1 0 0 0 -1 0 0
# 60	0 -1 1 0 0 1 -1 0 0 0	0 0 1 1 1 1 0 0 0 0	0 0 1 0 0 0 -1 0 0 0
# 61	-1 1 0 0 1 -1 0 0 0 0	0 1 1 1 1 0 0 0 0 0	0 1 0 0 0 -1 0 0 0 0
# 62	1 0 0 1 -1 0 0 0 0 0	1 1 1 1 0 0 0 0 0 0	1 0 0 0 -1 0 0 0 0 0
# 63	0 0 0 0 0 0 -1 0 1 1	0 0 0 0 0 0 0 1 1 1	0 0 0 0 0 0 0 1 1 0
# 64	0 0 0 0 0 -1 1 0 -1 1	0 0 0 0 0 0 1 1 0 1	0 0 0 0 0 0 1 0 0 1
# 65	0 0 0 0 0 -1 1 0 1 -1	0 0 0 0 0 0 1 1 1 0	0 0 0 0 0 0 1 0 0 -1
# 66	0 0 0 0 -1 1 0 1 -1 -1	0 0 0 0 0 1 1 1 0 0	0 0 0 0 0 1 0 0 -1 0
# 67	0 0 0 -1 1 0 1 -1 0 0	0 0 0 0 1 1 1 0 0 0	0 0 0 0 1 0 0 -1 0 0
# 68	0 0 -1 1 0 1 -1 0 0 0	0 0 0 1 1 1 0 0 0 0	0 0 0 1 0 0 -1 0 0 0
# 69	0 -1 1 0 1 -1 0 0 0 0	0 0 1 1 1 0 0 0 0 0	0 0 1 0 0 -1 0 0 0 0
# 70	-1 1 0 1 -1 0 0 0 0 0	0 1 1 1 0 0 0 0 0 0	0 1 0 0 -1 0 0 0 0 0
# 71	1 0 1 -1 0 0 0 0 0 0	1 1 1 0 0 0 0 0 0 0	1 0 0 -1 0 0 0 0 0 0
# 72	0 0 0 0 0 0 -1 1 -1 1	0 0 0 0 0 0 0 1 0 1	0 0 0 0 0 0 0 1 0 1
# 73	0 0 0 0 0 0 -1 1 1 -1	0 0 0 0 0 0 0 1 1 0	0 0 0 0 0 0 0 1 0 -1
# 74	0 0 0 0 0 -1 1 1 -1 -1	0 0 0 0 0 0 1 1 0 0	0 0 0 0 0 0 1 0 -1 0
# 75	0 0 0 0 -1 1 1 -1 0 0	0 0 0 0 0 1 1 0 0 0	0 0 0 0 0 1 0 -1 0 0
# 76	0 0 0 -1 1 1 -1 0 0 0	0 0 0 0 1 1 0 0 0 0	0 0 0 0 1 0 -1 0 0 0
# 77	0 0 -1 1 1 -1 0 0 0 0	0 0 0 1 1 0 0 0 0 0	0 0 0 1 0 -1 0 0 0 0
# 78	0 -1 1 1 -1 0 0 0 0 0	0 0 1 1 0 0 0 0 0 0	0 0 1 0 -1 0 0 0 0 0
# 79	-1 1 1 -1 0 0 0 0 0 0	0 1 1 0 0 0 0 0 0 0	0 1 0 -1 0 0 0 0 0 0
# 80	1 1 -1 0 0 0 0 0 0 0	1 1 0 0 0 0 0 0 0 0	1 0 -1 0 0 0 0 0 0 0
# 81	0 0 0 0 0 0 0 -1 0 2	0 0 0 0 0 0 0 0 0 1	0 0 0 0 0 0 0 0 1 1
# 82	0 0 0 0 0 0 -1 2 -1 -1	0 0 0 0 0 0 0 1 0 0	0 0 0 0 0 0 0 1 -1 0
# 83	0 0 0 0 0 0 0 -1 2 0	0 0 0 0 0 0 0 0 1 0	0 0 0 0 0 0 0 0 1 -1
# 84	0 0 0 0 0 -1 2 -1 0 0	0 0 0 0 0 0 1 0 0 0	0 0 0 0 0 0 1 -1 0 0
# 85	0 0 0 0 -1 2 -1 0 0 0	0 0 0 0 0 1 0 0 0 0	0 0 0 0 0 1 -1 0 0 0
# 86	0 0 0 -1 2 -1 0 0 0 0	0 0 0 0 1 0 0 0 0 0	0 0 0 0 1 -1 0 0 0 0
# 87	0 0 -1 2 -1 0 0 0 0 0	0 0 0 1 0 0 0 0 0 0	0 0 0 1 -1 0 0 0 0 0
# 88	0 -1 2 -1 0 0 0 0 0 0	0 0 1 0 0 0 0 0 0 0	0 0 1 -1 0 0 0 0 0 0
# 89	-1 2 -1 0 0 0 0 0 0 0	0 1 0 0 0 0 0 0 0 0	0 1 -1 0 0 0 0 0 0 0
# 90	2 -1 0 0 0 0 0 0 0 0	1 0 0 0 0 0 0 0 0 0	1 -1 0 0 0 0 0 0 0 0

Table 2. *(continued)*

ALGEBRA D11

#	OMEGA											ALPHA											EPSILON										
# 1	0	1	0	0	0	0	0	0	0	0	0	1	2	2	2	2	2	2	2	2	1	1	1	1	0	0	0	0	0	0	0	0	0
# 2	1	-1	1	0	0	0	0	0	0	0	0	1	1	2	2	2	2	2	2	2	1	1	1	0	1	0	0	0	0	0	0	0	0
# 3	-1	0	1	0	0	0	0	0	0	0	0	0	1	2	2	2	2	2	2	2	1	1	0	1	1	0	0	0	0	0	0	0	0
# 4	1	0	-1	1	0	0	0	0	0	0	0	1	1	1	2	2	2	2	2	2	1	1	1	0	0	1	0	0	0	0	0	0	0
# 5	-1	1	-1	1	0	0	0	0	0	0	0	0	1	1	2	2	2	2	2	2	1	1	0	1	0	1	0	0	0	0	0	0	0
# 6	1	0	0	-1	1	0	0	0	0	0	0	1	1	1	1	2	2	2	2	2	1	1	1	0	0	0	1	0	0	0	0	0	0
# 7	0	-1	0	1	0	0	0	0	0	0	0	0	0	1	2	2	2	2	2	2	1	1	0	0	1	1	0	0	0	0	0	0	0
# 8	-1	1	0	-1	1	0	0	0	0	0	0	0	1	1	1	2	2	2	2	2	1	1	0	1	0	0	1	0	0	0	0	0	0
# 9	1	0	0	0	-1	1	0	0	0	0	0	1	1	1	1	1	2	2	2	2	1	1	1	0	0	0	0	1	0	0	0	0	0
# 10	0	-1	1	-1	1	0	0	0	0	0	0	0	0	1	1	2	2	2	2	2	1	1	0	0	1	0	1	0	0	0	0	0	0
# 11	-1	1	0	0	-1	1	0	0	0	0	0	0	1	1	1	1	2	2	2	2	1	1	0	1	0	0	0	1	0	0	0	0	0
# 12	1	0	0	0	0	-1	1	0	0	0	0	1	1	1	1	1	1	2	2	2	1	1	1	0	0	0	0	0	1	0	0	0	0
# 13	0	0	-1	0	1	0	0	0	0	0	0	0	0	0	1	2	2	2	2	2	1	1	0	0	0	1	1	0	0	0	0	0	0
# 14	0	-1	1	0	-1	1	0	0	0	0	0	0	0	1	1	1	2	2	2	2	1	1	0	0	1	0	0	1	0	0	0	0	0
# 15	-1	1	0	0	0	-1	1	0	0	0	0	0	1	1	1	1	1	2	2	2	1	1	0	1	0	0	0	0	1	0	0	0	0
# 16	1	0	0	0	0	0	-1	1	0	0	0	1	1	1	1	1	1	1	2	2	1	1	1	0	0	0	0	0	0	1	0	0	0
# 17	0	0	-1	1	-1	1	0	0	0	0	0	0	0	0	1	1	2	2	2	2	1	1	0	0	0	1	0	1	0	0	0	0	0
# 18	0	-1	1	0	0	-1	1	0	0	0	0	0	0	1	1	1	1	2	2	2	1	1	0	0	1	0	0	0	1	0	0	0	0
# 19	-1	1	0	0	0	0	-1	1	0	0	0	0	1	1	1	1	1	1	2	2	1	1	0	1	0	0	0	0	0	1	0	0	0
# 20	1	0	0	0	0	0	0	-1	1	0	0	1	1	1	1	1	1	1	1	2	1	1	1	0	0	0	0	0	0	0	1	0	0
# 21	0	0	0	-1	0	1	0	0	0	0	0	0	0	0	0	1	2	2	2	2	1	1	0	0	0	0	1	1	0	0	0	0	0
# 22	0	0	-1	1	0	-1	1	0	0	0	0	0	0	0	1	1	1	2	2	2	1	1	0	0	0	1	0	0	1	0	0	0	0
# 23	0	-1	1	0	0	0	-1	1	0	0	0	0	0	1	1	1	1	1	2	2	1	1	0	0	1	0	0	0	0	1	0	0	0
# 24	-1	1	0	0	0	0	0	-1	1	0	0	0	1	1	1	1	1	1	1	2	1	1	0	1	0	0	0	0	0	0	1	0	0
# 25	1	0	0	0	0	0	0	0	-1	1	1	1	1	1	1	1	1	1	1	1	1	1	1	0	0	0	0	0	0	0	0	1	0
# 26	0	0	0	-1	1	-1	1	0	0	0	0	0	0	0	0	1	1	2	2	2	1	1	0	0	0	0	1	0	1	0	0	0	0
# 27	0	0	-1	1	0	0	-1	1	0	0	0	0	0	0	1	1	1	1	2	2	1	1	0	0	0	1	0	0	0	1	0	0	0
# 28	0	-1	1	0	0	0	0	-1	1	0	0	0	0	1	1	1	1	1	1	2	1	1	0	0	1	0	0	0	0	0	1	0	0
# 29	-1	1	0	0	0	0	0	0	-1	1	1	0	1	1	1	1	1	1	1	1	1	1	0	1	0	0	0	0	0	0	0	1	0
# 30	1	0	0	0	0	0	0	0	0	-1	1	1	1	1	1	1	1	1	1	1	0	1	1	0	0	0	0	0	0	0	0	0	1
# 31	1	0	0	0	0	0	0	0	0	1	-1	1	1	1	1	1	1	1	1	1	1	0	1	0	0	0	0	0	0	0	0	0	-1
# 32	0	0	0	0	-1	0	1	0	0	0	0	0	0	0	0	0	1	2	2	2	1	1	0	0	0	0	0	1	1	0	0	0	0
# 33	0	0	0	-1	1	0	-1	1	0	0	0	0	0	0	0	1	1	1	2	2	1	1	0	0	0	0	1	0	0	1	0	0	0
# 34	0	0	-1	1	0	0	0	-1	1	0	0	0	0	0	1	1	1	1	1	2	1	1	0	0	0	1	0	0	0	0	1	0	0
# 35	0	-1	1	0	0	0	0	0	-1	1	1	0	0	1	1	1	1	1	1	1	1	1	0	0	1	0	0	0	0	0	0	1	0
# 36	-1	1	0	0	0	0	0	0	0	-1	1	0	1	1	1	1	1	1	1	1	0	1	0	1	0	0	0	0	0	0	0	0	1
# 37	-1	1	0	0	0	0	0	0	0	1	-1	0	1	1	1	1	1	1	1	1	1	0	0	1	0	0	0	0	0	0	0	0	-1
# 38	1	0	0	0	0	0	0	0	1	-1	-1	1	1	1	1	1	1	1	1	1	0	0	1	0	0	0	0	0	0	0	0	-1	0
# 39	0	0	0	0	-1	1	-1	1	0	0	0	0	0	0	0	0	1	1	2	2	1	1	0	0	0	0	0	1	0	1	0	0	0
# 40	0	0	0	-1	1	0	0	-1	1	0	0	0	0	0	0	1	1	1	1	2	1	1	0	0	0	0	1	0	0	0	1	0	0
# 41	0	0	-1	1	0	0	0	0	-1	1	1	0	0	0	1	1	1	1	1	1	1	1	0	0	0	1	0	0	0	0	0	1	0
# 42	0	-1	1	0	0	0	0	0	0	-1	1	0	0	1	1	1	1	1	1	1	0	1	0	0	1	0	0	0	0	0	0	0	1
# 43	0	-1	1	0	0	0	0	0	0	1	-1	0	0	1	1	1	1	1	1	1	1	0	0	0	1	0	0	0	0	0	0	0	-1
# 44	-1	1	0	0	0	0	0	0	1	-1	-1	0	1	1	1	1	1	1	1	1	0	0	0	1	0	0	0	0	0	0	0	-1	0
# 45	1	0	0	0	0	0	0	1	-1	0	0	1	1	1	1	1	1	1	1	0	0	0	1	0	0	0	0	0	0	0	-1	0	0
# 46	0	0	0	0	0	-1	0	1	0	0	0	0	0	0	0	0	0	1	2	2	1	1	0	0	0	0	0	0	1	1	0	0	0
# 47	0	0	0	0	-1	1	0	-1	1	0	0	0	0	0	0	0	1	1	1	2	1	1	0	0	0	0	0	1	0	0	1	0	0
# 48	0	0	0	-1	1	0	0	0	-1	1	1	0	0	0	0	1	1	1	1	1	1	1	0	0	0	0	1	0	0	0	0	1	0
# 49	0	0	-1	1	0	0	0	0	0	-1	1	0	0	0	1	1	1	1	1	1	0	1	0	0	0	1	0	0	0	0	0	0	1
# 50	0	0	-1	1	0	0	0	0	0	1	-1	0	0	0	1	1	1	1	1	1	1	0	0	0	0	1	0	0	0	0	0	0	-1
# 51	0	-1	1	0	0	0	0	0	1	-1	-1	0	0	1	1	1	1	1	1	1	0	0	0	0	1	0	0	0	0	0	0	-1	0

#																																	
# 52	-1	1	0	0	0	0	0	1	-1	0	0	0	1	1	1	1	1	1	1	0	0	0	0	1	0	0	0	0	0	0	-1	0	0
# 53	1	0	0	0	0	0	1	-1	0	0	0	1	1	1	1	1	1	1	0	0	0	0	1	0	0	0	0	0	0	-1	0	0	0
# 54	0	0	0	0	0	-1	1	-1	1	0	0	0	0	0	0	0	0	1	1	2	1	1	0	0	0	0	0	0	1	0	1	0	0
# 55	0	0	0	0	-1	1	0	0	-1	1	1	0	0	0	0	0	1	1	1	1	1	1	0	0	0	0	0	1	0	0	0	1	0
# 56	0	0	0	-1	1	0	0	0	0	-1	1	0	0	0	0	1	1	1	1	1	0	1	0	0	0	0	1	0	0	0	0	0	1
# 57	0	0	0	-1	1	0	0	0	0	1	-1	0	0	0	0	1	1	1	1	1	1	0	0	0	0	0	1	0	0	0	0	0	-1
# 58	0	0	-1	1	0	0	0	0	1	-1	-1	0	0	0	1	1	1	1	1	1	0	0	0	0	0	1	0	0	0	0	0	-1	0
# 59	0	-1	1	0	0	0	0	1	-1	0	0	0	0	1	1	1	1	1	1	0	0	0	0	0	1	0	0	0	0	0	-1	0	0
# 60	-1	1	0	0	0	0	1	-1	0	0	0	0	1	1	1	1	1	1	0	0	0	0	0	1	0	0	0	0	0	-1	0	0	0
# 61	1	0	0	0	0	1	-1	0	0	0	0	1	1	1	1	1	1	0	0	0	0	0	1	0	0	0	0	0	-1	0	0	0	0
# 62	0	0	0	0	0	0	-1	0	1	0	0	0	0	0	0	0	0	0	1	2	1	1	0	0	0	0	0	0	0	1	1	0	0
# 63	0	0	0	0	0	-1	1	0	-1	1	1	0	0	0	0	0	0	1	1	1	1	1	0	0	0	0	0	0	1	0	0	1	0
# 64	0	0	0	0	-1	1	0	0	0	-1	1	0	0	0	0	0	1	1	1	1	0	1	0	0	0	0	0	1	0	0	0	0	1
# 65	0	0	0	0	-1	1	0	0	0	1	-1	0	0	0	0	0	1	1	1	1	1	0	0	0	0	0	0	1	0	0	0	0	-1
# 66	0	0	0	-1	1	0	0	0	1	-1	-1	0	0	0	0	1	1	1	1	1	0	0	0	0	0	0	1	0	0	0	0	-1	0
# 67	0	0	-1	1	0	0	0	1	-1	0	0	0	0	0	1	1	1	1	1	0	0	0	0	0	0	1	0	0	0	0	-1	0	0
# 68	0	-1	1	0	0	0	1	-1	0	0	0	0	0	1	1	1	1	1	0	0	0	0	0	0	1	0	0	0	0	-1	0	0	0
# 69	-1	1	0	0	0	1	-1	0	0	0	0	0	1	1	1	1	1	0	0	0	0	0	0	1	0	0	0	0	-1	0	0	0	0
# 70	1	0	0	0	1	-1	0	0	0	0	0	1	1	1	1	1	0	0	0	0	0	0	1	0	0	0	0	-1	0	0	0	0	0
# 71	0	0	0	0	0	0	-1	1	-1	1	1	0	0	0	0	0	0	0	1	1	1	1	0	0	0	0	0	0	0	1	0	1	0
# 72	0	0	0	0	0	-1	1	0	0	-1	1	0	0	0	0	0	0	1	1	1	0	1	0	0	0	0	0	0	1	0	0	0	1
# 73	0	0	0	0	0	-1	1	0	0	1	-1	0	0	0	0	0	0	1	1	1	1	0	0	0	0	0	0	0	1	0	0	0	-1
# 74	0	0	0	0	-1	1	0	0	1	-1	-1	0	0	0	0	0	1	1	1	1	0	0	0	0	0	0	0	1	0	0	0	-1	0
# 75	0	0	0	-1	1	0	0	1	-1	0	0	0	0	0	0	1	1	1	1	0	0	0	0	0	0	0	1	0	0	0	-1	0	0
# 76	0	0	-1	1	0	0	1	-1	0	0	0	0	0	0	1	1	1	1	0	0	0	0	0	0	0	1	0	0	0	-1	0	0	0
# 77	0	-1	1	0	0	1	-1	0	0	0	0	0	0	1	1	1	1	0	0	0	0	0	0	0	1	0	0	0	-1	0	0	0	0
# 78	-1	1	0	0	1	-1	0	0	0	0	0	0	1	1	1	1	0	0	0	0	0	0	0	1	0	0	0	-1	0	0	0	0	0
# 79	1	0	0	1	-1	0	0	0	0	0	0	1	1	1	1	0	0	0	0	0	0	0	1	0	0	0	-1	0	0	0	0	0	0
# 80	0	0	0	0	0	0	0	-1	0	1	1	0	0	0	0	0	0	0	0	1	1	1	0	0	0	0	0	0	0	0	1	1	0
# 81	0	0	0	0	0	0	-1	1	0	-1	1	0	0	0	0	0	0	0	1	1	0	1	0	0	0	0	0	0	0	1	0	0	1
# 82	0	0	0	0	0	0	-1	1	0	1	-1	0	0	0	0	0	0	0	1	1	1	0	0	0	0	0	0	0	0	1	0	0	-1
# 83	0	0	0	0	0	-1	1	0	1	-1	-1	0	0	0	0	0	0	1	1	1	0	0	0	0	0	0	0	0	1	0	0	-1	0
# 84	0	0	0	0	-1	1	0	1	-1	0	0	0	0	0	0	0	1	1	1	0	0	0	0	0	0	0	0	1	0	0	-1	0	0
# 85	0	0	0	-1	1	0	1	-1	0	0	0	0	0	0	0	1	1	1	0	0	0	0	0	0	0	0	1	0	0	-1	0	0	0
# 86	0	0	-1	1	0	1	-1	0	0	0	0	0	0	0	1	1	1	0	0	0	0	0	0	0	0	1	0	0	-1	0	0	0	0
# 87	0	-1	1	0	1	-1	0	0	0	0	0	0	0	1	1	1	0	0	0	0	0	0	0	0	1	0	0	-1	0	0	0	0	0
# 88	-1	1	0	1	-1	0	0	0	0	0	0	0	1	1	1	0	0	0	0	0	0	0	0	1	0	0	-1	0	0	0	0	0	0
# 89	1	0	1	-1	0	0	0	0	0	0	0	1	1	1	0	0	0	0	0	0	0	0	1	0	0	-1	0	0	0	0	0	0	0
# 90	0	0	0	0	0	0	0	-1	1	-1	1	0	0	0	0	0	0	0	0	1	0	1	0	0	0	0	0	0	0	0	1	0	1
# 91	0	0	0	0	0	0	0	-1	1	1	-1	0	0	0	0	0	0	0	0	1	1	0	0	0	0	0	0	0	0	0	1	0	-1
# 92	0	0	0	0	0	0	-1	1	1	-1	-1	0	0	0	0	0	0	0	1	1	0	0	0	0	0	0	0	0	0	1	0	-1	0
# 93	0	0	0	0	0	-1	1	1	-1	0	0	0	0	0	0	0	0	1	1	0	0	0	0	0	0	0	0	0	1	0	-1	0	0
# 94	0	0	0	0	-1	1	1	-1	0	0	0	0	0	0	0	0	1	1	0	0	0	0	0	0	0	0	0	1	0	-1	0	0	0
# 95	0	0	0	-1	1	1	-1	0	0	0	0	0	0	0	0	1	1	0	0	0	0	0	0	0	0	0	1	0	-1	0	0	0	0
# 96	0	0	-1	1	1	-1	0	0	0	0	0	0	0	0	1	1	0	0	0	0	0	0	0	0	0	1	0	-1	0	0	0	0	0
# 97	0	-1	1	1	-1	0	0	0	0	0	0	0	0	1	1	0	0	0	0	0	0	0	0	0	1	0	-1	0	0	0	0	0	0
# 98	-1	1	1	-1	0	0	0	0	0	0	0	0	1	1	0	0	0	0	0	0	0	0	0	1	0	-1	0	0	0	0	0	0	0
# 99	1	1	-1	0	0	0	0	0	0	0	0	1	1	0	0	0	0	0	0	0	0	0	1	0	-1	0	0	0	0	0	0	0	0
#100	0	0	0	0	0	0	0	0	-1	0	2	0	0	0	0	0	0	0	0	0	0	1	0	0	0	0	0	0	0	0	0	1	1
#101	0	0	0	0	0	0	0	-1	2	-1	-1	0	0	0	0	0	0	0	0	1	0	0	0	0	0	0	0	0	0	0	1	-1	0
#102	0	0	0	0	0	0	0	0	-1	2	0	0	0	0	0	0	0	0	0	0	1	0	0	0	0	0	0	0	0	0	0	1	-1
#103	0	0	0	0	0	0	-1	2	-1	0	0	0	0	0	0	0	0	0	1	0	0	0	0	0	0	0	0	0	0	1	-1	0	0
#104	0	0	0	0	0	-1	2	-1	0	0	0	0	0	0	0	0	0	1	0	0	0	0	0	0	0	0	0	0	1	-1	0	0	0
#105	0	0	0	0	-1	2	-1	0	0	0	0	0	0	0	0	0	1	0	0	0	0	0	0	0	0	0	0	1	-1	0	0	0	0
#106	0	0	0	-1	2	-1	0	0	0	0	0	0	0	0	0	1	0	0	0	0	0	0	0	0	0	0	1	-1	0	0	0	0	0
#107	0	0	-1	2	-1	0	0	0	0	0	0	0	0	0	1	0	0	0	0	0	0	0	0	0	0	1	-1	0	0	0	0	0	0
#108	0	-1	2	-1	0	0	0	0	0	0	0	0	0	1	0	0	0	0	0	0	0	0	0	0	1	-1	0	0	0	0	0	0	0
#109	-1	2	-1	0	0	0	0	0	0	0	0	0	1	0	0	0	0	0	0	0	0	0	0	1	-1	0	0	0	0	0	0	0	0
#110	2	-1	0	0	0	0	0	0	0	0	0	1	0	0	0	0	0	0	0	0	0	0	1	-1	0	0	0	0	0	0	0	0	0

Table 2. *(continued)*

ALGEBRA D12

#	OMEGA											
# 1	0	1	0	0	0	0	0	0	0	0	0	0
# 2	1	-1	1	0	0	0	0	0	0	0	0	0
# 3	-1	0	1	0	0	0	0	0	0	0	0	0
# 4	1	0	-1	1	0	0	0	0	0	0	0	0
# 5	-1	1	-1	1	0	0	0	0	0	0	0	0
# 6	1	0	0	-1	1	0	0	0	0	0	0	0
# 7	0	-1	0	1	0	0	0	0	0	0	0	0
# 8	-1	1	0	-1	1	0	0	0	0	0	0	0
# 9	1	0	0	0	-1	1	0	0	0	0	0	0
# 10	0	-1	1	-1	1	0	0	0	0	0	0	0
# 11	-1	1	0	0	-1	1	0	0	0	0	0	0
# 12	1	0	0	0	0	-1	1	0	0	0	0	0
# 13	0	0	-1	0	1	0	0	0	0	0	0	0
# 14	0	-1	1	0	-1	1	0	0	0	0	0	0
# 15	-1	1	0	0	0	-1	1	0	0	0	0	0
# 16	1	0	0	0	0	0	-1	1	0	0	0	0
# 17	0	0	-1	1	-1	1	0	0	0	0	0	0
# 18	0	-1	1	0	0	-1	1	0	0	0	0	0
# 19	-1	1	0	0	0	0	-1	1	0	0	0	0
# 20	1	0	0	0	0	0	0	-1	1	0	0	0
# 21	0	0	0	-1	0	1	0	0	0	0	0	0
# 22	0	0	-1	1	0	-1	1	0	0	0	0	0
# 23	0	-1	1	0	0	0	-1	1	0	0	0	0
# 24	-1	1	0	0	0	0	0	-1	1	0	0	0
# 25	1	0	0	0	0	0	0	0	-1	1	0	0
# 26	0	0	0	-1	1	-1	1	0	0	0	0	0
# 27	0	0	-1	1	0	0	-1	1	0	0	0	0
# 28	0	-1	1	0	0	0	0	-1	1	0	0	0
# 29	-1	1	0	0	0	0	0	0	-1	1	0	0
# 30	1	0	0	0	0	0	0	0	0	-1	1	1
# 31	0	0	0	0	-1	0	1	0	0	0	0	0
# 32	0	0	0	-1	1	0	-1	1	0	0	0	0
# 33	0	0	-1	1	0	0	0	-1	1	0	0	0
# 34	0	-1	1	0	0	0	0	0	-1	1	0	0
# 35	-1	1	0	0	0	0	0	0	0	-1	1	1
# 36	1	0	0	0	0	0	0	0	0	0	-1	1
# 37	1	0	0	0	0	0	0	0	0	0	1	-1
# 38	0	0	0	0	-1	1	-1	1	0	0	0	0
# 39	0	0	0	-1	1	0	0	-1	1	0	0	0
# 40	0	0	-1	1	0	0	0	0	-1	1	0	0
# 41	0	-1	1	0	0	0	0	0	0	-1	1	1
# 42	-1	1	0	0	0	0	0	0	0	0	-1	1
# 43	-1	1	0	0	0	0	0	0	0	0	1	-1
# 44	1	0	0	0	0	0	0	0	0	1	-1	-1
# 45	0	0	0	0	0	-1	0	1	0	0	0	0
# 46	0	0	0	0	-1	1	0	-1	1	0	0	0
# 47	0	0	0	-1	1	0	0	0	-1	1	0	0
# 48	0	0	-1	1	0	0	0	0	0	-1	1	1
# 49	0	-1	1	0	0	0	0	0	0	0	-1	1
# 50	0	-1	1	0	0	0	0	0	0	0	1	-1
# 51	-1	1	0	0	0	0	0	0	0	1	-1	-1
# 52	1	0	0	0	0	0	0	0	1	-1	0	0
# 53	0	0	0	0	0	-1	1	-1	1	0	0	0
# 54	0	0	0	0	-1	1	0	0	-1	1	0	0
# 55	0	0	0	-1	1	0	0	0	0	-1	1	1
# 56	0	0	-1	1	0	0	0	0	0	0	-1	1
# 57	0	0	-1	1	0	0	0	0	0	0	1	-1
# 58	0	-1	1	0	0	0	0	0	0	1	-1	-1
# 59	-1	1	0	0	0	0	0	0	1	-1	0	0
# 60	1	0	0	0	0	0	0	1	-1	0	0	0
# 61	0	0	0	0	0	0	-1	0	1	0	0	0
# 62	0	0	0	0	0	-1	1	0	-1	1	0	0

#	ALPHA											
# 1	1	2	2	2	2	2	2	2	2	2	1	1
# 2	1	1	2	2	2	2	2	2	2	2	1	1
# 3	0	1	2	2	2	2	2	2	2	2	1	1
# 4	1	1	1	2	2	2	2	2	2	2	1	1
# 5	0	1	1	2	2	2	2	2	2	2	1	1
# 6	1	1	1	1	2	2	2	2	2	2	1	1
# 7	0	0	1	2	2	2	2	2	2	2	1	1
# 8	0	1	1	1	2	2	2	2	2	2	1	1
# 9	1	1	1	1	1	2	2	2	2	2	1	1
# 10	0	0	1	1	2	2	2	2	2	2	1	1
# 11	0	1	1	1	1	2	2	2	2	2	1	1
# 12	1	1	1	1	1	1	2	2	2	2	1	1
# 13	0	0	0	1	2	2	2	2	2	2	1	1
# 14	0	0	1	1	1	2	2	2	2	2	1	1
# 15	0	1	1	1	1	1	2	2	2	2	1	1
# 16	1	1	1	1	1	1	1	2	2	2	1	1
# 17	0	0	0	1	1	2	2	2	2	2	1	1
# 18	0	0	1	1	1	1	2	2	2	2	1	1
# 19	0	1	1	1	1	1	1	2	2	2	1	1
# 20	1	1	1	1	1	1	1	1	2	2	1	1
# 21	0	0	0	0	1	2	2	2	2	2	1	1
# 22	0	0	0	1	1	1	2	2	2	2	1	1
# 23	0	0	1	1	1	1	1	2	2	2	1	1
# 24	0	1	1	1	1	1	1	1	2	2	1	1
# 25	1	1	1	1	1	1	1	1	1	2	1	1
# 26	0	0	0	0	1	1	2	2	2	2	1	1
# 27	0	0	0	1	1	1	1	2	2	2	1	1
# 28	0	0	1	1	1	1	1	1	2	2	1	1
# 29	0	1	1	1	1	1	1	1	1	2	1	1
# 30	1	1	1	1	1	1	1	1	1	1	1	1
# 31	0	0	0	0	0	1	2	2	2	2	1	1
# 32	0	0	0	0	1	1	1	2	2	2	1	1
# 33	0	0	0	1	1	1	1	1	2	2	1	1
# 34	0	0	1	1	1	1	1	1	1	2	1	1
# 35	0	1	1	1	1	1	1	1	1	1	1	1
# 36	1	1	1	1	1	1	1	1	1	1	0	1
# 37	1	1	1	1	1	1	1	1	1	1	1	0
# 38	0	0	0	0	0	1	1	2	2	2	1	1
# 39	0	0	0	0	1	1	1	1	2	2	1	1
# 40	0	0	0	1	1	1	1	1	1	2	1	1
# 41	0	0	1	1	1	1	1	1	1	1	1	1
# 42	0	1	1	1	1	1	1	1	1	1	0	1
# 43	0	1	1	1	1	1	1	1	1	1	1	0
# 44	1	1	1	1	1	1	1	1	1	1	0	0
# 45	0	0	0	0	0	0	1	2	2	2	1	1
# 46	0	0	0	0	0	1	1	1	2	2	1	1
# 47	0	0	0	0	1	1	1	1	1	2	1	1
# 48	0	0	0	1	1	1	1	1	1	1	1	1
# 49	0	0	1	1	1	1	1	1	1	1	0	1
# 50	0	0	1	1	1	1	1	1	1	1	1	0
# 51	0	1	1	1	1	1	1	1	1	1	0	0
# 52	1	1	1	1	1	1	1	1	1	0	0	0
# 53	0	0	0	0	0	0	1	1	2	2	1	1
# 54	0	0	0	0	0	1	1	1	1	2	1	1
# 55	0	0	0	0	1	1	1	1	1	1	1	1
# 56	0	0	0	1	1	1	1	1	1	1	0	1
# 57	0	0	0	1	1	1	1	1	1	1	1	0
# 58	0	0	1	1	1	1	1	1	1	1	0	0
# 59	0	1	1	1	1	1	1	1	1	0	0	0
# 60	1	1	1	1	1	1	1	1	0	0	0	0
# 61	0	0	0	0	0	0	0	1	2	2	1	1
# 62	0	0	0	0	0	0	1	1	1	2	1	1

#	EPSILON											
# 1	1	1	0	0	0	0	0	0	0	0	0	0
# 2	1	0	1	0	0	0	0	0	0	0	0	0
# 3	0	1	1	0	0	0	0	0	0	0	0	0
# 4	1	0	0	1	0	0	0	0	0	0	0	0
# 5	0	1	0	1	0	0	0	0	0	0	0	0
# 6	1	0	0	0	1	0	0	0	0	0	0	0
# 7	0	0	1	1	0	0	0	0	0	0	0	0
# 8	0	1	0	0	1	0	0	0	0	0	0	0
# 9	1	0	0	0	0	1	0	0	0	0	0	0
# 10	0	0	1	0	1	0	0	0	0	0	0	0
# 11	0	1	0	0	0	1	0	0	0	0	0	0
# 12	1	0	0	0	0	0	1	0	0	0	0	0
# 13	0	0	0	1	1	0	0	0	0	0	0	0
# 14	0	0	1	0	0	1	0	0	0	0	0	0
# 15	0	1	0	0	0	0	1	0	0	0	0	0
# 16	1	0	0	0	0	0	0	1	0	0	0	0
# 17	0	0	0	1	0	1	0	0	0	0	0	0
# 18	0	0	1	0	0	0	1	0	0	0	0	0
# 19	0	1	0	0	0	0	0	1	0	0	0	0
# 20	1	0	0	0	0	0	0	0	1	0	0	0
# 21	0	0	0	0	1	1	0	0	0	0	0	0
# 22	0	0	0	1	0	0	1	0	0	0	0	0
# 23	0	0	1	0	0	0	0	1	0	0	0	0
# 24	0	1	0	0	0	0	0	0	1	0	0	0
# 25	1	0	0	0	0	0	0	0	0	1	0	0
# 26	0	0	0	0	1	0	1	0	0	0	0	0
# 27	0	0	0	1	0	0	0	1	0	0	0	0
# 28	0	0	1	0	0	0	0	0	1	0	0	0
# 29	0	1	0	0	0	0	0	0	0	1	0	0
# 30	1	0	0	0	0	0	0	0	0	0	1	0
# 31	0	0	0	0	0	1	1	0	0	0	0	0
# 32	0	0	0	0	1	0	0	1	0	0	0	0
# 33	0	0	0	1	0	0	0	0	1	0	0	0
# 34	0	0	1	0	0	0	0	0	0	1	0	0
# 35	0	1	0	0	0	0	0	0	0	0	1	0
# 36	1	0	0	0	0	0	0	0	0	0	0	1
# 37	1	0	0	0	0	0	0	0	0	0	0	-1
# 38	0	0	0	0	0	1	0	1	0	0	0	0
# 39	0	0	0	0	1	0	0	0	1	0	0	0
# 40	0	0	0	1	0	0	0	0	0	1	0	0
# 41	0	0	1	0	0	0	0	0	0	0	1	0
# 42	0	1	0	0	0	0	0	0	0	0	0	1
# 43	0	1	0	0	0	0	0	0	0	0	0	-1
# 44	1	0	0	0	0	0	0	0	0	0	-1	0
# 45	0	0	0	0	0	0	1	1	0	0	0	0
# 46	0	0	0	0	0	1	0	0	1	0	0	0
# 47	0	0	0	0	1	0	0	0	0	1	0	0
# 48	0	0	0	1	0	0	0	0	0	0	1	0
# 49	0	0	1	0	0	0	0	0	0	0	0	1
# 50	0	0	1	0	0	0	0	0	0	0	0	-1
# 51	0	1	0	0	0	0	0	0	0	0	-1	0
# 52	1	0	0	0	0	0	0	0	0	-1	0	0
# 53	0	0	0	0	0	0	1	0	1	0	0	0
# 54	0	0	0	0	0	1	0	0	0	1	0	0
# 55	0	0	0	0	1	0	0	0	0	0	1	0
# 56	0	0	0	1	0	0	0	0	0	0	0	1
# 57	0	0	0	1	0	0	0	0	0	0	0	-1
# 58	0	0	1	0	0	0	0	0	0	0	-1	0
# 59	0	1	0	0	0	0	0	0	0	-1	0	0
# 60	1	0	0	0	0	0	0	0	-1	0	0	0
# 61	0	0	0	0	0	0	0	1	1	0	0	0
# 62	0	0	0	0	0	0	1	0	0	1	0	0

# 63	0	0	0	0	-1	1	0	0	0	-1	1	1	0	0	0	0	0	1	1	1	1	1	1	1	0	0	0	0	0	1	0	0	0	0	1	0
# 64	0	0	0	-1	1	0	0	0	0	0	-1	1	0	0	0	0	1	1	1	1	1	1	0	1	0	0	0	0	1	0	0	0	0	0	0	1
# 65	0	0	0	-1	1	0	0	0	0	0	1	-1	0	0	0	0	1	1	1	1	1	1	1	0	0	0	0	0	1	0	0	0	0	0	0	-1
# 66	0	0	-1	1	0	0	0	0	0	1	-1	-1	0	0	0	1	1	1	1	1	1	1	0	0	0	0	0	1	0	0	0	0	0	0	-1	0
# 67	0	-1	1	0	0	0	0	0	1	-1	0	0	0	0	1	1	1	1	1	1	1	0	0	0	0	0	1	0	0	0	0	0	0	-1	0	0
# 68	-1	1	0	0	0	0	0	1	-1	0	0	0	0	1	1	1	1	1	1	1	0	0	0	0	0	1	0	0	0	0	0	0	-1	0	0	0
# 69	1	0	0	0	0	0	1	-1	0	0	0	0	1	1	1	1	1	1	1	0	0	0	0	0	1	0	0	0	0	0	0	-1	0	0	0	0
# 70	0	0	0	0	0	0	-1	1	-1	1	0	0	0	0	0	0	0	0	0	1	1	2	1	1	0	0	0	0	0	0	0	1	0	1	0	0
# 71	0	0	0	0	0	-1	1	0	0	-1	1	1	0	0	0	0	0	0	1	1	1	1	1	1	0	0	0	0	0	0	1	0	0	0	1	0
# 72	0	0	0	0	-1	1	0	0	0	0	-1	1	0	0	0	0	0	1	1	1	1	1	0	1	0	0	0	0	0	1	0	0	0	0	0	1
# 73	0	0	0	0	-1	1	0	0	0	0	1	-1	0	0	0	0	0	1	1	1	1	1	1	0	0	0	0	0	0	1	0	0	0	0	0	-1
# 74	0	0	0	-1	1	0	0	0	0	1	-1	-1	0	0	0	0	1	1	1	1	1	1	0	0	0	0	0	0	1	0	0	0	0	0	-1	0
# 75	0	0	-1	1	0	0	0	0	1	-1	0	0	0	0	0	1	1	1	1	1	1	0	0	0	0	0	0	1	0	0	0	0	0	-1	0	0
# 76	0	-1	1	0	0	0	0	1	-1	0	0	0	0	0	1	1	1	1	1	1	0	0	0	0	0	0	1	0	0	0	0	0	-1	0	0	0
# 77	-1	1	0	0	0	0	1	-1	0	0	0	0	0	1	1	1	1	1	1	0	0	0	0	0	0	1	0	0	0	0	0	-1	0	0	0	0
# 78	1	0	0	0	0	1	-1	0	0	0	0	0	1	1	1	1	1	1	0	0	0	0	0	0	1	0	0	0	0	0	-1	0	0	0	0	0
# 79	0	0	0	0	0	0	0	-1	0	1	0	0	0	0	0	0	0	0	0	0	1	2	1	1	0	0	0	0	0	0	0	0	1	1	0	0
# 80	0	0	0	0	0	0	-1	1	0	-1	1	1	0	0	0	0	0	0	0	1	1	1	1	1	0	0	0	0	0	0	1	0	0	0	1	0
# 81	0	0	0	0	0	-1	1	0	0	0	-1	1	0	0	0	0	0	0	1	1	1	1	0	1	0	0	0	0	0	0	1	0	0	0	0	1
# 82	0	0	0	0	0	-1	1	0	0	0	1	-1	0	0	0	0	0	0	1	1	1	1	1	0	0	0	0	0	0	0	1	0	0	0	0	-1
# 83	0	0	0	0	-1	1	0	0	0	1	-1	-1	0	0	0	0	0	1	1	1	1	1	0	0	0	0	0	0	0	1	0	0	0	0	-1	0
# 84	0	0	0	-1	1	0	0	0	1	-1	0	0	0	0	0	0	1	1	1	1	1	0	0	0	0	0	0	0	1	0	0	0	0	-1	0	0
# 85	0	0	-1	1	0	0	0	1	-1	0	0	0	0	0	0	1	1	1	1	1	0	0	0	0	0	0	0	1	0	0	0	0	-1	0	0	0
# 86	0	-1	1	0	0	0	1	-1	0	0	0	0	0	0	1	1	1	1	1	0	0	0	0	0	0	0	1	0	0	0	0	-1	0	0	0	0
# 87	-1	1	0	0	0	1	-1	0	0	0	0	0	0	1	1	1	1	1	0	0	0	0	0	0	0	1	0	0	0	0	-1	0	0	0	0	0
# 88	1	0	0	0	1	-1	0	0	0	0	0	0	1	1	1	1	1	0	0	0	0	0	0	0	1	0	0	0	0	-1	0	0	0	0	0	0
# 89	0	0	0	0	0	0	0	-1	1	-1	1	1	0	0	0	0	0	0	0	0	1	1	1	1	0	0	0	0	0	0	0	0	1	0	1	0
# 90	0	0	0	0	0	0	-1	1	0	0	-1	1	0	0	0	0	0	0	0	1	1	1	0	1	0	0	0	0	0	0	0	1	0	0	0	1
# 91	0	0	0	0	0	0	-1	1	0	0	1	-1	0	0	0	0	0	0	0	1	1	1	1	0	0	0	0	0	0	0	0	1	0	0	0	-1
# 92	0	0	0	0	0	-1	1	0	0	1	-1	-1	0	0	0	0	0	0	1	1	1	1	0	0	0	0	0	0	0	0	1	0	0	0	-1	0
# 93	0	0	0	0	-1	1	0	0	1	-1	0	0	0	0	0	0	0	1	1	1	1	0	0	0	0	0	0	0	0	1	0	0	0	-1	0	0
# 94	0	0	0	-1	1	0	0	1	-1	0	0	0	0	0	0	0	1	1	1	1	0	0	0	0	0	0	0	0	1	0	0	0	-1	0	0	0
# 95	0	0	-1	1	0	0	1	-1	0	0	0	0	0	0	0	1	1	1	1	0	0	0	0	0	0	0	0	1	0	0	0	-1	0	0	0	0
# 96	0	-1	1	0	0	1	-1	0	0	0	0	0	0	0	1	1	1	1	0	0	0	0	0	0	0	0	1	0	0	0	-1	0	0	0	0	0
# 97	-1	1	0	0	1	-1	0	0	0	0	0	0	0	1	1	1	1	0	0	0	0	0	0	0	0	1	0	0	0	-1	0	0	0	0	0	0
# 98	1	0	0	1	-1	0	0	0	0	0	0	0	1	1	1	1	0	0	0	0	0	0	0	0	1	0	0	0	-1	0	0	0	0	0	0	0
# 99	0	0	0	0	0	0	0	0	-1	0	1	1	0	0	0	0	0	0	0	0	0	1	1	1	0	0	0	0	0	0	0	0	0	1	1	0
#100	0	0	0	0	0	0	0	-1	1	0	-1	1	0	0	0	0	0	0	0	0	1	1	0	1	0	0	0	0	0	0	0	0	1	0	0	1
#101	0	0	0	0	0	0	0	-1	1	0	1	-1	0	0	0	0	0	0	0	0	1	1	1	0	0	0	0	0	0	0	0	0	1	0	0	-1
#102	0	0	0	0	0	0	-1	1	0	1	-1	-1	0	0	0	0	0	0	0	1	1	1	0	0	0	0	0	0	0	0	0	1	0	0	-1	0
#103	0	0	0	0	0	-1	1	0	1	-1	0	0	0	0	0	0	0	0	1	1	1	0	0	0	0	0	0	0	0	0	1	0	0	-1	0	0
#104	0	0	0	0	-1	1	0	1	-1	0	0	0	0	0	0	0	0	1	1	1	0	0	0	0	0	0	0	0	0	1	0	0	-1	0	0	0
#105	0	0	0	-1	1	0	1	-1	0	0	0	0	0	0	0	0	1	1	1	0	0	0	0	0	0	0	0	0	1	0	0	-1	0	0	0	0
#106	0	0	-1	1	0	1	-1	0	0	0	0	0	0	0	0	1	1	1	0	0	0	0	0	0	0	0	0	1	0	0	-1	0	0	0	0	0
#107	0	-1	1	0	1	-1	0	0	0	0	0	0	0	0	1	1	1	0	0	0	0	0	0	0	0	0	1	0	0	-1	0	0	0	0	0	0
#108	-1	1	0	1	-1	0	0	0	0	0	0	0	0	1	1	1	0	0	0	0	0	0	0	0	0	1	0	0	-1	0	0	0	0	0	0	0
#109	1	0	1	-1	0	0	0	0	0	0	0	0	1	1	1	0	0	0	0	0	0	0	0	0	1	0	0	-1	0	0	0	0	0	0	0	0
#110	0	0	0	0	0	0	0	0	-1	1	-1	1	0	0	0	0	0	0	0	0	0	1	0	1	0	0	0	0	0	0	0	0	0	1	0	1
#111	0	0	0	0	0	0	0	0	-1	1	1	-1	0	0	0	0	0	0	0	0	0	1	1	0	0	0	0	0	0	0	0	0	0	1	0	-1
#112	0	0	0	0	0	0	0	-1	1	1	-1	-1	0	0	0	0	0	0	0	0	1	1	0	0	0	0	0	0	0	0	0	0	1	0	-1	0
#113	0	0	0	0	0	0	-1	1	1	-1	0	0	0	0	0	0	0	0	0	1	1	0	0	0	0	0	0	0	0	0	0	1	0	-1	0	0
#114	0	0	0	0	0	-1	1	1	-1	0	0	0	0	0	0	0	0	0	1	1	0	0	0	0	0	0	0	0	0	0	1	0	-1	0	0	0
#115	0	0	0	0	-1	1	1	-1	0	0	0	0	0	0	0	0	0	1	1	0	0	0	0	0	0	0	0	0	0	1	0	-1	0	0	0	0
#116	0	0	0	-1	1	1	-1	0	0	0	0	0	0	0	0	0	1	1	0	0	0	0	0	0	0	0	0	0	1	0	-1	0	0	0	0	0
#117	0	0	-1	1	1	-1	0	0	0	0	0	0	0	0	0	1	1	0	0	0	0	0	0	0	0	0	0	1	0	-1	0	0	0	0	0	0
#118	0	-1	1	1	-1	0	0	0	0	0	0	0	0	0	1	1	0	0	0	0	0	0	0	0	0	0	1	0	-1	0	0	0	0	0	0	0
#119	-1	1	1	-1	0	0	0	0	0	0	0	0	0	1	1	0	0	0	0	0	0	0	0	0	0	1	0	-1	0	0	0	0	0	0	0	0
#120	1	1	-1	0	0	0	0	0	0	0	0	0	1	1	0	0	0	0	0	0	0	0	0	0	1	0	-1	0	0	0	0	0	0	0	0	0
#121	0	0	0	0	0	0	0	0	0	-1	0	2	0	0	0	0	0	0	0	0	0	0	0	1	0	0	0	0	0	0	0	0	0	0	1	1
#122	0	0	0	0	0	0	0	0	-1	2	-1	-1	0	0	0	0	0	0	0	0	0	1	0	0	0	0	0	0	0	0	0	0	0	1	-1	0
#123	0	0	0	0	0	0	0	0	0	-1	2	0	0	0	0	0	0	0	0	0	0	0	1	0	0	0	0	0	0	0	0	0	0	0	1	-1
#124	0	0	0	0	0	0	0	-1	2	-1	0	0	0	0	0	0	0	0	0	0	1	0	0	0	0	0	0	0	0	0	0	0	1	-1	0	0
#125	0	0	0	0	0	0	-1	2	-1	0	0	0	0	0	0	0	0	0	0	1	0	0	0	0	0	0	0	0	0	0	0	1	-1	0	0	0
#126	0	0	0	0	0	-1	2	-1	0	0	0	0	0	0	0	0	0	0	1	0	0	0	0	0	0	0	0	0	0	0	1	-1	0	0	0	0
#127	0	0	0	0	-1	2	-1	0	0	0	0	0	0	0	0	0	0	1	0	0	0	0	0	0	0	0	0	0	0	1	-1	0	0	0	0	0
#128	0	0	0	-1	2	-1	0	0	0	0	0	0	0	0	0	0	1	0	0	0	0	0	0	0	0	0	0	0	1	-1	0	0	0	0	0	0
#129	0	0	-1	2	-1	0	0	0	0	0	0	0	0	0	0	1	0	0	0	0	0	0	0	0	0	0	0	1	-1	0	0	0	0	0	0	0
#130	0	-1	2	-1	0	0	0	0	0	0	0	0	0	0	1	0	0	0	0	0	0	0	0	0	0	0	1	-1	0	0	0	0	0	0	0	0
#131	-1	2	-1	0	0	0	0	0	0	0	0	0	0	1	0	0	0	0	0	0	0	0	0	0	0	1	-1	0	0	0	0	0	0	0	0	0
#132	2	-1	0	0	0	0	0	0	0	0	0	0	1	0	0	0	0	0	0	0	0	0	0	0	1	-1	0	0	0	0	0	0	0	0	0	0

Table 2. *(continued)*

ALGEBRA E6

#	OMEGA						ALPHA						EPSILON (*2)							
# 1	0	0	0	0	0	1	1	2	3	2	1	2	1	1	1	1	1	-1	-1	1
# 2	0	0	1	0	0	-1	1	2	3	2	1	1	-1	-1	1	1	1	-1	-1	1
# 3	0	1	-1	1	0	0	1	2	2	2	1	1	-1	1	-1	1	1	-1	-1	1
# 4	1	-1	0	1	0	0	1	1	2	2	1	1	1	-1	-1	1	1	-1	-1	1
# 5	0	1	0	-1	1	0	1	2	2	1	1	1	-1	1	1	-1	1	-1	-1	1
# 6	-1	0	0	1	0	0	0	1	2	2	1	1	0	0	0	2	2	0	0	0
# 7	1	-1	1	-1	1	0	1	1	2	1	1	1	1	-1	1	-1	1	-1	-1	1
# 8	0	1	0	0	-1	0	1	2	2	1	0	1	-1	1	1	1	-1	-1	-1	1
# 9	-1	0	1	-1	1	0	0	1	2	1	1	1	0	0	2	0	2	0	0	0
# 10	1	0	-1	0	1	1	1	1	1	1	1	1	1	1	-1	-1	1	-1	-1	1
# 11	1	-1	1	0	-1	0	1	1	2	1	0	1	1	-1	1	1	-1	-1	-1	1
# 12	-1	1	-1	0	1	1	0	1	1	1	1	1	0	2	0	0	2	0	0	0
# 13	-1	0	1	0	-1	0	0	1	2	1	0	1	0	0	2	2	0	0	0	0
# 14	1	0	0	0	1	-1	1	1	1	1	1	0	-1	-1	-1	-1	1	-1	-1	1
# 15	1	0	-1	1	-1	1	1	1	1	1	0	1	1	1	-1	1	-1	-1	-1	1
# 16	-1	1	0	0	1	-1	0	1	1	1	1	0	-2	0	0	0	2	0	0	0
# 17	0	-1	0	0	1	1	0	0	1	1	1	1	2	0	0	0	2	0	0	0
# 18	-1	1	-1	1	-1	1	0	1	1	1	0	1	0	2	0	2	0	0	0	0
# 19	1	0	0	1	-1	-1	1	1	1	1	0	0	-1	-1	-1	1	-1	-1	-1	1
# 20	1	0	0	-1	0	1	1	1	1	0	0	1	1	1	1	-1	-1	-1	-1	1
# 21	0	-1	1	0	1	-1	0	0	1	1	1	0	0	-2	0	0	2	0	0	0
# 22	-1	1	0	1	-1	-1	0	1	1	1	0	0	-2	0	0	2	0	0	0	0
# 23	0	-1	0	1	-1	1	0	0	1	1	0	1	2	0	0	2	0	0	0	0
# 24	-1	1	0	-1	0	1	0	1	1	0	0	1	0	2	2	0	0	0	0	0
# 25	1	0	1	-1	0	-1	1	1	1	0	0	0	-1	-1	1	-1	-1	-1	-1	1
# 26	0	0	-1	1	1	0	0	0	0	1	1	0	0	0	-2	0	2	0	0	0
# 27	0	-1	1	1	-1	-1	0	0	1	1	0	0	0	-2	0	2	0	0	0	0
# 28	-1	1	1	-1	0	-1	0	1	1	0	0	0	-2	0	2	0	0	0	0	0
# 29	0	-1	1	-1	0	1	0	0	1	0	0	1	2	0	2	0	0	0	0	0
# 30	1	1	-1	0	0	0	1	1	0	0	0	0	-1	1	-1	-1	-1	-1	-1	1
# 31	0	0	0	-1	2	0	0	0	0	0	1	0	0	0	0	-2	2	0	0	0
# 32	0	0	-1	2	-1	0	0	0	0	1	0	0	0	0	-2	2	0	0	0	0
# 33	0	-1	2	-1	0	-1	0	0	1	0	0	0	0	-2	2	0	0	0	0	0
# 34	-1	2	-1	0	0	0	0	1	0	0	0	0	-2	2	0	0	0	0	0	0
# 35	0	0	-1	0	0	2	0	0	0	0	0	1	2	2	0	0	0	0	0	0
# 36	2	-1	0	0	0	0	1	0	0	0	0	0	1	-1	-1	-1	-1	-1	-1	1

ALGEBRA E7

#	OMEGA							ALPHA							EPSILON (*2)							
# 1	1	0	0	0	0	0	0	2	3	4	3	2	1	2	0	0	0	0	0	0	-2	2
# 2	-1	1	0	0	0	0	0	1	3	4	3	2	1	2	-1	1	1	1	1	1	-1	1
# 3	0	-1	1	0	0	0	0	1	2	4	3	2	1	2	1	-1	1	1	1	1	-1	1
# 4	0	0	-1	1	0	0	1	1	2	3	3	2	1	2	1	1	-1	1	1	1	-1	1
# 5	0	0	0	1	0	0	-1	1	2	3	3	2	1	1	-1	-1	-1	1	1	1	-1	1
# 6	0	0	0	-1	1	0	1	1	2	3	2	2	1	2	1	1	1	-1	1	1	-1	1
# 7	0	0	1	-1	1	0	-1	1	2	3	2	2	1	1	-1	-1	1	-1	1	1	-1	1
# 8	0	0	0	0	-1	1	1	1	2	3	2	1	1	2	1	1	1	1	-1	1	-1	1
# 9	0	1	-1	0	1	0	0	1	2	2	2	2	1	1	-1	1	-1	-1	1	1	-1	1
# 10	0	0	1	0	-1	1	-1	1	2	3	2	1	1	1	-1	-1	1	1	-1	1	-1	1
# 11	0	0	0	0	0	-1	1	1	2	3	2	1	0	2	1	1	1	1	1	-1	-1	1
# 12	1	-1	0	0	1	0	0	1	1	2	2	2	1	1	1	-1	-1	-1	1	1	-1	1
# 13	0	1	-1	1	-1	1	0	1	2	2	2	1	1	1	-1	1	-1	1	-1	1	-1	1
# 14	0	0	1	0	0	-1	-1	1	2	3	2	1	0	1	-1	-1	1	1	1	-1	-1	1
# 15	-1	0	0	0	1	0	0	0	1	2	2	2	1	1	0	0	0	0	2	2	0	0
# 16	1	-1	0	1	-1	1	0	1	1	2	2	1	1	1	1	-1	-1	1	-1	1	-1	1
# 17	0	1	0	-1	0	1	0	1	2	2	1	1	1	1	-1	1	1	-1	-1	1	-1	1
# 18	0	1	-1	1	0	-1	0	1	2	2	2	1	0	1	-1	1	-1	1	1	-1	-1	1
# 19	-1	0	0	1	-1	1	0	0	1	2	2	1	1	1	0	0	0	2	0	2	0	0
# 20	1	-1	1	-1	0	1	0	1	1	2	1	1	1	1	1	-1	1	-1	-1	1	-1	1
# 21	1	-1	0	1	0	-1	0	1	1	2	2	1	0	1	1	-1	-1	1	1	-1	-1	1
# 22	0	1	0	-1	1	-1	0	1	2	2	1	1	0	1	-1	1	1	-1	1	-1	-1	1
# 23	-1	0	1	-1	0	1	0	0	1	2	1	1	1	1	0	0	2	0	0	2	0	0
# 24	-1	0	0	1	0	-1	0	0	1	2	2	1	0	1	0	0	0	2	2	0	0	0
# 25	1	0	-1	0	0	1	1	1	1	1	1	1	1	1	1	1	-1	-1	-1	1	-1	1
# 26	1	-1	1	-1	1	-1	0	1	1	2	1	1	0	1	1	-1	1	-1	1	-1	-1	1
# 27	0	1	0	0	-1	0	0	1	2	2	1	0	0	1	-1	1	1	1	-1	-1	-1	1
# 28	-1	1	-1	0	0	1	1	0	1	1	1	1	1	1	0	2	0	0	0	2	0	0
# 29	-1	0	1	-1	1	-1	0	0	1	2	1	1	0	1	0	0	2	0	2	0	0	0
# 30	1	0	0	0	0	1	-1	1	1	1	1	1	1	0	-1	-1	-1	-1	-1	1	-1	1
# 31	1	0	-1	0	1	-1	1	1	1	1	1	1	0	1	1	1	-1	-1	1	-1	-1	1
# 32	1	-1	1	0	-1	0	0	1	1	2	1	0	0	1	1	-1	1	1	-1	-1	-1	1
# 33	-1	1	0	0	0	1	-1	0	1	1	1	1	1	0	-2	0	0	0	0	2	0	0
# 34	0	-1	0	0	0	1	1	0	0	1	1	1	1	1	2	0	0	0	0	2	0	0
# 35	-1	1	-1	0	1	-1	1	0	1	1	1	1	0	1	0	2	0	0	2	0	0	0
# 36	-1	0	1	0	-1	0	0	0	1	2	1	0	0	1	0	0	2	2	0	0	0	0
# 37	1	0	0	0	1	-1	-1	1	1	1	1	1	0	0	-1	-1	-1	-1	1	-1	-1	1
# 38	1	0	-1	1	-1	0	1	1	1	1	1	0	0	1	1	1	-1	1	-1	-1	-1	1
# 39	0	-1	1	0	0	1	-1	0	0	1	1	1	1	0	0	-2	0	0	0	2	0	0
# 40	-1	1	0	0	1	-1	-1	0	1	1	1	1	0	0	-2	0	0	0	2	0	0	0
# 41	0	-1	0	0	1	-1	1	0	0	1	1	1	0	1	2	0	0	0	2	0	0	0
# 42	-1	1	-1	1	-1	0	1	0	1	1	1	0	0	1	0	2	0	2	0	0	0	0
# 43	1	0	0	1	-1	0	-1	1	1	1	1	0	0	0	-1	-1	-1	1	-1	-1	-1	1
# 44	1	0	0	-1	0	0	1	1	1	1	0	0	0	1	1	1	1	-1	-1	-1	-1	1
# 45	0	0	-1	1	0	1	0	0	0	0	1	1	1	0	0	0	-2	0	0	2	0	0
# 46	0	-1	1	0	1	-1	-1	0	0	1	1	1	0	0	0	-2	0	0	2	0	0	0
# 47	-1	1	0	1	-1	0	-1	0	1	1	1	0	0	0	-2	0	0	2	0	0	0	0
# 48	0	-1	0	1	-1	0	1	0	0	1	1	0	0	1	2	0	0	2	0	0	0	0
# 49	-1	1	0	-1	0	0	1	0	1	1	0	0	0	1	0	2	2	0	0	0	0	0
# 50	1	0	1	-1	0	0	-1	1	1	1	0	0	0	0	-1	-1	1	-1	-1	-1	-1	1
# 51	0	0	0	-1	1	1	0	0	0	0	0	1	1	0	0	0	0	-2	0	2	0	0
# 52	0	0	-1	1	1	-1	0	0	0	0	1	1	0	0	0	0	-2	0	2	0	0	0
# 53	0	-1	1	1	-1	0	-1	0	0	1	1	0	0	0	0	-2	0	2	0	0	0	0
# 54	-1	1	1	-1	0	0	-1	0	1	1	0	0	0	0	-2	0	2	0	0	0	0	0
# 55	0	-1	1	-1	0	0	1	0	0	1	0	0	0	1	2	0	2	0	0	0	0	0
# 56	1	1	-1	0	0	0	0	1	1	0	0	0	0	0	-1	1	-1	-1	-1	-1	-1	1
# 57	0	0	0	0	-1	2	0	0	0	0	0	0	1	0	0	0	0	0	-2	2	0	0
# 58	0	0	0	-1	2	-1	0	0	0	0	0	1	0	0	0	0	0	-2	2	0	0	0
# 59	0	0	-1	2	-1	0	0	0	0	0	1	0	0	0	0	0	-2	2	0	0	0	0
# 60	0	-1	2	-1	0	0	-1	0	0	1	0	0	0	0	0	-2	2	0	0	0	0	0
# 61	-1	2	-1	0	0	0	0	0	1	0	0	0	0	0	-2	2	0	0	0	0	0	0
# 62	0	0	-1	0	0	0	2	0	0	0	0	0	0	1	2	2	0	0	0	0	0	0
# 63	2	-1	0	0	0	0	0	1	0	0	0	0	0	0	1	-1	-1	-1	-1	-1	-1	1

Table 2. *(continued)*

ALGEBRA E8

	OMEGA								ALPHA								EPSILON (*2)							
# 1	1	0	0	0	0	0	0	0	2	3	4	5	6	4	2	3	0	0	0	0	0	0	2	2
# 2	-1	1	0	0	0	0	0	0	1	3	4	5	6	4	2	3	0	0	0	0	0	2	0	2
# 3	0	-1	1	0	0	0	0	0	1	2	4	5	6	4	2	3	0	0	0	0	2	0	0	2
# 4	0	0	-1	1	0	0	0	0	1	2	3	5	6	4	2	3	0	0	0	2	0	0	0	2
# 5	0	0	0	-1	1	0	0	0	1	2	3	4	6	4	2	3	0	0	2	0	0	0	0	2
# 6	0	0	0	0	-1	1	0	1	1	2	3	4	5	4	2	3	0	2	0	0	0	0	0	2
# 7	0	0	0	0	0	1	0	-1	1	2	3	4	5	4	2	2	-2	0	0	0	0	0	0	2
# 8	0	0	0	0	0	-1	1	1	1	2	3	4	5	3	2	3	2	0	0	0	0	0	0	2
# 9	0	0	0	0	1	-1	1	-1	1	2	3	4	5	3	2	2	0	-2	0	0	0	0	0	2
# 10	0	0	0	0	0	0	-1	1	1	2	3	4	5	3	1	3	1	1	1	1	1	1	1	1
# 11	0	0	0	0	1	0	-1	-1	1	2	3	4	5	3	1	2	-1	-1	1	1	1	1	1	1
# 12	0	0	0	1	-1	0	1	0	1	2	3	4	4	3	2	2	0	0	-2	0	0	0	0	2
# 13	0	0	0	1	-1	1	-1	0	1	2	3	4	4	3	1	2	-1	1	-1	1	1	1	1	1
# 14	0	0	1	-1	0	0	1	0	1	2	3	3	4	3	2	2	0	0	0	-2	0	0	0	2
# 15	0	0	0	1	0	-1	0	0	1	2	3	4	4	2	1	2	1	-1	-1	1	1	1	1	1
# 16	0	0	1	-1	0	1	-1	0	1	2	3	3	4	3	1	2	-1	1	1	-1	1	1	1	1
# 17	0	1	-1	0	0	0	1	0	1	2	2	3	4	3	2	2	0	0	0	0	-2	0	0	2
# 18	0	0	1	-1	1	-1	0	0	1	2	3	3	4	2	1	2	1	-1	1	-1	1	1	1	1
# 19	0	1	-1	0	0	1	-1	0	1	2	2	3	4	3	1	2	-1	1	1	1	-1	1	1	1
# 20	1	-1	0	0	0	0	1	0	1	1	2	3	4	3	2	2	0	0	0	0	0	-2	0	2
# 21	0	0	1	0	-1	0	0	1	1	2	3	3	3	2	1	2	1	1	-1	-1	1	1	1	1
# 22	0	1	-1	0	1	-1	0	0	1	2	2	3	4	2	1	2	1	-1	1	1	-1	1	1	1
# 23	1	-1	0	0	0	1	-1	0	1	1	2	3	4	3	1	2	-1	1	1	1	1	-1	1	1
# 24	-1	0	0	0	0	0	1	0	0	1	2	3	4	3	2	2	0	0	0	0	0	0	-2	2
# 25	0	0	1	0	0	0	0	-1	1	2	3	3	3	2	1	1	-1	-1	-1	-1	1	1	1	1
# 26	0	1	-1	1	-1	0	0	1	1	2	2	3	3	2	1	2	1	1	-1	1	-1	1	1	1
# 27	1	-1	0	0	1	-1	0	0	1	1	2	3	4	2	1	2	1	-1	1	1	1	-1	1	1
# 28	-1	0	0	0	0	1	-1	0	0	1	2	3	4	3	1	2	-1	1	1	1	1	1	-1	1
# 29	0	1	-1	1	0	0	0	-1	1	2	2	3	3	2	1	1	-1	-1	-1	1	-1	1	1	1
# 30	0	1	0	-1	0	0	0	1	1	2	2	2	3	2	1	2	1	1	1	-1	-1	1	1	1
# 31	1	-1	0	1	-1	0	0	1	1	1	2	3	3	2	1	2	1	1	-1	1	1	-1	1	1
# 32	-1	0	0	0	1	-1	0	0	0	1	2	3	4	2	1	2	1	-1	1	1	1	1	-1	1
# 33	0	1	0	-1	1	0	0	-1	1	2	2	2	3	2	1	1	-1	-1	1	-1	-1	1	1	1
# 34	1	-1	0	1	0	0	0	-1	1	1	2	3	3	2	1	1	-1	-1	-1	1	1	-1	1	1
# 35	1	-1	1	-1	0	0	0	1	1	1	2	2	3	2	1	2	1	1	1	-1	1	-1	1	1
# 36	-1	0	0	1	-1	0	0	1	0	1	2	3	3	2	1	2	1	1	-1	1	1	1	-1	1
# 37	0	1	0	0	-1	1	0	0	1	2	2	2	2	2	1	1	-1	1	-1	-1	-1	1	1	1
# 38	1	-1	1	-1	1	0	0	-1	1	1	2	2	3	2	1	1	-1	-1	1	-1	1	-1	1	1
# 39	-1	0	0	1	0	0	0	-1	0	1	2	3	3	2	1	1	-1	-1	-1	1	1	1	-1	1
# 40	1	0	-1	0	0	0	0	1	1	1	1	2	3	2	1	2	1	1	1	1	-1	-1	1	1
# 41	-1	0	1	-1	0	0	0	1	0	1	2	2	3	2	1	2	1	1	1	-1	1	1	-1	1
# 42	0	1	0	0	0	-1	1	0	1	2	2	2	2	1	1	1	1	-1	-1	-1	-1	1	1	1
# 43	1	-1	1	0	-1	1	0	0	1	1	2	2	2	2	1	1	-1	1	-1	-1	1	-1	1	1
# 44	1	0	-1	0	1	0	0	-1	1	1	1	2	3	2	1	1	-1	-1	1	1	-1	-1	1	1
# 45	-1	0	1	-1	1	0	0	-1	0	1	2	2	3	2	1	1	-1	-1	1	-1	1	1	-1	1
# 46	-1	1	-1	0	0	0	0	1	0	1	1	2	3	2	1	2	1	1	1	1	-1	1	-1	1
# 47	0	1	0	0	0	0	-1	0	1	2	2	2	2	1	0	1	0	0	0	0	0	2	2	0
# 48	1	-1	1	0	0	-1	1	0	1	1	2	2	2	1	1	1	1	-1	-1	-1	1	-1	1	1
# 49	1	0	-1	1	-1	1	0	0	1	1	1	2	2	2	1	1	-1	1	-1	1	-1	-1	1	1
# 50	-1	0	1	0	-1	1	0	0	0	1	2	2	2	2	1	1	-1	1	-1	-1	1	1	-1	1
# 51	-1	1	-1	0	1	0	0	-1	0	1	1	2	3	2	1	1	-1	-1	1	1	-1	1	-1	1
# 52	0	-1	0	0	0	0	0	1	0	0	1	2	3	2	1	2	1	1	1	1	1	-1	-1	1
# 53	1	-1	1	0	0	0	-1	0	1	1	2	2	2	1	0	1	0	0	0	0	2	0	2	0
# 54	1	0	-1	1	0	-1	1	0	1	1	1	2	2	1	1	1	1	-1	-1	1	-1	-1	1	1
# 55	-1	0	1	0	0	-1	1	0	0	1	2	2	2	1	1	1	1	-1	-1	-1	1	1	-1	1
# 56	1	0	0	-1	0	1	0	0	1	1	1	1	2	2	1	1	-1	1	1	-1	-1	-1	1	1
# 57	-1	1	-1	1	-1	1	0	0	0	1	1	2	2	2	1	1	-1	1	-1	1	-1	1	-1	1

# 58	0	-1	0	0	1	0	0	-1	0	0	1	2	3	2	1	1	-1	-1	1	1	1	-1	-1	1
# 59	1	0	-1	1	0	0	-1	0	1	1	1	2	2	1	0	1	0	0	0	2	0	0	2	0
# 60	-1	0	1	0	0	0	-1	0	0	1	2	2	2	1	0	1	0	0	0	0	2	2	0	0
# 61	1	0	0	-1	1	-1	1	0	1	1	1	1	2	1	1	1	1	-1	1	-1	-1	-1	1	1
# 62	-1	1	-1	1	0	-1	1	0	0	1	1	2	2	1	1	1	1	-1	-1	1	-1	1	-1	1
# 63	-1	1	0	-1	0	1	0	0	0	1	1	1	2	2	1	1	-1	1	1	-1	-1	1	-1	1
# 64	0	-1	0	1	-1	1	0	0	0	0	1	2	2	2	1	1	-1	1	-1	1	1	-1	-1	1
# 65	1	0	0	-1	1	0	-1	0	1	1	1	1	2	1	0	1	0	0	2	0	0	0	2	0
# 66	-1	1	-1	1	0	0	-1	0	0	1	1	2	2	1	0	1	0	0	0	2	0	2	0	0
# 67	1	0	0	0	-1	0	1	1	1	1	1	1	1	1	1	1	1	1	-1	-1	-1	-1	1	1
# 68	-1	1	0	-1	1	-1	1	0	0	1	1	1	2	1	1	1	1	-1	1	-1	-1	1	-1	1
# 69	0	-1	0	1	0	-1	1	0	0	0	1	2	2	1	1	1	1	-1	-1	1	1	-1	-1	1
# 70	0	-1	1	-1	0	1	0	0	0	0	1	1	2	2	1	1	-1	1	1	-1	1	-1	-1	1
# 71	1	0	0	0	-1	1	-1	1	1	1	1	1	1	1	0	1	0	2	0	0	0	0	2	0
# 72	-1	1	0	-1	1	0	-1	0	0	1	1	1	2	1	0	1	0	0	2	0	0	2	0	0
# 73	0	-1	0	1	0	0	-1	0	0	0	1	2	2	1	0	1	0	0	0	2	2	0	0	0
# 74	1	0	0	0	0	0	1	-1	1	1	1	1	1	1	1	0	-1	-1	-1	-1	-1	-1	1	1
# 75	-1	1	0	0	-1	0	1	1	0	1	1	1	1	1	1	1	1	1	-1	-1	-1	1	-1	1
# 76	0	-1	1	-1	1	-1	1	0	0	0	1	1	2	1	1	1	1	-1	1	-1	1	-1	-1	1
# 77	0	0	-1	0	0	1	0	0	0	0	0	1	2	2	1	1	-1	1	1	1	-1	-1	-1	1
# 78	1	0	0	0	0	1	-1	-1	1	1	1	1	1	1	0	0	-2	0	0	0	0	0	2	0
# 79	1	0	0	0	0	-1	0	1	1	1	1	1	1	0	0	1	2	0	0	0	0	0	2	0
# 80	-1	1	0	0	-1	1	-1	1	0	1	1	1	1	1	0	1	0	2	0	0	0	2	0	0
# 81	0	-1	1	-1	1	0	-1	0	0	0	1	1	2	1	0	1	0	0	2	0	2	0	0	0
# 82	-1	1	0	0	0	0	1	-1	0	1	1	1	1	1	1	0	-1	-1	-1	-1	-1	1	-1	1
# 83	0	-1	1	0	-1	0	1	1	0	0	1	1	1	1	1	1	1	1	-1	-1	1	-1	-1	1
# 84	0	0	-1	0	1	-1	1	0	0	0	0	1	2	1	1	1	1	-1	1	1	-1	-1	-1	1
# 85	1	0	0	0	1	-1	0	-1	1	1	1	1	1	0	0	0	0	-2	0	0	0	0	2	0
# 86	-1	1	0	0	0	1	-1	-1	0	1	1	1	1	1	0	0	-2	0	0	0	0	2	0	0
# 87	-1	1	0	0	0	-1	0	1	0	1	1	1	1	0	0	1	2	0	0	0	0	2	0	0
# 88	0	-1	1	0	-1	1	-1	1	0	0	1	1	1	1	0	1	0	2	0	0	2	0	0	0
# 89	0	0	-1	0	1	0	-1	0	0	0	0	1	2	1	0	1	0	0	2	2	0	0	0	0
# 90	0	-1	1	0	0	0	1	-1	0	0	1	1	1	1	1	0	-1	-1	-1	-1	1	-1	-1	1
# 91	0	0	-1	1	-1	0	1	1	0	0	0	1	1	1	1	1	1	1	-1	1	-1	-1	-1	1
# 92	1	0	0	1	-1	0	0	0	1	1	1	1	0	0	0	0	0	0	-2	0	0	0	2	0
# 93	-1	1	0	0	1	-1	0	-1	0	1	1	1	1	0	0	0	0	-2	0	0	0	2	0	0
# 94	0	-1	1	0	0	1	-1	-1	0	0	1	1	1	1	0	0	-2	0	0	0	2	0	0	0
# 95	0	-1	1	0	0	-1	0	1	0	0	1	1	1	0	0	1	2	0	0	0	2	0	0	0
# 96	0	0	-1	1	-1	1	-1	1	0	0	0	1	1	1	0	1	0	2	0	2	0	0	0	0
# 97	0	0	-1	1	0	0	1	-1	0	0	0	1	1	1	1	0	-1	-1	-1	1	-1	-1	-1	1
# 98	0	0	0	-1	0	0	1	1	0	0	0	0	1	1	1	1	1	1	1	-1	-1	-1	-1	1
# 99	1	0	1	-1	0	0	0	0	1	1	1	0	0	0	0	0	0	0	0	-2	0	0	2	0
#100	-1	1	0	1	-1	0	0	0	0	1	1	1	0	0	0	0	0	0	-2	0	0	2	0	0
#101	0	-1	1	0	1	-1	0	-1	0	0	1	1	1	0	0	0	0	-2	0	0	2	0	0	0
#102	0	0	-1	1	0	1	-1	-1	0	0	0	1	1	1	0	0	-2	0	0	2	0	0	0	0
#103	0	0	-1	1	0	-1	0	1	0	0	0	1	1	0	0	1	2	0	0	2	0	0	0	0
#104	0	0	0	-1	0	1	-1	1	0	0	0	0	1	1	0	1	0	2	2	0	0	0	0	0
#105	0	0	0	-1	1	0	1	-1	0	0	0	0	1	1	1	0	-1	-1	1	-1	-1	-1	-1	1
#106	1	1	-1	0	0	0	0	0	1	1	0	0	0	0	0	0	0	0	0	0	-2	0	2	0
#107	-1	1	1	-1	0	0	0	0	0	1	1	0	0	0	0	0	0	0	0	-2	0	2	0	0
#108	0	-1	1	1	-1	0	0	0	0	0	1	1	0	0	0	0	0	0	-2	0	2	0	0	0
#109	0	0	-1	1	1	-1	0	-1	0	0	0	1	1	0	0	0	0	-2	0	2	0	0	0	0
#110	0	0	0	-1	1	1	-1	-1	0	0	0	0	1	1	0	0	-2	0	2	0	0	0	0	0
#111	0	0	0	-1	1	-1	0	1	0	0	0	0	1	0	0	1	2	0	2	0	0	0	0	0
#112	0	0	0	0	-1	1	1	0	0	0	0	0	0	1	1	0	-1	1	-1	-1	-1	-1	-1	1
#113	2	-1	0	0	0	0	0	0	1	0	0	0	0	0	0	0	0	0	0	0	0	-2	2	0
#114	-1	2	-1	0	0	0	0	0	0	1	0	0	0	0	0	0	0	0	0	0	-2	2	0	0
#115	0	-1	2	-1	0	0	0	0	0	0	1	0	0	0	0	0	0	0	0	-2	2	0	0	0
#116	0	0	-1	2	-1	0	0	0	0	0	0	1	0	0	0	0	0	0	-2	2	0	0	0	0
#117	0	0	0	-1	2	-1	0	-1	0	0	0	0	1	0	0	0	0	-2	2	0	0	0	0	0
#118	0	0	0	0	-1	2	-1	0	0	0	0	0	0	1	0	0	-2	2	0	0	0	0	0	0
#119	0	0	0	0	-1	0	0	2	0	0	0	0	0	0	0	1	2	2	0	0	0	0	0	0
#120	0	0	0	0	0	-1	2	0	0	0	0	0	0	0	1	0	1	-1	-1	-1	-1	-1	-1	1

Table 2. *(continued)*

ALGEBRA F4

#	OMEGA				ALPHA				EPSILON (*2)			
# 1	1	0	0	0	2	3	4	2	2	2	0	0
# 2	-1	1	0	0	1	3	4	2	2	0	2	0
# 3	0	-1	2	0	1	2	4	2	2	0	0	2
# 4	0	0	0	1	1	2	3	2	2	0	0	0
# 5	0	1	-2	2	1	2	2	2	2	0	0	-2
# 6	0	0	1	-1	1	2	3	1	1	1	1	1
# 7	1	-1	0	2	1	1	2	2	2	0	-2	0
# 8	0	1	-1	0	1	2	2	1	1	1	1	-1
# 9	0	1	0	-2	1	2	2	0	0	2	2	0
# 10	-1	0	0	2	0	1	2	2	2	-2	0	0
# 11	1	-1	1	0	1	1	2	1	1	1	-1	1
# 12	1	-1	2	-2	1	1	2	0	0	2	0	2
# 13	-1	0	1	0	0	1	2	1	1	-1	1	1
# 14	1	0	-1	1	1	1	1	1	1	1	-1	-1
# 15	-1	0	2	-2	0	1	2	0	0	0	2	2
# 16	1	0	0	-1	1	1	1	0	0	2	0	0
# 17	-1	1	-1	1	0	1	1	1	1	-1	1	-1
# 18	1	1	-2	0	1	1	0	0	0	2	0	-2
# 19	-1	1	0	-1	0	1	1	0	0	0	2	0
# 20	0	-1	1	1	0	0	1	1	1	-1	-1	1
# 21	-1	2	-2	0	0	1	0	0	0	0	2	-2
# 22	2	-1	0	0	1	0	0	0	0	2	-2	0
# 23	0	-1	2	-1	0	0	1	0	0	0	0	2
# 24	0	0	-1	2	0	0	0	1	1	-1	-1	-1

ALGEBRA G2

#	OMEGA		ALPHA		EPSILON		
# 1	1	0	2	3	-1	-1	2
# 2	-1	3	1	3	1	-2	1
# 3	0	1	1	2	0	-1	1
# 4	1	-1	1	1	-1	0	1
# 5	2	-3	1	0	-2	1	1
# 6	-1	2	0	1	1	-1	0

4

Multiplicity Tables

The multiplicity tables (Table 3) are listed by Lie algebra and within each Lie algebra by congruence class. Since diagram automorphisms (corresponding to outer automorphisms of the Lie algebra or group) permute the congruence classes but otherwise do not affect the multiplicities, we have omitted any table which is simply obtained by permuting the labels of another. For example in the cases D_n, $n = 4, 5, \cdots$ interchanging labels on the fork converts the multiplicity table for one congruence class into another (and of course in D_4 the three nonzero congruence classes are permuted among each other by the various diagram automorphisms). We have *not*, however, removed the few symmetries of label permutations within a given congruence class.

The highest weight labels appear in vertical format across the page. The labeling is built around the Dynkin diagram and thus is free from any indexing conventions. In B_n, C_n, F_4, and G_2 the insertion of an inequality ($<$ or $>$) indicates the orientation of the Dynkin diagram (short roots $\longleftrightarrow$ black nodes are on the small side of the inequality).

The weights are numbered by increasing level and lexicographically within a given level. With the possible exception of the highest level $L_{\max}$ the weight lists are complete by level, that is, any $\lambda \in P^{++}$ in the given congruence class and with $L(\lambda) < L_{\max}$ appears in the table. The ordering by level guarantees the triangular form of the multiplicity tables. Under a given highest weight appear the multiplicities of the dominant weights of the corresponding representation; e.g., C_3 class 1, representation #15, highest weight 3 2 0 (diagram 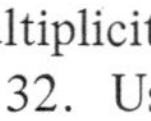): dominant weights #15, #14, #13 occur with multiplicity 1, #12 with multiplicity 0, #11 with multiplicity 2, etc., all the way down to #0 with multiplicity 32. Usually the most convenient way to read the table is to move left from a given multiplicity to the diagonal which then selects the column of the appropriate weight. The quantities above the weight labels λ are

DIM'N: the dimension of the representation with highest weight λ;
LEVEL: the *number* of levels in the weight system $\Omega(\lambda)$;
O.S.: orbit size, the number of weights on the Weyl group orbit of λ;
S.P.: the scalar product (λ,λ). For the purposes of keeping everything integral, we use here the normalization $(\alpha,\alpha) = 2 \det$ (Cartan matrix) for short roots α.

It should be noted that dominant weights which do not appear in the tables do not necessarily determine representations whose dimensions are at least as large as those in the table.

Table 3. Dominant Weight Multiplicity Table

ALGEBRA A2, REPRESENTATIONS OF CLASS 0

0	6	18	18	24	42	42	72	54	72	78	78	114	96	114	162	126	126	162	168	150	168	222	186	186	222	288	234	216	234	288	294	258	258	294	366	312	294	312	366	450	378	342	342	378	450	456	402	384	402	456	546	S.P.
1	6	3	3	6	6	6	3	6	3	6	6	6	6	6	3	6	6	3	6	6	6	6	6	6	6	3	6	6	6	3	6	6	6	6	6	6	6	6	6	3	6	6	6	6	3	6	6	6	6	6	6	O.S.
1	5	7	7	9	11	11	13	13	13	15	15	17	17	17	19	19	19	19	21	21	21	23	23	23	23	25	25	25	25	25	27	27	27	27	29	29	29	29	29	31	31	31	31	31	31	33	33	33	33	33	35	LEVEL
1	8	10	10	27	35	35	28	64	28	81	81	80	125	80	55	154	154	55	162	216	162	143	260	260	143	91	280	343	280	91	270	405	405	270	224	440	512	440	224	136	442	595	595	442	136	405	648	729	648	405	323	DIM'N
0 0	1 1	3 0	0 3	2 2	4 1	1 4	6 0	3 3	0 6	5 2	2 5	7 1	4 4	1 7	9 0	6 3	3 6	0 9	8 2	5 5	2 8	10 1	7 4	4 7	1 10	12 0	9 3	6 6	3 9	0 12	11 2	8 5	5 8	2 11	13 1	10 4	7 7	4 10	1 13	15 0	12 3	9 6	6 9	3 12	0 15	14 2	11 5	8 8	5 11	2 14	16 1	WEIGHT
0	1	2	3	4	5	6	7	8	9	10	11	12	13	14	15	16	17	18	19	20	21	22	23	24	25	26	27	28	29	30	31	32	33	34	35	36	37	38	39	40	41	42	43	44	45	46	47	48	49	50	51	NUMBER
1	2	1	1	3	2	2	1	4	1	3	3	2	5	2	1	4	4	1	3	6	3	2	5	5	2	1	4	7	4	1	3	6	6	3	2	5	8	5	2	1	4	7	7	4	1	3	6	9	6	3	2	0
	1	1	1	2	2	2	1	3	1	3	3	2	4	2	1	4	4	1	3	5	3	2	5	5	2	1	4	6	4	1	3	6	6	3	2	5	7	5	2	1	4	7	7	4	1	3	6	8	6	3	2	1
		1	.	1	2	1	1	2	1	3	2	2	3	2	1	4	3	1	3	4	3	2	5	4	2	1	4	5	4	1	3	6	5	3	2	5	6	5	2	1	4	7	6	4	1	3	6	7	6	3	2	2
			1	1	1	2	1	2	1	2	3	2	3	2	1	3	4	1	3	4	3	2	4	5	2	1	4	5	4	1	3	5	6	3	2	5	6	5	2	1	4	6	7	4	1	3	6	7	6	3	2	3
				1	1	1	1	2	1	2	2	2	3	2	1	3	3	1	3	4	3	2	4	4	2	1	4	5	4	1	3	5	5	3	2	5	6	5	2	1	4	6	6	4	1	3	6	7	6	3	2	4
					1	.	1	1	.	2	1	2	2	1	1	3	2	1	3	3	2	2	4	3	2	1	4	4	3	1	3	5	4	3	2	5	5	4	2	1	4	6	5	4	1	3	6	6	5	3	2	5
						1	.	1	1	1	2	1	2	2	1	2	3	1	2	3	3	2	3	4	2	1	3	4	4	1	3	4	5	3	2	4	5	5	2	1	4	5	6	4	1	3	5	6	6	3	2	6
							1	.	.	1	.	2	1	.	1	2	1	.	3	2	1	2	3	2	1	1	4	3	2	1	3	4	3	2	2	5	4	3	2	1	4	5	4	3	1	3	6	5	4	3	2	7
								1	.	1	1	1	2	1	1	2	2	1	2	3	2	2	3	3	2	1	3	4	3	1	3	4	4	3	2	4	5	4	2	1	4	5	5	4	1	3	5	6	5	3	2	8
									1	.	1	.	1	2	.	1	2	1	1	2	3	1	2	3	2	1	2	3	4	1	2	3	4	3	2	3	4	5	2	1	3	4	5	4	1	3	4	5	6	3	2	9
										1	.	1	1	.	1	2	1	.	2	2	1	2	3	2	1	1	3	3	2	1	3	4	3	2	2	4	4	3	2	1	4	5	4	3	1	3	5	5	4	3	2	10
											1	.	1	1	.	1	2	1	1	2	2	1	2	3	2	1	2	3	3	1	2	3	4	3	2	3	4	4	2	1	3	4	5	4	1	3	4	5	5	3	2	11
												1	.	.	1	1	.	.	2	1	.	2	2	1	.	1	3	2	1	.	3	3	2	1	2	4	3	2	1	1	4	4	3	2	1	3	5	4	3	2	2	12
													1	.	.	1	1	.	1	2	1	1	2	2	1	1	2	3	2	1	2	3	3	2	2	3	4	3	2	1	3	4	4	3	1	3	4	5	4	3	2	13
														1	.	.	1	1	.	1	2	.	1	2	2	.	1	2	3	1	1	2	3	3	1	2	3	4	2	1	2	3	4	4	1	2	3	4	5	3	2	14
															1	.	.	.	1	.	.	2	1	.	.	1	2	1	.	.	3	2	1	.	2	3	2	1	.	1	4	3	2	1	.	3	4	3	2	1	2	15
																1	.	.	1	1	.	1	2	1	.	1	2	2	1	.	2	3	2	1	2	3	3	2	1	1	3	4	3	2	1	3	4	4	3	2	2	16
																	1	.	.	1	1	.	1	2	1	.	1	2	2	1	1	2	3	2	1	2	3	3	2	1	2	3	4	3	1	2	3	4	4	3	2	17
																		1	.	.	1	.	.	1	2	.	.	1	2	1	.	1	2	3	.	1	2	3	2	.	1	2	3	4	1	1	2	3	4	3	1	18

1	.	.	1	1	.	.	1	2	1	.	.	2	2	1	.	2	3	2	1	.	1	3	3	2	1	.	3	4	3	2	1	2	19
	1	.	.	1	1	.	.	1	2	1	.	1	2	2	1	1	2	3	2	1	1	2	3	3	2	1	2	3	4	3	2	2	20
		1	.	.	1	1	.	.	1	2	1	.	1	2	2	.	1	2	3	2	.	1	2	3	3	1	1	2	3	4	3	1	21
			1	.	.	.	1	1	.	.	.	2	1	.	.	2	2	1	.	.	1	3	2	1	.	.	3	3	2	1	.	2	22
				1	.	.	.	1	1	.	.	1	2	1	.	1	2	2	1	.	1	2	3	2	1	.	2	3	3	2	1	2	23
					1	.	.	.	1	1	.	.	1	2	1	.	1	2	2	1	.	1	2	3	2	1	1	2	3	3	2	1	24
						1	.	.	.	1	1	.	.	1	2	.	.	1	2	2	.	.	1	2	3	1	.	1	2	3	3	.	25
							1	.	.	.	.	1	.	.	.	2	1	.	.	.	1	2	1	.	.	.	3	2	1	.	.	2	26
								1	.	.	.	1	1	.	.	1	2	1	.	.	1	2	2	1	.	.	2	3	2	1	.	2	27
									1	.	.	.	1	1	.	.	1	2	1	.	.	1	2	2	1	.	1	2	3	2	1	1	28
										1	.	.	.	1	1	.	.	1	2	1	.	.	1	2	2	1	.	1	2	3	2	.	29
											1	.	.	.	1	.	.	.	1	2	.	.	.	1	2	1	.	.	1	2	3	.	30
												1	.	.	.	1	1	.	.	.	1	2	1	.	.	.	2	2	1	.	.	2	31
													1	.	.	.	1	1	.	.	.	1	2	1	.	.	1	2	2	1	.	1	32
														1	.	.	.	1	1	.	.	.	1	2	1	.	.	1	2	2	1	.	33
															1	.	.	.	1	1	.	.	.	1	2	1	.	.	1	2	2	.	34
																1	.	.	.	.	1	1	.	.	.	.	2	1	.	.	.	2	35
																	1	.	.	.	.	1	1	.	.	.	1	2	1	.	.	1	36
																		1	.	.	.	.	1	1	.	.	.	1	2	1	.	.	37
																			1	.	.	.	.	1	1	.	.	.	1	2	1	.	38
																				1	.	.	.	.	1	1	.	.	.	1	2	.	39
																					1	.	.	.	.	.	1	.	.	.	.	2	40
																						1	.	.	.	.	1	1	.	.	.	1	41
																							1	.	.	.	.	1	1	.	.	.	42
																								1	.	.	.	.	1	1	.	.	43
																									1	.	.	.	.	1	1	.	44
																										1	.	.	.	.	1	.	45
																											1	.	.	.	.	1	46
																												1	.	.	.	.	47
																													1	.	.	.	48
																														1	.	.	49
																															1	.	50
																																1	51

Table 3. *(continued)*

ALGEBRA A2, REPRESENTATIONS OF CLASS 1

2	8	14	32	26	38	50	62	56	98	74	86	104	98	128	146	122	134	200	158	152	182	206	182	194	242	266	224	218	248	338	278	254	266	314	344	302	296	326	392	422	362	338	350	398	512	434	392	386	416	482	518	S.P.
3	3	6	3	6	6	3	6	6	3	6	6	6	6	3	6	6	6	3	6	6	6	6	6	6	3	6	6	6	6	3	6	6	6	6	6	6	6	6	3	6	6	6	6	6	3	6	6	6	6	6	6	O.S.
3	5	7	9	9	11	11	13	13	15	15	15	17	17	17	19	19	19	21	21	21	21	23	23	23	23	25	25	25	25	27	27	27	27	27	29	29	29	29	29	31	31	31	31	31	33	33	33	33	33	33	35	LEVEL
3	6	15	15	24	42	21	48	60	36	90	63	105	120	45	99	165	132	66	192	210	120	195	273	234	78	168	315	336	231	105	330	420	375	195	312	480	504	384	120	255	510	612	561	357	153	504	693	720	585	288	456	DIM'N
1	0	2	4	1	3	0	5	2	7	4	1	6	3	0	8	5	2	10	7	4	1	9	6	3	0	11	8	5	2	13	10	7	4	1	12	9	6	3	0	14	11	8	5	2	16	13	10	7	4	1	15	
0	2	1	0	3	2	5	1	4	0	3	6	2	5	8	1	4	7	0	3	6	9	2	5	8	11	1	4	7	10	0	3	6	9	12	2	5	8	11	14	1	4	7	10	13	0	3	6	9	12	15	2	WEIGHT
0	1	2	3	4	5	6	7	8	9	10	11	12	13	14	15	16	17	18	19	20	21	22	23	24	25	26	27	28	29	30	31	32	33	34	35	36	37	38	39	40	41	42	43	44	45	46	47	48	49	50	51	NUMBER
1	1	2	1	2	3	1	2	3	1	4	2	3	4	1	2	5	3	1	4	5	2	3	6	4	1	2	5	6	3	1	4	7	5	2	3	6	7	4	1	2	5	8	6	3	1	4	7	8	5	2	3	0
	1	1	1	2	2	1	2	3	1	3	2	3	4	1	2	4	3	1	4	5	2	3	5	4	1	2	5	6	3	1	4	6	5	2	3	6	7	4	1	2	5	7	6	3	1	4	7	8	5	2	3	1
		1	1	1	2	1	2	2	1	3	2	3	3	1	2	4	3	1	4	4	2	3	5	4	1	2	5	5	3	1	4	6	5	2	3	6	6	4	1	2	5	7	6	3	1	4	7	7	5	2	3	2
			1	.	1	.	2	1	1	2	1	3	2	1	2	3	2	1	4	3	2	3	4	3	1	2	5	4	3	1	4	5	4	2	3	6	5	4	1	2	5	6	5	3	1	4	7	6	5	2	3	3
				1	1	1	1	2	1	2	2	2	3	1	2	3	3	1	3	4	2	3	4	4	1	2	4	5	3	1	4	5	5	2	3	5	6	4	1	2	5	6	6	3	1	4	6	7	5	2	3	4
					1	.	1	1	1	2	1	2	2	1	2	3	2	1	3	3	2	3	4	3	1	2	4	4	3	1	4	5	4	2	3	5	5	4	1	2	5	6	5	3	1	4	6	6	5	2	3	5
						1	.	1	.	1	2	1	2	1	1	2	3	1	2	3	2	2	3	4	1	2	3	4	3	1	3	4	5	2	3	4	5	4	1	2	4	5	6	3	1	4	5	6	5	2	3	6
							1	.	1	1	.	2	1	.	2	2	1	1	3	2	1	3	3	2	1	2	4	3	2	1	4	4	3	2	3	5	4	3	1	2	5	5	4	3	1	4	6	5	4	2	3	7
								1	.	1	1	1	2	1	1	2	2	1	2	3	2	2	3	3	1	2	3	4	3	1	3	4	4	2	3	4	5	4	1	2	4	5	5	3	1	4	5	6	5	2	3	8
									1	.	.	1	.	.	2	1	.	1	2	1	.	3	2	1	.	2	3	2	1	1	4	3	2	1	3	4	3	2	1	2	5	4	3	2	1	4	5	4	3	2	3	9
										1	.	1	1	.	1	2	1	1	2	2	1	2	3	2	1	2	3	3	2	1	3	4	3	2	3	4	4	3	1	2	4	5	4	3	1	4	5	5	4	2	3	10
											1	.	1	1	.	1	2	.	1	2	2	1	2	3	1	1	2	3	3	1	2	3	4	2	2	3	4	4	1	2	3	4	5	3	1	3	4	5	5	2	3	11
												1	.	.	1	1	.	1	2	1	.	2	2	1	.	2	3	2	1	1	3	3	2	1	3	4	3	2	1	2	4	4	3	2	1	4	5	4	3	2	3	12
													1	.	.	1	1	.	1	2	1	1	2	2	1	1	2	3	2	1	2	3	3	2	2	3	4	3	1	2	3	4	4	3	1	3	4	5	4	2	3	13
														1	.	.	1	.	.	1	2	.	1	2	1	.	1	2	3	.	1	2	3	2	1	2	3	4	1	1	2	3	4	3	1	2	3	4	5	2	2	14
															1	.	.	1	1	.	.	2	1	.	.	2	2	1	.	1	3	2	1	.	3	3	2	1	.	2	4	3	2	1	1	4	4	3	2	1	3	15
																1	.	.	1	1	.	1	2	1	.	1	2	2	1	1	2	3	2	1	2	3	3	2	1	2	3	4	3	2	1	3	4	4	3	2	3	16
																	1	.	.	1	1	.	1	2	1	.	1	2	2	.	1	2	3	2	1	2	3	3	1	1	2	3	4	3	1	2	3	4	4	2	2	17
																		1	.	.	.	1	.	.	.	2	1	.	.	1	2	1	.	.	3	2	1	.	.	2	3	2	1	.	1	4	3	2	1	.	3	18

1	.	.	1	1	.	.	1	2	1	.	1	2	2	1	.	2	3	2	1	.	2	3	3	2	1	1	3	4	3	2	1	3	19
	1	.	.	1	1	.	.	1	2	1	.	1	2	2	1	1	2	3	2	1	1	2	3	3	2	1	2	3	4	3	2	2	20
		1	.	.	1	1	.	.	1	2	.	.	1	2	2	.	1	2	3	1	.	1	2	3	3	.	1	2	3	4	2	1	21
			1	.	.	.	1	1	.	.	1	2	1	.	.	2	2	1	.	.	2	3	2	1	.	1	3	3	2	1	.	3	22
				1	.	.	.	1	1	.	.	1	2	1	.	1	2	2	1	.	1	2	3	2	1	1	2	3	3	2	1	2	23
					1	.	.	.	1	1	.	.	1	2	1	.	1	2	2	1	.	1	2	3	2	.	1	2	3	3	2	1	24
						1	.	.	.	1	.	.	.	1	2	.	.	1	2	1	.	.	1	2	3	.	.	1	2	3	2	.	25
							1	.	.	.	1	1	.	.	.	2	1	.	.	.	2	2	1	.	.	1	3	2	1	.	.	3	26
								1	.	.	.	1	1	.	.	1	2	1	.	.	1	2	2	1	.	1	2	3	2	1	.	2	27
									1	.	.	.	1	1	.	.	1	2	1	.	.	1	2	2	1	.	1	2	3	2	1	1	28
										1	.	.	.	1	1	.	.	1	2	1	.	.	1	2	2	.	.	1	2	3	2	.	29
											1	.	.	.	.	1	.	.	.	.	2	1	.	.	.	1	2	1	.	.	.	3	30
												1	.	.	.	1	1	.	.	.	1	2	1	.	.	1	2	2	1	.	.	2	31
													1	.	.	.	1	1	.	.	.	1	2	1	.	.	1	2	2	1	.	1	32
														1	.	.	.	1	1	.	.	.	1	2	1	.	.	1	2	2	1	.	33
															1	.	.	.	1	1	.	.	.	1	2	.	.	.	1	2	2	.	34
																1	.	.	.	.	1	1	.	.	.	1	2	1	.	.	.	2	35
																	1	.	.	.	.	1	1	.	.	.	1	2	1	.	.	1	36
																		1	.	.	.	.	1	1	.	.	.	1	2	1	.	.	37
																			1	.	.	.	.	1	1	.	.	.	1	2	1	.	38
																				1	.	.	.	.	1	.	.	.	.	1	2	.	39
																					1	.	.	.	.	1	1	.	.	.	.	2	40
																						1	.	.	.	.	1	1	.	.	.	1	41
																							1	.	.	.	.	1	1	.	.	.	42
																								1	.	.	.	.	1	1	.	.	43
																									1	.	.	.	.	1	1	.	44
																										1	.	.	.	.	.	1	45
																											1	.	.	.	.	1	46
																												1	.	.	.	.	47
																													1	.	.	.	48
																														1	.	.	49
																															1	.	50
																																1	51

Table 3. *(continued)*

ALGEBRA A3, REPRESENTATIONS OF CLASS 0

0	8	16	24	24	48	32	48	40	64	56	56	72	88	72	72	88	96	80	96	136	104	104	104	136	192	S.P.
1	12	6	12	12	4	12	4	24	6	24	24	12	12	12	12	12	12	24	12	12	24	24	24	12	4	O.S.
1	7	9	11	11	13	13	13	15	17	17	17	19	19	19	19	19	21	21	21	23	23	23	23	23	25	LEVEL
1	15	20	45	45	35	84	35	175	105	256	256	280	189	280	300	189	360	729	360	315	735	875	875	315	165	DIM'N
0 0 0	1 0 1	0 2 0	2 1 0	0 1 2	4 0 0	2 0 2	0 0 4	1 2 1	0 4 0	3 1 1	1 1 3	2 3 0	5 0 1	0 3 2	3 0 3	1 0 5	4 2 0	2 2 2	0 2 4	6 1 0	1 4 1	4 1 2	2 1 4	0 1 6	8 0 0	WEIGHT
0	1	2	3	4	5	6	7	8	9	10	11	12	13	14	15	16	17	18	19	20	21	22	23	24	25	NUMBER
1	3	2	3	3	1	6	1	7	3	8	8	6	3	6	10	3	6	15	6	3	11	15	15	3	1	0
	1	1	2	2	1	3	1	5	2	6	6	5	3	5	6	3	5	12	5	3	9	12	12	3	1	1
		1	1	1	1	1	1	4	2	4	4	4	3	4	3	3	4	10	4	3	8	9	9	3	1	2
			1	.	1	1	.	2	1	4	2	3	3	3	3	1	4	7	3	3	6	9	6	2	1	3
				1	.	1	1	2	1	2	4	3	1	3	3	3	3	7	4	2	6	6	9	3	1	4
					1	.	.	.	.	2	.	1	3	.	1	.	3	3	.	3	2	6	2	.	1	5
						1	.	1	1	2	2	2	1	2	3	1	3	5	3	2	5	6	6	2	1	6
							1	.	.	.	2	.	.	1	1	3	.	3	3	.	2	2	6	3	.	7
								1	1	1	1	2	1	2	1	1	2	4	2	2	5	4	4	2	1	8
									1	.	.	1	.	1	.	.	1	1	1	1	4	1	1	1	1	9
										1	.	1	1	.	1	.	2	2	.	2	2	4	2	.	1	10
											1	.	.	1	1	1	.	2	2	.	2	2	4	2	.	11
												1	.	.	.	.	1	1	.	1	2	1	1	.	1	12
													1	.	.	.	1	.	.	2	.	2	.	.	1	13
														1	.	.	.	1	1	.	2	1	1	1	.	14
															1	.	.	1	.	.	1	2	2	.	.	15
																1	.	.	1	.	.	.	2	2	.	16
																	1	.	.	1	.	1	.	.	1	17
																		1	.	.	1	1	1	.	.	18

144	120	144	120	128	144	192	152	152	152	136	152	176	200	160	168	176	168	200	216	264	200	184	216	184	200	S.P.
6	24	12	24	12	12	4	12	24	12	24	24	12	24	24	24	12	24	24	12	12	24	24	12	24	12	O.S.
25	25	25	25	25	25	25	27	27	27	27	27	29	29	29	29	29	29	29	31	31	31	31	31	31	31	LEVEL
336	1280	616	1280	825	616	165	945	1485	945	2156	1485	1375	1280	2640	2304	1375	2304	1280	1540	715	2079	3780	1560	3780	1911	DIM'N
0 6 0	3 3 1	6 0 2	1 3 3	4 0 4	2 0 6	0 0 8	2 5 0	5 2 1	0 5 2	3 2 3	1 2 5	4 4 0	7 1 1	2 4 2	5 1 3	0 4 4	3 1 5	1 1 7	6 3 0	9 0 1	1 6 1	4 3 2	7 0 3	2 3 4	5 0 5	WEIGHT
26	27	28	29	30	31	32	33	34	35	36	37	38	39	40	41	42	43	44	45	46	47	48	49	50	51	NUMBER
4	16	6	16	15	6	1	9	15	9	26	15	11	8	24	24	11	24	8	10	3	15	30	10	30	21	0
3	14	6	14	10	6	1	8	13	8	22	13	10	8	21	20	10	20	8	9	3	13	27	10	27	15	1
3	12	6	12	6	6	1	7	11	7	19	11	9	8	19	16	9	16	8	8	3	12	24	10	24	10	2
2	10	6	10	6	3	1	6	11	6	15	9	8	8	16	16	8	12	6	8	3	10	21	10	21	10	3
2	10	3	10	6	6	1	6	9	6	15	11	8	6	16	12	8	16	8	7	3	10	21	6	21	10	4
1	6	6	4	3	1	.	3	9	3	9	3	6	8	9	12	5	6	2	7	3	6	15	10	12	6	5
2	8	3	8	6	3	1	5	9	5	12	9	7	6	14	12	7	12	6	7	3	9	18	6	18	10	6
1	4	1	6	3	6	1	3	3	3	9	9	5	2	9	6	6	12	8	4	1	6	12	3	15	6	7
2	7	3	7	3	3	1	5	7	5	10	7	6	6	13	9	6	9	6	6	3	9	16	6	16	6	8
2	4	1	4	1	1	1	4	4	4	4	4	4	4	10	4	4	4	4	4	3	8	10	3	10	3	9
1	5	3	3	3	1	.	3	7	3	7	3	5	6	8	9	4	6	2	6	3	6	13	6	10	6	10
1	3	1	5	3	3	1	3	3	3	7	7	4	2	8	6	5	9	6	4	1	6	10	3	13	6	11
1	4	1	2	1	1	.	3	4	3	4	2	4	4	7	4	3	4	2	4	3	6	10	3	7	3	12
.	2	3	.	1	.	.	1	5	.	3	.	3	6	3	6	.	2	.	5	3	2	8	6	4	3	13
1	2	1	4	1	1	1	3	2	3	4	4	3	2	7	4	4	4	4	3	1	6	7	3	10	3	14
1	2	1	2	3	1	.	2	3	2	5	3	3	2	5	6	3	6	2	4	1	5	8	3	8	6	15
.	.	.	2	1	3	1	.	.	1	3	5	.	.	3	2	3	6	6	.	.	2	4	1	8	3	16
.	2	1	.	1	.	.	1	4	.	2	.	3	4	3	4	.	2	.	4	3	2	7	3	3	3	17
1	2	1	2	1	1	.	2	2	2	4	2	3	2	5	4	3	4	2	3	1	5	7	3	7	3	18

1	.	.	.	1	1	.	.	.	.	2	1	1	1	.	.	1	2	4	.	.	3	2	3	4	4	.	.	2	3	1	7	3	19
	1	.	.	.	.	1	.	.	1	.	.	.	.	.	2	.	.	.	1	4	.	2	.	.	.	3	3	.	3	3	.	1	20
		1	.	.	.	.	1	1	.	1	.	.	.	2	1	2	1	1	2	1	4	1	2	1	1	2	1	5	4	1	4	1	21
			1	.	.	.	.	1	1	.	1	.	.	1	2	.	2	.	2	2	2	4	.	2	.	3	1	2	5	3	3	3	22
				1	.	.	.	.	.	1	1	1	.	.	.	1	2	2	.	.	2	2	2	4	2	.	.	2	3	1	5	3	23
					1	.	.	.	.	.	.	1	1	.	.	.	.	2	.	.	.	.	1	2	4	.	.	.	.	.	3	1	24
						1	.	.	.	.	.	.	.	.	.	.	.	.	.	2	.	.	.	.	.	1	3	.	.	1	.	.	25
							1	.	.	.	.	.	.	1	.	1	.	.	1	.	1	.	1	.	.	1	.	4	1	.	1	.	26
								1	.	.	.	.	.	1	1	.	1	.	2	1	2	1	.	1	.	2	1	2	4	1	2	1	27
									1	.	.	.	.	.	1	.	.	.	1	2	.	2	.	.	.	2	1	.	2	3	.	1	28
										1	.	.	.	.	.	1	1	1	.	.	2	1	2	1	1	.	.	2	2	1	4	1	29
											1	.	.	.	.	.	1	.	.	.	1	2	.	2	.	.	.	1	2	1	2	3	30
												1	.	.	.	.	.	1	.	.	.	.	1	2	2	.	.	.	.	.	2	1	31
													1	.	.	.	.	.	.	.	.	.	.	.	2	.	.	.	.	.	.	.	32
														1	.	.	.	.	1	.	1	.	.	.	.	1	.	2	1	.	1	.	33
															1	.	.	.	1	1	.	1	.	.	.	2	1	.	2	1	.	1	34
																1	.	.	.	.	1	.	1	.	.	.	.	2	1	.	1	.	35
																	1	.	.	.	1	1	.	1	.	.	.	1	2	1	2	1	36
																		1	.	.	.	.	1	1	1	.	.	.	.	.	2	1	37
																			1	.	.	.	.	.	.	1	.	.	1	.	.	.	38
																				1	.	.	.	.	.	1	1	.	.	1	.	.	39
																					1	.	.	.	.	.	.	1	1	.	1	.	40
																						1	.	.	.	.	.	.	1	1	.	1	41
																							1	.	.	.	.	.	.	.	1	.	42
																								1	.	.	.	.	.	.	1	1	43
																									1	.	.	.	.	.	.	.	44
																										1	.	.	.	.	.	.	45
																											1	.	.	.	.	.	46
																												1	.	.	.	.	47
																													1	.	.	.	48
																														1	.	.	49
																															1	.	50
																																1	51

Table 3. *(continued)*

ALGEBRA A3, REPRESENTATIONS OF CLASS 1

0	1	2	3	4	5	6	7	8	9	10	11	12	13	14	15	16	17	18	19	20	21	22	23	24	25	26	27	28	29	30	31	32	33	34	35	36	37	38	39	40	41	42	43	44	45	46	47	48	49	50	51	NUMBER
3	11	19	27	27	43	35	75	51	51	59	59	67	83	83	75	99	99	123	91	99	107	147	131	123	107	115	179	131	139	139	131	155	243	171	155	187	147	163	171	171	211	187	195	179	171	179	219	219	251	227	195	S.P.
4	12	12	4	12	12	24	4	12	12	12	24	12	12	24	24	12	12	12	24	12	12	4	12	12	24	24	12	24	24	12	24	24	4	12	24	12	24	12	12	12	12	12	24	24	24	24	12	12	24	12	24	O.S.
4	8	10	10	12	14	14	16	16	16	16	18	18	20	20	20	20	22	22	22	22	22	22	24	24	24	24	26	26	26	26	26	26	28	28	28	28	28	28	28	28	28	30	30	30	30	30	30	32	32	32	32	LEVEL
4	20	36	20	60	84	140	56	140	160	120	360	224	280	420	540	216	480	280	756	500	420	120	540	504	1 260	1 000	440	1 400	1 320	900	1 540	924	220	840	1 980	864	2 464	1 260	1 100	1 120	540	1 540	2 100	2 376	3 360	2 916	1 056	1 980	1 716	1 320	4 320	DIM'N
1 0 0	0 1 1	2 0 1	0 0 3	1 2 0	3 1 0	1 1 2	5 0 0	0 3 1	3 0 2	1 0 4	2 2 1	0 2 3	1 4 0	4 1 1	2 1 3	0 1 5	3 3 0	6 0 1	1 3 2	4 0 3	2 0 5	0 0 7	5 2 0	0 5 1	3 2 2	1 2 4	7 1 0	2 4 1	5 1 2	0 4 3	3 1 4	1 1 6	9 0 0	1 6 0	4 3 1	7 0 2	2 3 3	5 0 4	0 3 5	3 0 6	1 0 8	3 5 0	6 2 1	1 5 2	4 2 3	2 2 5	0 2 7	5 4 0	8 1 1	0 7 1	3 4 2	WEIGHT
1	2	3	1	3	3	6	1	4	6	3	9	5	5	8	12	3	7	3	12	10	6	1	6	6	18	13	3	15	15	9	20	8	1	7	18	6	24	15	9	10	3	11	15	18	30	24	6	12	8	8	30	0
	1	1	1	2	2	4	1	3	3	3	7	4	4	6	9	3	6	3	10	6	6	1	5	5	15	11	3	13	12	8	16	8	1	6	16	6	21	10	8	10	3	10	13	16	26	21	6	11	8	7	27	1
		1	.	1	2	2	1	2	3	1	5	3	3	6	6	2	5	3	8	6	3	1	5	4	12	9	3	11	12	7	12	6	1	5	14	6	18	10	7	6	3	9	13	14	22	18	5	10	8	6	24	2
			1	.	.	2	.	1	1	3	3	3	2	2	6	3	4	1	6	3	6	1	3	3	9	9	2	9	6	6	12	8	1	4	12	3	15	6	7	10	3	8	9	12	18	18	6	9	6	5	21	3
				1	1	1	1	2	1	1	4	2	3	4	4	2	4	3	7	3	3	1	4	4	10	7	3	10	9	6	9	6	1	5	12	6	16	6	6	6	3	8	11	13	19	15	5	9	8	6	22	4
					1	.	1	.	1	.	2	.	1	4	2	.	3	3	3	3	1	.	4	2	7	3	3	6	9	4	6	2	1	3	10	6	10	6	4	3	1	6	11	9	15	9	3	8	8	4	16	5
						1	.	1	1	1	2	2	2	2	4	2	3	1	5	3	3	1	3	3	7	7	2	8	6	5	9	6	1	4	10	3	13	6	6	6	3	7	9	11	15	15	5	8	6	5	19	6
							1	.	.	.	.	.	.	2	.	.	1	3	.	1	.	.	3	.	3	.	3	2	6	.	2	.	1	1	6	6	4	3	.	1	.	3	9	3	9	3	.	6	8	2	9	7
								1	.	.	1	1	2	1	1	1	2	1	4	1	1	1	2	3	4	4	2	7	4	4	4	4	1	4	7	3	10	3	4	3	3	6	7	10	10	10	4	6	6	5	16	8
									1	.	1	.	1	2	2	.	2	1	2	3	1	.	3	2	5	3	2	5	6	3	6	2	1	3	8	3	8	6	4	3	1	5	9	8	12	9	3	7	6	4	14	9
										1	.	1	.	.	2	2	.	.	2	1	3	1	.	1	3	5	.	3	2	3	6	6	.	2	4	1	8	3	5	6	3	4	3	6	9	12	5	5	2	3	11	10
											1	.	1	1	1	.	2	1	2	1	1	.	2	2	4	2	2	5	4	3	4	2	1	3	7	3	7	3	3	3	1	5	7	8	10	7	3	6	6	4	13	11
												1	.	.	1	1	.	.	2	1	1	1	.	1	2	4	.	3	2	3	4	4	.	2	3	1	7	3	4	3	3	4	3	6	7	10	4	4	2	3	10	12
													1	.	.	.	1	.	1	.	.	.	1	2	1	1	1	4	1	2	1	1	1	3	4	1	4	1	2	1	1	4	4	7	4	4	2	4	4	4	10	13
														1	.	.	1	1	.	1	.	.	2	.	2	.	2	2	4	.	2	.	1	1	5	3	3	3	.	1	.	3	7	3	7	3	.	5	6	2	8	14
															1	.	.	.	1	1	1	.	.	1	2	2	.	2	2	2	4	2	.	2	3	1	5	3	3	3	1	3	3	5	7	7	3	4	2	3	8	15
																1	.	.	.	.	1	1	.	.	.	2	.	.	.	1	2	4	.	.	.	.	3	1	3	3	3	.	.	2	3	7	4	.	.	1	4	16
																	1	.	.	.	.	.	1	.	1	.	1	2	1	.	1	.	1	1	4	1	2	1	.	1	.	3	4	3	4	2	.	4	4	2	7	17

	18	19	20	21	22	23	24	25	26	27	28	29	30	31	32	33	34	35	36	37	38	39	40	41	42	43	44	45	46	47	48	49	50	51
18	1	.	.	.	.	1	.	.	.	2	.	2	.	.	.	1	.	2	3	.	1	.	.	.	1	5	.	3	.	.	3	6	.	3
19		1	.	.	.	.	1	1	1	.	2	1	2	1	1	.	2	2	1	4	1	2	1	1	3	2	5	4	4	2	3	2	3	7
20			1	.	.	.	.	1	.	.	1	2	.	2	.	.	1	2	1	2	3	.	1	.	2	3	2	5	3	.	3	2	2	5
21				1	.	.	.	.	1	.	.	.	1	2	2	.	.	.	.	2	1	2	3	1	.	.	2	3	5	3	.	.	1	3
22					1	.	.	.	.	.	.	.	.	.	2	.	.	.	.	.	.	1	1	3	.	.	.	.	3	3	.	.	.	.
23						1	.	.	.	1	.	1	.	.	.	1	.	2	1	.	1	.	.	.	1	4	.	2	.	.	3	4	.	3
24							1	.	.	.	1	.	1	.	.	.	2	1	.	1	.	1	.	.	2	1	4	1	1	1	2	1	3	4
25								1	.	.	1	1	.	1	.	.	1	2	1	2	1	.	1	.	2	2	2	4	2	.	3	2	2	5
26									1	.	.	.	1	1	1	.	.	.	.	2	1	2	1	1	.	.	2	2	4	2	.	.	1	3
27										1	.	.	.	.	.	1	.	.	1	.	.	.	.	.	.	2	.	.	.	.	1	4	.	.
28											1	.	.	.	.	.	1	1	.	1	.	.	.	.	2	1	2	1	1	.	2	1	2	4
29												1	.	.	.	.	.	1	1	.	1	.	.	.	1	2	.	2	.	.	2	2	.	2
30													1	.	.	.	.	.	.	1	.	1	.	.	.	.	2	1	1	1	.	.	1	2
31														1	.	.	.	.	.	1	1	.	1	.	.	.	1	2	2	.	.	.	1	2
32															1	.	.	.	.	.	.	1	1	1	.	.	.	.	2	2	.	.	.	.
33																1	.	.	.	.	.	.	.	.	.	.	.	.	.	.	.	2	.	.
34																	1	.	.	.	.	.	.	.	1	.	1	.	.	.	1	.	2	1
35																		1	.	.	.	.	.	.	1	1	.	1	.	.	2	1	.	2
36																			1	.	.	.	.	.	.	1	.	.	.	.	1	2	.	.
37																				1	.	.	.	.	.	.	1	1	1	.	.	.	1	2
38																					1	.	.	.	.	.	.	1	.	.	.	.	.	1
39																						1	.	.	.	.	.	.	1	1	.	.	.	.
40																							1	.	.	.	.	.	1	.	.	.	.	.
41																								1	.	.	.	.	.	1	.	.	.	.
42																									1	.	.	.	.	.	1	.	.	1
43																										1	.	.	.	.	1	1	.	.
44																											1	.	.	.	.	.	1	1
45																												1	.	.	.	.	.	1
46																													1	.	.	.	.	.
47																														1	.	.	.	.
48																															1	.	.	.
49																																1	.	.
50																																	1	.
51																																		1

Table 3. *(continued)*

ALGEBRA A3, REPRESENTATIONS OF CLASS 2

4	12	12	20	36	36	36	44	44	68	52	68	108	68	76	76	108	100	84	84	108	116	108	100	116	132	164	116	132	132	132	164	172	148	140	140	172	228	196	164	180	164	164	180	228	300	204	196	236	204	180	204	S.P.
6	4	4	24	6	12	12	12	12	12	24	12	4	24	12	12	4	6	24	24	12	24	12	24	24	12	12	24	12	12	12	12	12	24	24	24	12	12	6	24	24	24	24	24	12	4	12	24	12	12	24	12	O.S.
5	7	7	11	13	13	13	15	15	17	17	17	19	19	19	19	19	21	21	21	23	23	23	23	23	25	25	25	25	25	25	27	27	27	27	27	27	29	29	29	29	29	29	29	29	31	31	31	31	31	31	31	LEVEL
6	10	10	64	50	70	70	126	126	140	300	140	84	384	270	270	84	196	630	630	540	640	540	960	640	750	396	1470	770	750	770	396	770	1280	1980	1980	770	594	540	2310	1890	2310	2450	1890	594	286	1540	2880	1170	1540	4096	1820	DIM'N
0 1 0	2 0 0	0 0 2	1 1 1	0 3 0	3 0 1	1 0 3	2 2 0	0 2 2	4 1 0	2 1 2	0 1 4	6 0 0	1 3 1	4 0 2	2 0 4	0 0 6	0 5 0	3 2 1	1 2 3	2 4 0	5 1 1	0 4 2	3 1 3	1 1 5	4 3 0	7 0 1	2 3 2	5 0 3	0 3 4	3 0 5	1 0 7	6 2 0	1 5 1	4 2 2	2 2 4	0 2 6	8 1 0	0 7 0	3 4 1	6 1 2	1 4 3	4 1 4	2 1 6	0 1 8	10 0 0	2 6 0	5 3 1	8 0 2	0 6 2	3 3 3	6 0 4	WEIGHT
0	1	2	3	4	5	6	7	8	9	10	11	12	13	14	15	16	17	18	19	20	21	22	23	24	25	26	27	28	29	30	31	32	33	34	35	36	37	38	39	40	41	42	43	44	45	46	47	48	49	50	51	NUMBER
1	1	1	4	2	3	3	4	4	3	9	3	1	8	6	6	1	3	11	11	7	8	7	16	8	8	3	18	10	8	10	3	6	12	21	21	6	3	4	19	15	19	25	15	3	1	10	20	6	10	32	15	0
	1	.	2	1	3	1	3	3	3	6	2	1	6	6	3	1	2	9	9	6	8	6	12	6	7	3	15	10	7	6	3	6	10	18	18	5	3	3	17	15	17	20	12	3	1	9	18	6	9	28	15	1
		1	2	1	1	3	3	3	2	6	3	1	6	3	6	1	2	9	9	6	6	6	12	8	7	3	15	6	7	10	3	5	10	18	18	6	3	3	17	12	17	20	15	3	1	9	18	6	9	28	10	2
			1	1	1	1	2	2	2	4	2	1	5	3	3	1	2	7	7	5	6	5	9	6	6	3	13	6	6	6	3	5	9	15	15	5	3	3	15	12	15	16	12	3	1	8	16	6	8	25	10	3
				1	.	.	1	1	1	1	1	1	4	1	1	1	2	4	4	4	4	4	4	4	4	3	10	3	4	3	3	4	8	10	10	4	3	3	12	9	12	9	9	3	1	7	12	6	7	20	6	4
					1	.	1	.	2	2	.	1	2	3	1	.	1	5	3	3	6	3	6	2	5	3	8	6	4	3	1	5	6	12	9	3	3	2	11	12	11	12	6	2	1	6	14	6	6	18	10	5
						1	.	1	.	2	2	.	2	1	3	1	1	3	5	3	2	3	6	6	4	1	8	3	5	6	3	3	6	9	12	5	2	2	11	6	11	12	12	3	1	6	12	3	6	18	6	6
							1	.	1	1	.	1	2	1	1	.	1	4	2	3	4	3	4	2	4	3	7	3	3	3	1	4	6	10	7	3	3	2	10	9	10	9	6	2	1	6	12	6	6	16	6	7
								1	.	1	1	.	2	1	1	1	1	2	4	3	2	3	4	4	3	1	7	3	4	3	3	3	6	7	10	4	2	2	10	6	10	9	9	3	1	6	10	3	6	16	6	8
									1	.	.	1	.	1	.	.	.	2	.	1	4	.	2	.	3	3	3	3	.	1	.	4	2	7	3	.	3	1	6	9	4	6	2	.	1	3	10	6	3	10	6	9
										1	.	.	1	1	1	.	1	2	2	2	2	2	4	2	3	1	5	3	3	3	1	3	5	7	7	3	2	2	8	6	8	9	6	2	1	5	10	3	5	13	6	10
											1	.	.	.	1	1	.	.	2	.	.	1	2	4	.	.	3	1	3	3	3	.	2	3	7	4	.	1	4	2	6	6	9	3	.	3	4	1	3	10	3	11
												1	.	.	.	.	.	.	.	.	2	.	.	.	1	3	.	1	.	.	.	3	.	3	.	.	3	.	2	6	.	2	.	.	1	1	6	6	.	4	3	12
													1	.	.	.	1	1	1	2	1	2	1	1	2	1	4	1	2	1	1	2	5	4	4	2	2	2	7	4	7	4	4	2	1	5	7	3	5	10	3	13
														1	.	.	.	1	.	1	2	.	2	.	2	1	2	3	.	1	.	3	2	5	3	.	2	1	5	6	3	6	2	.	1	3	8	3	3	8	6	14
															1	.	.	.	1	.	.	1	2	2	.	.	2	1	2	3	1	.	2	3	5	3	.	1	3	2	5	6	6	2	.	3	4	1	3	8	3	15
																1	.	.	.	.	.	.	.	2	.	.	.	.	1	1	3	.	.	.	3	3	.	.	.	.	2	2	6	3	.	.	.	.	1	4	1	16
																	1	.	.	1	.	1	.	.	1	.	1	.	1	.	.	1	4	1	1	1	1	2	4	1	4	1	1	1	1	4	4	1	4	4	1	17
																		1	.	1	1	.	1	.	2	1	2	1	.	1	.	2	2	4	2	.	2	1	5	4	3	4	2	.	1	3	7	3	3	7	3	18

1	.	.	1	1	1	.	.	2	1	2	1	1	.	2	2	4	2	.	1	3	2	5	4	4	2	.	3	3	1	3	7	3	19
	1	.	.	.	.	1	.	1	.	.	.	.	1	2	1	1	.	1	1	4	1	2	1	1	.	1	3	4	1	3	4	1	20
		1	.	.	.	1	1	.	1	.	.	.	2	.	2	.	.	2	.	2	4	.	2	.	.	1	1	5	3	.	3	3	21
			1	.	.	.	.	1	.	1	.	.	.	2	1	1	1	.	1	2	1	4	1	1	1	.	3	2	1	3	4	1	22
				1	.	.	.	1	1	.	1	.	.	1	2	2	.	.	1	2	2	2	4	2	.	.	2	3	1	2	5	3	23
					1	.	.	.	.	1	1	1	.	.	.	2	2	.	.	.	.	2	2	4	2	.	.	.	.	1	3	1	24
						1	.	.	.	.	.	.	1	.	1	.	.	1	.	2	1	.	1	.	.	1	1	4	1	.	2	1	25
							1	.	.	.	.	.	1	.	.	.	.	2	.	.	2	.	.	.	.	1	.	2	3	.	.	1	26
								1	.	.	.	.	.	1	1	1	.	.	1	2	1	2	1	1	.	.	2	2	1	2	4	1	27
									1	.	.	.	.	.	1	.	.	.	.	1	2	.	2	.	.	.	1	2	1	.	2	3	28
										1	.	.	.	.	.	1	1	.	.	.	.	2	1	1	1	.	.	.	.	1	2	1	29
											1	.	.	.	.	1	.	.	.	.	.	1	2	2	.	.	.	.	.	1	2	1	30
												1	.	.	.	.	1	.	.	.	.	.	.	2	2	.	.	.	.	.	.	.	31
													1	.	.	.	.	1	.	.	1	.	.	.	.	1	.	2	1	.	.	1	32
														1	.	.	.	.	1	1	.	1	.	.	.	.	2	1	.	2	1	.	33
															1	.	.	.	.	1	1	.	1	.	.	.	1	2	1	.	2	1	34
																1	.	.	.	.	.	1	1	1	.	.	.	.	.	1	2	1	35
																	1	.	.	.	.	.	.	1	1	.	.	.	.	.	.	.	36
																		1	.	.	.	.	.	.	.	1	.	.	1	.	.	.	37
																			1	.	.	.	.	.	.	.	1	.	.	1	.	.	38
																				1	.	.	.	.	.	.	1	1	.	.	1	.	39
																					1	.	.	.	.	.	.	1	1	.	.	1	40
																						1	.	.	.	.	.	.	.	1	1	.	41
																							1	.	.	.	.	.	.	.	1	1	42
																								1	.	.	.	.	.	.	.	.	43
																									1	.	.	.	.	.	.	.	44
																										1	.	.	.	.	.	.	45
																											1	.	.	.	.	.	46
																												1	.	.	.	.	47
																													1	.	.	.	48
																														1	.	.	49
																															1	.	50
																																1	51

Table 3. *(continued)*

ALGEBRA A4, REPRESENTATIONS OF CLASS 0

0	10	20	30	30	40	40	40	60	60	100	50	100	70	70	70	70	80	80	80	90	90	110	90	110	110	S.P.
1	20	30	30	30	20	20	20	20	20	5	120	5	30	30	60	60	30	60	60	20	60	60	60	60	60	O.S.
1	9	13	15	15	17	17	17	19	19	21	21	21	23	23	23	23	25	25	25	25	27	27	27	27	29	LEVEL
1	24	75	126	126	175	175	200	224	224	126	1 024	126	700	700	1 050	1 050	1 176	1 701	1 701	1 000	2 700	1 750	2 700	1 750	3 024	DIM'N
0	1	0	2	0	1	0	2	3	0	5	1	0	1	0	3	1	0	2	1	3	2	4	0	1	1	
0	0	1	0	1	2	0	0	1	0	0	1	0	0	3	0	1	2	2	0	0	1	1	2	0	3	
0	0	1	1	0	0	2	0	0	1	0	1	0	3	0	1	0	2	0	2	0	2	0	1	1	1	
0	1	0	0	2	0	1	2	0	3	0	1	5	0	1	1	3	0	1	2	3	0	1	2	4	0	WEIGHT
0	1	2	3	4	5	6	7	8	9	10	11	12	13	14	15	16	17	18	19	20	21	22	23	24	25	NUMBER
1	4	5	6	6	5	5	10	4	4	1	24	1	10	10	20	20	16	21	21	20	30	15	30	15	24	0
	1	2	3	3	3	3	4	3	3	1	14	1	7	7	12	12	10	15	15	10	21	12	21	12	18	1
		1	1	1	2	2	1	2	2	1	8	1	5	5	6	6	7	11	11	4	14	9	14	9	14	2
			1	.	1	.	1	2	.	1	4	.	2	3	6	3	4	8	4	4	10	9	9	3	10	3
				1	.	1	1	.	2	.	4	1	3	2	3	6	4	4	8	4	9	3	10	9	8	4
					1	.	.	1	.	1	2	.	1	3	2	1	3	6	3	1	6	6	6	2	9	5
						1	.	.	1	.	2	1	3	1	1	2	3	3	6	1	6	2	6	6	6	6
							1	.	.	.	2	.	1	1	3	3	3	3	3	4	6	3	6	3	6	7
								1	.	1	.	.	.	.	2	.	.	3	.	1	3	6	.	.	3	8
									1	.	.	1	.	.	.	2	.	.	3	1	.	.	3	6	.	9
										1	.	.	.	.	.	.	.	.	.	.	.	3	.	.	.	10
											1	.	1	1	1	1	2	2	2	1	4	2	4	2	5	11
												1	.	.	.	.	.	.	.	.	.	.	.	3	.	12
													1	.	.	.	1	.	1	.	2	.	1	1	2	13
														1	.	.	1	1	.	.	1	1	2	.	4	14
															1	.	.	1	.	1	2	2	.	.	2	15
																1	.	.	1	1	.	.	2	2	.	16
																	1	.	.	.	1	.	1	.	2	17

120	160	110	100	120	160	150	130	150	120	120	130	130	130	160	170	130	140	160	140	160	170	190	230	160	160	S.P.
30	20	60	120	30	20	10	60	10	60	60	60	60	60	20	60	120	60	20	60	20	60	20	30	60	60	O.S.
29	29	29	29	29	29	31	31	31	31	31	31	31	31	33	33	33	33	33	33	33	33	35	35	35	35	LEVEL
2 250	924	3 024	6 125	2 250	924	1 176	4 725	1 176	5 600	5 600	4 950	4 725	4 950	3 150	4 400	12 000	8 624	3 150	8 624	3 675	4 400	4 125	2 376	8 400	8 400	DIM'N
4	6	0	2	0	1	0	3	0	2	1	4	0	2	2	5	1	3	0	2	4	0	4	7	1	0	
0	0	1	1	2	0	5	2	0	0	3	0	1	1	4	1	2	2	0	0	0	1	3	0	1	4	
2	0	3	1	0	0	0	1	5	3	0	1	2	0	0	1	2	0	4	2	0	1	0	1	4	1	
0	1	1	2	4	6	0	0	0	1	2	2	3	4	0	0	1	2	2	3	4	5	0	0	0	1	WEIGHT
26	27	28	29	30	31	32	33	34	35	36	37	38	39	40	41	42	43	44	45	46	47	48	49	50	51	NUMBER
20	4	24	65	20	4	6	30	6	40	40	45	30	45	15	20	80	54	15	54	35	20	15	6	40	40	0
14	4	18	44	14	4	5	24	5	31	31	30	24	30	12	17	59	42	12	42	20	17	13	6	32	32	1
9	4	14	29	9	4	4	19	4	24	24	18	19	18	10	14	44	33	10	33	10	14	11	6	26	26	2
9	4	8	20	6	1	3	16	3	14	18	18	11	12	8	14	32	27	7	18	10	8	10	6	18	20	3
6	1	10	20	9	4	3	11	3	18	14	12	16	18	7	8	32	18	8	27	10	14	8	3	20	18	4
5	4	6	12	3	1	3	12	2	10	15	9	9	6	7	11	24	21	6	14	4	6	8	6	14	18	5
3	1	9	12	5	4	2	9	3	15	10	6	12	9	6	6	24	14	7	21	4	11	7	3	18	14	6
6	1	6	14	6	1	2	9	2	10	10	12	9	12	6	8	24	15	6	15	10	8	7	3	14	14	7
5	4	.	6	.	.	1	9	.	4	6	9	.	3	4	11	9	15	.	5	4	.	7	6	5	8	8
.	.	3	6	5	4	.	.	1	6	4	3	9	9	.	.	9	5	4	15	4	11	.	.	8	5	9
2	4	.	.	.	.	.	3	.	.	.	3	.	.	1	8	.	6	.	.	1	.	4	6	.	.	10
3	1	5	8	3	1	2	7	2	8	8	6	7	6	5	6	18	11	5	11	4	6	6	3	12	12	11
.	.	.	.	2	4	.	.	.	.	.	.	3	3	.	.	.	.	1	6	1	8	.	.	.	.	12
1	.	4	2	1	1	1	3	2	6	2	1	4	2	3	2	10	3	4	6	1	4	4	1	10	6	13
1	1	2	2	1	.	2	4	1	2	6	2	3	1	4	4	10	6	3	3	1	2	4	3	6	10	14
3	1	.	4	.	.	1	5	.	3	3	6	.	3	3	6	7	8	.	4	4	.	5	3	4	6	15
.	.	2	4	3	1	.	.	1	3	3	3	5	6	.	.	7	4	3	8	4	6	.	.	6	4	16
1	.	2	1	1	.	1	2	1	2	2	1	2	1	3	2	8	2	3	2	1	2	3	1	6	6	17

1	.	.	1	1	.	.	2	1	1	.	2	.	.	1	4	.	2	3	2	.	1	3	4	5	6	.	3	1	.	4	3	3	6	18
	1	.	.	.	1	1	.	.	.	2	2	1	1	.	.	1	3	2	1	4	2	.	.	5	3	3	6	1	4	.	.	6	3	19
		1	.	.	.	.	.	.	.	.	2	.	.	.	.	.	1	1	3	.	3	.	.	3	3	.	3	4	.	.	.	2	2	20
			1	.	.	.	1	1	.	.	1	.	.	1	2	.	2	1	1	.	1	2	2	4	2	.	2	1	.	3	1	3	4	21
				1	.	.	.	1	1	.	.	.	.	.	2	.	.	.	2	.	.	1	4	.	3	.	.	1	.	3	3	.	.	22
					1	.	.	.	.	1	1	1	.	.	.	1	1	2	1	2	1	.	.	4	2	2	2	1	2	.	.	4	3	23
						1	.	.	.	.	.	1	1	.	.	.	.	.	.	2	2	.	.	.	.	1	3	1	4	.	.	.	.	24
							1	.	.	.	.	.	.	1	1	.	.	1	.	.	.	2	1	2	1	.	.	.	.	2	1	1	4	25
								1	.	.	.	.	.	.	1	.	.	.	1	.	.	1	2	.	1	.	.	1	.	2	1	.	.	26
									1	.	.	.	.	.	.	.	.	.	.	.	.	.	2	.	.	.	.	.	.	1	3	.	.	27
										1	.	.	.	.	.	1	1	.	.	1	.	.	.	2	.	2	1	.	1	.	.	4	1	28
											1	.	.	.	.	.	1	1	1	.	1	.	.	2	2	.	2	1	.	.	.	2	2	29
												1	.	.	.	.	.	.	.	1	1	.	.	.	.	1	1	1	2	.	.	.	.	30
													1	.	.	.	.	.	.	.	.	.	.	.	.	.	.	.	2	.	.	.	.	31
														1	.	.	.	.	.	.	.	1	.	.	.	.	.	.	.	1	.	.	2	32
															1	.	.	.	.	.	.	1	1	.	1	.	.	.	.	2	1	.	.	33
																1	.	.	.	.	.	.	.	.	.	1	.	.	.	.	.	2	.	34
																	1	.	.	.	.	.	.	1	.	.	1	.	.	.	.	2	1	35
																		1	.	.	.	.	.	1	1	.	.	.	.	.	.	1	2	36
																			1	.	.	.	.	.	1	.	.	1	.	.	.	.	.	37
																				1	.	.	.	.	.	1	1	.	1	.	.	.	.	38
																					1	.	.	.	.	.	1	1	.	.	.	.	.	39
																						1	.	.	.	.	.	.	.	1	.	.	.	40
																							1	.	.	.	.	.	.	1	1	.	.	41
																								1	.	.	.	.	.	.	.	1	1	42
																									1	.	.	.	.	.	.	.	.	43
																										1	.	.	.	.	.	.	.	44
																											1	.	.	.	.	.	.	45
																												1	.	.	.	.	.	46
																													1	.	.	.	.	47
																														1	.	.	.	48
																															1	.	.	49
																																1	.	50
																																	1	51

Table 3. *(continued)*

ALGEBRA A4, REPRESENTATIONS OF CLASS 1

4	14	24	24	34	34	64	54	54	44	64	54	64	94	64	74	144	84	74	84	114	94	94	104	94	94	S.P.
5	30	10	20	20	60	5	10	30	60	20	60	20	20	60	60	5	30	120	30	20	30	60	60	60	60	O.S.
5	11	13	13	15	17	17	19	19	19	21	21	21	23	23	23	25	25	25	25	25	27	27	27	27	27	LEVEL
5	45	50	70	105	280	70	175	280	480	420	720	450	420	1 120	945	210	980	2 520	1 200	560	1 800	2 205	2 000	2 430	2 625	DIM'N
1000	0101	0020	2001	0012	1110	0004	0300	3010	1102	2200	1021	3002	4100	0211	1013	6000	0130	2111	0203	1005	2030	1301	4011	0122	2103	WEIGHT
0	1	2	3	4	5	6	7	8	9	10	11	12	13	14	15	16	17	18	19	20	21	22	23	24	25	NUMBER
1	3	2	4	3	8	1	3	6	12	6	12	10	4	16	12	1	10	32	14	4	16	19	20	21	30	0
	1	1	1	2	4	1	2	3	6	4	8	4	3	10	9	1	7	20	9	4	12	14	12	16	18	1
		1	.	1	2	1	1	1	2	3	6	1	2	6	6	1	6	12	5	4	9	10	6	12	9	2
			1	.	2	.	1	3	3	3	3	4	3	6	3	1	4	14	6	1	7	10	12	9	12	3
				1	.	1	.	.	2	.	3	1	.	3	6	.	3	6	5	4	6	4	3	9	9	4
					1	.	1	1	1	2	2	1	2	4	2	1	3	8	3	1	5	8	6	7	6	5
						1	.	.	.	.	.	.	.	.	3	.	.	.	2	4	.	.	.	3	3	6
							1	.	.	1	.	.	1	2	.	1	1	2	1	.	1	6	2	3	1	7
								1	.	1	.	1	2	.	.	1	.	4	.	.	2	3	6	.	3	8
									1	.	1	1	.	2	2	.	2	4	3	1	3	3	3	5	6	9
										1	.	.	1	.	.	1	.	2	.	.	1	3	2	.	1	10
											1	.	.	1	1	.	2	2	1	1	3	2	1	4	2	11
												1	.	.	.	.	.	2	.	.	1	1	3	.	3	12
													1	.	.	1	.	.	.	.	.	.	2	.	.	13
														1	.	.	1	1	1	.	1	2	1	2	1	14
															1	.	.	.	1	1	.	.	.	2	2	15
																1	.	.	.	.	.	.	.	.	.	16
																	1	.	.	.	1	.	.	1	.	17
																		1	.	.	1	1	1	.	1	18

104	114	124	104	124	124	134	124	154	114	144	134	174	144	164	214	134	144	134	144	184	184	184	174	174	154	S.P.
60	60	20	60	20	60	30	60	60	120	20	60	30	60	30	20	120	30	120	60	20	20	20	60	30	60	O.S.
29	29	29	29	29	29	31	31	31	31	31	31	31	33	33	33	33	33	33	33	33	33	35	35	35	35	LEVEL
3 780	3 360	1 470	4 410	1 925	2 625	2 520	5 400	2 970	8 750	2 400	4 620	1 540	6 615	3 850	1 440	10 240	4 725	12 320	7 425	2 475	2 520	3 360	8 250	4 410	10 500	DIM'N
1220	3201	0041	2022	4003	0114	0410	3120	5101	1212	0033	2014	0106	2310	5020	7001	1131	0402	3112	1204	0025	2006	1500	4210	1050	0321	WEIGHT
26	27	28	29	30	31	32	33	34	35	36	37	38	39	40	41	42	43	44	45	46	47	48	49	50	51	NUMBER
30	24	9	36	20	17	14	36	15	63	12	30	6	33	20	4	56	24	80	45	9	10	12	33	17	50	0
21	18	7	27	10	14	11	26	12	45	10	24	6	26	14	4	44	19	56	32	8	10	10	27	14	38	1
15	14	6	21	4	11	8	18	9	31	8	18	6	21	9	4	36	15	38	21	7	10	8	22	12	29	2
15	15	5	15	10	8	8	21	12	33	7	12	3	21	14	4	30	15	44	24	6	4	8	24	11	29	3
9	5	4	15	4	11	5	12	3	22	7	18	6	12	6	1	26	9	26	21	7	10	6	13	9	21	4
11	11	4	11	4	6	7	14	9	23	6	9	3	17	9	4	24	12	29	15	5	4	7	19	9	23	5
.	.	1	6	1	8	.	.	.	7	4	12	6	.	.	.	8	3	8	12	6	10	.	.	3	6	6
6	6	2	3	1	2	6	6	6	12	4	2	1	12	5	4	12	9	12	6	3	1	6	12	5	15	7
4	8	.	4	4	.	3	10	9	9	.	3	.	10	9	4	8	6	20	6	.	1	4	16	3	11	8
7	4	3	8	4	6	4	9	3	16	5	9	3	10	6	1	18	7	20	15	5	4	5	11	7	17	9
3	6	.	3	1	.	3	6	6	6	.	2	.	9	5	4	6	6	12	3	.	1	4	12	2	9	10
5	3	3	6	1	4	3	6	2	10	4	6	3	8	3	1	16	5	12	8	4	4	4	9	7	13	11
3	3	.	3	4	.	2	6	3	6	.	3	.	6	6	1	6	3	14	6	.	1	3	9	2	9	12
.	3	.	.	1	.	.	3	6	.	.	.	.	3	5	4	.	.	6	.	.	.	1	9	.	.	13
4	2	2	2	1	2	3	4	2	8	3	2	1	7	3	1	10	5	8	6	3	1	4	7	5	11	14
.	.	1	3	1	4	.	.	.	5	3	6	3	.	.	.	6	2	6	8	4	4	.	.	3	5	15
.	.	.	.	.	.	.	.	3	.	.	.	.	.	2	4	.	.	.	.	.	.	.	3	.	.	16
2	.	2	1	.	1	1	2	.	2	2	1	1	3	1	.	8	1	2	2	2	1	2	3	5	6	17
2	2	.	2	1	.	2	4	2	4	.	2	.	5	3	1	5	3	8	3	.	1	3	7	2	7	18

1	.	.	.	.	1	1	.	.	1	1	1	2	.	.	.	4	2	2	1	.	.	.	4	2	4	6	3	1	.	.	3	4	19
	1	.	.	.	.	.	.	.	.	.	.	2	.	.	.	.	1	3	3	.	.	.	.	.	.	3	3	4	.	.	.	.	20
		1	.	.	.	.	1	.	.	1	.	.	1	2	.	1	.	1	.	2	1	.	4	1	2	1	.	1	2	3	2	4	21
			1	.	.	.	1	1	.	.	.	.	2	1	1	2	.	.	.	4	1	1	2	3	2	1	.	.	3	4	1	5	22
				1	.	.	.	1	.	.	1	.	.	2	2	.	.	.	.	2	3	1	.	.	4	.	.	.	1	5	.	.	23
					1	.	.	.	1	1	.	1	.	.	.	2	2	1	1	.	.	.	4	1	2	2	2	1	.	.	3	3	24
						1	.	.	.	1	1	.	.	.	.	2	.	2	.	.	.	.	2	1	4	3	.	1	.	.	1	3	25
							1	.	.	.	.	.	1	1	.	1	.	.	.	2	1	.	2	1	1	1	.	.	2	2	1	4	26
								1	.	.	.	.	.	1	1	.	.	.	.	2	1	1	.	.	2	.	.	.	1	4	.	.	27
									1	.	.	.	.	.	.	.	1	.	.	.	.	.	2	.	.	.	1	.	.	.	3	1	28
										1	.	.	.	.	.	1	.	1	.	.	.	.	2	1	2	1	.	1	.	.	1	2	29
											1	.	.	.	.	.	.	.	.	.	.	.	.	2	.	.	.	.	.	.	.	30
												1	.	.	.	.	1	1	1	.	.	.	.	.	.	2	2	1	.	.	.	.	31
													1	.	.	.	.	.	.	1	.	.	.	1	.	.	.	.	2	1	.	2	32
														1	.	.	.	.	.	1	1	.	.	.	1	.	.	.	1	2	.	.	33
															1	.	.	.	.	.	1	1	.	.	.	.	.	.	.	2	.	.	34
																1	.	.	.	.	.	.	1	1	1	1	.	.	.	.	1	2	35
																	1	.	.	.	.	.	.	.	.	1	.	.	.	.	.	36
																		1	.	.	.	.	.	.	.	1	.	1	.	.	.	.	37
																			1	.	.	.	.	.	.	.	1	1	.	.	.	.	38
																				1	.	.	.	.	.	.	.	.	1	1	.	.	39
																					1	.	.	.	.	.	.	.	.	1	.	.	40
																						1	.	.	.	.	.	.	.	.	.	.	41
																							1	.	.	.	.	.	.	.	1	1	42
																								1	.	.	.	.	.	.	.	1	43
																									1	.	.	.	.	.	.	.	44
																										1	.	.	.	.	.	.	45
																											1	.	.	.	.	.	46
																												1	.	.	.	.	47
																													1	.	.	.	48
																														1	.	.	49
																															1	.	50
																																1	51

Table 3. *(continued)*

ALGEBRA A4, REPRESENTATIONS OF CLASS 2

NUMBER	0	1	2	3	4	5	6	7	8	9	10	11	12	13	14	15	16	17	18	19	20	21	22	23	24	25
S.P.	6	16	16	36	26	36	46	46	46	56	56	76	76	86	66	66	96	96	76	96	86	136	86	106	96	126
O.S.	10	5	20	5	60	30	20	30	60	60	30	20	20	30	60	60	20	10	60	20	60	20	120	20	60	30
LEVEL	7	9	11	13	15	17	17	19	19	21	21	21	23	23	23	23	25	25	25	25	25	27	27	27	27	27
DIM'N	10	15	40	35	175	210	160	315	450	700	560	315	560	540	1 050	1 215	840	490	1 890	875	1 440	720	3 675	1 260	2 520	945
WEIGHT	0 1 0 0	2 0 0 0	0 0 1 1	0 0 0 3	1 1 0 1	1 0 2 0	3 0 0 1	0 2 1 0	1 0 1 2	2 1 1 0	0 2 0 2	1 0 0 4	1 3 0 0	4 0 1 0	0 1 2 1	2 1 0 2	3 2 0 0	0 0 4 0	2 0 2 1	4 0 0 2	0 1 1 3	5 1 0 0	1 2 1 1	0 0 3 2	2 0 1 3	0 1 0 5
0	1	1	2	1	6	5	4	6	9	11	9	4	6	6	13	18	7	4	21	10	14	4	34	9	24	6
1		1	.	.	3	2	4	3	3	8	6	1	4	6	7	12	6	3	12	10	8	4	24	6	12	3
2			1	1	2	3	1	3	6	6	5	4	4	3	9	9	5	3	15	4	11	3	22	7	18	6
3				1	.	.	.	.	3	.	2	4	.	.	3	3	.	1	6	1	8	.	7	4	12	6
4					1	1	1	2	2	4	3	1	3	3	5	6	4	2	8	4	6	3	16	5	9	3
5						1	.	1	1	2	1	1	2	1	4	2	3	2	6	1	4	2	10	4	6	3
6							1	.	.	2	.	.	1	3	.	3	3	.	3	4	.	3	6	.	3	.
7								1	.	1	1	.	2	1	2	1	2	1	2	1	2	2	8	3	2	1
8									1	.	1	1	.	.	2	2	.	1	3	1	4	.	5	3	6	3
9										1	.	.	1	1	.	1	2	.	2	1	.	2	4	.	2	.
10											1	.	.	.	1	1	.	1	1	1	2	.	4	2	2	1
11												1	.	.	.	.	.	.	.	.	2	.	.	1	3	3
12													1	.	.	.	1	.	.	.	.	1	2	.	.	.
13														1	.	.	1	.	.	1	.	2	.	.	.	.
14															1	.	.	1	1	.	1	.	2	2	1	1
15																1	.	.	1	1	.	.	2	.	2	.
16																	1	.	.	.	.	1	.	.	.	.
17																		1	.	.	.	.	.	1	.	.
18																			1	.	.	.	1	.	1	.

NUMBER	26	27	28	29	30	31	32	33	34	35	36	37	38	39	40	41	42	43	44	45	46	47	48	49	50	51
S.P.	196	106	116	106	106	136	136	126	126	126	146	116	126	186	136	156	146	136	136	166	146	256	166	166	206	156
O.S.	5	60	30	120	60	20	20	30	30	60	60	120	60	20	60	60	60	60	60	20	120	5	60	60	60	60
LEVEL	29	29	29	29	29	29	29	31	31	31	31	31	31	31	33	33	33	33	33	33	33	33	35	35	35	35
DIM'N	330	3 360	1 890	5 120	4 000	1 500	1 540	3 360	3 750	4 900	3 465	8 400	5 390	1 155	8 505	5 940	5 600	7 875	9 240	3 360	9 240	495	7 350	9 625	4 725	10 080
WEIGHT	7 0 0 0	1 1 3 0	0 4 0 1	3 1 1 1	1 2 0 3	0 0 2 4	2 0 0 5	0 3 2 0	3 0 3 0	2 3 0 1	5 0 1 1	1 1 2 2	3 1 0 3	0 0 1 6	2 2 2 0	4 2 0 1	1 0 4 1	0 3 1 2	3 0 2 2	5 0 0 3	1 1 1 4	0 0 0 8	1 4 1 0	4 1 2 0	6 1 0 1	0 2 3 1
0	1	24	12	40	32	8	10	20	23	28	20	54	40	4	45	27	26	42	54	20	48	1	29	42	15	44
1	1	16	9	32	24	6	4	14	16	22	20	36	30	3	36	24	20	33	36	20	32	1	23	36	15	32
2	1	18	9	26	21	7	10	14	18	21	12	42	24	4	33	21	21	31	42	10	40	1	23	31	12	35
3	.	8	3	8	12	6	10	6	10	6	3	24	12	4	15	6	12	16	24	4	32	1	11	15	3	20
4	1	12	7	20	15	5	4	11	12	17	12	28	18	3	26	18	16	24	27	10	26	1	19	26	12	26
5	1	10	5	12	8	4	4	8	9	13	6	22	9	3	19	14	14	17	21	4	20	1	15	18	9	22
6	1	4	3	14	6	.	1	5	7	10	12	9	12	.	15	15	5	12	15	10	8	.	11	21	12	12
7	1	6	5	8	6	3	1	7	5	11	6	14	6	2	15	11	10	14	11	4	12	1	14	14	9	16
8	.	6	2	6	8	4	4	5	6	5	3	16	9	3	12	5	10	12	15	4	20	1	9	12	3	16
9	1	3	3	8	3	.	1	4	5	8	6	7	6	.	11	11	4	9	11	4	6	.	10	14	9	10
10	.	4	2	4	6	3	1	4	3	4	3	10	6	2	10	4	8	10	8	4	12	1	8	9	3	12
11	.	.	.	.	3	3	4	.	.	.	.	6	3	3	.	.	3	4	6	1	14	1	.	.	.	5
12	1	1	3	2	1	.	.	3	1	6	2	3	1	.	6	6	2	6	3	1	2	.	9	6	6	6
13	1	.	.	4	.	.	.	.	2	3	6	.	3	.	4	8	.	.	4	4	.	.	3	10	9	.
14	.	4	1	2	2	2	1	3	3	3	1	8	2	2	7	3	8	6	6	1	8	1	6	6	2	11
15	.	2	1	4	3	.	1	3	3	3	3	5	6	.	7	4	3	6	8	4	6	.	6	9	3	8
16	1	.	.	2	.	.	.	.	1	3	2	.	1	.	3	6	.	.	3	1	.	.	3	6	6	.
17	.	2	.	.	.	1	.	1	1	.	.	2	.	1	3	.	6	1	1	.	2	1	2	2	.	6
18	.	2	1	2	1	.	1	2	3	2	1	4	2	.	5	3	3	4	6	1	4	.	5	6	2	7

1	.	.	.	.	.	.	.	.	.	2	.	.	.	.	1	1	3	.	3	.	3	3	.	.	3	4	.	.	2	6	3	.	19
	1	.	.	1	1	1	.	.	.	.	2	2	1	.	.	.	.	4	2	2	.	.	3	3	3	1	8	1	.	.	.	4	20
		1	.	.	.	.	1	.	.	.	.	.	.	.	.	.	2	.	.	.	.	3	.	.	.	1	.	.	.	3	6	.	21
			1	.	.	.	.	1	1	1	1	.	.	2	1	2	1	2	1	.	4	2	2	4	2	1	2	.	5	4	2	5	22
				1	.	.	.	.	.	.	.	1	.	.	.	.	.	2	.	1	.	.	3	1	1	.	2	1	.	.	.	3	23
					1	.	.	.	.	.	1	.	1	.	.	.	.	2	2	.	.	.	1	2	3	1	4	.	.	.	.	3	24
						1	.	.	.	.	.	1	1	.	.	.	.	.	.	2	.	.	.	.	.	.	4	1	.	.	.	.	25
							1	.	.	.	.	.	.	.	.	.	.	.	.	.	.	.	.	.	.	.	.	.	.	.	3	.	26
								1	.	.	.	.	.	1	1	.	.	1	.	.	2	.	2	1	1	.	1	.	2	2	.	4	27
									1	.	.	.	.	1	.	1	.	.	.	.	1	1	.	2	.	.	.	.	4	1	1	2	28
										1	.	.	.	.	1	1	1	.	1	.	2	2	.	.	2	1	.	.	2	4	2	.	29
											1	.	.	.	.	.	.	1	1	.	.	.	1	2	1	1	2	.	.	.	.	2	30
												1	.	.	.	.	.	.	.	1	.	.	.	.	.	.	2	1	.	.	.	.	31
													1	.	.	.	.	.	.	.	.	.	.	.	.	.	2	.	.	.	.	.	32
														1	.	.	.	.	.	.	1	.	.	1	.	.	.	.	2	1	.	2	33
															1	.	.	.	.	.	1	.	.	.	1	.	.	.	1	2	.	.	34
																1	.	.	.	.	1	1	.	.	.	.	.	.	2	1	1	.	35
																	1	.	.	.	.	1	.	.	.	1	.	.	.	2	2	.	36
																		1	.	.	.	.	1	1	1	.	1	.	.	.	.	2	37
																			1	.	.	.	.	.	1	1	.	.	.	.	.	.	38
																				1	.	.	.	.	.	.	.	1	.	.	.	.	39
																					1	.	.	.	.	.	.	.	1	1	.	.	40
																						1	.	.	.	.	.	.	.	1	1	.	41
																							1	.	.	.	.	.	.	.	.	1	42
																								1	.	.	.	.	.	.	.	1	43
																									1	.	.	.	.	.	.	.	44
																										1	.	.	.	.	.	.	45
																											1	.	.	.	.	.	46
																												1	.	.	.	.	47
																													1	.	.	.	48
																														1	.	.	49
																															1	.	50
																																1	51

Table 3. *(continued)*

ALGEBRA A5, REPRESENTATIONS OF CLASS 0

0	12	24	36	36	36	48	48	48	72	72	72	72	84	60	84	120	72	84	84	120	180	84	84	108	180	96	96	96	108	108	120	132	120	108	132	144	120	144	144	120	120	144	144	132	192	132	132	156	156	192	156	S.P.
1	30	90	20	60	60	30	120	120	15	60	15	60	30	360	30	30	180	120	120	30	6	180	180	30	6	90	360	360	180	180	120	180	120	180	180	20	180	60	120	180	360	120	60	360	120	360	180	120	120	120	120	O.S.
1	11	17	19	19	19	21	23	23	25	25	25	25	27	27	27	29	29	29	29	29	31	31	31	31	31	33	33	33	35	35	35	35	35	35	35	37	37	37	37	37	37	37	37	39	39	39	39	39	39	39	41	LEVEL
1	35	189	175	280	280	405	896	896	490	840	490	840	1 134	3 675	1 134	1 050	3 969	3 200	3 200	1 050	462	5 670	5 670	2 695	462	6 720	12 250	12 250	11 340	14 175	8 960	8 910	8 960	14 175	8 910	4 116	19 845	9 625	12 936	19 845	29 700	12 936	9 625	36 750	10 368	36 750	34 496	19 845	19 845	10 368	26 880	DIM'N
0	1	0	0	2	0	2	1	0	0	3	0	0	2	1	0	4	1	3	1	0	6	1	0	3	0	0	2	1	0	2	1	4	1	0	1	0	2	4	3	0	2	1	0	1	5	0	2	4	2	1	1	
0	0	1	0	0	1	0	1	0	3	0	0	0	2	1	0	1	0	0	1	0	0	0	2	0	0	2	1	0	1	1	3	0	0	2	0	0	0	0	2	1	1	0	2	2	1	1	0	0	1	0	1	
0	0	0	2	0	0	0	1	1	0	1	0	1	0	0	0	0	2	0	0	0	0	1	1	0	0	0	1	1	2	0	0	1	0	0	1	4	2	0	0	2	0	0	0	1	0	1	2	0	0	0	3	
0	0	1	0	1	0	0	0	1	0	0	3	0	0	1	2	0	0	1	0	1	0	2	0	0	0	2	0	1	1	2	0	0	3	1	0	0	1	2	0	0	1	2	0	1	0	2	0	1	0	1	0	
0	1	0	0	0	2	2	0	1	0	0	0	3	0	1	2	0	1	1	3	4	0	0	1	3	6	0	1	2	0	0	1	1	1	2	4	0	0	0	1	2	2	3	4	0	1	1	2	2	4	5	0	WEIGHT
0	1	2	3	4	5	6	7	8	9	10	11	12	13	14	15	16	17	18	19	20	21	22	23	24	25	26	27	28	29	30	31	32	33	34	35	36	37	38	39	40	41	42	43	44	45	46	47	48	49	50	51	NUMBER
1	5	9	5	10	10	15	16	16	5	10	5	10	9	55	9	5	39	40	40	5	1	45	45	35	1	50	85	85	60	90	40	45	40	90	45	16	90	50	46	90	180	46	50	145	24	145	146	105	105	24	80	0
	1	3	2	4	4	5	8	8	3	6	3	6	6	27	6	4	22	20	20	4	1	27	27	15	1	28	52	52	38	54	28	30	28	54	30	10	60	30	34	60	105	34	30	97	20	97	96	60	60	20	56	1
		1	1	1	1	1	4	4	2	3	2	3	4	12	4	3	13	8	8	3	1	16	16	5	1	16	31	31	25	30	20	18	20	30	18	7	40	16	25	40	57	25	16	65	16	65	64	30	30	16	40	2
			1	.	.	.	2	2	1	1	1	1	3	4	3	2	9	2	2	2	1	9	9	1	1	8	18	18	18	15	14	9	14	15	9	6	27	7	19	27	27	19	7	43	12	43	45	12	12	12	30	3
				1	.	1	2	.	1	3	.	.	3	6	.	3	5	8	4	.	1	6	9	5	.	9	22	11	13	21	14	18	6	18	6	4	24	16	21	21	39	7	10	45	16	34	37	30	20	4	26	4
					1	1	.	2	.	.	1	3	.	6	3	.	5	4	8	3	.	9	6	5	1	9	11	22	13	18	6	6	14	21	18	4	21	10	7	24	39	21	16	34	4	45	37	20	30	16	24	5
						1	.	.	.	.	.	.	.	3	.	.	2	4	4	.	.	3	3	5	.	6	8	8	7	12	4	6	4	12	6	3	12	10	6	12	27	6	10	24	4	24	22	20	20	4	16	6
							1	.	1	1	.	.	2	2	.	2	3	2	1	.	1	3	6	1	.	5	12	6	9	10	11	9	4	9	3	3	15	7	15	15	18	5	4	31	12	22	24	12	8	3	19	7
								1	.	.	1	1	.	2	2	.	3	1	2	2	.	6	3	1	1	5	6	12	9	9	4	3	11	10	9	3	15	4	5	15	18	15	7	22	3	31	24	8	12	12	18	8
									1	.	.	.	1	.	.	1	.	.	.	.	1	.	3	.	.	2	3	.	3	2	8	3	.	3	.	1	3	2	8	6	3	.	1	16	8	7	6	3	1	.	8	9
										1	.	.	1	.	.	2	.	2	.	.	1	.	.	1	.	.	6	.	.	5	4	9	.	.	.	.	6	7	11	.	9	.	.	11	12	.	9	12	4	.	6	10
											1	.	.	.	1	.	.	.	.	1	.	3	.	.	1	2	.	3	3	3	.	.	8	2	3	1	6	1	.	3	3	8	2	7	.	16	6	1	3	8	8	11
												1	.	.	1	.	.	.	2	2	.	.	.	1	1	.	.	6	.	.	.	.	4	5	9	.	.	.	.	6	9	11	7	.	.	11	9	4	12	12	.	12
													1	.	.	1	.	.	.	.	1	.	.	.	.	.	3	.	.	2	4	3	.	.	.	.	3	2	8	.	3	.	.	8	8	.	6	3	1	.	4	13
														1	.	.	1	1	1	.	.	2	2	1	.	3	4	4	5	6	3	3	3	6	3	2	8	4	4	8	12	4	4	16	3	16	13	8	8	3	12	14
															1	.	.	.	.	1	.	.	.	.	1	.	.	3	.	.	.	.	4	2	3	.	.	.	.	3	3	8	2	.	.	6	6	1	3	8	.	15
																1	.	.	.	.	1	.	.	.	.	.	.	.	.	.	.	3	.	.	.	.	.	2	4	.	.	.	.	.	8	.	.	3	.	.	.	16
																	1	.	.	.	.	1	1	.	.	1	2	2	4	2	2	1	2	2	1	2	6	1	3	6	4	3	1	10	2	10	9	2	2	2	10	17

	18	19	20	21	22	23	24	25	26	27	28	29	30	31	32	33	34	35	36	37	38	39	40	41	42	43	44	45	46	47	48	49	50	51
18	1	.	.	.	.	.	1	.	.	2	.	.	3	1	3	.	.	.	.	3	4	3	.	6	.	.	6	3	.	5	8	4	.	4
19		1	.	.	.	.	1	.	.	.	2	.	.	.	.	1	3	3	.	.	.	.	3	6	3	4	.	.	6	5	4	8	3	.
20			1	.	.	.	.	1	.	.	.	.	.	.	.	.	.	3	.	.	.	.	.	.	4	2	.	.	.	.	.	3	8	.
21				1	.	.	.	.	.	.	.	.	.	.	.	.	.	.	.	.	.	.	.	.	.	.	.	4	.	.	.	.	.	.
22					1	.	.	.	1	.	1	2	2	.	.	2	1	1	1	3	1	.	2	2	2	1	5	.	8	3	1	2	2	6
23						1	.	.	1	1	.	2	1	2	1	.	2	.	1	2	1	2	3	2	.	1	8	2	5	3	2	1	.	6
24							1	.	.	.	.	.	.	.	.	.	.	.	.	.	.	.	.	3	.	.	.	.	.	2	4	4	.	.
25								1	.	.	.	.	.	.	.	.	.	.	.	.	.	.	.	.	.	.	.	.	.	.	.	.	4	.
26									1	.	.	1	1	.	.	.	1	.	1	1	1	.	1	1	.	1	4	.	4	1	1	1	.	4
27										1	.	.	1	1	1	.	.	.	.	2	1	2	.	2	.	.	4	2	.	3	2	1	.	3
28											1	.	.	.	.	1	1	1	.	.	.	.	2	2	2	1	.	.	4	3	1	2	2	.
29												1	.	.	.	.	.	.	1	1	.	.	1	.	.	.	2	.	2	1	.	.	.	4
30													1	.	.	.	.	.	.	1	1	.	.	1	.	.	2	.	.	1	1	1	.	2
31														1	.	.	.	.	.	.	.	1	.	.	.	.	2	1	.	.	.	.	.	1
32															1	.	.	.	.	.	1	1	.	.	.	.	.	2	.	.	2	.	.	.
33																1	.	.	.	.	.	.	.	.	1	.	.	.	2	.	.	.	1	.
34																	1	.	.	.	.	.	1	1	.	1	.	.	2	1	1	1	.	.
35																		1	.	.	.	.	.	.	1	1	.	.	.	.	.	2	2	.
36																			1	.	.	.	.	.	.	.	.	.	.	.	.	.	.	2
37																				1	.	.	.	.	.	.	1	.	.	1	.	.	.	2
38																					1	.	.	.	.	.	.	.	.	.	1	.	.	.
39																						1	.	.	.	.	.	1	.	.	.	.	.	.
40																							1	.	.	.	.	.	1	1	.	.	.	.
41																								1	.	.	.	.	.	1	1	1	.	.
42																									1	.	.	.	.	.	.	.	1	.
43																										1	.	.	.	.	.	1	.	.
44																											1	.	.	.	.	.	.	1
45																												1	.	.	.	.	.	.
46																													1	.	.	.	.	.
47																														1	.	.	.	.
48																															1	.	.	.
49																																1	.	.
50																																	1	.
51																																		1

Table 3. *(continued)*

ALGEBRA A5, REPRESENTATIONS OF CLASS 1

5	17	29	29	41	41	53	53	65	53	77	65	77	125	77	65	101	113	89	77	89	125	89	89	101	101	173	101	101	125	113	137	245	125	113	113	113	149	197	125	149	137	137	137	125	149	137	149	161	185	149	149	S.P.
6	60	30	60	60	180	30	60	60	120	30	60	30	6	120	360	30	60	60	180	180	30	180	360	120	60	30	180	180	120	120	120	6	60	180	360	360	30	30	180	60	360	180	60	360	180	360	180	120	180	120	120	O.S.
6	14	16	18	20	22	22	24	24	24	24	26	26	26	28	28	30	30	30	30	30	32	32	32	32	32	34	34	34	34	34	34	36	36	36	36	36	36	36	38	38	38	38	38	38	38	40	40	40	40	40	40	LEVEL
6	84	120	210	336	840	420	840	720	1 260	504	1 176	1 050	252	2 520	4 410	1 764	1 800	2 520	5 040	4 200	2 520	6 804	10 500	5 880	4 500	1 980	10 080	11 760	6 930	9 240	5 544	792	7 056	17 010	21 000	22 050	5 880	2 520	25 200	8 820	27 720	17 640	12 600	36 960	16 500	44 100	32 340	22 176	17 010	25 200	27 720	DIM'N
1	0	2	0	0	1	0	1	3	1	0	0	3	0	2	1	1	4	1	0	1	3	0	2	1	0	5	0	2	4	2	1	7	0	2	1	0	4	1	1	0	3	0	0	2	0	1	3	2	5	0	2	
0	1	0	0	0	1	0	0	0	1	0	2	0	0	1	0	3	0	0	2	0	2	1	1	0	2	1	1	0	0	1	0	0	0	0	2	1	0	0	2	4	1	1	0	0	1	1	1	3	0	0	0	
0	0	0	1	1	0	0	2	0	0	0	1	0	0	1	1	0	1	0	0	1	0	2	0	0	0	0	1	2	0	0	0	0	3	1	1	1	0	0	0	0	1	0	3	1	1	2	0	0	1	2	0	
0	0	0	1	0	1	2	0	1	0	1	0	0	0	0	1	0	0	3	1	0	0	0	1	2	0	0	2	0	1	0	1	0	1	2	0	1	0	0	2	0	0	3	0	1	0	1	2	0	0	2	3	
0	1	1	0	2	0	1	0	0	2	3	0	2	5	0	1	0	0	0	1	3	0	1	1	2	3	0	0	1	1	3	4	0	0	0	1	2	3	6	0	1	1	1	2	2	4	0	0	1	1	1	1	WEIGHT
0	1	2	3	4	5	6	7	8	9	10	11	12	13	14	15	16	17	18	19	20	21	22	23	24	25	26	27	28	29	30	31	32	33	34	35	36	37	38	39	40	41	42	43	44	45	46	47	48	49	50	51	NUMBER
1	4	5	5	6	15	5	10	10	20	4	11	15	1	20	40	9	10	15	40	30	10	39	75	30	30	5	50	60	40	60	20	1	26	75	90	95	35	5	100	25	100	55	40	150	55	135	110	56	45	70	80	0
	1	1	2	3	6	3	5	4	8	3	6	5	1	11	22	6	6	10	21	18	7	24	39	21	16	4	31	37	20	30	16	1	17	48	57	62	15	5	61	18	64	40	28	96	38	92	68	41	30	49	59	1
		1	.	.	3	.	2	4	4	.	3	5	.	8	8	4	6	3	12	6	6	12	27	6	10	4	16	22	20	20	4	1	10	27	39	32	15	1	42	13	52	19	16	52	20	59	54	32	30	30	28	2
			1	1	2	2	3	1	2	2	3	1	1	6	12	4	3	7	10	9	5	15	18	15	7	3	20	24	8	12	12	1	12	30	35	39	5	5	36	13	40	30	19	60	24	64	39	30	18	35	44	3
				1	.	1	.	.	2	2	.	1	1	.	6	.	.	4	5	9	.	6	9	11	7	.	11	9	4	12	12	.	6	18	13	28	5	5	21	4	14	21	12	42	24	34	24	9	6	24	32	4
					1	.	1	1	1	.	2	1	.	4	4	3	3	2	6	3	4	8	12	4	4	3	10	13	8	8	3	1	7	16	25	20	5	1	25	10	31	14	12	31	12	41	30	24	18	21	20	5
						1	.	.	.	1	.	.	1	.	3	.	.	4	2	3	.	3	3	8	2	.	8	6	1	3	8	.	4	12	7	16	1	5	12	3	8	18	7	24	13	24	12	6	3	18	26	6
							1	.	.	.	1	.	.	2	2	2	1	1	2	1	3	6	4	3	1	2	6	9	2	2	2	1	6	9	15	12	1	1	13	7	18	10	9	18	6	30	15	18	9	15	14	7
								1	.	.	.	1	.	2	.	1	3	.	.	.	3	.	6	.	.	3	.	5	8	4	.	1	.	6	9	.	5	.	9	4	22	.	.	11	.	13	21	14	18	.	6	8
									1	.	.	1	.	.	2	.	.	1	3	3	.	3	6	3	4	.	6	5	4	8	3	.	4	9	9	14	5	1	15	3	11	10	7	22	12	22	18	7	6	15	14	9
										1	.	.	1	.	.	.	.	.	.	3	.	.	.	4	2	.	.	.	.	3	8	.	.	.	.	8	1	5	.	.	.	6	3	12	13	.	.	.	.	6	10	10
											1	.	.	1	.	2	1	.	2	.	2	3	2	.	1	2	3	3	2	1	.	1	3	3	12	6	1	.	10	7	12	4	6	6	3	18	10	15	9	9	4	11
												1	.	.	.	.	.	.	.	.	.	.	3	.	.	.	.	2	4	4	.	.	.	3	3	.	5	.	6	1	8	.	.	8	.	7	12	4	6	.	4	12
													1	.	.	.	.	.	.	.	.	.	.	.	.	.	.	.	.	.	4	.	.	.	.	.	.	5	.	.	.	.	.	.	5	.	.	.	.	.	.	13
														1	.	1	1	.	.	.	2	.	2	.	.	2	.	3	2	1	.	1	.	3	6	.	1	.	5	4	12	.	.	6	.	9	10	11	9	.	4	14
															1	.	.	1	1	1	.	2	2	2	1	.	4	3	1	2	2	.	3	6	5	8	1	1	8	2	6	8	5	12	6	16	9	5	3	11	11	15
																1	.	.	.	.	1	.	.	.	.	1	.	.	.	.	.	1	.	.	3	.	.	.	2	4	3	.	.	.	.	3	2	8	3	.	.	16
																	1	.	.	.	1	.	.	.	.	2	.	.	2	.	.	1	.	.	.	.	1	.	.	.	6	.	.	.	.	.	5	4	9	.	.	17

1	.	.	.	.	.	1	.	.	2	.	.	.	1	.	1	3	.	2	.	1	3	.	.	6	1	3	2	6	3	.	.	6	8	18
	1	.	.	1	1	.	1	.	2	1	1	1	.	.	2	2	4	4	1	.	6	2	4	3	3	4	3	10	6	4	3	7	3	19
		1	.	.	.	1	1	.	.	.	.	2	2	.	.	.	.	4	1	1	.	.	.	3	2	6	6	.	.	.	.	4	4	20
			1	.	.	.	.	1	.	.	.	.	.	1	.	.	.	.	.	.	.	.	3	.	.	.	.	.	2	4	3	.	.	21
				1	.	.	.	.	1	1	.	.	.	.	2	1	2	2	.	.	2	1	2	2	3	2	1	8	2	3	1	5	2	22
					1	.	.	.	.	1	1	1	.	.	.	2	2	.	1	.	3	1	4	.	.	4	.	5	6	3	3	.	3	23
						1	.	.	.	.	.	.	1	.	.	.	.	2	.	1	.	.	.	3	1	3	2	.	.	.	.	3	4	24
							1	.	.	.	.	1	.	.	.	.	.	2	1	.	.	.	.	1	1	2	3	.	.	.	.	3	1	25
								1	.	.	.	.	.	1	.	.	.	.	.	.	.	.	.	.	.	.	.	.	.	.	3	.	.	26
									1	.	.	.	.	.	1	1	.	1	.	.	2	.	.	2	1	1	1	4	2	.	.	4	2	27
										1	.	.	.	.	.	1	1	.	.	.	1	1	2	.	.	2	.	4	2	2	1	.	2	28
											1	.	.	.	.	.	.	.	1	.	.	.	2	.	.	.	.	.	3	1	3	.	.	29
												1	.	.	.	.	.	.	1	.	.	.	.	.	.	2	.	.	.	.	.	.	1	30
													1	.	.	.	.	.	.	1	.	.	.	.	.	.	2	.	.	.	.	.	.	31
														1	.	.	.	.	.	.	.	.	.	.	.	.	.	.	.	.	.	.	.	32
															1	.	.	.	.	.	.	.	.	.	1	.	.	2	.	.	.	2	.	33
																1	.	.	.	.	1	.	.	.	.	1	.	2	2	.	.	.	2	34
																	1	.	.	.	1	1	1	.	.	.	.	2	1	2	1	.	.	35
																		1	.	.	.	.	.	1	1	1	1	.	.	.	.	2	1	36
																			1	.	.	.	.	.	.	.	.	.	.	.	.	.	.	37
																				1	.	.	.	.	.	.	.	.	.	.	.	.	.	38
																					1	.	.	.	.	.	.	1	1	.	.	.	.	39
																						1	.	.	.	.	.	.	.	1	.	.	.	40
																							1	.	.	.	.	.	1	1	1	.	.	41
																								1	.	.	.	.	.	.	.	1	1	42
																									1	.	.	.	.	.	.	1	.	43
																										1	.	.	.	.	.	.	1	44
																											1	.	.	.	.	.	.	45
																												1	.	.	.	.	.	46
																													1	.	.	.	.	47
																														1	.	.	.	48
																															1	.	.	49
																																1	.	50
																																	1	51

Table 3. *(continued)*

ALGEBRA A5, REPRESENTATIONS OF CLASS 2

NUMBER	0	1	2	3	4	5	6	7	8	9	10	11	12	13	14	15	16	17	18	19	20	21	22	23	24	25
S.P.	8	20	20	32	32	44	56	80	44	56	56	68	68	68	68	80	104	80	92	92	80	116	140	128	116	104
O.S.	15	6	60	15	120	30	30	6	180	90	180	60	180	120	60	60	60	120	120	120	360	30	30	15	120	90
LEVEL	9	11	15	17	19	19	21	21	23	25	25	27	27	27	27	29	29	29	29	31	31	31	31	33	33	33
DIM'N	15	21	105	105	384	210	315	126	1050	1176	1701	1470	2430	2205	1800	2520	1575	3675	2688	4410	8064	2310	1386	1764	5760	5040
	0	2	0	0	1	0	3	0	1	0	1	0	2	1	0	2	4	2	1	1	0	4	1	0	3	0
	1	0	0	0	1	0	0	0	0	2	0	1	1	0	2	0	0	1	0	2	1	0	0	4	1	1
	0	0	1	0	0	0	0	0	1	0	1	2	0	0	0	2	0	0	0	1	1	0	0	0	1	0
	0	0	0	2	0	1	0	0	1	1	0	0	1	2	0	0	1	0	1	0	1	0	0	0	0	3
WEIGHT	0	0	1	0	1	2	1	4	0	0	2	0	0	1	2	0	0	2	3	0	1	2	5	0	0	0
0	1	1	3	2	8	3	5	1	13	12	18	11	21	17	16	16	10	30	16	23	48	15	5	6	24	22
1		1	.	.	4	.	5	.	5	6	6	5	15	5	10	10	10	20	4	15	24	15	1	4	20	10
2			1	1	2	2	1	1	6	5	9	6	9	11	7	9	4	12	12	13	28	5	5	4	14	15
3				1	.	1	.	1	3	2	3	3	3	8	2	6	1	3	8	7	16	1	5	3	8	12
4					1	.	1	.	2	3	3	3	6	3	4	5	4	8	3	9	14	5	1	3	11	7
5						1	.	1	.	.	3	.	.	4	2	.	.	3	8	.	8	1	5	.	.	6
6							1	.	.	.	.	.	3	.	.	2	4	4	.	3	.	5	.	1	8	.
7								1	.	.	.	.	.	.	.	.	.	.	4	.	.	.	5	.	.	.
8									1	1	1	2	2	2	1	3	1	2	2	5	8	1	1	2	6	5
9										1	.	1	1	.	1	1	1	1	.	4	4	1	.	2	4	2
10											1	.	.	1	1	.	.	2	2	.	4	1	1	.	.	3
11												1	.	.	.	1	.	.	.	2	2	.	.	1	2	1
12													1	.	.	1	1	1	.	2	.	1	.	1	4	.
13														1	.	.	.	.	1	.	2	.	1	.	.	3
14															1	.	.	1	.	.	2	1	.	.	.	1
15																1	.	.	.	1	.	.	.	1	2	.
16																	1	.	.	.	.	1	.	.	2	.
17																		1	.	.	.	1	.	.	.	.

NUMBER	26	27	28	29	30	31	32	33	34	35	36	37	38	39	40	41	42	43	44	45	46	47	48	49	50	51
S.P.	104	92	104	140	164	116	116	104	116	116	176	116	128	128	128	128	152	236	128	140	140	176	152	152	164	164
O.S.	60	360	180	30	60	60	60	360	180	180	30	360	360	120	120	120	180	30	360	180	180	120	120	120	60	120
LEVEL	33	33	33	35	35	35	35	35	35	35	37	37	37	37	37	37	37	39	39	39	39	39	39	39	39	39
DIM'N	4410	13440	7875	4410	3465	7056	7875	19200	11907	12936	4950	26880	24255	15750	18480	17325	11550	3465	40320	22050	27720	13440	16128	21504	12375	17640
	0	2	0	2	5	0	2	1	0	2	4	1	3	0	2	1	0	6	1	0	3	5	0	3	0	2
	0	0	1	3	0	0	0	2	1	0	2	1	1	0	0	2	1	1	1	3	0	0	0	1	0	0
	3	1	1	0	1	2	0	0	0	1	0	2	0	2	0	0	0	0	1	1	2	0	1	0	2	0
	0	1	0	0	0	2	3	1	2	0	0	0	1	1	2	0	1	0	2	0	0	1	3	0	0	1
WEIGHT	1	1	3	0	0	0	0	1	2	3	0	1	1	2	2	3	4	0	0	1	1	1	1	3	4	4
0	19	72	38	13	10	26	31	88	45	60	11	96	95	51	66	70	33	5	128	63	84	40	40	80	31	50
1	10	40	20	10	10	16	15	60	21	30	10	60	75	30	30	50	15	5	80	45	60	40	24	60	20	20
2	12	42	24	9	6	17	22	48	33	36	6	62	51	35	48	38	27	4	84	41	55	20	30	40	23	40
3	7	24	13	6	3	12	16	24	24	18	6	40	24	23	36	17	21	3	56	26	38	8	24	16	16	30
4	7	22	12	7	6	10	10	33	15	18	7	39	39	21	21	26	12	4	52	30	37	20	18	30	14	16
5	3	12	13	.	.	6	10	12	18	18	.	16	12	17	26	17	21	.	32	10	14	4	16	16	16	30
6	.	8	.	4	6	.	3	12	.	6	6	12	27	.	6	10	.	4	16	12	22	20	.	20	.	4
7	.	.	5	.	.	.	.	.	6	6	.	.	.	5	10	5	15	.	.	.	.	.	4	4	10	20
8	5	12	6	5	3	7	7	16	11	9	5	26	18	14	15	11	9	3	34	19	24	8	14	12	9	12
9	3	4	3	4	3	4	2	12	4	3	4	16	12	9	4	8	3	3	20	16	13	8	8	8	6	3
10	2	6	6	.	.	4	4	8	8	9	.	10	9	10	11	11	9	.	20	7	9	4	10	12	9	12
11	3	2	1	3	1	3	1	4	3	1	3	12	4	6	3	2	2	2	12	9	9	2	6	2	3	2
12	.	4	.	3	3	.	2	6	.	3	4	8	12	.	4	4	.	3	10	9	13	8	.	8	.	3
13	1	3	2	.	.	3	4	3	6	3	.	6	3	6	8	3	6	.	14	4	6	1	9	3	5	8
14	1	2	3	.	.	3	1	6	3	3	.	6	6	6	3	8	3	.	12	6	5	4	6	8	6	3
15	.	2	.	2	1	.	1	2	.	1	3	6	4	.	3	1	.	2	6	6	9	2	.	2	.	2
16	.	.	.	1	3	.	.	.	.	.	3	.	6	.	.	.	.	3	.	.	5	8	.	4	.	.
17	.	2	.	.	.	.	1	3	.	3	.	3	6	.	3	4	.	.	6	3	5	4	.	8	.	3

18	19	20	21	22	23	24	25	26	27	28	29	30	31	32	33	34	35	36	37	38	39	40	41	42	43	44	45	46	47	48	49	50	51
8	·	2	·	4	·	·	1	·	2	2	·	·	·	1	·	1	2	·	·	·	·	1	·	1	·	·	·	·	·	·	·	·	1
5	·	3	·	2	·	·	1	1	·	3	·	·	1	·	·	1	·	·	·	·	1	·	·	1	·	·	·	·	·	·	·	1	
3	1	2	4	·	·	1	·	·	2	2	·	·	·	·	1	·	2	·	·	1	·	·	1	·	·	·	·	·	·	·	1		
3	·	5	·	·	·	·	4	2	·	2	·	·	2	·	·	2	·	·	·	·	1	·	·	·	·	·	·	·	·	1			
·	2	1	4	·	·	2	·	·	1	·	·	2	·	·	1	·	·	·	·	1	·	·	·	·	·	·	·	·	1				
·	3	3	2	·	·	3	·	1	3	·	·	·	·	·	1	·	·	·	1	1	·	·	·	·	·	·	·	1					
·	6	3	·	·	3	·	·	1	2	·	·	·	·	·	2	·	·	·	1	·	·	·	·	·	·	·	1						
·	3	8	·	·	·	·	4	2	4	·	·	·	2	2	2	·	·	·	1	·	·	·	·	·	·	1							
·	2	·	·	·	1	2	·	·	·	·	1	2	·	·	·	·	·	1	·	·	·	·	·	·	1								
6	·	2	·	3	·	·	1	·	·	2	·	·	·	·	·	1	·	·	·	·	·	·	·	1									
3	1	2	·	·	·	·	·	·	1	2	·	·	·	·	1	·	1	·	·	·	·	·	1										
4	·	2	·	·	·	·	1	·	2	1	·	·	·	1	·	1	1	·	·	·	·	1											
3	·	4	·	·	·	·	1	2	·	2	·	·	1	·	·	1	·	·	·	·	1												
·	2	2	3	·	·	2	·	·	2	·	·	·	·	·	1	·	·	·	·	1													
·	3	4	·	·	·	·	·	2	2	·	·	·	·	·	1	·	·	·	1														
·	2	·	·	·	1	2	·	·	·	·	1	1	·	·	·	·	·	1															
3	·	1	·	·	·	·	·	·	1	1	·	·	·	·	·	·	1																
3	·	2	·	·	·	·	1	·	·	1	·	·	·	·	·	1																	
·	2	2	·	·	·	·	·	·	1	·	·	·	·	·	1																		
·	·	1	·	·	·	·	1	·	1	·	·	·	·	1																			
·	·	2	·	·	·	·	1	1	·	·	·	·	1																				
·	1	·	·	·	·	1	·	·	·	·	·	1																					
·	2	·	·	·	1	1	·	·	·	·	1																						
2	·	1	·	·	·	·	·	·	·	1																							
·	·	1	·	·	·	·	·	·	1																								
·	·	1	·	·	·	·	·	1																									
·	·	1	·	·	·	·	1																										
·	1	·	·	·	·	1																											
·	1	·	·	·	1																												
1	·	·	·	1																													
·	·	·	1																														
·	·	1																															
·	1																																
1																																	

Table 3. *(continued)*

ALGEBRA A5, REPRESENTATIONS OF CLASS 3

NUMBER	0	1	2	3	4	5	6	7	8	9	10	11	12	13	14	15	16	17	18	19	20	21	22	23	24	25
S.P.	9	21	21	45	45	33	45	45	57	57	57	93	93	81	69	69	81	81	81	93	105	93	93	93	105	117
O.S.	20	30	30	6	6	180	60	60	120	120	180	30	30	20	180	180	180	180	180	120	180	120	120	120	180	60
LEVEL	10	14	14	16	16	20	22	22	24	24	26	26	26	28	28	28	30	30	30	32	32	32	32	32	32	34
DIM'N	20	70	70	56	56	540	560	560	1 134	1 134	1 960	700	700	980	3 240	3 240	4 536	4 536	5 292	5 880	5 670	5 880	7 000	7 000	5 670	4 704
	0	1	0	3	0	1	1	0	2	1	0	4	1	0	2	0	1	0	2	1	3	0	2	1	0	0
	0	1	0	0	0	0	0	2	1	0	1	0	0	0	0	1	2	1	0	1	1	0	0	2	1	3
	1	0	0	0	0	1	0	0	0	0	1	0	0	3	1	1	0	0	1	2	0	2	0	0	0	1
	0	0	1	0	0	0	2	0	0	1	1	0	0	0	1	0	1	2	0	0	1	1	2	0	1	0
WEIGHT	0	0	1	0	3	1	0	1	1	2	0	1	4	0	0	2	0	1	2	0	0	1	1	2	3	0
0	1	2	2	1	1	9	7	7	12	12	17	5	5	6	24	24	27	27	36	28	27	28	38	38	27	17
1		1	.	1	.	3	2	4	8	3	8	5	1	3	13	12	18	12	18	17	21	16	17	26	12	12
2			1	.	1	3	4	2	3	8	8	1	5	3	12	13	12	18	18	16	12	17	26	17	21	10
3				1	.	.	.	.	4	.	.	5	.	.	5	.	6	.	6	5	15	.	5	10	.	4
4					1	.	.	.	.	4	.	.	5	.	.	5	.	6	6	.	.	5	10	5	15	.
5						1	1	1	2	2	4	1	1	2	6	6	8	8	9	10	9	10	11	11	9	7
6							1	.	.	1	2	.	1	1	3	2	3	6	3	6	3	6	8	3	6	4
7								1	1	.	2	1	.	1	2	3	6	3	3	6	6	6	3	8	3	6
8									1	.	.	1	.	.	2	.	3	.	3	3	6	.	3	4	.	3
9										1	.	.	1	.	.	2	.	3	3	.	.	3	4	3	6	.
10											1	.	.	1	1	1	2	2	1	4	2	4	2	2	2	3
11												1	.	.	.	.	.	.	.	.	3	.	.	.	.	.
12													1	.	.	.	.	.	.	.	.	.	.	.	3	.
13														1	.	.	.	.	.	2	.	2	.	.	.	1
14															1	.	1	.	1	2	2	.	2	1	.	2
15																1	.	1	1	.	.	2	1	2	2	.
16																	1	.	.	1	1	.	.	1	.	2
17																		1	.	.	.	1	1	.	1	.

NUMBER	26	27	28	29	30	31	32	33	34	35	36	37	38	39	40	41	42	43	44	45	46	47	48	49	50	51
S.P.	117	153	117	117	117	117	153	129	105	165	129	165	165	165	129	129	129	129	165	129	165	189	225	141	153	141
O.S.	60	60	60	120	60	120	60	120	720	30	120	30	30	120	180	180	180	360	30	360	120	30	60	180	60	180
LEVEL	34	34	34	34	34	34	34	36	36	36	36	36	38	38	38	38	38	38	38	38	38	40	40	40	40	40
DIM'N	6 000	3 080	4 704	8 624	6 000	8 624	3 080	11 340	32 768	4 536	11 340	4 536	5 880	11 550	21 000	19 600	21 000	32 340	5 880	32 340	11 550	9 240	6 160	31 752	19 250	31 752
	3	5	0	3	0	2	0	2	1	5	0	2	1	4	1	1	0	3	0	1	0	3	6	1	3	0
	0	0	0	1	0	0	1	2	1	0	0	0	4	1	1	0	3	0	0	1	0	3	0	0	0	2
	2	0	1	0	2	0	0	1	1	0	1	0	0	1	0	3	0	1	0	1	1	0	1	2	0	2
	0	1	3	0	0	1	0	0	1	0	2	0	0	0	3	0	1	1	4	0	1	0	0	2	3	0
WEIGHT	0	0	0	2	3	3	5	0	1	2	2	5	0	0	0	1	1	1	1	3	4	0	0	0	0	1
0	23	10	17	40	23	40	10	36	132	15	36	15	13	28	67	59	67	112	13	112	28	17	10	84	54	84
1	16	10	10	30	14	16	4	27	80	15	21	5	10	24	38	38	48	72	9	68	17	14	10	56	31	57
2	14	4	12	16	16	30	10	21	80	5	27	15	9	17	48	38	38	68	10	72	24	12	6	57	40	56
3	10	10	.	20	.	4	.	15	24	15	.	1	5	20	10	10	20	40	.	20	.	10	10	16	15	21
4	.	.	4	4	10	20	10	.	24	1	15	15	.	.	20	10	10	20	5	40	20	.	.	21	20	16
5	9	4	7	12	9	12	4	16	48	5	16	5	7	14	27	26	27	42	7	42	14	10	6	38	22	38
6	6	1	6	3	5	8	4	9	28	1	12	5	5	8	21	17	14	24	6	22	11	7	3	27	16	25
7	5	4	4	8	6	3	1	12	28	5	9	1	6	11	14	17	21	22	5	24	8	8	6	25	10	27
8	5	4	.	8	.	3	.	9	14	5	.	1	4	11	7	7	12	22	.	12	.	7	6	10	10	15
9	.	.	3	3	5	8	4	.	14	1	9	5	.	.	12	7	7	12	4	22	11	.	.	15	10	10
10	3	1	3	2	3	2	1	7	16	1	7	1	4	6	10	13	10	12	4	12	6	6	3	18	7	18
11	2	4	.	4	.	.	.	3	.	5	.	.	1	8	.	.	.	8	.	.	.	4	6	.	3	.
12	.	.	.	.	2	4	4	.	.	.	3	5	.	.	.	.	.	.	1	8	8	.	.	.	.	.
13	1	.	1	.	1	.	.	3	4	.	3	.	2	2	2	9	2	2	2	2	2	4	1	9	1	9
14	3	1	.	2	.	2	.	5	8	1	.	1	3	6	5	5	6	12	.	6	.	5	3	7	7	11
15	.	.	2	2	3	2	1	.	8	1	5	1	.	.	6	5	5	6	3	12	6	.	.	11	4	7
16	1	1	.	1	.	.	.	4	4	1	.	.	3	4	2	3	6	4	.	3	.	4	3	4	2	8
17	.	.	2	.	1	1	1	.	4	.	4	1	.	.	6	3	2	3	3	4	4	.	.	8	4	4

1	.	.	.	1	1	.	.	.	.	.	2	.	2	.	.	4	1	.	1	.	.	3	2	3	6	.	6	.	.	.	4	4	4	18
	1	.	.	.	.	.	1	1	.	.	.	.	.	.	2	2	.	.	.	2	2	1	3	2	2	.	1	.	3	1	3	1	6	19
		1	.	.	.	.	.	1	1	.	1	.	.	.	2	.	1	.	.	1	4	.	.	.	4	.	.	.	3	3	.	2	.	20
			1	.	.	.	.	.	.	1	.	1	.	.	.	2	.	2	.	.	.	2	3	1	1	2	2	2	.	.	6	1	3	21
				1	.	.	.	.	.	.	.	.	1	.	.	2	.	.	1	.	.	3	1	1	3	.	2	.	.	.	3	4	3	22
					1	.	.	.	.	.	1	.	.	.	.	2	1	.	.	.	.	1	1	3	2	.	3	.	.	.	3	1	3	23
						1	.	.	.	.	.	1	1	1	.	.	.	2	1	.	.	.	.	.	.	1	4	4	.	.	.	.	.	24
							1	.	.	.	.	.	.	.	1	.	.	.	.	2	1	.	.	2	.	.	.	.	2	1	.	.	3	25
								1	.	.	.	.	.	.	1	.	.	.	.	1	2	.	.	.	2	.	.	.	2	1	.	1	.	26
									1	.	.	.	.	.	.	.	1	.	.	.	2	.	.	.	.	.	.	.	1	3	.	.	.	27
										1	.	.	.	.	.	.	.	1	.	.	.	2	.	.	.	2	.	1	.	.	3	1	.	28
											1	.	.	.	.	.	1	.	.	.	.	.	.	.	2	.	.	.	.	.	.	1	.	29
												1	.	.	.	.	.	1	.	.	.	.	.	.	.	1	2	2	.	.	.	.	.	30
													1	.	.	.	.	.	1	.	.	.	.	.	.	.	2	.	.	.	.	.	.	31
														1	.	.	.	.	1	.	.	.	.	.	.	.	.	2	.	.	.	.	.	32
															1	.	.	.	.	1	1	.	.	.	.	.	.	.	2	1	.	.	.	33
																1	.	.	.	.	.	1	1	1	1	.	1	.	.	.	2	1	2	34
																	1	.	.	.	.	.	.	.	.	.	.	.	.	.	.	.	.	35
																		1	.	.	.	.	.	.	.	1	.	1	.	.	.	.	.	36
																			1	.	.	.	.	.	.	.	.	.	.	.	.	.	.	37
																				1	.	.	.	.	.	.	.	.	1	.	.	.	.	38
																					1	.	.	.	.	.	.	.	1	1	.	.	.	39
																						1	.	.	.	.	.	.	.	.	1	1	.	40
																							1	.	.	.	.	.	.	.	1	.	1	41
																								1	.	.	.	.	.	.	.	.	1	42
																									1	.	.	.	.	.	.	1	.	43
																										1	.	.	.	.	.	.	.	44
																											1	.	.	.	.	.	.	45
																												1	.	.	.	.	.	46
																													1	.	.	.	.	47
																														1	.	.	.	48
																															1	.	.	49
																																1	.	50
																																	1	51

Table 3. *(continued)*

ALGEBRA A6, REPRESENTATIONS OF CLASS 0

0	14	28	42	42	42	56	56	56	70	84	70	84	84	84	70	98	98	98	98	126	140	84	126	126	140	154	98	98	154	210	112	112	112	112	112	210	294	126	126	154	126	126	154	294	126	168	140	140	140	168	154	S.P.
1	42	210	105	105	140	42	420	420	105	140	105	140	105	105	840	210	210	210	210	42	105	1260	42	42	105	42	630	630	42	42	420	210	420	1260	1260	42	7	420	420	420	420	420	420	7	1260	105	840	840	840	105	210	O.S.
1	13	21	23	23	25	25	29	29	31	31	31	31	33	33	33	35	35	35	35	37	37	37	37	37	37	39	39	39	39	41	41	41	41	41	41	41	43	43	43	43	43	43	43	43	45	45	45	45	45	45	47	LEVEL
1	48	392	540	540	784	735	2940	2940	2646	2400	2646	2400	3528	3528	10240	7350	8085	7350	8085	4704	4950	24500	4704	6272	4950	6468	26460	26460	6468	4752	30870	27000	30870	53460	53460	4752	1716	53900	51744	33075	51744	53900	33075	1716	105840	32340	87318	87318	107163	32340	52920	DIM'N
0	1	0	2	0	0	2	1	0	1	3	0	0	0	0	1	2	3	0	1	1	4	1	0	3	0	3	1	0	0	5	1	0	0	2	1	0	7	2	2	4	1	0	1	0	0	4	1	1	2	0	0	
0	0	1	0	1	0	0	1	0	0	0	0	0	2	0	1	1	0	0	1	3	0	0	0	0	0	2	0	2	0	1	0	2	1	1	0	0	0	1	0	0	0	2	0	0	1	0	2	0	1	2	1	
0	0	0	0	0	1	0	0	1	2	0	0	1	1	0	0	1	0	0	0	0	1	1	0	0	0	0	1	0	0	0	0	0	2	0	1	0	0	0	2	0	0	0	1	0	1	0	1	0	0	0	0	
0	0	0	0	0	1	0	1	0	0	1	2	0	0	1	0	0	0	1	0	0	0	1	0	0	1	0	0	1	0	0	2	0	0	1	0	0	0	0	0	1	2	0	0	0	1	0	0	1	0	0	3	
0	0	1	1	0	0	0	0	1	0	0	0	0	0	2	1	0	1	1	0	0	0	0	3	0	0	0	2	0	2	0	1	2	0	0	1	1	0	2	0	0	0	1	0	0	1	2	0	2	1	0	0	
0	1	0	0	2	0	2	0	1	0	0	1	3	0	0	1	0	1	2	3	0	0	1	1	3	4	0	0	1	3	0	0	0	1	1	2	5	0	0	1	1	2	2	4	7	0	0	1	1	2	4	0	WEIGHT
0	1	2	3	4	5	6	7	8	9	10	11	12	13	14	15	16	17	18	19	20	21	22	23	24	25	26	27	28	29	30	31	32	33	34	35	36	37	38	39	40	41	42	43	44	45	46	47	48	49	50	51	NUMBER
1	6	14	15	15	14	21	35	35	21	20	21	20	21	21	104	35	70	35	70	14	15	140	14	56	15	14	126	126	14	6	105	120	105	225	225	6	1	210	154	105	154	210	105	1	294	105	210	210	399	105	105	0
	1	4	5	5	5	6	15	15	10	10	10	10	11	11	44	20	30	20	30	9	10	70	9	21	10	10	66	66	10	5	60	60	60	120	120	5	1	110	90	60	90	110	60	1	169	55	130	130	204	55	65	1
		1	1	1	2	1	6	6	5	4	5	4	6	6	16	11	10	11	10	6	6	35	6	6	6	7	33	33	7	4	35	29	35	60	60	4	1	52	53	30	53	52	30	1	98	25	80	80	94	25	41	2
			1	.	.	1	3	.	2	4	.	.	3	.	8	8	10	.	5	4	6	13	.	6	.	6	12	18	.	4	11	16	17	42	21	.	1	36	32	30	16	30	10	.	48	25	54	23	64	15	19	3
				1	.	1	.	3	.	.	2	4	.	3	8	.	5	8	10	.	.	13	4	6	6	.	18	12	6	.	17	16	11	21	42	4	.	30	16	10	32	36	30	1	48	15	23	54	64	25	22	4
					1	.	2	2	3	1	3	1	3	3	4	6	2	6	2	4	3	18	4	1	3	5	15	15	5	3	21	12	21	27	27	3	1	21	33	12	33	21	12	1	57	9	48	48	36	9	27	5
						1	.	.	.	.	.	.	.	.	4	.	5	.	5	.	.	5	.	6	.	.	6	6	.	.	5	10	5	15	15	.	.	20	10	10	10	20	10	.	24	15	15	15	44	15	10	6
							1	.	1	1	.	.	2	.	2	4	2	.	1	3	3	6	.	1	.	4	5	9	.	3	6	7	11	18	9	.	1	14	18	12	9	12	4	.	28	9	34	13	24	5	12	7
								1	.	.	1	1	.	2	2	.	1	4	2	.	.	6	3	1	3	.	9	5	4	.	11	7	6	9	18	3	.	12	9	4	18	14	12	1	28	5	13	34	24	9	15	8
									1	.	.	.	1	.	.	2	.	.	.	2	1	3	.	.	.	3	2	3	.	2	3	2	8	6	3	.	1	4	12	3	6	3	1	.	16	2	20	7	6	1	7	9
										1	.	.	.	.	.	2	2	.	.	1	3	.	.	1	.	3	.	.	.	3	.	.	.	9	.	.	1	7	7	12	.	.	.	.	.	9	12	.	12	.	.	10
											1	.	.	1	.	.	.	2	.	.	.	3	2	.	1	.	3	2	3	.	8	2	3	3	6	2	.	3	6	1	12	4	3	1	16	1	7	20	6	2	12	11
												1	.	.	.	.	.	2	2	.	.	.	1	1	3	.	.	.	3	.	.	.	.	.	9	3	.	.	.	.	7	7	12	1	.	.	.	12	12	9	.	12
													1	.	.	1	.	.	.	2	1	.	.	.	.	2	.	3	.	2	.	2	4	3	.	.	1	2	4	3	.	3	.	.	8	2	16	.	3	1	3	13
														1	.	.	.	1	.	.	.	.	2	.	1	.	3	.	2	.	4	2	.	.	3	2	.	3	.	.	4	2	3	1	8	1	.	16	3	2	6	14
															1	.	1	.	1	.	.	2	.	1	.	.	3	3	.	.	3	4	3	6	6	.	.	8	5	4	5	8	4	.	14	5	9	9	16	5	7	15
																1	.	.	.	1	1	.	.	.	.	2	.	.	.	2	.	.	.	3	.	.	1	2	4	3	.	.	.	.	.	2	8	.	3	.	.	16
																	1	.	.	.	.	.	.	1	.	.	.	.	.	.	.	.	.	3	.	.	.	4	2	4	.	.	.	.	.	5	3	.	8	.	.	17

1	.	.	.	.	1	.	1	.	.	.	2	.	.	.	.	.	3	2	.	.	.	.	4	2	3	1	.	.	.	8	3	2	.	18
	1	.	.	.	.	1	.	.	.	.	.	.	.	.	.	.	3	.	.	.	.	.	2	4	4	.	.	.	.	3	8	5	.	19
		1	.	.	.	.	.	1	.	.	.	1	.	.	.	.	.	.	1	.	.	.	.	.	.	.	.	.	4	.	.	.	.	20
			1	.	.	.	.	1	.	.	.	2	.	.	.	.	.	.	1	.	.	3	.	.	.	.	.	2	.	.	.	.	.	21
				1	.	.	.	.	1	1	.	.	2	1	2	2	2	.	.	2	3	1	3	2	1	.	8	1	5	5	4	1	5	22
					1	.	.	.	.	.	1	.	.	.	.	.	.	1	.	.	.	.	.	.	.	1	.	.	.	4	.	.	.	23
						1	.	.	.	.	.	.	.	.	.	.	.	.	.	.	.	.	.	.	.	.	.	.	.	.	4	.	.	24
							1	.	.	.	1	.	.	.	.	.	.	2	.	.	.	.	.	.	3	1	.	.	.	.	.	2	.	25
								1	.	.	.	1	.	.	.	.	.	.	1	.	.	.	.	.	.	.	.	.	.	.	.	.	.	26
									1	.	.	.	1	1	.	.	1	.	.	2	.	.	1	1	1	.	4	1	.	4	2	1	3	27
										1	.	.	.	1	1	1	.	.	.	1	1	1	.	2	.	.	4	1	4	.	2	1	2	28
											1	.	.	.	.	.	.	1	.	.	.	.	.	.	.	1	.	.	.	.	.	.	.	29
												1	.	.	.	.	.	.	1	.	.	.	.	.	.	.	.	.	.	.	.	.	.	30
													1	.	.	.	.	.	.	.	.	.	1	.	.	.	2	.	.	2	.	.	3	31
														1	.	.	.	.	.	1	.	.	.	1	.	.	2	1	.	.	1	1	1	32
															1	.	.	.	.	.	1	.	.	.	.	.	2	.	2	.	.	.	1	33
																1	.	.	.	1	1	1	.	.	.	.	.	1	2	.	2	.	.	34
																	1	.	.	.	.	.	1	1	1	.	.	.	.	2	2	1	.	35
																		1	.	.	.	.	.	.	.	1	.	.	.	.	.	.	.	36
																			1	.	.	.	.	.	.	.	.	.	.	.	.	.	.	37
																				1	.	.	.	.	.	.	.	1	.	.	1	.	.	38
																					1	.	.	.	.	.	.	.	1	.	.	.	.	39
																						1	.	.	.	.	.	1	.	.	.	.	.	40
																							1	.	.	.	.	.	.	1	.	.	.	41
																								1	.	.	.	.	.	.	1	1	.	42
																									1	.	.	.	.	.	.	1	.	43
																										1	.	.	.	.	.	.	.	44
																											1	.	.	.	.	.	1	45
																												1	.	.	.	.	.	46
																													1	.	.	.	.	47
																														1	.	.	.	48
																															1	.	.	49
																																1	.	50
																																	1	51

Table 3. *(continued)*

ALGEBRA A6, REPRESENTATIONS OF CLASS 1

6	20	34	34	48	48	48	76	62	62	62	90	90	90	76	104	90	90	76	146	104	132	90	90	104	216	S.P.
7	105	42	210	35	140	420	105	210	210	420	21	42	105	210	42	105	420	1260	42	105	140	420	420	420	7	O.S.
7	17	19	23	25	25	27	29	29	29	31	31	31	31	33	33	35	35	35	35	37	37	37	37	37	37	LEVEL
7	140	189	588	490	840	2016	1575	2352	2800	4410	1176	2156	2100	4704	2646	5292	9450	16800	2310	8820	5775	15680	16800	13860	924	DIM'N
1	0	2	0	0	0	1	3	0	1	1	0	3	0	0	0	0	2	1	0	2	4	1	0	1	0	
0	1	0	0	0	0	1	0	0	1	0	0	0	0	2	0	1	1	0	0	0	0	0	2	0	0	
0	0	0	1	0	1	0	0	0	0	1	0	0	0	0	0	2	0	1	0	2	0	0	0	1	0	
0	0	0	0	2	0	0	0	1	0	1	0	0	1	1	0	0	1	0	0	0	1	2	0	0	0	
0	0	0	1	0	0	1	1	1	0	0	3	0	0	0	2	0	0	1	1	0	0	0	1	0	0	
0	1	1	0	0	2	0	0	1	2	0	0	2	3	0	2	0	0	1	4	0	0	1	1	3	6	WEIGHT
0	1	2	3	4	5	6	7	8	9	10	11	12	13	14	15	16	17	18	19	20	21	22	23	24	25	NUMBER
1	5	6	9	5	10	24	15	16	30	30	5	21	10	26	9	21	45	90	5	30	20	60	80	60	1	0
	1	1	3	2	4	8	5	8	10	13	3	6	6	12	6	11	21	42	4	16	10	32	36	30	1	1
		1	.	.	.	4	5	.	5	5	.	6	.	6	.	5	15	15	.	10	10	10	20	10	.	2
			1	1	1	2	1	4	2	6	2	1	3	5	4	6	9	18	3	9	4	18	14	12	1	3
				1	.	.	.	2	.	3	1	.	1	2	3	3	3	6	2	6	1	12	4	3	1	4
					1	.	.	2	2	.	1	1	3	.	3	.	.	9	3	.	.	7	7	12	1	5
						1	1	.	1	2	.	1	.	3	.	3	6	6	.	5	4	5	8	4	.	6
							1	.	.	.	.	1	.	.	.	.	3	.	.	2	4	.	.	.	.	7
								1	.	.	1	.	1	.	2	.	.	3	2	.	.	4	2	3	1	8
									1	.	.	1	.	.	.	.	.	3	.	.	.	2	4	4	.	9
										1	.	.	.	1	.	2	2	2	.	3	1	3	2	1	.	10
											1	.	.	.	1	.	.	.	1	.	.	.	.	.	1	11
												1	.	.	.	.	.	.	.	.	.	.	.	.	.	12
													1	.	1	.	.	.	2	.	.	.	.	3	1	13
														1	.	1	1	.	.	1	1	.	2	.	.	14
															1	.	.	.	1	.	.	.	.	.	1	15
																1	.	.	.	1	.	.	.	.	.	16

118	104	104	118	160	146	104	146	118	132	174	202	118	132	118	146	174	160	216	132	132	132	132	174	286	146	S.P.
210	420	840	105	21	210	1260	210	840	210	42	105	630	140	1260	210	42	420	42	630	630	1260	1260	210	42	420	O.S.
39	39	39	39	41	41	41	41	41	41	43	43	43	43	43	43	43	43	45	45	45	45	45	45	47	47	LEVEL
14700	21168	33264	13475	5292	18480	51450	19600	44352	26460	12936	10395	56700	23520	83160	29106	15288	30240	13860	71442	90552	112266	115500	40320	9009	88200	DIM'N
1	1	2	0	0	3	0	4	1	2	2	5	0	0	2	1	4	1	4	0	2	1	0	1	6	0	
2	0	1	2	4	1	1	0	0	1	3	0	1	0	0	0	0	0	2	1	0	2	1	0	1	0	
1	0	0	0	0	1	1	0	0	0	0	1	1	3	1	0	0	0	0	0	1	0	1	0	0	2	
0	1	0	0	0	0	1	0	1	0	0	0	0	0	1	0	0	1	0	2	0	1	0	0	0	1	
0	2	1	0	0	0	0	1	1	0	0	0	2	0	0	3	0	0	0	1	2	0	1	2	0	1	
0	0	1	3	0	0	1	1	2	3	0	0	0	1	1	1	3	4	0	0	0	1	2	3	0	0	WEIGHT
26	27	28	29	30	31	32	33	34	35	36	37	38	39	40	41	42	43	44	45	46	47	48	49	50	51	NUMBER
40	66	144	55	9	40	155	70	120	105	19	15	150	50	225	54	56	60	15	144	216	261	270	60	6	155	0
23	38	64	25	6	24	84	30	72	45	13	10	82	30	122	37	21	40	11	87	120	144	153	44	5	93	1
15	11	44	15	4	20	40	30	20	30	10	10	40	15	70	9	21	10	10	39	66	96	75	10	5	50	2
13	22	24	9	4	14	45	10	42	15	9	6	44	18	66	26	6	24	8	54	63	75	81	32	4	57	3
7	12	6	2	3	8	24	2	24	3	6	3	21	10	36	18	1	12	6	36	30	36	39	24	3	35	4
.	12	12	9	.	.	17	5	30	15	.	.	24	6	24	18	6	24	.	27	36	27	57	27	.	28	5
9	6	16	5	3	11	22	10	11	10	7	6	21	10	35	6	6	6	7	24	33	51	39	7	4	31	6
3	.	8	.	1	8	.	10	.	5	4	6	.	.	13	.	6	.	6	.	12	18	.	.	4	.	7
.	8	3	2	.	.	8	1	16	3	.	.	13	3	12	14	1	12	.	18	18	12	26	19	.	17	8
.	3	8	5	.	.	8	5	8	10	.	.	12	3	13	4	6	6	.	12	18	18	27	6	.	16	9
5	3	4	1	2	6	12	2	6	2	5	3	10	7	18	4	1	3	5	15	15	24	18	5	3	20	10
.	4	.	.	.	.	.	.	4	.	.	.	5	.	.	10	.	4	.	6	6	.	5	10	.	5	11
.	.	4	.	.	.	.	5	.	5	.	.	.	.	5	.	6	.	.	.	6	6	.	.	.	.	12
.	.	.	2	.	.	.	.	8	3	.	.	.	.	.	5	1	12	.	.	.	.	13	14	.	.	13
4	.	2	1	2	4	6	2	.	1	4	3	5	4	6	.	1	.	4	6	5	18	9	.	3	11	14
.	.	.	.	.	.	.	.	4	.	.	.	.	.	.	5	.	4	.	.	.	.	5	10	.	.	15
2	.	.	.	1	2	3	.	.	.	3	1	2	4	3	.	.	.	3	3	2	6	3	.	2	8	16

1	.	.	1	1	.	.	.	.	2	.	2	.	1	4	.	2	.	1	3	3	.	.	6	.	1	.	4	.	5	9	.	.	3	.	17
	1	.	.	.	1	1	1	.	.	2	2	1	.	.	4	1	4	2	.	.	6	2	6	3	1	3	.	8	9	8	12	4	.	10	18
		1	.	.	.	.	.	1	.	.	.	.	.	.	.	.	.	.	.	.	.	.	.	.	.	4	.	.	.	.	.	5	.	.	19
			1	.	.	.	.	.	1	.	.	.	1	2	.	.	.	.	2	1	.	.	3	.	.	.	3	.	2	3	.	.	2	.	20
				1	.	.	.	.	.	.	.	.	.	2	.	2	.	.	1	3	.	.	.	.	1	.	3	.	.	.	.	.	3	.	21
					1	.	.	.	.	1	.	.	.	.	2	.	2	.	.	.	2	1	3	2	.	1	.	6	3	3	4	3	.	6	22
						1	.	.	.	.	1	1	.	.	2	1	.	1	.	.	3	1	2	.	1	.	.	3	3	6	6	.	.	6	23
							1	.	.	.	.	1	.	.	.	.	2	2	.	.	.	.	.	1	1	3	.	.	.	.	6	3	.	.	24
								1	.	.	.	.	.	.	.	.	.	.	.	.	.	.	.	.	.	.	.	.	.	.	.	.	.	.	25
									1	.	.	.	1	1	.	.	.	.	2	1	.	.	.	.	.	.	2	.	.	3	.	.	2	.	26
										1	.	.	.	.	.	.	1	.	.	.	2	.	.	2	.	1	.	3	3	.	2	2	.	3	27
											1	.	.	.	.	1	.	1	.	.	.	.	2	.	1	.	.	.	3	3	.	.	.	.	28
												1	.	.	.	.	.	1	.	.	.	.	.	.	1	.	.	.	.	.	3	.	.	.	29
													1	.	.	.	.	.	1	.	.	.	.	.	.	.	1	.	.	.	.	.	1	.	30
														1	.	.	.	.	1	1	.	.	.	.	.	.	2	.	.	.	.	.	2	.	31
															1	.	.	.	.	.	1	1	1	.	.	.	.	2	1	2	2	.	.	4	32
																1	.	.	.	.	.	.	.	.	1	.	.	.	.	.	.	.	.	.	33
																	1	.	.	.	.	.	.	1	.	1	.	.	.	.	2	2	.	.	34
																		1	.	.	.	.	.	.	1	.	.	.	.	.	.	.	.	.	35
																			1	.	.	.	.	.	.	.	1	.	.	.	.	.	1	.	36
																				1	.	.	.	.	.	.	1	.	.	.	.	.	2	.	37
																					1	.	.	.	.	.	.	1	1	.	1	.	.	2	38
																						1	.	.	.	.	.	.	.	.	.	.	.	2	39
																							1	.	.	.	.	.	1	1	.	.	.	.	40
																								1	.	.	.	.	.	.	.	1	.	.	41
																									1	.	.	.	.	.	.	.	.	.	42
																										1	.	.	.	.	.	1	.	.	43
																											1	.	.	.	.	.	1	.	44
																												1	.	.	.	.	.	1	45
																													1	.	.	.	.	.	46
																														1	.	.	.	.	47
																															1	.	.	.	48
																																1	.	.	49
																																	1	.	50
																																		1	51

Table 3. *(continued)*

ALGEBRA A6, REPRESENTATIONS OF CLASS 2

0	1	2	3	4	5	6	7	8	9	10	11	12	13	14	15	16	17	18	19	20	21	22	23	24	25	NUMBER
10	24	24	38	38	66	52	66	52	94	66	66	66	150	80	80	80	122	80	94	108	94	108	136	108	108	S.P.
21	7	140	105	210	42	105	42	630	42	140	210	420	7	420	105	420	105	840	210	35	420	105	42	420	420	O.S.
11	13	19	23	23	25	25	27	29	29	31	31	31	31	33	33	35	35	35	35	37	37	37	37	37	39	LEVEL
21	28	224	490	735	560	756	882	3 402	1 008	2 940	3 360	4 900	462	6 615	4 725	9 408	3 850	14 336	9 240	4 116	15 120	7 350	5 292	12 936	20 160	DIM'N
0 1 0 0 0 0	2 0 0 0 0 0	0 0 1 0 0 1	0 0 0 1 1 0	1 1 0 0 0 1	3 0 0 0 0 1	0 0 0 1 0 2	0 0 0 0 2 1	1 0 1 0 1 0	0 0 0 0 1 3	1 0 0 2 0 0	0 2 0 0 1 0	1 0 1 0 0 2	0 0 0 0 0 5	2 1 0 0 1 0	0 2 0 0 0 2	0 1 1 1 0 0	4 0 0 0 1 0	1 0 0 1 1 1	2 1 0 0 0 2	0 0 3 0 0 0	2 0 1 1 0 0	1 0 0 0 3 0	4 0 0 0 0 2	1 0 0 1 0 3	1 2 0 1 0 0	WEIGHT
1	1	4	5	10	6	6	5	24	4	15	20	30	1	34	25	36	15	56	45	11	50	20	21	40	56	0
	1	.	.	5	6	.	.	9	.	5	10	10	.	24	15	16	15	16	30	5	30	5	21	10	36	1
		1	2	2	1	3	3	9	3	7	7	12	1	12	9	17	5	30	15	6	24	13	6	24	27	2
			1	.	.	1	2	3	2	4	2	3	1	3	2	8	1	16	3	3	12	9	1	12	12	3
				1	1	.	.	3	.	2	4	4	.	8	5	8	5	8	10	3	13	3	6	6	18	4
					1	.	.	.	.	.	.	.	.	4	.	.	5	.	5	.	5	.	6	.	6	5
						1	1	.	2	.	.	3	1	.	2	.	.	8	3	.	.	5	1	12	.	6
							1	.	1	.	.	.	1	.	.	.	.	4	.	.	.	5	.	4	.	7
								1	.	1	1	1	.	2	1	4	1	4	2	2	6	2	1	3	8	8
									1	.	.	.	1	.	.	.	.	.	.	.	.	.	.	4	.	9
										1	.	.	.	.	.	2	.	2	.	1	3	1	.	1	3	10
											1	.	.	1	1	2	1	.	1	1	2	.	1	.	6	11
												1	.	.	1	.	.	2	2	.	.	1	1	3	.	12
													1	.	.	.	.	.	.	.	.	.	.	.	.	13
														1	.	.	1	.	1	.	2	.	1	.	3	14
															1	.	.	.	1	.	.	.	1	.	.	15
																1	.	.	.	1	1	.	.	.	2	16

26	27	28	29	30	31	32	33	34	35	36	37	38	39	40	41	42	43	44	45	46	47	48	49	50	51	NUMBER
94	122	122	136	108	108	122	164	150	150	192	122	122	122	122	136	234	164	136	136	150	136	150	206	206	150	S.P.
1 260	210	210	420	630	1 260	420	210	105	105	140	630	420	420	840	420	42	210	210	420	840	1 260	210	42	210	420	O.S.
39	39	41	41	41	41	41	41	43	43	43	43	43	43	43	43	43	45	45	45	45	45	45	47	47	47	LEVEL
36 750	16 632	23 520	24 255	37 044	58 212	30 800	14 700	17 640	23 100	12 320	51 030	44 100	55 440	72 765	48 384	6 006	41 580	50 400	75 460	86 016	107 800	58 800	19 404	40 425	79 380	DIM'N
0 1 1 0 1 1	1 0 0 0 2 2	1 1 2 0 0 0	3 1 0 1 0 0	0 1 0 2 0 1	2 0 1 0 1 1	0 1 1 0 0 3	1 0 0 0 1 4	0 3 1 0 0 0	3 0 2 0 0 0	5 0 0 1 0 0	0 1 0 1 2 0	0 0 2 1 0 1	2 0 0 2 0 1	1 2 0 0 1 1	2 0 1 0 0 3	1 0 0 0 0 6	2 2 1 0 0 0	0 0 2 0 2 0	2 0 0 1 2 0	3 1 0 0 1 1	0 1 0 1 1 2	1 2 0 0 0 3	1 4 0 0 0 0	4 1 1 0 0 0	0 0 1 2 1 0	WEIGHT
115	39	50	55	89	165	80	25	29	40	20	105	90	120	180	120	6	60	90	141	184	195	130	19	45	115	0
55	9	30	45	39	90	40	5	19	30	20	45	45	60	120	60	1	45	50	66	144	85	90	14	40	60	1
57	27	28	27	51	81	42	20	17	23	10	63	51	69	84	60	6	36	52	86	84	123	60	13	28	72	2
26	19	16	12	30	36	18	15	10	14	4	39	28	42	34	24	6	21	30	52	32	75	22	9	17	47	3
27	6	17	21	22	42	20	4	12	16	10	27	27	32	56	30	1	27	28	38	64	52	40	10	24	38	4
.	.	5	15	.	15	.	.	4	10	10	.	.	10	20	10	.	15	.	11	44	.	15	5	20	.	5
13	14	.	.	12	18	18	15	.	.	.	21	11	16	17	24	6	.	16	30	16	54	22	.	.	23	6
5	10	.	.	6	6	5	10	.	.	.	15	5	10	5	6	6	.	10	20	4	30	5	.	.	15	7
12	4	10	9	13	18	8	3	7	9	4	16	16	18	22	12	1	16	16	22	24	31	14	7	14	25	8
.	5	.	.	.	.	5	10	.	.	.	.	.	.	.	6	6	.	.	.	.	15	5	.	.	.	9
4	3	6	3	9	6	2	2	4	6	1	9	9	12	6	3	1	9	8	12	6	18	3	5	8	18	10
6	.	6	6	5	6	4	.	6	5	4	6	9	5	16	4	.	12	9	6	16	11	10	6	11	13	11
6	3	.	.	5	9	8	3	.	.	.	9	6	7	11	12	1	.	9	12	12	22	14	.	.	13	12
.	.	.	.	.	.	.	5	.	.	.	.	.	.	.	.	6	.	.	.	.	.	.	.	.	.	13
.	.	3	6	.	6	.	.	3	5	4	.	.	5	8	4	.	9	.	6	16	.	5	4	11	.	14
.	.	.	.	2	3	4	.	.	.	.	3	3	2	8	4	.	.	6	3	8	8	10	.	.	7	15
2	.	4	2	3	2	1	.	3	3	1	3	6	3	4	1	.	7	5	3	4	6	2	4	6	9	16

1	.	.	.	.	.	1	.	.	.	.	.	3	.	.	.	.	.	2	4	.	.	.	.	.	.	3	.	.	8	.	.	1	8	.	17
	1	.	.	.	1	.	1	.	2	2	.	.	3	3	2	2	.	.	.	6	3	4	3	3	1	.	5	8	3	12	3	.	.	9	18
		1	.	.	.	1	.	.	.	.	.	.	.	3	.	.	.	.	.	.	.	2	4	4	.	.	.	3	8	.	5	.	.	.	19
			1	.	.	.	.	.	.	.	2	.	.	.	.	.	1	1	.	.	3	.	.	.	.	3	2	.	.	.	.	2	2	3	20
				1	.	.	.	1	.	.	2	2	.	2	.	.	2	3	1	.	.	3	2	1	.	5	.	3	4	.	1	3	6	.	21
					1	.	.	.	.	1	.	.	.	.	.	1	.	.	.	3	.	.	.	.	1	.	2	4	.	3	.	.	.	3	22
						1	.	.	.	.	.	.	.	.	.	.	.	.	.	.	.	.	.	.	.	.	.	.	4	.	.	.	.	.	23
							1	.	.	1	.	.	.	.	2	2	.	.	.	.	.	.	.	3	1	.	.	.	.	6	3	.	.	.	24
								1	.	.	1	1	.	.	.	.	2	1	1	.	.	.	2	.	.	4	.	.	2	.	1	3	4	.	25
									1	.	.	.	1	1	1	.	.	.	.	2	2	1	2	1	.	.	3	2	2	4	2	.	.	5	26
										1	.	.	.	.	.	1	.	.	.	.	.	.	.	.	1	.	.	.	.	3	.	.	.	.	27
											1	.	.	.	.	.	1	1	.	.	.	.	.	.	.	2	.	.	.	.	.	2	2	.	28
												1	.	.	.	.	.	1	1	.	.	.	.	.	.	2	.	.	2	.	.	1	4	.	29
													1	.	.	.	.	.	.	1	1	1	.	.	.	.	1	1	.	2	.	.	.	4	30
														1	.	.	.	.	.	.	.	1	1	1	.	.	.	2	2	.	1	.	.	.	31
															1	.	.	.	.	.	.	.	.	1	.	.	.	.	.	2	2	.	.	.	32
																1	.	.	.	.	.	.	.	.	1	.	.	.	.	.	.	.	.	.	33
																	1	.	.	.	.	.	.	.	.	1	.	.	.	.	.	2	1	.	34
																		1	.	.	.	.	.	.	.	1	.	.	.	.	.	1	2	.	35
																			1	.	.	.	.	.	.	.	.	.	.	.	.	.	2	.	36
																				1	.	.	.	.	.	.	1	1	.	1	.	.	.	2	37
																					1	.	.	.	.	.	1	.	.	.	.	.	.	2	38
																						1	.	.	.	.	.	1	.	.	.	.	.	.	39
																							1	.	.	.	.	.	1	.	1	.	.	.	40
																								1	.	.	.	.	.	.	1	.	.	.	41
																									1	.	.	.	.	.	.	.	.	.	42
																										1	.	.	.	.	.	1	1	.	43
																											1	.	.	.	.	.	.	1	44
																												1	.	.	.	.	.	.	45
																													1	.	.	.	.	.	46
																														1	.	.	.	.	47
																															1	.	.	.	48
																																1	.	.	49
																																	1	.	50
																																		1	51

Table 3. *(continued)*

ALGEBRA A6, REPRESENTATIONS OF CLASS 3

NUMBER	0	1	2	3	4	5	6	7	8	9	10	11	12	13	14	15	16	17	18	19	20	21	22	23	24	25
S.P.	12	26	54	26	40	54	40	96	54	54	68	110	68	68	82	82	82	82	110	96	96	96	96	166	124	96
O.S.	35	42	7	105	21	42	420	7	105	420	210	42	420	630	210	210	630	420	210	140	140	420	420	42	420	1260
LEVEL	13	17	19	19	21	23	25	25	27	29	29	31	31	33	33	35	35	35	35	37	37	37	37	37	39	39
DIM'N	35	112	84	210	196	378	1323	210	1260	3024	2450	1386	4704	7560	5600	7056	11907	11025	6468	8232	10500	14700	17248	3024	17248	35280
WEIGHT	0 0 1 0 0 0	1 1 0 0 0 0	3 0 0 0 0 0	0 0 0 1 0 1	0 0 0 0 2 0	0 0 0 0 1 2	1 0 1 0 0 1	0 0 0 0 0 4	0 2 0 0 0 1	1 0 0 1 1 0	2 1 0 0 0 1	4 0 0 0 0 1	1 0 0 1 0 2	0 1 1 0 1 0	1 0 0 0 2 1	0 1 0 2 0 0	2 0 1 0 1 0	0 1 1 0 0 2	1 0 0 0 1 3	0 0 2 1 0 0	2 0 0 2 0 0	1 2 0 0 1 0	2 0 1 0 0 2	1 0 0 0 0 5	3 1 0 0 1 0	0 1 0 1 1 1
0	1	2	1	3	2	3	12	1	9	18	15	6	24	33	22	24	45	42	20	23	31	46	60	6	44	96
1		1	1	.	.	.	4	.	5	5	10	6	6	15	5	10	24	20	4	11	15	30	30	1	34	40
2			1	.	.	.	.	.	.	.	5	6	.	.	.	.	9	.	.	.	5	10	10	.	24	.
3				1	1	2	3	1	2	8	3	1	12	13	14	12	18	18	15	11	16	17	24	6	16	54
4					1	1	.	1	.	4	.	.	4	5	10	6	6	5	10	5	10	5	6	6	4	30
5						1	.	1	.	.	.	.	4	.	5	.	.	5	10	.	.	.	6	6	.	15
6							1	.	1	2	2	1	3	6	3	5	9	8	3	6	7	11	12	1	12	22
7								1	.	.	.	.	.	.	.	.	.	.	5	.	.	.	.	6	.	.
8									1	.	1	1	.	3	.	2	3	4	.	3	2	8	4	.	8	8
9										1	.	.	1	2	2	3	3	2	2	3	4	3	3	1	3	12
10											1	1	.	.	.	.	3	.	.	.	2	4	4	.	8	.
11												1	.	.	.	.	.	.	.	.	.	.	.	.	4	.
12													1	.	1	.	.	2	2	.	.	.	3	1	.	6
13														1	.	1	1	1	.	2	1	2	1	.	2	4
14															1	.	.	.	1	.	.	.	.	1	.	3
15																1	.	.	.	1	1	.	.	.	.	2
16																	1	.	.	.	1	1	1	.	2	.
17																		1	.	.	.	.	1	.	.	2
18																			1	.	.	.	.	1	.	.

NUMBER	26	27	28	29	30	31	32	33	34	35	36	37	38	39	40	41	42	43	44	45	46	47	48	49	50	51
S.P.	110	110	180	124	110	110	138	124	138	138	138	138	124	194	138	138	138	152	152	138	124	152	180	166	194	166
O.S.	210	840	105	210	420	840	210	630	105	210	420	105	420	42	420	140	420	105	420	420	2520	210	420	210	420	105
LEVEL	39	41	41	41	41	41	41	41	43	43	43	43	43	43	43	43	43	45	45	45	45	45	45	47	47	47
DIM'N	20790	43008	8316	18900	33600	51744	24192	32340	20580	25200	40425	26950	42336	11466	43120	28875	47040	31752	58212	60480	169785	61236	38808	69300	53900	37044
WEIGHT	1 2 0 0 0 2	1 1 1 1 0 0	5 0 0 0 1 0	0 1 0 0 3 0	0 0 2 0 1 1	2 0 0 1 1 1	3 1 0 0 0 2	0 1 0 1 0 3	1 0 3 0 0 0	0 3 0 1 0 0	3 0 1 1 0 0	2 0 0 0 3 0	0 0 1 2 0 1	5 0 0 0 0 2	0 1 0 0 2 2	0 0 2 0 0 3	2 0 0 1 0 3	0 2 2 0 0 0	2 2 0 1 0 0	0 0 1 1 2 0	1 1 1 0 1 1	2 0 0 0 2 2	0 1 0 0 1 4	2 1 2 0 0 0	4 1 0 1 0 0	0 0 0 3 1 0
0	60	96	15	39	76	128	60	72	34	44	74	51	78	21	78	54	100	45	90	96	326	105	54	87	65	45
1	40	56	15	15	40	56	45	30	20	30	50	20	39	21	30	30	40	29	66	50	190	39	20	60	55	26
2	15	16	15	.	.	16	30	.	5	10	30	5	.	21	.	.	10	9	36	.	55	9	.	30	45	.
3	22	48	5	26	40	72	20	45	19	22	38	35	46	6	56	31	60	26	45	60	168	75	44	51	33	29
4	5	24	1	20	20	40	4	24	10	10	20	25	26	1	40	15	30	16	21	40	80	55	34	30	15	19
5	5	.	.	10	10	20	4	24	.	.	.	15	11	1	30	15	30	.	.	20	40	40	34	.	.	9
6	14	28	5	10	21	30	15	18	12	15	24	13	24	6	21	16	24	17	33	31	96	27	16	35	27	17
7	.	.	.	.	.	.	.	9	.	.	.	.	.	.	10	5	10	.	.	.	.	15	24	.	.	.
8	10	16	5	3	12	8	10	6	7	12	13	3	12	6	6	10	6	12	24	16	54	6	4	23	21	10
9	3	14	1	7	10	16	3	9	7	7	12	9	15	1	15	7	12	10	15	20	44	19	12	21	12	12
10	5	8	5	.	.	8	10	.	3	6	13	3	.	6	.	.	6	6	18	.	27	6	.	17	21	.
11	.	.	5	.	.	.	5	.	.	.	5	.	.	6	.	.	.	.	6	.	.	.	.	5	15	.
12	3	.	.	4	5	8	3	9	.	.	.	5	6	1	11	7	12	.	.	11	22	14	12	.	.	6
13	2	8	1	2	6	4	2	3	5	5	6	2	8	1	4	4	3	7	11	10	24	4	3	14	9	7
14	.	.	.	4	2	4	.	3	.	.	.	5	3	.	8	2	4	.	.	8	8	10	8	.	.	4
15	.	4	.	1	2	2	.	1	3	2	3	1	6	.	3	1	1	4	4	6	8	3	2	9	3	6
16	1	4	1	.	.	4	2	.	2	3	6	2	.	1	.	.	3	4	8	.	12	4	.	10	9	.
17	2	.	.	1	3	2	2	3	.	.	.	1	3	1	3	4	3	.	.	6	12	3	3	.	.	4
18	.	.	.	.	.	.	.	3	.	.	.	.	.	.	4	2	4	.	.	.	.	5	8	.	.	.

19	1	.	.	.	.	.	.	.	2	.	.	2	.	.	.	3	1	1	.	3	.	.	1	.	3	3	3	4	.	.	6	2	3
20		1	.	.	.	.	.	.	2	.	.	.	2	.	.	1	1	3	1	.	.	.	.	1	3	3	.	4	3	.	6	3	.
21			1	.	.	1	.	1	2	1	.	.	.	1	.	1	3	2	.	.	1	.	.	.	3	6	.	6	.	.	6	6	.
22				1	.	.	.	1	.	.	.	.	2	2	.	.	.	.	1	.	1	.	.	3	.	.	.	6	3	.	.	.	.
23					1	.	.	.	.	.	.	.	.	.	.	.	.	.	.	.	.	.	.	.	.	.	.	.	.	4	.	.	.
24						1	.	.	.	1	.	.	.	1	.	.	.	2	.	.	1	.	.	.	.	3	.	.	.	.	3	6	.
25							1	.	.	.	1	1	1	.	1	.	.	.	1	2	.	2	1	1	.	.	4	4	2	2	.	.	3
26								1	.	.	.	.	.	1	.	.	.	.	.	.	1	.	.	.	.	.	.	3	.	.	.	.	.
27									1	.	.	.	.	.	.	1	1	1	.	.	.	.	.	.	2	2	.	2	.	.	4	2	.
28										1	.	.	.	.	.	.	.	.	.	.	1	.	.	.	.	.	.	.	.	.	.	3	.
29											1	.	.	.	.	.	.	.	1	.	.	1	.	.	.	.	2	.	1	1	.	.	1
30												1	.	.	.	.	.	.	.	1	.	.	1	.	.	.	2	2	.	.	.	.	2
31													1	.	.	.	.	.	1	.	.	.	.	1	.	.	.	2	2	.	.	.	.
32														1	.	.	.	.	.	.	1	.	.	.	.	.	.	.	.	.	.	.	.
33															1	.	.	.	.	.	.	1	1	1	.	.	.	.	1	2	.	.	.
34																1	.	.	.	.	.	.	.	.	1	.	.	.	.	.	2	.	.
35																	1	.	.	.	.	.	.	.	1	1	.	.	.	.	1	1	.
36																		1	.	.	.	.	.	.	.	1	.	.	.	.	2	2	.
37																			1	.	.	.	.	.	.	.	.	.	1	.	.	.	.
38																				1	.	.	.	.	.	.	1	.	.	.	.	.	2
39																					1	.	.	.	.	.	.	.	.	.	.	.	.
40																						1	.	.	.	.	.	.	1	1	.	.	.
41																							1	.	.	.	.	.	.	.	.	.	.
42																								1	.	.	.	.	1	.	.	.	.
43																									1	.	.	.	.	.	1	.	.
44																										1	.	.	.	.	1	1	.
45																											1	.	.	.	.	.	1
46																												1	.	.	.	.	.
47																													1	.	.	.	.
48																														1	.	.	.
49																															1	.	.
50																																1	.
51																																	1

Table 3. *(continued)*

ALGEBRA A7, REPRESENTATIONS OF CLASS 0

0	16	32	48	48	64	48	64	64	64	96	96	80	S.P.
1	56	420	168	168	56	560	70	1120	1120	280	280	840	O.S.
1	15	25	27	27	29	31	33	35	35	37	37	39	LEVEL
1	63	720	945	945	1232	2352	1764	7680	7680	5775	5775	14700	DIM'N
0	1	0	2	0	2	0	0	1	0	3	0	1	
0	0	1	0	1	0	0	0	1	0	0	0	0	
0	0	0	0	0	0	1	0	0	1	0	1	1	
0	0	0	0	0	0	0	2	0	0	0	0	1	
0	0	0	0	0	0	1	0	1	0	1	0	0	
0	0	1	1	0	0	0	0	0	1	0	0	0	
0	1	0	0	2	2	0	0	0	1	0	3	0	WEIGHT
0	1	2	3	4	5	6	7	8	9	10	11	12	NUMBER
1	7	20	21	21	28	28	14	64	64	35	35	70	0
	1	5	6	6	7	9	5	24	24	15	15	30	1
		1	1	1	1	3	2	8	8	5	5	13	2
			1	.	1	.	.	4	.	5	.	5	3
				1	1	.	.	.	4	.	5	.	4
					1	.	.	.	.	.	.	.	5
						1	1	2	2	1	1	6	6
							1	.	.	.	.	3	7
								1	.	1	.	2	8
									1	.	1	.	9
										1	.	.	10
											1	.	11
												1	12

80	80	96	96	112	112	112	112	112	112	144	128	160	S.P.
840	1680	420	420	336	336	168	840	168	840	56	168	280	O.S.
39	39	41	41	41	41	43	43	43	43	43	45	45	LEVEL
14700	24255	15120	15120	17920	17920	15876	29700	15876	29700	13104	25872	17325	DIM'N
0	1	0	0	3	1	0	2	0	0	3	2	4	
0	1	2	0	0	1	1	1	0	0	0	0	0	
0	0	0	0	0	0	2	0	0	0	0	2	0	
1	0	1	1	0	0	0	1	0	1	0	0	1	
1	0	0	0	0	0	0	0	2	0	0	0	0	
0	1	0	2	1	0	0	0	1	1	0	0	0	
1	1	0	0	1	3	0	0	0	2	3	0	0	WEIGHT
13	14	15	16	17	18	19	20	21	22	23	24	25	NUMBER
70	175	56	56	112	112	42	90	42	90	84	56	35	0
30	65	26	26	42	42	21	45	21	45	28	30	20	1
13	20	12	12	12	12	11	21	11	21	7	16	10	2
.	10	6	.	12	6	5	15	.	.	7	10	10	3
5	10	.	6	6	12	.	.	5	15	7	.	.	4
.	5	.	.	6	6	.	.	.	.	7	.	.	5
6	4	5	5	2	2	6	9	6	9	1	9	4	6
3	.	2	2	.	.	3	3	3	3	.	6	1	7
.	2	3	.	2	1	3	6	.	.	1	5	4	8
2	2	.	3	1	2	.	.	3	6	1	.	.	9
.	.	.	.	2	.	.	3	.	.	1	2	4	10
.	.	.	.	.	2	.	.	.	3	1	.	.	11
.	.	1	.	.	.	2	2	.	.	.	3	1	12
1	.	.	1	.	.	.	.	2	2	.	.	.	13
	1	.	.	1	1	.	.	.	.	1	.	.	14
		1	.	.	.	1	1	.	.	.	1	1	15
			1	.	.	.	.	1	1	.	.	.	16

96	128	160	144	112	112	112	144	192	176	128	192	128	S.P.
5040	168	280	336	1680	1120	1680	336	28	336	420	28	3360	O.S.
45	45	45	47	47	47	47	47	49	49	49	49	49	LEVEL
96228	25872	17325	41580	92400	77616	92400	41580	13860	50688	86625	13860	178605	DIM'N
1	0	0	1	1	1	0	0	0	3	0	0	2	
0	0	0	2	0	0	2	0	4	1	2	0	1	
1	0	0	1	1	0	0	0	0	1	0	0	0	
0	0	1	0	0	2	0	0	0	0	0	0	0	
1	2	0	0	0	0	1	1	0	0	0	0	1	
0	0	0	0	2	0	0	2	0	0	2	4	0	
1	2	4	0	0	1	1	1	0	0	0	0	1	WEIGHT
26	27	28	29	30	31	32	33	34	35	36	37	38	NUMBER
342	56	35	70	280	196	280	70	14	64	245	14	483	0
153	30	20	40	130	95	130	40	9	40	110	9	228	1
66	16	10	23	56	47	56	23	6	24	46	6	98	2
24	.	.	15	20	15	30	.	4	20	25	.	68	3
24	10	10	.	30	15	20	15	.	.	25	4	34	4
9	.	.	.	10	5	10	.	.	.	15	.	24	5
27	9	4	13	21	25	21	13	4	14	16	4	36	6
9	6	1	7	6	16	6	7	3	8	4	3	9	7
9	.	.	9	7	7	12	.	3	11	9	.	24	8
9	5	4	.	12	7	7	9	.	.	9	3	12	9
.	.	.	3	.	.	.	.	1	8	.	.	12	10
.	2	4	.	.	.	.	3	.	.	.	1	.	11
3	.	.	5	2	4	3	.	2	6	2	.	6	12
3	3	1	.	3	4	2	5	.	.	2	2	3	13
3	.	.	.	4	2	4	.	.	.	5	.	8	14
.	.	.	4	.	.	3	.	2	4	2	.	3	15
.	1	1	.	3	.	.	4	.	.	2	2	.	16

128	176	208	240	128	144	128	176	144	208	176	240	256	S.P.
3360	336	56	168	3360	840	3360	840	840	56	840	168	56	O.S.
49	49	51	51	51	51	51	51	51	51	51	51	53	LEVEL
178605	50688	33264	27027	236544	166320	236544	100100	166320	33264	100100	27027	34320	DIM'N
1	0	2	5	1	2	0	4	0	0	1	0	4	
0	0	3	0	0	1	1	0	2	0	0	0	2	
1	0	0	1	0	0	1	0	0	0	1	0	0	
0	0	0	0	1	0	1	0	0	0	0	0	0	
0	1	0	0	1	0	0	1	0	0	0	1	0	
1	1	0	0	1	2	0	0	1	3	0	0	0	
2	3	0	0	0	0	1	1	2	2	4	5	0	WEIGHT
39	40	41	42	43	44	45	46	47	48	49	50	51	NUMBER
483	64	28	21	448	420	448	210	420	28	210	21	20	0
228	40	19	15	232	195	232	105	195	19	105	15	15	1
98	24	13	10	120	80	120	45	80	13	45	10	11	2
34	.	10	10	36	55	56	45	45	.	15	.	10	3
68	20	.	.	56	45	36	15	55	10	45	10	.	4
24	.	.	.	16	30	16	15	30	.	15	.	.	5
36	14	9	6	62	27	62	15	27	9	15	6	8	6
9	8	6	3	32	6	32	3	6	6	3	3	6	7
12	.	7	6	17	18	30	15	15	.	5	.	7	8
24	11	.	.	30	15	17	5	18	7	15	6	.	9
.	.	4	6	.	9	.	15	.	.	.	.	6	10
12	8	.	.	.	.	.	.	9	4	15	6	.	11
3	.	5	3	8	4	16	3	3	.	1	.	5	12
6	6	.	.	16	3	8	1	4	5	3	3	.	13
8	.	.	.	8	10	8	5	10	.	5	.	.	14
.	.	4	3	.	2	8	3	3	.	.	.	4	15
3	4	.	.	8	3	.	.	2	4	3	3	.	16

1	.	.	.	.	.	1	.	.	.	.	.	.	.	.	.	.	.	.	.	.	4	.	.	.	.	.	5	.	5	.	.	.	.	.	17
	1	.	.	.	.	1	.	.	.	.	.	.	.	.	.	.	.	.	.	.	4	.	.	.	.	.	.	.	5	.	5	.	.	18
		1	.	.	.	.	1	.	.	.	.	2	.	.	.	.	1	2	.	.	.	.	.	3	1	.	.	4	.	.	.	.	.	3	19
			1	.	.	.	1	1	.	.	.	2	.	.	.	.	1	4	.	.	3	.	.	3	3	.	2	.	3	.	.	.	.	4	20
				1	.	.	.	.	.	1	.	.	.	.	.	2	.	.	.	1	.	.	2	.	.	.	4	.	.	.	.	3	1	.	21
					1	.	.	.	.	1	1	.	.	.	.	2	.	.	.	1	.	3	4	.	.	.	.	.	.	2	3	3	3	.	22
						1	.	.	.	.	.	.	.	.	.	.	.	.	.	.	.	.	.	.	.	.	.	.	.	.	.	.	.	.	23
							1	.	.	.	.	1	.	.	.	.	1	2	.	.	.	.	.	2	1	.	.	.	.	.	.	.	.	3	24
								1	.	.	.	.	.	.	.	.	.	2	.	.	.	.	.	1	3	.	.	.	3	.	.	.	.	3	25
									1	.	.	.	1	1	1	.	.	.	1	.	2	2	.	.	.	4	2	4	1	2	.	1	.	.	26
										1	.	.	.	.	.	1	.	.	.	1	.	.	2	.	.	.	.	.	.	.	2	.	1	.	27
											1	.	.	.	.	.	.	.	.	.	.	.	2	.	.	.	.	.	.	.	1	3	3	.	28
												1	.	.	.	.	1	1	.	.	.	.	.	2	1	.	.	.	.	.	.	.	.	2	29
													1	.	.	.	.	.	1	.	.	1	.	.	.	2	2	.	.	1	.	1	.	.	30
														1	.	.	.	.	.	.	.	.	.	.	.	2	.	2	.	.	.	.	.	.	31
															1	.	.	.	1	.	1	.	.	.	.	.	1	2	1	2	.	.	.	.	32
																1	.	.	.	1	.	.	1	.	.	.	.	.	.	.	2	.	1	.	33
																	1	.	.	.	.	.	.	1	.	.	.	.	.	.	.	.	.	1	34
																		1	.	.	.	.	.	1	1	.	.	.	.	.	.	.	.	2	35
																			1	.	.	.	.	.	.	.	1	.	.	1	.	.	.	.	36
																				1	.	.	.	.	.	.	.	.	.	.	1	.	.	.	37
																					1	.	.	.	.	.	1	.	1	.	.	.	.	.	38
																						1	.	.	.	.	.	.	.	1	.	1	.	.	39
																							1	.	.	.	.	.	.	.	1	.	1	.	40
																								1	.	.	.	.	.	.	.	.	.	1	41
																									1	.	.	.	.	.	.	.	.	1	42
																										1	.	.	.	.	.	.	.	.	43
																											1	.	.	.	.	.	.	.	44
																												1	.	.	.	.	.	.	45
																													1	.	.	.	.	.	46
																														1	.	.	.	.	47
																															1	.	.	.	48
																																1	.	.	49
																																	1	.	50
																																		1	51

Table 3. *(continued)*

ALGEBRA A7, REPRESENTATIONS OF CLASS 1

NUMBER	0	1	2	3	4	5	6	7	8	9	10	11	12	13	14	15	16	17	18	19	20	21	22	23	24	25
S.P.	7	23	39	39	55	55	55	87	71	71	103	71	87	103	87	87	103	103	87	119	103	151	103	151	119	167
O.S.	8	168	56	560	280	280	840	168	336	840	56	1680	168	280	280	560	168	1120	3360	336	840	280	840	56	840	168
LEVEL	8	20	22	28	30	32	32	34	34	36	36	38	38	38	40	40	40	42	42	42	44	44	44	44	44	44
DIM'N	8	216	280	1344	1800	2352	4200	3080	5544	8400	4032	15120	7056	6600	11760	14400	9072	27720	49896	18480	35280	15400	46200	11088	37800	11880
WEIGHT	1000000	0100001	2000001	0010010	0010002	0001100	1100010	3000010	1100002	0001011	3000002	1010100	0000201	0001003	1002000	0200100	0000120	2100100	1010011	0000112	0111000	4000100	0200011	0000031	1010003	0000104
0	1	6	7	14	15	14	35	21	42	35	28	63	21	20	35	50	21	84	168	35	77	35	140	14	105	15
1		1	1	4	5	5	10	6	12	15	7	24	10	10	15	20	11	34	68	20	36	15	55	9	45	10
2			1	.	.	.	5	6	6	.	7	9	.	.	5	10	.	24	24	.	16	15	30	.	15	.
3				1	1	2	2	1	2	6	1	9	5	4	7	7	6	12	24	11	17	5	18	6	15	6
4					1	.	.	.	2	3	1	.	2	4	.	.	3	.	12	8	.	.	9	4	15	6
5						1	.	.	.	2	.	3	3	1	4	2	3	3	6	6	8	1	4	4	3	3
6							1	1	1	.	1	3	.	.	2	4	.	8	8	.	8	5	10	.	5	.
7								1	.	.	1	.	.	.	.	.	.	4	.	.	.	5	.	.	.	.
8									1	.	1	.	.	.	.	.	.	.	4	.	.	.	5	.	5	.
9										1	.	.	1	1	.	.	2	.	3	4	.	.	2	3	3	3
10											1	.	.	.	.	.	.	.	.	.	.	.	.	.	.	.
11												1	.	.	1	1	.	2	2	.	4	1	2	.	1	.
12													1	.	.	.	1	.	.	2	.	.	.	2	.	1
13														1	.	.	.	.	.	2	.	.	.	1	3	3
14															1	.	.	.	.	.	2	.	.	.	.	.
15																1	.	1	.	.	2	1	2	.	.	.
16																	1	.	.	1	.	.	.	2	.	1

NUMBER	26	27	28	29	30	31	32	33	34	35	36	37	38	39	40	41	42	43	44	45	46	47	48	49	50	51
S.P.	135	119	103	119	135	183	135	119	167	151	247	151	167	135	119	135	199	343	183	183	231	135	135	135	151	183
O.S.	56	840	3360	1680	168	56	840	1680	336	336	56	336	840	840	5040	3360	56	8	168	168	280	1680	1680	5040	840	1120
LEVEL	46	46	46	46	46	46	48	48	48	48	48	50	50	50	50	50	50	50	52	52	52	52	52	52	52	52
DIM'N	14112	55440	99792	88200	34496	14784	71280	103488	48048	65520	10296	77616	83160	110880	237600	204120	35280	3432	55440	73920	40040	231000	206976	367416	181440	120120
WEIGHT	0030000	2011000	1001101	2100011	0200003	0000023	1201000	1001020	4000011	2100003	0000015	1120000	3101000	1000210	0110101	1001012	4000003	0000007	0310000	3020000	5001000	0110020	0102001	2010101	1000202	1001004
0	21	105	210	245	91	14	111	182	112	168	6	91	105	147	405	315	84	1	49	70	35	350	266	567	210	140
1	11	50	100	95	36	10	56	92	42	63	5	50	55	81	195	165	28	1	29	40	20	170	140	273	120	80
2	5	30	30	65	21	.	36	26	42	42	.	30	45	21	90	45	28	.	19	30	20	80	60	153	30	20
3	6	24	48	30	11	7	27	45	12	18	4	28	27	46	90	81	7	1	17	23	10	78	75	126	69	40
4	.	.	18	15	11	6	.	24	6	18	4	.	.	22	33	57	7	1	.	.	.	42	24	45	42	40
5	3	12	24	6	2	5	12	20	2	3	3	16	12	27	39	36	1	1	10	14	4	31	42	54	42	16
6	3	13	13	20	6	.	18	12	12	12	.	17	21	11	42	21	7	.	12	16	10	36	32	66	16	10
7	.	5	.	10	.	.	6	.	12	6	.	5	15	.	.	.	7	.	4	10	10	.	.	24	.	.
8	.	.	5	10	6	.	.	6	6	12	.	.	.	5	15	15	7	.	.	.	.	20	10	24	10	10
9	.	.	8	3	2	4	.	12	1	3	3	.	.	14	13	24	1	1	.	.	.	18	12	18	23	16
10	.	.	.	5	.	.	.	.	6	6	.	.	.	.	.	.	7	.	.	.	.	.	.	9	.	.
11	2	6	6	4	1	.	8	5	2	2	.	10	9	6	18	9	1	.	7	9	4	14	18	27	9	4
12	.	.	4	.	.	3	.	4	.	.	2	.	.	10	5	8	.	1	.	.	.	5	6	6	15	4
13	.	.	.	.	2	3	.	.	.	3	3	.	.	.	.	12	1	1	.	.	.	.	.	.	9	16
14	1	3	3	.	.	.	3	2	.	.	.	6	3	3	6	3	.	.	4	6	1	4	12	9	6	1
15	1	2	.	2	1	.	6	.	2	1	.	6	6	.	9	.	1	.	6	5	4	7	7	9	.	.
16	.	.	.	.	.	2	.	4	.	.	2	.	.	5	.	4	.	1	.	.	.	5	.	.	5	4

1	.	.	.	1	.	.	.	.	.	2	.	2	.	.	3	.	2	1	.	3	6	.	.	.	1	.	3	5	4	.	.	9	.	.	17
	1	.	.	.	1	.	1	.	.	.	2	2	1	.	.	3	1	2	.	.	.	3	6	6	1	.	.	.	.	8	5	9	5	4	18
		1	.	.	.	1	.	1	.	.	.	.	.	2	.	.	.	.	2	.	.	.	.	4	.	1	.	.	.	.	.	.	5	4	19
			1	.	.	.	.	.	1	1	.	.	.	.	2	.	.	.	.	4	2	.	3	.	.	.	3	3	1	2	4	3	.	.	20
				1	.	.	.	.	.	.	.	.	.	.	.	.	2	.	.	.	3	.	.	.	1	.	.	2	4	.	.	.	.	.	21
					1	.	.	.	.	.	.	1	1	.	.	.	1	1	.	.	.	.	3	.	1	.	.	.	.	4	2	3	.	.	22
						1	.	.	.	.	.	.	.	1	.	.	.	.	1	.	.	.	.	.	.	1	.	.	.	.	.	.	.	.	23
							1	.	.	.	.	.	1	.	.	.	.	2	.	.	.	.	.	3	1	.	.	.	.	.	.	.	2	4	24
								1	.	.	.	.	.	1	.	.	.	.	2	.	.	.	.	.	.	1	.	.	.	.	.	.	.	4	25
									1	.	.	.	.	.	.	.	.	.	.	2	.	.	.	.	.	.	1	1	.	.	.	.	.	.	26
										1	.	.	.	.	1	.	.	.	.	2	2	.	.	.	.	.	2	3	1	.	.	3	.	.	27
											1	.	.	.	.	1	.	.	.	.	.	2	2	2	.	.	.	.	.	2	3	3	3	1	28
												1	.	.	.	.	1	1	.	.	.	.	.	.	1	.	.	.	.	.	.	3	.	.	29
													1	.	.	.	.	1	.	.	.	.	.	.	1	.	.	.	.	.	.	.	.	.	30
														1	.	.	.	.	1	.	.	.	.	.	.	1	.	.	.	.	.	.	.	.	31
															1	.	.	.	.	1	1	.	.	.	.	.	2	1	1	.	.	.	.	.	32
																1	.	.	.	.	.	1	.	1	.	.	.	.	.	2	.	.	1	1	33
																	1	.	.	.	.	.	.	.	1	.	.	.	.	.	.	.	.	.	34
																		1	.	.	.	.	.	.	1	.	.	.	.	.	.	.	.	.	35
																			1	.	.	.	.	.	.	1	.	.	.	.	.	.	.	.	36
																				1	.	.	.	.	.	.	1	1	.	.	.	.	.	.	37
																					1	.	.	.	.	.	.	1	1	.	.	.	.	.	38
																						1	.	.	.	.	.	.	.	.	.	.	1	.	39
																							1	.	.	.	.	.	.	1	1	1	.	.	40
																								1	.	.	.	.	.	.	.	.	1	1	41
																									1	.	.	.	.	.	.	.	.	.	42
																										1	.	.	.	.	.	.	.	.	43
																											1	.	.	.	.	.	.	.	44
																												1	.	.	.	.	.	.	45
																													1	.	.	.	.	.	46
																														1	.	.	.	.	47
																															1	.	.	.	48
																																1	.	.	49
																																	1	.	50
																																		1	51

Table 3. *(continued)*

ALGEBRA A7, REPRESENTATIONS OF CLASS 2

NUMBER	0	1	2	3	4	5	6	7	8	9	10	11	12
S.P.	12	28	28	44	44	76	60	60	60	76	76	108	76
O.S.	28	8	280	336	420	56	56	280	1680	336	420	28	840
LEVEL	13	15	23	27	29	29	31	31	35	35	37	37	37
DIM'N	28	36	420	1280	1512	924	1176	2100	8820	5376	8100	2520	11880
	0	2	0	1	0	3	0	0	1	0	0	0	1
	1	0	0	1	0	0	0	0	0	0	2	0	0
	0	0	1	0	0	0	0	0	1	0	0	0	1
	0	0	0	0	1	0	0	1	0	0	0	0	0
	0	0	0	0	0	0	2	0	0	1	0	0	0
	0	0	0	0	1	0	0	0	1	1	1	3	0
WEIGHT	0	0	1	1	0	1	0	2	0	1	0	0	2
0	1	1	5	12	9	7	5	10	38	16	30	5	45
1		1	.	6	.	7	.	.	14	.	15	.	15
2			1	2	3	1	2	4	12	8	9	3	15
3				1	.	1	.	.	4	.	5	.	5
4					1	.	1	1	3	4	2	2	3
5						1	.	.	.	.	.	.	.
6							1	.	.	2	.	1	.
7								1	.	2	.	1	3
8									1	.	1	.	1
9										1	.	1	.
10											1	.	.
11												1	.
12													1
13													
14													
15													
16													

NUMBER	13	14	15	16	17	18	19	20	21	22	23	24	25
S.P.	108	76	92	92	124	140	108	172	92	92	156	252	108
O.S.	168	1120	840	168	56	168	336	56	1680	3360	56	8	420
LEVEL	37	39	39	39	39	41	41	41	43	43	43	43	45
DIM'N	4620	16128	15400	10780	5544	8316	20412	4620	37800	58212	10920	1716	31752
	0	1	2	0	0	4	2	0	0	1	4	0	0
	0	0	1	2	0	0	1	0	1	0	0	0	1
	0	0	0	0	0	0	0	0	1	0	0	0	0
	0	1	0	0	0	0	0	0	0	1	0	0	2
	1	1	0	0	0	0	0	0	1	0	0	0	0
	0	0	1	0	2	1	0	1	0	1	0	0	0
WEIGHT	3	0	0	2	2	0	2	4	0	1	2	6	0
0	10	44	50	36	9	21	63	5	80	125	28	1	51
1	.	14	35	21	.	21	42	.	35	35	28	.	21
2	6	18	15	11	6	6	18	4	33	57	7	1	24
3	.	5	10	6	.	6	12	.	15	15	7	.	10
4	3	8	3	2	4	1	3	3	13	24	1	1	12
5	.	.	5	.	.	6	6	.	.	.	7	.	.
6	1	4	.	.	3	.	.	2	5	8	.	1	6
7	3	.	.	2	3	.	3	3	.	12	1	1	.
8	.	2	2	1	.	1	2	.	6	6	1	.	5
9	1	.	.	.	2	.	.	2	.	4	.	1	.
10	.	.	1	1	.	1	1	.	3	.	1	.	2
11	.	.	.	.	1	.	.	1	.	.	.	1	.
12	.	.	.	1	.	.	2	.	.	3	1	.	.
13	1	.	.	.	1	.	.	2	.	.	.	1	.
14		1	.	.	.	.	.	.	2	2	.	.	3
15			1	.	.	1	1	.	.	.	1	.	.
16				1	.	.	1	.	.	.	1	.	.

NUMBER	26	27	28	29	30	31	32	33	34	35	36	37	38
S.P.	108	108	124	124	124	124	124	108	156	124	140	140	140
O.S.	1680	840	1120	280	280	1120	840	3360	1120	3360	1680	840	1680
LEVEL	45	45	45	47	47	47	47	47	49	49	49	49	51
DIM'N	58212	49896	46080	35280	46200	69300	64680	126720	78848	193536	132300	97020	177408
	2	1	1	0	2	1	1	0	3	2	1	0	1
	0	0	0	0	0	2	0	1	1	0	0	1	1
	1	0	0	2	0	0	0	1	0	1	0	1	1
	0	0	1	1	2	0	0	0	0	0	0	0	1
	1	2	0	0	0	1	1	0	1	0	1	0	0
	0	0	0	0	0	0	2	1	0	1	1	0	0
WEIGHT	0	1	3	0	0	0	0	1	0	1	2	3	0
0	108	81	80	47	65	110	87	224	104	312	155	145	192
1	63	21	20	21	35	70	21	104	84	168	35	70	112
2	45	42	40	23	31	46	49	96	44	132	92	65	96
3	24	10	10	11	15	30	11	44	34	68	20	30	56
4	18	23	16	11	16	17	28	36	16	48	53	23	48
5	9	.	.	.	5	10	.	.	24	24	.	.	16
6	6	15	4	5	10	5	15	10	4	12	30	5	24
7	.	9	16	.	.	.	15	18	.	24	38	23	.
8	9	5	4	6	7	11	6	16	12	24	11	10	28
9	.	5	4	.	.	.	10	5	.	6	20	5	.
10	3	.	.	3	2	8	.	8	8	8	.	5	16
11	.	.	.	.	.	.	5	.	.	.	5	.	.
12	.	2	4	.	.	.	3	8	.	12	8	10	.
13	.	.	4	.	.	.	.	.	.	.	10	5	.
14	3	3	1	3	4	3	3	4	3	6	6	2	14
15	3	.	.	.	2	4	.	.	8	8	.	.	8
16	.	.	.	.	.	.	.	4	.	4	.	5	.

NUMBER	39	40	41	42	43	44	45	46	47	48	49	50	51
S.P.	220	124	140	172	156	188	172	172	172	140	140	140	172
O.S.	280	5040	1680	336	640	840	168	420	840	2520	1680	3360	1680
LEVEL	51	51	51	51	51	51	53	53	53	53	53	53	53
DIM'N	36036	277200	224000	80640	147420	85800	77616	99000	161700	291060	249480	396900	252252
	5	0	1	1	2	1	1	0	3	0	0	2	3
	0	1	2	0	0	0	0	3	0	1	0	0	1
	0	0	0	0	1	0	3	0	1	0	2	0	0
	0	1	0	0	0	0	0	1	1	1	0	1	0
	1	1	0	0	0	1	0	0	0	0	1	1	0
	0	0	1	3	0	0	0	0	0	2	0	0	1
WEIGHT	0	1	1	1	3	4	0	0	0	0	1	1	1
0	35	335	320	68	210	75	61	85	145	306	261	435	315
1	35	140	210	14	105	15	35	55	105	126	126	210	245
2	15	171	130	46	90	50	34	44	74	162	132	222	125
3	15	70	85	9	45	10	20	30	50	66	66	100	95
4	5	87	44	32	30	30	19	22	38	84	64	114	40
5	15	.	30	.	15	.	5	10	30	.	.	30	65
6	1	45	10	22	6	15	10	10	20	39	30	60	8
7	.	33	22	22	30	30	.	.	.	45	24	42	20
8	5	35	28	6	15	6	12	15	24	33	33	48	30
9	.	15	5	17	6	15	.	.	.	24	10	20	4
10	5	13	20	.	5	.	7	12	13	12	18	13	20
11	.	.	.	12	.	5	.	.	.	9	.	.	.
12	.	13	14	4	15	6	.	.	.	18	12	18	15
13	.	.	.	6	6	15	.	.	.	.	.	.	.
14	1	18	6	4	3	3	7	7	12	15	15	24	6
15	5	.	10	.	5	.	3	6	13	.	.	13	20
16	.	5	10	.	5	.	.	.	.	6	6	5	10

1	.	.	1	.	.	.	1	.	.	.	.	.	.	.	.	.	.	.	5	.	.	.	.	.	6	.	5	.	.	.	.	.	.	.	17
	1	.	.	.	.	1	.	.	.	.	.	.	.	.	.	.	4	.	.	.	.	5	.	.	.	.	.	.	.	5	.	.	.	10	18
		1	.	.	.	1	.	.	.	.	.	.	.	.	.	.	.	4	.	.	.	.	.	5	.	5	.	.	.	.	.	.	5	10	19
			1	.	.	.	1	.	.	.	.	.	.	.	.	.	.	.	.	.	.	.	.	.	.	.	5	.	.	.	.	.	.	.	20
				1	.	.	.	1	1	.	.	2	1	2	.	2	2	2	.	1	8	1	6	4	.	1	.	5	5	6	5	9	6	4	21
					1	.	.	.	.	1	1	.	.	.	2	2	.	3	4	2	.	.	6	3	3	3	3	.	.	.	9	5	8	3	22
						1	.	.	.	.	.	.	.	.	.	.	.	.	.	.	.	.	.	.	.	.	.	.	.	.	.	.	.	5	23
							1	.	.	.	.	.	.	.	.	.	.	.	.	.	.	.	.	.	.	.	.	.	.	.	.	.	.	.	24
								1	.	.	.	1	1	.	.	.	.	.	.	.	4	.	3	.	.	.	.	3	2	3	2	3	3	.	25
									1	.	.	.	1	1	.	.	2	2	.	.	4	1	.	2	.	1	.	2	3	6	.	.	6	4	26
										1	.	.	.	.	1	.	.	.	2	.	.	.	3	.	2	.	1	.	.	.	3	2	4	.	27
											1	.	.	.	.	.	.	.	2	2	.	.	.	.	1	3	3	.	.	.	.	.	.	.	28
												1	.	.	.	.	.	.	.	.	2	.	.	.	.	.	.	3	1	1	.	3	.	.	29
													1	.	.	.	.	.	.	.	2	.	.	.	.	.	.	1	1	3	.	.	3	.	30
														1	.	.	1	.	.	.	2	1	.	2	.	.	.	1	3	2	.	.	.	2	31
															1	.	.	.	1	.	.	.	.	.	2	.	1	.	.	.	3	.	.	.	32
																1	.	1	.	1	.	.	2	2	.	1	.	.	.	.	3	3	2	2	33
																	1	.	.	.	.	1	.	.	.	.	.	.	.	2	.	.	.	2	34
																		1	.	.	.	.	.	1	.	1	.	.	.	.	.	.	2	2	35
																			1	.	.	.	.	.	1	.	1	.	.	.	.	.	.	.	36
																				1	.	.	.	.	.	1	.	.	.	.	.	.	.	.	37
																					1	.	.	.	.	.	.	1	1	1	.	.	.	.	38
																						1	.	.	.	.	.	.	.	.	.	.	.	.	39
																							1	.	.	.	.	.	.	.	1	1	1	.	40
																								1	.	.	.	.	.	.	.	.	.	1	41
																									1	.	.	.	.	.	.	.	.	.	42
																										1	.	.	.	.	.	.	.	.	43
																											1	.	.	.	.	.	.	.	44
																												1	.	.	.	.	.	.	45
																													1	.	.	.	.	.	46
																														1	.	.	.	.	47
																															1	.	.	.	48
																																1	.	.	49
																																	1	.	50
																																		1	51

Table 3. *(continued)*

ALGEBRA A7, REPRESENTATIONS OF CLASS 3

NUMBER	0	1	2	3	4	5	6	7	8	9	10	11	12	13	14	15	16	17
S.P.	15	31	63	31	47	47	63	63	79	79	111	63	127	175	79	79	79	95
O.S.	56	56	8	280	168	840	168	168	56	336	56	1 680	56	8	280	1 120	1 680	1 680
LEVEL	16	20	22	24	28	30	30	32	32	34	34	36	36	36	38	38	40	42
DIM'N	56	168	120	504	1 008	2 800	1 512	2 520	1 680	4 752	1 848	10 584	2 520	792	8 400	14 784	22 680	34 496
WEIGHT	0 0 1 0 0 0 0	1 1 0 0 0 0 0	3 0 0 0 0 0 0	0 0 0 1 0 0 1	0 0 0 0 1 1 0	1 0 1 0 0 0 1	0 0 0 0 1 0 2	0 2 0 0 0 0 1	0 0 0 0 0 2 1	2 1 0 0 0 0 1	0 0 0 0 0 1 3	1 0 0 1 0 1 0	4 0 0 0 0 0 1	0 0 0 0 0 0 5	1 0 0 0 2 0 0	1 0 0 1 0 0 2	0 1 1 0 0 1 0	2 0 1 0 0 1 0
0	1	2	1	4	5	15	6	11	5	18	4	33	7	1	20	40	54	72
1		1	1	.	.	5	.	6	.	12	.	9	7	.	5	10	24	38
2			1	.	.	.	.	.	.	6	.	.	7	.	.	.	.	14
3				1	2	3	3	2	3	3	3	12	1	1	9	16	18	24
4					1	.	1	.	2	.	2	4	.	1	5	4	5	6
5						1	.	1	.	2	.	3	1	.	2	4	8	12
6							1	.	1	.	2	.	.	1	.	4	.	.
7								1	.	1	.	.	1	.	.	.	4	4
8									1	.	1	.	.	1	.	.	.	.
9										1	.	.	1	.	.	.	.	4
10											1	.	.	1	.	.	.	.
11												1	.	.	1	1	2	3
12													1	.	.	.	.	.
13														1	.	.	.	.
14															1	.	.	.
15																1	.	.
16																	1	1

NUMBER	18	19	20	21	22	23	24	25	26	27	28	29	30	31	32	33	34	35
S.P.	95	95	95	111	127	111	127	111	111	143	127	143	127	207	111	159	191	127
O.S.	1 680	840	1 680	840	168	840	840	560	1 120	840	336	336	280	168	5 040	336	336	3 360
LEVEL	42	42	44	44	44	44	44	46	46	46	46	46	48	48	48	48	48	50
DIM'N	38 808	30 800	45 360	39 600	18 480	46 656	33 600	40 320	64 680	44 352	52 920	40 824	42 336	19 656	166 320	58 968	34 320	194 040
WEIGHT	1 0 0 0 1 1 1	0 1 1 0 0 0 2	0 1 0 1 1 0 0	1 2 0 0 0 1 0	1 0 0 0 0 3 0	2 0 1 0 0 0 2	1 0 0 0 1 0 3	0 0 2 0 1 0 0	2 0 0 1 1 0 0	3 1 0 0 0 1 0	1 2 0 0 0 0 2	1 0 0 0 0 2 2	0 0 1 2 0 0 0	5 0 0 0 0 1 0	0 1 0 1 0 1 1	3 1 0 0 0 0 2	1 0 0 0 0 1 4	1 1 1 0 1 0 0
0	72	65	75	70	25	90	50	56	94	65	87	48	47	21	225	84	30	220
1	16	30	30	45	5	45	10	26	44	50	57	9	21	21	90	63	5	125
2	.	.	.	15	.	15	.	.	14	35	21	.	.	21	.	42	.	35
3	38	23	33	22	16	30	30	24	42	20	27	33	23	6	108	24	24	96
4	20	5	15	5	11	6	15	10	20	4	5	23	11	1	48	4	18	40
5	8	10	13	14	3	15	6	12	18	15	17	6	11	6	42	18	4	54
6	10	5	.	.	6	6	15	.	.	.	5	17	.	.	24	4	18	.
7	.	5	5	10	.	5	.	6	5	10	12	.	5	6	15	12	.	30
8	5	.	.	.	6	.	5	.	.	.	.	12	.	.	9	.	12	.
9	.	.	.	5	.	5	.	.	5	10	6	.	.	6	.	12	.	15
10	.	.	.	.	.	.	5	.	.	.	.	6	.	.	.	.	12	.
11	4	2	6	3	2	3	3	5	8	3	3	4	6	1	18	3	3	22
12	.	.	.	.	.	.	.	.	.	5	.	.	.	6	.	6	.	.
13	.	.	.	.	.	.	.	.	.	.	.	.	.	.	.	.	6	.
14	2	.	3	.	1	.	1	2	4	.	.	3	3	.	6	.	2	8
15	2	2	.	.	1	3	3	.	.	.	3	3	.	.	9	3	3	.
16	.	1	2	2	.	1	.	3	2	2	2	.	3	1	6	2	.	12

NUMBER	36	37	38	39	40	41	42	43	44	45	46	47	48	49	50	51
S.P.	127	127	127	223	143	271	143	159	159	143	143	159	159	159	175	175
O.S.	1 680	1 120	3 360	56	1 680	56	840	560	1 680	1 680	840	280	1 120	840	280	1 120
LEVEL	50	50	50	50	50	50	52	52	52	52	52	52	52	54	54	54
DIM'N	147 840	138 600	235 200	25 872	133 056	12 936	166 320	99 000	172 480	194 040	204 120	107 800	187 200	199 584	138 600	221 760
WEIGHT	0 1 0 0 2 0 1	0 0 2 0 0 1 1	2 0 0 1 0 1 1	5 0 0 0 0 0 2	0 1 0 1 0 0 3	1 0 0 0 0 0 6	1 1 0 2 0 0 0	0 3 0 0 1 0 0	3 0 1 0 1 0 0	0 1 0 0 1 2 0	2 0 0 0 2 0 1	0 0 2 0 0 0 3	2 0 0 1 0 0 3	1 0 2 1 0 0 0	3 0 0 2 0 0 0	2 2 0 0 1 0 0
0	160	160	290	28	150	7	151	90	163	181	201	105	200	153	105	180
1	60	80	125	28	60	1	81	60	108	66	81	55	80	87	65	130
2	.	.	35	28	.	.	21	20	63	.	21	.	20	21	35	70
3	88	73	138	7	78	7	76	39	72	106	111	51	100	78	54	78
4	50	30	60	1	33	7	40	15	30	64	65	20	40	39	30	30
5	32	36	57	7	30	1	40	26	45	38	42	25	40	46	31	56
6	20	15	30	1	33	7	.	.	.	34	25	20	40	.	.	.
7	10	20	15	7	10	.	20	20	24	11	10	15	10	26	15	40
8	10	5	10	.	9	7	.	.	.	24	15	5	10	.	.	.
9	.	.	15	7	.	.	10	10	24	.	10	.	10	11	15	30
10	.	.	.	.	9	7	.	.	.	.	.	5	10	.	.	.
11	18	14	24	1	12	1	21	10	18	22	23	9	16	24	16	21
12	.	.	.	7	.	.	.	.	9	.	.	.	.	.	5	10
13	.	.	.	.	.	7	.	.	.	.	.	.	.	.	.	.
14	12	4	8	.	3	1	12	3	6	12	15	2	4	12	10	6
15	7	7	12	1	12	1	.	.	.	12	9	9	16	.	.	.
16	5	8	6	1	4	.	10	7	9	6	5	5	4	15	7	15

1	.	.	.	1	.	1	.	.	2	2	1	.	.	1	.	2	.	6	.	.	6	1	.	.	5	4	9	.	5	.	4	6	7	11	17
	1	.	.	.	1	.	1	.	.	.	.	2	.	.	3	.	2	.	4	2	4	.	3	1	.	.	.	8	5	2	4	.	.	.	18
		1	.	.	.	1	.	.	.	.	2	.	.	.	3	2	.	.	2	4	3	1	4	.	.	.	.	3	2	5	4	.	.	.	19
			1	.	.	.	.	1	1	.	.	.	2	.	2	.	.	4	3	2	2	.	1	.	6	2	3	3	3	1	1	8	4	4	20
				1	.	.	.	.	.	1	1	.	.	1	.	1	.	3	.	.	.	1	.	.	2	4	3	.	.	.	.	3	2	8	21
					1	.	.	.	.	.	.	1	.	.	.	.	1	.	.	.	.	.	.	1	.	.	.	4	.	.	.	.	.	.	22
						1	.	.	.	.	1	.	.	.	.	2	.	.	.	.	3	1	.	.	.	.	.	.	2	.	4	.	.	.	23
							1	.	.	.	.	1	.	.	.	.	2	.	.	.	.	.	3	1	.	.	.	.	.	2	4	.	.	.	24
								1	.	.	.	.	1	.	.	.	.	2	.	2	.	.	.	.	2	1	1	.	.	1	.	6	1	3	25
									1	.	.	.	.	.	.	.	.	2	.	.	2	.	.	.	3	1	3	.	3	.	1	3	4	3	26
										1	.	.	.	1	.	1	.	.	.	.	.	1	.	.	.	.	3	.	.	.	.	.	2	4	27
											1	.	.	.	.	1	.	.	.	.	.	1	.	.	.	.	.	.	.	.	.	.	.	.	28
												1	.	.	.	.	1	.	.	.	.	.	.	1	.	.	.	.	.	.	.	.	.	.	29
													1	.	.	.	.	.	.	.	.	.	.	.	2	.	.	.	.	.	.	3	1	.	30
														1	.	.	.	.	.	.	.	1	.	.	.	.	.	.	.	.	.	.	.	.	31
															1	.	.	.	1	1	1	.	1	.	.	.	.	2	1	1	1	.	.	.	32
																1	.	.	.	.	.	1	.	.	.	.	.	.	.	.	.	.	.	.	33
																	1	.	.	.	.	.	.	1	.	.	.	.	.	.	.	.	.	.	34
																		1	.	.	.	.	.	.	1	1	1	.	.	.	.	2	1	2	35
																			1	.	.	.	.	.	.	.	.	1	1	.	.	.	.	.	36
																				1	.	.	.	.	.	.	.	.	.	1	.	.	.	.	37
																					1	.	.	.	.	.	.	.	1	.	1	.	.	.	38
																						1	.	.	.	.	.	.	.	.	.	.	.	.	39
																							1	.	.	.	.	.	.	1	1	.	.	.	40
																								1	.	.	.	.	.	.	.	.	.	.	41
																									1	.	.	.	.	.	.	1	1	.	42
																										1	.	.	.	.	.	.	.	1	43
																											1	.	.	.	.	.	1	1	44
																												1	.	.	.	.	.	.	45
																													1	.	.	.	.	.	46
																														1	.	.	.	.	47
																															1	.	.	.	48
																																1	.	.	49
																																	1	.	50
																																		1	51

Table 3. *(continued)*

ALGEBRA A7, REPRESENTATIONS OF CLASS 4

16	32	32	48	48	64	64	112	112	48	64	64	80	80	96	96	80	128	128	96	96	96	192	96	192	112	112	112	112	112	112	144	144	144	144	128	128	144	128	160	160	128	160	160	160	160	144	160	208	144	176	176	S.P.
70	168	168	28	28	56	56	8	8	1 120	840	840	840	840	336	336	2 520	336	336	560	560	1 680	56	1 680	56	280	280	1 120	1 120	3 360	3 360	70	420	1 680	420	1 680	1 680	1 680	3 360	840	168	3 360	168	840	840	840	1 680	1 120	840.	1 680	336	336	O.S.
17	23	23	25	25	27	27	29	29	31	35	35	37	37	39	39	41	41	41	43	43	43	43	43	43	45	45	45	47	47	47	49	49	49	49	49	49	49	51	51	51	51	51	51	51	51	53	53	53	53	53	53	LEVEL
70	378	378	336	336	630	630	330	330	3 584	7 350	7 350	11 088	11 088	12 474	12 474	30 618	13 824	13 824	25 200	25 200	43 120	6 006	43 120	6 006	34 650	34 650	60 480	64 512	118 272	118 272	24 696	57 750	103 488	57 750	161 280	161 280	103 488	242 550	128 304	77 616	242 550	77 616	140 400	128 304	140 400	232 848	197 120	109 200	232 848	171 990	171 990	DIM'N
0	1	0	0	0	2	0	4	0	1	1	0	2	1	1	1	0	3	1	0	0	2	5	0	1	2	0	2	0	1	0	0	0	3	0	2	1	0	1	2	2	0	0	3	0	2	1	3	4	0	2	2	
0	0	0	2	0	1	0	0	0	0	0	1	0	0	2	0	1	1	0	1	0	0	0	1	0	0	0	0	0	1	1	0	3	0	1	0	1	1	1	2	0	0	3	0	1	0	0	0	1	0	2	0	
0	1	0	0	0	0	0	0	0	0	0	1	1	0	0	0	0	0	0	0	2	0	0	0	0	0	2	0	1	1	0	0	0	1	0	0	1	0	0	0	0	1	0	1	0	0	2	0	0	1	0	0	
1	0	0	0	0	0	0	0	0	1	0	0	0	0	0	0	1	0	0	0	0	1	0	1	0	0	0	1	1	0	0	3	0	0	0	0	0	0	1	0	0	1	0	0	0	0	0	1	0	0	0	0	
0	0	1	0	0	0	0	0	0	0	1	0	0	1	0	0	0	0	0	2	0	0	0	0	0	2	0	0	1	0	1	0	0	0	0	1	0	1	1	0	0	0	0	0	0	1	1	1	0	2	0	0	
0	0	0	0	2	0	1	0	0	0	1	0	0	0	0	2	1	0	1	0	1	1	0	0	0	0	0	0	0	1	1	0	1	1	3	1	0	0	0	1	3	1	0	0	2	0	0	0	1	0	0	2	
0	0	1	0	0	0	2	0	4	1	0	1	1	2	1	1	0	1	3	0	0	0	1	2	5	0	2	2	0	0	1	0	0	0	0	1	2	3	0	0	0	1	2	2	2	3	0	0	0	1	2	2	WEIGHT
0	1	2	3	4	5	6	7	8	9	10	11	12	13	14	15	16	17	18	19	20	21	22	23	24	25	26	27	28	29	30	31	32	33	34	35	36	37	38	39	40	41	42	43	44	45	46	47	48	49	50	51	NUMBER
1	3	3	2	2	3	3	1	1	16	23	23	30	30	27	27	63	24	24	42	42	78	7	78	7	51	51	100	80	160	160	23	61	116	61	200	200	116	247	120	76	247	76	150	120	150	198	168	80	198	153	153	0
	1	.	1	.	2	.	1	.	4	5	10	15	6	17	5	24	18	4	15	20	33	7	30	1	20	25	40	36	88	56	11	40	72	20	72	110	40	125	85	25	115	51	90	39	50	112	94	65	89	108	48	1
		1	.	1	.	2	.	1	4	10	5	6	15	5	17	24	4	18	20	15	30	1	33	7	25	20	40	36	56	88	11	20	40	40	110	72	72	115	39	51	125	25	50	85	90	89	80	25	112	48	108	2
			1	.	1	.	1	.	.	.	5	5	.	12	.	9	12	.	5	10	9	7	10	.	5	15	10	16	48	16	5	30	38	5	16	60	10	60	60	5	55	36	45	9	10	62	44	50	39	78	9	3
				1	.	1	.	1	.	5	.	.	5	.	12	9	.	12	10	5	10	.	9	7	15	5	10	16	16	48	5	5	10	30	60	16	38	55	9	36	60	5	10	60	45	39	40	5	62	9	78	4
					1	.	1	.	.	.	.	5	.	6	.	.	12	.	.	.	9	7	.	.	5	.	10	.	24	.	.	15	38	.	16	30	.	30	45	5	.	21	45	.	10	26	44	50	.	57	9	5
						1	.	1	.	.	.	.	5	.	6	.	.	12	.	.	.	.	9	7	.	5	10	.	.	24	.	.	.	15	30	16	38	.	.	21	30	5	10	45	45	.	.	.	26	9	57	6
							1	.	.	.	.	.	.	.	.	.	6	.	.	.	.	7	.	.	.	.	.	.	.	.	.	.	14	.	.	.	.	.	15	.	.	.	15	.	.	.	14	35	.	21	.	7
								1	.	.	.	.	.	.	.	.	.	6	.	.	.	.	.	7	.	.	.	.	.	.	.	.	.	.	.	.	14	.	.	.	.	.	.	15	15	.	.	.	.	.	21	8
									1	2	2	3	3	3	3	9	3	3	7	7	12	1	12	1	9	9	16	17	30	30	6	13	24	13	38	38	24	57	27	16	57	16	30	27	30	51	42	20	51	33	33	9
										1	.	.	1	.	2	3	.	2	4	2	4	.	3	1	5	2	4	8	8	16	3	3	6	9	20	8	12	27	6	11	26	3	6	19	15	22	20	4	30	6	23	10
											1	1	.	2	.	3	2	.	2	4	3	1	4	.	2	5	4	8	16	8	3	9	12	3	8	20	6	26	19	3	27	11	15	6	6	30	18	15	22	23	6	11
												1	.	1	.	.	2	.	.	.	3	1	.	.	2	.	4	.	8	.	.	5	12	.	8	10	.	13	14	3	.	6	15	.	6	12	18	15	.	17	6	12
													1	.	1	.	.	2	.	.	.	.	3	1	.	2	4	.	.	8	.	.	.	5	10	8	12	.	.	6	13	3	6	14	15	.	.	.	12	6	17	13
														1	.	.	1	.	.	.	.	1	.	.	.	.	.	.	4	.	.	5	4	.	.	5	.	5	10	.	.	6	5	.	.	6	5	10	.	12	.	14
															1	.	.	1	.	.	.	.	.	1	.	.	.	.	.	4	.	.	.	5	5	.	4	.	.	6	5	.	.	10	5	.	.	.	6	.	12	15
																1	.	.	1	1	1	.	1	.	1	1	1	4	4	4	2	2	3	2	4	4	3	12	4	2	12	2	3	4	3	13	8	3	13	4	4	16

1	.	.	.	.	1	.	.	.	.	.	.	.	.	.	.	4	.	.	.	.	.	5	.	.	.	5	.	.	.	5	10	.	6	.	17
	1	.	.	.	.	.	.	1	.	.	.	.	.	.	.	.	.	.	.	4	.	.	.	.	.	.	5	5	.	.	.	.	.	6	18
		1	.	.	.	.	.	1	.	.	2	.	2	1	.	.	1	2	.	1	6	.	1	4	.	.	3	1	5	4	.	9	.	3	19
			1	.	.	.	.	.	1	.	2	2	.	1	1	1	.	.	2	.	4	3	.	6	1	1	.	.	9	2	2	5	3	.	20
				1	.	.	.	1	.	1	.	2	.	.	1	3	.	4	2	.	6	3	2	.	1	3	.	3	5	8	3	.	3	4	21
					1	.	.	.	.	.	.	.	.	.	.	.	.	.	.	.	.	.	.	.	.	.	.	.	.	.	5	.	.	.	22
						1	.	.	1	1	.	.	2	.	.	.	1	2	4	3	.	.	1	6	2	3	3	3	.	.	.	5	4	3	23
							1	.	.	.	.	.	.	.	.	.	.	.	.	.	.	.	.	.	.	.	.	.	.	.	.	.	.	.	24
								1	.	.	.	.	.	.	.	.	.	2	.	.	3	.	1	.	.	.	.	1	2	4	.	.	.	3	25
									1	.	.	.	.	.	.	.	.	.	2	.	.	.	.	3	1	1	.	.	.	.	.	2	3	.	26
										1	.	.	.	.	.	.	.	2	2	.	.	.	1	.	1	3	.	3	.	.	.	.	3	3	27
											1	.	.	1	.	.	.	.	.	.	2	.	.	2	.	.	.	.	3	1	.	3	.	.	28
												1	.	.	1	1	.	.	1	.	2	2	.	.	1	1	.	.	3	2	2	.	2	.	29
													1	.	.	.	1	1	.	1	.	.	1	2	.	.	2	1	.	.	.	3	.	2	30
														1	.	.	.	.	.	.	.	.	.	.	.	.	.	.	.	.	.	.	.	.	31
															1	.	.	.	.	.	.	1	.	.	1	.	.	.	.	.	1	.	1	.	32
																1	.	.	.	.	.	1	.	.	.	1	.	.	.	2	2	.	1	.	33
																	1	.	.	.	.	.	1	.	.	.	1	.	.	.	.	.	.	1	34
																		1	.	.	.	.	1	.	.	.	.	1	.	.	.	.	.	2	35
																			1	.	.	.	.	.	1	1	.	.	.	.	.	.	2	.	36
																				1	.	.	.	.	.	.	1	1	.	.	.	.	.	1	37
																					1	.	.	.	.	.	.	.	1	1	.	.	.	.	38
																						1	.	.	.	.	.	.	.	.	1	.	1	.	39
																							1	.	.	.	.	.	.	.	.	.	.	1	40
																								1	.	.	.	.	.	.	.	1	.	.	41
																									1	.	.	.	.	.	.	.	1	.	42
																										1	.	.	.	.	.	.	1	.	43
																											1	.	.	.	.	.	.	1	44
																												1	.	.	.	.	.	1	45
																													1	.	.	.	.	.	46
																														1	.	.	.	.	47
																															1	.	.	.	48
																																1	.	.	49
																																	1	.	50
																																		1	51

Table 3. *(continued)*

ALGEBRA A8, REPRESENTATIONS OF CLASS 0

0	18	36	54	54	72	54	72	72	72	108	108	90	90	90	126	126	108	108	108	108	162	126	126	126	180	126	108	180	162	144	162	126	126	144	162	144	126	144	162	144	180	198	162	198	180	162	198	198	216	216	270	S.P.
1	72	756	252	252	72	1680	630	2520	2520	504	504	3024	3780	3780	504	504	504	1260	1260	504	72	2520	2520	1512	630	1512	15120	630	84	1512	84	3780	3780	1512	1512	756	10080	7560	1512	7560	504	1512	1512	1512	504	1512	1512	1512	252	252	504	O.S.
1	17	29	31	31	33	37	41	41	41	43	43	45	47	47	47	47	49	49	49	49	49	51	51	53	53	53	53	53	55	55	55	55	55	55	57	57	57	57	57	57	59	59	59	59	59	59	59	59	61	61	61	LEVEL
1	80	1215	1540	1540	1944	5760	8820	17325	17325	12320	12320	51200	53460	53460	36036	36036	38808	49500	49500	38808	25200	93555	93555	110880	50050	110880	297675	50050	41580	171072	41580	267300	267300	171072	213840	237160	497664	500500	213840	500500	221760	243243	442260	261954	221760	442260	243243	261954	152460	205920	112112	DIM'N
0	1	0	2	0	2	0	0	1	0	3	0	1	1	0	3	1	1	0	0	0	3	2	0	0	4	0	1	0	0	2	0	1	0	0	1	0	1	2	0	1	1	3	2	4	0	0	0	1	0	3	5	
0	0	1	0	1	0	0	0	1	0	0	0	1	0	0	0	1	0	2	0	0	0	1	0	1	0	0	0	0	0	0	0	0	2	0	2	2	0	1	0	0	1	1	1	0	0	2	0	0	3	0	0	
0	0	0	0	0	0	1	0	0	1	0	1	0	1	0	0	0	0	0	0	0	0	0	0	1	0	0	1	0	3	1	0	1	0	0	0	0	0	0	0	1	2	0	0	0	0	0	0	1	1	2	0	
0	0	0	0	0	0	0	1	0	0	0	0	0	0	1	0	0	2	0	1	0	0	0	1	1	0	0	0	1	0	1	0	0	0	0	1	0	1	0	0	0	0	1	0	0	0	0	0	0	0	0	1	
0	0	0	0	0	0	0	1	0	0	0	0	0	1	0	0	0	0	1	0	2	0	1	0	0	1	1	0	0	0	0	0	0	0	1	0	0	1	0	1	0	0	0	0	0	0	0	1	0	0	0	0	
0	0	0	0	0	0	1	0	1	0	1	0	0	0	1	0	0	0	0	0	0	0	0	0	0	0	1	1	0	0	0	3	0	1	1	0	0	0	1	0	0	0	0	0	1	2	0	0	0	0	0	0	
0	0	1	1	0	0	0	0	0	1	0	0	1	0	0	1	0	0	0	2	0	0	0	1	0	0	1	0	0	0	0	0	2	0	0	0	2	0	0	2	1	0	0	2	0	1	1	1	0	0	0	0	
0	1	0	0	2	2	0	0	0	1	0	3	1	0	1	1	3	0	0	0	1	3	0	2	0	0	0	1	4	0	0	0	0	1	2	0	0	1	1	1	2	0	0	0	1	1	2	3	4	0	0	0	WEIGHT
0	1	2	3	4	5	6	7	8	9	10	11	12	13	14	15	16	17	18	19	20	21	22	23	24	25	26	27	28	29	30	31	32	33	34	35	36	37	38	39	40	41	42	43	44	45	46	47	48	49	50	51	NUMBER
1	8	27	28	28	36	48	42	105	105	56	56	272	162	162	168	168	84	120	120	84	120	189	189	168	70	168	693	70	42	216	42	540	540	216	216	448	720	910	216	910	168	189	756	378	168	756	189	378	84	120	56	0
	1	6	7	7	8	14	14	35	35	21	21	90	63	63	56	56	35	50	50	35	36	84	84	77	35	77	280	35	21	105	21	225	225	105	111	182	322	385	111	385	91	105	315	168	91	315	105	168	49	70	35	1
		1	1	1	1	4	5	10	10	6	6	24	24	24	14	14	15	20	20	15	8	34	34	36	15	36	106	15	11	50	11	85	85	50	56	67	144	145	56	145	50	55	114	63	50	114	55	63	29	40	20	2
			1	.	1	.	.	5	.	6	.	12	9	.	14	7	5	10	.	.	8	24	.	16	15	.	38	.	5	30	.	30	45	.	36	36	44	100	.	50	30	45	78	63	.	63	.	21	19	30	20	3
				1	1	.	.	.	5	.	6	12	.	9	7	14	.	.	10	5	8	.	24	.	.	16	38	15	.	.	5	45	30	30	.	36	44	50	36	100	.	.	63	21	30	78	45	63	.	.	.	4
					1	.	.	.	.	.	.	6	.	.	7	7	.	.	.	.	8	.	.	.	.	.	14	.	.	.	.	15	15	.	.	21	14	35	.	35	.	.	42	21	.	42	.	21	.	.	.	5
						1	2	2	2	1	1	4	9	9	2	2	7	7	7	7	1	12	12	17	5	17	36	5	6	24	6	27	27	24	27	20	66	45	27	45	28	27	33	18	28	33	27	18	17	23	10	6
							1	.	.	.	.	.	3	3	.	.	4	2	2	4	.	3	3	8	1	8	9	1	3	12	3	6	6	12	12	4	32	9	12	9	16	12	6	3	16	6	12	3	10	14	4	7
								1	.	1	.	2	3	.	2	1	2	4	.	.	1	8	.	8	5	.	12	.	3	13	.	9	15	.	18	11	18	30	.	15	17	21	22	18	.	18	.	6	12	16	10	8
									1	.	1	2	.	3	1	2	.	.	4	2	1	.	8	.	.	8	12	5	.	.	3	15	9	13	.	11	18	15	18	30	.	.	18	6	17	22	21	18	.	.	.	9
										1	.	.	.	.	2	.	.	.	.	.	1	4	.	.	5	.	.	.	.	5	.	.	.	.	6	.	.	15	.	.	5	15	11	18	.	.	.	.	4	10	10	10
											1	.	.	.	.	2	.	.	.	.	1	.	4	.	.	.	.	5	.	.	.	.	.	5	.	.	.	.	6	15	.	.	.	.	5	11	15	18	.	.	.	11
												1	.	.	1	1	.	.	.	.	1	.	.	.	.	.	4	.	.	.	.	5	5	.	.	6	5	10	.	10	.	.	12	6	.	12	.	6	.	.	.	12
													1	.	.	.	1	1	.	.	.	2	.	4	1	.	3	.	2	6	.	2	3	.	8	2	8	6	.	3	10	9	4	3	.	3	.	1	7	9	4	13
														1	.	.	.	.	1	1	.	.	2	.	.	4	3	1	.	.	2	3	2	6	.	2	8	3	8	6	.	.	3	1	10	4	9	3	.	.	.	14
															1	.	.	.	.	.	1	.	.	.	.	.	.	.	.	.	.	.	.	.	.	.	.	5	.	.	.	.	6	6	.	.	.	.	.	.	.	15
																1	.	.	.	.	1	.	.	.	.	.	.	.	.	.	.	.	.	.	.	.	.	.	.	5	.	.	.	.	.	6	.	6	.	.	.	16

1	.	.	.	.	.	.	2	.	.	.	.	1	3	.	.	.	.	3	.	4	.	.	.	6	3	.	.	.	.	.	.	4	6	1	17
	1	.	.	.	1	.	2	1	.	.	.	1	2	.	.	3	.	6	2	.	3	.	.	6	6	2	3	.	3	.	.	6	5	4	18
		1	.	.	.	1	.	.	2	.	1	.	.	1	3	.	2	.	2	.	.	6	3	.	.	3	.	6	2	6	3	.	.	.	19
			1	.	.	.	.	.	2	.	.	.	.	1	.	.	3	.	.	4	.	3	.	.	.	.	.	6	.	3	.	.	.	.	20
				1	.	.	.	.	.	.	.	.	.	.	.	.	.	.	.	.	.	.	.	.	.	.	.	.	.	.	.	.	.	.	21
					1	.	.	1	.	.	.	.	2	.	.	.	.	3	.	.	3	.	.	3	6	2	3	.	.	.	.	3	5	4	22
						1	.	.	.	.	1	.	.	.	.	.	2	.	.	.	.	3	3	.	.	.	.	3	2	6	3	.	.	.	23
							1	.	.	.	.	1	1	.	.	.	.	2	.	.	.	.	.	4	2	.	.	.	.	.	.	3	3	1	24
								1	.	.	.	.	.	.	.	.	.	.	.	.	.	.	.	.	3	.	3	.	.	.	.	.	2	4	25
									1	.	.	.	.	1	.	.	1	.	.	.	.	2	.	.	.	.	.	4	.	2	.	.	.	.	26
										1	.	.	.	.	1	1	.	.	1	2	2	.	2	.	.	2	1	.	2	.	1	.	.	.	27
											1	.	.	.	.	.	.	.	.	.	.	.	.	.	.	.	.	.	.	3	3	.	.	.	28
												1	.	.	.	.	.	.	.	.	.	.	.	2	.	.	.	.	.	.	.	1	1	.	29
													1	.	.	.	.	1	.	.	.	.	.	2	2	.	.	.	.	.	.	2	3	1	30
														1	.	.	.	.	.	.	.	.	.	.	.	.	.	2	.	.	.	.	.	.	31
															1	.	.	.	1	.	.	.	1	.	.	2	.	.	1	.	1	.	.	.	32
																1	.	.	1	.	1	.	.	.	.	1	1	.	2	.	.	.	.	.	33
																	1	.	.	.	.	1	.	.	.	.	.	2	.	2	.	.	.	.	34
																		1	.	.	.	.	.	1	1	.	.	.	.	.	.	2	1	1	35
																			1	.	.	.	.	.	.	1	.	.	1	.	.	.	.	.	36
																				1	.	.	.	.	.	.	.	.	.	.	.	.	.	.	37
																					1	.	.	.	.	1	1	.	.	.	.	.	.	.	38
																						1	.	.	.	.	.	1	.	1	.	.	.	.	39
																							1	.	.	.	.	.	1	.	1	.	.	.	40
																								1	.	.	.	.	.	.	.	1	1	.	41
																									1	.	.	.	.	.	.	.	1	1	42
																										1	.	.	.	.	.	.	.	.	43
																											1	.	.	.	.	.	.	.	44
																												1	.	.	.	.	.	.	45
																													1	.	.	.	.	.	46
																														1	.	.	.	.	47
																															1	.	.	.	48
																																1	.	.	49
																																	1	.	50
																																		1	51

Table 3. *(continued)*

ALGEBRA A8, REPRESENTATIONS OF CLASS 1

S.P.	8	26	44	44	62	62	62	98	80	80	116	80	80	116	98	98	98	116	116	98	170	134	116	134	134	116
O.S.	9	252	72	1 260	504	1 512	1 260	252	504	126	72	2 520	5 040	630	1 260	1 512	2 520	2 520	756	7 560	504	252	1 512	1 512	1 512	3 780
LEVEL	9	23	25	33	35	37	39	39	39	41	41	43	45	45	47	47	49	49	49	49	51	51	51	51	51	53
DIM'N	9	315	396	2 700	3 465	7 920	7 560	5 544	10 080	5 292	7 020	23 760	41 580	17 325	37 125	41 580	66 528	69 300	41 580	126 000	36 036	41 580	110 880	80 190	90 090	148 500
	1	0	2	0	0	1	0	3	1	0	3	0	1	0	0	0	1	2	0	1	4	0	0	0	1	0
	0	1	0	0	0	1	0	0	1	0	0	0	0	0	2	0	0	1	0	0	0	0	2	0	0	1
	0	0	0	1	1	0	0	0	0	0	0	0	1	0	0	0	0	0	0	1	0	0	0	0	1	1
	0	0	0	0	0	0	1	0	0	0	0	1	0	1	0	0	1	0	0	0	0	0	0	0	0	0
	0	0	0	0	0	0	0	0	0	2	0	0	0	0	0	1	1	0	1	0	0	0	0	1	0	1
	0	0	0	0	0	0	1	0	0	0	0	0	1	0	1	1	0	1	0	0	1	2	0	0	0	0
	0	0	0	1	0	1	0	1	0	0	0	1	0	0	0	0	0	0	2	1	0	1	1	1	0	0
WEIGHT	0	1	1	0	2	0	0	0	2	0	2	1	0	3	0	1	0	0	0	1	0	0	1	2	3	0
NUMBER	0	1	2	3	4	5	6	7	8	9	10	11	12	13	14	15	16	17	18	19	20	21	22	23	24	25
0	1	7	8	20	21	48	28	28	56	14	36	64	112	35	85	70	112	140	56	280	56	42	224	90	168	190
1		1	1	5	6	12	9	7	14	5	8	24	38	15	30	30	44	50	26	100	21	21	78	45	63	80
2			1	.	.	6	.	7	7	.	8	.	14	.	15	.	14	35	.	35	21	.	42	.	21	35
3				1	1	2	3	1	2	2	1	8	12	5	9	13	18	15	12	30	6	11	22	21	18	33
4					1	.	.	.	2	.	1	4	.	5	.	5	.	.	6	15	.	5	11	15	18	.
5						1	.	1	1	.	1	.	4	.	5	.	5	10	.	10	6	.	12	.	6	15
6							1	.	.	1	.	2	3	1	2	6	8	3	5	6	1	6	4	9	3	13
7								1	.	.	1	.	.	.	.	.	.	5	.	.	6	.	.	.	.	.
8									1	.	1	.	.	.	.	.	.	.	.	5	.	.	6	.	6	.
9										1	.	.	.	.	.	3	4	.	2	.	.	3	.	3	.	5
10											1	.	.	.	.	.	.	.	.	.	.	.	.	.	.	.
11												1	.	1	.	2	.	.	3	3	.	3	2	6	3	.
12													1	.	1	.	2	2	.	2	1	.	2	.	1	6
13														1	.	.	.	.	.	.	.	.	.	3	3	.
14															1	.	.	1	.	.	1	.	2	.	.	3
15																1	.	.	1	.	.	2	.	2	.	.
16																	1	.	.	.	.	.	.	.	.	2

S.P.	134	152	152	188	134	134	116	188	170	170	152	152	152	134	224	134	224	206	188	134	152	242	278	170	260	152
O.S.	3 024	252	252	504	756	3 780	15 120	504	504	504	504	504	2 520	5 040	36	2 520	72	504	2 520	15 120	10 080	72	252	3 024	630	7 560
LEVEL	53	53	53	53	55	55	55	55	55	55	57	57	57	57	57	57	57	57	59	59	59	59	59	61	61	61
DIM'N	205 920	66 528	78 624	45 045	116 424	224 532	408 240	105 840	103 950	145 530	124 740	166 320	259 875	380 160	32 670	302 400	74 250	123 552	288 288	841 995	720 720	77 220	63 063	608 256	126 126	873 180
	2	0	0	0	0	2	1	4	0	2	0	2	1	1	0	1	4	0	3	0	1	0	0	1	5	1
	1	0	2	0	1	0	0	0	0	1	0	0	2	0	0	0	0	0	1	1	0	0	0	1	0	0
	0	0	0	0	0	1	0	0	0	0	2	0	0	0	0	0	0	0	0	1	0	0	0	1	0	0
	0	0	0	0	2	0	1	0	0	0	1	2	0	1	0	0	0	0	0	0	1	0	0	1	0	0
	0	0	0	1	0	1	0	0	0	0	0	0	1	0	0	2	0	0	1	0	0	0	0	0	1	1
	0	2	0	0	0	0	1	0	1	0	0	0	0	0	0	0	0	1	0	1	0	0	1	0	0	1
	1	0	0	0	0	0	0	1	2	0	0	0	0	2	4	0	0	1	0	0	1	3	0	0	0	1
WEIGHT	1	2	3	4	0	0	1	1	1	3	0	0	0	0	0	1	3	3	0	1	2	2	5	0	0	0
NUMBER	26	27	28	29	30	31	32	33	34	35	36	37	38	39	40	41	42	43	44	45	46	47	48	49	50	51
0	384	56	140	35	112	252	504	168	70	252	98	140	245	400	14	280	120	64	224	861	672	28	21	384	70	616
1	132	30	49	20	51	108	216	56	40	84	47	65	110	180	9	130	36	40	104	372	312	19	15	192	35	309
2	90	.	28	.	21	63	63	56	.	56	21	35	70	50	.	35	36	.	84	168	84	.	.	112	35	77
3	36	16	13	10	24	45	90	14	23	21	23	31	46	76	6	62	8	24	44	150	132	13	10	96	15	156
4	18	10	13	10	.	.	33	7	15	21	.	.	.	40	4	20	8	20	.	54	92	10	10	.	.	72
5	24	.	7	.	10	24	24	14	.	14	11	15	30	20	.	15	8	.	34	68	34	.	.	56	15	36
6	6	9	2	4	12	18	36	2	13	3	11	16	17	28	4	32	1	14	16	54	48	9	6	48	5	79
7	12	.	.	.	.	9	.	14	.	7	.	5	10	.	.	.	8	.	24	.	.	.	.	16	15	.
8	12	.	7	.	.	.	9	7	.	14	.	.	.	10	.	5	8	.	.	24	24	.	.	.	.	16
9	.	6	.	1	6	6	12	.	7	.	5	10	5	8	3	20	.	8	4	15	12	6	3	24	1	40
10	6	.	.	.	.	.	.	7	.	7	.	.	.	.	.	.	8	.	.	.	.	.	.	.	.	.
11	3	5	2	4	.	.	12	1	9	3	.	.	.	16	3	9	1	11	.	18	32	7	6	.	.	38
12	4	.	1	.	5	9	9	2	.	2	6	7	11	7	.	7	1	.	12	24	12	.	.	28	5	17
13	.	2	2	4	.	.	.	.	3	3	.	.	.	.	1	.	1	8	.	.	16	4	6	.	.	.
14	2	.	1	.	2	3	.	2	.	1	3	2	8	.	.	.	1	.	8	12	.	.	.	16	5	.
15	.	3	.	1	.	.	4	.	5	.	.	.	.	4	2	5	.	6	.	5	8	5	3	.	.	20
16	.	.	.	.	3	3	3	.	.	.	3	4	3	2	.	4	.	.	3	6	3	.	.	14	1	8

1	.	.	1	.	.	.	.	.	2	.	.	.	.	3	.	2	.	1	.	2	4	.	.	.	1	.	8	.	.	.	.	8	5	.	17
	1	.	.	1	.	1	.	.	.	1	.	1	.	.	.	.	4	.	.	.	.	4	2	.	.	4	.	.	4	4	3	.	.	10	18
		1	.	.	1	.	1	.	2	.	1	.	.	.	3	1	.	2	.	.	.	4	.	2	1	.	.	8	8	.	.	.	.	8	19
			1	.	.	.	.	.	.	.	.	.	.	.	.	2	.	.	.	.	.	.	.	.	1	.	4	.	.	.	.	.	5	.	20
				1	.	.	.	.	.	1	.	.	.	.	.	.	2	.	.	.	.	.	1	.	.	2	.	.	.	3	1	.	.	5	21
					1	.	.	.	1	.	1	.	.	.	.	1	.	1	.	.	.	.	.	.	1	.	.	4	.	.	.	.	.	.	22
						1	.	.	.	1	.	1	.	.	.	.	2	.	.	.	.	.	1	.	.	4	.	.	4	3	3	.	.	.	23
							1	.	.	.	1	.	.	.	.	.	.	2	.	.	.	.	.	.	1	.	.	.	4	.	.	.	.	.	24
								1	.	.	.	.	1	1	.	.	.	.	2	1	2	.	.	.	.	.	2	3	.	.	.	8	1	.	25
									1	.	.	.	.	.	.	1	.	1	.	.	.	.	.	.	1	.	.	.	.	.	.	.	.	.	26
										1	.	.	.	.	.	.	1	.	.	.	.	.	1	.	.	2	.	.	.	2	1	.	.	.	27
											1	.	.	.	.	.	.	1	.	.	.	.	.	.	1	.	.	.	.	.	.	.	.	.	28
												1	.	.	.	.	.	.	.	.	.	.	.	.	.	2	.	.	.	1	3	.	.	.	29
													1	.	.	.	.	.	1	1	.	.	.	.	.	.	.	.	.	.	.	4	.	.	30
														1	.	.	.	.	.	1	1	.	.	.	.	.	2	.	.	.	.	4	1	.	31
															1	.	.	.	.	.	.	1	.	1	.	.	.	2	2	.	.	.	.	4	32
																1	.	.	.	.	.	.	.	.	1	.	.	.	.	.	.	.	.	.	33
																	1	.	.	.	.	.	1	.	.	1	.	.	.	2	1	.	.	.	34
																		1	.	.	.	.	.	.	1	.	.	.	.	.	.	.	.	.	35
																			1	.	.	.	.	.	.	.	.	.	.	.	.	2	.	.	36
																				1	.	.	.	.	.	.	.	.	.	.	.	2	.	.	37
																					1	.	.	.	.	.	1	.	.	.	.	2	1	.	38
																						1	.	.	.	.	.	.	1	.	.	.	.	2	39
																							1	.	.	.	.	.	.	1	.	.	.	.	40
																								1	.	.	.	.	.	.	.	.	.	2	41
																									1	.	.	.	.	.	.	.	.	.	42
																										1	.	.	.	1	1	.	.	.	43
																											1	.	.	.	.	.	1	.	44
																												1	.	.	.	.	.	.	45
																													1	.	.	.	.	.	46
																														1	.	.	.	.	47
																															1	.	.	.	48
																																1	.	.	49
																																	1	.	50
																																		1	51

Table 3. *(continued)*

ALGEBRA A8, REPRESENTATIONS OF CLASS 2

14	32	32	50	86	50	68	68	68	86	86	86	104	104	104	122	86	158	122	122	104	176	140	104	104	176	194	122	140	212	122	140	122	122	284	140	140	176	140	140	158	392	158	248	158	158	158	176	158	140	194	176	S.P.
36	9	504	504	72	1 260	630	504	3 780	756	1 512	1 512	1 512	252	252	504	5 040	252	252	504	630	72	504	5 040	10 080	72	252	5 040	2 520	72	3 780	2 520	7 560	7 560	72	1 260	2 520	2 520	3 780	7 560	1 512	9	504	504	1 512	3 024	7 560	1 512	7 560	22 680	3 024	1 512	O.S.
15	17	27	31	33	35	37	39	41	43	43	43	45	45	45	45	47	47	47	47	49	49	49	51	51	51	51	53	53	53	55	55	55	55	55	57	57	57	57	57	57	57	59	59	59	59	59	59	61	61	61	61	LEVEL
36	45	720	2 079	1 440	3 780	4 950	6 048	19 800	17 325	20 790	25 515	32 076	16 632	22 176	15 840	58 212	16 380	20 790	40 950	41 580	20 790	41 580	118 800	184 320	23 760	25 740	177 408	135 135	30 888	207 900	198 000	330 750	363 825	20 592	178 200	291 060	216 216	332 640	540 540	262 080	6 435	174 636	91 728	336 798	594 594	643 500	388 080	831 600	1 299 078	645 120	540 540	DIM'N
0	2	0	1	3	0	0	0	1	0	0	1	2	0	0	0	1	4	0	2	1	4	0	0	1	0	0	2	1	0	0	1	1	0	0	0	2	3	1	2	0	0	0	5	1	1	1	2	1	0	3	1	
1	0	0	1	0	0	0	0	0	2	0	0	1	0	2	0	0	0	0	1	0	0	0	1	0	0	0	0	0	0	1	2	0	1	0	0	0	1	0	0	1	0	0	0	0	2	0	0	1	1	1	0	
0	0	1	0	0	0	0	0	1	0	0	1	0	0	0	0	0	0	0	0	0	0	0	1	0	0	0	1	0	0	0	0	0	1	0	2	0	0	0	1	1	0	1	0	0	0	0	1	1	0	0	0	
0	0	0	0	0	1	1	0	0	0	0	0	0	0	0	0	1	0	0	0	0	0	0	0	1	0	0	0	1	0	1	0	0	0	0	0	1	0	0	0	0	0	2	0	0	0	0	0	0	1	0	0	
0	0	0	0	0	0	0	1	0	0	1	0	0	0	0	1	0	0	0	0	2	0	0	0	0	0	0	0	0	0	1	0	1	0	0	1	1	0	1	0	0	0	0	0	0	0	1	0	1	0	0	0	
0	0	0	0	0	0	0	1	0	0	0	0	0	2	0	0	1	0	1	0	0	0	1	1	0	0	1	1	0	0	0	1	1	0	0	0	0	1	0	0	0	0	0	1	2	0	0	0	0	1	0	2	
0	0	0	0	0	1	0	0	1	1	1	0	1	0	0	0	0	1	2	0	0	0	1	0	1	3	0	0	0	2	0	0	0	1	1	0	0	0	2	1	0	0	0	0	1	1	1	0	0	0	1	0	
0	0	1	1	1	0	2	0	0	0	1	2	0	1	2	3	0	0	0	2	0	2	2	0	1	1	4	0	3	3	0	0	1	1	5	0	0	0	0	1	3	7	0	0	0	1	2	3	0	1	1	2	WEIGHT
0	1	2	3	4	5	6	7	8	9	10	11	12	13	14	15	16	17	18	19	20	21	22	23	24	25	26	27	28	29	30	31	32	33	34	35	36	37	38	39	40	41	42	43	44	45	46	47	48	49	50	51	NUMBER
1	1	6	14	8	14	15	14	55	42	35	63	69	21	49	20	91	28	21	84	49	36	35	148	232	14	15	196	140	14	175	190	280	385	6	127	217	175	238	525	238	1	98	56	189	518	405	336	490	846	496	266	0
	1	.	7	8	.	.	.	20	21	.	21	48	.	28	.	28	28	.	56	14	36	.	64	64	.	.	112	35	.	70	120	70	175	.	56	112	140	56	280	112	.	42	56	42	336	90	168	280	342	384	56	1
		1	2	1	4	5	5	15	11	15	18	18	10	13	10	33	7	11	21	20	8	20	54	92	9	10	72	60	10	75	70	130	145	5	56	94	65	118	195	93	1	47	21	102	186	210	126	220	387	174	150	2
			1	1	.	.	.	5	6	.	6	12	.	7	.	9	7	.	14	5	8	.	24	24	.	.	38	15	.	30	45	30	65	.	26	44	50	26	100	42	.	21	21	21	120	45	63	125	153	132	30	3
				1	.	.	.	.	.	.	.	6	.	.	.	.	7	.	7	.	8	.	.	.	.	.	14	.	.	.	15	.	.	.	.	14	35	.	35	.	.	.	21	.	42	.	21	35	.	90	.	4
					1	1	2	3	2	6	3	3	5	2	4	12	1	6	3	9	1	11	18	32	6	6	24	20	7	33	22	61	46	4	24	42	20	57	60	28	1	23	6	57	54	102	36	96	171	48	85	5
						1	.	.	.	3	3	.	2	2	4	.	.	3	3	.	1	8	.	16	4	6	.	20	6	.	.	23	23	4	.	.	.	30	30	28	1	.	.	27	27	72	36	.	63	24	52	6
							1	.	.	2	.	.	3	.	1	4	.	3	.	5	.	6	5	8	4	3	6	4	5	15	5	30	10	3	10	20	4	25	12	5	1	11	1	33	10	45	6	40	72	8	51	7
								1	1	.	1	2	.	1	.	3	1	.	2	2	1	.	8	8	.	.	12	5	.	13	14	13	20	.	12	18	15	12	30	12	.	11	6	11	34	21	18	54	66	36	16	8
									1	.	.	1	.	1	.	.	1	.	1	.	1	.	4	.	.	.	4	.	.	5	10	.	10	.	6	5	10	.	10	6	.	5	6	.	24	.	6	30	24	24	.	9
										1	.	.	1	.	1	.	.	2	.	.	.	4	.	4	3	3	.	4	4	.	.	10	5	3	.	.	.	15	6	5	1	.	.	17	5	30	6	.	24	4	28	10
											1	.	.	1	.	.	.	.	2	.	1	.	.	4	.	.	.	5	.	.	.	5	10	.	.	.	.	6	15	12	.	.	.	5	17	15	18	.	24	18	10	11
												1	.	.	.	.	1	.	1	.	1	.	.	.	.	.	4	.	.	.	5	.	.	.	.	5	10	.	10	.	.	.	6	.	12	.	6	15	.	24	.	12
													1	.	.	.	.	1	.	.	.	2	.	.	2	1	.	.	3	.	.	5	.	2	.	.	.	5	.	.	1	.	.	12	.	10	.	.	9	.	18	13
														1	.	.	.	.	1	.	1	.	.	.	.	.	.	.	.	.	.	.	5	.	.	.	.	.	5	6	.	.	.	.	12	.	6	.	9	12	.	14
															1	.	.	.	.	.	.	2	.	.	1	3	.	4	3	.	.	.	.	3	.	.	.	.	.	5	1	.	.	.	.	15	6	.	.	.	11	15
																1	.	.	.	1	.	.	2	2	.	.	3	1	.	6	3	6	4	.	5	8	3	5	6	2	.	6	1	6	6	9	3	22	27	6	9	16

17	1	.	.	.	1	.	.	.	.	.	.	.	.	.	.	.	.	.	.	.	5	.	.	.	.	.	6	.	.	.	.	.	.	12	.
18		1	.	.	.	1	.	.	2	1	.	.	2	.	.	.	.	2	.	.	.	5	.	.	1	.	.	6	.	5	.	.	.	.	6
19			1	.	1	.	.	.	.	.	.	.	.	.	.	.	.	.	.	.	.	.	5	.	.	.	.	.	6	.	6	.	.	12	.
20				1	.	.	.	.	.	.	.	.	.	3	.	3	.	.	2	4	.	2	.	.	.	3	.	3	.	3	.	8	9	.	6
21					1	.	.	.	.	.	.	.	.	.	.	.	.	.	.	.	.	.	.	.	.	.	.	.	.	.	.	.	.	6	.
22						1	.	.	1	1	.	.	2	.	.	.	.	2	.	.	.	.	.	.	1	.	.	.	.	5	.	.	.	.	6
23							1	.	.	.	1	.	.	2	2	.	2	.	3	2	2	.	2	1	.	3	1	.	4	.	1	12	9	4	.
24								1	.	.	.	1	.	.	.	2	2	.	.	.	.	3	3	2	.	.	.	3	3	6	3	.	9	3	5
25									1	.	.	.	1	.	.	.	.	1	.	.	.	.	.	.	1	.	.	.	.	.	.	.	.	.	.
26										1	.	.	1	.	.	.	.	2	.	.	.	.	.	.	1	.	.	.	.	.	.	.	.	.	.
27											1	.	.	.	1	.	.	.	.	2	2	.	2	.	.	.	1	.	2	.	1	6	.	4	.
28												1	.	.	.	.	.	.	.	.	.	.	.	2	.	.	.	.	.	3	3	.	.	.	2
29													1	.	.	.	.	1	.	.	.	.	.	.	1	.	.	.	.	.	.	.	.	.	.
30														1	.	.	.	.	1	1	.	.	.	.	.	2	.	.	.	.	.	4	3	.	.
31															1	.	.	.	.	.	1	.	.	.	.	.	1	.	2	.	.	3	.	2	.
32																1	.	.	.	.	.	1	.	.	.	.	.	2	.	2	.	.	3	.	3
33																	1	.	.	.	.	.	1	1	.	.	.	.	2	.	1	.	3	2	.
34																		1	.	.	.	.	.	.	1	.	.	.	.	.	.	.	.	.	.
35																			1	.	.	.	.	.	.	1	.	.	.	.	.	2	.	.	.
36																				1	.	.	.	.	.	.	.	.	.	.	.	2	.	.	.
37																					1	.	.	.	.	.	1	.	.	.	.	.	.	2	.
38																						1	.	.	.	.	.	1	.	1	.	.	.	.	1
39																							1	.	.	.	.	.	1	.	1	.	.	2	.
40																								1	.	.	.	.	.	.	1	.	.	.	.
41																									1	.	.	.	.	.	.	.	.	.	.
42																										1	.	.	.	.	.	.	.	.	.
43																											1	.	.	.	.	.	.	.	.
44																												1	.	.	.	.	.	.	1
45																													1	.	.	.	.	1	.
46																														1	.	.	.	.	1
47																															1	.	.	.	.
48																																1	.	.	.
49																																	1	.	.
50																																		1	.
51																																			1

Table 3. *(continued)*

ALGEBRA A8, REPRESENTATIONS OF CLASS 3

18	36	72	36	54	54	72	72	72	90	144	90	72	126	126	90	144	90	90	198	108	108	288	126	108	126	108	162	126	144	144	126	126	126	234	144	180	144	126	252	162	144	144	144	144	198	162	216	180	180	180	144	S.P.
84	72	9	630	756	1 512	84	252	504	504	72	504	5 040	36	252	2 520	72	2 520	3 780	72	3 780	1 512	9	1 512	7 560	1 512	7 560	1 512	1 512	504	2 520	1 260	1 680	5 040	252	1 512	504	630	15 120	72	3 024	2 520	10 080	2 520	10 080	504	3 780	1 512	126	1 260	5 040	15 120	O.S.
19	23	25	29	35	35	37	37	37	39	41	41	43	43	43	45	45	47	47	47	49	49	49	51	51	51	53	53	53	53	53	55	55	55	55	55	55	57	57	57	57	59	59	59	59	59	59	59	61	61	61	61	LEVEL
84	240	165	1 050	3 402	5 346	2 520	4 620	4 620	8 505	4 290	11 088	29 568	4 950	9 240	38 880	10 692	48 510	57 750	8 580	85 536	74 844	3 003	93 555	168 000	110 565	187 110	101 088	136 080	120 120	128 700	138 600	148 500	258 720	42 042	171 072	129 360	187 110	598 752	53 460	343 200	332 640	675 840	462 000	823 680	198 198	442 260	213 840	116 424	317 625	576 576	1 122 660	DIM'N
0	1	3	0	0	1	0	0	0	2	4	0	1	0	0	1	0	1	0	0	2	0	0	1	1	2	0	3	1	1	1	0	0	2	5	1	3	2	0	5	1	0	1	0	2	1	0	1	0	0	3	0	
0	1	0	0	0	0	0	2	0	1	0	0	0	0	0	0	0	0	1	0	0	1	0	2	0	0	1	1	0	2	0	1	0	0	0	0	1	0	1	0	0	0	1	0	0	0	1	0	0	3	0	1	
1	0	0	0	0	1	0	0	0	0	0	0	0	0	0	0	0	0	1	0	1	1	0	0	0	1	0	0	0	0	0	0	2	0	0	0	0	0	0	0	0	1	1	2	0	0	0	0	0	0	1	0	
0	0	0	1	0	0	0	0	0	0	0	0	1	0	0	1	0	0	0	0	0	0	0	0	0	0	1	0	0	0	0	0	0	1	0	0	0	0	1	0	0	1	0	0	1	0	1	0	3	0	0	0	
0	0	0	0	1	0	0	0	1	0	0	0	0	0	0	0	0	1	0	0	0	0	0	0	1	0	0	0	0	0	1	2	0	0	0	0	0	2	0	0	0	1	0	0	0	0	0	0	0	0	0	1	
0	0	0	0	0	0	2	0	0	0	0	1	0	0	1	0	0	1	0	0	0	0	0	0	0	0	1	0	2	0	0	0	1	1	0	1	0	0	0	0	1	0	1	0	0	0	0	1	0	1	1	1	
0	0	0	0	1	0	0	0	0	0	0	1	1	3	0	0	2	0	1	1	1	0	0	1	1	0	0	1	0	0	0	0	0	0	1	2	0	0	1	0	1	0	0	1	1	3	0	0	0	0	0	0	
0	0	0	1	0	1	0	1	2	1	1	1	0	0	3	2	2	0	0	4	0	2	6	0	1	2	0	0	1	2	3	0	0	0	0	0	2	0	1	2	2	0	0	1	1	1	3	4	0	0	0	1	WEIGHT
0	1	2	3	4	5	6	7	8	9	10	11	12	13	14	15	16	17	18	19	20	21	22	23	24	25	26	27	28	29	30	31	32	33	34	35	36	37	38	39	40	41	42	43	44	45	46	47	48	49	50	51	NUMBER
1	2	1	5	9	18	5	13	10	21	8	16	52	5	10	60	9	58	80	5	105	93	1	99	160	126	162	90	102	119	100	95	110	199	28	108	112	114	432	36	190	178	416	290	544	82	270	90	47	160	300	585	0
	1	1	.	.	6	.	7	.	14	8	.	14	.	.	15	.	14	35	.	55	42	.	63	35	63	63	69	21	77	20	35	50	91	28	21	84	49	168	36	35	77	232	140	232	14	105	15	21	105	196	210	1
		1	.	.	.	.	.	.	7	8	.	.	.	.	.	.	.	.	.	20	.	.	21	.	21	.	48	.	28	.	.	.	28	28	.	56	14	.	36	.	.	64	.	64	.	.	.	.	35	112	.	2
			1	3	3	2	2	4	3	1	8	16	3	6	20	6	23	23	4	30	28	1	27	72	36	63	24	52	32	50	42	42	78	7	60	28	51	180	8	112	80	160	116	224	55	120	60	23	61	116	288	3
				1	.	1	.	1	.	.	4	4	2	3	4	4	10	5	3	6	5	1	5	30	6	24	4	28	5	20	20	15	30	1	34	4	25	66	1	64	36	56	40	80	38	42	36	11	20	40	142	4
					1	.	1	.	2	1	.	4	.	.	5	.	5	10	.	15	12	.	17	15	18	24	18	10	20	10	15	20	33	7	11	21	20	68	8	20	36	88	55	92	9	45	10	11	40	72	100	5
						1	.	.	.	.	2	.	1	1	.	3	5	.	2	.	.	1	.	10	.	9	.	18	.	5	10	5	10	.	18	.	15	18	.	36	16	16	10	20	26	9	18	5	5	10	72	6
							1	.	1	1	.	.	.	.	.	.	.	5	.	5	6	.	12	.	6	9	12	.	14	.	5	10	9	7	.	14	5	24	8	.	16	48	30	24	.	15	.	5	30	38	30	7
								1	.	.	2	.	1	3	4	3	.	.	3	.	5	1	.	15	6	.	.	11	5	20	.	.	.	.	18	4	.	33	1	46	.	.	20	40	26	42	36	.	.	.	54	8
									1	1	.	.	.	.	.	.	.	.	.	5	.	.	6	.	6	.	12	.	7	.	.	.	9	7	.	14	5	.	8	.	.	24	.	24	.	.	.	.	15	38	.	9
										1	.	.	.	.	.	.	.	.	.	.	.	.	.	.	.	.	6	.	.	.	.	.	.	7	.	7	.	.	8	.	.	.	.	.	.	.	.	.	.	14	.	10
											1	.	1	1	.	2	.	.	2	.	.	1	.	5	.	.	.	6	.	5	.	.	.	.	12	.	.	9	.	24	.	.	5	10	20	9	18	.	.	.	24	11
												1	.	.	1	.	2	2	.	3	2	.	3	6	3	9	3	5	3	4	7	7	12	1	6	3	9	24	1	11	17	30	18	32	6	15	6	6	13	24	48	12
													1	.	.	1	.	.	1	.	.	1	.	.	.	.	.	.	.	.	.	.	.	.	6	.	.	.	.	6	.	.	.	.	14	.	6	.	.	.	.	13
														1	.	1	.	.	2	.	.	1	.	.	.	.	.	.	.	5	.	.	.	.	.	.	.	.	.	12	.	.	.	.	7	9	18	.	.	.	.	14
															1	.	.	.	.	.	2	.	.	3	3	.	.	2	3	4	.	.	.	.	3	3	.	12	1	8	.	.	9	16	4	15	6	.	.	.	18	15
																1	.	.	1	.	.	1	.	.	.	.	.	.	.	.	.	.	.	.	.	.	.	.	.	6	.	.	.	.	7	.	6	.	.	.	.	16

1	.	.	.	.	.	.	2	.	3	.	3	.	1	4	2	4	.	3	.	5	6	.	6	8	8	4	8	4	3	3	3	3	6	24	17
	1	.	1	1	.	2	.	1	3	2	.	2	.	2	4	3	1	.	2	2	8	1	.	8	16	10	8	.	5	.	3	9	12	13	18
		1	.	.	1	.	.	.	.	.	.	.	.	.	.	.	.	.	.	.	.	.	.	.	.	.	.	.	.	6	.	.	.	.	19
			1	.	.	1	.	1	.	2	.	1	.	.	.	3	1	.	2	2	.	1	.	.	8	.	8	.	.	.	.	5	12	.	20
				1	.	.	.	1	.	.	.	2	.	.	.	.	.	.	2	.	4	1	.	.	.	5	4	.	5	.	.	.	.	5	21
					1	.	.	.	.	.	.	.	.	.	.	.	.	.	.	.	.	.	.	.	.	.	.	.	.	.	.	.	.	.	22
						1	.	.	.	1	.	1	.	.	.	.	1	.	1	.	.	1	.	.	4	.	.	.	.	.	.	5	4	.	23
							1	.	.	.	1	.	1	.	.	.	.	2	.	.	3	.	4	.	.	2	4	3	3	3	.	.	.	8	24
								1	.	.	.	1	.	.	.	.	.	.	2	.	.	1	.	.	.	.	4	.	.	.	.	.	.	.	25
									1	.	.	.	.	1	1	1	.	.	.	1	2	.	.	4	4	2	2	.	1	.	2	2	3	6	26
										1	.	.	.	.	.	.	1	.	1	.	.	1	.	.	.	.	.	.	.	.	.	.	4	.	27
											1	.	.	.	.	.	.	1	.	.	.	.	2	.	.	.	.	2	.	1	.	.	.	4	28
												1	.	.	.	.	.	.	1	.	.	1	.	.	.	.	.	.	.	.	.	.	.	.	29
													1	.	.	.	.	.	.	.	.	.	2	.	.	.	.	1	3	3	.	.	.	.	30
														1	.	.	.	.	.	1	.	.	.	2	.	.	.	.	.	.	1	.	.	3	31
															1	.	.	.	.	.	.	.	.	2	2	2	.	.	.	.	1	1	1	.	32
																1	.	.	.	1	.	.	.	.	2	.	2	.	.	.	.	1	3	.	33
																	1	.	.	.	.	1	.	.	.	.	.	.	.	.	.	.	.	.	34
																		1	.	.	.	.	1	.	.	.	.	2	.	1	.	.	.	.	35
																			1	.	.	1	.	.	.	.	.	.	.	.	.	.	.	.	36
																				1	.	.	.	.	.	.	.	.	.	.	.	.	.	.	37
																					1	.	.	.	.	1	1	.	1	.	.	.	.	2	38
																						1	.	.	.	.	.	.	.	.	.	.	.	.	39
																							1	.	.	.	.	1	.	1	.	.	.	.	40
																								1	.	.	.	.	.	.	1	.	.	.	41
																									1	.	.	.	.	.	.	1	1	.	42
																										1	.	.	.	.	.	.	.	.	43
																											1	.	.	.	.	.	.	.	44
																												1	.	.	.	.	.	.	45
																													1	.	.	.	.	.	46
																														1	.	.	.	.	47
																															1	.	.	.	48
																																1	.	.	49
																																	1	.	50
																																		1	51

Table 3. *(continued)*

ALGEBRA A8, REPRESENTATIONS OF CLASS 4

20	38	56	38	74	128	56	74	56	92	128	74	200	74	92	92	110	92	146	92	218	110	110	110	146	110	S.P.
126	252	36	504	72	9	252	252	2 520	72	72	1 512	9	3 780	1 512	504	504	2 520	504	7 560	72	3 024	1 260	5 040	252	3 780	O.S.
21	27	29	29	31	33	33	35	37	37	39	41	41	43	43	45	45	45	47	49	49	49	51	51	51	51	LEVEL
126	630	540	1 008	990	495	1 890	2 772	8 316	2 970	3 168	15 840	1 287	27 720	23 328	20 790	25 200	37 800	27 027	97 020	11 088	92 160	74 250	133 056	41 580	128 304	DIM'N
0	1	0	0	2	4	0	0	1	0	0	0	0	1	2	1	1	1	3	0	5	1	0	2	1	0	
0	0	2	0	1	0	0	0	0	0	0	1	0	0	0	0	2	0	1	1	0	0	0	0	0	1	
0	1	0	0	0	0	0	0	0	0	0	1	0	0	1	0	0	0	0	0	0	0	2	0	0	0	
1	0	0	0	0	0	0	0	1	0	0	0	0	0	0	0	0	0	0	1	0	0	0	1	0	1	
0	0	0	1	0	0	0	0	0	0	0	0	0	1	0	0	0	1	0	0	0	0	0	0	0	0	
0	0	0	0	0	0	1	1	0	0	0	0	0	0	0	2	0	0	0	0	0	1	0	0	0	0	
0	0	0	0	0	0	1	0	0	2	1	0	0	1	0	0	0	0	0	1	0	1	1	1	3	0	
0	0	0	1	0	0	0	2	1	1	3	1	5	0	1	0	1	2	1	0	1	1	0	0	0	2	WEIGHT
0	1	2	3	4	5	6	7	8	9	10	11	12	13	14	15	16	17	18	19	20	21	22	23	24	25	NUMBER
1	3	2	4	3	1	5	6	20	5	4	28	1	42	36	25	32	50	28	102	8	88	65	124	30	120	0
	1	1	.	2	1	.	.	5	.	.	12	.	9	18	5	20	10	21	38	8	16	30	52	5	45	1
		1	.	1	1	.	.	.	.	.	6	.	.	6	.	14	.	14	14	8	.	15	14	.	15	2
			1	.	.	2	3	4	3	3	5	1	15	6	11	5	20	4	33	1	46	20	40	19	42	3
				1	1	.	.	.	.	.	.	.	.	6	.	7	.	14	.	8	.	.	14	.	.	4
					1	.	.	.	.	.	.	.	.	.	.	.	.	7	.	8	.	.	.	.	.	5
						1	1	.	2	2	.	1	5	.	6	.	5	.	9	.	24	5	10	13	9	6
							1	.	1	2	.	1	.	.	.	.	5	.	.	.	12	.	.	7	9	7
								1	.	.	2	.	3	3	2	3	4	3	12	1	8	9	16	3	15	8
									1	1	.	1	.	.	.	.	.	.	.	.	6	.	.	7	.	9
										1	.	1	.	.	.	.	.	.	.	.	.	.	.	.	.	10
											1	.	.	1	.	2	.	2	4	1	.	5	4	.	5	11
												1	.	.	.	.	.	.	.	.	.	.	.	.	.	12
													1	.	1	.	1	.	3	.	4	2	4	2	3	13
														1	.	1	.	2	.	1	.	.	4	.	.	14
															1	.	.	.	.	.	2	.	.	1	.	15
																1	.	1	.	1	.	.	.	.	.	16

146	110	128	128	164	128	128	218	128	164	164	128	146	308	146	182	146	146	182	182	164	164	146	236	164	164	S.P.
1 512	5 040	504	2 520	504	2 520	7 560	504	5 040	756	3 780	15 120	3 024	72	1 260	1 512	3 780	7 560	252	1 512	5 040	630	15 120	1 512	3 780	1 512	O.S.
51	53	53	53	53	55	55	55	57	57	57	57	57	57	59	59	59	59	59	59	59	61	61	61	61	61	LEVEL
77 220	166 320	97 020	175 500	90 090	221 760	332 640	72 765	311 850	152 460	280 800	582 120	432 432	25 740	249 480	331 695	481 140	772 200	196 560	363 825	449 280	249 480	1 108 800	270 270	609 840	630 630	DIM'N
1	0	0	2	1	2	1	1	0	0	3	0	1	1	0	2	0	2	0	3	0	0	1	4	0	2	
0	1	0	0	0	0	1	0	0	3	0	1	1	0	0	2	1	0	3	0	1	0	1	1	1	0	
0	0	2	0	0	0	1	0	1	0	1	0	1	0	1	0	0	0	0	1	0	0	0	0	0	0	
0	0	0	1	0	0	0	0	1	0	0	0	0	0	0	0	0	0	0	0	0	2	1	0	0	0	
0	1	0	0	0	1	0	0	0	0	0	1	0	0	2	0	0	1	0	0	1	1	0	0	0	0	
1	1	0	0	0	1	0	0	1	0	0	0	0	0	0	0	2	0	0	0	0	0	1	0	1	2	
0	0	0	0	2	0	1	1	0	1	1	1	0	0	0	1	0	1	0	0	0	0	0	1	2	0	
3	0	2	2	2	0	0	4	0	0	0	1	2	6	0	0	1	1	2	2	3	0	0	0	0	1	WEIGHT
26	27	28	29	30	31	32	33	34	35	36	37	38	39	40	41	42	43	44	45	46	47	48	49	50	51	NUMBER
60	128	76	150	57	152	240	35	185	88	170	370	288	8	122	171	252	450	106	210	240	103	548	111	278	303	0
10	44	36	60	9	58	130	5	80	57	105	125	156	1	51	120	81	160	70	126	80	47	270	90	87	102	1
.	14	21	15	.	14	70	.	35	42	55	35	84	.	21	84	21	35	49	63	20	21	126	69	21	21	2
36	54	25	50	39	65	72	28	75	25	50	174	88	8	56	48	135	210	30	60	124	47	225	30	160	163	3
.	.	.	15	.	14	35	.	.	21	55	.	42	.	.	63	.	35	28	63	.	.	63	69	.	21	4
.	.	.	.	.	.	.	.	.	.	20	.	.	.	.	21	.	.	.	21	.	.	.	48	.	.	5
18	24	5	10	27	30	16	21	30	5	10	76	16	8	26	9	75	90	5	10	52	21	90	5	95	93	6
18	.	5	10	20	.	.	21	.	.	.	38	16	8	.	.	30	45	5	10	52	.	.	.	50	36	7
6	18	11	20	6	23	38	4	33	16	30	57	46	1	24	33	42	72	19	36	40	23	108	24	49	52	8
6	.	.	.	14	.	.	14	.	.	.	14	.	8	.	.	15	15	.	.	14	.	.	.	35	21	9
6	.	.	.	7	.	.	14	.	.	.	.	.	8	.	.	.	.	.	.	14	.	.	.	.	.	10
.	5	6	5	.	5	20	.	15	11	15	15	24	.	10	23	10	15	13	18	10	11	48	18	11	10	11
.	.	.	.	.	.	.	7	.	.	.	.	.	8	.	.	.	.	.	.	.	.	.	.	.	.	12
3	8	2	4	4	10	8	3	13	3	6	24	8	1	12	6	23	30	3	6	16	11	42	4	28	28	13
.	.	.	5	.	5	10	.	.	6	15	.	12	.	.	17	.	15	7	18	.	.	24	18	.	10	14
1	4	.	.	3	5	.	2	5	.	.	8	.	1	6	.	15	10	.	.	4	5	15	.	15	18	15
.	.	.	.	.	.	5	.	.	6	5	.	6	.	.	12	.	.	7	6	.	.	9	12	.	.	16

1	.	.	.	2	.	.	1	3	3	.	2	4	3	.	.	3	.	.	.	12	8	1	.	.	9	15	3	6	16	.	.	.	15	11	17
	1	.	1	.	.	.	.	.	.	.	.	.	.	.	.	.	.	.	5	.	.	.	.	6	.	.	.	6	.	.	.	12	.	.	18
		1	.	.	1	1	.	1	.	2	1	1	.	2	4	.	6	2	3	6	4	.	5	4	5	6	2	3	4	6	18	3	6	5	19
			1	.	.	.	.	.	.	.	.	.	.	.	.	.	.	.	.	.	.	.	.	.	.	.	.	.	.	.	.	6	.	.	20
				1	.	.	1	.	1	.	.	.	2	.	.	2	.	.	.	4	.	1	.	.	5	5	.	.	4	.	.	.	10	6	21
					1	.	.	.	.	.	1	.	.	.	2	.	3	1	1	.	2	.	2	3	.	.	1	1	.	3	6	2	.	.	22
						1	.	.	.	.	.	1	.	2	2	.	.	1	3	.	2	.	.	3	.	6	1	3	.	.	9	3	.	5	23
							1	.	.	.	.	.	1	.	.	1	.	.	.	.	.	1	.	.	.	.	.	.	.	.	.	.	5	.	24
								1	.	.	1	1	.	.	.	.	.	.	.	3	4	.	.	.	2	3	2	3	4	.	.	.	3	2	25
									1	.	.	.	1	.	.	2	.	.	.	.	.	1	.	.	.	.	.	.	4	.	.	.	.	.	26
										1	.	.	.	1	.	.	2	.	.	2	.	.	3	.	3	2	.	.	1	3	6	.	3	3	27
											1	.	.	.	.	.	.	.	.	.	2	.	.	.	.	.	1	1	.	.	.	.	.	.	28
												1	.	.	.	.	.	.	.	.	2	.	.	.	.	3	1	3	.	.	.	.	.	2	29
													1	.	.	1	.	.	.	.	.	1	.	.	.	.	.	.	.	.	.	.	.	.	30
														1	.	.	.	.	.	.	.	.	.	.	.	2	.	.	.	.	3	.	.	3	31
															1	.	.	1	1	.	1	.	.	2	.	.	1	1	.	.	3	2	.	.	32
																1	.	.	.	.	.	1	.	.	.	.	.	.	.	.	.	.	.	.	33
																	1	.	.	.	.	.	1	.	.	.	.	.	.	2	2	.	.	.	34
																		1	.	.	.	.	.	1	.	.	1	.	.	.	.	1	.	.	35
																			1	.	.	.	.	1	.	.	.	1	.	.	.	2	.	.	36
																				1	.	.	.	.	1	1	.	.	1	.	.	.	2	1	37
																					1	.	.	.	.	.	1	1	.	.	.	.	.	.	38
																						1	.	.	.	.	.	.	.	.	.	.	.	.	39
																							1	.	.	.	.	.	.	1	.	.	.	.	40
																								1	.	.	.	.	.	.	.	1	.	.	41
																									1	.	.	.	.	.	.	.	1	1	42
																										1	.	.	.	.	.	.	.	1	43
																											1	.	.	.	.	.	.	.	44
																												1	.	.	.	.	.	.	45
																													1	.	.	.	.	.	46
																														1	.	.	.	.	47
																															1	.	.	.	48
																																1	.	.	49
																																	1	.	50
																																		1	51

Table 3. *(continued)*

ALGEBRA A9, REPRESENTATIONS OF CLASS 0

NUMBER	0	1	2	3	4	5	6	7	8	9	10	11	12	13	14	15	16	17
S.P.	0	20	40	60	60	80	60	80	80	80	120	120	100	100	140	140	100	100
O.S.	1	90	1260	360	360	90	4200	5040	5040	3150	840	840	252	5040	720	720	12600	12600
LEVEL	1	19	33	35	35	37	43	47	47	49	49	49	51	51	53	53	55	55
DIM'N	1	99	1925	2376	2376	2925	12375	35200	35200	29700	24024	24024	19404	99099	67200	67200	155925	155925
	0	1	0	0	2	2	0	0	1	0	0	3	0	1	1	3	0	1
	0	0	1	1	0	0	0	0	1	0	0	0	0	1	1	0	0	0
	0	0	0	0	0	0	1	1	0	0	1	0	0	0	0	0	0	1
	0	0	0	0	0	0	0	0	0	1	0	0	0	0	0	0	1	0
	0	0	0	0	0	0	0	0	0	0	0	0	2	0	0	0	0	0
	0	0	0	0	0	0	0	0	0	1	0	0	0	0	0	0	0	1
	0	0	0	0	0	0	1	0	1	0	0	1	0	0	0	0	1	0
	0	0	1	0	1	0	0	1	0	0	0	0	0	1	0	1	0	0
WEIGHT	0	1	0	2	0	2	0	1	0	0	3	0	0	1	3	1	1	0
0	1	9	35	36	36	45	75	160	160	90	84	84	42	399	240	240	315	315
1		1	7	8	8	9	20	48	48	28	28	28	14	119	72	72	112	112
2			1	1	1	1	5	12	12	9	7	7	5	28	16	16	38	38
3				1	.	1	.	6	.	.	7	.	.	14	16	8	14	.
4					1	1	.	.	6	.	.	7	.	14	8	16	.	14
5						1	.	.	.	.	.	.	.	7	8	8	.	.
6							1	2	2	3	1	1	2	4	2	2	12	12
7								1	.	.	1	.	.	2	2	1	4	.
8									1	.	.	1	.	2	1	2	.	4
9										1	.	.	1	.	.	.	3	3
10											1	.	.	.	2	.	.	.
11												1	.	.	.	2	.	.
12													1	.	.	.	.	.
13														1	1	1	.	.
14															1	.	.	.
15																1	.	.
16																	1	.

NUMBER	18	19	20	21	22	23	24	25	26	27	28	29	30	31	32	33	34
S.P.	180	120	120	120	140	120	140	200	120	200	140	140	140	140	160	160	160
O.S.	90	3150	3150	5040	6300	5040	6300	1260	37800	1260	7560	7560	7560	7560	1260	7560	1260
LEVEL	55	57	57	59	59	59	59	61	61	61	63	63	63	63	65	65	65
DIM'N	45375	136125	136125	228096	250250	228096	250250	126126	786500	126126	490050	490050	675675	675675	365904	729729	365904
	3	0	0	0	0	1	2	0	1	4	0	0	0	1	0	0	0
	0	0	2	0	0	0	1	0	0	0	0	1	2	0	0	0	1
	0	0	0	0	0	0	0	0	1	0	0	1	0	1	0	0	0
	0	1	0	0	1	1	0	1	0	0	0	0	0	0	0	0	2
	0	0	0	1	0	1	0	0	0	0	1	1	0	0	0	1	0
	0	0	1	1	0	0	1	0	0	1	0	0	0	0	2	0	0
	0	0	0	0	0	0	0	0	1	0	1	0	1	0	0	1	0
	0	2	0	0	1	0	0	0	0	0	1	0	0	2	1	0	0
WEIGHT	3	0	0	1	2	0	0	4	1	0	0	0	1	0	0	2	0
0	165	225	225	288	350	288	350	126	1250	126	450	450	945	945	252	567	252
1	45	85	85	112	140	112	140	56	460	56	190	190	357	357	112	252	112
2	9	30	30	44	50	44	50	21	155	21	80	80	120	120	51	108	51
3	9	15	.	14	35	.	.	21	55	.	35	.	42	63	21	63	.
4	9	.	15	.	.	14	35	.	55	21	.	35	63	42	.	.	21
5	9	.	.	.	.	.	.	.	20	.	.	.	21	21	.	.	.
6	1	9	9	18	15	18	15	6	45	6	33	33	33	33	24	45	24
7	1	5	.	5	10	.	.	6	15	.	15	.	11	18	10	24	.
8	1	.	5	.	.	5	10	.	15	6	.	15	18	11	.	.	10
9	.	2	2	8	3	8	3	1	9	1	13	13	6	6	12	18	12
10	1	.	.	.	5	.	.	6	.	.	.	.	.	.	.	9	.
11	1	.	.	.	.	.	5	.	.	6	.	.	.	.	.	.	.
12	.	.	.	4	.	4	.	.	.	.	5	5	.	.	6	6	6
13	1	.	.	.	.	.	.	.	5	.	.	.	6	6	.	.	.
14	1	.	.	.	.	.	.	.	.	.	.	.	.	.	.	.	.
15	1	.	.	.	.	.	.	.	.	.	.	.	.	.	.	.	.
16	.	1	.	2	2	.	.	1	3	.	6	.	2	3	5	9	.

NUMBER	35	36	37	38	39	40	41	42	43	44	45	46	47	48	49	50	51
S.P.	160	160	160	160	180	180	180	180	180	140	220	180	180	180	220	220	240
O.S.	1260	15120	7560	15120	840	840	5040	840	2520	50400	2520	5040	840	2520	2520	5040	360
LEVEL	65	65	65	65	67	67	67	67	67	67	67	67	67	67	67	69	69
DIM'N	576576	1234926	729729	1234926	381150	514800	825825	381150	1051050	2081079	614250	825825	514800	1051050	614250	896896	519750
	0	1	2	2	0	0	0	0	0	1	1	1	2	2	4	0	0
	2	0	0	1	0	0	0	0	2	0	0	2	0	1	0	0	2
	0	1	1	0	0	0	0	2	0	0	1	0	0	0	0	0	0
	0	0	0	0	0	0	0	1	0	1	0	0	2	0	0	0	0
	0	0	1	0	0	0	1	0	0	0	0	1	0	0	0	1	0
	0	0	0	0	1	2	0	0	0	1	0	0	0	0	0	0	0
	0	0	0	1	2	0	0	0	0	0	0	0	0	0	1	0	0
	2	1	0	0	0	0	2	0	1	0	0	0	0	2	0	1	0
WEIGHT	0	2	0	1	0	2	1	0	2	1	4	0	0	0	1	3	4
0	756	1566	567	1566	210	300	525	210	1260	1827	630	525	300	1260	630	448	540
1	280	600	252	600	98	140	245	98	476	756	252	245	140	476	252	224	204
2	92	201	108	201	47	65	110	47	154	307	84	110	65	154	84	104	64
3	49	138	.	69	21	35	70	.	105	91	84	.	.	84	28	84	64
4	49	69	63	138	.	.	.	21	84	91	28	70	35	105	84	.	36
5	28	48	.	48	.	.	.	.	56	28	28	.	.	56	28	.	36
6	24	54	45	54	23	31	46	23	39	123	21	46	31	39	21	44	15
7	13	36	.	18	11	15	30	.	26	33	21	.	.	21	7	34	15
8	13	18	24	36	.	.	.	11	21	33	7	30	15	26	21	.	8
9	4	9	18	9	11	16	17	11	6	48	3	17	16	6	3	16	2
10	.	18	.	.	.	5	10	.	13	.	21	.	.	.	.	24	15
11	.	.	9	18	.	.	.	.	.	.	.	10	5	13	21	.	.
12	.	.	6	.	5	10	5	5	.	16	.	5	10	.	.	4	.
13	7	12	.	12	.	.	.	.	14	9	7	.	.	14	7	.	8
14	.	6	.	.	.	.	.	.	7	.	7	.	.	.	.	.	8
15	.	.	.	6	.	.	.	.	.	.	.	.	.	7	7	.	.
16	2	6	.	3	6	7	11	.	4	12	3	.	.	3	1	12	2

1	.	.	1	.	.	2	2	.	3	1	.	6	3	2	.	.	5	2	3	9	6	.	.	.	6	3	12	1	11	7	4	3	.	1	17
	1	.	.	.	.	.	.	.	.	.	.	.	.	.	.	.	.	.	.	.	.	.	.	.	.	.	.	.	.	.	.	.	.	.	18
		1	.	.	1	.	.	1	.	.	3	.	.	3	2	3	.	2	3	.	.	3	2	8	.	2	.	3	.	.	3	.	8	2	19
			1	.	.	.	1	.	.	1	.	3	3	.	.	.	2	2	.	3	3	.	.	.	3	3	.	.	8	2	2	3	.	1	20
				1	.	.	.	.	.	.	2	.	.	.	3	3	.	.	.	.	.	3	4	3	.	.	4	.	.	.	.	.	3	.	21
					1	.	.	1	.	.	.	.	.	.	.	3	.	.	3	.	.	.	2	4	.	2	.	3	.	.	.	.	8	2	22
						1	.	.	.	.	.	2	.	.	.	.	3	.	.	3	.	.	.	.	3	.	4	.	3	4	.	.	.	.	23
							1	.	.	1	.	.	.	.	.	.	.	.	.	3	3	.	.	.	.	.	.	.	4	2	2	3	.	.	24
								1	.	.	.	.	.	.	.	.	.	.	.	.	.	.	.	.	.	.	.	3	.	.	.	.	4	2	25
									1	.	.	.	1	1	.	.	.	1	2	.	2	.	.	.	.	2	3	1	.	.	2	1	.	1	26
										1	.	.	.	.	.	.	.	.	.	.	.	.	.	.	.	.	.	.	.	.	.	3	.	.	27
											1	.	.	.	1	1	.	.	.	.	.	2	1	2	.	.	.	.	.	.	.	.	2	.	28
												1	.	.	.	.	1	.	.	1	.	.	.	.	2	.	.	.	2	1	.	.	.	.	29
													1	.	.	.	.	1	.	.	1	.	.	.	.	2	.	.	.	.	1	1	.	1	30
														1	.	.	.	1	1	.	.	.	.	.	.	1	.	1	.	.	2	.	.	1	31
															1	.	.	.	.	.	.	1	1	.	.	.	.	.	.	.	.	.	.	.	32
																1	.	.	.	.	.	.	1	1	.	.	.	.	.	.	.	.	2	.	33
																	1	.	.	.	.	.	.	.	1	.	.	.	.	1	.	.	.	.	34
																		1	.	.	.	.	.	.	.	1	.	.	.	.	1	.	.	1	35
																			1	.	.	.	.	.	.	1	.	1	.	.	.	.	.	1	36
																				1	.	.	.	.	.	.	.	.	1	1	.	.	.	.	37
																					1	.	.	.	.	.	.	.	.	.	1	1	.	.	38
																						1	.	.	.	.	.	.	.	.	.	.	.	.	39
																							1	.	.	.	.	.	.	.	.	.	.	.	40
																								1	.	.	.	.	.	.	.	.	1	.	41
																									1	.	.	.	.	.	.	.	.	.	42
																										1	.	.	.	.	.	.	.	1	43
																											1	.	.	.	.	.	.	.	44
																												1	.	.	.	.	.	1	45
																													1	.	.	.	.	.	46
																														1	.	.	.	.	47
																															1	.	.	.	48
																																1	.	.	49
																																	1	.	50
																																		1	51

Table 3. *(continued)*

ALGEBRA A9, REPRESENTATIONS OF CLASS 1

9	29	49	49	69	69	89	109	69	129	89	89	129	89	109	109	109	129	129	129	129	109	149	189	149	169	129	149	149	209	129	189	209	189	169	129	149	249	189	149	149	169	209	229	169	149	149	169	169	209	249	249	S.P.
10	360	90	2520	840	2520	720	360	4200	90	1260	6300	1260	12600	2520	7560	15120	5040	840	2520	2520	12600	2520	840	5040	360	1260	5040	2520	1260	12600	720	720	120	2520	50400	12600	90	2520	7560	12600	6300	720	2520	2520	37800	25200	25200	5040	6300	360	360	O.S.
10	26	28	38	40	42	44	44	46	46	50	50	52	52	54	56	56	56	58	58	58	58	58	58	60	60	60	60	62	62	62	62	62	64	64	64	64	64	66	66	66	66	68	68	68	68	68	68	68	68	70	70	LEVEL
10	440	540	4950	6160	13860	17160	9360	19800	11550	27720	57750	40040	99000	84700	160380	283140	154440	110880	145200	240240	249480	194040	76440	270270	163800	166320	436590	304920	140140	490050	297000	214500	108900	463320	1321320	720720	145200	566280	784080	1158300	786500	566280	630630	653400	2502500	1981980	2134440	1081080	840840	377520	514800	DIM'N
1	0	2	0	0	1	1	3	0	3	0	0	0	1	0	0	1	2	0	0	0	1	1	4	0	0	1	2	0	0	0	2	4	0	0	1	2	4	0	0	1	1	0	0	0	0	1	1	2	3	0	0	
0	1	0	0	0	1	1	0	0	0	0	0	0	0	2	0	0	1	0	0	2	0	0	0	0	2	0	1	0	0	1	1	0	0	0	0	0	0	0	1	0	2	0	0	0	1	0	0	0	1	0	0	
0	0	0	1	1	0	0	0	0	0	0	0	0	1	0	0	1	0	0	0	0	0	1	0	0	0	0	0	0	0	1	0	0	0	0	0	1	0	0	0	0	0	0	0	2	1	0	0	0	0	0	0	
0	0	0	0	0	0	0	0	1	0	0	1	1	0	0	0	0	0	0	0	0	1	0	0	0	0	0	0	0	0	0	0	0	0	0	1	0	0	0	1	1	0	0	0	0	0	0	1	1	0	0	0	
0	0	0	0	0	0	0	0	0	0	1	0	0	0	0	1	0	0	0	1	0	0	0	0	1	0	2	0	0	1	0	0	0	0	0	0	0	0	0	1	0	0	0	0	1	0	1	0	1	0	0	0	
0	0	0	0	0	0	0	0	0	0	1	0	0	0	0	0	0	0	2	0	0	1	0	0	0	0	0	0	1	0	1	0	0	0	1	0	1	0	1	0	0	1	0	1	0	0	1	0	0	1	0	0	
0	0	0	0	0	0	0	0	1	0	0	0	0	1	1	1	0	1	0	0	0	0	0	1	0	0	0	0	1	0	0	0	0	3	1	1	0	0	0	0	0	0	2	0	0	1	0	0	0	0	1	2	
0	0	0	1	0	1	0	1	0	0	0	1	0	0	0	0	1	0	0	2	1	0	0	0	1	0	0	1	1	0	0	0	1	0	0	0	0	0	2	0	2	0	1	1	0	0	0	1	0	0	3	0	
0	1	1	0	2	0	2	0	0	2	0	1	3	0	0	1	1	0	1	0	1	0	3	0	2	3	0	1	0	4	0	3	1	0	2	1	0	3	1	0	0	0	1	3	0	1	1	2	0	0	0	3	WEIGHT
0	1	2	3	4	5	6	7	8	9	10	11	12	13	14	15	16	17	18	19	20	21	22	23	24	25	26	27	28	29	30	31	32	33	34	35	36	37	38	39	40	41	42	43	44	45	46	47	48	49	50	51	NUMBER
1	8	9	27	28	63	72	36	48	45	42	105	56	180	133	162	432	216	84	120	336	252	252	84	189	204	126	567	168	70	387	360	240	42	216	1008	504	165	216	414	765	470	168	189	288	1610	1008	1260	504	420	84	120	0
	1	1	6	7	14	16	8	14	9	14	35	21	55	42	63	138	69	35	50	105	91	84	28	84	64	49	175	77	35	148	108	72	21	105	392	196	45	111	175	310	190	91	105	127	630	434	525	217	175	49	70	1
		1	.	.	7	8	8	.	9	.	.	.	20	21	.	48	48	.	.	56	28	28	28	.	36	14	119	.	.	64	72	72	.	.	112	112	45	.	70	85	120	.	.	56	280	112	140	112	140	.	.	2
			1	1	2	2	1	4	1	5	10	6	15	11	24	36	18	15	20	26	33	21	7	34	15	20	42	36	15	54	24	16	11	50	144	72	9	56	75	115	70	50	55	56	225	188	195	94	65	29	40	3
				1	.	2	.	.	1	.	5	6	.	.	9	18	.	5	10	13	.	21	.	24	15	.	21	16	15	.	24	8	5	30	52	.	9	36	.	60	.	30	45	.	80	58	135	.	.	19	30	4
					1	1	1	.	1	.	.	.	5	6	.	12	12	.	.	14	9	7	7	.	8	5	28	.	.	24	16	16	.	.	38	38	9	.	30	30	45	.	.	26	100	44	50	44	50	.	.	5
						1	.	.	1	.	.	.	.	.	.	6	.	.	.	7	.	7	.	.	8	.	14	.	.	.	16	8	.	.	14	.	9	.	.	15	.	.	.	.	35	14	35	.	.	.	.	6
							1	.	1	.	.	.	.	.	.	.	6	.	.	.	.	.	7	.	.	.	14	.	.	.	8	16	.	.	.	14	9	.	.	.	15	.	.	.	.	.	.	14	35	.	.	7
								1	.	2	2	1	3	2	9	6	3	7	7	4	12	3	1	12	2	9	6	17	5	18	3	2	6	24	48	24	1	27	33	36	22	28	27	24	69	84	60	42	20	17	23	8
									1	.	.	.	.	.	.	.	.	.	.	.	.	.	.	.	.	.	7	.	.	.	8	8	.	.	.	.	9	.	.	.	.	.	.	.	.	.	.	.	.	.	.	9
										1	.	.	.	.	3	.	.	4	2	.	4	.	.	3	.	5	.	8	1	5	.	.	3	12	12	6	.	12	15	8	5	16	12	10	15	40	12	20	4	10	14	10
											1	1	.	.	3	3	.	2	4	2	.	3	.	8	2	.	3	8	5	.	3	1	3	13	16	.	1	18	.	20	.	17	21	.	23	23	40	.	.	12	16	11
												1	.	.	.	.	.	.	.	.	.	3	.	4	2	.	.	.	5	.	3	.	.	5	.	.	1	6	.	.	.	5	15	.	.	.	20	.	.	4	10	12
													1	1	.	2	2	.	.	2	3	1	1	.	1	2	4	.	.	8	2	2	.	.	12	12	1	.	13	9	14	.	.	12	30	18	15	18	15	.	.	13
														1	.	.	1	.	.	2	.	.	1	.	1	.	2	.	.	4	1	2	.	.	.	4	1	.	5	.	10	.	.	6	15	.	.	5	10	.	.	14
															1	.	.	1	1	.	.	.	.	2	.	.	.	4	1	.	.	.	2	6	4	.	.	8	.	4	.	10	9	.	5	10	8	.	.	7	9	15
																1	.	.	.	1	.	1	.	.	1	.	2	.	.	.	2	1	.	.	4	.	1	.	.	5	.	.	.	.	10	5	10	.	.	.	.	16

1	.	.	.	.	.	1	.	.	.	2	.	.	.	1	2	.	.	.	4	1	.	.	.	5	.	.	.	.	.	.	5	10	.	.	17
	1	.	.	.	.	.	.	.	.	.	2	.	.	.	.	1	3	.	.	.	3	.	.	.	6	3	.	.	5	.	.	.	4	6	18
		1	.	.	.	.	1	.	.	.	2	1	.	.	.	1	2	.	.	.	6	.	4	.	6	6	.	.	.	4	.	.	6	5	19
			1	.	.	.	.	1	.	1	.	.	.	1	1	.	.	.	.	1	.	.	.	.	.	.	.	5	.	.	.	.	.	.	20
				1	.	.	.	.	1	.	.	.	2	.	.	.	.	3	3	.	.	6	2	3	.	.	5	6	8	3	8	3	.	.	21
					1	.	.	1	.	.	.	.	.	2	.	.	.	.	.	1	.	.	.	.	.	.	.	.	.	5	.	.	.	.	22
						1	.	.	.	.	.	.	.	.	2	.	.	.	.	1	.	.	.	.	.	.	.	.	.	.	.	5	.	.	23
							1	.	.	.	.	1	.	.	.	.	2	.	.	.	3	.	.	.	3	6	.	.	.	4	.	.	3	5	24
								1	.	.	.	.	.	1	.	.	.	.	.	1	.	.	.	.	.	.	.	.	.	.	.	.	.	.	25
									1	.	.	.	.	.	.	.	.	.	.	.	.	3	.	.	.	.	2	.	4	.	4	.	.	.	26
										1	.	.	.	1	1	.	.	.	.	1	.	.	.	.	.	.	.	.	.	.	.	.	.	.	27
											1	.	.	.	.	1	1	.	.	.	2	.	.	.	4	2	.	.	.	.	.	.	3	3	28
												1	.	.	.	.	.	.	.	.	.	.	.	.	.	3	.	.	.	.	.	.	.	2	29
													1	.	.	.	.	.	1	.	.	2	.	2	.	.	3	3	.	.	2	2	.	.	30
														1	.	.	.	.	.	1	.	.	.	.	.	.	.	.	.	.	.	.	.	.	31
															1	.	.	.	.	1	.	.	.	.	.	.	.	.	.	.	.	.	.	.	32
																1	.	.	.	.	.	.	.	.	2	.	.	.	.	.	.	.	1	1	33
																	1	.	.	.	1	.	.	.	2	2	.	.	.	.	.	.	2	3	34
																		1	.	.	.	.	1	.	.	.	.	2	2	2	.	.	.	.	35
																			1	.	.	.	.	1	.	.	.	.	.	.	2	2	.	.	36
																				1	.	.	.	.	.	.	.	.	.	.	.	.	.	.	37
																					1	.	.	.	1	1	.	.	.	.	.	.	2	1	38
																						1	.	.	.	.	1	.	.	.	1	.	.	.	39
																							1	.	.	.	.	.	.	1	.	.	.	.	40
																								1	.	.	.	.	.	.	.	1	.	.	41
																									1	.	.	.	.	.	.	.	1	1	42
																										1	.	.	.	.	.	.	.	1	43
																											1	.	.	.	.	.	.	.	44
																												1	.	.	.	.	.	.	45
																													1	.	.	.	.	.	46
																														1	.	.	.	.	47
																															1	.	.	.	48
																																1	.	.	49
																																	1	.	50
																																		1	51

Table 3. *(continued)*

ALGEBRA A9, REPRESENTATIONS OF CLASS 2

16	36	36	56	96	56	76	76	76	96	96	96	96	116	116	136	136	176	116	96	196	136	156	156	116	116	S.P.
45	10	840	720	90	3 150	1 260	2 520	7 560	210	1 260	2 520	5 040	360	2 520	1 260	720	360	2 520	16 800	90	1 260	360	2 520	12 600	6 300	O.S.
17	19	31	35	37	41	43	47	47	49	49	49	51	51	51	53	53	53	55	55	55	57	59	59	59	59	LEVEL
45	55	1 155	3 200	2 145	8 250	10 395	20 790	40 095	13 860	33 880	50 050	63 360	42 120	61 425	45 045	76 230	30 030	103 950	168 960	37 125	101 640	98 010	193 050	317 625	242 550	DIM'N
0	2	0	1	3	0	0	0	1	0	0	1	0	0	2	0	2	4	0	1	4	0	0	0	0	1	
1	0	0	1	0	0	0	0	0	0	2	0	0	2	1	0	1	0	0	0	0	0	0	0	1	0	
0	0	1	0	0	0	0	0	1	0	0	1	0	0	0	0	0	0	0	0	0	0	0	0	1	0	
0	0	0	0	0	1	1	0	0	0	0	0	0	0	0	0	0	0	0	1	0	0	0	0	0	0	
0	0	0	0	0	0	0	1	0	0	0	0	1	0	0	1	0	0	0	0	0	0	0	0	0	1	
0	0	0	0	0	0	0	0	0	2	0	0	0	0	0	0	0	0	1	0	0	1	0	1	0	1	
0	0	0	0	0	0	0	1	0	0	0	0	0	0	0	0	0	0	1	1	0	0	2	0	1	0	
0	0	0	0	0	1	0	0	1	0	1	0	1	0	1	0	0	1	0	0	0	2	1	1	0	0	
0	0	1	1	1	0	2	0	0	0	0	2	1	2	0	3	2	0	1	0	2	0	0	2	0	0	WEIGHT
0	1	2	3	4	5	6	7	8	9	10	11	12	13	14	15	16	17	18	19	20	21	22	23	24	25	NUMBER
1	1	7	16	9	20	21	28	75	14	56	84	64	64	91	35	108	36	70	160	45	56	42	90	245	154	0
	1	.	8	9	.	.	.	27	.	28	28	.	36	63	.	72	36	.	48	45	.	.	.	105	42	1
		1	2	1	5	6	9	18	5	13	21	24	15	21	15	24	8	30	52	9	26	21	45	80	58	2
			1	1	.	.	.	6	.	7	7	.	8	14	.	16	8	.	14	9	.	.	.	35	14	3
				1	.	.	.	.	.	.	.	.	.	7	.	8	8	.	.	9	.	.	.	.	.	4
					1	1	3	3	2	2	3	8	2	3	5	3	1	13	16	1	12	11	21	23	23	5
						1	.	.	.	.	3	4	2	.	5	3	.	5	.	1	6	5	15	.	.	6
							1	.	1	.	.	2	.	.	1	.	.	6	4	.	5	6	9	5	10	7
								1	.	1	1	.	1	2	.	2	1	.	4	1	.	.	.	10	5	8
									1	.	.	.	.	.	.	.	.	3	.	.	2	3	3	.	5	9
										1	.	.	1	1	.	1	1	.	.	1	.	.	.	5	.	10
											1	.	1	.	.	2	.	.	.	1	.	.	.	.	.	11
												1	.	.	1	.	.	2	.	.	3	3	6	.	.	12
													1	.	.	1	.	.	.	1	.	.	.	.	.	13
														1	.	1	1	.	.	1	.	.	.	.	.	14
															1	.	.	.	.	.	.	.	3	.	.	15
																1	.	.	.	1	.	.	.	.	.	16

116	176	216	156	136	196	136	156	256	236	136	176	136	156	196	276	316	156	156	156	156	176	156	196	276	336	S.P.
25 200	360	840	5 040	12 600	720	15 120	5 040	45	720	18 900	2 520	37 800	15 120	5 040	90	360	4 200	2 520	5 040	12 600	5 040	12 600	2 520	840	90	O.S.
59	61	61	61	61	63	63	63	65	65	65	65	65	65	65	67	67	67	67	67	67	67	67	67	67	69	LEVEL
495 495	154 440	105 105	344 960	463 320	235 950	915 200	495 495	70 785	274 560	882 090	630 630	1 415 700	1 330 560	524 160	165 165	135 135	680 625	609 840	990 990	1 287 000	1 411 200	1 201 200	914 760	210 210	160 875	DIM'N
1	0	0	1	2	0	0	1	0	0	0	0	1	2	3	0	0	0	0	1	1	1	2	2	5	0	
0	0	0	0	0	0	1	2	0	0	1	1	0	0	1	0	0	0	1	0	0	2	0	0	0	0	
0	0	0	0	1	0	1	0	0	0	0	1	0	1	0	0	0	2	0	0	0	0	0	1	0	0	
1	0	0	1	0	0	0	0	0	0	1	0	0	0	0	0	0	0	0	0	0	0	1	0	0	0	
0	0	0	0	0	0	0	0	0	0	0	0	1	0	0	0	0	0	2	0	1	0	0	0	0	0	
0	0	1	0	0	0	0	0	0	0	1	0	0	0	0	0	0	1	0	2	0	0	1	0	0	0	
0	2	0	0	1	1	0	1	0	1	0	0	1	0	1	0	1	0	0	0	0	0	0	0	1	0	
1	0	0	0	0	2	1	0	4	1	0	0	0	1	0	3	0	0	0	0	2	1	0	0	0	2	
1	2	4	3	0	1	1	0	0	3	0	3	1	1	0	2	5	0	0	1	0	1	0	3	0	4	WEIGHT
26	27	28	29	30	31	32	33	34	35	36	37	38	39	40	41	42	43	44	45	46	47	48	49	50	51	NUMBER
385	56	35	224	320	70	608	301	14	64	414	364	666	816	272	28	21	275	224	364	520	784	504	504	84	20	0
105	.	.	56	180	.	272	189	.	.	162	168	162	432	216	.	.	120	84	84	120	504	252	252	84	.	1
135	30	20	84	105	40	204	99	9	40	162	126	279	270	90	19	15	110	95	165	230	252	199	168	28	15	2
35	.	.	21	55	.	90	63	.	.	63	56	63	138	69	.	.	50	35	35	50	161	91	84	28	.	3
.	.	.	.	20	.	.	21	.	.	.	.	.	48	48	.	.	.	.	.	.	56	28	28	28	.	4
40	16	10	24	30	23	56	27	6	24	63	33	114	72	24	13	10	42	42	77	96	64	78	42	7	11	5
20	10	10	24	.	15	28	.	4	20	.	33	42	36	.	10	10	.	.	25	50	32	.	42	.	10	6
8	9	4	4	6	13	10	5	4	14	24	5	45	12	4	9	6	15	20	39	35	10	30	6	1	8	7
10	.	.	6	15	.	24	17	.	.	24	14	24	36	18	.	.	20	15	15	20	40	33	21	7	.	8
.	6	1	.	.	7	.	.	3	8	9	.	15	.	.	6	3	5	10	24	10	.	10	.	.	6	9
.	.	.	.	5	.	12	12	.	.	9	7	.	12	12	.	.	10	5	.	.	28	9	7	7	.	10
5	.	.	6	.	.	12	.	.	.	.	14	9	18	.	.	.	.	.	5	10	20	.	21	.	.	11
4	5	4	4	.	9	5	.	3	11	.	5	15	6	.	7	6	.	.	11	20	5	.	6	.	7	12
.	.	.	.	.	.	6	.	.	.	.	7	.	6	.	.	.	.	.	.	.	14	.	7	.	.	13
.	.	.	.	5	.	.	6	.	.	.	.	.	12	12	.	.	.	.	.	.	14	9	7	7	.	14
.	2	4	4	.	3	.	.	1	8	.	5	.	.	.	4	6	.	.	.	.	.	.	6	.	6	15
.	.	.	.	.	.	.	.	.	.	.	.	.	6	.	.	.	.	.	.	.	7	.	7	.	.	16

1	.	.	1	.	.	.	.	.	.	.	.	.	.	.	.	.	.	.	.	.	.	.	6	.	.	.	.	.	.	.	.	.	7	.	17
	1	.	.	1	2	2	.	.	.	3	1	.	.	5	.	.	2	6	.	.	5	.	.	5	3	.	.	6	5	.	.	.	.	5	18
		1	.	.	.	.	2	2	2	.	.	1	3	.	4	3	.	.	9	2	9	6	3	.	.	7	7	7	7	6	12	3	1	.	19
			1	.	.	.	.	.	.	.	.	.	.	.	.	.	.	.	.	.	.	.	.	.	.	.	.	.	.	.	.	.	.	.	20
				1	1	1	.	.	.	1	1	.	.	4	.	.	2	4	.	.	.	.	.	4	3	.	.	.	5	.	.	.	.	4	21
					1	.	.	.	.	1	.	.	.	2	.	.	1	2	.	.	.	.	.	3	1	.	.	.	.	.	.	.	.	3	22
						1	.	.	.	1	1	.	.	2	.	.	1	4	.	.	.	.	.	3	3	.	.	.	.	.	.	.	.	4	23
							1	.	.	.	.	.	1	.	2	2	.	.	3	1	.	2	2	.	.	4	2	.	.	4	3	1	1	.	24
								1	.	.	.	.	.	.	.	.	.	.	3	.	3	.	.	.	.	2	4	4	2	.	4	.	.	.	25
									1	.	.	1	.	.	2	.	.	.	.	2	3	3	.	.	.	.	.	2	4	3	.	3	.	.	26
										1	.	.	.	1	.	.	1	2	.	.	.	.	.	2	1	.	.	.	.	.	.	.	.	3	27
											1	.	.	.	.	.	.	2	.	.	.	.	.	1	3	.	.	.	.	.	.	.	.	3	28
												1	.	.	.	.	.	.	.	2	.	.	.	.	.	.	.	.	.	.	.	3	.	.	29
													1	.	.	1	.	.	.	.	.	2	2	.	.	.	.	.	.	2	3	1	1	.	30
														1	.	.	1	1	.	.	.	.	.	2	1	.	.	.	.	.	.	.	.	2	31
															1	.	.	.	.	1	.	1	.	.	.	.	.	.	.	2	.	1	.	.	32
																1	.	.	.	.	.	.	1	.	.	.	.	.	.	2	.	.	1	.	33
																	1	.	.	.	.	.	.	1	.	.	.	.	.	.	.	.	.	1	34
																		1	.	.	.	.	.	1	1	.	.	.	.	.	.	.	.	2	35
																			1	.	.	.	.	.	.	1	1	.	.	.	1	.	.	.	36
																				1	.	.	.	.	.	.	.	.	.	.	.	1	.	.	37
																					1	.	.	.	.	.	.	1	1	.	.	.	.	.	38
																						1	.	.	.	.	.	.	.	1	.	1	.	.	39
																							1	.	.	.	.	.	.	.	.	.	1	.	40
																								1	.	.	.	.	.	.	.	.	.	1	41
																									1	.	.	.	.	.	.	.	.	1	42
																										1	.	.	.	.	.	.	.	.	43
																											1	.	.	.	.	.	.	.	44
																												1	.	.	.	.	.	.	45
																													1	.	.	.	.	.	46
																														1	.	.	.	.	47
																															1	.	.	.	48
																																1	.	.	49
																																	1	.	50
																																		1	51

Table 3. *(continued)*

ALGEBRA A9, REPRESENTATIONS OF CLASS 3

21	41	81	41	61	61	81	81	101	81	161	101	81	121	141	101	141	101	161	121	101	121	201	221	121	141	141	241	121	161	181	321	121	161	201	261	441	141	141	141	281	141	141	161	161	181	181	181	161	161	161	161	S.P.
120	90	10	1260	2520	2520	360	1260	720	840	90	2520	12600	360	840	5040	360	7560	720	2520	12600	7560	90	360	1260	2520	2520	90	25200	720	2520	90	25200	6300	720	360	10	4200	15120	16800	90	12600	37800	7560	5040	7560	2520	15120	25200	6300	25200	12600	O.S.
22	26	28	34	40	42	42	44	44	46	46	50	50	52	52	52	54	54	56	56	56	56	58	58	58	58	58	60	60	60	60	62	62	62	62	62	64	64	64	64	64	66	66	66	68	68	68	68	68	68	68	70	LEVEL
120	330	220	1980	9450	9240	7920	11880	14300	13860	6930	46200	71280	35640	34320	90090	43560	130680	85800	163800	184800	189540	47190	51480	124740	200200	237160	60060	566280	249480	210210	38610	609840	404250	261360	83160	11440	453750	943800	823680	102960	914760	1801800	926640	1321320	1261260	900900	1764180	1981980	1201200	2425500	1633500	DIM'N
0	1	3	0	1	0	0	0	2	0	4	0	1	0	0	1	0	0	0	0	1	2	0	0	1	1	2	0	1	1	3	0	0	1	3	5	0	0	1	2	5	0	0	1	0	0	1	1	1	2	2	0	
0	1	0	0	0	0	2	0	1	0	0	0	0	0	0	0	0	1	0	1	0	0	0	0	0	2	0	0	0	2	1	0	1	0	1	0	0	0	0	0	0	1	1	0	0	1	0	0	1	0	0	0	
1	0	0	0	1	0	0	0	0	0	0	0	0	0	0	0	0	1	0	1	0	1	0	0	0	0	1	0	0	0	0	0	0	0	0	0	0	2	0	0	0	0	0	0	2	0	0	0	1	0	0	1	
0	0	0	1	0	0	0	0	0	0	0	0	1	0	0	1	0	0	0	0	0	0	0	0	0	0	0	0	0	0	0	0	1	0	0	0	0	0	0	1	0	0	1	0	0	1	0	0	0	0	1	1	
0	0	0	0	0	1	0	1	0	0	0	0	0	0	0	0	0	0	0	0	1	0	0	0	0	0	0	0	1	0	0	0	0	1	0	0	0	0	0	0	0	1	0	0	0	0	0	0	0	1	0	0	
0	0	0	0	0	0	0	0	0	1	0	1	0	0	1	0	0	0	0	0	0	0	0	0	2	0	0	0	0	0	0	0	0	0	0	0	0	0	1	0	0	1	0	1	0	0	0	1	0	1	0	1	
0	0	0	0	0	0	0	0	0	1	0	0	0	2	0	0	1	0	1	0	1	0	0	1	0	0	0	0	0	0	0	0	1	0	0	0	0	1	1	1	0	0	0	0	0	0	2	0	1	0	0	0	
0	0	0	0	0	1	0	0	0	0	0	1	1	0	0	0	2	1	1	0	0	1	3	0	0	1	0	2	1	0	1	1	0	0	0	1	0	0	0	0	0	0	1	2	1	0	1	1	0	0	1	0	
0	0	0	1	1	0	1	2	1	0	1	1	0	1	3	2	0	0	2	2	0	0	1	4	0	0	2	3	1	2	0	5	0	3	2	0	7	0	1	0	2	0	1	0	1	3	0	2	0	0	1	0	WEIGHT
0	1	2	3	4	5	6	7	8	9	10	11	12	13	14	15	16	17	18	19	20	21	22	23	24	25	26	27	28	29	30	31	32	33	34	35	36	37	38	39	40	41	42	43	44	45	46	47	48	49	50	51	NUMBER
1	2	1	6	21	14	15	15	24	14	9	35	75	21	20	84	21	111	35	126	119	144	14	15	63	133	168	14	296	156	119	6	294	175	144	36	1	190	350	356	45	315	735	294	476	441	231	495	700	371	910	450	0
	1	1	.	7	.	8	.	16	.	9	.	20	.	.	21	.	48	.	56	28	75	.	.	14	84	84	.	64	100	91	.	112	35	108	36	.	85	70	160	45	112	280	56	224	168	42	90	385	154	385	190	1
		1	.	.	.	.	.	8	.	9	.	.	.	.	.	.	.	.	.	.	27	.	.	.	28	28	.	.	36	63	.	.	.	72	36	.	.	.	48	45	.	.	.	.	.	.	.	105	42	105	.	2
			1	3	4	2	5	3	5	1	15	20	10	10	24	11	28	20	33	42	36	9	10	25	32	42	10	116	37	28	5	102	75	32	8	1	65	160	124	9	128	270	144	169	171	123	255	240	152	330	185	3
				1	.	1	.	2	.	1	.	5	.	.	6	.	12	.	14	9	18	.	.	5	20	21	.	24	23	21	.	38	15	24	8	.	30	30	52	9	44	100	26	78	63	21	45	130	58	135	80	4
					1	.	1	.	2	.	6	4	5	4	4	6	5	11	5	15	6	6	6	11	5	6	7	40	5	4	4	33	25	4	1	1	20	74	40	1	54	84	69	50	51	68	123	72	65	100	75	5
						1	.	1	.	1	.	.	.	.	.	.	6	.	7	.	6	.	.	.	14	7	.	.	16	14	.	14	.	16	8	.	15	.	14	9	14	35	.	42	21	.	.	70	14	35	35	6
							1	.	.	.	3	.	2	4	4	3	.	8	5	.	.	4	6	.	.	6	6	20	5	.	4	.	25	4	.	1	.	28	.	1	.	42	36	25	51	32	87	.	.	50	.	7
								1	.	1	.	.	.	.	.	.	.	.	.	.	6	.	.	.	7	7	.	.	8	14	.	.	.	16	8	.	.	.	14	9	.	.	.	.	.	.	.	35	14	35	.	8
									1	.	2	.	3	1	.	3	.	6	.	5	.	4	3	6	.	.	5	10	.	.	3	9	5	.	.	1	5	36	10	.	24	18	30	10	9	39	54	16	30	20	30	9
										1	.	.	.	.	.	.	.	.	.	.	.	.	.	.	.	.	.	.	.	7	.	.	.	8	8	.	.	.	.	9	.	.	.	.	.	.	.	.	.	.	.	10
											1	.	1	1	.	2	.	4	.	.	.	3	3	.	.	.	4	5	.	.	3	.	5	.	.	1	.	12	.	.	.	9	18	5	9	20	36	.	.	10	.	11
												1	.	.	1	.	2	.	2	3	3	.	.	2	3	3	.	8	3	3	.	12	5	3	1	.	9	13	16	1	18	30	12	22	18	11	21	38	23	40	33	12
													1	.	.	1	.	2	.	.	.	2	1	.	.	.	3	.	.	.	2	.	.	.	.	1	.	6	.	.	.	.	6	.	.	14	12	.	.	.	.	13
														1	.	.	.	2	.	.	.	1	3	.	.	.	3	.	.	.	3	.	5	.	.	1	.	.	.	.	.	.	.	.	9	.	18	.	.	.	.	14
															1	.	.	.	2	.	.	.	.	.	.	3	.	4	3	.	.	.	5	3	.	.	.	5	.	1	.	15	6	11	18	5	15	.	.	20	.	15
																1	.	1	.	.	.	2	1	.	.	.	2	.	.	.	2	.	.	.	.	1	.	.	.	.	.	.	6	.	.	7	6	.	.	.	.	16
																	1	.	1	.	1	.	.	.	2	1	.	.	2	2	.	4	.	2	1	.	5	.	4	1	5	10	.	12	6	.	.	20	5	10	15	17

18	1	.	.	.	1	1	.	.	.	2	.	.	.	2	.	.	.	.	1	.	.	.	.	.	.	.	.	.	.	6	.	.	.	.
19		1	.	.	.	.	.	.	1	.	.	2	.	.	.	.	2	.	.	.	.	.	1	.	5	.	6	6	.	.	.	.	5	.
20			1	.	.	.	1	.	.	.	2	.	.	.	3	1	.	.	.	2	6	4	.	8	6	5	4	3	6	9	8	10	8	13
21				1	.	.	.	1	1	.	.	1	2	.	.	.	2	1	.	.	.	4	1	.	.	.	.	.	.	.	10	5	10	.
22					1	.	.	.	.	1	.	.	.	1	.	.	.	.	1	.	.	.	.	.	.	.	.	.	.	.	.	.	.	.
23						1	.	.	.	1	.	.	.	2	.	.	.	.	1	.	.	.	.	.	.	.	.	.	.	.	.	.	.	.
24							1	.	.	.	.	.	.	.	.	.	.	.	.	.	3	.	.	4	.	2	.	.	3	3	.	5	.	5
25								1	.	.	.	1	1	.	.	.	1	1	.	.	.	.	1	.	.	.	.	.	.	.	5	.	.	.
26									1	.	.	1	.	.	.	.	2	.	.	.	.	.	1	.	.	.	.	.	.	.	.	.	5	.
27										1	.	.	.	1	.	.	.	.	1	.	.	.	.	.	.	.	.	.	.	.	.	.	.	.
28											1	.	.	.	.	1	.	.	.	.	2	.	.	.	3	3	2	3	3	6	.	.	4	.
29												1	.	.	.	.	1	.	.	.	.	.	1	.	.	.	.	.	.	.	.	.	.	.
30													1	.	.	.	1	1	.	.	.	.	1	.	.	.	.	.	.	.	.	.	.	.
31														1	.	.	.	.	1	.	.	.	.	.	.	.	.	.	.	.	.	.	.	.
32															1	.	.	.	.	1	.	1	.	2	2	.	2	1	.	.	4	2	2	6
33																1	.	.	.	.	.	.	.	.	.	.	.	3	.	3	.	.	.	.
34																	1	.	.	.	.	.	1	.	.	.	.	.	.	.	.	.	.	.
35																		1	.	.	.	.	1	.	.	.	.	.	.	.	.	.	.	.
36																			1	.	.	.	.	.	.	.	.	.	.	.	.	.	.	.
37																				1	.	.	.	.	.	.	2	.	.	.	2	.	.	3
38																					1	.	.	.	.	1	.	.	2	2	.	.	.	.
39																						1	.	.	.	.	.	.	.	.	2	2	2	.
40																							1	.	.	.	.	.	.	.	.	.	.	.
41																								1	.	.	.	.	.	.	.	1	.	2
42																									1	.	1	1	.	.	.	.	1	.
43																										1	.	.	1	1	.	.	.	.
44																											1	.	.	.	.	.	.	.
45																												1	.	.	.	.	.	.
46																													1	.	.	.	.	.
47																														1	.	.	.	.
48																															1	.	.	.
49																																1	.	.
50																																	1	.
51																																		1

Table 3. *(continued)*

ALGEBRA A9, REPRESENTATIONS OF CLASS 4

S.P.	24	44	64	44	84	144	64	84	84	64	104	84	144	144	104	164	84	124	224	104	164	324	104	244	104	124
O.S.	210	360	45	1 260	90	10	1 260	120	840	5 040	720	2 520	45	360	2 520	90	12 600	720	90	6 300	720	10	5 040	90	18 900	2 520
LEVEL	25	31	33	35	35	37	41	43	43	43	47	47	49	49	49	51	51	51	53	53	53	55	55	55	57	59
DIM'N	210	990	825	2 310	1 485	715	6 930	4 950	9 240	17 280	21 120	31 185	9 075	17 160	45 045	19 305	83 160	47 190	15 015	107 250	49 280	5 005	126 720	19 305	261 360	190 575
	0	1	0	0	2	4	0	0	0	1	0	0	0	0	2	0	1	1	0	1	3	0	1	5	0	0
	0	0	2	0	1	0	0	0	0	0	0	1	0	0	0	0	0	2	0	0	1	0	0	0	1	0
	0	1	0	0	0	0	0	0	0	0	0	1	0	0	1	0	0	0	0	0	0	0	0	0	0	2
	1	0	0	0	0	0	0	0	0	1	0	0	0	0	0	0	0	0	0	0	0	0	0	0	1	0
	0	0	0	1	0	0	0	0	0	0	0	0	0	0	0	0	1	0	0	1	0	0	0	0	0	0
	0	0	0	0	0	0	1	0	1	0	0	0	0	0	0	0	0	0	0	0	0	0	1	0	0	0
	0	0	0	0	0	0	0	2	0	0	1	0	0	1	0	0	0	0	0	0	0	0	1	0	0	0
	0	0	0	0	0	0	1	0	0	0	1	0	3	0	0	2	1	0	1	0	0	0	0	0	1	1
WEIGHT	0	0	0	1	0	0	0	0	2	1	1	1	0	3	1	2	0	1	4	2	1	6	0	1	0	0
NUMBER	0	1	2	3	4	5	6	7	8	9	10	11	12	13	14	15	16	17	18	19	20	21	22	23	24	25
0	1	3	2	5	3	1	9	5	10	24	16	33	5	10	42	9	66	37	5	75	32	1	72	9	150	93
1		1	1	.	2	1	.	.	.	6	.	14	.	.	21	.	14	23	.	15	24	.	14	9	55	42
2			1	.	1	1	.	.	.	.	.	7	.	.	7	.	.	16	.	.	16	.	.	9	20	21
3				1	.	.	3	2	4	4	8	5	3	6	6	6	20	5	4	25	4	1	28	1	42	25
4					1	1	.	.	.	.	.	.	.	.	7	.	.	8	.	.	16	.	.	9	.	.
5						1	.	.	.	.	.	.	.	.	.	.	.	.	.	.	8	.	.	9	.	.
6							1	1	1	.	4	.	2	3	.	4	5	.	3	5	.	1	12	.	9	5
7								1	.	.	2	.	1	1	.	3	.	.	2	.	.	1	6	.	.	.
8									1	.	2	.	1	3	.	3	.	.	3	5	.	1	.	.	.	.
9										1	.	2	.	.	3	.	4	3	.	5	3	.	5	1	15	11
10											1	.	1	1	.	2	.	.	2	.	.	1	.	.	.	.
11												1	.	.	1	.	.	2	.	.	2	.	.	1	5	6
12													1	.	.	1	.	.	1	.	.	1	.	.	.	.
13														1	.	1	.	.	2	.	.	1	.	.	.	.
14															1	.	.	1	.	.	2	.	.	1	.	.
15																1	.	.	1	.	.	1	.	.	.	.
16																	1	.	.	1	.	.	2	.	3	2

S.P.	124	124	124	144	144	164	144	124	164	144	144	184	184	164	144	184	144	144	204	224	244	164	204	204	184	264
O.S.	7 560	15 120	12 600	840	2 520	5 040	5 040	25 200	2 520	15 120	3 150	1 260	5 040	5 040	12 600	7 560	16 800	50 400	360	720	2 520	1 260	2 520	2 520	12 600	720
LEVEL	59	59	59	61	61	61	61	63	63	63	65	65	65	65	65	65	67	67	67	67	67	67	67	67	69	69
DIM'N	331 695	424 710	351 000	240 240	330 330	316 800	444 675	711 480	405 405	823 680	490 050	360 360	800 415	1 034 880	926 640	675 675	1 161 600	2 196 480	450 450	443 520	482 625	630 630	770 770	847 000	1 576 575	566 280
	0	1	2	0	1	1	2	0	1	1	0	0	1	1	2	3	0	0	0	1	1	2	2	3	0	1
	1	0	0	0	0	0	0	1	0	1	1	3	0	1	0	0	0	1	3	0	0	0	2	0	1	0
	0	0	0	2	0	0	0	0	0	1	0	0	0	1	0	1	1	0	0	0	0	0	0	1	0	0
	1	0	1	0	0	0	1	0	0	0	0	0	0	0	0	0	1	0	0	0	0	0	0	0	0	0
	0	0	0	0	0	0	0	1	0	0	0	0	0	0	1	0	0	1	0	0	0	0	0	0	1	0
	0	1	0	0	0	1	0	0	0	0	2	0	0	0	0	0	0	0	0	0	0	2	0	0	0	0
	0	0	0	0	2	0	0	1	1	0	0	0	1	0	1	0	1	0	0	0	1	0	0	0	0	0
	0	1	1	0	0	0	0	0	2	1	0	1	1	0	0	1	0	1	0	3	0	0	1	0	0	2
WEIGHT	2	1	0	2	1	3	2	0	0	0	0	0	2	2	0	0	0	1	2	1	4	0	0	2	3	3
NUMBER	26	27	28	29	30	31	32	33	34	35	36	37	38	39	40	41	42	43	44	45	46	47	48	49	50	51
0	171	195	180	106	123	120	210	272	129	336	153	120	225	392	318	234	352	704	141	96	105	177	231	280	430	102
1	63	35	75	49	21	20	84	91	21	180	49	77	35	210	119	144	148	232	92	14	15	63	161	168	140	14
2	21	.	20	28	.	.	21	28	.	96	14	56	.	112	28	75	64	64	64	.	.	14	112	84	35	.
3	51	87	50	30	62	60	60	102	71	88	65	30	132	104	120	60	128	288	35	64	70	76	57	70	190	74
4	.	.	20	.	.	.	21	.	.	48	.	28	.	56	28	75	.	.	36	.	.	14	84	84	.	.
5	.	.	.	.	.	.	.	.	.	.	.	.	.	.	.	27	.	.	.	.	.	.	28	28	.	.
6	9	36	10	5	33	24	10	38	40	16	30	5	75	16	45	10	44	104	5	44	42	36	9	10	66	53
7	.	12	.	.	21	6	.	14	21	.	15	.	42	.	15	.	14	28	.	30	21	21	.	.	14	39
8	9	18	.	5	13	24	10	.	21	.	.	.	54	16	.	.	.	52	5	30	42	.	.	10	66	45
9	18	15	20	13	10	10	24	33	11	46	20	19	20	54	42	36	54	92	22	9	10	25	39	42	60	10
10	.	6	.	.	7	6	.	.	14	.	.	.	28	.	.	.	.	14	.	23	21	.	.	.	14	31
11	6	.	5	7	.	.	6	9	.	24	5	13	.	28	9	18	24	24	15	.	.	5	27	21	15	.
12	.	.	.	.	.	.	.	.	7	.	.	.	7	.	.	.	.	.	.	16	7	.	.	.	.	16
13	.	.	.	.	.	6	.	.	.	.	.	.	14	.	.	.	.	.	.	8	21	.	.	.	14	23
14	.	.	5	.	.	.	6	.	.	12	.	7	.	14	9	18	.	.	8	.	.	5	20	21	.	.
15	.	.	.	.	.	.	.	.	.	.	.	.	7	.	.	.	.	.	.	8	7	.	.	.	.	16
16	3	6	4	2	5	4	4	12	6	8	9	3	11	8	15	6	18	32	3	6	6	11	6	6	20	7

1	.	.	1	.	.	1	.	.	.	.	.	.	.	.	.	.	.	6	.	7	.	7	.	6	.	.	8	.	.	.	14	7	.	.	17
	1	.	.	1	.	.	.	.	.	.	.	.	.	.	.	.	.	.	.	.	.	.	.	.	.	.	.	.	7	.	.	.	.	8	18
		1	.	.	.	.	.	.	3	3	.	2	2	4	4	.	3	.	.	.	8	8	.	.	.	16	3	4	6	.	.	6	20	6	19
			1	.	.	1	.	.	.	.	.	.	.	.	.	.	.	.	.	.	.	.	.	6	.	.	.	.	.	.	7	7	.	.	20
				1	.	.	.	.	.	.	.	.	.	.	.	.	.	.	.	.	.	.	.	.	.	.	.	.	.	.	.	.	.	.	21
					1	.	.	.	.	2	.	.	3	1	.	4	3	.	5	.	6	.	5	.	5	8	.	4	3	6	.	.	4	5	22
						1	.	.	.	.	.	.	.	.	.	.	.	.	.	.	.	.	.	.	.	.	.	.	.	.	.	.	.	.	23
							1	1	1	.	1	1	.	.	1	3	.	4	2	2	.	4	3	3	8	8	2	.	.	2	4	3	5	.	24
								1	.	.	.	1	.	.	.	.	.	2	.	1	.	2	.	1	4	.	1	.	.	.	3	1	.	.	25
									1	.	.	1	.	.	1	.	.	.	.	.	.	4	.	.	.	4	2	.	.	.	.	3	5	.	26
										1	.	.	1	1	.	.	2	.	.	.	4	.	.	.	.	4	.	3	3	.	.	.	4	4	27
											1	.	.	.	1	.	.	2	.	1	.	2	3	3	.	.	1	.	.	2	3	3	.	.	28
												1	.	.	.	.	.	.	.	.	.	2	.	.	.	.	1	.	.	.	.	1	.	.	29
													1	.	.	.	1	.	.	.	2	.	.	.	.	.	.	2	1	.	.	.	.	3	30
														1	.	.	.	.	.	.	2	.	.	.	.	.	.	1	3	.	.	.	4	3	31
															1	.	.	.	.	.	.	2	.	.	.	.	1	.	.	.	.	3	.	.	32
																1	.	.	1	.	.	.	1	.	2	2	.	.	.	1	.	.	1	.	33
																	1	.	.	.	1	.	.	.	.	.	.	2	1	.	.	.	.	2	34
																		1	.	1	.	1	.	1	.	.	1	.	.	.	2	1	.	.	35
																			1	.	.	.	.	.	.	.	.	.	.	1	.	.	.	.	36
																				1	.	.	.	.	.	.	1	.	.	.	1	.	.	.	37
																					1	.	.	.	.	.	.	1	1	.	.	.	.	2	38
																						1	.	.	.	.	1	.	.	.	.	1	.	.	39
																							1	.	.	.	.	.	.	1	.	.	.	.	40
																								1	.	.	.	.	.	.	1	1	.	.	41
																									1	.	.	.	.	.	.	.	.	.	42
																										1	.	.	.	.	.	.	1	.	43
																											1	.	.	.	.	.	.	.	44
																												1	.	.	.	.	.	1	45
																													1	.	.	.	.	1	46
																														1	.	.	.	.	47
																															1	.	.	.	48
																																1	.	.	49
																																	1	.	50
																																		1	51

Table 3. *(continued)*

ALGEBRA A9, REPRESENTATIONS OF CLASS 5

25	45	45	65	65	85	85	105	105	145	65	145	225	225	85	85	105	105	105	105	125	125	105	165	165	165	S.P.
252	840	840	360	360	360	360	90	90	90	6 300	90	10	10	7 560	7 560	840	840	5 040	5 040	5 040	5 040	25 200	360	360	2 520	O.S.
26	34	34	38	38	40	40	42	42	44	44	44	46	46	50	50	52	52	52	52	56	56	58	58	58	58	LEVEL
252	1 848	1 848	3 300	3 300	4 752	4 752	4 950	4 950	5 148	21 000	5 148	2 002	2 002	64 152	64 152	46 200	46 200	85 800	85 800	198 198	198 198	325 248	85 800	85 800	161 700	DIM'N
0	0	1	0	0	0	2	0	1	0	1	3	0	5	0	1	0	1	1	2	1	1	0	0	1	1	
0	0	0	0	1	0	0	0	2	0	0	1	0	0	1	0	0	0	0	0	0	1	1	3	0	0	
0	0	0	0	1	0	1	0	0	0	0	0	0	0	0	0	2	0	0	0	0	1	0	0	0	0	
0	0	1	0	0	0	0	0	0	0	0	0	0	0	1	0	0	0	0	1	0	0	0	0	0	0	
1	0	0	0	0	0	0	0	0	0	1	0	0	0	0	0	0	0	0	0	0	0	1	0	0	0	
0	1	0	0	0	0	0	0	0	0	0	0	0	0	0	1	0	0	1	0	0	0	0	0	0	0	
0	0	0	1	0	1	0	0	0	0	0	0	0	0	0	0	0	2	0	0	1	0	0	0	0	1	
0	0	0	1	0	0	0	2	0	1	0	0	0	0	0	1	0	0	0	0	1	0	1	0	3	0	
0	1	0	0	0	2	0	1	0	3	1	0	5	0	1	0	1	0	2	1	1	1	0	1	0	3	WEIGHT
0	1	2	3	4	5	6	7	8	9	10	11	12	13	14	15	16	17	18	19	20	21	22	23	24	25	NUMBER
1	4	4	5	5	6	6	5	5	4	25	4	1	1	51	51	30	30	60	60	104	104	164	35	35	70	0
	1	.	2	.	3	.	3	.	3	5	.	1	.	9	18	5	13	24	10	54	16	52	5	22	42	1
		1	.	2	.	3	.	3	.	5	3	.	1	18	9	13	5	10	24	16	54	52	22	5	10	2
			1	.	1	.	2	.	2	.	.	1	.	.	6	.	7	6	.	28	.	14	.	15	21	3
				1	.	1	.	2	.	.	2	.	1	6	.	7	.	.	6	.	28	14	15	.	.	4
					1	.	1	.	2	.	.	1	.	.	.	.	.	6	.	14	.	.	.	8	21	5
						1	.	1	.	.	2	.	1	.	.	.	.	.	6	.	14	.	8	.	.	6
							1	.	1	.	.	1	.	.	.	.	.	.	.	7	.	.	.	8	7	7
								1	.	.	1	.	1	.	.	.	.	.	.	.	7	.	8	.	.	8
									1	.	.	1	.	.	.	.	.	.	.	.	.	.	.	.	7	9
										1	.	.	.	3	3	2	2	4	4	8	8	16	3	3	6	10
											1	.	1	.	.	.	.	.	.	.	.	.	.	.	.	11
												1	.	.	.	.	.	.	.	.	.	.	.	.	.	12
													1	.	.	.	.	.	.	.	.	.	.	.	.	13
														1	.	1	.	.	1	.	4	4	2	.	.	14
															1	.	1	1	.	4	.	4	.	2	3	15
																1	.	.	.	.	2	.	1	.	.	16

165	125	185	125	185	125	125	245	145	245	145	345	145	345	145	145	145	165	185	165	165	165	185	165	165	185	S.P.
2 520	12 600	720	12 600	720	12 600	12 600	720	6 300	720	5 040	90	5 040	90	37 800	37 800	25 200	7 560	12 600	7 560	15 120	15 120	12 600	4 200	4 200	7 560	O.S.
58	60	60	60	60	62	62	62	62	62	64	64	64	64	66	66	68	68	68	68	68	68	68	70	70	70	LEVEL
161 700	421 200	182 952	421 200	182 952	508 200	508 200	143 000	544 500	143 000	648 648	48 048	648 648	48 048	1 716 000	1 716 000	1 524 600	1 351 350	1 287 000	1 351 350	2 182 950	2 182 950	1 287 000	1 089 000	1 089 000	1 667 952	DIM'N
3	0	1	2	2	0	0	1	2	4	0	1	2	6	0	1	0	0	0	1	1	2	3	0	0	0	
0	1	0	0	2	0	1	0	0	1	0	0	0	0	1	1	0	1	1	0	1	0	0	0	0	1	
1	0	0	0	0	1	0	0	0	0	1	0	0	0	0	0	1	0	0	2	0	0	0	0	1	0	
0	0	0	0	0	1	0	0	0	0	1	0	0	0	0	1	0	0	0	0	1	0	1	2	0	0	
0	1	0	1	0	0	0	0	1	0	0	0	0	0	0	0	1	0	0	0	0	0	0	0	0	0	
0	0	0	0	0	0	1	0	0	0	0	0	1	0	1	0	0	0	1	0	0	1	0	0	2	0	
0	0	0	0	0	0	1	0	0	0	0	0	1	0	0	0	1	2	0	0	0	0	0	1	0	1	
0	0	2	1	0	1	0	1	0	0	0	0	0	0	1	1	0	0	0	1	0	1	1	0	0	2	
1	2	2	0	1	0	0	4	2	1	2	6	0	1	1	0	0	1	3	0	2	1	0	0	0	0	WEIGHT
26	27	28	29	30	31	32	33	34	35	36	37	38	39	40	41	42	43	44	45	46	47	48	49	50	51	NUMBER
70	190	66	190	66	195	195	40	225	40	224	9	224	9	550	550	415	365	350	365	645	645	350	250	250	396	0
10	66	45	60	9	58	80	32	75	5	72	9	93	1	255	160	162	192	180	102	195	297	100	95	110	225	1
42	60	9	66	45	80	58	5	75	32	93	1	72	9	160	255	162	102	100	192	297	195	180	110	95	108	2
.	14	31	15	.	14	35	24	15	.	14	9	42	.	110	35	63	105	75	21	35	126	20	35	50	132	3
21	15	.	14	31	35	14	.	15	24	42	.	14	9	35	110	63	21	20	105	126	35	75	50	35	21	4
.	14	23	.	.	.	.	24	15	.	14	9	.	.	55	.	.	42	75	.	35	63	.	.	.	69	5
21	.	.	14	23	.	.	.	15	24	.	.	14	9	.	55	.	.	.	42	63	35	75	.	.	.	6
.	.	16	.	.	.	.	16	.	.	.	9	.	.	20	.	.	21	20	.	.	21	.	.	.	48	7
7	.	.	.	16	.	.	.	.	16	.	.	.	9	.	20	.	.	.	21	21	.	20	.	.	.	8
.	.	8	.	.	.	.	16	.	.	.	9	.	.	.	.	.	.	20	.	.	.	.	.	.	.	9
6	20	6	20	6	23	23	4	25	4	28	1	28	1	72	72	63	52	50	52	87	87	50	42	42	60	10
7	.	.	.	8	.	.	.	.	16	.	.	.	9	.	.	.	.	.	.	.	.	20	.	.	.	11
.	.	.	.	.	.	.	8	.	.	.	9	.	.	.	.	.	.	.	.	.	.	.	.	.	.	12
.	.	.	.	.	.	.	.	.	8	.	.	.	9	.	.	.	.	.	.	.	.	.	.	.	.	13
3	5	.	4	4	10	5	.	5	3	12	.	5	1	15	30	24	10	10	28	36	15	20	20	15	11	14
.	4	4	5	.	5	10	3	5	.	5	1	12	.	30	15	24	28	20	10	15	36	10	15	20	34	15
1	.	.	.	3	5	.	.	.	2	6	.	.	1	.	10	9	.	.	18	12	.	5	10	5	.	16

1	.	.	2	.	.	.	1	1	.	.	3	.	.	.	5	2	.	.	.	1	6	.	10	.	9	18	5	.	.	12	.	5	10	18	17
	1	.	2	.	.	.	1	3	.	4	3	.	.	.	.	3	5	.	5	1	.	.	15	.	.	11	20	.	15	18	.	.	.	18	18
		1	.	2	.	1	.	.	3	.	.	4	3	.	.	.	5	3	.	.	5	1	.	15	.	.	.	11	18	15	20	.	.	.	19
			1	.	.	.	1	1	.	.	2	.	.	.	.	2	.	.	.	1	.	.	5	.	.	6	5	.	.	6	.	.	.	12	20
				1	.	1	.	.	1	.	.	.	2	.	.	.	.	2	.	.	.	1	.	5	.	.	.	6	6	.	5	.	.	.	21
					1	.	.	.	.	1	.	1	.	2	2	.	1	.	2	.	2	.	6	6	9	5	4	5	6	6	4	7	7	6	22
						1	.	.	.	.	.	.	1	.	.	.	.	1	.	.	.	1	.	.	.	.	.	.	.	.	.	.	.	.	23
							1	.	.	.	1	.	.	.	.	1	.	.	.	1	.	.	.	.	.	.	.	.	.	.	.	.	.	6	24
								1	.	.	1	.	.	.	.	2	.	.	.	1	.	.	.	.	.	.	5	.	.	.	.	.	.	.	25
									1	.	.	.	1	.	.	.	.	2	.	.	.	1	.	.	.	.	.	.	.	.	5	.	.	.	26
										1	.	.	.	.	.	.	1	.	2	.	.	.	3	.	.	2	4	.	6	3	.	.	.	3	27
											1	.	.	.	.	1	.	.	.	1	.	.	.	.	.	.	.	.	.	.	.	.	.	.	28
												1	.	.	.	.	1	.	.	.	2	.	.	3	.	.	.	2	3	6	4	.	.	.	29
													1	.	.	.	.	1	.	.	.	1	.	.	.	.	.	.	.	.	.	.	.	.	30
														1	.	.	.	.	1	.	.	.	.	2	3	.	.	3	2	.	1	4	2	.	31
															1	.	.	.	.	.	1	.	2	.	3	3	1	.	.	2	.	2	4	3	32
																1	.	.	.	1	.	.	.	.	.	.	.	.	.	.	.	.	.	.	33
																	1	.	.	.	.	.	.	.	.	.	.	.	3	3	.	.	.	.	34
																		1	.	.	.	1	.	.	.	.	.	.	.	.	.	.	.	.	35
																			1	.	.	.	.	.	.	.	.	.	2	.	.	.	.	.	36
																				1	.	.	.	.	.	.	.	.	.	.	.	.	.	.	37
																					1	.	.	.	.	.	.	.	.	2	.	.	.	.	38
																						1	.	.	.	.	.	.	.	.	.	.	.	.	39
																							1	.	.	1	1	.	.	1	.	.	.	2	40
																								1	.	.	.	1	1	.	1	.	.	.	41
																									1	.	.	.	.	.	.	1	1	.	42
																										1	.	.	.	.	.	.	.	1	43
																											1	.	.	.	.	.	.	.	44
																												1	.	.	.	.	.	.	45
																													1	.	.	.	.	.	46
																														1	.	.	.	.	47
																															1	.	.	.	48
																																1	.	.	49
																																	1	.	50
																																		1	51

Table 3. *(continued)*

ALGEBRA A10, REPRESENTATIONS OF CLASS 0

0	22	44	66	66	88	66	88	88	132	132	88	110	154	154	110	198	110	110	132	132	154	154	132	220	132	S.P.
1	110	1 980	495	495	110	9 240	9 240	9 240	1 320	1 320	11 550	7 920	990	990	2 772	110	34 650	34 650	6 930	6 930	13 860	13 860	27 720	2 310	27 720	O.S.
1	21	37	39	39	41	49	53	53	55	55	57	57	59	59	61	61	63	63	65	65	67	67	69	69	69	LEVEL
1	120	2 904	3 510	3 510	4 235	24 200	66 066	66 066	43 680	43 680	81 675	179 200	117 975	117 975	104 544	77 440	393 250	393 250	330 330	330 330	594 594	594 594	891 891	286 650	891 891	DIM'N
0	1	0	0	2	2	0	0	1	0	3	0	1	1	3	0	3	0	1	0	0	0	2	0	0	1	
0	0	1	1	0	0	0	0	1	0	0	0	1	1	0	0	0	0	0	0	2	0	1	0	0	0	
0	0	0	0	0	0	1	1	0	1	0	0	0	0	0	0	0	0	1	0	0	0	0	0	0	0	
0	0	0	0	0	0	0	0	0	0	0	1	0	0	0	0	0	1	0	1	0	1	0	0	1	1	
0	0	0	0	0	0	0	0	0	0	0	0	0	0	0	1	0	0	0	0	0	0	0	1	0	0	
0	0	0	0	0	0	0	0	0	0	0	0	0	0	0	1	0	0	0	0	0	0	0	0	0	1	
0	0	0	0	0	0	0	0	0	0	0	1	0	0	0	0	0	0	1	0	1	0	1	1	0	0	
0	0	0	0	0	0	1	0	1	0	1	0	0	0	0	0	0	1	0	0	0	0	0	0	0	0	
0	0	1	0	1	0	0	1	0	0	0	0	1	0	1	0	0	0	0	2	0	1	0	0	0	0	
0	1	0	2	0	2	0	1	0	3	0	0	1	3	1	0	3	1	0	0	0	2	0	1	4	0	WEIGHT
0	1	2	3	4	5	6	7	8	9	10	11	12	13	14	15	16	17	18	19	20	21	22	23	24	25	NUMBER
1	10	44	45	45	55	110	231	231	120	120	165	560	330	330	132	220	550	550	385	385	594	594	693	210	693	0
	1	8	9	9	10	27	63	63	36	36	48	152	90	90	42	55	180	180	133	133	216	216	252	84	252	1
		1	1	1	1	6	14	14	8	8	14	32	18	18	14	10	55	55	42	42	69	69	91	28	91	2
			1	.	1	.	7	.	8	.	.	16	18	9	.	10	20	.	21	.	48	.	28	28	.	3
				1	1	.	.	7	.	8	.	16	9	18	.	10	.	20	.	21	.	48	.	.	28	4
					1	.	.	.	.	.	.	8	9	9	.	10	.	.	.	.	.	.	.	.	.	5
						1	2	2	1	1	4	4	2	2	5	1	15	15	11	11	18	18	33	7	33	6
							1	.	1	.	.	2	2	1	.	1	5	.	6	.	12	.	9	7	.	7
								1	.	1	.	2	1	2	.	1	.	5	.	6	.	12	.	.	9	8
									1	.	.	.	2	.	.	1	.	.	.	.	6	.	.	7	.	9
										1	.	.	.	2	.	1	.	.	.	.	.	6	.	.	.	10
											1	.	.	.	2	.	3	3	2	2	3	3	12	1	12	11
												1	1	1	.	1	.	.	.	.	.	.	.	.	.	12
													1	.	.	1	.	.	.	.	.	.	.	.	.	13
														1	.	1	.	.	.	.	.	.	.	.	.	14
															1	.	.	.	.	.	.	.	4	.	4	15

132	220	154	154	154	154	154	154	176	176	176	176	198	242	176	198	242	176	198	176	264	154	198	220	264	198	S.P.
83 160	2 310	2 310	13 860	2 310	13 860	27 720	27 720	1 980	27 720	27 720	27 720	3 960	3 960	27 720	3 960	3 960	13 860	13 860	13 860	495	104 800	13 860	7 920	495	4 620	O.S.
69	69	71	71	71	71	73	73	73	73	73	75	75	75	75	75	75	77	77	77	77	77	77	77	77	79	LEVEL
1 852 389	286 650	566 280	1 541 540	566 280	1 541 540	1 698 840	1 698 840	1 277 640	2 764 125	2 764 125	2 466 464	2 286 900	1 321 320	2 466 464	2 286 900	1 321 320	2 548 260	2 642 640	2 548 260	1 089 000	6 938 624	2 642 640	4 089 800	1 089 000	2 076 360	DIM'N
1	4	0	0	1	1	0	0	0	1	2	0	0	1	2	2	4	0	0	0	0	1	1	2	4	0	
0	0	0	2	0	0	0	1	2	0	1	0	2	0	0	1	0	0	0	1	2	0	2	1	0	0	
1	0	0	0	0	1	0	1	0	1	0	0	0	1	1	0	0	0	0	0	0	0	0	0	0	0	
0	0	0	0	0	0	0	0	0	0	0	0	0	0	0	0	0	0	0	1	0	1	0	0	0	0	
0	0	0	0	2	0	1	0	0	0	0	1	0	0	0	0	0	0	1	1	0	0	0	0	0	0	
0	0	2	0	0	0	0	1	0	0	0	0	0	0	1	0	0	1	0	0	0	0	1	0	0	1	
0	1	0	0	0	0	0	0	0	0	0	0	0	0	0	0	0	1	0	0	0	1	0	0	0	0	
1	0	0	1	0	0	1	0	0	0	1	1	0	0	0	0	1	0	0	0	0	0	0	0	0	2	
0	0	0	0	0	2	1	0	2	1	0	0	1	0	0	2	0	1	2	0	0	0	0	1	2	0	
1	0	1	1	0	0	0	0	0	2	1	2	2	4	0	0	1	0	1	0	4	1	0	2	0	0	WEIGHT
26	27	28	29	30	31	32	33	34	35	36	37	38	39	40	41	42	43	44	45	46	47	48	49	50	51	NUMBER
2 079	210	330	1 540	330	1 540	990	990	1 200	2 520	2 520	1 232	1 980	990	1 232	1 980	990	990	1 100	990	825	3 872	1 100	3 410	825	660	0
702	84	126	532	126	532	387	387	408	882	882	504	684	360	504	684	360	414	470	414	285	1 488	470	1 160	285	288	1
213	28	49	161	49	161	148	148	121	266	266	196	200	108	196	200	108	175	190	175	81	552	190	332	81	127	2
75	.	14	56	.	84	64	.	64	182	91	112	136	108	.	108	36	70	120	.	81	160	.	224	45	56	3
75	28	.	84	14	56	.	64	64	91	182	.	108	36	112	136	108	.	.	70	45	160	120	224	81	.	4
27	.	.	28	.	28	.	.	36	63	63	.	72	36	.	72	36	.	.	.	45	48	.	152	45	.	5
54	7	20	39	20	39	54	54	28	63	63	72	45	24	72	45	24	75	70	75	17	196	70	72	17	56	6
18	.	5	13	.	21	24	.	15	42	21	38	30	24	.	24	8	30	45	.	17	52	.	48	9	26	7
18	7	.	21	5	13	.	24	15	21	42	.	24	8	38	30	24	.	.	30	9	52	45	48	17	.	8
.	.	.	.	.	.	.	.	.	21	.	14	15	24	.	.	.	.	15	.	17	.	.	24	.	.	9
.	7	.	.	.	.	.	.	.	.	21	.	.	.	14	15	24	.	.	.	.	.	15	24	17	.	10
9	1	9	6	9	6	18	18	4	9	9	24	6	3	24	6	3	33	22	33	2	64	22	9	2	24	11
6	.	.	7	.	7	.	.	8	14	14	.	16	8	.	16	8	.	.	.	9	14	.	32	9	.	12
.	.	.	.	.	.	.	.	.	7	.	.	8	8	.	.	.	.	.	.	9	.	.	16	.	.	13
.	.	.	.	.	.	.	.	.	.	7	.	.	.	.	8	8	.	.	.	.	.	.	16	9	.	14
.	.	5	.	5	.	5	5	.	.	.	6	.	.	6	.	.	15	5	15	.	16	5	.	.	10	15

1	.	.	.	.	.	.	.	.	.	.	.	.	.	.	.	.	.	.	.	.	.	.	.	.	.	.	.	.	.	.	.	.	8	.	.	16
	1	.	1	.	2	.	3	1	.	3	.	2	2	.	3	8	.	2	6	3	12	4	3	.	3	1	13	14	.	2	16	.	6	1	12	17
		1	.	1	.	2	.	.	3	3	1	.	3	2	2	.	8	2	3	6	.	3	1	12	4	3	.	.	13	1	16	14	6	2	.	18
			1	.	1	.	.	1	.	.	.	.	.	.	3	4	.	2	3	.	4	2	3	.	3	.	5	10	.	2	.	.	3	1	6	19
				1	.	1	.	.	.	.	1	.	3	.	.	.	4	2	.	3	.	3	.	4	2	3	.	.	5	1	.	10	3	2	.	20
					1	.	.	1	.	.	.	.	.	.	.	.	.	.	3	.	4	2	3	.	.	.	.	5	.	2	.	.	3	.	.	21
						1	.	.	.	.	1	.	.	.	.	.	.	.	.	3	.	.	.	4	2	3	.	.	.	.	.	5	3	2	.	22
							1	.	.	.	.	1	.	.	.	2	.	.	.	.	3	.	.	.	.	.	6	3	.	.	4	.	.	.	5	23
								1	.	.	.	.	.	.	.	.	.	.	.	.	.	.	3	.	.	.	.	.	.	2	.	.	.	.	.	24
									1	.	.	.	.	1	.	.	2	.	.	.	.	.	.	3	.	.	.	.	6	.	4	3	.	.	.	25
										1	.	.	1	.	1	.	.	1	2	2	.	2	1	.	2	1	.	.	.	1	4	.	4	1	.	26
											1	.	.	.	.	.	.	.	.	.	.	.	.	.	.	3	.	.	.	.	.	.	.	2	.	27
												1	.	.	.	.	.	.	.	.	.	.	.	.	.	.	3	.	.	.	.	.	.	.	2	28
													1	.	.	.	.	1	.	1	.	2	.	.	1	1	.	.	.	1	.	.	2	1	.	29
														1	.	.	.	.	.	.	.	.	.	.	.	.	.	.	3	.	.	.	.	.	.	30
															1	.	.	1	1	.	.	1	1	.	2	.	.	.	.	1	.	.	2	1	.	31
																1	.	.	.	.	1	.	.	.	.	.	2	2	.	.	.	.	.	.	3	32
																	1	.	.	.	.	.	.	1	.	.	.	.	2	.	.	2	.	.	.	33
																		1	.	.	.	1	.	.	1	.	.	.	.	1	.	.	1	1	.	34
																			1	.	.	1	1	.	.	.	.	.	.	1	.	.	2	.	.	35
																				1	.	.	.	.	1	1	.	.	.	.	.	.	2	1	.	36
																					1	.	.	.	.	.	.	1	.	.	.	.	.	.	.	37
																						1	.	.	.	.	.	.	.	1	.	.	1	.	.	38
																							1	.	.	.	.	.	.	1	.	.	.	.	.	39
																								1	.	.	.	.	.	.	.	1	.	.	.	40
																									1	.	.	.	.	.	.	.	1	1	.	41
																										1	.	.	.	.	.	.	.	1	.	42
																											1	.	.	.	.	.	.	.	1	43
																												1	.	.	.	.	.	.	.	44
																													1	.	.	.	.	.	.	45
																														1	.	.	.	.	.	46
																															1	.	.	.	.	47
																																1	.	.	.	48
																																	1	.	.	49
																																		1	.	50
																																			1	51

Table 3. *(continued)*

ALGEBRA A10, REPRESENTATIONS OF CLASS 1

10	32	54	54	76	76	98	120	142	76	98	98	142	98	120	120	120	142	120	142	164	208	142	186	120	164	142	164	208	230	230	142	142	274	164	142	164	186	186	164	186	208	208	208	164	164	186	230	252	186	186	186	S.P.
			4	1	3				11	13	6	2	27		4	27	9	27	3	3	1	6		46	7	9	13			2	34	13		13	138	34	1	13	27	13	1	1	9	41	83	55	13	9	9	4	13	
11	495	110	620	320	960	990	495	110	550	860	930	310	720	462	620	720	240	720	960	960	320	930	495	200	920	240	860	990	990	772	650	860	110	860	600	650	980	860	720	860	320	320	240	580	160	440	860	240	240	620	860	O.S.
11	29	31	43	45	47	49	49	51	53	57	59	59	59	61	61	63	63	65	65	65	65	67	67	67	67	69	69	69	69	71	71	71	71	73	73	73	75	75	75	75	77	77	77	77	77	77	77	79	79	79	79	LEVEL
																															1			1	3	1	1	2	3	2	1	1	2	3	6	5	2	2	2	2	5	
			8	10	22	27	15	18	45	125	98	84	212	60	176	582	315	495	480	386	150	424	317	755	858	679	770	566	406	378	376	990	267	415	659	981	019	081	083	081	038	415	312	440	540	559	162	466	595	265	465	
11	594	715	470	296	880	720	015	150	375	840	010	084	355	984	176	120	315	495	480	100	150	710	625	040	000	536	770	280	560	378	375	990	410	700	040	980	304	079	080	079	180	700	310	151	534	840	160	464	450	120	460	DIM'N
1	0	2	0	0	1	1	3	3	0	0	0	0	1	0	0	1	2	0	0	1	4	0	0	1	2	0	0	2	4	0	0	1	4	0	1	2	0	0	1	1	0	0	0	0	0	1	3	0	0	0	0	
0	1	0	0	0	1	1	0	0	0	0	0	0	0	0	2	0	1	0	2	0	0	0	2	0	1	0	0	1	0	0	1	0	0	0	0	0	0	0	0	2	0	0	0	1	1	0	1	0	0	1	1	
0	0	0	1	1	0	0	0	0	0	0	0	0	1	0	0	1	0	0	0	1	0	0	0	0	0	0	0	0	0	0	1	0	0	0	0	1	0	0	0	0	0	0	0	0	1	0	0	0	2	0	1	
0	0	0	0	0	0	0	0	0	1	1	0	1	0	0	0	0	0	0	0	0	0	0	0	1	0	0	0	0	0	0	0	0	0	0	1	0	0	0	1	0	0	0	0	1	0	1	0	0	0	0	0	
0	0	0	0	0	0	0	0	0	0	0	1	0	0	0	0	0	0	1	0	0	0	1	0	0	0	0	1	0	0	1	0	1	0	0	0	0	0	0	0	0	0	0	0	0	0	0	0	0	0	2	0	
0	0	0	0	0	0	0	0	0	0	0	0	0	0	2	0	0	0	0	0	0	0	0	0	0	0	1	0	0	0	0	0	1	0	1	0	0	0	1	0	0	0	0	1	1	0	0	0	1	1	0	0	
0	0	0	0	0	0	0	0	0	0	0	1	0	0	0	0	0	0	0	0	0	0	0	0	1	0	1	0	0	0	0	1	0	0	0	0	1	2	0	0	1	1	2	0	0	0	0	1	0	0	0	0	
0	0	0	0	0	0	0	0	0	1	0	0	0	1	0	1	0	1	1	0	0	1	0	0	0	0	0	0	0	0	0	0	0	0	1	1	0	0	1	0	0	2	0	0	0	1	0	0	0	0	0	0	
0	0	0	1	0	1	0	1	0	0	1	0	0	0	0	0	1	0	0	1	0	0	2	0	0	1	0	1	0	1	0	0	0	0	1	0	0	1	0	2	0	0	0	2	0	0	1	0	1	0	0	2	
0	1	1	0	2	0	2	0	2	0	1	0	3	0	0	0	1	0	1	1	3	0	0	3	0	1	1	2	3	1	4	0	0	3	0	1	0	0	2	0	0	0	2	1	0	1	2	0	3	0	0	0	WEIGHT
0	1	2	3	4	5	6	7	8	9	10	11	12	13	14	15	16	17	18	19	20	21	22	23	24	25	26	27	28	29	30	31	32	33	34	35	36	37	38	39	40	41	42	43	44	45	46	47	48	49	50	51	NUMBER
																																			1				1					1	2	2					2	
1	9	10	35	36	80	90	45	55	75	160	90	84	270	42	196	630	315	315	480	360	120	225	285	480	800	288	350	495	330	126	700	420	220	450	800	900	252	567	330	819	210	300	525	053	754	160	720	448	675	540	100	0
	1	1	7	8	16	18	9	10	20	48	28	28	75	14	56	182	91	112	136	108	36	85	81	160	224	112	140	135	90	56	245	154	55	190	640	320	112	252	490	301	98	140	245	414	984	816	272	224	275	224	756	1
		1	.	.	8	9	9	10	.	.	.	.	27	.	28	63	63	.	72	36	36	.	45	48	152	.	.	90	90	.	105	42	55	.	180	180	.	.	133	189	.	.	.	162	432	216	216	.	120	84	336	2
			1	1	2	2	1	1	5	12	9	7	18	5	13	42	21	38	30	24	8	30	17	52	48	44	50	27	18	21	80	58	10	80	210	105	51	108	162	99	47	65	110	162	315	270	90	104	110	95	240	3
				1	.	2	.	1	.	6	.	7	.	.	.	21	.	14	15	24	.	15	17	.	24	14	35	27	9	21	.	.	10	35	75	.	21	63	84	.	21	35	70	.	111	186	.	84	.	.	126	4
					1	1	1	1	.	.	.	.	6	.	7	14	14	.	16	8	8	.	9	14	32	.	.	18	18	.	35	14	10	.	55	55	.	.	42	63	.	.	.	63	138	69	69	.	50	35	105	5
						1	.	1	.	.	.	.	.	.	.	7	.	.	8	8	.	.	9	.	16	.	.	18	9	.	.	.	10	.	20	.	.	.	21	.	.	.	.	.	48	48	.	.	.	.	56	6
							1	1	.	.	.	.	.	.	.	.	7	.	.	.	8	.	.	.	16	.	.	9	18	.	.	.	10	.	.	20	.	.	.	21	.	.	.	.	.	.	48	.	.	.	.	7
								1	.	.	.	.	.	.	.	.	.	.	.	.	.	.	.	.	8	.	.	9	9	.	.	.	10	.	.	.	.	.	.	.	.	.	.	.	.	.	.	.	.	.	.	8
									1	2	3	1	3	2	2	6	3	12	4	3	1	9	2	16	6	18	15	3	2	6	23	23	1	33	60	30	24	45	44	27	23	31	46	63	84	72	24	44	42	42	61	9
										1	.	1	.	.	.	3	.	4	2	3	.	5	2	.	3	5	10	3	1	6	.	.	1	15	20	.	10	24	24	.	11	15	30	.	28	48	.	34	.	.	33	10
											1	.	.	1	.	.	.	3	.	.	.	2	.	4	.	8	3	.	.	1	5	10	.	13	12	6	12	18	8	5	11	16	17	24	15	12	4	16	15	20	10	11
												1	.	.	.	.	.	.	.	3	.	.	2	.	.	.	5	3	.	6	.	.	1	.	.	.	.	9	.	.	.	5	10	.	.	24	.	24	.	.	.	12
													1	.	1	2	2	.	2	1	1	.	1	4	4	.	.	2	2	.	10	5	1	.	15	15	.	.	11	17	.	.	.	24	36	18	18	.	20	15	26	13
														1	.	.	.	.	.	.	.	.	.	.	.	4	.	.	.	.	.	5	.	5	.	.	6	6	.	.	5	10	5	9	.	.	.	4	5	10	.	14
															1	.	1	.	2	.	1	.	1	.	2	.	.	1	2	.	5	.	1	.	.	5	.	.	.	12	.	.	.	9	18	.	12	.	10	5	13	15

1	.	.	1	1	.	.	1	.	2	.	.	2	1	.	.	.	1	.	5	.	.	.	6	.	.	.	.	.	12	12	.	.	.	.	14	16
	1	.	.	.	1	.	.	.	2	.	.	1	2	.	.	.	1	.	.	5	.	.	.	6	.	.	.	.	.	.	12	.	.	.	.	17
		1	.	.	.	1	.	.	.	2	2	.	.	1	.	.	.	6	4	.	5	9	4	.	6	7	11	.	5	8	.	12	.	.	5	18
			1	.	.	.	1	.	1	.	.	1	1	.	.	.	1	.	.	.	.	.	.	.	.	.	.	.	6	.	.	.	.	.	7	19
				1	.	.	1	.	.	.	.	2	.	.	.	.	1	.	.	.	.	.	.	.	.	.	.	.	.	6	.	.	.	.	.	20
					1	.	.	.	.	.	.	.	2	.	.	.	1	.	.	.	.	.	.	.	.	.	.	.	.	.	6	.	.	.	.	21
						1	.	.	.	.	1	.	.	1	.	.	.	3	.	.	2	3	4	.	3	2	8	.	.	4	.	8	.	.	5	22
							1	.	.	.	.	1	.	.	.	.	1	.	.	.	.	.	.	.	.	.	.	.	.	.	.	.	.	.	.	23
								1	.	.	.	.	.	.	2	2	.	.	3	3	.	.	2	3	.	.	.	9	6	3	3	.	7	7	4	24
										.	.	1	1	.	.	.	1	.	.	.	.	.	.	.	.	.	.	.	.	.	.	.	.	.	.	25
										1	.	.	.	.	.	.	.	2	.	.	3	3	.	.	3	4	3	.	.	.	.	3	.	.	.	26
											1	.	.	1	.	.	.	.	.	.	.	3	.	.	.	2	4	.	.	4	.	8	.	.	.	27
												1	.	.	.	.	1	.	.	.	.	.	.	.	.	.	.	.	.	.	.	.	.	.	.	28
													1	.	.	.	1	.	.	.	.	.	.	.	.	.	.	.	.	.	.	.	.	.	.	29
														1	.	.	.	.	.	.	.	.	.	.	.	.	.	.	.	.	.	4	.	.	.	30
															1	.	.	.	.	1	.	.	.	2	.	.	.	3	3	.	2	.	4	2	2	31
																1	.	.	.	.	.	.	.	.	.	.	.	3	.	.	.	.	2	4	.	32
																	1	.	.	.	.	.	.	.	.	.	.	.	.	.	.	.	.	.	.	33
																		1	.	.	1	1	.	.	2	1	2	.	.	.	.	2	.	.	.	34
																			1	.	.	.	1	.	.	.	.	.	2	2	.	.	.	.	2	35
																				1	.	.	.	1	.	.	.	.	.	.	2	.	.	.	.	36
																					1	.	.	.	1	1	.	.	.	.	.	.	.	.	.	37
																						1	.	.	.	1	1	.	.	.	.	2	.	.	.	38
																							1	.	.	.	.	.	.	1	.	.	.	.	2	39
																								1	.	.	.	.	.	.	1	.	.	.	.	40
																									1	.	.	.	.	.	.	.	.	.	.	41
																										1	.	.	.	.	.	.	.	.	.	42
																											1	.	.	.	.	1	.	.	.	43
																												1	.	.	.	.	1	1	.	44
																													1	.	.	.	.	.	1	45
																														1	.	.	.	.	.	46
																															1	.	.	.	.	47
																																1	.	.	.	48
																																	1	.	.	49
																																		1	.	50
																																			1	51

Table 3. *(continued)*

ALGEBRA A10, REPRESENTATIONS OF CLASS 2

18	40	40	62	106	62	84	84	84	106	106	128	128	106	106	150	194	150	216	106	128	150	150	128	128	172	128	172	150	172	238	150	150	172	216	194	194	172	216	216	150	150	194	216	304	238	260	172	172	238	172	238	S.P.
55	11	1 320	990	110	6 930	2 310	13 860	9 240	1 980	3 960	495	3 960	2 310	13 860	990	495	2 772	110	46 200	13 860	1 320	4 620	27 720	55 440	9 240	34 650	9 240	27 720	3 960	2 310	27 720	2 772	9 240	165	3 960	3 960	27 720	9 240	3 960	69 300	138 600	7 920	3 960	1 320	990	3 960	11 550	34 650	990	46 200	7 920	O.S.
19	21	35	39	41	47	49	53	55	55	55	57	57	59	59	59	59	61	61	63	65	67	67	67	67	69	69	69	69	71	71	71	71	71	73	73	73	73	73	75	75	75	75	75	75	77	77	77	77	77	77	77	LEVEL
55	66	1 760	4 719	3 080	16 335	20 020	75 075	58 080	61 776	91 476	75 075	110 110	76 230	165 165	133 650	51 975	112 112	62 920	424 710	424 710	283 140	377 520	755 040	1 182 720	693 693	943 800	792 792	1 081 080	755 040	350 350	2 081 079	594 594	1 121 120	259 545	1 132 560	1 386 000	2 972 970	1 156 155	1 359 072	2 972 970	4 802 490	3 067 350	1 976 832	443 520	1 321 320	1 486 485	2 162 875	4 129 125	2 032 800	3 963 960	3 153 920	DIM'N
0	2	0	1	3	0	0	1	0	0	1	0	2	0	0	2	4	0	4	1	0	0	0	0	1	0	1	1	2	0	0	0	1	1	0	0	0	2	3	0	0	1	1	2	5	0	0	0	1	1	2	3	
1	0	0	1	0	0	0	0	0	2	0	2	1	0	0	1	0	0	0	0	0	0	0	1	0	0	0	0	0	0	0	1	0	2	0	0	1	0	1	0	1	0	2	0	0	0	0	0	0	2	0	1	
0	0	1	0	0	0	0	1	0	0	1	0	0	0	0	0	0	0	0	0	0	0	0	1	0	0	0	0	1	0	0	1	0	0	0	0	1	1	0	0	0	0	0	1	0	0	0	2	0	0	0	0	
0	0	0	0	0	1	1	0	0	0	0	0	0	0	0	0	0	0	0	1	0	0	0	0	1	0	0	1	0	0	0	0	0	0	0	0	0	0	0	0	1	0	0	0	0	0	0	0	0	0	1	0	
0	0	0	0	0	0	0	0	1	0	0	0	0	0	1	0	0	1	0	0	0	0	0	0	0	0	1	0	0	0	0	0	0	0	0	0	0	0	0	0	0	1	0	0	0	0	0	0	1	0	0	0	
0	0	0	0	0	0	0	0	0	0	0	0	0	1	0	0	0	0	0	0	1	0	1	0	0	1	0	0	0	0	1	0	2	0	0	0	0	0	0	0	0	0	0	0	0	0	0	0	0	0	0	0	
0	0	0	0	0	0	0	0	0	0	0	0	0	1	0	0	0	0	0	0	0	2	0	0	0	0	1	0	0	1	0	0	0	0	0	1	0	0	0	1	1	0	0	0	0	0	1	1	0	0	1	0	
0	0	0	0	0	0	0	0	1	0	0	0	0	0	0	0	0	0	0	1	1	0	0	1	0	0	0	0	1	1	0	0	0	1	3	1	0	0	1	0	0	1	0	0	1	2	0	0	0	0	0	0	
0	0	0	0	0	1	0	1	0	1	0	0	1	0	1	0	1	0	0	0	0	0	2	0	1	1	0	0	0	1	0	1	0	0	0	0	0	1	0	2	0	0	1	0	0	1	1	0	2	0	0	1	
0	0	1	1	1	0	2	0	0	0	2	2	0	0	1	2	0	3	2	0	1	1	0	0	1	2	0	3	0	0	4	1	0	0	0	2	3	1	0	1	0	1	1	3	0	1	3	0	0	3	0	1	WEIGHT
0	1	2	3	4	5	6	7	8	9	10	11	12	13	14	15	16	17	18	19	20	21	22	23	24	25	26	27	28	29	30	31	32	33	34	35	36	37	38	39	40	41	42	43	44	45	46	47	48	49	50	51	NUMBER
1	1	8	18	10	27	28	98	48	72	108	81	116	42	105	135	45	56	55	255	162	84	120	376	592	189	342	336	486	168	70	903	168	448	42	216	528	1 197	399	216	819	1 323	1 128	720	120	168	189	520	990	690	984	1 040	0
	1	.	9	10	.	.	35	.	36	36	45	80	.	.	90	45	.	55	75	.	.	.	160	160	.	90	84	270	.	.	399	42	280	.	.	240	630	315	.	315	315	720	360	120	.	.	225	225	450	480	800	1
		1	2	1	6	7	21	14	15	24	17	24	14	35	27	9	21	10	75	63	35	50	111	186	84	119	112	144	77	35	273	63	133	21	105	164	357	119	111	294	504	328	216	36	91	105	190	395	198	356	296	2
			1	1	.	.	7	.	8	8	9	16	.	.	18	9	.	10	20	.	.	.	48	48	.	28	28	75	.	.	119	14	84	.	.	72	182	91	.	112	112	208	108	36	.	.	85	85	126	160	224	3
				1	.	.	.	.	.	.	.	8	.	.	9	9	.	10	.	.	.	.	.	.	.	.	.	27	.	.	.	.	28	.	.	.	63	63	.	.	.	72	36	36	.	.	.	.	45	48	152	4
					1	1	3	4	2	3	2	3	5	10	3	1	6	1	20	24	15	20	28	48	34	42	28	36	36	15	66	25	32	11	50	38	84	28	56	102	182	74	48	8	50	55	65	145	42	124	64	5
						1	.	.	.	3	2	.	.	5	3	.	6	1	.	9	5	10	.	24	24	.	28	.	16	15	33	.	.	5	30	38	42	.	36	.	66	37	48	.	30	45	.	75	42	.	32	6
							1	.	1	1	1	2	.	.	2	1	.	1	5	.	.	.	12	12	.	9	7	18	.	.	28	5	20	.	.	16	42	21	.	38	38	46	24	8	.	.	30	30	26	52	48	7
								1	.	.	.	.	2	2	.	.	1	.	4	9	7	7	5	8	12	15	4	6	17	5	10	11	5	6	24	5	12	4	27	33	60	10	6	1	28	27	20	45	5	40	8	8
									1	.	1	1	.	.	1	1	.	1	.	.	.	.	6	.	.	.	.	6	.	.	14	.	14	.	.	8	14	14	.	14	.	32	8	8	.	.	15	.	18	14	32	9
										1	1	.	.	.	2	.	.	1	.	.	.	.	.	6	.	.	7	.	.	.	14	.	.	.	.	16	21	.	.	.	14	23	24	.	.	.	.	15	26	.	24	10
											1	.	.	.	1	.	.	1	.	.	.	.	.	.	.	.	.	.	.	.	7	.	.	.	.	8	7	.	.	.	.	16	8	.	.	.	.	.	18	.	16	11
												1	.	.	1	1	.	1	.	.	.	.	.	.	.	.	.	6	.	.	.	.	7	.	.	.	14	14	.	.	.	16	8	8	.	.	.	.	9	14	32	12
													1	.	.	.	.	.	.	3	4	2	.	.	3	5	.	.	8	1	.	6	.	3	12	.	.	.	12	9	15	.	.	.	16	12	5	10	.	10	.	13
														1	.	.	1	.	.	3	2	4	.	4	8	.	4	.	8	5	5	.	.	3	13	5	6	.	18	.	20	5	6	.	17	21	.	25	5	.	4	14
															1	.	.	1	.	.	.	.	.	.	.	.	.	.	.	.	.	.	.	.	.	.	7	.	.	.	.	8	8	.	.	.	.	.	9	.	16	15

1	.	1	.	.	.	.	.	.	.	.	.	.	.	.	.	.	.	.	.	.	.	7	.	.	.	.	.	8	.	.	.	.	.	.	16	16
	1	.	.	.	.	.	.	.	4	.	4	.	.	5	.	.	.	.	5	5	.	.	6	.	.	.	6	.	5	15	.	.	5	.	.	17
		1	.	.	.	.	.	.	.	.	.	.	.	.	.	.	.	.	.	.	.	.	.	.	.	.	.	.	.	.	.	.	.	.	8	18
			1	.	.	.	2	2	.	3	1	3	.	.	4	2	3	.	.	2	6	3	.	12	12	6	3	1	.	.	9	9	3	16	6	19
				1	1	1	.	.	2	.	.	.	4	1	.	.	.	2	6	.	.	.	8	.	5	.	.	.	10	9	.	5	.	.	.	20
					1	.	.	.	.	.	.	.	2	.	.	.	.	1	3	.	.	.	3	.	.	.	.	.	6	3	.	.	.	.	.	21
						1	.	.	1	.	.	.	2	1	.	.	.	1	2	.	.	.	6	.	.	.	.	.	6	6	.	5	.	.	.	22
							1	.	.	.	.	1	.	.	2	.	2	.	.	1	2	2	.	4	.	4	1	1	.	.	5	.	2	4	4	23
								1	.	.	1	.	.	.	2	.	.	.	.	2	3	.	.	.	4	3	3	.	.	.	.	5	3	.	3	24
									1	.	.	.	.	1	.	.	.	.	2	.	.	.	3	.	.	.	.	.	3	6	.	.	.	.	.	25
										1	.	.	.	.	.	1	.	.	.	.	.	.	.	3	3	.	.	.	.	.	2	2	.	4	.	26
											1	.	.	.	.	.	.	.	.	2	.	.	.	.	.	.	3	.	.	.	.	.	3	.	.	27
												1	.	.	.	.	1	.	.	.	2	2	.	.	.	2	1	1	.	.	.	.	1	4	4	28
													1	.	.	.	.	1	1	.	.	.	2	.	.	.	.	.	4	2	.	.	.	.	.	29
														1	.	.	.	.	.	.	.	.	.	.	.	.	.	.	.	3	.	.	.	.	.	30
															1	.	.	.	.	1	1	.	.	.	.	2	1	.	.	.	.	.	2	.	2	31
																1	.	.	.	.	.	.	.	.	.	.	.	.	.	.	.	.	.	.	.	32
																	1	.	.	.	.	1	.	.	.	2	.	1	.	.	.	.	1	.	2	33
																		1	.	.	.	.	.	.	.	.	.	.	2	.	.	.	.	.	.	34
																			1	.	.	.	1	.	.	.	.	.	2	2	.	.	.	.	.	35
																				1	.	.	.	.	.	.	1	.	.	.	.	.	2	.	.	36
																					1	.	.	.	.	1	1	.	.	.	.	.	1	.	2	37
																						1	.	.	.	.	.	1	.	.	.	.	.	.	2	38
																							1	.	.	.	.	.	1	1	.	.	.	.	.	39
																								1	.	.	.	.	.	.	1	.	.	1	.	40
																									1	.	.	.	.	.	.	1	.	.	.	41
																										1	.	.	.	.	.	.	1	.	1	42
																											1	.	.	.	.	.	1	.	.	43
																												1	.	.	.	.	.	.	.	44
																													1	.	.	.	.	.	.	45
																														1	.	.	.	.	.	46
																															1	.	.	.	.	47
																																1	.	.	.	48
																																	1	.	.	49
																																		1	.	50
																																			1	51

Table 3. *(continued)*

ALGEBRA A10, REPRESENTATIONS OF CLASS 3

24	46	90	46	68	90	68	112	90	178	90	112	90	112	112	156	112	134	134	134	156	112	156	156	178	178	178	200	200	244	134	134	222	288	222	134	178	310	288	266	156	156	310	354	156	156	376	178	156	200	178	178	S.P.
165	110	11	2310	3960	495	6930	990	2772	110	4620	330	27720	9240	9240	2310	13860	3960	3960	13860	1980	46200	3960	3960	495	3960	990	3960	495	1320	13860	69300	990	495	990	69300	13860	110	55	990	9240	46200	110	495	83160	83160	110	9240	69300	13060	9240	27720	O.S.
25	29	31	39	45	47	49	49	51	51	55	57	57	59	59	61	61	63	63	63	65	65	65	65	67	67	67	67	69	69	69	69	69	69	71	71	71	71	73	73	73	73	75	75	75	75	77	77	77	77	77	77	LEVEL
165	440	286	3465	15730	12870	21780	22869	27027	10725	50820	32670	154440	151008	189728	105105	270270	235950	330330	385385	226512	566280	396396	470448	212355	424710	482625	406560	330330	225225	755040	1617000	494208	154440	495495	1698840	1101100	187187	143143	566280	1211210	2252250	330330	270270	4756752	4234032	314600	3363360	3963960	3201660	2845920	3775200	DIM'N
0	1	3	0	1	0	0	2	0	4	0	0	1	0	1	0	0	0	0	2	0	1	1	2	0	0	1	3	0	0	1	1	3	5	0	0	1	5	0	0	0	2	0	0	0	1	0	0	0	0	1	1	
0	1	0	0	0	2	0	1	0	0	0	0	0	0	0	0	1	0	1	0	0	0	2	0	0	0	2	1	0	0	0	0	1	0	0	1	0	0	0	0	0	0	0	0	1	0	0	0	1	1	0	0	
1	0	0	0	1	0	0	0	0	0	0	0	0	0	0	0	1	0	1	1	0	0	0	1	0	0	0	0	0	0	0	0	0	0	0	0	0	0	0	0	2	0	0	0	0	0	0	2	0	0	0	0	
0	0	0	1	0	0	0	0	0	0	0	0	1	0	1	0	0	0	0	0	0	0	0	0	0	0	0	0	0	0	0	0	0	0	0	1	0	0	0	0	0	1	0	0	1	0	0	0	0	1	0	0	
0	0	0	0	0	0	1	0	1	0	0	0	0	0	0	0	0	0	0	0	0	1	0	0	0	0	0	0	0	0	0	1	0	0	0	0	1	0	0	0	0	0	0	0	0	0	0	0	1	0	0	0	
0	0	0	0	0	0	0	0	0	0	1	0	0	1	0	1	0	0	0	0	0	0	0	0	0	0	0	0	0	0	1	0	0	0	0	0	0	0	0	0	0	0	0	0	0	1	0	0	0	0	0	1	
0	0	0	0	0	0	0	0	0	0	0	2	0	0	0	0	0	1	0	0	1	0	0	0	0	1	0	0	0	1	1	0	0	0	0	0	0	0	0	0	0	0	0	0	0	0	0	0	1	0	2	0	
0	0	0	0	0	0	0	0	0	0	1	0	0	0	0	0	0	1	0	0	0	1	0	0	2	0	0	0	2	0	0	0	0	0	1	1	0	0	0	1	1	1	0	1	0	1	0	0	0	0	0	0	
0	0	0	0	0	0	1	0	0	0	0	0	1	1	0	0	1	0	0	1	2	0	1	0	1	1	0	1	0	0	0	1	0	1	2	0	0	0	4	1	0	0	3	0	1	0	2	1	0	0	0	2	
0	0	0	1	1	1	0	1	2	1	0	0	0	1	2	3	0	1	2	0	0	0	0	2	0	2	2	0	2	4	0	1	2	0	1	0	3	2	0	3	0	0	2	5	1	1	4	1	0	3	1	0	WEIGHT
0	1	2	3	4	5	6	7	8	9	10	11	12	13	14	15	16	17	18	19	20	21	22	23	24	25	26	27	28	29	30	31	32	33	34	35	36	37	38	39	40	41	42	43	44	45	46	47	48	49	50	51	NUMBER
1	2	1	7	24	17	20	27	21	10	28	14	102	64	112	35	147	70	164	189	56	208	172	216	42	90	198	152	56	35	196	490	180	45	70	480	280	55	14	64	301	575	28	21	1152	828	20	728	728	672	448	640	0
	1	1	.	8	9	.	18	.	10	.	.	27	.	28	.	63	.	72	98	.	48	108	108	.	.	126	116	.	.	42	105	135	45	.	180	56	55	.	.	133	255	.	.	432	162	.	336	252	252	84	120	1
		1	.	.	.	.	9	.	10	.	.	.	.	.	.	.	.	.	35	.	.	36	36	.	.	45	80	.	.	.	.	90	45	.	.	.	55	.	.	.	75	.	.	.	.	.	.	.	.	.	.	2
			1	3	2	5	3	6	1	9	5	24	24	28	15	33	30	38	42	26	66	37	48	21	45	42	32	30	20	72	170	36	9	40	150	105	10	9	40	93	180	19	15	378	342	15	232	272	231	200	280	3
				1	1	.	2	.	1	.	.	6	.	7	.	14	.	16	21	.	14	23	24	.	.	26	24	.	.	14	35	27	9	.	55	21	10	.	.	42	75	.	.	138	63	.	105	91	84	35	50	4
					1	.	1	.	1	.	.	.	.	.	.	7	.	8	7	.	.	16	8	.	.	18	16	.	.	.	.	18	9	.	20	.	10	.	.	21	20	.	.	48	.	.	56	28	28	.	.	5
						1	.	1	.	3	2	4	8	4	5	5	13	5	6	12	20	5	6	11	21	5	4	16	10	28	50	4	1	23	42	30	1	6	24	25	50	13	10	102	138	11	60	102	60	92	116	6
							1	.	1	.	.	.	.	.	.	.	.	.	7	.	.	8	8	.	.	9	16	.	.	.	.	18	9	.	.	.	10	.	.	.	20	.	.	.	.	.	.	.	.	.	.	7
								1	.	.	.	.	4	4	5	.	5	5	.	6	.	.	6	5	15	5	.	10	10	.	25	4	.	15	.	30	1	4	20	.	.	10	10	51	51	10	30	.	60	30	60	8
									1	.	.	.	.	.	.	.	.	.	.	.	.	.	.	.	.	.	8	.	.	.	.	9	9	.	.	.	10	.	.	.	.	.	.	.	.	.	.	.	.	.	.	9
										1	1	.	2	.	1	.	6	.	.	5	5	.	.	6	9	.	.	9	4	12	10	.	.	13	9	5	.	4	14	5	10	9	6	18	54	8	10	38	9	46	42	10
											1	.	.	.	.	.	3	.	.	2	.	.	.	3	3	.	.	6	1	6	.	.	.	7	.	.	.	3	8	.	.	6	3	.	18	6	.	14	.	28	12	11
												1	.	1	.	2	.	2	3	.	4	3	3	.	.	3	3	.	.	5	10	3	1	.	15	6	1	.	.	11	20	.	.	36	24	.	26	33	21	15	20	12
													1	.	1	.	2	.	.	3	.	.	.	3	6	.	.	5	4	.	5	.	.	9	.	5	.	3	11	.	.	7	6	9	18	7	5	.	9	13	24	13
														1	.	.	2	.	.	.	.	.	3	.	.	3	.	.	.	.	5	3	.	.	.	6	1	.	.	.	.	.	.	18	9	.	13	.	21	5	10	14
															1	.	.	.	.	.	.	.	.	.	3	.	.	2	4	.	.	.	.	3	.	5	.	1	8	.	.	4	6	.	.	6	.	.	9	.	.	15
																1	.	1	1	.	.	2	1	.	.	2	2	.	.	.	.	2	1	.	5	.	1	.	.	6	5	.	.	12	.	.	14	9	7	.	.	16

1	.	.	1	.	.	.	2	2	.	.	3	1	.	.	.	.	5	.	.	.	2	6	.	.	5	3	.	6	5	.	.	.	7	6	17
	1	.	.	.	.	1	.	.	2	.	.	.	.	.	2	.	.	.	.	1	.	.	.	.	.	.	6	.	.	7	.	7	.	.	18
		1	.	.	1	1	.	.	1	2	.	.	.	.	2	1	.	.	.	1	.	.	.	5	.	.	.	.	.	.	.	.	.	.	19
			1	.	.	.	1	1	.	.	1	1	.	.	.	.	4	.	.	.	2	4	.	.	4	3	.	.	4	.	.	.	.	6	20
				1	.	.	.	.	.	.	.	.	2	2	.	.	.	3	1	.	.	.	2	4	.	.	6	9	.	4	12	3	7	7	21
					1	.	.	.	1	1	.	.	.	.	1	1	.	.	.	1	.	.	.	.	.	.	.	.	.	.	.	.	.	.	22
						1	.	.	1	.	.	.	.	.	2	.	.	.	.	1	.	.	.	.	.	.	.	.	.	.	.	.	.	.	23
							1	.	.	.	1	.	.	.	.	.	2	.	.	.	1	2	.	.	3	1	.	.	3	.	.	.	.	.	24
								1	.	.	1	1	.	.	.	.	2	.	.	.	1	4	.	.	3	3	.	.	4	.	.	.	.	.	25
									1	.	.	.	.	.	1	.	.	.	.	1	.	.	.	.	.	.	.	.	.	.	.	.	.	.	26
										1	.	.	.	.	1	1	.	.	.	1	.	.	.	.	.	.	.	.	.	.	.	.	.	.	27
											1	.	.	.	.	.	1	.	.	.	1	2	.	.	2	1	.	.	3	.	.	.	.	.	28
												1	.	.	.	.	.	.	.	.	.	2	.	.	1	3	.	.	3	.	.	.	.	.	29
													1	.	.	.	.	.	.	.	.	.	.	.	.	.	.	3	.	.	4	.	4	2	30
														1	.	.	.	.	1	.	.	.	.	.	.	.	3	3	.	2	.	3	2	4	31
															1	.	.	.	.	1	.	.	.	.	.	.	.	.	.	.	.	.	.	.	32
																1	.	.	.	1	.	.	.	.	.	.	.	.	.	.	.	.	.	.	33
																	1	.	.	.	1	1	.	.	2	1	.	.	2	.	.	.	.	.	34
																		1	.	.	.	.	1	1	.	.	2	.	.	2	3	1	.	.	35
																			1	.	.	.	.	.	.	.	.	.	.	.	.	3	.	.	36
																				1	.	.	.	.	.	.	.	.	.	.	.	.	.	.	37
																					1	.	.	.	1	.	.	.	1	.	.	.	.	.	38
																						1	.	.	1	1	.	.	2	.	.	.	.	.	39
																							1	.	.	.	.	.	.	2	.	.	.	.	40
																								1	.	.	.	.	.	.	.	.	.	.	41
																									1	.	.	.	1	.	.	.	.	.	42
																										1	.	.	1	.	.	.	.	.	43
																											1	.	.	1	.	1	.	.	44
																												1	.	.	.	.	1	1	45
																													1	.	.	.	.	.	46
																														1	.	.	.	.	47
																															1	.	.	.	48
																																1	.	.	49
																																	1	.	50
																																		1	51

Table 3. *(continued)*

ALGEBRA A10, REPRESENTATIONS OF CLASS 4

S.P.	28	50	72	94	50	160	72	72	94	94	94	116	116	138	138	160	94	182	160	116	270	182	226	248	116	116
O.S.	330	495	55	110	2 772	11	4 620	9 240	2 310	1 320	3 960	3 960	3 960	990	495	1 320	34 650	990	495	13 860	110	990	110	495	41 580	27 720
LEVEL	29	35	37	39	41	41	49	49	51	53	53	55	57	57	59	59	59	59	61	61	61	63	65	65	65	65
DIM'N	330	1 485	1 210	2 145	4 752	1 001	20 328	33 033	25 740	29 040	57 200	81 312	94 380	83 160	70 785	68 640	214 500	84 942	84 942	266 805	32 032	165 165	88 088	96 525	625 482	509 652
	0	1	0	2	0	4	0	1	0	0	0	2	0	1	0	0	1	3	0	1	5	0	0	0	0	1
	0	0	2	1	0	0	0	0	0	0	1	0	0	2	0	0	0	1	0	0	0	0	0	0	1	0
	0	1	0	0	0	0	0	0	0	0	1	1	0	0	0	0	0	0	0	0	0	0	0	0	0	0
	1	0	0	0	0	0	0	1	0	0	0	0	0	0	0	0	0	0	0	0	0	0	0	0	1	0
	0	0	0	0	1	0	0	0	0	0	0	0	0	0	0	0	1	0	0	1	0	0	0	0	0	0
	0	0	0	0	0	0	1	0	1	0	0	0	0	0	0	0	0	0	0	0	0	0	0	0	0	1
	0	0	0	0	0	0	0	0	0	1	0	0	1	0	0	1	0	0	0	0	0	0	0	0	0	0
	0	0	0	0	0	0	0	0	0	1	0	0	0	0	2	0	0	0	1	0	0	1	0	1	0	1
	0	0	0	0	0	0	1	0	0	0	0	0	1	0	0	0	1	0	2	0	0	1	3	0	1	0
WEIGHT	0	0	0	0	1	0	0	1	2	0	1	1	1	1	1	3	0	1	0	2	1	2	1	4	0	0
NUMBER	0	1	2	3	4	5	6	7	8	9	10	11	12	13	14	15	16	17	18	19	20	21	22	23	24	25
0	1	3	2	3	6	1	14	28	15	14	38	48	35	42	21	20	95	36	21	105	10	35	14	15	207	147
1		1	1	2	.	1	.	7	.	.	16	24	.	26	.	.	20	27	.	21	10	.	.	.	75	28
2			1	1	.	1	.	.	.	.	8	8	.	18	.	.	.	18	.	.	10	.	.	.	27	.
3				1	.	1	.	.	.	.	.	8	.	9	.	.	.	18	.	.	10	.	.	.	.	.
4					1	.	4	4	5	5	5	6	15	5	10	10	25	4	11	30	1	20	9	10	51	51
5						1	.	.	.	.	.	.	.	.	.	.	.	9	.	.	10	.	.	.	.	.
6							1	.	1	2	.	.	6	.	5	4	5	.	6	5	.	11	6	6	9	18
7								1	.	.	2	3	.	3	.	.	5	3	.	6	1	.	.	.	18	9
8									1	.	.	.	3	.	2	4	.	.	3	5	.	8	4	6	.	.
9										1	.	.	2	.	3	1	.	.	3	.	.	6	4	3	.	6
10											1	1	.	2	.	.	.	2	.	.	1	.	.	.	6	.
11												1	.	1	.	.	.	2	.	.	1	.	.	.	.	.
12													1	.	1	1	.	.	2	.	.	4	3	3	.	.
13														1	.	.	.	1	.	.	1	.	.	.	.	.
14															1	.	.	.	1	.	.	2	2	1	.	.
15																1	.	.	.	.	.	2	1	3	.	.

S.P.	270	138	138	138	138	358	160	138	160	490	182	160	138	204	160	182	204	226	182	160	226	226	160	160	160	204
O.S.	110	4 620	13 860	2 310	27 720	110	1 320	55 440	9 240	11	13 860	27 720	92 400	1 980	27 720	7 920	13 860	495	13 860	46 200	3 960	3 960	46 200	34 650	138 600	3 960
LEVEL	67	67	67	67	67	69	69	69	69	71	71	71	73	73	73	73	73	75	75	75	75	75	77	77	77	77
DIM'N	110 110	440 440	770 770	330 330	825 825	68 640	540 540	1 520 640	1 016 400	19 448	1 061 775	1 849 848	2 416 128	780 780	2 401 245	2 265 120	1 482 250	952 875	2 312 310	3 088 800	1 646 568	1 812 096	3 633 630	3 303 300	6 936 930	2 180 178
	0	0	0	1	2	0	0	1	2	0	1	1	0	0	1	1	3	0	1	2	2	3	0	0	0	1
	0	0	1	0	0	0	0	0	0	0	0	1	1	3	0	1	0	3	0	0	2	0	0	1	1	0
	0	2	0	0	0	0	2	0	0	0	0	1	0	0	0	1	1	0	0	0	0	1	1	0	0	0
	0	0	1	0	1	0	0	0	1	0	0	0	0	0	0	0	0	0	0	0	0	0	1	0	0	0
	0	0	0	0	0	0	0	0	0	0	0	0	1	0	0	0	0	0	0	1	0	0	0	0	1	0
	0	0	0	0	0	0	0	1	0	0	1	0	0	0	0	0	0	0	0	0	0	0	0	1	0	0
	0	0	0	2	0	0	0	0	0	0	0	0	0	0	1	0	0	0	1	0	0	0	0	1	0	0
	0	0	0	0	0	0	0	0	0	0	0	0	1	0	1	0	0	0	0	1	0	0	1	0	0	2
	2	1	0	0	1	1	0	1	0	0	0	1	0	1	0	0	1	0	2	0	1	0	0	0	1	1
WEIGHT	3	0	2	0	0	5	2	1	2	7	3	0	0	0	1	2	0	2	0	0	0	2	0	0	1	0
NUMBER	26	27	28	29	30	31	32	33	34	35	36	37	38	39	40	41	42	43	44	45	46	47	48	49	50	51
0	14	126	231	77	246	6	141	360	280	1 210	448	488	157	420	512	308	181	350	564	300	360	595	497	1 190	273	
1	.	56	84	14	102	.	64	64	112	.	35	238	160	100	70	272	189	117	56	208	208	216	245	154	385	42
2	.	28	28	.	27	.	36	.	28	.	.	126	48	72	.	144	98	81	.	48	144	108	105	42	105	.
3	.	.	.	.	27	.	.	.	28	.	.	63	.	36	.	72	98	45	.	48	108	108	.	.	.	.
4	10	30	60	30	60	5	35	140	70	1	90	104	164	35	190	120	70	40	170	190	66	80	195	195	430	144
5	.	.	.	.	.	.	.	.	.	.	.	.	.	.	.	.	35	.	.	.	36	36	.	.	.	.
6	7	5	9	13	10	4	5	48	10	1	30	16	52	5	87	16	10	5	81	60	9	10	58	80	132	79
7	.	13	21	5	24	.	15	24	28	.	15	54	52	22	30	62	42	25	26	66	45	48	80	58	135	21
8	6	.	9	.	.	4	5	24	10	1	30	.	.	.	33	16	.	5	42	.	.	10	.	.	66	37
9	5	.	.	7	.	3	.	12	.	1	6	.	14	.	42	.	.	.	35	15	.	.	14	35	28	45
10	.	7	7	.	6	.	8	.	7	.	.	28	14	15	.	32	21	17	.	14	31	24	35	14	35	.
11	.	.	.	.	6	.	.	.	7	.	.	14	.	8	.	16	21	9	.	14	23	24	.	.	.	.
12	4	.	.	.	.	3	.	6	.	1	6	.	.	.	14	.	.	.	21	.	.	.	.	.	14	23
13	.	.	.	.	.	.	.	.	.	.	.	7	.	8	.	8	7	9	.	.	16	8	.	.	.	.
14	3	.	.	.	.	2	.	.	.	1	.	.	.	.	7	.	.	.	7	.	.	.	.	.	.	16
15	3	.	.	.	.	3	.	.	.	1	6	.	.	.	.	.	.	.	.	.	.	.	.	.	.	.

1	.	.	1	.	.	.	.	3	3	.	2	3	2	4	.	2	8	4	.	5	8	16	3	13	8	6	3	12	20	6	6	23	23	40	11	16
	1	.	.	1	.	.	.	.	.	.	.	.	.	.	.	.	.	.	.	.	.	.	.	.	7	.	.	.	8	8	.	.	.	.	.	17
		1	.	.	1	2	1	.	.	2	.	.	.	.	2	.	.	.	1	.	.	.	.	.	.	.	.	7	.	.	.	.	.	.	8	18
			1	.	.	.	.	.	.	.	.	3	.	.	.	2	4	4	.	5	.	.	.	5	8	.	3	6	.	.	6	.	.	20	5	19
				1	.	.	.	.	.	.	.	.	.	.	.	.	.	.	.	.	.	.	.	.	.	.	.	.	.	.	.	.	.	.	.	20
					1	1	1	.	.	2	.	.	.	.	2	.	.	.	1	.	.	.	.	.	.	.	.	.	.	.	.	.	.	.	.	21
						1	.	.	.	1	.	.	.	.	1	.	.	.	1	.	.	.	.	.	.	.	.	.	.	.	.	.	.	.	.	22
							1	.	.	1	.	.	.	.	2	.	.	.	1	.	.	.	.	.	.	.	.	.	.	.	.	.	.	.	.	23
								1	.	.	1	1	.	1	.	1	.	1	.	.	4	4	2	.	4	3	2	.	4	4	3	10	5	10	.	24
									1	.	.	.	1	.	.	.	2	.	.	1	.	4	.	6	.	.	.	5	5	.	.	5	10	8	6	25
										1	.	.	.	.	1	.	.	.	1	.	.	.	.	.	.	.	.	.	.	.	.	.	.	.	.	26
											1	.	.	.	.	1	.	.	.	.	2	.	1	.	2	1	1	.	.	3	1	5	.	.	.	27
												1	.	.	.	1	.	1	.	.	.	.	.	.	4	.	2	.	.	.	3	.	.	5	.	28
													1	.	.	.	.	.	.	.	.	.	.	3	.	.	.	2	.	.	.	.	5	.	3	29
														1	.	.	.	1	.	.	2	.	1	.	2	3	1	.	4	3	3	.	.	.	.	30
															1	.	.	.	1	.	.	.	.	.	.	.	.	.	.	.	.	.	.	.	.	31
																1	.	.	.	.	.	.	.	.	2	.	1	.	.	.	1	.	.	.	.	32
																	1	.	.	1	.	.	.	2	.	.	.	3	.	.	.	.	.	4	3	33
																		1	.	.	.	.	.	.	2	.	1	.	.	.	3	.	.	.	.	34
																			1	.	.	.	.	.	.	.	.	.	.	.	.	.	.	.	.	35
																				1	.	.	.	.	.	.	.	.	.	.	.	.	.	.	.	36
																					1	.	1	.	1	1	1	.	.	2	1	.	.	.	.	37
																						1	.	.	.	.	.	.	1	.	.	2	2	2	.	38
																							1	.	.	.	1	.	.	1	.	.	.	.	.	39
																								1	.	.	.	1	.	.	.	.	.	.	2	40
																									1	.	1	.	.	.	1	.	.	.	.	41
																										1	.	.	.	1	1	.	.	.	.	42
																											1	.	.	.	.	.	.	.	.	43
																												1	.	.	.	.	.	.	1	44
																													1	.	.	.	.	.	.	45
																														1	.	.	.	.	.	46
																															1	.	.	.	.	47
																																1	.	.	.	48
																																	1	.	.	49
																																		1	.	50
																																			1	51

Table 3. *(continued)*

ALGEBRA A10, REPRESENTATIONS OF CLASS 5

30	52	52	74	96	74	118	96	96	162	74	250	118	162	162	184	96	250	118	96	118	360	118	118	140	184	S.P.
462	1 320	2 310	495	495	1 980	110	165	1 320	110	13 860	11	990	55	495	110	13 860	110	1 320	27 720	9 240	11	13 860	9 240	7 920	495	O.S.
31	39	41	43	45	47	47	49	49	49	51	51	53	55	55	57	57	59	59	59	59	61	61	63	63	65	LEVEL
462	3 168	4 620	5 445	7 722	13 068	7 865	9 075	17 160	8 008	47 190	3 003	37 752	15 730	30 030	33 033	135 135	25 025	94 380	205 920	177 870	8 008	261 360	297 297	394 240	165 165	DIM'N
0	1	0	0	2	0	1	0	0	3	1	5	0	0	0	0	0	0	0	1	2	0	1	1	1	0	
0	0	0	1	0	0	2	0	0	1	0	0	0	0	0	0	1	0	0	0	0	0	0	0	1	3	
0	0	0	1	1	0	0	0	0	0	0	0	0	0	0	0	0	0	2	0	0	0	0	0	1	0	
0	1	0	0	0	0	0	0	0	0	0	0	0	0	0	0	1	0	0	0	1	0	0	0	0	0	
1	0	0	0	0	0	0	0	0	0	1	0	0	0	0	0	0	0	0	0	0	0	0	0	0	0	
0	0	1	0	0	0	0	0	0	0	0	0	0	0	0	0	0	0	0	1	0	0	1	0	0	0	
0	0	0	0	0	1	0	0	1	0	0	0	0	0	0	0	0	0	0	0	0	0	0	1	0	0	
0	0	0	0	0	0	0	2	0	0	0	0	1	0	1	0	0	0	0	0	0	0	0	1	0	0	
0	0	0	0	0	1	0	0	0	0	0	0	1	3	0	2	0	1	0	1	0	0	0	0	0	0	
0	0	1	0	0	0	0	0	2	0	1	0	1	0	3	2	1	4	1	0	1	6	2	0	1	1	WEIGHT
0	1	2	3	4	5	6	7	8	9	10	11	12	13	14	15	16	17	18	19	20	21	22	23	24	25	NUMBER
1	4	5	5	6	9	5	5	10	4	30	1	16	5	10	9	60	5	35	80	70	1	90	86	120	40	0
	1	.	2	3	.	3	.	.	3	6	1	.	.	.	.	21	.	15	14	28	.	15	14	62	25	1
		1	.	.	3	.	2	4	.	5	.	8	3	6	6	9	4	5	24	10	1	30	33	16	5	2
			1	1	.	2	.	.	2	.	1	.	.	.	.	7	.	8	.	7	.	.	.	32	17	3
				1	.	1	.	.	2	.	1	.	.	.	.	.	.	.	.	7	.	.	.	16	9	4
					1	.	1	1	.	.	.	4	2	3	4	.	3	.	6	.	1	6	14	.	.	5
						1	.	.	1	.	1	.	.	.	.	.	.	.	.	.	.	.	.	8	9	6
							1	.	.	.	.	2	1	1	3	.	2	.	.	.	1	.	7	.	.	7
								1	.	.	.	2	1	3	3	.	3	.	.	.	1	6	.	.	.	8
									1	.	1	.	.	.	.	.	.	.	.	.	.	.	.	.	.	9
										1	.	.	.	.	.	3	.	2	4	4	.	5	5	8	3	10
											1	.	.	.	.	.	.	.	.	.	.	.	.	.	.	11
												1	1	1	2	.	2	.	.	.	1	.	.	.	.	12
													1	.	1	.	1	.	.	.	1	.	.	.	.	13
														1	1	.	2	.	.	.	1	.	.	.	.	14
															1	.	1	.	.	.	1	.	.	.	.	15

184	118	140	206	140	162	184	140	272	140	184	162	382	162	140	206	162	250	272	162	162	162	294	184	184	184	S.P.
3 960	69 300	27 720	990	27 720	3 960	9 240	34 650	990	27 720	3 960	13 860	110	9 240	69 300	7 920	6 930	990	3 960	83 160	27 720	138 600	990	13 860	27 720	2 310	O.S.
65	67	67	67	69	69	69	69	69	71	71	71	71	73	73	73	75	75	75	75	75	77	77	77	77	77	LEVEL
314 600	926 640	970 200	347 490	1 156 155	731 808	707 850	1 179 750	264 264	1 359 072	880 880	1 470 150	85 085	1 681 680	2 265 120	1 716 000	1 486 485	920 205	1 006 720	4 459 455	2 837 835	6 795 360	1 153 152	3 391 388	5 505 500	1 849 848	DIM'N
3	0	1	2	0	1	1	2	4	0	1	2	6	0	0	1	0	1	1	1	2	0	1	1	1	2	
0	1	0	2	1	0	0	0	1	0	0	0	0	0	1	0	1	0	0	1	0	1	0	0	1	0	
1	0	0	0	0	0	0	0	0	1	0	0	0	1	0	0	0	0	0	0	0	0	0	2	0	0	
0	0	0	0	0	0	0	0	0	1	0	0	0	1	0	0	0	0	0	1	0	0	0	0	1	0	
0	1	0	0	1	0	0	1	0	0	0	1	0	0	0	0	0	0	0	0	0	0	0	0	0	0	
0	0	0	0	0	0	0	0	0	0	0	0	0	0	1	0	0	0	0	0	1	1	0	0	0	0	
0	0	1	0	0	0	1	0	0	0	0	0	0	0	0	0	2	0	0	0	0	0	0	0	0	2	
0	0	0	0	0	2	0	0	0	0	1	0	0	0	1	1	0	0	1	0	1	0	0	0	0	0	
0	1	1	0	0	0	0	1	0	1	2	0	0	0	0	1	0	3	0	1	0	1	2	1	0	0	
1	0	1	1	2	1	3	0	1	0	0	2	1	2	0	2	0	1	4	0	0	1	3	0	2	0	WEIGHT
26	27	28	29	30	31	32	33	34	35	36	37	38	39	40	41	42	43	44	45	46	47	48	49	50	51	NUMBER
80	240	230	75	270	144	140	275	45	276	150	315	10	310	410	260	225	110	120	765	465	1 040	116	499	875	254	0
48	75	35	51	84	21	20	95	36	111	21	105	10	126	119	35	63	14	15	351	147	296	14	259	399	77	1
10	66	102	9	80	72	70	75	5	72	82	90	1	86	150	152	93	73	80	195	171	420	84	123	230	106	2
24	20	.	35	21	.	.	20	27	48	.	21	10	56	28	.	14	.	.	150	28	64	.	140	168	14	3
24	.	.	26	.	.	.	20	27	.	.	21	10	.	.	.	.	.	.	75	28	.	.	56	84	14	4
.	14	42	.	14	38	28	15	.	14	46	15	.	14	55	86	42	50	48	35	63	150	60	21	35	49	5
8	.	.	18	.	.	.	.	18	.	.	.	10	.	.	.	.	.	.	27	.	.	.	28	28	.	6
.	.	14	.	.	24	7	.	.	.	24	.	.	.	20	48	21	34	24	.	21	40	44	.	.	28	7
.	.	21	.	14	15	28	.	.	.	24	15	.	14	.	62	.	34	48	.	.	75	51	.	35	.	8
8	.	.	9	.	.	.	.	18	.	.	.	10	.	.	.	.	.	.	.	.	.	.	.	.	.	9
6	20	15	6	24	10	10	25	4	28	11	30	1	33	42	20	25	9	10	87	51	116	10	62	102	30	10
.	.	.	.	.	.	.	.	9	.	.	.	10	.	.	.	.	.	.	.	.	.	.	.	.	.	11
.	.	7	.	.	8	7	.	.	.	16	.	.	.	.	32	.	26	24	.	.	20	35	.	.	.	12
.	.	.	.	.	.	.	.	.	.	8	.	.	.	.	8	.	18	8	.	.	.	18	.	.	.	13
.	.	.	.	.	.	7	.	.	.	.	.	.	.	.	16	.	9	24	.	.	.	26	.	.	.	14
.	.	.	.	.	.	.	.	.	.	.	.	.	.	.	8	.	9	8	.	.	.	18	.	.	.	15

1	.	1	.	1	.	.	.	4	2	3	5	.	4	6	.	.	5	3	12	.	6	1	14	9	.	5	.	.	36	9	24	.	33	42	5	16
	1	.	.	.	1	.	.	.	.	.	.	.	.	.	.	.	.	.	.	.	.	.	.	.	.	.	.	8	.	.	.	9	.	.	.	17
		1	.	.	.	.	.	2	1	1	.	.	3	.	.	.	.	2	6	.	.	1	7	.	.	.	.	.	12	.	.	.	21	14	.	18
			1	.	.	1	2	.	.	.	4	6	.	4	5	4	5	.	5	6	5	.	5	15	11	11	6	6	15	18	40	7	10	15	13	19
				1	.	.	.	2	1	3	.	.	3	.	.	.	5	3	.	.	6	1	.	.	.	.	.	.	18	9	.	.	13	21	5	20
					1	.	.	.	.	.	.	.	.	.	.	.	.	.	.	.	.	.	.	.	.	.	.	.	.	.	.	.	.	.	.	21
						1	.	.	.	.	.	3	.	4	2	4	.	.	.	3	5	.	5	.	8	.	4	6	.	.	20	6	.	15	.	22
							1	.	.	.	.	2	.	.	3	1	.	.	.	3	.	.	.	5	6	6	4	3	.	6	10	5	.	.	7	23
								1	1	1	.	.	2	.	.	.	.	2	.	.	.	1	.	.	.	.	.	.	6	.	.	.	7	7	.	24
									1	.	.	.	1	.	.	.	.	1	.	.	.	1	.	.	.	.	.	.	.	.	.	.	.	.	.	25
										1	.	.	1	.	.	.	.	2	.	.	.	1	.	.	.	.	.	.	.	.	.	.	.	.	.	26
											1	.	.	1	.	.	1	.	2	.	1	.	2	3	.	2	.	.	6	3	8	.	5	6	2	27
												1	.	.	1	1	.	.	.	2	.	.	.	.	4	.	3	3	.	.	5	4	.	.	.	28
													1	.	.	.	.	1	.	.	.	1	.	.	.	.	.	.	.	.	.	.	.	.	.	29
														1	.	.	.	.	.	.	1	.	2	.	.	.	.	.	.	.	4	.	.	6	.	30
															1	.	.	.	.	1	.	.	.	.	2	.	2	1	.	.	.	3	.	.	.	31
																1	.	.	.	.	.	.	.	.	2	.	1	3	.	.	.	3	.	.	.	32
																	1	.	.	.	1	.	.	.	.	.	.	.	3	3	.	.	2	3	2	33
																		1	.	.	.	1	.	.	.	.	.	.	.	.	.	.	.	.	.	34
																			1	.	.	.	1	.	.	.	.	.	2	.	.	.	3	2	.	35
																				1	.	.	.	.	1	.	2	1	.	.	.	2	.	.	.	36
																					1	.	.	.	.	.	.	.	.	.	.	.	.	3	.	37
																						1	.	.	.	.	.	.	.	.	.	.	.	.	.	38
																							1	.	.	.	.	.	.	.	.	.	.	2	.	39
																								1	.	1	.	.	.	1	2	.	.	.	1	40
																									1	.	1	1	.	.	.	2	.	.	.	41
																										1	.	.	.	.	.	.	.	.	1	42
																											1	.	.	.	.	1	.	.	.	43
																												1	.	.	.	1	.	.	.	44
																													1	.	.	.	1	1	.	45
																														1	.	.	.	.	1	46
																															1	.	.	.	.	47
																																1	.	.	.	48
																																	1	.	.	49
																																		1	.	50
																																			1	51

Table 3. *(continued)*

ALGEBRA A11, REPRESENTATIONS OF CLASS 0

0	24	48	72	72	96	72	96	96	144	144	120	96	168	168	216	120	120	120	144	144	144	168	168	240	144	240	144	168	144	168	192	192	192	168	168	168	216	168	264	216	264	192	288	192	240	288	216	168	216	312	312	S.P.
1	132	2970	660	660	132	18480	15840	15840	1980	1980	11880	34650	1320	1320	132	16632	83160	83160	924	13860	13860	27720	27720	3960	166320	3960	110880	23760	110880	23760	2970	47520	47520	27720	83160	83160	5940	27720	5940	5940	5940	83160	660	83160	11880	660	33264	554400	33264	1320	1320	O.S.
1	23	41	43	43	45	55	59	59	61	61	63	65	65	65	67	71	71	71	73	73	73	75	75	77	77	77	79	79	79	79	81	81	81	83	83	83	83	83	83	83	83	85	85	85	85	85	87	87	87	87	87	LEVEL
1	143	4212	5005	5005	5940	44044	116480	116480	75075	75075	306735	196625	197120	197120	126412	382239	891891	891891	226512	728728	728728	1289925	1289925	600600	3992625	600600	2818816	3243240	2818816	3243240	2627625	5725720	5725720	3468465	5010005	5010005	4633200	3468465	2650725	4633200	2650725	7135128	2139280	7135128	8164233	2139280	7378371	19876428	7378371	3764475	3764475	DIM'N
0	1	0	0	2	2	0	0	1	0	3	1	0	1	3	3	0	0	1	0	0	0	0	2	0	1	4	0	0	1	1	0	1	2	0	0	0	0	1	1	2	4	0	0	2	2	4	0	1	1	2	4	
0	0	1	1	0	0	0	0	1	0	0	1	0	1	0	0	0	0	0	0	0	2	0	1	0	0	0	0	2	0	0	2	0	1	0	0	1	2	0	0	1	0	0	2	0	1	0	0	0	2	1	0	
0	0	0	0	0	0	1	1	0	1	0	0	0	0	0	0	0	0	1	0	0	0	0	0	0	1	0	0	0	0	1	0	1	0	0	0	1	0	0	1	0	0	0	0	1	0	0	0	0	0	0	0	
0	0	0	0	0	0	0	0	0	0	0	0	1	0	0	0	0	1	0	0	1	0	1	0	1	0	0	0	0	1	0	0	0	0	0	0	0	0	0	0	0	0	0	0	0	0	0	0	1	0	0	0	
0	0	0	0	0	0	0	0	0	0	0	0	0	0	0	0	1	0	0	0	0	0	0	0	0	0	0	1	0	0	0	0	0	0	0	1	0	0	1	0	0	0	1	0	0	0	0	1	0	0	0	0	
0	0	0	0	0	0	0	0	0	0	0	0	0	0	0	0	0	0	0	2	0	0	0	0	0	0	0	0	0	0	0	0	0	0	1	0	0	0	1	0	0	0	0	0	0	0	0	0	0	0	0	0	
0	0	0	0	0	0	0	0	0	0	0	0	0	0	0	0	1	0	0	0	0	0	0	0	0	0	0	0	0	1	0	0	0	0	1	0	1	0	0	0	0	0	0	0	1	0	0	0	0	1	0	0	
0	0	0	0	0	0	0	0	0	0	0	0	1	0	0	0	0	0	1	0	0	1	0	1	0	0	1	1	0	0	0	0	0	0	0	0	0	0	0	0	0	0	0	0	0	0	0	0	1	0	0	0	
0	0	0	0	0	0	1	0	1	0	1	0	0	0	0	0	0	1	0	0	0	0	0	0	0	1	0	0	1	0	0	0	0	1	0	1	0	0	0	0	0	1	1	0	0	0	0	0	0	0	0	0	
0	0	1	0	1	0	0	1	0	0	0	1	0	0	1	0	0	0	0	0	2	0	1	0	0	0	0	0	0	0	2	2	1	0	0	1	0	1	0	0	2	0	0	0	0	1	2	2	0	0	0	1	
0	1	0	2	0	2	0	1	0	3	0	1	0	3	1	3	0	1	0	0	0	0	2	0	4	1	0	1	1	0	0	0	2	1	1	0	0	2	0	4	0	1	2	4	0	2	0	1	1	0	4	2	WEIGHT
0	1	2	3	4	5	6	7	8	9	10	11	12	13	14	15	16	17	18	19	20	21	22	23	24	25	26	27	28	29	30	31	32	33	34	35	36	37	38	39	40	41	42	43	44	45	46	47	48	49	50	51	NUMBER
1	11	54	55	55	66	154	320	320	165	165	759	275	440	440	286	297	891	891	132	616	616	945	945	330	3255	330	1408	2376	1408	2376	1815	3850	3850	1155	1925	1925	2970	1155	1485	2970	1485	2376	1210	2376	5049	1210	2079	7326	2079	2145	2145	0
	1	9	10	10	11	35	80	80	45	45	189	75	110	110	66	90	270	270	42	196	196	315	315	120	1015	120	480	756	480	756	570	1240	1240	420	700	700	945	420	495	945	495	900	385	900	1584	385	819	2625	819	660	660	1
		1	1	1	1	7	16	16	9	9	36	20	20	20	11	28	75	75	14	56	56	91	91	36	280	36	160	208	160	208	154	340	340	154	245	245	252	154	135	252	135	320	100	320	414	100	301	895	301	165	165	2
			1	.	1	.	8	.	9	.	18	.	20	10	11	.	27	.	.	28	.	63	.	36	98	.	48	72	.	108	81	232	116	42	105	.	171	.	135	135	45	180	100	.	279	55	189	255	.	165	110	3
				1	1	.	.	8	.	9	18	.	10	20	11	.	.	27	.	.	28	.	63	.	98	36	.	108	48	72	81	116	232	.	.	105	135	42	45	171	135	.	55	180	279	100	.	255	189	110	165	4
					1	.	.	.	.	.	9	.	10	10	11	.	.	.	.	.	.	.	.	.	35	.	.	36	.	36	45	80	80	.	.	.	90	.	45	90	45	.	55	.	189	55	.	75	.	110	110	5
						1	2	2	1	1	4	5	2	2	1	9	18	18	5	13	13	21	21	8	63	8	52	45	52	45	32	72	72	58	80	80	51	58	27	51	27	105	19	105	81	19	99	285	99	30	30	6
							1	.	1	.	2	.	2	1	1	.	6	.	.	7	.	14	.	8	21	.	14	15	.	24	17	48	24	14	35	.	34	.	27	27	9	55	19	.	54	10	63	75	.	30	20	7
								1	.	1	2	.	1	2	1	.	.	6	.	.	7	.	14	.	21	8	.	24	14	15	17	24	48	.	.	35	27	14	9	34	27	.	10	55	54	19	.	75	63	20	30	8
									1	.	.	.	2	.	1	.	.	.	.	.	.	7	.	8	.	.	.	.	.	.	.	24	.	.	.	.	17	.	27	.	.	20	19	.	27	.	21	.	.	30	10	9
										1	.	.	.	2	1	.	.	.	.	.	.	.	7	.	.	8	.	.	.	.	.	.	24	.	.	.	.	.	.	17	27	.	.	20	27	19	.	.	21	10	30	10
											1	.	1	1	1	.	.	.	.	.	.	.	.	.	7	.	.	8	.	8	9	16	16	.	.	.	18	.	9	18	9	.	10	.	36	10	.	20	.	20	20	11
												1	.	.	.	3	3	3	2	2	2	3	3	1	9	1	16	6	16	6	4	9	9	23	23	23	6	23	3	6	3	30	2	30	9	2	27	80	27	3	3	12
													1	.	1	.	.	.	.	.	.	.	.	.	.	.	.	.	.	.	.	8	.	.	.	.	9	.	9	.	.	.	10	.	18	.	.	.	.	20	10	13
														1	1	.	.	.	.	.	.	.	.	.	.	.	.	.	.	.	.	.	8	.	.	.	.	.	.	9	9	.	.	.	18	10	.	.	.	10	20	14
															1	.	.	.	.	.	.	.	.	.	.	.	.	.	.	.	.	.	.	.	.	.	.	.	.	.	.	.	.	.	9	.	.	.	.	10	10	15

1	.	.	1	.	.	.	.	.	.	.	4	.	4	.	.	.	.	10	5	5	.	10	.	.	.	6	.	6	.	.	5	16	5	.	.	16
	1	.	.	1	.	2	.	1	3	.	4	2	.	3	2	6	3	5	10	.	4	.	3	3	1	15	2	.	6	1	17	20	.	3	2	17
		1	.	.	1	.	2	.	3	1	.	3	4	2	2	3	6	.	.	10	3	5	1	4	3	.	1	15	6	2	.	20	17	2	3	18
			1	.	.	.	.	.	.	.	.	.	.	.	.	.	.	5	.	.	.	5	.	.	.	.	.	.	.	.	.	.	.	.	.	19
				1	.	1	.	1	.	.	.	.	.	3	2	3	.	.	5	.	2	.	3	3	.	5	2	.	3	1	12	.	.	3	1	20
					1	.	1	.	.	1	.	3	.	.	2	.	3	.	.	5	3	.	.	2	3	.	1	5	3	2	.	.	12	1	3	21
						1	.	1	.	.	.	.	.	.	.	3	.	.	.	.	2	.	3	.	.	5	2	.	3	.	6	.	.	3	1	22
							1	.	.	1	.	.	.	.	.	.	3	.	.	.	.	.	.	2	3	.	.	5	3	2	.	.	6	1	3	23
								1	.	.	.	.	.	.	.	.	.	.	.	.	.	.	3	.	.	.	2	.	.	.	.	.	.	3	.	24
									1	.	.	1	.	1	1	2	2	.	.	.	2	.	1	2	1	.	1	.	4	1	.	5	.	2	2	25
										1	.	.	.	.	.	.	.	.	.	.	.	.	.	.	3	.	.	.	.	2	.	.	.	.	3	26
											1	.	.	.	.	.	.	2	2	.	.	.	.	.	.	3	.	.	.	.	3	4	.	.	.	27
												1	.	.	1	.	1	.	.	.	2	.	.	1	1	.	1	.	2	1	.	.	.	1	2	28
													1	.	.	.	.	.	.	2	.	2	.	.	.	.	.	3	.	.	.	4	3	.	.	29
														1	1	1	.	.	.	.	1	.	1	2	.	.	1	.	2	1	.	.	.	2	1	30
															1	.	.	.	.	.	1	.	.	1	.	.	1	.	1	1	.	.	.	1	1	31
																1	.	.	.	.	1	.	1	.	.	.	1	.	2	.	.	.	.	2	1	32
																	1	.	.	.	.	.	.	1	1	.	.	.	2	1	.	.	.	1	2	33
																		1	.	.	.	.	.	.	.	.	.	.	.	.	.	.	.	.	.	34
																			1	.	.	.	.	.	.	1	.	.	.	.	2	.	.	.	.	35
																				1	.	.	.	.	.	.	.	1	.	.	.	.	2	.	.	36
																					1	.	.	.	.	.	1	.	1	.	.	.	.	1	1	37
																						1	.	.	.	.	.	.	.	.	.	.	.	.	.	38
																							1	.	.	.	1	.	.	.	.	.	.	2	.	39
																								1	.	.	.	.	1	1	.	.	.	1	1	40
																									1	.	.	.	.	1	.	.	.	.	2	41
																										1	.	.	.	.	1	.	.	.	.	42
																											1	.	.	.	.	.	.	1	.	43
																												1	.	.	.	.	1	.	.	44
																													1	.	.	.	.	1	1	45
																														1	.	.	.	.	1	46
																															1	.	.	.	.	47
																																1	.	.	.	48
																																	1	.	.	49
																																		1	.	50
																																			1	51

Table 3. *(continued)*

ALGEBRA A11, REPRESENTATIONS OF CLASS 1

NUMBER	0	1	2	3	4	5	6	7	8	9	10	11	12	13	14	15	16	17	18	19	20	21	22	23	24	25
S.P.	11	35	59	59	83	83	107	131	155	83	107	155	107	107	131	131	155	131	155	179	227	131	203	179	155	131
O.S.	12	660	132	7920	1980	5940	1320	660	132	27720	27720	3960	55440	27720	7920	47520	15840	5544	5940	5940	1980	83160	660	11880	16632	138600
LEVEL	12	32	34	48	50	52	54	54	56	60	64	66	66	68	68	70	70	72	72	72	72	74	74	74	76	76
DIM'N	12	780	924	13728	16380	36036	42900	23100	27456	94380	252252	163800	420420	283140	340704	1115400	600600	339768	900900	720720	277200	1321320	580800	1585584	1093092	1981980
	1	0	2	0	0	1	1	3	3	0	0	0	1	0	0	1	2	0	0	1	4	0	0	2	0	1
	0	1	0	0	0	1	1	0	0	0	0	0	0	0	2	0	1	0	2	0	0	0	2	1	0	0
	0	0	0	1	1	0	0	0	0	0	0	0	1	0	0	1	0	0	0	1	0	0	0	0	0	0
	0	0	0	0	0	0	0	0	0	1	1	1	0	0	0	0	0	0	0	0	0	0	0	0	0	1
	0	0	0	0	0	0	0	0	0	0	0	0	0	1	0	0	0	0	0	0	0	1	0	0	1	0
	0	0	0	0	0	0	0	0	0	0	0	0	0	0	0	0	0	1	0	0	0	0	0	0	0	0
	0	0	0	0	0	0	0	0	0	0	0	0	0	0	0	0	0	1	0	0	0	0	0	0	0	0
	0	0	0	0	0	0	0	0	0	0	0	0	0	1	0	0	0	0	0	0	0	0	0	0	0	1
	0	0	0	0	0	0	0	0	0	1	0	0	1	0	1	0	1	0	0	0	1	1	0	0	0	0
	0	0	0	1	0	1	0	1	0	0	1	0	0	0	0	1	0	0	1	0	0	0	0	1	2	0
WEIGHT	0	1	1	0	2	0	2	0	2	0	1	3	0	0	0	1	0	0	1	3	0	1	3	1	0	0
0	1	10	11	44	45	99	110	55	66	110	231	120	385	165	276	880	440	132	660	495	165	550	385	1089	385	825
1		1	1	8	9	18	20	10	11	27	63	36	98	48	72	232	116	42	171	135	45	180	100	279	133	255
2			1	.	.	9	10	10	11	.	.	.	35	.	36	80	80	.	90	45	45	.	55	189	.	75
3				1	1	2	2	1	1	6	14	8	21	14	15	48	24	14	34	27	9	55	19	54	42	75
4					1	.	2	.	1	.	7	8	.	.	.	24	.	.	17	27	.	20	19	27	21	.
5						1	1	1	1	.	.	.	7	.	8	16	16	.	18	9	9	.	10	36	.	20
6							1	.	1	.	.	.	.	.	.	8	.	.	9	9	.	.	10	18	.	.
7								1	1	.	.	.	.	.	.	.	8	.	.	.	9	.	.	18	.	.
8									1	.	.	.	.	.	.	.	.	.	.	.	.	.	.	9	.	.
9										1	2	1	3	4	2	6	3	5	4	3	1	15	2	6	11	20
10											1	1	.	.	.	3	.	.	2	3	.	5	2	3	6	.
11												1	.	.	.	.	.	.	.	3	.	.	2	.	.	.
12													1	.	1	2	2	.	2	1	1	.	1	4	.	5
13														1	.	.	.	2	.	.	.	3	.	.	2	4
14															1	.	1	.	2	.	1	.	1	2	.	.

NUMBER	26	27	28	29	30	31	32	33	34	35	36	37	38	39	40	41	42	43	44	45	46	47	48	49	50	51
S.P.	227	251	179	299	155	251	155	179	155	155	179	179	179	179	203	203	179	203	251	203	227	179	203	275	203	347
O.S.	1320	1320	33264	132	55440	5544	83160	3960	83160	332640	83160	55440	5544	55440	27720	55440	166320	110880	27720	23760	27720	166320	23760	15840	166320	3960
LEVEL	76	76	78	78	80	80	80	82	82	82	82	84	84	84	84	86	86	86	86	88	88	88	88	88	88	88
DIM'N	1021020	729300	1945944	468468	2774772	917280	3435432	1698840	3963960	9034740	4864860	5153148	2378376	7399392	4968600	7399392	15479100	13117104	5045040	7361640	7807800	12024012	12612600	6126120	21861840	1884960
	2	4	0	4	0	0	0	0	1	1	2	0	1	1	1	0	0	1	3	0	0	0	0	1	2	5
	1	0	0	0	0	0	1	0	0	0	0	0	0	0	2	0	1	0	1	0	0	1	1	0	0	0
	0	0	0	0	0	0	1	0	0	0	1	0	0	0	0	0	1	0	0	0	0	0	1	0	1	0
	0	0	0	0	0	0	0	0	0	1	0	0	0	1	0	0	0	1	0	0	0	1	0	1	0	0
	0	0	1	0	0	1	0	0	1	0	0	0	0	0	0	0	0	0	0	0	0	0	0	0	0	0
	0	0	0	0	1	0	0	0	0	0	0	1	2	0	0	1	0	0	0	0	1	0	0	0	0	0
	0	0	0	0	0	0	0	2	1	0	0	0	0	0	0	0	0	0	0	1	0	1	0	0	0	0
	0	0	0	0	1	0	1	0	0	0	1	0	0	0	1	0	0	0	1	1	0	0	0	0	0	1
	0	0	0	0	0	0	0	0	0	1	0	1	0	0	0	1	1	0	0	0	0	0	0	0	1	0
	0	1	1	0	0	0	0	0	0	0	0	1	0	2	0	0	0	1	0	1	2	0	2	0	0	0
WEIGHT	3	1	2	3	1	4	0	1	0	1	0	0	0	0	0	2	1	2	0	0	1	0	0	4	1	0
0	660	440	594	286	693	210	1166	330	990	2970	1485	990	462	2156	1330	1232	4410	3465	1155	990	1100	2233	3300	1320	5775	330
1	165	110	216	66	252	84	376	126	342	972	486	387	168	728	448	504	1449	1197	399	414	470	819	1092	480	1897	120
2	110	110	.	66	.	.	160	.	90	270	270	.	42	196	280	.	630	315	315	.	.	315	480	120	1015	120
3	30	20	69	11	91	28	111	49	119	288	144	148	63	217	133	196	420	357	119	175	190	294	314	144	546	36
4	30	10	48	11	28	28	.	14	.	102	.	64	.	112	.	112	147	245	.	70	120	.	164	144	189	.
5	20	20	.	11	.	.	48	.	28	75	75	.	14	56	84	.	182	91	91	.	.	112	136	36	280	36
6	20	10	.	11	.	.	.	.	.	27	.	.	.	28	.	.	63	63	.	.	.	.	72	36	98	.
7	10	20	.	11	.	.	.	.	.	.	27	.	.	.	28	.	.	.	63	.	.	.	.	.	98	36
8	10	10	.	11	.	.	.	.	.	.	.	.	.	.	.	.	.	.	.	.	.	.	.	.	35	.
9	3	2	18	1	33	7	28	20	42	72	36	54	25	52	32	72	99	84	28	75	70	102	71	32	126	8
10	3	1	12	1	9	7	.	5	.	24	.	24	.	28	.	38	33	56	.	30	45	.	38	32	42	.
11	3	.	6	1	.	7	.	.	.	.	.	.	.	.	.	14	.	28	.	.	15	.	.	32	.	.
12	2	2	.	1	.	.	12	.	9	18	18	.	5	13	20	.	42	21	21	.	.	38	30	8	63	8
13	.	.	3	.	12	1	5	9	15	12	6	18	11	8	5	24	15	12	4	33	22	33	10	4	18	1
14	1	2	.	1	.	.	6	.	.	.	6	.	.	.	14	.	21	.	14	.	.	14	15	.	21	8

1	.	.	1	1	.	.	1	2	.	.	2	1	.	1	.	.	.	.	.	6	.	.	.	7	.	.	14	14	.	.	.	.	16	8	21	.	15
	1	.	.	.	1	.	.	2	.	.	1	2	.	1	.	.	.	.	.	.	6	.	.	.	7	.	.	.	14	.	.	.	.	.	21	8	16
		1	.	.	.	.	.	.	.	.	.	.	.	.	4	.	.	5	5	.	.	5	6	.	.	6	.	.	.	15	5	9	.	.	.	.	17
			1	.	.	.	1	1	.	.	1	1	.	1	.	.	.	.	.	.	.	.	.	.	.	.	7	.	.	.	.	.	8	.	7	.	18
				1	.	.	1	.	.	.	2	.	.	1	.	.	.	.	.	.	.	.	.	.	.	.	.	7	.	.	.	.	.	8	.	.	19
					1	.	.	.	.	.	.	2	.	1	.	.	.	.	.	.	.	.	.	.	.	.	.	.	7	.	.	.	.	.	.	8	20
						1	.	.	1	.	.	.	2	.	3	1	.	2	.	4	.	8	.	4	.	12	5	8	.	13	14	.	5	4	6	.	21
							1	.	.	.	1	.	.	1	.	.	.	.	.	.	.	.	.	.	.	.	.	.	.	.	.	.	.	.	.	.	22
								1	.	.	1	1	.	1	.	.	.	.	.	.	.	.	.	.	.	.	.	.	.	.	.	.	.	.	7	.	23
									1	.	.	.	1	.	.	1	.	.	.	.	.	4	.	4	.	4	.	4	.	5	10	.	5	4	.	.	24
										1	.	.	.	.	.	.	2	.	3	3	3	.	2	2	3	.	6	3	3	.	.	12	4	1	9	1	25
											1	.	.	1	.	.	.	.	.	.	.	.	.	.	.	.	.	.	.	.	.	.	.	.	.	.	26
												1	.	1	.	.	.	.	.	.	.	.	.	.	.	.	.	.	.	.	.	.	.	.	.	.	27
													1	.	.	1	.	.	.	.	.	.	.	.	.	4	.	4	.	.	5	.	.	4	.	.	28
														1	.	.	.	.	.	.	.	.	.	.	.	.	.	.	.	.	.	.	.	.	.	.	29
															1	.	.	1	.	.	.	2	.	.	.	3	.	.	.	6	3	.	.	.	.	.	30
																1	.	.	.	.	.	.	.	.	.	.	.	.	.	.	.	.	.	4	.	.	31
																	1	.	.	.	1	.	.	.	2	.	3	.	2	.	.	4	2	.	3	1	32
																		1	.	.	.	.	.	.	.	.	.	.	.	3	.	.	.	.	.	.	33
																			1	.	.	.	1	.	.	.	.	.	.	.	.	3	.	.	.	.	34
																				1	.	.	.	1	.	.	2	2	.	.	.	.	2	1	3	.	35
																					1	.	.	.	1	.	.	.	2	.	.	.	.	.	3	1	36
																						1	.	.	.	1	.	.	.	2	2	.	.	.	.	.	37
																							1	.	.	.	.	.	.	.	.	.	.	.	.	.	38
																								1	.	.	.	1	.	.	.	.	2	1	.	.	39
																									1	.	.	.	1	.	.	.	.	.	.	1	40
																										1	.	.	.	.	1	.	.	.	.	.	41
																											1	.	.	.	.	.	1	.	1	.	42
																												1	.	.	.	.	.	1	.	.	43
																													1	.	.	.	.	.	.	1	44
																														1	.	.	.	.	.	.	45
																															1	.	.	.	.	.	46
																																1	.	.	.	.	47
																																	1	.	.	.	48
																																		1	.	.	49
																																			1	.	50
																																				1	51

Table 3. *(continued)*

ALGEBRA A11, REPRESENTATIONS OF CLASS 2

NUMBER	0	1	2	3	4	5	6	7	8	9	10	11	12	13	14	15	16	17	18	19	20	21	22	23	24	25
S.P.	20	44	44	68	116	68	92	92	116	116	92	140	140	164	212	116	236	116	164	140	116	140	140	140	164	188
O.S.	66	12	1 980	1 320	132	13 860	3 960	23 760	2 970	5 940	27 720	660	5 940	1 320	660	33 264	132	13 860	5 544	792	110 880	55 440	55 440	110 880	13 860	15 840
LEVEL	21	23	39	43	45	53	55	59	61	61	63	63	63	65	65	67	67	69	69	71	71	75	75	75	77	77
DIM'N	66	78	2 574	6 720	4 290	30 030	36 036	132 132	106 470	157 950	141 570	127 050	187 110	223 080	85 800	384 384	102 102	283 140	252 252	169 884	960 960	1 387 386	1 639 638	2 576 574	1 171 170	1 677 312
	0	2	0	1	3	0	0	1	0	1	0	0	2	2	4	0	4	0	0	0	1	0	0	1	0	1
	1	0	0	1	0	0	0	0	2	0	0	2	1	1	0	0	0	0	0	0	0	0	1	0	0	0
	0	0	1	0	0	0	0	1	0	1	0	0	0	0	0	0	0	0	0	0	0	0	1	0	0	0
	0	0	0	0	0	1	1	0	0	0	0	0	0	0	0	0	0	0	0	0	1	0	0	1	0	1
	0	0	0	0	0	0	0	0	0	0	1	0	0	0	0	1	0	0	1	0	0	0	0	0	0	0
	0	0	0	0	0	0	0	0	0	0	0	0	0	0	0	0	0	1	0	0	0	1	0	0	1	0
	0	0	0	0	0	0	0	0	0	0	0	0	0	0	0	0	0	0	0	2	0	0	0	0	0	0
	0	0	0	0	0	0	0	0	0	0	0	0	0	0	0	0	0	1	0	0	0	0	0	0	0	0
	0	0	0	0	0	0	0	0	0	0	1	0	0	0	0	0	0	0	0	0	1	1	1	0	0	0
	0	0	0	0	0	1	0	1	1	0	0	0	1	0	1	1	0	0	0	0	0	0	0	1	2	0
WEIGHT	0	0	1	1	1	0	2	0	0	2	0	2	0	2	0	1	2	0	3	0	0	1	0	1	0	3
0	1	1	9	20	11	35	36	124	90	135	75	100	144	165	55	160	66	90	84	42	380	315	546	861	225	480
1		1	.	10	11	.	.	44	45	45	.	55	99	110	55	.	66	.	.	.	110	.	231	231	.	120
2			1	2	1	7	8	24	17	27	20	19	27	30	10	48	11	28	28	14	102	112	147	245	85	144
3				1	1	.	.	8	9	9	.	10	18	20	10	.	11	.	.	.	27	.	63	63	.	36
4					1	.	.	.	.	.	.	.	9	10	10	.	11	.	.	.	.	.	.	.	.	.
5						1	1	3	2	3	5	2	3	3	1	12	1	9	7	5	24	38	33	56	30	32
6							1	.	.	3	.	2	.	3	.	6	1	.	7	.	.	14	.	28	15	32
7								1	1	1	.	1	2	2	1	.	1	.	.	.	6	.	14	14	.	8
8									1	.	.	1	1	1	1	.	1	.	.	.	.	.	7	.	.	.
9										1	.	1	.	2	.	.	1	.	.	.	.	.	.	7	.	8
10											1	.	.	.	.	2	.	3	1	2	4	12	5	8	9	4
11												1	.	1	.	.	1	.	.	.	.	.	.	.	.	.
12													1	1	1	.	1	.	.	.	.	.	.	.	.	.
13														1	.	.	1	.	.	.	.	.	.	.	.	.
14															1	.	1	.	.	.	.	.	.	.	.	.
15																1	.	.	1	.	.	4	.	4	5	4

NUMBER	26	27	28	29	30	31	32	33	34	35	36	37	38	39	40	41	42	43	44	45	46	47	48	49	50	51
S.P.	164	164	188	164	140	188	260	212	188	236	188	164	212	236	332	212	212	164	164	260	260	236	236	236	188	188
O.S.	55 440	15 840	27 720	47 520	138 600	15 840	5 544	5 940	47 520	15 840	23 760	33 264	11 880	5 940	1 980	2 970	23 760	207 900	415 800	1 320	11 880	1 980	1 980	15 840	27 720	83 160
LEVEL	77	79	79	79	79	79	81	81	81	81	83	83	83	83	83	85	85	85	85	85	85	87	87	87	87	87
DIM'N	2 312 310	1 812 096	2 102 100	4 368 000	2 972 970	2 342 340	1 009 008	2 831 400	6 150 144	2 365 440	3 680 820	3 624 192	6 209 280	3 981 978	875 160	2 576 574	5 351 346	8 588 580	13 942 500	4 011 150	6 257 394	2 576 574	3 539 250	5 855 850	6 012 006	11 561 550
	2	0	0	0	1	1	0	0	2	3	0	1	1	2	5	0	0	0	1	1	3	0	0	0	0	1
	0	0	0	1	0	2	0	1	0	1	0	0	2	0	0	0	0	1	0	2	1	0	0	0	0	0
	1	0	0	1	0	0	0	1	1	0	0	0	0	1	0	0	0	0	0	0	0	0	0	0	2	0
	0	0	0	0	0	0	0	0	0	0	0	0	0	0	0	0	0	1	0	0	0	0	0	0	0	0
	0	0	0	0	1	0	0	0	0	0	0	0	0	0	0	0	0	0	1	0	0	0	0	0	0	1
	0	0	1	0	0	0	1	0	0	0	0	1	0	0	0	0	0	0	0	0	0	0	0	0	0	0
	0	1	0	0	0	0	0	0	0	0	1	1	0	0	0	0	1	0	0	0	0	0	0	1	0	0
	0	1	0	0	1	0	0	0	0	0	0	0	0	0	0	2	0	1	0	0	0	1	2	0	1	0
	1	0	0	0	0	1	0	0	0	1	1	0	0	0	1	0	1	0	1	0	0	2	0	0	0	0
	0	0	1	1	0	0	0	0	1	0	1	0	1	0	0	1	0	0	0	0	1	0	0	2	0	2
WEIGHT	0	1	2	1	0	0	4	3	1	0	0	0	1	3	0	0	2	0	1	3	1	0	2	1	0	0
0	700	288	350	1 280	645	636	126	735	1 680	560	450	552	1 560	990	165	252	567	1 450	2 350	935	1 419	210	300	525	896	1 715
1	385	.	.	560	165	396	.	330	880	440	.	132	990	495	165	.	.	550	550	605	1 089	.	.	.	385	385
2	189	112	140	352	208	172	56	207	456	152	190	196	414	270	45	112	252	480	820	245	369	98	140	245	301	623
3	98	.	.	152	48	108	.	90	232	116	.	42	261	135	45	.	.	180	180	155	279	.	.	.	133	133
4	35	.	.	.	.	36	.	.	80	80	.	.	90	45	45	.	.	.	.	55	189	.	.	.	.	.
5	42	44	50	76	66	37	21	43	96	32	80	72	84	54	9	51	108	150	265	47	72	47	65	110	93	204
6	.	14	35	38	.	.	21	43	48	.	35	.	42	54	.	21	63	.	95	47	36	21	35	70	.	105
7	21	.	.	32	14	23	.	18	48	24	.	14	52	27	9	.	.	55	55	29	54	.	.	.	42	42
8	7	.	.	16	.	16	.	9	16	16	.	.	36	9	9	.	.	20	.	20	36	.	.	.	21	.
9	.	.	.	16	.	.	.	18	24	.	.	.	26	27	.	.	.	.	20	29	27	.	.	.	.	21
10	6	18	15	10	20	5	6	5	12	4	33	28	10	6	1	24	45	42	75	5	8	23	31	46	25	55
11	.	.	.	8	.	.	.	9	8	.	.	.	18	9	.	.	.	.	.	20	18	.	.	.	.	.
12	7	.	.	.	.	8	.	.	16	16	.	.	18	9	9	.	.	.	.	10	36	.	.	.	.	.
13	.	.	.	.	.	.	.	.	8	.	.	.	9	9	.	.	.	.	.	10	18	.	.	.	.	.
14	.	.	.	.	.	.	.	.	.	8	.	.	.	.	9	.	.	.	.	.	18	.	.	.	.	.
15	.	5	10	5	.	.	6	5	6	.	15	.	5	6	.	10	24	.	25	5	4	11	15	30	.	30

1	.	.	.	.	.	.	.	.	.	.	.	.	.	.	.	.	.	.	.	.	.	.	.	.	.	.	.	.	.	9	.	.	.	.	.	16
	1	.	1	.	3	.	.	2	.	.	8	3	.	5	.	1	.	.	.	13	12	.	.	.	12	18	9	15	.	.	11	16	17	5	10	17
		1	.	.	.	.	.	.	4	.	.	5	.	.	.	6	5	.	.	.	.	.	6	.	.	9	.	.	5	.	.	5	10	.	.	18
			1	.	.	.	.	.	.	.	4	.	.	.	.	.	.	.	.	5	6	.	.	.	6	6	.	.	.	.	5	10	5	.	.	19
				1	.	2	2	.	1	3	.	.	4	4	3	.	2	6	3	.	5	6	3	1	.	.	15	15	3	6	.	.	.	11	11	20
					1	.	.	1	.	.	2	2	.	.	.	1	.	.	.	6	.	.	.	.	5	9	.	5	.	.	6	7	11	.	5	21
						1	.	.	.	1	.	.	2	.	2	.	1	2	2	.	.	4	1	1	.	.	5	.	2	4	.	.	.	6	.	22
							1	.	1	.	.	.	2	.	.	.	2	3	.	.	.	3	3	.	.	.	.	5	3	3	.	.	.	.	6	23
								1	.	.	.	1	.	.	.	1	.	.	.	3	.	.	.	.	2	3	.	.	.	.	3	2	8	.	5	24
									1	.	.	.	.	.	.	.	2	.	.	.	.	.	3	.	.	.	.	.	3	.	.	.	.	.	.	25
										1	.	.	.	.	1	.	.	2	2	.	.	2	1	1	.	.	.	.	1	4	.	.	.	.	.	26
											1	.	.	.	.	.	.	.	.	2	.	.	.	.	3	3	.	.	.	.	3	4	3	.	.	27
												1	.	.	.	1	.	.	.	.	.	.	.	.	.	3	.	.	.	.	.	2	4	.	.	28
													1	.	.	.	1	1	.	.	.	2	1	.	.	.	.	.	2	2	.	.	.	.	.	29
														1	.	.	.	.	.	.	2	.	.	.	.	.	3	3	.	.	.	.	.	2	2	30
															1	.	.	.	1	.	.	2	.	1	.	.	.	.	1	2	.	.	.	.	.	31
																1	.	.	.	.	.	.	.	.	.	.	.	.	.	.	.	.	.	.	.	32
																	1	.	.	.	.	.	1	.	.	.	.	.	2	.	.	.	.	.	.	33
																		1	.	.	.	1	1	.	.	.	.	.	1	2	.	.	.	.	.	34
																			1	.	.	.	.	1	.	.	.	.	.	2	.	.	.	.	.	35
																				1	.	.	.	.	1	1	.	.	.	.	2	1	2	.	.	36
																					1	.	.	.	.	.	.	.	.	.	.	.	.	.	.	37
																						1	.	.	.	.	.	.	1	1	.	.	.	.	.	38
																							1	.	.	.	.	.	1	.	.	.	.	.	.	39
																								1	.	.	.	.	.	.	.	.	.	.	.	40
																									1	.	.	.	.	.	1	1	.	.	.	41
																										1	.	.	.	.	.	1	1	.	.	42
																											1	.	.	.	.	.	.	1	.	43
																												1	.	.	.	.	.	.	1	44
																													1	.	.	.	.	.	.	45
																														1	.	.	.	.	.	46
																															1	.	.	.	.	47
																																1	.	.	.	48
																																	1	.	.	49
																																		1	.	50
																																			1	51

Table 3. *(continued)*

ALGEBRA A11, REPRESENTATIONS OF CLASS 3

27	51	99	51	75	99	123	75	195	99	99	99	123	123	123	123	171	147	147	171	171	147	123	195	219	171	171	243	315	195	147	339	195	267	147	147	195	243	219	171	171	171	243	171	267	291	195	219	171	195	195	315	S.P.
220	132	12	3 960	5 940	660	1 320	16 632	132	5 544	18 480	55 440	15 840	3 960	27 720	23 760	5 544	5 940	23 760	5 940	5 940	23 760	138 600	1 320	5 940	1 980	7 920	1 320	660	15 840	166 320	132	5 940	3 960	166 320	83 160	27 720	220	5 940	18 480	5 544	110 880	5 940	166 320	1 320	5 940	15 840	23 760	332 640	110 880	110 880	660	O.S.
28	32	34	44	50	52	54	56	56	58	64	64	66	68	68	68	70	70	70	72	72	74	74	74	74	76	76	76	76	78	78	78	80	80	80	80	80	82	82	82	82	82	84	84	86	86	86	86	86	86	86	88	LEVEL
220	572	364	5 720	24 948	20 020	35 100	46 332	16 016	56 056	151 008	308 308	370 656	188 760	420 420	520 520	280 280	623 700	731 808	737 100	876 096	1 019 304	1 501 500	880 880	741 312	660 660	892 320	884 884	272 272	1 621 620	4 089 800	324 324	1 717 716	800 800	4 216 212	3 057 912	2 690 688	572 572	2 548 260	2 914 912	1 849 848	5 505 500	3 011 580	11 351 340	2 862 860	3 243 240	7 807 800	7 413 120	15 057 900	12 024 012	14 798 784	1 821 820	DIM'N
0	1	3	0	1	0	2	0	4	0	0	1	1	0	0	0	0	0	2	1	2	0	1	1	3	0	0	3	5	0	1	5	0	0	0	1	1	0	0	0	1	2	0	0	0	0	0	0	1	1	2	0	
0	1	0	0	0	2	1	0	0	0	0	0	0	0	0	1	0	1	0	2	0	0	0	2	1	0	0	1	0	0	0	0	0	0	1	0	0	0	0	0	0	0	0	1	0	0	0	1	0	1	0	0	
1	0	0	0	1	0	0	0	0	0	0	0	0	0	0	1	0	1	1	0	1	0	0	0	0	0	0	0	0	0	0	0	0	0	0	0	0	0	0	2	0	0	0	0	0	0	2	0	0	1	0	0	
0	0	0	1	0	0	0	0	0	0	0	1	1	0	0	0	0	0	0	0	0	0	0	0	0	0	0	0	0	0	0	0	0	0	1	0	0	0	0	0	0	1	0	1	0	0	0	1	0	0	1	0	
0	0	0	0	0	0	0	1	0	1	0	0	0	0	0	0	0	0	0	0	0	0	1	0	0	0	0	0	0	0	1	0	0	0	0	0	1	0	0	0	0	0	0	0	0	0	0	0	0	0	0	0	
0	0	0	0	0	0	0	0	0	0	1	0	0	0	1	0	1	0	0	0	0	0	0	0	0	0	0	0	0	0	0	0	0	0	0	1	0	0	0	0	0	0	0	0	0	0	0	0	1	0	0	0	
0	0	0	0	0	0	0	0	0	0	0	0	0	1	0	0	0	0	0	0	0	1	0	0	0	0	1	0	0	1	0	0	0	1	0	0	0	0	0	0	2	0	0	0	0	0	0	0	0	0	0	0	
0	0	0	0	0	0	0	0	0	0	0	0	0	1	0	0	0	0	0	0	0	0	0	0	0	2	0	0	0	0	0	0	1	0	0	1	0	0	1	0	0	0	1	0	0	1	0	0	0	0	0	0	
0	0	0	0	0	0	0	0	0	0	1	0	0	0	0	0	0	0	0	0	0	1	1	0	0	0	0	0	0	0	0	0	1	0	1	0	0	3	1	1	0	1	0	0	2	0	0	0	1	1	0	1	
0	0	0	0	0	0	0	1	0	0	0	1	0	0	1	1	0	0	1	1	0	0	0	0	1	0	2	0	1	1	1	0	1	0	0	0	0	0	0	0	0	0	2	1	1	1	1	0	0	0	1	3	
0	0	0	1	1	1	1	0	1	2	0	0	2	0	1	0	3	2	0	0	2	1	0	2	0	1	0	2	0	2	1	2	0	4	0	0	3	0	2	0	0	0	1	1	1	3	1	3	1	0	1	0	WEIGHT
0	1	2	3	4	5	6	7	8	9	10	11	12	13	14	15	16	17	18	19	20	21	22	23	24	25	26	27	28	29	30	31	32	33	34	35	36	37	38	39	40	41	42	43	44	45	46	47	48	49	50	51	NUMBER
1	2	1	8	27	19	30	27	11	28	48	133	144	42	105	188	56	207	240	216	270	162	330	245	189	84	120	220	55	189	752	66	168	70	729	432	420	42	216	448	210	866	216	1 701	168	189	1 056	972	1 638	1 596	2 058	84	0
	1	1	.	9	10	20	.	11	.	.	35	36	.	.	80	.	90	124	135	135	.	75	155	144	.	.	165	55	.	160	66	.	.	270	90	84	.	.	196	42	380	.	630	.	.	480	360	315	861	861	.	1
		1	.	.	.	10	.	11	.	.	.	.	.	.	.	.	.	44	45	45	.	.	55	99	.	.	110	55	.	.	66	.	.	.	.	.	.	.	.	.	110	.	.	.	.	.	.	.	231	231	.	2
			1	3	2	3	6	1	7	14	28	32	14	35	38	21	43	48	42	54	63	95	47	36	35	50	40	10	84	234	11	77	35	207	147	140	21	105	126	77	246	111	504	91	105	305	300	616	448	602	49	3
				1	1	2	.	1	.	.	7	8	.	.	16	.	18	24	26	27	.	20	29	27	.	.	30	10	.	48	11	.	.	75	28	28	.	.	56	14	102	.	182	.	.	136	108	112	238	245	.	4
					1	1	.	1	.	.	.	.	.	.	8	.	9	8	18	9	.	.	20	18	.	.	20	10	.	.	11	.	.	27	.	.	.	.	28	.	27	.	63	.	.	72	36	.	126	63	.	5
						1	.	1	.	.	.	.	.	.	.	.	.	8	9	9	.	.	10	18	.	.	20	10	.	.	11	.	.	.	.	.	.	.	.	.	27	.	.	.	.	.	.	.	63	63	.	6
							1	.	1	4	4	4	5	10	5	6	5	6	5	6	24	25	5	4	15	20	4	1	34	60	1	36	15	51	51	35	11	50	30	30	60	56	120	50	55	70	69	220	104	140	29	7
								1	.	.	.	.	.	.	.	.	.	.	.	.	.	.	.	9	.	.	10	10	.	.	11	.	.	.	.	.	.	.	.	.	.	.	.	.	.	.	.	.	.	.	.	8
									1	.	.	4	.	5	.	6	5	.	.	6	9	.	5	.	5	10	4	.	24	30	1	16	15	.	.	35	5	30	.	.	.	36	60	30	45	35	69	80	.	70	19	9
										1	.	.	2	2	.	1	.	.	.	.	9	5	.	.	7	7	.	.	12	10	.	17	5	9	18	5	6	24	5	13	10	27	18	28	27	10	9	72	16	20	17	10
											1	1	.	.	2	.	2	3	3	3	.	5	3	3	.	.	3	1	.	12	1	.	.	18	9	7	.	.	13	5	24	.	42	.	.	30	24	38	54	56	.	11
												1	.	.	.	.	2	.	.	3	.	.	3	.	.	.	3	.	.	6	1	.	.	.	.	7	.	.	.	.	.	.	21	.	.	15	24	14	.	28	.	12
													1	.	.	.	.	.	.	.	3	.	.	.	4	2	.	.	3	.	.	8	1	.	6	.	3	12	.	7	.	12	.	16	12	.	.	18	.	.	10	13
														1	.	1	.	.	.	.	3	.	.	.	2	4	.	.	8	5	.	8	5	.	.	5	3	13	.	.	.	18	9	17	21	5	9	24	.	10	12	14
															1	.	1	1	2	1	.	.	2	2	.	.	2	1	.	.	1	.	.	6	.	.	.	.	7	.	6	.	14	.	.	16	8	.	28	14	.	15
																1	.	.	.	.	.	.	.	.	.	.	.	.	4	.	.	.	5	.	.	5	.	5	.	.	.	6	.	5	15	.	9	.	.	.	4	16

1	.	.	1	.	.	2	.	.	.	2	.	.	.	1	.	.	.	.	.	.	.	.	.	.	.	7	.	.	8	8	.	.	7	.	17
	1	1	1	.	.	1	2	.	.	2	1	.	.	1	.	.	.	.	.	.	.	.	.	6	.	.	.	.	.	.	.	14	14	.	18
		1	.	.	.	1	1	.	.	1	1	.	.	1	.	.	.	.	.	.	.	.	.	.	.	.	.	.	.	.	.	7	.	.	19
			1	.	.	1	.	.	.	2	.	.	.	1	.	.	.	.	.	.	.	.	.	.	.	.	.	.	.	.	.	.	7	.	20
				1	.	.	.	1	1	.	.	2	.	.	4	1	.	.	.	2	6	.	.	.	8	.	10	9	.	.	6	.	.	7	21
					1	.	.	.	.	.	.	.	2	.	.	.	3	3	1	.	.	2	2	4	.	6	.	.	4	3	12	8	8	.	22
						1	.	.	.	1	.	.	.	1	.	.	.	.	.	.	.	.	.	.	.	.	.	.	.	.	.	.	.	.	23
							1	.	.	1	1	.	.	1	.	.	.	.	.	.	.	.	.	.	.	.	.	.	.	.	.	.	.	.	24
								1	.	.	.	.	.	.	2	.	.	.	.	1	3	.	.	.	3	.	6	3	.	.	.	.	.	4	25
									1	.	.	1	.	.	2	1	.	.	.	1	2	.	.	.	6	.	6	6	.	.	.	.	.	6	26
										1	.	.	.	1	.	.	.	.	.	.	.	.	.	.	.	.	.	.	.	.	.	.	.	.	27
											1	.	.	1	.	.	.	.	.	.	.	.	.	.	.	.	.	.	.	.	.	.	.	.	28
												1	.	.	.	1	.	.	.	.	2	.	.	.	3	.	3	6	.	.	.	.	.	3	29
													1	.	.	.	.	.	1	.	.	.	.	.	.	3	.	.	2	3	4	.	4	.	30
														1	.	.	.	.	.	.	.	.	.	.	.	.	.	.	.	.	.	.	.	.	31
															1	.	.	.	.	1	1	.	.	.	2	.	4	2	.	.	.	.	.	3	32
																1	.	.	.	.	.	.	.	.	.	.	.	3	.	.	.	.	.	.	33
																	1	.	.	.	.	1	.	1	.	2	.	.	2	1	.	4	2	.	34
																		1	.	.	.	.	1	.	.	.	.	.	.	.	3	.	.	.	35
																			1	.	.	.	.	.	.	.	.	.	.	3	.	.	.	.	36
																				1	.	.	.	.	.	.	2	.	.	.	.	.	.	1	37
																					1	.	.	.	1	.	2	2	.	.	.	.	.	2	38
																						1	.	.	.	.	.	.	2	.	.	2	.	.	39
																							1	.	.	.	.	.	.	.	.	.	.	.	40
																								1	.	.	.	.	.	.	.	2	2	.	41
																									1	.	1	1	.	.	.	.	.	2	42
																										1	.	.	1	1	.	.	1	.	43
																											1	.	.	.	.	.	.	1	44
																												1	.	.	.	.	.	.	45
																													1	.	.	.	.	.	46
																														1	.	.	.	.	47
																															1	.	.	.	48
																																1	.	.	49
																																	1	.	50
																																		1	51

Table 3. *(continued)*

ALGEBRA A11, REPRESENTATIONS OF CLASS 4

NUMBER	0	1	2	3	4	5	6	7	8	9	10	11	12	13	14	15	16	17	18	19	20	21	22	23	24	25
S.P.	32	56	80	104	176	56	80	80	104	104	128	104	152	128	200	128	104	296	176	128	152	176	128	200	200	152
O.S.	495	660	66	132	12	5544	15840	13860	5544	5940	5940	7920	1320	495	1320	15840	83160	132	3960	27720	5940	2970	83160	660	5940	7920
LEVEL	33	39	41	43	45	47	55	57	59	59	61	63	63	65	65	67	67	67	69	69	71	73	73	75	75	75
DIM'N	495	2145	1716	3003	1365	9009	59136	51480	63063	99099	138996	113256	139425	70785	139776	329472	495495	51051	225225	600600	495495	468468	1366365	429429	868725	936936
WEIGHT	0 0 0 1 0 0 0 0 0 0 0	1 0 1 0 0 0 0 0 0 0 0	0 2 0 0 0 0 0 0 0 0 0	2 1 0 0 0 0 0 0 0 0 0	4 0 0 0 0 0 0 0 0 0 0	0 0 0 0 1 0 0 0 0 0 1	1 0 0 1 0 0 0 0 0 0 1	0 0 0 0 0 1 0 0 0 1 0	0 0 0 0 0 1 0 0 0 0 2	0 1 1 0 0 0 0 0 0 0 1	2 0 1 0 0 0 0 0 0 0 1	0 0 0 0 0 0 1 0 1 0 0	1 2 0 0 0 0 0 0 0 0 1	0 0 0 0 0 0 0 2 0 0 0	3 1 0 0 0 0 0 0 0 0 1	0 0 0 0 0 0 1 0 0 1 1	1 0 0 0 1 0 0 0 0 1 0	5 0 0 0 0 0 0 0 0 0 1	0 0 0 0 0 0 1 0 0 0 3	1 0 0 0 1 0 0 0 0 0 2	0 0 0 0 0 0 0 1 1 0 1	0 0 0 0 0 0 0 1 0 2 0	0 1 0 1 0 0 0 0 0 1 0	0 0 0 0 0 0 0 0 2 1 0	0 0 0 0 0 0 0 1 0 1 2	0 0 2 0 0 0 0 0 0 1 0
0	1	3	2	3	1	7	32	20	21	43	54	28	47	14	40	64	129	11	35	140	70	56	273	42	90	164
1		1	1	2	1	.	8	.	.	18	27	.	29	.	30	.	27	11	.	28	.	.	98	.	.	72
2			1	1	1	.	.	.	.	9	9	.	20	.	20	.	.	11	.	.	.	.	35	.	.	36
3				1	1	.	.	.	.	.	9	.	10	.	20	.	.	11	.	.	.	.	.	.	.	.
4					1	.	.	.	.	.	.	.	.	.	10	.	.	11	.	.	.	.	.	.	.	.
5						1	4	5	6	5	6	9	5	5	4	24	30	1	15	35	30	26	60	21	45	35
6							1	.	.	2	3	.	3	.	3	.	6	1	.	7	.	.	21	.	.	15
7								1	1	.	.	3	.	2	.	8	5	.	5	5	13	12	9	11	21	5
8									1	.	.	.	.	.	.	4	.	.	5	5	5	6	.	5	15	.
9										1	1	.	2	.	2	.	.	1	.	.	.	.	7	.	.	8
10											1	.	1	.	2	.	.	1	.	.	.	.	.	.	.	.
11												1	.	1	.	2	.	.	1	.	6	5	.	6	9	.
12													1	.	1	.	.	1	.	.	.	.	.	.	.	.
13														1	.	.	.	.	.	.	3	2	.	3	3	.
14															1	.	.	1	.	.	.	.	.	.	.	.

NUMBER	26	27	28	29	30	31	32	33	34	35	36	37	38	39	40	41	42	43	44	45	46	47	48	49	50	51
S.P.	152	128	152	224	272	176	176	248	152	152	176	320	296	224	200	200	224	344	392	152	248	248	248	416	176	176
O.S.	23760	110880	55440	660	1980	1980	15840	1320	27720	166320	47520	66	1320	2970	33264	11880	23760	132	660	277200	660	5940	5940	132	166320	138600
LEVEL	75	75	75	77	77	77	77	79	79	79	79	81	81	81	81	81	81	83	83	83	83	83	83	85	85	85
DIM'N	1646568	1647360	1778700	660660	450450	1126125	2141568	975975	2081079	4601025	3843840	273273	1098240	1576575	3075072	4612608	3020160	624624	510510	7027020	1887600	3285360	3619980	583440	11293425	8848125
WEIGHT	0 1 0 1 0 0 0 0 0 0 2	1 0 0 0 0 1 0 0 1 0 0	2 0 0 1 0 0 0 0 0 1 0	0 0 0 0 0 0 0 0 2 0 2	0 0 0 0 0 0 0 1 0 0 4	0 0 2 0 0 0 0 0 0 0 2	2 0 0 1 0 0 0 0 0 0 2	0 0 0 0 0 0 0 0 1 2 1	1 0 0 0 0 0 1 1 0 0 0	1 0 0 0 0 1 0 0 0 1 1	1 1 1 0 0 0 0 0 0 1 0	0 0 0 0 0 0 0 0 0 4 0	0 0 0 0 0 0 0 0 1 1 3	0 3 0 0 0 0 0 0 0 1 0	1 0 0 0 0 1 0 0 0 0 3	1 1 1 0 0 0 0 0 0 0 2	3 0 1 0 0 0 0 0 0 1 0	0 0 0 0 0 0 0 0 0 3 2	0 0 0 0 0 0 0 0 1 0 5	0 1 0 0 1 0 0 0 1 0 0	0 3 0 0 0 0 0 0 0 0 2	2 2 0 0 0 0 0 0 0 1 0	3 0 1 0 0 0 0 0 0 0 2	0 0 0 0 0 0 0 0 0 2 4	1 0 0 0 0 0 1 0 1 0 1	2 0 0 0 1 0 0 0 0 1 0
0	300	256	322	56	35	181	360	70	238	595	576	14	64	199	336	648	392	28	21	790	226	378	450	20	990	905
1	108	48	133	.	.	81	144	.	42	105	304	.	.	126	56	342	240	.	.	255	145	261	270	.	162	330
2	36	.	35	.	.	45	36	.	.	.	160	.	.	90	.	180	124	.	.	75	100	180	135	.	.	75
3	.	.	35	.	.	.	36	.	.	.	80	.	.	45	.	90	124	.	.	.	55	135	135	.	.	75
4	.	.	.	.	.	.	.	.	.	.	.	.	.	.	.	.	44	.	.	.	.	45	45	.	.	.
5	69	80	70	30	20	40	80	40	86	205	120	9	40	40	126	136	80	19	15	240	45	75	90	15	405	275
6	24	14	28	.	.	17	32	.	14	35	62	.	.	25	21	70	48	.	.	75	28	51	54	.	63	95
7	9	24	10	16	10	5	10	23	33	60	16	6	24	5	36	16	10	13	10	66	5	9	10	11	162	75
8	9	.	.	10	10	5	10	15	.	30	.	4	20	.	36	16	.	10	10	.	5	.	10	10	60	.
9	8	.	7	.	.	9	8	.	.	.	32	.	.	17	.	36	24	.	.	20	19	35	27	.	.	20
10	.	.	7	.	.	.	8	.	.	.	16	.	.	9	.	18	24	.	.	.	10	26	27	.	.	20
11	.	6	.	9	4	.	.	13	14	12	.	4	14	.	6	.	.	9	6	14	.	.	.	8	63	15
12	.	.	.	.	.	.	.	.	.	.	8	.	.	9	.	9	8	.	.	.	10	18	9	.	.	.
13	.	.	.	6	1	.	.	7	7	.	.	3	8	.	.	.	.	6	3	.	.	.	.	6	21	.
14	.	.	.	.	.	.	.	.	.	.	.	.	.	.	.	.	8	.	.	.	.	9	9	.	.	.

1	.	.	1	.	2	3	.	3	6	.	.	.	.	5	4	.	.	9	.	6	.	3	11	.	6	.	.	7	6	.	.	.	.	7	21	.	15
	1	.	.	1	.	.	3	.	.	2	3	4	4	.	.	2	4	.	5	10	8	.	.	3	6	8	6	.	.	20	3	6	6	.	24	25	16
		1	.	.	.	.	.	.	.	.	.	.	.	.	.	.	.	.	.	.	.	.	.	.	.	.	.	.	.	.	.	.	.	.	.	.	17
			1	.	.	.	.	.	3	.	.	.	.	2	4	.	.	3	.	.	.	1	8	.	6	.	.	4	6	.	.	.	.	6	.	.	18
				1	.	.	.	.	.	.	3	.	.	.	.	2	4	.	.	5	.	.	.	.	6	8	.	.	.	.	3	.	6	.	9	.	19
					1	1	.	2	2	.	.	.	.	3	1	.	.	5	.	.	.	2	6	.	.	.	.	5	3	.	.	.	.	5	7	.	20
						1	.	1	1	.	.	.	.	1	1	.	.	4	.	.	.	2	4	.	.	.	.	4	3	.	.	.	.	4	.	.	21
							1	.	.	1	1	.	1	.	.	1	1	.	.	.	4	.	.	2	.	4	3	.	.	5	2	4	3	.	.	5	22
								1	.	.	.	.	.	1	.	.	.	2	.	.	.	1	2	.	.	.	.	3	1	.	.	.	.	3	.	.	23
									1	.	.	.	.	1	1	.	.	2	.	.	.	1	4	.	.	.	.	3	3	.	.	.	.	4	.	.	24
										1	.	.	.	.	.	1	.	.	.	.	2	.	.	1	.	2	1	.	.	.	1	3	1	.	.	.	25
											1	.	.	.	.	1	1	.	.	.	.	.	.	.	.	4	.	.	.	.	2	.	3	.	.	.	26
												1	.	.	.	.	.	.	2	2	.	.	.	.	1	.	.	.	.	4	.	.	.	.	9	5	27
													1	.	.	.	1	.	.	.	2	.	.	1	.	2	3	.	.	.	1	3	3	.	.	5	28
														1	.	.	.	1	.	.	.	1	2	.	.	.	.	2	1	.	.	.	.	3	.	.	29
															1	.	.	.	.	.	.	.	2	.	.	.	.	1	3	.	.	.	.	3	.	.	30
																1	.	.	.	.	.	.	.	.	.	2	.	.	.	.	1	.	1	.	.	.	31
																	1	.	.	.	.	.	.	.	.	2	.	.	.	.	1	.	3	.	.	.	32
																		1	.	.	.	1	1	.	.	.	.	2	1	.	.	.	.	2	.	.	33
																			1	.	.	.	.	.	.	.	.	.	.	.	.	.	.	.	3	.	34
																				1	.	.	.	.	1	.	.	.	.	.	.	.	.	.	3	.	35
																					1	.	.	1	.	1	1	.	.	.	1	2	1	.	.	.	36
																						1	.	.	.	.	.	1	.	.	.	.	.	1	.	.	37
																							1	.	.	.	.	1	1	.	.	.	.	2	.	.	38
																								1	.	.	.	.	.	.	1	1	.	.	.	.	39
																									1	.	.	.	.	.	.	.	.	.	.	.	40
																										1	.	.	.	.	1	.	1	.	.	.	41
																											1	.	.	.	.	1	1	.	.	.	42
																												1	.	.	.	.	.	1	.	.	43
																													1	.	.	.	.	1	.	.	44
																														1	.	.	.	.	.	1	45
																															1	.	.	.	.	.	46
																																1	.	.	.	.	47
																																	1	.	.	.	48
																																		1	.	.	49
																																			1	.	50
																																				1	51

Table 3. *(continued)*

ALGEBRA A11, REPRESENTATIONS OF CLASS 5

35	59	59	83	107	131	179	83	275	107	83	107	131	107	155	179	131	131	179	107	203	131	155	251	275	203	203	299	131	227	395	131	155	299	539	155	155	155	419	155	203	179	179	179	155	203	179	227	227	203	203	179	S.P.
792	1 980	5 544	660	660	132	132	7 920	12	3 960	27 720	1 900	5 940	23 760	660	1 980	1 980	15 840	660	83 160	1 320	33 264	11 880	132	660	660	5 940	132	55 440	1 320	132	166 320	3 960	1 320	12	55 440	110 880	83 160	132	55 440	27 720	27 720	15 840	47 520	277 200	23 760	166 320	5 940	47 520	23 760	47 520	110 880	O.S.
36	44	48	48	50	52	54	56	56	58	58	60	64	64	66	66	66	66	68	68	70	70	70	72	72	72	72	74	74	74	76	76	76	76	78	78	78	78	78	80	80	80	82	82	84	84	84	86	86	86	86	86	LEVEL
792	5 148	10 296	8 580	12 012	12 012	12 012	41 184	4 368	51 480	97 020	56 628	180 180	264 264	132 132	128 700	180 180	343 200	156 156	566 280	300 300	694 980	736 164	156 156	171 600	300 300	576 576	192 192	1 261 260	624 624	116 688	2 342 340	792 792	464 100	31 824	2 845 920	3 680 820	2 940 300	144 144	3 279 276	2 522 520	3 569 280	3 963 960	5 577 000	8 030 880	5 285 280	10 510 500	4 864 860	9 815 520	7 783 776	12 684 672	9 909 900	DIM'N
0	1	0	0	2	1	3	0	5	0	1	0	0	0	0	0	0	2	0	1	0	1	1	0	0	0	3	0	1	2	0	0	1	4	0	0	1	2	6	0	1	2	0	1	0	1	1	1	1	1	1	2	
0	0	0	1	0	2	1	0	0	0	0	0	0	1	0	0	0	0	0	0	0	0	1	0	0	3	0	0	0	2	0	1	0	1	0	1	0	0	0	0	0	0	0	0	1	0	1	0	0	0	1	0	
0	0	0	1	1	0	0	0	0	0	0	0	0	0	0	0	2	0	0	0	0	0	1	0	0	0	1	0	0	0	0	0	0	0	0	0	0	0	0	1	0	0	1	0	0	0	0	0	0	2	0	0	
0	1	0	0	0	0	0	0	0	0	0	0	0	1	0	0	0	1	0	0	0	0	0	0	0	0	0	0	0	0	0	0	0	0	0	0	0	0	0	1	0	0	1	0	0	0	1	0	0	0	1	0	
1	0	0	0	0	0	0	0	0	0	1	0	0	0	0	0	0	0	0	0	0	0	0	0	0	0	0	0	0	0	0	1	0	0	0	1	0	1	0	0	0	1	0	0	0	0	0	0	0	0	0	0	
0	0	1	0	0	0	0	0	0	0	0	0	0	0	0	0	0	0	0	1	0	1	0	0	0	0	0	0	0	0	0	0	0	0	0	0	0	0	0	0	0	0	0	0	1	0	0	0	0	0	0	1	
0	0	0	0	0	0	0	1	0	1	0	0	0	0	0	0	0	0	0	0	0	0	0	0	0	0	0	0	1	0	0	0	0	0	0	0	1	0	0	0	1	0	0	0	0	0	0	0	0	0	0	0	
0	0	0	0	0	0	0	0	0	0	0	1	1	0	0	1	0	0	0	0	0	0	0	0	0	0	0	0	0	0	0	0	2	0	0	0	0	0	0	0	0	0	0	1	0	1	0	0	1	0	0	0	
0	0	0	0	0	0	0	0	0	0	0	1	0	0	2	0	0	0	1	0	1	0	0	0	1	0	0	0	1	0	0	0	0	0	0	0	0	0	0	0	0	0	0	1	1	0	0	2	0	0	0	1	
0	0	0	0	0	0	0	1	0	0	0	0	1	0	0	0	0	0	2	1	1	0	0	3	0	0	0	2	0	0	1	1	0	0	0	0	1	1	0	1	0	0	0	0	0	2	1	1	1	1	0	0	
0	0	1	0	0	0	0	0	0	2	1	0	1	1	1	3	1	1	0	0	2	2	1	1	4	1	1	3	0	1	5	0	0	1	7	2	1	0	1	0	3	2	2	1	0	0	0	0	2	0	2	0	WEIGHT
0	1	2	3	4	5	6	7	8	9	10	11	12	13	14	15	16	17	18	19	20	21	22	23	24	25	26	27	28	29	30	31	32	33	34	35	36	37	38	39	40	41	42	43	44	45	46	47	48	49	50	51	NUMBER
1	4	6	5	6	5	4	14	1	15	35	14	35	69	21	20	40	80	21	115	35	126	136	14	15	45	90	14	175	84	6	330	91	50	1	364	424	375	11	371	245	420	410	490	730	406	1 015	315	675	654	1 140	820	0
	1	.	2	3	3	3	.	1	.	7	.	.	24	.	.	17	32	.	20	.	21	70	.	.	28	54	.	28	57	.	102	14	40	.	112	64	129	11	147	35	140	164	70	208	56	462	42	90	336	516	256	1
		1	.	.	.	.	4	.	5	5	5	15	9	10	10	5	10	11	30	20	36	16	9	10	5	10	10	60	9	5	80	35	5	1	94	164	90	1	86	105	105	100	220	240	196	230	165	345	144	265	270	2
			1	1	2	2	.	1	.	.	.	.	8	.	.	9	8	.	.	.	.	36	.	.	19	27	.	.	39	.	27	.	30	.	28	.	27	11	63	.	28	72	.	48	.	196	.	.	180	216	48	3
				1	1	2	.	1	.	.	.	.	.	.	.	.	8	.	.	.	.	18	.	.	10	27	.	.	29	.	.	.	30	.	.	.	27	11	.	.	28	.	.	.	.	98	.	.	72	108	48	4
					1	1	.	1	.	.	.	.	.	.	.	.	.	.	.	.	.	9	.	.	10	9	.	.	20	.	.	.	20	.	.	.	.	11	.	.	.	.	.	.	.	35	.	.	36	36	.	5
						1	.	1	.	.	.	.	.	.	.	.	.	.	.	.	.	.	.	.	.	9	.	.	10	.	.	.	20	.	.	.	.	11	.	.	.	.	.	.	.	.	.	.	.	.	.	6
							1	.	1	.	2	6	.	5	4	.	.	6	6	11	6	.	6	6	.	.	7	21	.	4	14	15	.	1	14	56	15	.	14	35	15	14	100	75	93	35	90	165	21	35	84	7
								1	.	.	.	.	.	.	.	.	.	.	.	.	.	.	.	.	.	.	.	.	.	.	.	.	10	.	.	.	.	11	.	.	.	.	.	.	.	.	.	.	.	.	.	8
									1	.	.	3	.	2	4	.	.	3	.	8	6	.	4	6	.	.	6	.	.	4	.	.	.	1	14	28	.	.	.	35	15	14	38	.	48	.	42	117	.	35	.	9
										1	.	.	3	.	.	2	4	.	5	.	6	8	.	.	3	6	.	9	6	.	24	5	4	.	28	24	30	1	33	15	35	38	30	66	26	102	21	45	72	117	80	10
											1	2	.	3	1	.	.	3	.	6	.	.	4	3	.	.	5	7	.	3	.	8	.	1	.	14	.	.	.	7	.	.	48	20	40	.	51	72	.	.	21	11
												1	.	1	1	.	.	2	.	4	.	.	3	3	.	.	4	.	.	3	.	.	.	1	.	7	.	.	.	7	.	.	16	.	24	.	26	48	.	.	.	12
													1	.	.	1	1	.	.	.	.	4	.	.	2	3	.	.	4	.	6	.	3	.	7	.	6	1	14	.	7	16	.	14	.	42	.	.	38	48	14	13
														1	.	.	.	1	.	2	.	.	2	1	.	.	3	.	.	2	.	.	.	1	.	.	.	.	.	.	.	.	8	.	8	.	18	16	.	.	.	14
															1	.	.	.	.	2	.	.	1	3	.	.	3	.	.	3	.	.	.	1	.	.	.	.	.	7	.	.	.	.	.	.	.	24	.	.	.	15
																1	.	.	.	.	.	2	.	.	1	1	.	.	3	.	.	.	2	.	.	.	.	1	7	.	.	8	.	.	.	14	.	.	24	16	.	16

1	.	.	.	.	2	.	.	1	3	.	.	3	.	.	.	3	.	.	.	6	1	.	.	7	.	.	.	.	21	.	.	15	24	14	17
	1	.	1	.	.	2	1	.	.	2	.	.	2	.	.	.	1	.	.	.	.	.	.	.	.	.	.	8	.	9	8	.	.	.	18
		1	.	1	.	.	.	.	.	.	3	.	.	4	2	.	.	4	8	5	.	5	5	5	5	13	20	12	15	11	21	10	15	24	19
			1	.	.	1	1	.	.	2	.	.	2	.	.	.	1	.	.	.	.	.	.	.	.	.	.	.	.	.	8	.	.	.	20
				1	.	.	.	.	.	.	.	.	.	.	.	.	.	4	4	.	.	.	5	5	5	5	.	6	.	5	15	.	15	.	21
					1	.	.	1	1	.	.	2	.	.	.	2	.	.	.	.	1	.	.	.	.	.	.	.	7	.	.	8	8	.	22
						1	.	.	.	1	.	.	1	.	.	.	1	.	.	.	.	.	.	.	.	.	.	.	.	.	.	.	.	.	23
							1	.	.	1	.	.	2	.	.	.	1	.	.	.	.	.	.	.	.	.	.	.	.	.	.	.	.	.	24
								1	.	.	.	1	.	.	.	1	.	.	.	.	1	.	.	.	.	.	.	.	.	.	.	.	.	.	25
									1	.	.	1	.	.	.	2	.	.	.	.	1	.	.	.	.	.	.	.	.	.	.	.	.	.	26
										1	.	.	1	.	.	.	1	.	.	.	.	.	.	.	.	.	.	.	.	.	.	.	.	.	27
											1	.	.	.	1	.	.	.	2	.	.	.	1	.	.	6	5	5	.	6	9	.	.	6	28
												1	.	.	.	1	.	.	.	.	1	.	.	.	.	.	.	.	.	.	.	.	.	.	29
													1	.	.	.	1	.	.	.	.	.	.	.	.	.	.	.	.	.	.	.	.	.	30
														1	.	.	.	1	.	1	.	2	.	1	2	.	4	.	6	.	.	5	6	4	31
															1	.	.	.	.	.	.	.	.	.	.	3	.	2	.	3	3	.	.	.	32
																1	.	.	.	.	1	.	.	.	.	.	.	.	.	.	.	.	.	.	33
																	1	.	.	.	.	.	.	.	.	.	.	.	.	.	.	.	.	.	34
																		1	.	.	.	.	.	1	2	.	.	.	.	.	.	.	6	.	35
																			1	.	.	.	1	.	.	2	.	3	.	3	6	.	.	.	36
																				1	.	.	.	1	.	.	.	.	3	.	.	2	3	4	37
																					1	.	.	.	.	.	.	.	.	.	.	.	.	.	38
																						1	.	.	1	.	.	.	2	.	.	3	2	.	39
																							1	.	.	.	.	.	.	.	3	.	.	.	40
																								1	.	.	.	.	.	.	.	.	3	.	41
																									1	.	.	.	.	.	.	.	2	.	42
																										1	.	1	.	2	2	.	.	.	43
																											1	.	.	.	.	.	.	1	44
																												1	.	1	1	.	.	.	45
																													1	.	.	1	1	.	46
																														1	.	.	.	.	47
																															1	.	.	.	48
																																1	.	.	49
																																	1	.	50
																																		1	51

Table 3. *(continued)*

ALGEBRA A11, REPRESENTATIONS OF CLASS 6

0	1	2	3	4	5	6	7	8	9	10	11	12	NUMBER
36	60	60	84	84	108	108	108	108	132	84	132	180	S.P.
924	3 960	3 960	2 970	2 970	220	1 980	220	1 980	1 320	33 264	1 320	66	O.S.
37	47	47	53	53	55	55	55	55	59	59	59	61	LEVEL
924	8 580	8 580	23 166	23 166	15 730	30 030	15 730	30 030	64 064	114 048	64 064	26 026	DIM'N
0	0	1	0	0	0	0	0	2	0	1	1	0	
0	0	0	0	1	0	0	0	0	0	0	1	0	
0	0	0	0	0	0	0	2	0	0	0	1	0	
0	0	0	0	1	0	0	0	1	0	0	0	0	
0	0	1	0	0	0	0	0	0	0	0	0	0	
1	0	0	0	0	0	0	0	0	0	1	0	0	
0	1	0	0	0	0	0	0	0	0	0	0	0	
0	0	0	1	0	0	1	0	0	0	0	0	0	
0	0	0	0	0	2	0	0	0	1	0	0	0	
0	0	0	1	0	0	0	0	0	1	0	0	3	
0	1	0	0	0	0	2	0	0	1	1	0	0	WEIGHT
1	5	5	9	9	5	10	5	10	16	36	16	5	0
	1	.	3	.	2	4	.	.	8	6	.	3	1
		1	.	3	.	.	2	4	.	6	8	.	2
			1	.	1	1	.	.	4	.	.	2	3
				1	.	.	1	1	.	.	4	.	4
					1	.	.	.	2	.	.	1	5
						1	.	.	2	.	.	1	6
							1	.	.	.	2	.	7
								1	.	.	2	.	8
									1	.	.	1	9
										1	.	.	10
											1	.	11
												1	12
													13
													14
													15
													16

13	14	15	16	17	18	19	20	21	22	23	24	25	NUMBER
180	180	180	204	204	276	276	396	108	108	396	132	132	S.P.
660	66	660	132	132	132	132	12	55 440	55 440	12	27 720	27 720	O.S.
61	61	61	63	63	65	65	67	67	67	67	69	69	LEVEL
50 050	26 026	50 050	54 054	54 054	40 040	40 040	12 376	462 462	462 462	12 376	579 150	579 150	DIM'N
0	0	3	0	2	0	4	0	0	1	6	1	2	
0	3	0	0	2	0	1	0	1	0	0	0	0	
0	0	1	0	0	0	0	0	0	0	0	0	0	
0	0	0	0	0	0	0	0	0	0	0	0	0	
0	0	0	0	0	0	0	0	1	0	0	0	1	
0	0	0	0	0	0	0	0	0	0	0	0	0	
0	0	0	0	0	0	0	0	0	1	0	1	0	
0	0	0	0	0	0	0	0	0	0	0	0	0	
1	0	0	0	0	0	0	0	0	0	0	0	0	
0	0	0	2	0	1	0	0	0	1	0	0	0	
3	0	0	2	0	4	0	6	1	0	0	2	1	WEIGHT
10	5	10	9	9	5	5	1	94	94	1	105	105	0
6	.	.	6	.	4	.	1	14	28	.	35	15	1
.	3	6	.	6	.	4	.	28	14	1	15	35	2
3	.	.	4	.	3	.	1	.	7	.	7	.	3
.	2	3	.	4	.	3	.	7	.	1	.	7	4
1	.	.	3	.	2	.	1	.	.	.	.	.	5
3	.	.	3	.	3	.	1	.	.	.	7	.	6
.	1	1	.	3	.	2	.	.	.	1	.	.	7
.	1	3	.	3	.	3	.	.	.	1	.	7	8
1	.	.	2	.	2	.	1	.	.	.	.	.	9
.	.	.	.	.	.	.	.	4	4	.	5	5	10
.	1	1	.	2	.	2	.	.	.	1	.	.	11
.	.	.	1	.	1	.	1	.	.	.	.	.	12
1	.	.	1	.	2	.	1	.	.	.	.	.	13
	1	.	.	1	.	1	.	.	.	1	.	.	14
		1	.	1	.	2	.	.	.	1	.	.	15
			1	.	1	.	1	.	.	.	.	.	16

26	27	28	29	30	31	32	33	34	35	36	37	38	NUMBER
132	132	156	156	132	180	204	180	204	156	204	204	156	S.P.
15 840	15 840	47 520	47 520	207 900	5 940	15 840	5 940	15 840	83 160	5 940	5 940	83 160	O.S.
71	71	75	75	77	77	77	77	77	79	79	79	79	LEVEL
640 640	640 640	2 044 900	2 044 900	2 788 500	1 505 790	1 464 320	1 505 790	1 464 320	3 430 350	1 783 782	1 783 782	3 430 350	DIM'N
0	1	1	1	0	1	1	1	3	0	0	1	2	
0	0	0	1	1	0	0	0	0	1	2	0	0	
1	0	0	0	0	0	0	2	0	0	1	0	0	
1	0	0	1	0	0	0	0	1	0	0	0	0	
0	0	0	0	0	0	0	0	0	0	0	0	0	
0	0	0	0	1	0	0	0	0	1	0	0	1	
0	0	0	0	0	0	0	0	0	0	0	0	0	
0	1	1	0	0	0	1	0	0	0	0	0	0	
0	1	0	0	0	2	0	0	0	0	0	1	0	
0	0	1	0	1	0	0	0	0	0	0	2	1	
1	0	1	1	0	1	3	1	1	2	1	0	0	WEIGHT
100	100	265	265	350	165	160	165	160	390	171	171	390	0
14	38	117	35	95	82	80	21	20	115	21	93	105	1
38	14	35	117	95	21	20	82	80	105	93	21	115	2
.	16	48	.	20	43	32	.	.	20	.	52	21	3
16	.	.	48	20	.	.	43	32	21	52	.	20	4
.	8	16	.	.	27	8	.	.	.	.	27	.	5
.	.	24	.	.	17	32	.	.	20	.	27	.	6
8	.	.	16	.	.	.	27	8	.	27	.	.	7
.	.	.	24	.	.	.	17	32	.	27	.	20	8
.	.	8	.	.	9	8	.	.	.	.	18	.	9
5	5	15	15	25	10	10	10	10	30	11	11	30	10
.	.	.	8	.	.	.	9	8	.	18	.	.	11
.	.	.	.	.	.	.	.	.	.	.	9	.	12
.	.	.	.	.	.	8	.	.	.	.	.	.	13
.	.	.	.	.	.	.	.	.	.	9	.	.	14
.	.	.	.	.	.	.	.	8	.	.	.	.	15
.	.	.	.	.	.	.	.	.	.	.	.	.	16

39	40	41	42	43	44	45	46	47	48	49	50	51	NUMBER
228	180	228	156	156	276	300	276	300	180	180	180	324	S.P.
11 880	33 264	11 880	166 320	166 320	1 320	5 940	1 320	5 940	13 060	55 440	13 860	1 320	O.S.
81	81	81	83	83	83	83	83	83	85	85	85	85	LEVEL
3 435 432	4 216 212	3 435 432	6 324 318	6 324 318	1 793 792	1 969 110	1 793 792	1 969 110	4 008 004	7 707 700	4 008 004	2 212 210	DIM'N
1	2	2	0	0	1	1	1	4	0	0	0	1	
0	0	1	0	1	0	0	3	0	0	0	1	0	
0	0	1	1	0	0	0	0	1	0	1	0	0	
0	0	0	0	0	0	0	0	0	2	0	0	0	
0	0	0	1	0	0	0	0	0	0	1	0	0	
0	1	0	0	0	0	0	0	0	0	0	0	0	
0	0	0	0	1	0	0	0	0	0	0	0	0	
0	0	0	0	0	0	0	0	0	0	0	2	0	
1	0	0	0	1	0	1	0	0	0	0	0	0	
1	0	0	0	0	3	0	0	0	1	0	0	2	
2	2	1	0	0	1	4	1	1	0	2	0	3	WEIGHT
295	441	295	576	576	124	135	124	135	311	640	311	130	0
172	126	35	147	207	82	90	14	15	77	175	126	94	1
35	126	172	207	147	14	15	82	90	126	231	77	14	2
97	21	.	28	75	56	54	.	.	14	28	56	67	3
.	21	97	75	28	.	.	56	54	56	84	14	.	4
54	.	.	.	27	38	27	.	.	.	.	28	49	5
70	21	.	.	.	38	54	.	.	.	28	.	57	6
.	.	54	27	.	.	.	38	27	28	28	.	.	7
.	21	70	.	.	.	.	38	54	.	.	.	.	8
36	.	.	.	.	29	27	.	.	.	.	.	39	9
20	36	20	51	51	9	10	9	10	30	60	30	10	10
.	.	36	.	.	.	.	29	27	.	.	.	.	11
9	.	.	.	.	20	9	.	.	.	.	.	20	12
18	.	.	.	.	10	27	.	.	.	.	.	29	13
.	.	9	.	.	.	.	20	9	.	.	.	.	14
.	.	18	.	.	.	.	10	27	.	.	.	.	15
9	.	.	.	.	10	9	.	.	.	.	.	20	16

1	.	1	.	.	.	1	.	.	.	.	.	.	.	.	.	.	.	.	.	.	.	.	.	9	.	.	.	.	10	9	.	.	.	.	17
	1	.	1	.	.	.	.	.	.	.	.	.	.	.	.	.	.	.	.	.	.	.	.	.	.	.	.	9	.	.	.	.	.	10	18
		1	.	.	.	1	.	.	.	.	.	.	.	.	.	.	.	.	.	.	.	.	.	.	.	.	.	.	.	9	.	.	.	.	19
			1	.	.	.	.	.	.	.	.	.	.	.	.	.	.	.	.	.	.	.	.	.	.	.	.	.	.	.	.	.	.	.	20
				1	.	.	.	1	2	.	.	6	5	.	.	5	4	6	6	.	5	.	6	11	18	9	.	.	6	6	13	21	5	.	21
					1	.	1	.	.	2	6	.	5	5	4	.	.	5	.	6	6	11	6	.	9	18	6	6	.	.	5	9	13	7	22
						1	.	.	.	.	.	.	.	.	.	.	.	.	.	.	.	.	.	.	.	.	.	.	.	.	.	.	.	.	23
							1	.	.	.	3	.	.	2	4	.	.	5	.	3	.	8	6	.	.	.	4	6	.	.	.	9	.	6	24
								1	.	.	.	3	.	.	.	2	4	.	3	.	5	.	6	8	.	.	.	.	4	6	.	.	.	.	25
									1	.	.	2	.	.	.	3	1	.	3	.	.	.	.	6	6	.	.	.	4	3	7	7	.	.	26
										1	2	.	.	3	1	.	.	.	.	3	.	6	.	.	.	6	4	3	.	.	.	.	7	5	27
											1	.	.	1	1	.	.	.	.	2	.	4	.	.	.	.	3	3	.	.	.	.	.	4	28
												1	.	.	.	1	1	.	2	.	.	.	.	4	.	.	.	.	3	3	.	.	.	.	29
													1	.	.	.	.	1	.	.	1	.	1	.	3	3	.	.	.	.	2	3	2	.	30
														1	.	.	.	.	.	1	.	2	.	.	.	.	2	1	.	.	.	.	.	3	31
															1	.	.	.	.	.	.	2	.	.	.	.	1	3	.	.	.	.	.	3	32
																1	.	.	1	.	.	.	.	2	.	.	.	.	2	1	.	.	.	.	33
																	1	.	.	.	.	.	.	2	.	.	.	.	1	3	.	.	.	.	34
																		1	.	.	.	.	1	.	.	.	.	.	.	.	.	3	.	.	35
																			1	.	.	.	.	1	.	.	.	.	2	1	.	.	.	.	36
																				1	.	1	.	.	.	.	2	1	.	.	.	.	.	2	37
																					1	.	1	.	.	.	.	.	.	.	.	.	.	.	38
																						1	.	.	.	.	1	1	.	.	.	.	.	2	39
																							1	.	.	.	.	.	.	.	.	.	.	.	40
																								1	.	.	.	.	1	1	.	.	.	.	41
																									1	.	.	.	.	.	1	1	.	.	42
																										1	.	.	.	.	.	.	1	.	43
																											1	.	.	.	.	.	.	1	44
																												1	.	.	.	.	.	1	45
																													1	.	.	.	.	.	46
																														1	.	.	.	.	47
																															1	.	.	.	48
																																1	.	.	49
																																	1	.	50
																																		1	51

Table 3. *(continued)*

ALGEBRA A12, REPRESENTATIONS OF CLASS 0

S.P.	0	26	52	78	78	104	78	104	104	156	156	130	182	182	104	234	130	130	130	156	156	182	182	156	260	156
O.S.	1	156	4290	858	858	156	34320	25740	25740	2860	2860	17160	1716	1716	90090	156	180180	180180	72072	25740	25740	51480	51480	12012	6435	308880
LEVEL	1	25	45	47	47	49	61	65	65	67	67	69	71	71	73	73	79	79	81	81	81	83	83	85	85	85
DIM'N	1	168	5915	6930	6930	8112	75712	195195	195195	123200	123200	501760	316030	316030	429429	198744	1863225	1863225	1145144	1490580	1490580	2602600	2602600	1288287	1178100	8016008
	0	1	0	0	2	2	0	0	1	0	3	1	1	3	0	3	0	1	0	0	0	0	2	0	0	1
	0	0	1	1	0	0	0	0	1	0	0	1	1	0	0	0	0	0	0	0	2	0	1	0	0	0
	0	0	0	0	0	0	1	1	0	1	0	0	0	0	0	0	0	1	0	0	0	0	0	0	0	1
	0	0	0	0	0	0	0	0	0	0	0	0	0	0	1	0	1	0	0	1	0	1	0	0	1	0
	0	0	0	0	0	0	0	0	0	0	0	0	0	0	0	0	0	0	1	0	0	0	0	0	0	0
	0	0	0	0	0	0	0	0	0	0	0	0	0	0	0	0	0	0	0	0	0	0	0	1	0	0
	0	0	0	0	0	0	0	0	0	0	0	0	0	0	0	0	0	0	0	0	0	0	0	1	0	0
	0	0	0	0	0	0	0	0	0	0	0	0	0	0	0	0	0	0	1	0	0	0	0	0	0	0
	0	0	0	0	0	0	0	0	0	0	0	0	0	0	1	0	0	1	0	0	1	0	1	0	0	0
	0	0	0	0	0	0	1	0	1	0	1	0	0	0	0	0	1	0	0	0	0	0	0	0	0	1
	0	0	1	0	1	0	0	1	0	0	0	1	0	1	0	0	0	0	0	2	0	1	0	0	0	0
WEIGHT	0	1	0	2	0	2	0	1	0	3	0	1	3	1	0	3	1	0	0	0	0	2	0	0	4	1
NUMBER	0	1	2	3	4	5	6	7	8	9	10	11	12	13	14	15	16	17	18	19	20	21	22	23	24	25
0	1	12	65	66	66	78	208	429	429	220	220	1000	572	572	429	364	1365	1365	572	936	936	1430	1430	429	495	4862
1		1	10	11	11	12	44	99	99	55	55	230	132	132	110	78	385	385	165	276	276	440	440	132	165	1408
2			1	1	1	1	8	18	18	10	10	40	22	22	27	12	98	98	48	72	72	116	116	42	45	356
3				1	.	1	.	9	.	10	.	20	22	11	.	12	35	.	.	36	.	80	.	.	45	124
4					1	1	.	.	9	.	10	20	11	22	.	12	.	35	.	.	36	.	80	.	.	124
5						1	.	.	.	.	.	10	11	11	.	12	.	.	.	.	.	.	.	.	.	44
6							1	2	2	1	1	4	2	2	6	1	21	21	14	15	15	24	24	14	9	72
7								1	.	1	.	2	2	1	.	1	7	.	.	8	.	16	.	.	9	24
8									1	.	1	2	1	2	.	1	.	7	.	.	8	.	16	.	.	24
9										1	.	.	2	.	.	1	.	.	.	.	.	8	.	.	9	.
10											1	.	.	2	.	1	.	.	.	.	.	.	8	.	.	.
11												1	1	1	.	1	.	.	.	.	.	.	.	.	.	8
12													1	.	.	1	.	.	.	.	.	.	.	.	.	.
13														1	.	1	.	.	.	.	.	.	.	.	.	.
14															1	.	3	3	4	2	2	3	3	5	1	9

S.P.	260	182	182	156	208	156	208	208	234	286	234	286	182	182	312	260	312	182	208	182	208	338	338	208	234	208
O.S.	6435	38610	38610	360360	4290	360360	77220	77220	8580	8580	8580	8580	216216	216216	858	17160	858	180180	216216	180180	216216	1716	1716	10296	72072	10296
LEVEL	85	87	87	89	89	89	89	89	91	91	91	91	93	93	93	93	93	95	95	95	95	95	95	97	97	97
DIM'N	1178100	6388200	6388200	7729722	5082000	7729722	11133045	11133045	8848840	5019300	8848840	5019300	13117104	13117104	3981978	15400125	3981978	14314300	18398016	14314300	18398016	6922916	6922916	8281845	18549440	8281845
	4	0	1	0	0	1	1	2	0	1	2	4	0	0	0	2	4	0	0	1	2	2	4	0	0	1
	0	2	0	0	2	0	0	1	2	0	1	0	0	1	2	1	0	0	0	0	0	1	0	0	0	0
	0	0	1	0	0	0	1	0	0	1	0	0	0	1	0	0	0	0	0	0	1	0	0	0	0	0
	0	0	0	0	0	1	0	0	0	0	0	0	0	0	0	0	0	0	0	0	0	0	0	0	0	0
	0	0	0	1	0	0	0	0	0	0	0	0	1	0	0	0	0	0	1	1	0	0	0	0	1	0
	0	0	0	0	0	0	0	0	0	0	0	0	0	0	0	0	0	1	0	0	0	0	0	0	0	2
	0	0	0	0	0	0	0	0	0	0	0	0	0	0	0	0	0	0	0	1	0	0	0	2	0	0
	0	0	0	0	0	1	0	0	0	0	0	0	0	1	0	0	0	1	0	0	1	0	0	0	0	0
	1	0	0	1	0	0	0	0	0	0	0	0	0	0	0	0	0	0	0	0	0	0	0	0	0	0
	0	1	0	0	0	0	0	1	0	0	0	1	1	0	0	0	0	0	1	0	0	0	0	0	0	0
	0	0	2	0	2	0	1	0	1	0	2	0	1	0	0	1	2	0	0	0	0	0	1	0	2	0
WEIGHT	0	1	0	1	0	0	2	1	2	4	0	1	0	0	4	2	0	1	2	0	0	4	2	1	1	0
NUMBER	26	27	28	29	30	31	32	33	34	35	36	37	38	39	40	41	42	43	44	45	46	47	48	49	50	51
0	495	3510	3510	2574	2640	2574	5643	5643	4290	2145	4290	2145	3432	3432	1716	7215	1716	2860	4212	2860	4212	3003	3003	1287	3640	1287
1	165	1035	1035	825	770	825	1683	1683	1265	660	1265	660	1166	1166	506	2100	506	990	1485	990	1485	858	858	462	1330	462
2	45	261	261	255	191	255	423	423	310	165	310	165	376	376	121	505	121	342	486	342	486	198	198	168	448	168
3	.	90	135	75	100	.	288	144	210	165	165	55	160	.	121	340	66	90	270	.	.	198	132	42	280	.
4	45	135	90	.	100	75	144	288	165	55	210	165	.	160	66	340	121	.	.	90	270	132	198	.	.	42
5	.	45	45	.	55	.	99	99	110	55	110	55	.	.	66	230	66	.	.	.	.	132	132	.	.	.
6	9	51	51	75	36	75	81	81	57	30	57	30	111	111	21	90	21	119	144	119	144	33	33	63	133	63
7	.	17	27	20	19	.	54	27	38	30	30	10	48	.	21	60	11	28	75	.	.	33	22	14	84	.
8	9	27	17	.	19	20	27	54	30	10	38	30	.	48	11	60	21	.	.	28	75	22	33	.	.	14
9	.	.	.	.	.	.	27	.	19	30	.	.	.	.	21	30	.	.	27	.	.	33	11	.	28	.
10	9	.	.	.	.	.	.	27	.	.	19	30	.	.	.	30	21	.	.	.	27	11	33	.	.	.
11	.	9	9	.	10	.	18	18	20	10	20	10	.	.	11	40	11	.	.	.	.	22	22	.	.	.
12	.	.	.	.	.	.	9	.	10	10	.	.	.	.	11	20	.	.	.	.	.	22	11	.	.	.
13	.	.	.	.	.	.	.	9	.	.	10	10	.	.	.	20	11	.	.	.	.	11	22	.	.	.
14	1	6	6	20	4	20	9	9	6	3	6	3	28	28	2	9	2	42	36	42	36	3	3	25	32	25

15	16	17	18	19	20	21	22	23	24	25	26	27	28	29	30	31	32	33	34	35	36	37	38	39	40	41	42	43	44	45	46	47	48	49	50	51
.	.	5	11	.	.	.	.	6	.	.	.	.	.	.	.	2	.	.	.	.	.	.	.	.	.	.	.	.	.	1	.	.	.	.	.	1
.	20	.	5	14	.	7	.	.	.	.	.	.	.	3	.	.	.	.	.	.	.	.	2	.	.	.	.	.	1	.	.	.	.	.	1	
.	5	.	11	.	.	.	.	6	.	.	.	.	.	2	.	.	.	.	.	.	.	.	.	.	.	.	.	1	.	.	.	.	.	1		
11	2	3	.	1	3	1	3	.	.	2	3	2	1	.	1	.	1	2	1	.	1	2	.	.	.	1	1	.	.	.	.	.	1			
11	3	2	.	3	1	3	1	.	3	2	.	1	2	.	1	.	2	1	1	2	1	.	.	.	1	1	.	.	.	.	.	1				
.	.	18	6	.	6	.	6	.	.	.	.	.	.	.	.	3	.	.	.	.	.	.	.	1	.	.	.	.	.	.	1					
.	.	9	15	.	.	.	.	5	.	.	.	.	.	.	.	3	.	.	.	.	.	.	.	.	.	.	.	.	.	1						
.	18	.	6	6	.	6	.	.	.	.	.	.	.	3	.	.	.	.	.	.	.	.	1	.	.	.	.	.	1							
.	9	.	15	.	.	.	.	5	.	.	.	.	.	3	.	.	.	.	.	.	.	.	.	.	.	.	.	1								
.	1	2	.	1	2	.	2	.	.	1	2	1	1	.	1	.	.	1	.	.	1	1	.	.	.	.	1									
10	6	6	.	3	3	3	3	.	.	4	.	2	2	.	1	.	2	2	1	.	1	.	.	.	.	1										
.	2	1	.	2	1	2	.	.	2	1	.	1	1	.	1	.	1	.	1	1	.	.	.	.	1											
.	.	12	5	.	6	.	.	.	.	.	.	.	.	.	.	2	.	.	.	.	.	.	.	1												
.	12	.	5	6	.	.	.	.	.	.	.	.	.	2	.	.	.	.	.	.	.	.	1													
.	1	3	.	.	3	.	3	.	.	1	3	1	.	.	.	.	.	1	.	.	.	1														
.	3	4	.	3	2	.	2	.	.	2	.	1	2	.	1	.	.	1	.	.	1															
.	3	1	.	3	.	3	.	.	3	1	.	.	1	.	.	.	1	.	.	1																
.	4	3	.	2	3	2	.	.	.	2	.	2	1	.	1	.	1	.	1																	
.	3	6	.	.	3	.	3	.	.	2	.	1	.	.	.	.	.	1																		
.	6	3	.	3	.	3	.	.	.	2	.	.	1	.	.	.	1																			
.	.	5	4	.	.	.	.	.	.	.	.	.	.	.	.	1																				
.	2	2	.	2	2	.	.	.	.	1	.	1	1	.	1																					
.	5	.	4	.	.	.	.	.	.	.	.	.	.	1																						
.	3	2	.	3	.	.	.	.	.	1	.	.	1																							
.	2	3	.	.	3	.	.	.	.	1	.	1																								
.	.	1	.	.	1	.	1	.	.	.	1																									
.	3	3	.	.	.	.	.	.	.	1																										
.	1	.	.	1	.	1	.	.	1																											
.	.	.	2	.	.	.	.	1																												
.	.	2	.	.	1	.	1																													
.	2	.	.	1	.	1																														
.	.	1	.	.	1																															
.	1	.	.	1																																
.	.	.	1																																	
.	.	1																																		
.	1																																			
1																																				

Table 3. *(continued)*

ALGEBRA A12, REPRESENTATIONS OF CLASS 1

S.P.	12	38	64	64	90	90	116	142	168	90	116	168	116	142	116	142	168	168	194	246	220	194	142	142	246	272
O.S.	13	858	156	12 870	2 860	8 580	1 716	858	156	60 060	51 480	6 435	102 960	12 870	90 090	77 220	25 740	8 580	8 580	2 860	858	17 160	36 036	216 216	1 716	1 716
LEVEL	13	35	37	53	55	57	59	59	61	67	71	73	73	75	77	77	77	79	79	79	81	81	83	83	83	83
DIM'N	13	1 001	1 170	21 294	25 025	54 600	64 064	34 320	40 222	182 182	473 200	300 300	780 780	621 075	715 715	2 018 016	1 081 080	1 601 600	1 276 275	486 200	1 011 296	2 784 600	1 288 287	3 162 159	1 756 755	1 249 248
	1	0	2	0	0	1	1	3	3	0	0	0	1	0	0	1	2	0	1	4	0	2	0	0	2	4
	0	1	0	0	0	1	1	0	0	0	0	0	0	2	0	0	1	2	0	0	2	1	0	0	1	0
	0	0	0	1	1	0	0	0	0	0	0	0	1	0	0	1	0	0	1	0	0	0	0	0	0	0
	0	0	0	0	0	0	0	0	0	1	1	1	0	0	0	0	0	0	0	0	0	0	0	0	0	0
	0	0	0	0	0	0	0	0	0	0	0	0	0	0	1	0	0	0	0	0	0	0	0	1	0	0
	0	0	0	0	0	0	0	0	0	0	0	0	0	0	0	0	0	0	0	0	0	0	1	0	0	0
	0	0	0	0	0	0	0	0	0	0	0	0	0	0	0	0	0	0	0	0	0	0	0	0	0	0
	0	0	0	0	0	0	0	0	0	0	0	0	0	0	0	0	0	0	0	0	0	0	1	0	0	0
	0	0	0	0	0	0	0	0	0	0	0	0	0	0	1	0	0	0	0	0	0	0	0	0	0	0
	0	0	0	0	0	0	0	0	0	1	0	0	1	1	0	0	1	0	0	1	0	0	0	1	0	0
	0	0	0	1	0	1	0	1	0	0	1	0	0	0	0	1	0	1	0	0	0	1	0	0	0	1
WEIGHT	0	1	1	0	2	0	2	0	2	0	1	3	0	0	0	1	0	1	3	0	3	1	0	1	3	1
NUMBER	0	1	2	3	4	5	6	7	8	9	10	11	12	13	14	15	16	17	18	19	20	21	22	23	24	25
0	1	11	12	54	55	120	132	66	78	154	320	165	528	375	275	1 188	594	880	660	220	506	1 440	297	891	858	572
1		1	1	9	10	20	22	11	12	35	80	45	124	90	75	288	144	210	165	55	121	340	90	270	198	132
2			1	.	.	10	11	11	12	.	.	.	44	45	.	99	99	110	55	55	66	230	.	.	132	132
3				1	1	2	2	1	1	7	16	9	24	17	20	54	27	38	30	10	21	60	28	75	33	22
4					1	.	2	.	1	.	8	9	.	.	.	27	.	19	30	.	21	30	.	27	33	11
5						1	1	1	1	.	.	.	8	9	.	18	18	20	10	10	11	40	.	.	22	22
6							1	.	1	.	.	.	.	.	.	9	.	10	10	.	11	20	.	.	22	11
7								1	1	.	.	.	.	.	.	.	9	.	.	10	.	20	.	.	11	22
8									1	.	.	.	.	.	.	.	.	.	.	.	.	10	.	.	11	11
9										1	2	1	3	2	5	6	3	4	3	1	2	6	9	18	3	2
10											1	1	.	.	.	3	.	2	3	.	2	3	.	6	3	1
11												1	.	.	.	.	.	.	3	.	2	.	.	.	3	.
12													1	1	.	2	2	2	1	1	1	4	.	.	2	2
13														1	.	.	1	2	.	1	1	2	.	.	1	2
14															1	.	.	.	.	.	.	.	3	3	.	.

S.P.	168	168	142	324	194	272	168	168	168	194	168	194	220	194	194	194	220	272	220	220	194	298	220	376	246	194
O.S.	1 716	36 036	360 360	156	72 072	10 296	180 180	240 240	720 720	180 180	360 360	102 960	51 480	51 480	180 180	308 880	205 920	51 480	180 180	38 610	72 072	25 740	308 880	6 435	72 072	540 540
LEVEL	85	85	85	85	87	89	89	91	91	91	93	93	93	95	95	95	95	95	97	97	97	97	97	97	99	99
DIM'N	736 164	2 550 548	4 684 680	786 695	4 471 740	2 042 040	7 825 545	9 161 152	20 404 384	10 930 920	12 882 870	16 353 792	10 930 920	10 735 725	15 940 925	33 833 800	28 588 560	10 890 880	22 486 464	27 027 000	14 723 280	12 972 960	47 171 124	3 938 220	22 954 932	36 072 036
	0	0	1	4	0	0	0	0	1	2	1	1	1	0	0	0	1	3	0	0	1	1	2	5	0	0
	0	0	0	0	0	0	1	0	0	0	0	0	2	0	0	1	0	1	0	1	0	0	0	0	0	1
	0	0	0	0	0	0	1	0	0	1	0	0	0	0	0	1	0	0	0	1	0	0	1	0	0	0
	0	0	1	0	0	0	0	0	1	0	0	1	0	0	0	0	1	0	0	0	0	1	0	0	0	1
	0	1	0	0	1	1	0	0	0	0	1	0	0	0	0	0	0	0	0	0	0	0	0	0	0	0
	0	0	0	0	0	0	0	1	0	0	0	0	0	0	1	0	0	0	1	0	1	0	0	0	1	0
	2	0	0	0	0	0	0	0	0	0	0	0	0	1	0	0	0	0	0	0	1	0	0	0	0	0
	0	0	0	0	0	0	0	0	0	0	1	0	0	1	0	0	0	0	0	0	0	0	0	0	0	1
	0	0	1	0	0	0	1	1	0	1	0	0	1	0	0	0	0	1	0	0	0	0	0	1	0	0
	0	0	0	0	0	0	0	0	1	0	0	0	0	0	1	1	0	0	1	0	0	0	1	0	0	0
	0	2	0	0	1	0	0	0	0	0	0	2	0	0	1	0	1	0	0	2	0	0	0	0	2	0
WEIGHT	0	0	0	3	2	4	0	1	1	0	0	0	0	1	0	1	2	0	2	0	0	4	1	0	1	0
NUMBER	26	27	28	29	30	31	32	33	34	35	36	37	38	39	40	41	42	43	44	45	46	47	48	49	50	51
0	132	616	1 320	364	945	330	1 827	1 408	4 620	2 310	1 980	3 312	2 046	1 155	1 925	6 710	5 280	1 760	2 376	4 950	1 584	1 980	8 712	495	2 079	4 224
1	42	196	380	78	315	120	546	480	1 400	700	645	1 032	636	420	700	2 040	1 680	560	900	1 515	552	660	2 648	165	819	1 450
2	.	.	110	78	.	.	231	.	385	385	165	276	396	.	.	880	440	440	.	660	132	165	1 408	165	.	550
3	14	56	102	12	91	36	147	160	378	189	208	280	172	154	245	540	456	152	320	398	196	180	696	45	301	480
4	.	28	.	12	63	36	.	48	133	.	.	144	.	42	105	188	312	.	180	207	.	180	240	.	189	.
5	.	.	27	12	.	.	63	.	98	98	48	72	108	.	.	232	116	116	.	171	42	45	356	45	.	180
6	.	.	.	12	.	.	.	.	35	.	.	36	.	.	.	80	80	.	.	90	.	45	124	.	.	.
7	.	.	.	12	.	.	.	.	.	35	.	.	36	.	.	.	.	80	.	.	.	.	124	45	.	.
8	.	.	.	12	.	.	.	.	.	.	.	.	.	.	.	.	.	.	.	.	.	.	44	.	.	.
9	5	13	24	1	21	8	33	52	84	42	66	60	37	58	80	114	96	32	105	81	72	36	144	9	99	150
10	.	7	.	1	14	8	.	14	28	.	.	32	.	14	35	38	64	.	55	43	.	36	48	.	63	.
11	.	.	.	1	7	8	.	.	.	.	.	.	.	.	.	.	32	.	20	.	.	36	.	.	21	.
12	.	.	6	1	.	.	14	.	21	21	14	15	23	.	.	48	24	24	.	34	14	9	72	9	.	55
13	.	.	.	1	.	.	7	.	.	7	.	.	16	.	.	24	.	16	.	17	.	.	24	9	.	20
14	2	2	4	.	3	1	5	16	12	6	20	8	5	23	23	15	12	4	30	10	28	4	18	1	27	42

15	1	.	1	1	.	1	2	.	.	2	1	.	.	.	1	.	.	.	.	7	.	.	8	.	.	.	16	16	.	.	18	.	9	24	.	.	.
16		1	.	.	1	.	2	.	.	1	2	.	.	.	1	.	.	.	.	.	7	.	.	8	.	.	.	.	16	.	.	.	.	24	9	.	.
17			1	.	.	1	1	.	.	1	1	.	.	.	1	.	.	.	.	.	.	.	.	.	.	.	8	.	.	.	9	.	.	8	.	.	.
18				1	.	1	.	.	.	2	.	.	.	.	1	.	.	.	.	.	.	.	.	.	.	.	.	8	.	.	.	.	9	.	.	.	.
19					1	.	.	.	.	.	2	.	.	.	1	.	.	.	.	.	.	.	.	.	.	.	.	.	8	.	.	.	.	.	9	.	.
20						1	.	.	.	1	.	.	.	.	1	.	.	.	.	.	.	.	.	.	.	.	.	.	.	.	.	.	.	.	.	.	.
21							1	.	.	1	1	.	.	.	1	.	.	.	.	.	.	.	.	.	.	.	.	.	.	.	.	.	.	8	.	.	.
22								1	.	.	.	1	.	.	.	.	.	.	4	.	.	5	.	.	10	5	.	.	.	6	.	12	.	.	.	5	9
23									1	.	.	.	1	.	.	2	1	.	4	4	.	.	4	.	5	10	5	8	.	15	5	.	4	6	.	17	.
24										1	.	.	.	.	1	.	.	.	.	.	.	.	.	.	.	.	.	.	.	.	.	.	.	.	.	.	.
25											1	.	.	.	1	.	.	.	.	.	.	.	.	.	.	.	.	.	.	.	.	.	.	.	.	.	.
26												1	.	.	.	.	.	.	.	.	.	.	.	.	5	.	.	.	.	.	.	6	.	.	.	.	.
27													1	.	.	1	1	.	.	.	.	.	4	.	.	5	.	4	.	5	5	.	4	.	.	12	.
28														1	.	.	.	2	.	3	3	4	2	3	.	.	6	3	3	.	4	5	1	9	1	.	15
29															1	.	.	.	.	.	.	.	.	.	.	.	.	.	.	.	.	.	.	.	.	.	.
30																1	1	.	.	.	.	.	.	.	.	.	.	4	.	5	.	.	4	.	.	6	.
31																	1	.	.	.	.	.	.	.	.	.	.	.	.	.	.	.	4	.	.	.	.
32																		1	.	.	1	.	.	2	.	.	3	.	2	.	2	.	.	3	1	.	5
33																			1	.	.	.	.	.	2	2	.	.	.	3	.	.	.	.	.	3	.
34																				1	.	.	1	.	.	.	2	2	.	.	2	.	1	3	.	.	.
35																					1	.	.	1	.	.	.	.	2	.	.	.	.	3	1	.	.
36																						1	.	.	.	.	.	.	.	.	.	2	.	.	.	.	3
37																							1	.	.	.	.	1	.	.	2	.	1	.	.	.	.
38																								1	.	.	.	.	1	.	.	.	.	.	1	.	.
39																									1	.	.	.	.	.	.	.	.	.	.	.	.
40																										1	.	.	.	1	.	.	.	.	.	2	.
41																											1	.	.	.	1	.	.	1	.	.	.
42																												1	.	.	.	.	1	.	.	.	.
43																													1	.	.	.	.	.	1	.	.
44																														1	.	.	.	.	.	1	.
45																															1	.	.	.	.	.	.
46																																1	.	.	.	.	.
47																																	1	.	.	.	.
48																																		1	.	.	.
49																																			1	.	.
50																																				1	.
51																																					1

Table 3. *(continued)*

ALGEBRA A12, REPRESENTATIONS OF CLASS 2

S.P.	22	48	48	74	126	74	100	100	126	126	152	152	100	178	230	256	126	178	126	126	152	152	152	152	204	178
O.S.	78	13	2 860	1 716	156	25 740	6 435	38 610	4 290	8 580	858	8 580	72 072	1 716	858	156	72 072	10 296	60 060	240 240	10 296	102 960	205 920	180 180	25 740	102 960
LEVEL	23	25	43	47	49	59	61	65	67	67	69	69	71	71	71	73	75	77	79	79	83	83	83	85	85	85
DIM'N	78	91	3 640	9 295	5 824	52 052	61 425	221 130	175 175	260 260	205 920	304 200	312 312	357 357	136 136	159 705	819 819	524 160	858 858	2 004 002	981 552	3 312 400	5 218 304	3 903 900	3 318 315	4 612 608
	0	2	0	1	3	0	0	1	0	1	0	2	0	2	4	4	0	0	0	1	0	0	1	0	1	2
	1	0	0	1	0	0	0	0	2	0	2	1	0	1	0	0	0	0	0	0	0	1	0	0	0	0
	0	0	1	0	0	0	0	1	0	1	0	0	0	0	0	0	0	0	0	0	0	1	0	0	0	1
	0	0	0	0	0	1	1	0	0	0	0	0	0	0	0	0	0	0	0	1	0	0	1	0	1	0
	0	0	0	0	0	0	0	0	0	0	0	0	1	0	0	0	1	1	0	0	0	0	0	0	0	0
	0	0	0	0	0	0	0	0	0	0	0	0	0	0	0	0	0	0	1	0	0	0	0	1	0	0
	0	0	0	0	0	0	0	0	0	0	0	0	0	0	0	0	0	0	0	0	1	0	0	0	0	0
	0	0	0	0	0	0	0	0	0	0	0	0	0	0	0	0	0	0	0	0	1	0	0	0	0	0
	0	0	0	0	0	0	0	0	0	0	0	0	0	0	0	0	0	0	1	0	0	0	0	0	0	0
	0	0	0	0	0	0	0	0	0	0	0	0	1	0	0	0	0	0	0	1	0	1	0	1	0	1
	0	0	0	0	0	1	0	1	1	0	0	1	0	0	1	0	1	0	0	0	0	0	1	0	0	0
WEIGHT	0	0	1	1	1	0	2	0	0	2	2	0	0	2	0	2	1	3	0	0	0	0	1	1	3	0
NUMBER	0	1	2	3	4	5	6	7	8	9	10	11	12	13	14	15	16	17	18	19	20	21	22	23	24	25
0	1	1	10	22	12	44	45	153	110	165	121	175	110	198	66	78	231	120	165	539	132	760	1 200	550	660	968
1		1	.	11	12	.	.	54	55	55	66	120	.	132	66	78	.	.	.	154	.	320	320	.	165	528
2			1	2	1	8	9	27	19	30	21	30	27	33	11	12	63	36	48	133	42	188	312	180	180	240
3				1	1	.	.	9	10	10	11	20	.	22	11	12	.	.	.	35	.	80	80	.	45	124
4					1	.	.	.	.	.	.	10	.	11	11	12	.	.	.	.	.	.	.	.	.	44
5						1	1	3	2	3	2	3	6	3	1	1	14	8	14	28	14	38	64	55	36	48
6							1	.	.	3	2	.	.	3	.	1	7	8	.	.	.	.	32	20	36	.
7								1	1	1	1	2	.	2	1	1	.	.	.	7	.	16	16	.	9	24
8									1	.	1	1	.	1	1	1	.	.	.	.	.	8	.	.	.	8
9										1	1	.	.	2	.	1	.	.	.	.	.	.	8	.	9	.
10											1	.	.	1	.	1	.	.	.	.	.	.	.	.	.	.
11												1	.	1	1	1	.	.	.	.	.	.	.	.	.	8
12													1	.	.	.	2	1	4	4	5	5	8	15	4	6
13														1	.	1	.	.	.	.	.	.	.	.	.	.
14															1	1	.	.	.	.	.	.	.	.	.	.

S.P.	178	178	204	204	230	152	204	256	178	282	230	256	360	204	282	282	204	178	178	178	334	386	230	204	204	204
O.S.	36 036	77 220	25 740	72 072	8 580	450 450	77 220	25 740	102 960	12 012	17 160	8 580	2 860	6 435	1 716	17 160	102 960	540 540	216 216	1 081 080	1 716	1 716	102 960	60 060	12 012	180 180
LEVEL	87	87	87	89	89	89	89	89	91	91	91	91	91	93	93	93	95	95	95	95	95	95	97	97	97	97
DIM'N	3 188 185	8 588 580	4 586 400	5 621 616	5 445 440	8 133 125	11 945 934	4 550 832	7 729 722	2 598 960	11 851 125	7 567 560	1 633 632	4 601 025	7 495 488	11 741 184	14 054 040	22 135 113	14 907 321	36 072 036	7 417 410	4 203 199	19 987 968	15 071 420	8 588 580	29 149 120
	0	0	1	0	0	1	2	3	0	0	1	2	5	0	1	3	0	0	1	1	3	5	0	0	1	1
	0	1	2	0	1	0	0	1	0	0	2	0	0	0	2	1	0	1	0	0	1	0	0	0	0	0
	0	1	0	0	1	0	1	0	0	0	0	1	0	0	0	0	0	0	0	0	0	0	0	2	0	0
	0	0	0	0	0	0	0	0	0	0	0	0	0	0	0	0	0	1	0	0	0	0	0	0	0	0
	0	0	0	0	0	1	0	0	0	0	0	0	0	0	0	0	0	0	0	1	0	0	0	0	0	1
	1	0	0	1	0	0	0	0	0	1	0	0	0	0	0	0	0	0	1	0	0	0	0	0	0	0
	0	0	0	0	0	0	0	0	1	0	0	0	0	0	0	0	1	0	0	0	0	0	1	0	2	0
	0	0	0	0	0	0	0	0	0	0	0	0	0	2	0	0	0	0	1	0	0	0	0	0	0	0
	0	0	0	0	0	1	0	0	1	0	0	0	0	0	0	0	0	1	0	0	0	0	0	1	0	0
	0	0	1	0	0	0	0	1	0	0	0	0	1	0	0	0	1	0	0	1	0	0	1	0	0	0
	2	1	0	1	0	0	1	0	0	0	1	0	0	0	0	1	1	0	0	0	0	1	0	0	0	2
WEIGHT	0	1	0	2	3	0	1	0	1	4	1	3	0	1	3	1	0	0	0	1	3	1	2	0	0	0
NUMBER	26	27	28	29	30	31	32	33	34	35	36	37	38	39	40	41	42	43	44	45	46	47	48	49	50	51
0	385	1 749	870	594	990	1 100	2 277	759	693	210	2 090	1 320	220	330	1 232	1 880	990	2 376	1 287	3 861	1 144	572	1 232	1 442	594	2 772
1	.	759	540	.	440	275	1 188	594	.	.	1 320	660	220	.	792	1 440	.	891	297	891	858	572	.	616	132	616
2	133	441	216	216	255	330	567	189	252	84	510	330	55	126	297	450	387	729	432	1 242	264	132	504	448	210	924
3	.	189	135	.	110	75	288	144	.	.	320	165	55	.	187	340	.	270	90	270	198	132	.	196	42	196
4	.	.	45	.	.	.	99	99	.	.	110	55	55	.	66	230	.	.	.	.	132	132	.	.	.	.
5	42	86	42	69	48	95	108	36	91	28	94	60	10	49	52	80	148	207	147	363	44	22	196	126	77	273
6	21	43	.	48	48	.	54	.	28	28	47	60	.	14	52	40	64	.	.	129	44	11	112	.	.	140
7	.	36	26	.	20	20	54	27	.	.	58	30	10	.	32	60	.	75	28	75	33	22	.	56	14	56
8	.	18	18	.	10	.	18	18	.	.	40	10	10	.	22	40	.	27	.	.	22	22	.	28	.	.
9	.	18	.	.	20	.	27	.	.	.	29	30	.	.	32	30	.	.	.	27	33	11	.	.	.	28
10	.	9	.	.	10	.	9	.	.	.	20	10	.	.	22	20	.	.	.	.	22	11	.	.	.	.
11	.	.	9	.	.	.	18	18	.	.	20	10	10	.	11	40	.	.	.	.	22	22	.	.	.	.
12	11	10	5	18	5	25	12	4	33	7	10	6	1	20	5	8	54	51	51	90	4	2	72	30	30	65
13	.	.	.	.	.	.	9	.	.	.	10	10	.	.	11	20	.	.	.	.	22	11	.	.	.	.
14	.	.	.	.	.	.	.	9	.	.	.	.	10	.	.	20	.	.	.	.	11	22	.	.	.	.

1	.	.	.	.	.	.	.	.	.	.	.	.	.	.	.	.	.	.	.	.	.	.	.	.	.	10	.	.	.	.	11	11	.	.	.	.	15
	1	1	.	.	.	.	4	5	4	.	6	5	.	12	5	.	6	.	9	7	5	6	.	5	5	4	24	.	.	30	4	1	38	.	.	35	16
		1	.	.	.	.	.	.	4	.	.	.	.	6	5	.	.	.	.	7	.	6	.	.	5	.	.	.	.	.	4	.	14	.	.	.	17
			1	.	2	.	.	3	.	.	2	.	.	3	.	5	.	.	12	1	.	.	.	9	.	.	18	9	18	15	.	.	24	5	13	10	18
				1	.	2	2	.	1	3	.	4	3	.	2	5	6	3	.	.	6	3	1	.	3	6	.	18	9	18	3	2	.	13	5	13	19
					1	.	.	.	.	.	.	.	.	.	.	.	.	.	4	.	.	.	.	5	.	.	5	.	6	.	.	.	6	.	7	.	20
						1	.	.	.	1	.	2	2	.	1	.	2	2	.	.	4	1	1	.	2	4	.	6	.	.	2	2	.	7	.	.	21
							1	.	1	.	.	2	.	.	2	.	3	.	.	.	3	3	.	.	3	3	.	.	.	6	3	1	.	.	.	7	22
								1	.	.	1	.	.	2	.	.	.	.	3	1	.	.	.	2	.	.	8	.	.	5	.	.	12	.	.	5	23
									1	.	.	.	.	.	2	.	.	.	.	.	.	3	.	.	3	.	.	.	.	.	3	.	.	.	.	.	24
										1	.	.	1	.	.	.	2	2	.	.	2	1	1	.	1	4	.	.	.	.	2	2	.	.	.	.	25
											1	.	.	1	.	.	.	.	.	1	.	.	.	.	.	.	4	.	.	.	.	.	4	.	.	5	26
												1	.	.	1	.	1	.	.	.	2	1	.	.	2	2	.	.	.	.	2	1	.	.	.	.	27
													1	.	.	.	.	1	.	.	2	.	1	.	1	2	.	.	.	.	1	2	.	.	.	.	28
														1	.	.	.	.	.	1	.	.	.	.	.	.	.	.	.	.	.	.	4	.	.	.	29
															1	.	.	.	.	.	.	1	.	.	2	.	.	.	.	.	2	.	.	.	.	.	30
																1	.	.	.	.	.	.	.	.	.	.	.	3	3	3	.	.	.	2	2	2	31
																	1	.	.	.	1	1	.	.	1	2	.	.	.	.	2	1	.	.	.	.	32
																		1	.	.	.	.	1	.	.	2	.	.	.	.	1	2	.	.	.	.	33
																			1	.	.	.	.	1	.	.	2	.	.	.	.	.	3	.	.	.	34
																				1	.	.	.	.	.	.	.	.	.	.	.	.	.	.	.	.	35
																					1	.	.	.	1	1	.	.	.	.	1	1	.	.	.	.	36
																						1	.	.	1	.	.	.	.	.	2	.	.	.	.	.	37
																							1	.	.	.	.	.	.	.	.	2	.	.	.	.	38
																								1	.	.	.	.	.	.	.	.	.	.	.	.	39
																									1	.	.	.	.	.	1	.	.	.	.	.	40
																										1	.	.	.	.	1	1	.	.	.	.	41
																											1	.	.	.	.	.	1	.	.	.	42
																												1	.	.	.	.	.	1	.	.	43
																													1	.	.	.	.	.	1	.	44
																														1	.	.	.	.	.	1	45
																															1	.	.	.	.	.	46
																																1	.	.	.	.	47
																																	1	.	.	.	48
																																		1	.	.	49
																																			1	.	50
																																				1	51

Table 3. *(continued)*

ALGEBRA A12, REPRESENTATIONS OF CLASS 3

NUMBER	0	1	2	3	4	5	6	7	8	9	10	11	12	13	14	15	16	17	18	19	20	21	22	23	24	25
S.P.	30	56	108	56	82	108	134	212	82	108	108	108	134	134	134	160	160	134	186	186	186	160	212	238	134	264
O.S.	286	156	13	6 435	8 580	858	1 716	156	36 036	10 296	102 960	60 060	25 740	38 610	72 072	8 580	38 610	25 740	12 012	8 580	8 580	1 287	1 716	8 580	360 360	1 716
LEVEL	31	35	37	49	55	57	59	61	63	65	71	73	73	75	77	77	77	79	79	79	79	81	81	81	83	83
DIM'N	286	728	455	9 009	38 025	30 030	52 052	23 205	91 091	108 108	576 576	390 390	681 408	945 945	1 041 040	1 115 400	1 314 144	736 164	672 672	1 301 300	1 548 547	429 429	1 531 530	1 287 104	3 578 575	1 513 512
	0	1	3	0	1	0	2	4	0	0	1	0	1	0	0	0	2	0	0	1	2	0	1	3	1	3
	0	1	0	0	0	2	1	0	0	0	0	0	0	1	0	1	0	0	0	2	0	0	2	1	0	1
	1	0	0	0	1	0	0	0	0	0	0	0	0	1	0	1	1	0	0	0	1	0	0	0	0	0
	0	0	0	1	0	0	0	0	0	0	1	0	1	0	0	0	0	0	0	0	0	0	0	0	0	0
	0	0	0	0	0	0	0	0	1	1	0	0	0	0	0	0	0	0	0	0	0	0	0	0	1	0
	0	0	0	0	0	0	0	0	0	0	0	1	0	0	1	0	0	0	1	0	0	0	0	0	0	0
	0	0	0	0	0	0	0	0	0	0	0	0	0	0	0	0	0	1	0	0	0	0	0	0	0	0
	0	0	0	0	0	0	0	0	0	0	0	0	0	0	0	0	0	0	0	0	0	2	0	0	0	0
	0	0	0	0	0	0	0	0	0	0	0	0	0	0	0	0	0	1	0	0	0	0	0	0	0	0
	0	0	0	0	0	0	0	0	0	0	0	1	0	0	0	0	0	0	0	0	0	0	0	0	1	0
	0	0	0	0	0	0	0	0	1	0	1	0	0	1	1	0	1	0	0	1	0	0	0	1	0	0
WEIGHT	0	0	0	1	1	1	1	1	0	2	0	0	2	0	1	2	0	0	3	0	2	0	2	0	0	2
0	1	2	1	9	30	21	33	12	35	36	168	75	180	234	160	255	297	90	84	265	330	42	297	230	490	264
1		1	1	.	10	11	22	12	.	.	44	.	45	99	.	110	153	.	.	165	165	.	187	175	110	198
2			1	.	.	.	11	12	.	.	.	.	.	.	.	.	54	.	.	55	55	.	66	120	.	132
3				1	3	2	3	1	7	8	32	20	36	43	48	48	54	28	28	47	60	14	52	40	129	44
4					1	1	2	1	.	.	8	.	9	18	.	20	27	.	.	29	30	.	32	30	27	33
5						1	1	1	.	.	.	.	.	9	.	10	9	.	.	20	10	.	22	20	.	22
6							1	1	.	.	.	.	.	.	.	.	9	.	.	10	10	.	11	20	.	22
7								1	.	.	.	.	.	.	.	.	.	.	.	.	.	.	.	10	.	11
8									1	1	4	5	4	5	12	5	6	9	7	5	6	5	5	4	30	4
9										1	.	.	4	.	6	5	.	.	7	.	6	.	5	.	.	4
10											1	.	1	2	.	2	3	.	.	3	3	.	3	3	6	3
11												1	.	.	2	.	.	3	1	.	.	2	.	.	5	.
12													1	.	.	2	.	.	.	.	3	.	3	.	.	3
13														1	.	1	1	.	.	2	1	.	2	2	.	2
14															1	.	.	.	1	.	.	.	.	.	.	.
15																1	.	.	.	.	1	.	2	.	.	2

NUMBER	26	27	28	29	30	31	32	33	34	35	36	37	38	39	40	41	42	43	44	45	46	47	48	49	50	51
S.P.	342	160	368	186	160	186	212	160	212	290	186	160	186	212	186	238	238	212	238	186	212	212	264	264	264	264
O.S.	858	102 960	156	25 740	360 360	25 740	51 480	360 360	51 480	10 296	34 320	360 360	240 240	38 610	308 880	4 290	38 610	25 740	38 610	72 072	205 920	205 920	2 860	2 860	25 740	2 860
LEVEL	83	85	85	87	87	89	89	89	89	91	91	91	91	93	93	95	95	95	95	95	95	95	97	97	97	97
DIM'N	459 459	3 513 510	539 448	2 927 925	9 417 408	4 416 984	5 205 200	9 564 555	6 033 300	2 450 448	6 459 180	10 038 600	12 332 320	8 783 775	24 984 960	6 012 006	12 648 636	16 816 800	15 927 912	11 594 583	26 091 520	32 177 600	5 920 915	8 179 600	13 663 650	10 696 400
	5	0	5	0	1	0	0	0	1	0	0	1	2	0	0	0	0	0	0	1	1	2	0	0	0	0
	0	0	0	0	0	0	0	1	0	0	0	0	0	0	1	0	0	0	1	0	1	0	0	0	0	0
	0	0	0	0	0	0	0	0	0	0	2	0	0	0	0	0	0	2	0	0	1	0	0	0	0	2
	0	0	0	0	0	0	0	1	0	0	0	0	1	0	1	0	0	0	1	0	0	1	0	0	0	0
	0	0	0	0	1	0	0	0	1	0	0	0	0	0	0	0	0	0	0	0	0	0	0	0	0	0
	0	0	0	0	0	0	0	0	0	0	0	1	0	0	0	0	0	0	0	0	0	0	0	0	0	0
	0	1	0	1	0	0	1	0	0	1	0	0	0	0	0	0	0	0	0	1	0	0	0	0	0	0
	0	0	0	0	0	1	0	0	0	0	0	0	0	1	0	0	1	0	0	1	0	0	0	0	1	0
	0	0	0	0	0	1	0	0	0	0	0	1	0	0	0	2	0	0	0	0	0	0	1	2	0	0
	0	1	0	0	0	0	0	1	0	0	1	0	1	1	0	0	1	0	0	0	1	0	2	0	0	0
	1	0	0	2	1	0	1	0	0	0	0	0	0	1	1	1	0	1	0	0	0	1	0	0	2	0
WEIGHT	0	1	2	0	1	1	2	0	3	4	0	0	0	0	1	0	2	1	3	0	0	1	0	2	1	3
0	66	315	78	225	1 092	288	350	1 050	600	126	636	810	1 239	450	2 400	252	567	1 470	1 350	684	2 240	2 880	210	300	525	835
1	66	.	78	.	231	.	.	385	120	.	276	165	539	.	880	.	.	660	495	132	1 200	1 200	.	.	.	385
2	66	.	78	.	.	.	.	.	.	.	.	.	154	.	.	.	.	.	.	.	320	320	.	.	.	.
3	11	112	12	85	308	112	140	273	180	56	164	256	322	190	648	112	252	388	378	238	576	768	98	140	245	226
4	11	.	12	.	63	.	.	98	36	.	72	48	133	.	232	.	.	171	135	42	304	312	.	.	.	100
5	11	.	12	.	.	.	.	35	.	.	36	.	35	.	80	.	.	90	45	.	160	80	.	.	.	55
6	11	.	12	.	.	.	.	.	.	.	.	.	35	.	.	.	.	.	.	.	80	80	.	.	.	.
7	11	.	12	.	.	.	.	.	.	.	.	.	.	.	.	.	.	.	.	.	.	.	.	.	.	.
8	1	38	1	30	70	44	50	60	40	21	35	80	70	80	138	51	108	80	78	86	120	160	47	65	110	45
9	.	14	1	15	35	14	35	.	40	21	.	.	.	35	69	21	63	40	78	.	.	80	21	35	70	45
10	1	.	1	.	14	.	.	21	8	.	15	14	28	.	48	.	.	34	27	14	62	64	.	.	.	19
11	.	12	.	9	10	18	15	9	5	6	5	24	10	33	18	24	45	10	9	33	16	20	23	31	46	5
12	.	.	1	.	7	.	.	.	8	.	.	.	.	.	24	.	.	17	27	.	.	32	.	.	.	19
13	1	.	1	.	.	.	.	7	.	.	8	.	7	.	16	.	.	18	9	.	32	16	.	.	.	10
14	.	4	.	5	5	5	10	.	5	6	.	.	.	15	9	10	24	5	9	.	.	10	11	15	30	5
15	.	.	1	.	.	.	.	.	.	.	.	.	.	.	8	.	.	9	9	.	.	8	.	.	.	10

1	.	.	1	1	.	1	2	.	2	1	.	1	.	.	.	.	.	.	.	.	.	7	.	.	.	.	.	.	.	16	16	.	.	.	.	16
	1	.	.	.	1	.	.	.	.	.	3	.	2	.	8	3	.	.	1	.	6	.	13	.	12	18	.	.	14	.	.	11	16	17	.	17
		1	.	.	.	.	.	.	.	.	.	.	.	.	.	5	.	5	6	.	.	.	.	.	.	9	.	9	.	.	.	.	5	10	5	18
			1	.	.	1	1	.	1	1	.	1	.	.	.	.	.	.	.	.	.	.	.	.	.	.	.	.	.	8	.	.	.	.	.	19
				1	.	1	.	.	2	.	.	1	.	.	.	.	.	.	.	.	.	.	.	.	.	.	.	.	.	.	8	.	.	.	.	20
					1	.	.	.	.	.	.	.	.	.	4	.	.	.	.	.	.	.	5	.	6	6	.	.	7	.	.	5	10	5	.	21
						1	.	.	1	.	.	1	.	.	.	.	.	.	.	.	.	.	.	.	.	.	.	.	.	.	.	.	.	.	.	22
							1	.	1	1	.	1	.	.	.	.	.	.	.	.	.	.	.	.	.	.	.	.	.	.	.	.	.	.	.	23
								1	.	.	.	.	.	2	.	.	3	1	.	2	4	4	.	6	.	.	4	3	5	8	8	.	.	.	2	24
									1	.	.	1	.	.	.	.	.	.	.	.	.	.	.	.	.	.	.	.	.	.	.	.	.	.	.	25
										1	.	1	.	.	.	.	.	.	.	.	.	.	.	.	.	.	.	.	.	.	.	.	.	.	.	26
											1	.	1	.	2	2	.	.	1	.	.	.	6	.	5	9	.	.	.	.	.	6	7	11	.	27
												1	.	.	.	.	.	.	.	.	.	.	.	.	.	.	.	.	.	.	.	.	.	.	.	28
													1	.	.	1	.	.	1	.	.	.	3	.	2	3	.	.	.	.	.	3	2	8	.	29
														1	.	.	.	1	.	.	.	.	.	3	.	.	2	3	.	.	4	.	.	.	2	30
															1	.	.	.	.	.	.	.	2	.	3	3	.	.	.	.	.	3	4	3	.	31
																1	.	.	1	.	.	.	.	.	.	3	.	.	.	.	.	.	2	4	.	32
																	1	.	.	1	.	1	.	2	.	.	2	1	.	4	2	.	.	.	1	33
																		1	.	.	.	.	.	.	.	.	.	3	.	.	.	.	.	.	2	34
																			1	.	.	.	.	.	.	.	.	.	.	.	.	.	.	.	.	35
																				1	.	.	.	.	.	.	2	.	.	2	.	.	.	.	1	36
																					1	.	.	.	.	.	.	.	2	.	.	.	.	.	.	37
																						1	.	.	.	.	.	.	.	2	2	.	.	.	.	38
																							1	.	1	1	.	.	.	.	.	2	1	2	.	39
																								1	.	.	1	1	.	.	1	.	.	.	1	40
																									1	.	.	.	.	.	.	1	1	.	.	41
																										1	.	.	.	.	.	.	1	1	.	42
																											1	.	.	.	.	.	.	.	1	43
																												1	.	.	.	.	.	.	1	44
																													1	.	.	.	.	.	.	45
																														1	.	.	.	.	.	46
																															1	.	.	.	.	47
																																1	.	.	.	48
																																	1	.	.	49
																																		1	.	50
																																			1	51

Table 3. *(continued)*

ALGEBRA A12, REPRESENTATIONS OF CLASS 4

NUMBER	0	1	2	3	4	5	6	7	8	9	10	11	12	13	14	15	16	17	18	19	20	21	22	23	24	25
S.P.	36	62	88	114	192	62	88	88	114	114	140	166	218	114	322	114	140	140	140	192	140	166	166	166	166	192
O.S.	715	858	78	156	13	10 296	25 740	36 036	8 580	12 012	8 580	1 716	1 716	34 320	156	180 180	6 435	51 480	51 480	10 296	154 440	38 610	12 070	38 610	102 960	2 860
LEVEL	37	43	45	47	49	53	61	65	65	67	67	69	71	73	73	75	77	77	77	79	81	83	83	83	83	85
DIM'N	715	3 003	2 366	4 095	1 820	16 016	100 386	117 117	163 800	140 140	227 136	224 224	221 221	356 928	78 624	1 051 050	429 429	975 975	1 249 248	640 640	2 774 772	2 258 685	1 863 225	3 285 360	3 569 280	1 431 430
	0	1	0	2	4	0	1	0	0	0	2	1	3	0	5	1	0	0	1	0	0	0	0	0	2	0
	0	0	2	1	0	0	0	0	1	0	0	2	1	0	0	0	0	0	0	0	1	0	0	1	0	0
	0	1	0	0	0	0	0	0	1	0	1	0	0	0	0	0	0	0	0	0	0	0	2	0	0	0
	1	0	0	0	0	0	1	0	0	0	0	0	0	0	0	0	0	0	0	0	1	0	0	1	1	0
	0	0	0	0	0	1	0	0	0	0	0	0	0	0	0	1	0	0	1	0	0	0	0	0	0	0
	0	0	0	0	0	0	0	1	0	1	0	0	0	0	0	0	0	0	0	0	0	0	0	0	0	0
	0	0	0	0	0	0	0	0	0	0	0	0	0	1	0	0	0	1	0	1	0	0	0	0	0	0
	0	0	0	0	0	0	0	0	0	0	0	0	0	0	0	0	1	0	0	0	0	1	0	0	0	0
	0	0	0	0	0	0	0	0	0	0	0	0	0	0	0	0	1	0	0	0	0	0	0	0	0	2
	0	0	0	0	0	0	0	0	0	0	0	0	0	1	0	0	0	0	0	0	0	1	0	0	0	0
	0	0	0	0	0	0	0	1	0	0	0	0	0	0	0	1	0	1	0	0	1	0	1	0	1	0
WEIGHT	0	0	0	0	0	1	1	0	1	2	1	1	1	0	1	0	0	1	2	3	0	1	0	2	0	1
0	1	3	2	3	1	8	36	27	48	28	60	52	44	48	12	168	42	105	180	56	348	162	207	378	408	84
1		1	1	2	1	.	9	.	20	.	30	32	33	.	12	35	.	.	36	.	124	.	90	135	168	.
2			1	1	1	.	.	.	10	.	10	22	22	.	12	.	.	.	.	.	44	.	45	45	44	.
3				1	1	.	.	.	.	.	10	11	22	.	12	.	.	.	.	.	.	.	.	.	44	.
4					1	.	.	.	.	.	.	.	11	.	12	.	.	.	.	.	.	.	.	.	.	.
5						1	4	6	5	7	6	5	4	14	1	35	14	35	40	21	69	63	40	78	80	35
6							1	.	2	.	3	3	3	.	1	7	.	.	8	.	24	.	17	27	32	.
7								1	.	1	.	.	.	4	.	5	5	10	5	6	9	24	5	9	10	15
8									1	.	1	2	2	.	1	.	.	.	.	.	8	.	9	9	8	.
9										1	.	.	.	.	.	.	.	5	5	6	.	9	.	9	.	5
10											1	1	2	.	1	.	.	.	.	.	.	.	.	.	8	.
11												1	1	.	1	.	.	.	.	.	.	.	.	.	.	.
12													1	.	1	.	.	.	.	.	.	.	.	.	.	.
13														1	.	.	2	2	.	1	.	9	.	.	.	7
14															1	.	.	.	.	.	.	.	.	.	.	.
15																1	.	.	1	.	3	.	2	3	4	.

NUMBER	26	27	28	29	30	31	32	33	34	35	36	37	38	39	40	41	42	43	44	45	46	47	48	49	50	51
S.P.	192	192	140	192	218	192	218	296	244	166	218	244	270	244	270	166	218	270	270	270	166	192	296	348	296	322
O.S.	12 870	2 860	360 360	25 740	25 740	77 220	8 580	6 435	4 290	432 432	17 160	38 610	286	8 580	858	180 180	72 072	8 580	8 580	8 580	720 720	10 296	1 716	8 580	1 716	8 580
LEVEL	85	85	85	85	87	87	89	89	89	89	89	89	91	91	91	91	91	91	91	93	93	93	93	93	95	95
DIM'N	1 951 950	2 202 200	4 601 025	4 223 310	3 513 510	7 495 488	3 643 640	1 701 700	3 003 000	12 300 288	8 848 840	5 791 968	1 184 183	5 353 920	3 539 536	8 783 775	7 963 956	6 194 188	6 831 825	6 246 240	18 218 200	5 153 148	7 297 290	4 524 975	5 829 824	6 636 630
	0	0	1	2	0	1	0	0	0	1	1	3	0	0	0	1	1	2	3	0	0	1	2	4	0	0
	0	0	0	0	0	1	0	0	3	0	1	0	0	0	3	0	0	2	0	0	1	0	2	1	0	0
	0	2	0	0	0	1	0	0	0	0	1	1	0	0	0	0	0	0	1	0	0	0	0	0	0	0
	0	0	0	1	0	0	0	0	0	0	0	0	0	0	0	0	0	0	0	0	0	0	0	0	0	0
	0	0	0	0	0	0	0	0	0	0	0	0	0	0	0	0	0	0	0	0	1	0	0	0	0	0
	0	0	1	0	0	0	0	0	0	1	0	0	0	0	0	0	1	0	0	0	0	0	0	0	0	0
	0	0	0	0	0	0	0	0	0	0	0	0	0	0	0	1	0	0	0	0	0	0	0	0	0	0
	1	0	0	0	1	0	0	1	0	0	0	0	0	0	0	0	0	0	0	0	0	2	0	0	0	0
	0	0	0	0	0	0	1	0	0	0	0	0	0	1	0	1	0	0	0	1	0	0	0	0	0	1
	0	0	1	0	0	0	1	0	0	0	0	0	3	1	0	0	0	0	0	0	1	0	0	0	2	0
	2	0	0	0	1	1	1	0	1	1	0	1	0	0	0	0	0	1	0	2	0	0	0	1	1	1
WEIGHT	0	2	0	2	2	0	0	4	0	1	2	0	0	2	2	0	3	0	2	1	0	0	2	0	1	3
0	120	226	405	450	189	720	168	70	246	912	800	486	42	216	276	522	504	465	550	216	1 192	252	528	285	168	189
1	.	100	75	180	.	378	.	.	155	160	420	297	.	.	176	90	84	320	330	.	380	42	363	230	.	.
2	.	55	.	45	.	198	.	.	110	.	220	153	.	.	121	.	.	220	165	.	110	.	253	175	.	.
3	.	.	.	45	.	99	.	.	55	.	110	153	.	.	66	.	.	165	165	.	.	.	187	175	.	.
4	.	.	.	.	.	.	.	.	.	.	.	54	.	.	.	.	.	55	55	.	.	.	66	120	.	.
5	50	45	115	90	84	136	77	35	45	282	152	90	21	105	50	175	168	84	100	111	330	91	93	50	91	105
6	.	19	20	36	.	70	.	.	28	48	78	54	.	.	31	28	28	57	60	.	102	14	63	40	.	.
7	20	5	30	10	34	16	36	15	5	72	16	10	11	50	5	60	42	9	10	56	80	35	9	5	50	55
8	.	10	.	9	.	36	.	.	19	.	40	27	.	.	21	.	.	39	30	.	27	.	43	30	.	.
9	10	5	.	10	24	.	16	15	.	36	16	.	5	30	5	.	42	.	10	36	.	.	9	.	30	45
10	.	.	.	9	.	18	.	.	10	.	20	27	.	.	11	.	.	29	30	.	.	.	32	30	.	.
11	.	.	.	.	.	9	.	.	10	.	10	9	.	.	11	.	.	20	10	.	.	.	22	20	.	.
12	.	.	.	.	.	.	.	.	.	.	.	9	.	.	.	.	.	10	10	.	.	.	11	20	.	.
13	7	.	6	.	12	.	17	5	.	12	.	.	6	24	.	21	6	.	.	27	14	15	.	.	28	27
14	.	.	.	.	.	.	.	.	.	.	.	.	.	.	.	.	.	.	.	.	.	.	.	10	.	.
15	.	2	5	4	.	8	.	.	3	12	8	6	.	.	3	9	7	6	6	.	24	5	6	4	.	.

1	.	.	.	.	3	.	.	.	4	2	.	.	.	3	.	8	1	.	.	.	.	3	12	.	7	.	.	.	12	.	8	.	.	16	12	16
	1	.	1	.	3	.	.	.	2	4	.	.	.	8	.	8	5	.	6	.	.	3	13	.	.	6	.	.	18	.	.	.	.	17	21	17
		1	.	.	.	.	3	.	.	.	2	.	4	.	.	.	.	.	6	8	.	.	.	3	.	7	.	6	.	.	.	6	.	.	.	18
			1	.	.	.	.	.	.	.	.	.	.	4	.	.	5	.	.	.	.	.	5	.	.	6	.	.	6	.	.	.	.	5	15	19
				1	.	1	1	1	.	.	1	.	1	.	4	.	.	2	.	4	3	.	.	2	.	.	4	3	.	6	.	4	3	.	.	20
					1	.	.	.	1	1	.	.	.	2	.	4	1	.	.	.	.	2	6	.	.	.	.	.	8	.	.	.	.	10	9	21
						1	.	.	.	.	1	.	.	.	2	.	.	1	.	2	1	.	.	1	.	.	3	1	.	.	.	3	2	.	.	22
							1	.	.	.	1	.	1	.	.	.	.	.	.	4	.	.	.	2	.	.	.	3	.	.	.	4	.	.	.	23
								1	.	.	.	.	1	.	2	.	.	1	.	2	3	.	.	1	.	.	3	3	.	.	.	3	3	.	.	24
									1	.	.	.	.	.	.	2	.	.	.	.	.	1	3	.	.	.	.	.	3	.	.	.	.	6	3	25
										1	.	.	.	1	.	2	1	.	.	.	.	1	2	.	.	.	.	.	6	.	.	.	.	6	6	26
											1	.	.	.	.	.	.	.	.	2	.	.	.	1	.	.	.	1	.	.	.	3	.	.	.	27
												1	.	.	.	.	.	.	2	.	.	.	.	.	3	1	.	.	.	4	2	.	.	.	.	28
													1	.	.	.	.	.	.	2	.	.	.	1	.	.	.	3	.	.	.	3	.	.	.	29
														1	.	.	1	.	.	.	.	.	2	.	.	.	.	.	3	.	.	.	.	3	6	30
															1	.	.	1	.	1	1	.	.	1	.	.	2	1	.	.	.	2	2	.	.	31
																1	.	.	.	.	.	1	1	.	.	.	.	.	2	.	.	.	.	4	2	32
																	1	.	.	.	.	.	.	.	.	.	.	.	.	.	.	.	.	.	3	33
																		1	.	.	.	.	.	1	.	.	1	.	.	.	.	1	1	.	.	34
																			1	.	.	.	.	.	.	1	.	.	.	.	.	.	.	.	.	35
																				1	.	.	.	1	.	.	.	1	.	.	.	2	.	.	.	36
																					1	.	.	.	.	.	1	1	.	.	.	1	2	.	.	37
																						1	.	.	.	.	.	.	.	.	.	.	.	2	.	38
																							1	.	.	.	.	.	1	.	.	.	.	2	2	39
																								1	.	.	.	.	.	.	.	1	.	.	.	40
																									1	.	.	.	.	.	1	.	.	.	.	41
																										1	.	.	.	.	.	.	.	.	.	42
																											1	.	.	.	.	1	1	.	.	43
																												1	.	.	.	1	.	.	.	44
																													1	.	.	.	.	1	1	45
																														1	.	.	.	.	.	46
																															1	.	.	.	.	47
																																1	.	.	.	48
																																	1	.	.	49
																																		1	.	50
																																			1	51

Table 3. *(continued)*

ALGEBRA A12, REPRESENTATIONS OF CLASS 5

NUMBER	S.P.	O.S.	LEVEL	DIM'N	WEIGHT	0	1	2	3	4	5	6	7	8	9	10	11	12	13	14	15
0	40	1 287	41	1 287	0 0 0 0 1 0 0 0 0 0 0 0	1															
1	66	2 860	49	8 008	1 0 0 1 0 0 0 0 0 0 0 0	4	1														
2	92	858	53	13 013	0 1 1 0 0 0 0 0 0 0 0 0	5	2	1													
3	66	12 012	55	21 021	0 0 0 0 0 1 0 0 0 0 0 1	7	.	.	1												
4	118	858	55	18 018	2 0 1 0 0 0 0 0 0 0 0 0	6	3	1	.	1											
5	144	156	57	17 745	1 2 0 0 0 0 0 0 0 0 0 0	5	3	2	.	1	1										
6	196	156	59	17 472	3 1 0 0 0 0 0 0 0 0 0 0	4	3	2	.	2	1	1									
7	300	13	61	6 188	5 0 0 0 0 0 0 0 0 0 0 0	1	1	1	.	1	1	1	1								
8	92	25 740	65	111 540	0 0 0 0 0 0 1 0 0 0 1 0	20	.	.	5	.	.	.	.	1							
9	92	51 480	65	185 900	1 0 0 0 1 0 0 0 0 0 0 1	40	8	.	5	.	.	.	.	.	1						
10	118	10 296	67	135 135	0 0 0 0 0 0 1 0 0 0 0 2	21	.	.	6	.	.	.	.	1	.	1					
11	118	12 870	71	234 234	0 0 0 0 0 0 0 1 0 1 0 0	28	.	.	9	.	.	.	.	3	.	.	1				
12	118	38 610	71	486 486	0 1 0 1 0 0 0 0 0 0 0 1	78	27	9	9	.	.	.	.	.	3	.	.	1			
13	144	715	73	143 143	0 0 0 0 0 0 0 0 2 0 0 0	14	.	.	5	.	.	.	.	2	.	.	1	.	1		
14	144	2 860	73	325 325	0 0 2 0 0 0 0 0 0 0 0 1	45	19	10	5	.	.	.	.	.	2	.	.	1	.	1	
15	144	25 740	73	624 624	2 0 0 1 0 0 0 0 0 0 0 1	90	36	9	10	9	.	.	.	.	4	.	.	1	.	.	1
16	144	25 740	75	669 240	0 0 0 0 0 0 0 1 0 0 1 1	64	.	.	24	.	.	.	.	8	.	4	2	.	.	.	.
17	196	6 435	77	450 450	0 0 0 0 0 0 0 1 0 0 0 3	35	.	.	15	.	.	.	.	5	.	5	1	.	.	.	.
18	118	216 216	77	1 389 960	1 0 0 0 0 1 0 0 0 0 1 0	156	27	.	36	.	.	.	.	6	6	.	.	.	.	.	.
19	170	17 160	77	1 304 576	1 1 1 0 0 0 0 0 0 0 0 1	152	78	40	16	20	10	.	.	.	8	.	.	4	.	2	2
20	170	8 580	79	975 975	0 0 0 0 0 0 0 0 1 1 0 1	70	.	.	30	.	.	.	.	13	.	5	6	.	3	.	.
21	222	858	79	520 520	0 3 0 0 0 0 0 0 0 0 0 1	50	31	21	5	11	11	.	.	.	3	.	.	2	.	1	1
22	144	72 072	79	1 665 664	1 0 0 0 0 1 0 0 0 0 0 2	168	28	.	42	.	.	.	.	6	7	6	.	.	.	.	.
23	222	8 580	79	1 005 550	3 0 1 0 0 0 0 0 0 0 0 1	100	60	30	10	30	10	10	.	.	6	.	.	3	.	1	3
24	196	4 290	81	910 910	0 0 0 0 0 0 0 0 1 0 2 0	56	.	.	26	.	.	.	.	12	.	6	5	.	2	.	.
25	248	1 716	81	1 072 071	2 2 0 0 0 0 0 0 0 0 0 1	93	63	43	9	32	22	11	.	.	6	.	.	4	.	3	3
26	222	858	83	819 819	0 0 0 0 0 0 0 0 0 2 1 0	42	.	.	21	.	.	.	.	11	.	5	6	.	3	.	.
27	222	8 580	83	1 673 100	0 0 0 0 0 0 0 0 1 0 1 2	90	.	.	45	.	.	.	.	21	.	15	9	.	3	.	.
28	326	1 716	83	780 780	4 1 0 0 0 0 0 0 0 0 0 1	55	44	33	5	33	22	22	11	.	4	.	.	3	.	2	3
29	248	858	85	1 249 248	0 0 0 0 0 0 0 0 0 2 0 2	56	.	.	30	.	.	.	.	16	.	10	9	.	6	.	.
30	300	2 860	85	850 850	0 0 0 0 0 0 0 0 1 0 0 4	35	.	.	20	.	.	.	.	10	.	10	4	.	1	.	.
31	144	360 360	85	5 395 390	0 1 0 0 1 0 0 0 0 0 1 0	434	133	35	94	.	.	.	.	14	28	.	.	7	.	.	.
32	144	240 240	85	4 294 290	1 0 0 0 0 0 1 0 0 1 0 0	304	48	.	94	.	.	.	.	28	14	.	7	.	.	.	.
33	456	156	85	235 144	6 0 0 0 0 0 0 0 0 0 0 1	12	12	12	1	12	12	12	12	.	1	.	.	1	.	1	1
34	274	1 716	87	1 821 820	0 0 0 0 0 0 0 0 0 1 2 1	70	.	.	40	.	.	.	.	23	.	15	13	.	7	.	.
35	170	102 960	87	6 424 704	0 1 0 0 1 0 0 0 0 0 0 2	472	144	36	108	.	.	.	.	14	32	14	.	8	.	.	.
36	170	180 180	87	6 692 400	2 0 0 0 1 0 0 0 0 0 1 0	490	168	35	105	35	.	.	.	15	35	.	.	7	.	.	7
37	352	78	89	496 860	0 0 0 0 0 0 0 0 0 0 4 0	14	.	.	9	.	.	.	.	6	.	4	4	.	3	.	.
38	326	1 716	89	2 022 592	0 0 0 0 0 0 0 0 0 1 1 3	64	.	.	40	.	.	.	.	24	.	20	14	.	8	.	.
39	170	102 960	89	7 287 280	0 0 1 1 0 0 0 0 0 0 1 0	480	188	80	100	.	.	.	.	14	38	.	.	16	.	8	.
40	170	51 480	89	5 205 200	1 0 0 0 0 0 0 1 1 0 0 0	280	42	.	100	.	.	.	.	38	14	.	16	.	8	.	.
41	170	360 360	89	11 771 760	1 0 0 0 0 0 1 0 0 0 1 1	700	105	.	240	.	.	.	.	70	35	35	14	.	.	.	.
42	196	51 480	89	7 963 956	2 0 0 0 1 0 0 0 0 0 0 2	540	180	36	120	36	.	.	.	15	40	15	.	8	.	.	8
43	378	156	91	1 126 216	0 0 0 0 0 0 0 0 0 0 3 2	28	.	.	19	.	.	.	.	13	.	10	9	.	6	.	.
44	430	858	91	918 918	0 0 0 0 0 0 0 0 0 1 0 5	21	.	.	15	.	.	.	.	10	.	10	6	.	3	.	.
45	196	25 740	91	8 648 640	0 0 1 1 0 0 0 0 0 0 0 2	524	207	90	114	.	.	.	.	14	43	14	.	18	.	9	.
46	222	72 072	91	7 742 735	1 0 0 0 0 0 1 0 0 0 0 3	392	56	.	147	.	.	.	.	42	21	42	7	.	.	.	.
47	456	156	93	1 034 290	0 0 0 0 0 0 0 0 0 0 2 4	20	.	.	15	.	.	.	.	11	.	10	8	.	6	.	.
48	196	308 880	93	22 902 800	1 1 0 1 0 0 0 0 0 0 1 0	1 300	588	248	265	124	44	.	.	35	117	.	.	48	.	16	24
49	586	156	95	529 074	0 0 0 0 0 0 0 0 0 0 1 6	7	.	.	6	.	.	.	.	5	.	5	4	.	3	.	.
50	170	900 900	95	24 399 375	0 1 0 0 0 1 0 0 0 1 0 0	1 175	330	75	350	.	.	.	.	95	95	.	20	20	.	.	.
51	196	308 880	95	27 483 456	1 0 0 0 0 0 0 1 0 1 0 1	1 152	162	.	468	.	.	.	.	186	63	69	72	.	24	.	.

16	1	1	.	.	2	.	.	.	3	.	3	6	.	5	4	.	.	.	9	.	.	3	11	.	.	7	.	7	6	.	7	7	.	4	.	24
17		1	.	.	.	.	.	.	.	.	.	3	.	2	4	.	.	.	3	.	.	1	8	.	.	.	.	4	6	.	7	6	.	4	.	.
18			1	.	.	.	1	.	.	.	.	.	.	.	.	4	4	.	.	4	5	.	.	5	5	10	5	.	.	5	6	.	15	.	25	24
19				1	.	1	.	1	.	2	.	.	2	.	.	.	.	1	.	.	.	.	.	.	.	.	.	.	.	.	.	.	8	.	.	.
20					1	.	.	.	1	.	2	2	.	3	1	.	.	.	5	.	.	2	6	.	.	.	.	5	3	.	.	5	.	3	.	8
21						1	.	.	.	1	.	.	1	.	.	.	.	1	.	.	.	.	.	.	.	.	.	.	.	.	.	.	.	.	.	.
22							1	.	.	.	.	.	.	.	.	.	.	.	.	4	.	.	.	.	.	5	5	.	.	5	6	.	.	.	.	9
23								1	.	1	.	.	2	.	.	.	.	1	.	.	.	.	.	.	.	.	.	.	.	.	.	.	.	.	.	.
24									1	.	1	1	.	1	1	.	.	.	4	.	.	2	4	.	.	.	.	4	3	.	.	4	.	3	.	.
25										1	.	.	1	.	.	.	.	1	.	.	.	.	.	.	.	.	.	.	.	.	.	.	.	.	.	.
26											1	.	.	1	.	.	.	.	2	.	.	1	2	.	.	.	.	3	1	.	.	3	.	2	.	.
27												1	.	1	1	.	.	.	2	.	.	1	4	.	.	.	.	3	3	.	.	4	.	3	.	.
28													1	.	.	.	.	1	.	.	.	.	.	.	.	.	.	.	.	.	.	.	.	.	.	.
29														1	.	.	.	.	1	.	.	1	2	.	.	.	.	2	1	.	.	3	.	2	.	.
30															1	.	.	.	.	.	.	.	2	.	.	.	.	1	3	.	.	3	.	3	.	.
31																1	.	.	.	1	1	.	.	2	.	.	1	.	.	2	.	.	6	.	5	.
32																	1	.	.	.	.	.	.	.	2	2	.	.	.	.	1	.	.	.	5	9
33																		1	.	.	.	.	.	.	.	.	.	.	.	.	.	.	.	.	.	.
34																			1	.	.	1	1	.	.	.	.	2	1	.	.	2	.	2	.	.
35																				1	.	.	.	.	.	.	1	.	.	2	.	.	.	.	.	.
36																					1	.	.	.	.	.	1	.	.	.	.	.	3	.	.	.
37																						1	.	.	.	.	.	1	.	.	.	1	.	1	.	.
38																							1	.	.	.	.	1	1	.	.	2	.	2	.	.
39																								1	.	.	.	.	.	1	.	.	2	.	.	.
40																									1	.	.	.	.	.	.	.	.	.	.	3
41																										1	.	.	.	.	1	.	.	.	.	3
42																											1	.	.	.	.	.	.	.	.	.
43																												1	.	.	.	1	.	1	.	.
44																													1	.	.	1	.	2	.	.
45																														1	.	.	.	.	.	.
46																															1	.	.	.	.	.
47																																1	.	1	.	.
48																																	1	.	.	.
49																																		1	.	.
50																																			1	.
51																																				1

Table 3. *(continued)*

ALGEBRA A12, REPRESENTATIONS OF CLASS 6

S.P.	42	68	68	94	120	120	94	120	146	120	198	94	198	224	146	302	172	198	432	198	120	224	120	146	276	302
O.S.	1 716	6 435	10 296	4 290	286	2 860	12 870	6 435	1 716	2 860	78	72 072	858	156	8 580	156	858	2 860	13	858	102 960	1 716	180 180	51 480	156	858
LEVEL	43	53	55	59	61	61	63	65	65	67	67	67	67	69	71	71	73	73	73	75	75	77	77	77	79	79
DIM'N	1 716	15 015	20 592	39 039	26 026	50 050	78 078	96 525	104 104	104 104	41 405	250 965	80 080	85 176	325 325	61 880	234 234	228 800	18 564	273 273	960 960	520 520	1 351 350	1 189 760	264 992	291 720
	0	1	0	0	0	2	0	0	1	0	0	1	3	2	0	4	0	0	6	0	0	0	1	2	0	0
	0	0	0	1	0	0	0	0	1	0	3	0	0	2	0	1	0	0	0	0	1	0	0	0	0	0
	0	0	0	0	2	0	0	0	1	0	0	0	1	0	0	0	0	0	0	0	0	0	0	0	0	0
	0	0	0	1	0	1	0	0	0	0	0	0	0	0	0	0	0	0	0	0	0	0	0	0	0	0
	0	1	0	0	0	0	0	0	0	0	0	0	0	0	0	0	0	0	0	0	1	0	0	1	0	0
	1	0	0	0	0	0	0	0	0	0	0	1	0	0	0	0	0	0	0	0	0	0	0	0	0	0
	0	0	1	0	0	0	0	0	0	0	0	0	0	0	0	0	0	0	0	0	0	0	1	0	0	0
	0	0	0	0	0	0	1	1	0	0	0	0	0	0	0	0	0	0	0	0	0	0	0	0	0	0
	0	0	0	0	0	0	0	0	0	1	0	0	0	0	1	0	0	1	0	0	0	0	0	0	0	0
	0	0	0	0	0	0	0	0	0	1	0	0	0	0	0	0	2	0	0	1	0	1	0	0	0	1
	0	0	0	0	0	0	1	0	0	0	0	0	0	0	1	0	0	0	0	2	0	1	1	0	3	0
WEIGHT	0	0	1	0	0	0	0	2	0	0	0	1	0	0	1	0	1	3	0	0	1	2	0	1	1	4
NUMBER	0	1	2	3	4	5	6	7	8	9	10	11	12	13	14	15	16	17	18	19	20	21	22	23	24	25
0	1	5	6	9	5	10	14	15	16	14	5	42	10	9	35	5	21	20	1	21	108	35	135	120	14	15
1		1	.	3	2	4	.	.	8	.	3	7	6	6	.	4	.	.	1	.	32	.	20	40	.	.
2			1	.	.	.	4	5	.	5	.	6	.	.	15	.	10	10	.	11	14	20	35	15	9	10
3				1	1	1	.	.	4	.	2	.	3	4	.	3	.	.	1	.	8	.	.	8	.	.
4					1	.	.	.	2	.	1	.	1	3	.	2	.	.	1	.	.	.	.	.	.	.
5						1	.	.	2	.	1	.	3	3	.	3	.	.	1	.	.	.	.	8	.	.
6							1	1	.	2	.	.	.	.	6	.	5	4	.	6	.	11	7	.	6	6
7								1	.	.	.	.	.	.	3	.	2	4	.	3	.	8	.	.	4	6
8									1	.	1	.	1	2	.	2	.	.	1	.	.	.	.	.	.	.
9										1	.	.	.	.	2	.	3	1	.	3	.	6	.	.	4	3
10											1	.	.	1	.	1	.	.	1	.	.	.	.	.	.	.
11												1	.	.	.	.	.	.	.	.	4	.	5	5	.	.
12													1	1	.	2	.	.	1	.	.	.	.	.	.	.
13														1	.	1	.	.	1	.	.	.	.	.	.	.
14															1	.	1	1	.	2	.	4	.	.	3	3
15																1	.	.	1	.	.	.	.	.	.	.
16																	1	.	.	1	.	2	.	.	2	1

146	146	328	432	146	172	588	172	198	224	146	224	172	172	224	172	250	198	302	198	328	172	224	354	198	198	S.P.
25 740	72 072	156	156	102 960	77 220	13	6 435	8 580	25 740	540 540	8 580	205 920	180 180	51 480	216 216	17 160	77 220	1 716	72 072	8 580	360 360	38 610	1 716	25 740	102 960	O.S.
79	79	81	83	83	83	85	85	85	85	87	87	87	89	89	89	89	91	91	91	91	93	93	93	95	95	LEVEL
1 288 287	1 639 638	321 776	190 944	2 862 860	4 036 032	50 388	1 756 755	2 914 912	2 844 270	7 432 425	3 407 040	8 200 192	8 923 200	5 530 525	9 034 740	6 497 400	12 024 012	3 318 315	10 839 829	3 650 400	15 940 925	11 243 232	4 036 032	9 837 828	19 027 008	DIM'N
0	1	0	0	1	1	0	1	1	3	0	0	1	0	1	2	2	1	1	2	4	0	1	3	0	0	
0	0	0	0	0	1	0	0	0	0	1	2	0	1	0	0	1	0	3	0	0	0	0	2	0	0	
1	0	0	0	0	0	0	0	2	0	0	1	0	0	0	0	1	0	0	0	1	1	0	0	0	1	
1	0	0	0	0	1	0	0	0	1	0	0	0	0	0	0	0	0	0	0	0	0	0	0	2	0	
0	0	0	0	0	0	0	0	0	0	0	0	0	0	0	0	0	0	0	0	0	1	0	0	0	1	
0	0	0	0	0	0	0	0	0	0	1	0	0	1	0	1	0	0	0	1	0	0	0	0	0	0	
0	1	0	0	0	0	0	0	0	0	0	0	0	0	0	0	0	0	0	0	0	0	0	0	0	0	
0	0	0	0	1	0	0	0	0	0	0	0	1	0	1	0	0	0	0	0	0	0	0	0	0	0	
0	0	0	0	0	0	0	2	0	0	0	0	0	0	0	0	0	1	0	0	0	0	1	0	0	0	
0	0	0	0	1	0	0	0	0	0	0	0	0	0	0	0	0	1	0	0	0	0	0	0	0	0	
0	0	2	1	0	0	0	0	0	0	1	0	1	0	0	1	0	0	0	0	0	1	2	0	1	0	
1	2	3	5	0	1	7	0	1	1	0	1	1	2	3	0	1	1	1	2	1	0	0	1	0	2	WEIGHT
26	27	28	29	30	31	32	33	34	35	36	37	38	39	40	41	42	43	44	45	46	47	48	49	50	51	NUMBER
114	147	14	6	203	300	1	105	186	180	480	192	488	525	280	531	330	560	138	588	150	770	462	144	411	843	0
43	21	.	.	28	132	.	14	92	90	129	104	64	140	35	156	192	70	91	168	100	273	56	104	164	300	1
14	42	10	5	69	35	1	40	21	20	115	21	188	135	120	126	35	250	14	147	15	175	222	14	91	203	2
18	.	.	.	.	54	.	.	48	36	27	58	.	28	.	27	108	.	62	28	60	98	.	74	72	108	3
9	.	.	.	.	18	.	.	30	9	.	30	.	.	.	.	60	.	42	.	30	35	.	54	36	36	4
.	.	.	.	.	27	.	.	19	36	.	30	.	.	.	27	78	.	42	28	60	.	.	63	.	.	5
.	7	7	4	24	.	1	17	.	.	20	.	64	20	40	21	.	113	.	21	.	28	105	.	14	28	6
.	7	6	4	.	.	1	.	.	.	.	.	32	20	40	.	.	43	.	21	.	.	54	.	.	28	7
.	.	.	.	.	9	.	.	10	9	.	20	.	.	.	.	40	.	32	.	30	.	.	43	.	.	8
.	.	5	3	8	.	1	9	.	.	.	.	16	.	8	.	.	54	.	.	.	.	45	.	.	.	9
.	.	.	.	.	.	.	.	.	.	.	10	.	.	.	.	10	.	22	.	10	.	.	22	.	.	10
5	6	.	.	9	15	.	5	10	10	30	11	24	35	15	36	20	30	9	42	10	60	26	10	35	69	11
.	.	.	.	.	.	.	.	.	9	.	.	.	.	.	.	20	.	11	.	30	.	.	32	.	.	12
.	.	.	.	.	.	.	.	.	.	.	.	.	.	.	.	10	.	11	.	10	.	.	22	.	.	13
.	.	4	3	.	.	1	.	.	.	.	.	8	.	8	.	.	18	.	.	.	.	27	.	.	.	14
.	.	.	.	.	.	.	.	.	.	.	.	.	.	.	.	.	.	.	.	10	.	.	11	.	.	15
.	.	3	2	.	.	1	.	.	.	.	.	.	.	.	.	.	9	.	.	.	.	9	.	.	.	16

1	.	.	.	2	.	.	1	3	.	.	3	3	.	.	1	.	.	.	.	.	.	.	8	.	.	.	.	.	.	.	.	.	.	.	17
	1	.	.	.	.	.	.	.	.	.	.	.	.	.	.	.	.	.	.	.	.	.	.	.	.	.	.	.	.	.	.	.	.	.	18
		1	.	1	.	.	2	1	.	.	2	2	.	.	1	.	.	.	.	.	.	.	.	.	.	.	.	.	.	.	9	.	.	.	19
			1	.	.	1	.	.	2	.	.	.	.	6	.	.	5	4	6	6	.	7	.	6	11	.	6	7	6	21	.	7	15	24	20
				1	.	.	1	1	.	.	2	2	.	.	1	.	.	.	.	.	.	.	.	.	.	.	.	.	.	.	.	.	.	.	21
					1	.	.	.	.	1	.	.	3	.	.	2	.	.	5	.	8	5	5	6	.	13	.	6	.	9	12	.	5	9	22
						1	.	.	.	.	.	.	.	3	.	.	2	4	.	3	.	.	.	6	8	.	4	7	6	.	.	6	.	.	23
							1	.	.	.	1	1	.	.	1	.	.	.	.	.	.	.	.	.	.	.	.	.	.	.	.	.	.	.	24
								1	.	.	1	2	.	.	1	.	.	.	.	.	.	.	.	.	.	.	.	.	.	.	.	.	.	.	25
									1	.	.	.	.	2	.	.	3	1	.	3	.	.	.	.	6	.	4	.	3	7	.	5	8	8	26
										1	.	.	.	.	.	.	.	.	.	.	4	5	5	.	.	5	.	6	.	.	6	.	.	9	27
											1	1	.	.	1	.	.	.	.	.	.	.	.	.	.	.	.	.	.	.	.	.	.	.	28
												1	.	.	1	.	.	.	.	.	.	.	.	.	.	.	.	.	.	.	.	.	.	.	29
													1	.	.	1	.	.	.	.	2	.	1	.	.	6	.	.	.	.	5	.	.	.	30
														1	.	.	1	1	.	2	.	.	.	.	4	.	3	.	3	.	.	4	.	.	31
															1	.	.	.	.	.	.	.	.	.	.	.	.	.	.	.	.	.	.	.	32
																1	.	.	.	.	.	.	.	.	.	3	.	.	.	.	2	.	.	.	33
																	1	.	.	1	.	.	.	.	2	.	2	.	1	.	.	3	.	.	34
																		1	.	.	.	.	.	.	2	.	1	.	3	.	.	3	.	.	35
																			1	.	.	1	.	1	.	.	.	1	.	3	.	.	2	3	36
																				1	.	.	.	.	1	.	2	.	1	.	.	2	.	.	37
																					1	.	1	.	.	2	.	.	.	.	3	.	.	.	38
																						1	.	.	.	.	.	1	.	.	.	.	.	3	39
																							1	.	.	.	.	.	.	.	.	.	.	.	40
																								1	.	.	.	1	.	.	.	.	.	.	41
																									1	.	1	.	1	.	.	2	.	.	42
																										1	.	.	.	.	1	.	.	.	43
																											1	.	.	.	.	1	.	.	44
																												1	.	.	.	.	.	.	45
																													1	.	.	1	.	.	46
																														1	.	.	1	1	47
																															1	.	.	.	48
																																1	.	.	49
																																	1	.	50
																																		1	51

Table 3. *(continued)*

ALGEBRA B2, REPRESENTATIONS OF CLASS 0

NUMBER	0	1	2	3	4	5	6	7	8	9	10	11	12	13	14	15	16	17	18	19	20	21	22	23	24	25
S.P.	0	4	8	16	20	36	32	40	64	52	68	72	100	80	104	100	144	116	128	148	136	196	160	164	200	180
O.S.	1	4	4	4	8	4	4	8	4	8	8	4	4	8	8	8	4	8	4	8	8	4	8	8	8	8
LEVEL	1	5	7	9	11	13	13	15	17	17	19	19	21	21	23	23	25	25	25	27	27	29	29	29	31	31
DIM'N	1	5	10	14	35	30	35	81	55	105	154	84	91	220	260	231	140	390	165	405	455	204	625	429	595	770
WEIGHT	0>0	1>0	0>2	2>0	1>2	3>0	0>4	2>2	4>0	1>4	3>2	0>6	5>0	2>4	4>2	1>6	6>0	3>4	0>8	5>2	2>6	7>0	4>4	1>8	6>2	3>6
0	1	1	2	2	3	2	3	5	3	5	6	4	3	8	8	7	4	10	5	9	11	4	13	9	11	14
1		1	1	1	3	2	2	4	2	5	6	3	3	7	7	7	3	10	4	9	10	4	12	9	10	14
2			1	1	2	1	2	4	2	4	5	3	2	7	7	6	3	9	4	8	10	3	12	8	10	13
3				1	1	1	1	3	2	3	4	2	2	6	6	5	3	8	3	7	9	3	11	7	9	12
4					1	1	1	2	1	3	4	2	2	5	5	5	2	8	3	7	8	3	10	7	8	12
5						1	.	1	1	1	3	1	2	3	4	3	2	6	2	6	6	3	8	5	7	10
6							1	1	1	2	2	2	1	4	4	4	2	6	3	5	7	2	9	6	7	10
7								1	1	1	2	1	1	3	4	3	2	5	2	5	6	2	8	5	7	9
8									1	.	1	.	1	1	3	1	2	3	1	4	3	2	6	3	6	6
9										1	1	1	1	2	2	3	1	4	2	4	5	2	6	5	5	8
10											1	.	1	1	2	1	1	3	1	4	3	2	5	3	5	6
11												1	.	1	1	2	1	2	2	2	4	1	4	4	4	6
12													1	.	1	.	1	1	.	3	1	2	3	1	4	3
13														1	1	1	1	2	1	2	3	1	4	3	4	5
14															1	.	1	1	.	2	1	1	3	1	4	3
15																1	.	1	1	1	2	1	2	3	2	4
16																	1	.	.	1	.	1	1	.	3	1
17																		1	.	1	1	1	2	1	2	3
18																			1	.	1	.	1	2	1	2
19																				1	.	1	1	.	2	1
20																					1	.	1	1	1	2

NUMBER	26	27	28	29	30	31	32	33	34	35	36	37	38	39	40	41	42	43	44	45	46	47	48	49	50	51
S.P.	200	256	212	208	260	232	244	324	272	260	288	328	292	296	400	340	320	340	404	360	356	392	484	416	388	400
O.S.	4	4	8	8	8	8	8	4	8	8	4	8	8	8	4	8	8	8	8	8	8	4	4	8	8	8
LEVEL	31	33	33	33	35	35	35	37	37	37	37	39	39	39	41	41	41	41	43	43	43	43	45	45	45	45
DIM'N	286	285	935	810	836	1190	715	385	1330	1326	455	1134	1729	1309	506	1820	1995	1105	1495	2401	2090	680	650	2415	2835	1976
WEIGHT	0>10	8>0	5>4	2>8	7>2	4>6	1>10	9>0	6>4	3>8	0>12	8>2	5>6	2>10	10>0	7>4	4>8	1>12	9>2	6>6	3>10	0>14	11>0	8>4	5>8	2>12
0	6	5	15	14	12	18	11	5	18	18	7	14	21	17	6	20	23	13	15	25	22	8	6	23	27	20
1	5	4	15	13	12	17	11	5	17	18	6	13	21	16	5	20	22	13	15	24	22	7	6	22	27	19
2	5	4	14	13	11	17	10	4	17	17	6	13	20	16	5	19	22	12	14	24	21	7	5	22	26	19
3	4	4	13	12	10	16	9	4	16	16	5	12	19	15	5	18	21	11	13	23	20	6	5	21	25	18
4	4	3	13	11	10	15	9	4	15	16	5	11	19	14	4	18	20	11	13	22	20	6	5	20	25	17
5	3	3	11	9	9	13	7	4	13	14	4	10	17	12	4	16	18	9	12	20	18	5	5	18	23	15
6	4	3	11	10	8	14	8	3	14	14	5	10	17	13	4	16	19	10	11	21	18	6	4	19	23	16
7	3	3	10	9	8	13	7	3	13	13	4	10	16	12	4	15	18	9	11	20	17	5	4	18	22	15
8	2	3	8	6	7	10	5	3	11	10	3	9	13	9	4	13	15	7	10	17	14	4	4	16	19	12
9	3	2	9	8	7	11	7	3	11	12	4	8	15	11	3	14	16	9	10	18	16	5	4	16	21	14
10	2	2	8	6	7	9	5	3	10	10	3	8	13	9	3	13	14	7	10	16	14	4	4	15	19	12
11	3	2	6	7	5	9	6	2	9	10	4	7	12	10	3	11	14	8	8	16	14	5	3	14	18	13
12	1	2	6	3	6	6	3	3	8	6	2	7	10	6	3	11	10	5	9	13	10	3	4	13	15	9
13	2	2	6	6	5	8	5	2	9	9	3	7	11	9	3	11	13	7	8	15	13	4	3	14	17	12
14	1	2	5	3	5	6	3	2	8	6	2	7	9	6	3	10	10	5	8	13	10	3	3	13	14	9
15	2	1	4	5	4	6	5	2	6	8	3	5	9	8	2	9	11	7	7	12	12	4	3	11	15	11
16	.	2	3	1	4	3	1	2	6	3	1	6	6	3	3	8	6	3	7	10	6	2	3	11	10	6
17	1	1	4	3	4	5	3	2	6	6	2	5	8	6	2	9	9	5	7	11	10	3	3	11	13	9
18	2	1	2	4	2	4	4	1	4	6	3	4	6	7	2	6	9	6	5	9	10	4	2	9	12	10
19	.	1	3	1	4	3	1	2	5	3	1	5	6	3	2	8	6	3	7	9	6	2	3	10	10	6
20	1	1	2	3	2	4	3	1	4	5	2	4	6	6	2	6	8	5	5	9	9	3	2	9	11	9

21	22	23	24	25	26	27	28	29	30	31	32	33	34	35	36	37	38	39	40	41	42	43	44	45	46	47	48	49	50	51
3	6	8	3	6	7	1	3	6	1	3	5	.	1	3	4	.	1	3	.	.	1	2	.	.	1	1	.	.	.	1
6	9	9	6	8	6	3	6	6	3	5	4	1	3	4	2	1	3	2	.	1	2	1	.	1	1	.	.	.	1	
8	9	6	8	6	4	6	6	4	5	4	2	3	4	2	1	3	2	1	1	2	1	.	1	1	.	.	.	1		
3	2	2	2	2	1	2	2	1	2	1	1	2	1	1	.	1	1	.	1	1	.	.	1	.	.	.	1			
1	2	3	1	2	3	.	1	2	.	1	2	.	.	1	2	.	.	1	.	.	.	1	.	.	.	1				
3	6	8	3	6	6	1	3	5	1	3	4	.	1	3	2	.	1	2	.	.	1	1	.	.	1					
6	8	6	6	6	4	3	5	4	3	4	2	1	3	2	1	1	2	1	.	1	1	.	.	1						
6	5	4	5	4	2	4	4	2	4	2	1	3	2	1	.	2	1	.	1	1	.	.	1							
1	3	5	1	3	4	.	1	3	.	1	3	.	.	1	2	.	.	1	.	.	.	1								
3	6	6	3	5	4	1	3	4	1	3	2	.	1	2	1	.	1	1	.	.	1									
6	6	4	5	4	2	3	4	2	3	2	1	1	2	1	.	1	1	.	.	1										
2	2	1	2	1	1	2	1	1	1	1	.	1	1	.	.	1	.	.	1											
1	3	5	1	3	4	.	1	3	.	1	2	.	.	1	1	.	.	1												
3	5	4	3	4	2	1	3	2	1	2	1	.	1	1	.	.	1													
4	4	2	4	2	1	3	2	1	2	1	.	1	1	.	.	1														
.	1	2	.	1	2	.	.	1	.	.	1	.	.	.	1															
1	3	4	1	3	2	.	1	2	.	1	1	.	.	1																
3	4	2	3	2	1	1	2	1	1	1	.	.	1																	
2	1	1	1	1	.	1	1	.	1	.	.	1																		
.	1	3	.	1	2	.	.	1	.	.	1																			
1	3	2	1	2	1	.	1	1	.	1																				
3	2	1	2	1	.	1	1	.	1																					
.	1	2	.	1	1	.	.	1																						
1	2	1	1	1	.	.	1																							
1	1	.	1	.	.	1																								
.	.	1	.	.	1																									
.	1	1	.	1																										
1	1	.	1																											
.	.	1																												
.	1																													
1																														

Table 3. *(continued)*

ALGEBRA B2, REPRESENTATIONS OF CLASS 1

S.P.	2	10	18	26	34	50	50	58	82	74	90	98	122	106	130	130	170	146	162	178	170	226	194	202	234	218
O.S.	4	8	4	8	8	8	4	8	8	8	8	4	8	8	8	8	8	8	4	8	8	8	8	8	8	8
LEVEL	4	8	10	12	14	16	16	18	20	20	22	22	24	24	26	26	28	28	28	30	30	32	32	32	34	34
DIM'N	4	16	20	40	64	80	56	140	140	160	256	120	224	324	420	320	336	560	220	640	616	480	880	560	924	1 024
WEIGHT	0 > 1	1 > 1	0 > 3	2 > 1	1 > 3	3 > 1	0 > 5	2 > 3	4 > 1	1 > 5	3 > 3	0 > 7	5 > 1	2 > 5	4 > 3	1 > 7	6 > 1	3 > 5	0 > 9	5 > 3	2 > 7	7 > 1	4 > 5	1 > 9	6 > 3	3 > 7
NUMBER	0	1	2	3	4	5	6	7	8	9	10	11	12	13	14	15	16	17	18	19	20	21	22	23	24	25
0	1	2	2	3	4	4	3	6	5	6	8	4	6	9	10	8	7	12	5	12	12	8	15	10	14	16
1		1	1	2	3	3	2	5	4	5	7	3	5	8	9	7	6	11	4	11	11	7	14	9	13	15
2			1	1	2	2	2	4	3	4	6	3	4	7	8	6	5	10	4	10	10	6	13	8	12	14
3				1	1	2	1	3	3	3	5	2	4	6	7	5	5	9	3	9	9	6	12	7	11	13
4					1	1	1	2	2	3	4	2	3	5	6	5	4	8	3	8	8	5	11	7	10	12
5						1	.	1	2	1	3	1	3	3	5	3	4	6	2	7	6	5	9	5	9	10
6							1	1	1	2	2	2	2	4	4	4	3	6	3	6	7	4	9	6	8	10
7								1	1	1	2	1	2	3	4	3	3	5	2	6	6	4	8	5	8	9
8									1	.	1	.	2	1	3	1	3	3	1	5	3	4	6	3	7	6
9										1	1	1	1	2	2	3	2	4	2	4	5	3	6	5	6	8
10											1	.	1	1	2	1	2	3	1	4	3	3	5	3	6	6
11												1	.	1	1	2	1	2	2	2	4	2	4	4	4	6
12													1	.	1	.	2	1	.	3	1	3	3	1	5	3
13														1	1	1	1	2	1	2	3	2	4	3	4	5
14															1	.	1	1	.	2	1	2	3	1	4	3
15																1	.	1	1	1	2	1	2	3	2	4
16																	1	.	.	1	.	2	1	.	3	1
17																		1	.	1	1	1	2	1	2	3
18																			1	.	1	.	1	2	1	2
19																				1	.	1	1	.	2	1

S.P.	242	290	250	250	298	274	290	362	314	306	338	370	338	346	442	386	370	394	450	410	410	450	530	466	442	458
O.S.	4	8	8	8	8	8	8	8	8	8	4	8	8	8	8	8	8	8	8	8	8	4	8	8	8	8
LEVEL	34	36	36	36	38	38	38	40	40	40	40	42	42	42	44	44	44	44	46	46	46	46	48	48	48	48
DIM'N	364	660	1 296	1 040	1 280	1 560	896	880	1 820	1 680	560	1 716	2 240	1 620	1 144	2 464	2 500	1 344	2 240	3 080	2 560	816	1 456	3 240	3 520	2 380
WEIGHT	0 > 11	8 > 1	5 > 5	2 > 9	7 > 3	4 > 7	1 > 11	9 > 1	6 > 5	3 > 9	0 > 13	8 > 3	5 > 7	2 > 11	10 > 1	7 > 5	4 > 9	1 > 13	9 > 3	6 > 7	3 > 11	0 > 15	11 > 1	8 > 5	5 > 9	2 > 13
NUMBER	26	27	28	29	30	31	32	33	34	35	36	37	38	39	40	41	42	43	44	45	46	47	48	49	50	51
0	6	9	18	15	16	20	12	10	21	20	7	18	24	18	11	24	25	14	20	28	24	8	12	27	30	21
1	5	8	17	14	15	19	11	9	20	19	6	17	23	17	10	23	24	13	19	27	23	7	11	26	29	20
2	5	7	16	13	14	18	10	8	19	18	6	16	22	16	9	22	23	12	18	26	22	7	10	25	28	19
3	4	7	15	12	13	17	9	8	18	17	5	15	21	15	9	21	22	11	17	25	21	6	10	24	27	18
4	4	6	14	11	12	16	9	7	17	16	5	14	20	14	8	20	21	11	16	24	20	6	9	23	26	17
5	3	6	12	9	11	14	7	7	15	14	4	13	18	12	8	18	19	9	15	22	18	5	9	21	24	15
6	4	5	12	10	10	14	8	6	15	14	5	12	18	13	7	18	19	10	14	22	18	6	8	21	24	16
7	3	5	11	9	10	13	7	6	14	13	4	12	17	12	7	17	18	9	14	21	17	5	8	20	23	15
8	2	5	9	6	9	10	5	6	12	10	3	11	14	9	7	15	15	7	13	18	14	4	8	18	20	12
9	3	4	9	8	8	11	7	5	12	12	4	10	15	11	6	15	16	9	12	19	16	5	7	18	21	14
10	2	4	8	6	8	9	5	5	11	10	3	10	13	9	6	14	14	7	12	17	14	4	7	17	19	12
11	3	3	6	7	6	9	6	4	9	10	4	8	12	10	5	12	14	8	10	16	14	5	6	15	18	13
12	1	4	6	3	7	6	3	5	9	6	2	9	10	6	6	12	10	5	11	14	10	3	7	15	15	9
13	2	3	6	6	6	8	5	4	9	9	3	8	11	9	5	12	13	7	10	15	13	4	6	15	17	12
14	1	3	5	3	6	6	3	4	8	6	2	8	9	6	5	11	10	5	10	13	10	3	6	14	14	9
15	2	2	4	5	4	6	5	3	6	8	3	6	9	8	4	9	11	7	8	12	12	4	5	12	15	11
16	.	3	3	1	5	3	1	4	6	3	1	7	6	3	5	9	6	3	9	10	6	2	6	12	10	6
17	1	2	4	3	4	5	3	3	6	6	2	6	8	6	4	9	9	5	8	11	10	3	5	12	13	9
18	2	1	2	4	2	4	4	2	4	6	3	4	6	7	3	6	9	6	6	9	10	4	4	9	12	10
19	.	2	3	1	4	3	1	3	5	3	1	6	6	3	4	8	6	3	8	9	6	2	5	11	10	6

1	.	1	1	1	2	1	1	2	3	2	4	3	2	4	5	2	4	6	6	3	6	8	5	6	9	9	3	4	9	11	9	20
	1	.	.	1	.	.	2	1	.	3	1	.	3	3	1	.	5	3	1	4	6	3	1	7	6	3	1	5	9	6	3	21
		1	.	1	1	.	1	2	1	2	3	1	2	4	3	1	4	5	3	3	6	6	3	6	8	6	2	4	9	9	6	22
			1	.	1	1	.	1	2	1	2	3	1	2	4	2	2	4	5	2	4	6	5	4	6	8	3	3	6	9	8	23
				1	.	.	1	1	.	2	1	.	2	3	1	.	4	3	1	3	5	3	1	6	6	3	1	4	8	6	3	24
					1	.	.	1	1	1	2	1	1	2	3	1	2	4	3	2	4	5	3	4	6	6	2	3	6	8	6	25
						1	.	.	1	.	1	2	.	1	2	2	1	2	4	1	2	4	4	2	4	6	3	2	4	6	7	26
							1	.	.	1	.	.	2	1	.	.	3	1	.	3	3	1	.	5	3	1	.	4	6	3	1	27
								1	.	1	1	.	1	2	1	.	2	3	1	2	4	3	1	4	5	3	1	3	6	6	3	28
									1	.	1	1	.	1	2	1	1	2	3	1	2	4	3	2	4	5	2	2	4	6	6	29
										1	.	.	1	1	.	.	2	1	.	2	3	1	.	4	3	1	.	3	5	3	1	30
											1	.	.	1	1	.	1	2	1	1	2	3	1	2	4	3	1	2	4	5	3	31
												1	.	.	1	1	.	1	2	.	1	2	3	1	2	4	2	1	2	4	5	32
													1	.	.	.	1	.	.	2	1	.	.	3	1	.	.	3	3	1	.	33
														1	.	.	1	1	.	1	2	1	.	2	3	1	.	2	4	3	1	34
															1	.	.	1	1	.	1	2	1	1	2	3	1	1	2	4	3	35
																1	.	.	1	.	.	1	2	.	1	2	2	.	1	2	4	36
																	1	.	.	1	1	.	.	2	1	.	.	2	3	1	.	37
																		1	.	.	1	1	.	1	2	1	.	1	2	3	1	38
																			1	.	.	1	1	.	1	2	1	.	1	2	3	39
																				1	.	.	.	1	.	.	.	2	1	.	.	40
																					1	.	.	1	1	.	.	1	2	1	.	41
																						1	.	.	1	1	.	.	1	2	1	42
																							1	.	.	1	1	.	.	1	2	43
																								1	.	.	.	1	1	.	.	44
																									1	.	.	.	1	1	.	45
																										1	.	.	.	1	1	46
																											1	.	.	.	1	47
																												1	.	.	.	48
																													1	.	.	49
																														1	.	50
																															1	51

Table 3. *(continued)*

ALGEBRA B3, REPRESENTATIONS OF CLASS 0

0	1	2	3	4	5	6	7	8	9	10	11	12	13	14	15	16	17	18	19	20	21	22	23	24	25	NUMBER
0	4	8	16	12	20	36	24	32	40	36	64	44	48	52	68	56	100	72	72	68	80	76	104	84	88	S.P.
1	6	12	6	8	24	6	24	12	24	24	6	24	8	24	24	48	6	12	24	24	24	24	24	48	24	O.S.
1	7	11	13	13	17	19	19	21	23	23	25	25	25	27	29	29	31	31	31	31	33	33	35	35	35	LEVEL
1	7	21	27	35	105	77	189	168	330	378	182	616	294	693	819	1 617	378	825	1 560	1 386	1 911	2 079	1 750	4 550	2 310	DIM'N
0 0 > 0	1 0 > 0	0 1 > 0	2 0 > 0	0 0 > 2	1 1 > 0	3 0 > 0	1 0 > 2	0 2 > 0	2 1 > 0	0 1 > 2	4 0 > 0	2 0 > 2	0 0 > 4	1 2 > 0	3 1 > 0	1 1 > 2	5 0 > 0	0 3 > 0	3 0 > 2	1 0 > 4	2 2 > 0	0 2 > 2	4 1 > 0	2 1 > 2	0 1 > 4	WEIGHT
1	1	3	3	3	5	3	9	8	12	12	6	16	8	15	15	35	6	17	28	24	33	33	26	62	32	0
	1	1	1	2	5	3	6	4	7	10	3	14	6	15	15	27	6	11	22	21	23	29	18	57	27	1
		1	1	1	2	1	5	4	7	7	3	10	5	9	9	24	3	11	20	17	23	23	18	46	24	2
			1	.	1	1	2	2	5	3	3	6	3	6	7	15	3	7	14	11	18	15	15	33	17	3
				1	2	1	3	2	3	6	1	9	4	9	9	18	3	7	15	15	15	21	11	42	20	4
					1	1	1	1	2	3	1	5	2	6	7	11	3	5	10	10	11	14	9	30	14	5
						1	.	.	1	.	1	2	.	2	5	3	3	2	6	3	6	5	7	15	5	6
							1	1	2	2	1	3	2	3	3	10	1	5	9	8	11	11	9	23	13	7
								1	1	1	1	1	1	2	2	6	1	4	5	5	8	8	7	15	9	8
									1	.	1	1	.	1	2	3	1	2	5	2	6	4	7	11	5	9
										1	.	1	1	2	2	4	1	2	3	5	4	7	3	14	8	10
											1	.	.	.	1	.	1	.	2	.	2	.	5	3	.	11
												1	.	1	2	2	1	1	3	2	3	4	3	10	4	12
													1	.	.	2	.	1	1	3	3	3	2	6	6	13
														1	1	1	1	1	1	1	2	3	2	6	3	14
															1	.	1	.	1	.	1	.	2	3	.	15
																1	.	1	1	1	2	2	2	4	3	16
																	1	.	.	.	.	.	1	.	.	17
																		1	.	.	1	1	1	1	1	18

26	27	28	29	30	31	32	33	34	35	36	37	38	39	40	41	42	43	44	45	46	47	48	49	50	51	NUMBER
144	100	108	96	108	116	104	148	128	120	116	196	136	152	132	132	136	160	140	144	200	164	164	152	164	256	S.P.
6	24	24	24	8	24	48	24	12	48	48	6	24	24	24	24	24	24	48	24	24	24	48	48	24	6	O.S.
37	37	37	37	37	39	39	41	41	41	41	43	43	43	43	43	43	45	45	45	47	47	47	47	47	49	LEVEL
714	3 003	3 375	4 095	1 386	4 312	7 722	3 366	3 003	10 395	9 009	1 254	7 560	6 545	8 008	9 625	6 006	8 568	19 683	10 296	5 985	10 010	20 825	23 625	9 009	2 079	DIM'N
6 0 > 0	1 3 > 0	4 0 > 2	2 0 > 4	0 0 > 6	3 2 > 0	1 2 > 2	5 1 > 0	0 4 > 0	3 1 > 2	1 1 > 4	7 0 > 0	2 3 > 0	5 0 > 2	0 3 > 2	3 0 > 4	1 0 > 6	4 2 > 0	2 2 > 2	0 2 > 4	6 1 > 0	1 4 > 0	4 1 > 2	2 1 > 4	0 1 > 6	8 0 > 0	WEIGHT
10	35	39	31	16	44	90	30	33	105	89	10	72	57	72	83	52	72	159	84	45	70	147	177	63	15	0
6	35	36	43	14	44	75	30	24	89	81	10	56	48	66	77	46	56	149	74	34	70	139	157	58	10	1
6	25	29	39	12	32	69	21	24	83	70	6	56	45	56	67	42	56	128	68	34	55	120	146	52	10	2
6	19	22	29	9	26	51	18	18	65	53	6	46	36	42	53	33	48	102	54	30	45	98	118	42	10	3
3	25	27	33	11	32	57	21	18	69	64	6	42	37	52	62	37	42	120	60	24	55	113	129	48	6	4
3	19	20	24	8	26	42	18	14	53	49	6	34	29	40	49	29	35	95	47	21	45	92	104	39	6	5
3	10	14	11	4	18	21	15	8	33	25	6	22	22	22	29	16	26	59	26	18	29	65	63	24	6	6
3	13	15	22	7	17	39	11	14	49	42	3	34	27	34	42	27	35	80	44	21	35	77	97	35	6	7
3	10	10	16	5	13	29	9	12	36	32	3	27	20	27	32	21	29	62	35	18	29	60	78	28	6	8
3	7	10	10	3	11	20	9	8	30	21	3	22	20	19	24	15	26	49	25	18	23	53	59	21	6	9
1	10	9	13	5	13	23	9	8	28	30	3	19	15	25	30	19	19	58	31	11	29	56	68	27	3	10
3	2	6	3	.	6	5	7	3	15	5	3	10	14	6	11	4	18	21	8	15	11	33	25	7	6	11
1	7	9	8	3	11	16	9	6	23	20	3	15	15	18	22	13	17	46	22	11	23	49	51	20	3	12
1	4	3	9	4	5	15	3	6	18	20	1	15	9	15	19	15	15	36	24	9	17	33	51	21	3	13
1	6	5	5	2	8	12	7	5	15	15	3	12	10	15	16	10	13	35	18	9	20	36	40	16	3	14
1	2	5	2	.	6	4	7	2	11	5	3	7	10	6	10	3	11	20	7	9	11	30	21	7	3	15
1	3	3	5	2	4	11	3	5	14	13	1	12	9	12	13	10	13	28	17	9	14	28	38	15	3	16
1	.	2	.	.	2	.	5	.	3	.	3	2	6	.	3	.	6	5	.	7	3	15	5	.	3	17
1	2	1	1	1	2	6	2	4	6	6	1	8	5	8	5	5	8	16	10	7	10	15	21	9	3	18

1	.	1	.	2	2	.	1	1	3	2	.	3	4	3	2	10	4	1	7	9	5	8	3	11	16	7	9	8	23	20	6	3	19
	1	.	1	.	2	2	.	2	1	3	2	3	5	2	2	6	10	1	5	3	8	9	8	5	19	12	3	11	18	26	13	1	20
		1	.	1	1	.	1	1	1	1	.	2	3	2	2	6	3	1	6	5	4	5	2	8	12	6	7	7	15	15	5	3	21
			1	.	1	1	.	2	1	1	1	2	4	2	2	4	6	1	4	3	7	5	5	4	15	9	3	10	14	18	9	1	22
				1	.	.	1	.	1	.	.	1	.	2	.	3	.	1	2	5	.	2	.	6	4	.	7	2	11	5	.	3	23
					1	.	.	1	1	1	.	2	2	2	1	4	3	1	3	3	4	5	2	4	11	5	3	7	14	13	5	1	24
						1	.	.	.	1	1	.	2	.	1	2	4	.	3	1	3	3	5	3	7	7	2	4	6	14	8	1	25
							1	.	.	.	.	.	.	1	.	.	.	1	.	2	.	.	.	2	.	.	5	.	3	.	.	3	26
								1	.	.	.	1	1	1	1	1	1	1	2	1	3	1	1	2	6	3	2	6	6	6	3	1	27
									1	.	.	1	.	2	.	2	.	1	1	3	.	2	.	3	4	.	3	2	10	4	.	1	28
										1	.	.	1	.	1	2	2	.	2	1	2	3	2	3	5	4	2	3	6	10	4	1	29
											1	.	.	.	.	.	2	.	.	.	1	1	3	.	3	3	.	2	2	6	6	.	30
												1	.	1	.	1	.	1	1	1	.	1	.	2	3	.	2	2	6	3	.	1	31
													1	.	1	1	1	.	2	1	2	1	1	2	4	3	2	3	4	6	3	1	32
														1	.	.	.	1	.	1	.	.	.	1	.	.	2	.	3	.	.	1	33
															1	.	.	.	1	.	1	.	.	1	1	1	1	2	1	1	1	1	34
																1	.	.	1	1	.	1	.	2	2	.	2	1	4	3	.	1	35
																	1	.	.	.	1	1	1	.	2	2	.	2	2	4	3	.	36
																		1	.	.	.	.	.	.	.	.	1	.	.	.	.	1	37
																			1	.	.	.	.	1	1	.	1	1	1	1	.	1	38
																				1	.	.	.	1	.	.	2	.	2	.	.	1	39
																					1	.	.	.	1	1	.	2	1	1	1	.	40
																						1	.	.	1	.	.	1	2	2	.	.	41
																							1	.	.	1	.	.	.	2	2	.	42
																								1	.	.	1	.	1	.	.	1	43
																									1	.	.	1	1	1	.	.	44
																										1	.	.	.	1	1	.	45
																											1	.	.	.	.	1	46
																												1	.	.	.	.	47
																													1	.	.	.	48
																														1	.	.	49
																															1	.	50
																																1	51

Table 3. *(continued)*

ALGEBRA B3, REPRESENTATIONS OF CLASS 1

NUMBER	0	1	2	3	4	5	6	7	8	9	10	11	12	13	14	15	16	17	18	19	20	21	22	23	24	25
S.P.	3	11	19	27	27	35	51	43	51	59	59	83	67	75	75	91	83	123	99	99	99	107	107	131	115	123
O.S.	8	24	24	24	8	48	24	24	24	48	24	24	24	8	48	48	48	24	24	24	24	48	24	48	48	24
LEVEL	7	13	17	19	19	23	25	25	27	29	29	31	31	31	33	35	35	37	37	37	37	39	39	41	41	41
DIM'N	8	48	112	168	112	512	448	560	720	1512	1008	1008	1728	672	2800	3584	4096	2016	3080	4200	3024	7392	4928	7392	11088	4752
WEIGHT	0 0 > 1	1 0 > 1	0 1 > 1	2 0 > 1	0 0 > 3	1 1 > 1	3 0 > 1	1 0 > 3	0 2 > 1	2 1 > 1	0 1 > 3	4 0 > 1	2 0 > 3	0 0 > 5	1 2 > 1	3 1 > 1	1 1 > 3	5 0 > 1	0 3 > 1	3 0 > 3	1 0 > 5	2 2 > 1	0 2 > 3	4 1 > 1	2 1 > 3	0 1 > 5
0	1	3	5	6	4	14	10	13	15	27	18	15	27	10	40	44	52	21	35	46	33	75	50	65	102	42
1		1	2	3	2	8	6	8	9	18	12	10	19	7	28	32	38	15	25	35	25	57	38	50	80	33
2			1	1	1	4	3	5	6	11	8	6	13	5	19	22	28	10	19	26	19	42	29	37	62	26
3				1	.	2	3	2	3	8	4	6	8	3	13	18	18	10	13	19	13	33	20	32	46	19
4					1	2	1	3	3	6	6	3	9	4	13	14	20	6	13	19	15	30	23	26	48	21
5						1	1	1	2	4	3	3	5	2	9	11	13	6	10	13	10	23	16	22	35	15
6							1	.	.	2	.	3	2	.	3	8	4	6	4	8	3	13	6	18	18	6
7								1	1	2	2	1	3	2	5	6	10	3	7	9	8	16	12	14	26	13
8									1	1	1	1	1	1	4	4	6	3	6	5	5	12	9	11	18	9
9										1	.	1	1	.	2	4	3	3	3	5	2	9	5	11	13	5
10											1	.	1	1	2	2	4	1	3	3	5	7	7	6	14	8
11												1	.	.	.	2	.	3	.	2	.	3	.	8	4	.
12													1	.	1	2	2	1	2	3	2	5	4	6	10	4
13														1	.	.	2	.	1	1	3	3	3	2	6	6
14															1	1	1	1	2	1	1	4	3	4	6	3
15																1	.	1	.	1	.	2	.	4	3	.
16																	1	.	1	1	1	2	2	2	4	3
17																		1	.	.	.	.	.	2	.	.
18																			1	.	.	1	1	1	1	1

NUMBER	26	27	28	29	30	31	32	33	34	35	36	37	38	39	40	41	42	43	44	45	46	47	48	49	50	51
S.P.	171	131	139	131	147	147	139	179	163	155	155	227	171	187	171	171	179	195	179	187	235	203	203	195	211	291
O.S.	24	48	24	24	8	48	48	48	24	48	48	24	48	24	24	24	24	48	48	24	48	48	48	48	24	24
LEVEL	43	43	43	43	43	45	45	47	47	47	47	49	49	49	49	49	49	51	51	51	53	53	53	53	53	55
DIM'N	3696	10752	8800	8624	2640	16128	17640	13824	10192	24576	17920	6336	26208	16632	17472	19712	11088	31200	43680	19656	24024	32928	48048	45760	16016	10296
WEIGHT	6 0 > 1	1 3 > 1	4 0 > 3	2 0 > 5	0 0 > 7	3 2 > 1	1 2 > 3	5 1 > 1	0 4 > 1	3 1 > 3	1 1 > 5	7 0 > 1	2 3 > 1	5 0 > 3	0 3 > 3	3 0 > 5	1 0 > 7	4 2 > 1	2 2 > 3	0 2 > 5	6 1 > 1	1 4 > 1	4 1 > 3	2 1 > 5	0 1 > 7	8 0 > 1
0	28	90	70	70	20	120	135	90	70	168	122	36	165	99	110	121	66	175	255	111	119	175	250	242	80	45
1	21	70	56	56	16	96	108	72	55	138	100	28	135	82	90	101	55	145	213	93	98	145	212	205	68	36
2	15	54	44	45	13	75	87	56	45	112	82	21	109	67	74	84	46	118	177	78	79	120	178	174	58	28
3	15	42	35	34	10	63	66	50	35	90	64	21	91	56	58	68	37	103	144	63	72	100	150	143	48	28
4	10	42	34	36	11	57	69	42	35	90	68	15	87	54	62	70	39	94	147	66	62	100	148	147	50	21
5	10	33	26	27	8	47	53	37	29	71	53	15	72	44	49	56	31	81	119	53	56	84	123	121	41	21
6	10	18	19	13	4	33	28	32	17	46	28	15	48	35	28	34	18	63	76	31	50	56	90	76	26	21
7	6	24	19	22	7	34	43	26	23	56	44	10	57	34	40	46	27	62	97	46	42	68	100	103	36	15
8	6	20	13	16	5	27	33	22	20	42	34	10	47	26	33	36	21	52	78	37	37	59	80	84	29	15
9	6	14	13	10	3	23	23	22	14	35	23	10	38	26	24	27	15	47	62	27	37	47	71	64	22	15
10	3	14	9	13	5	19	25	14	14	32	30	6	35	19	27	30	19	38	64	32	26	47	64	71	27	10
11	6	4	8	3	.	13	6	18	5	18	6	10	18	19	8	13	4	33	28	9	32	23	46	28	8	15
12	3	10	9	8	3	16	18	14	11	26	20	6	28	19	20	22	13	34	51	23	26	38	56	54	20	10
13	1	6	3	9	4	9	15	6	8	18	20	3	21	9	16	19	15	22	39	24	14	29	38	51	21	6
14	3	9	5	5	2	12	14	11	10	18	15	6	24	13	17	16	10	27	40	19	22	34	42	43	16	10
15	3	3	5	2	.	9	5	11	4	13	5	6	14	13	7	10	3	23	23	8	22	19	35	23	7	10
16	1	5	3	5	2	7	11	6	7	14	13	3	17	9	13	13	10	19	31	17	14	25	32	38	15	6
17	3	.	2	.	.	3	.	8	.	4	.	6	4	8	.	3	.	13	6	.	18	5	18	6	.	10
18	1	4	1	1	1	4	6	4	6	6	6	3	12	5	9	5	5	12	19	10	11	20	18	21	9	6

1	.	1	.	2	2	.	1	2	3	2	.	5	4	6	3	10	4	3	10	9	6	8	3	16	18	7	14	15	26	20	6	6	19
	1	.	1	.	2	2	.	2	1	3	2	3	5	2	3	6	10	1	8	3	8	9	8	9	19	12	6	15	18	26	13	3	20
		1	.	1	1	.	1	2	1	1	.	4	3	4	3	6	3	3	9	5	5	5	2	12	14	6	11	14	18	15	5	6	21
			1	.	1	1	.	2	1	1	1	2	4	2	3	4	6	1	7	3	7	5	5	7	15	9	6	14	14	18	9	3	22
				1	.	.	1	.	1	.	.	2	.	4	.	3	.	3	3	5	.	2	.	9	5	.	11	4	13	5	.	6	23
					1	.	.	1	1	1	.	2	2	2	2	4	3	1	5	3	4	5	2	7	11	5	6	10	14	13	5	3	24
						1	.	.	.	1	1	.	2	.	1	2	4	.	3	1	3	3	5	3	7	7	2	6	6	14	8	1	25
							1	.	.	.	.	.	.	2	.	.	.	3	.	2	.	.	.	3	.	.	8	.	4	.	.	6	26
								1	.	.	.	1	1	1	2	1	1	1	4	1	3	1	1	4	6	3	4	9	6	6	3	3	27
									1	.	.	1	.	2	.	2	.	1	2	3	.	2	.	5	4	.	6	3	10	4	.	3	28
										1	.	.	1	.	1	2	2	.	2	1	2	3	2	3	5	4	2	5	6	10	4	1	29
											1	.	.	.	.	.	2	.	.	.	1	1	3	.	3	3	.	2	2	6	6	.	30
												1	.	1	.	1	.	1	2	1	.	1	.	4	3	.	4	3	6	3	.	3	31
													1	.	1	1	1	.	2	1	2	1	1	2	4	3	2	5	4	6	3	1	32
														1	.	.	.	1	.	1	.	.	.	2	.	.	4	.	3	.	.	3	33
															1	.	.	.	1	.	1	.	.	1	1	1	1	4	1	1	1	1	34
																1	.	.	1	1	.	1	.	2	2	.	2	2	4	3	.	1	35
																	1	.	.	.	1	1	1	.	2	2	.	2	2	4	3	.	36
																		1	.	.	.	.	.	.	.	.	2	.	.	.	.	3	37
																			1	.	.	.	.	1	1	.	1	2	1	1	.	1	38
																				1	.	.	.	1	.	.	2	.	2	.	.	1	39
																					1	.	.	.	1	1	.	2	1	1	1	.	40
																						1	.	.	1	.	.	1	2	2	.	.	41
																							1	.	.	1	.	.	.	2	2	.	42
																								1	.	.	1	.	1	.	.	1	43
																									1	.	.	1	1	1	.	.	44
																										1	.	.	.	1	1	.	45
																											1	.	.	.	.	1	46
																												1	.	.	.	.	47
																													1	.	.	.	48
																														1	.	.	49
																															1	.	50
																																1	51

Table 3. *(continued)*

ALGEBRA B4, REPRESENTATIONS OF CLASS 0

NUMBER	0	1	2	3	4	5	6	7	8	9	10	11	12	13	14	15	16	17	18	19	20	21	22	23	24	25
S.P.	0	4	8	16	12	16	20	36	24	32	28	40	64	36	44	40	52	48	48	68	52	100	56	64	72	72
O.S.	1	8	24	8	32	16	48	8	96	24	64	48	8	96	96	96	48	32	64	48	64	8	192	16	24	96
LEVEL	1	9	15	17	19	21	23	25	27	29	29	31	33	33	35	35	37	37	37	39	39	41	41	41	43	43
DIM'N	1	9	36	44	84	126	231	156	594	495	924	910	450	1650	2457	2772	2574	1980	3900	2772	4158	1122	9009	2772	4004	7700
	0	1	0	2	0	0	1	3	1	0	1	2	4	0	2	0	1	0	2	3	0	5	1	0	0	3
	0	0	1	0	0	0	1	0	0	2	0	1	0	1	0	1	2	0	0	1	0	0	1	0	3	0
	0	0	0	0	1	0	0	0	1	0	0	0	0	1	1	0	0	2	0	0	1	0	1	0	0	1
	>	>	>	>	>	>	>	>	>	>	>	>	>	>	>	>	>	>	>	>	>	>	>	>	>	>
WEIGHT	0	0	0	0	0	2	0	0	0	0	2	0	0	0	0	2	0	0	2	0	2	0	0	4	0	0
0	1	1	4	4	4	6	7	4	18	15	20	22	10	26	33	44	30	28	52	28	46	10	105	28	44	76
1		1	1	1	3	3	7	4	9	6	15	10	4	23	30	29	30	17	36	28	37	10	65	20	24	48
2			1	1	1	2	2	1	8	6	9	10	4	12	16	23	14	15	29	13	26	4	60	16	24	45
3				1	.	.	1	1	3	3	3	7	4	5	9	9	9	8	15	10	12	4	35	8	15	30
4					1	1	2	1	3	2	7	3	1	11	15	15	14	9	19	13	21	4	34	12	12	25
5						1	.	.	3	2	4	3	1	5	6	12	5	8	16	4	15	1	32	10	12	24
6							1	1	1	1	2	2	1	5	8	6	9	4	9	10	10	4	19	6	8	16
7								1	.	.	.	1	1	.	3	.	3	.	3	7	.	4	5	.	3	9
8									1	1	1	2	1	2	3	5	3	4	7	3	7	1	18	5	8	15
9										1	.	1	1	1	1	2	2	2	2	2	3	1	10	3	6	8
10											1	.	.	2	3	3	3	2	4	3	6	1	8	4	3	6
11												1	1	.	1	.	1	.	2	2	.	1	5	.	3	8
12													1	.	.	.	.	.	.	1	.	1	.	.	.	3
13														1	1	1	2	1	1	2	3	1	4	2	2	3
14															1	.	1	.	1	2	.	1	2	.	1	3
15																1	.	1	1	.	2	.	4	2	2	3
16																	1	.	.	1	.	1	1	.	1	1
17																		1	.	.	1	.	2	1	1	1

NUMBER	26	27	28	29	30	31	32	33	34	35	36	37	38	39	40	41	42	43	44	45	46	47	48	49	50	51
S.P.	60	80	68	76	104	76	72	144	84	80	84	100	108	88	88	116	96	112	92	148	104	108	100	104	196	128
O.S.	192	48	96	64	48	96	192	8	192	96	64	48	96	96	192	48	96	64	192	48	192	32	192	96	8	24
LEVEL	43	45	45	45	47	47	47	49	49	49	49	51	51	51	51	53	53	53	53	55	55	55	55	55	57	57
DIM'N	15444	8748	12012	12375	7140	15444	25740	2508	31500	27456	18018	18018	20196	27027	54675	23868	44352	32725	60060	16302	71500	23595	96228	46332	5148	22932
	1	2	1	3	4	0	1	6	2	0	1	1	4	0	2	3	2	4	0	5	1	0	2	0	7	0
	1	2	0	0	1	2	0	0	1	2	0	3	0	1	1	2	0	0	1	1	2	0	0	1	0	4
	0	0	2	0	0	1	1	0	1	0	0	0	1	2	0	0	2	0	1	0	1	3	1	0	0	0
	>	>	>	>	>	>	>	>	>	>	>	>	>	>	>	>	>	>	>	>	>	>	>	>	>	>
WEIGHT	2	0	0	2	0	0	2	0	0	2	4	0	0	0	2	0	0	2	2	0	0	0	2	4	0	0
0	140	84	92	95	60	108	196	20	196	192	114	98	108	171	339	116	256	173	324	70	388	107	468	228	20	116
1	116	48	83	80	34	99	148	10	184	144	95	98	102	123	261	116	184	134	276	70	274	96	405	183	20	74
2	83	48	55	58	34	64	124	10	120	122	74	58	66	112	224	69	170	116	216	40	258	72	320	156	10	74
3	45	36	32	36	28	39	72	10	80	72	44	42	48	72	144	54	116	80	135	34	178	48	210	99	10	54
4	69	24	50	49	16	60	93	4	113	91	62	58	63	78	168	69	119	87	185	40	175	66	276	126	10	46
5	48	24	32	34	16	36	78	4	68	78	49	31	36	72	144	36	110	76	144	19	166	48	216	108	4	46
6	37	17	30	29	13	37	53	4	75	54	37	42	45	49	105	54	78	58	116	34	119	44	181	81	10	34
7	9	9	8	15	10	12	12	4	35	16	8	21	30	15	45	36	32	36	35	28	54	17	72	24	10	18
8	25	17	18	19	13	22	45	4	43	46	29	22	25	46	89	27	73	49	90	16	113	32	139	70	4	34
9	12	12	10	9	10	14	25	4	26	28	17	16	16	29	51	20	48	29	56	13	77	21	86	46	4	27
10	21	6	17	16	4	21	33	1	41	33	25	22	24	30	63	27	47	34	78	16	70	30	120	57	4	19
11	6	9	4	9	10	7	10	4	19	14	6	11	16	15	37	17	30	29	27	13	52	11	53	21	4	18
12	.	3	.	3	7	.	.	4	5	.	.	3	9	.	9	9	8	15	.	10	12	.	12	.	4	6
13	10	4	10	7	3	13	18	1	25	19	15	16	15	19	35	20	29	19	49	13	46	20	75	37	4	14
14	5	3	4	7	3	7	7	1	18	10	5	11	15	9	25	17	18	19	24	13	31	11	45	17	4	10
15	6	4	5	4	3	6	16	1	11	16	12	6	6	18	30	7	28	16	37	4	44	14	56	33	1	14
16	2	2	2	2	2	5	3	1	10	6	3	9	8	6	12	12	10	9	15	10	21	7	25	11	4	8
17	2	3	3	1	2	3	9	1	6	8	7	4	3	12	15	5	18	7	23	3	28	10	33	21	1	10

1	.	.	.	2	.	1	3	3	3	2	4	3	3	6	1	8	9	4	4	6	9	21	6	17	16	18	4	30	7	33	15	1	10	18
	1	.	1	.	.	.	1	.	1	.	2	2	.	.	1	5	.	.	3	8	.	6	9	4	9	.	10	7	.	10	.	4	3	19
		1	.	.	1	.	.	2	.	3	1	.	3	6	.	6	5	7	4	3	6	9	5	9	4	20	3	13	9	30	18	1	4	20
			1	.	.	.	.	.	.	.	.	1	.	.	1	.	.	.	.	3	.	.	3	.	3	.	7	.	.	.	.	4	.	21
				1	.	1	1	1	2	1	1	2	2	3	1	4	5	2	3	3	6	10	4	10	7	11	3	20	5	18	10	1	8	22
					1	.	.	.	.	.	.	.	.	3	.	.	2	4	.	.	3	3	.	6	1	8	.	7	3	12	12	.	3	23
						1	.	.	1	.	.	1	1	.	1	1	2	.	2	1	2	2	2	2	2	3	2	10	2	3	3	1	6	24
							1	.	1	.	1	2	.	.	1	2	.	.	1	3	.	5	3	4	7	.	3	7	.	7	.	1	3	25
								1	.	1	1	.	2	2	.	4	3	2	3	3	3	6	4	5	4	10	3	9	5	16	8	1	3	26
									1	.	.	1	.	.	1	1	.	.	1	1	.	2	2	2	2	.	2	5	.	3	.	1	3	27
										1	.	.	1	1	.	2	1	1	2	1	2	2	3	3	1	6	2	5	4	9	5	1	2	28
											1	.	.	.	.	2	.	.	1	3	.	3	3	2	4	.	3	3	.	6	.	1	1	29
												1	.	.	1	.	.	.	.	1	.	.	1	.	2	.	2	.	.	.	.	1	.	30
													1	.	.	1	1	.	2	1	1	1	2	1	1	3	2	4	2	3	2	1	2	31
														1	.	.	1	1	.	.	2	2	.	3	1	4	.	5	2	6	5	.	2	32
															1	.	.	.	.	.	.	.	.	.	.	.	1	.	.	.	.	1	.	33
																1	.	.	1	1	.	1	2	1	1	.	2	2	.	3	.	1	1	34
																	1	.	.	.	1	1	.	1	1	2	.	4	1	2	2	.	2	35
																		1	.	.	.	.	.	.	.	2	.	.	1	3	3	.	.	36
																			1	.	.	.	1	.	.	.	1	1	.	.	.	1	1	37
																				1	.	.	1	.	1	.	2	.	.	.	.	1	.	38
																					1	.	.	1	.	1	.	2	1	1	1	.	1	39
																						1	.	1	1	.	.	2	.	2	.	.	1	40
																							1	.	.	.	1	.	.	.	.	1	.	41
																								1	.	.	.	1	.	1	.	.	1	42
																									1	.	.	.	.	.	.	.	.	43
																										1	.	.	1	1	1	.	.	44
																											1	.	.	.	.	1	.	45
																												1	.	.	.	.	1	46
																													1	.	.	.	.	47
																														1	.	.	.	48
																															1	.	.	49
																																1	.	50
																																	1	51

Table 3. *(continued)*

ALGEBRA B4, REPRESENTATIONS OF CLASS 1

NUMBER	0	1	2	3	4	5	6	7	8	9	10	11	12	13	14	15	16	17	18	19	20	21	22	23	24	25
S.P.	4	12	20	28	28	36	36	52	44	52	52	60	84	60	68	68	76	76	76	92	84	124	84	100	100	100
O.S.	16	64	96	64	64	16	192	64	192	96	64	192	64	192	192	96	192	64	64	192	64	64	384	16	96	192
LEVEL	11	19	25	27	29	31	33	35	37	39	39	41	43	43	45	45	47	47	47	49	49	51	51	51	53	53
DIM'N	16	128	432	576	768	672	2 560	1 920	5 040	4 928	4 608	9 504	5 280	12 672	19 712	12 672	24 192	13 200	18 480	27 648	16 896	12 672	65 536	9 504	34 944	59 136
WEIGHT	0 0 0 > 1	1 0 0 > 1	0 1 0 > 1	2 0 0 > 1	0 0 1 > 1	0 0 0 > 3	1 1 0 > 1	3 0 0 > 1	1 0 1 > 1	0 2 0 > 1	1 0 0 > 3	2 1 0 > 1	4 0 0 > 1	0 1 1 > 1	2 0 1 > 1	0 1 0 > 3	1 2 0 > 1	0 0 2 > 1	2 0 0 > 3	3 1 0 > 1	0 0 1 > 3	5 0 0 > 1	1 1 1 > 1	0 0 0 > 5	0 3 0 > 1	3 0 1 > 1
0	1	4	9	10	12	9	32	20	50	44	40	75	35	92	128	82	144	76	107	144	88	56	328	43	156	260
1		1	3	4	5	4	15	10	26	23	22	42	20	53	77	49	87	47	67	90	56	35	210	28	100	172
2			1	1	2	2	6	4	13	12	12	21	10	30	44	30	50	29	41	52	36	20	132	19	64	109
3				1	.	.	3	4	5	6	4	15	10	14	26	13	33	15	22	42	18	20	80	10	43	77
4					1	1	2	1	6	5	7	9	4	17	24	18	27	18	25	27	24	10	80	13	38	66
5						1	.	.	3	2	4	3	1	8	12	12	12	11	16	12	16	4	48	10	22	38
6							1	1	2	3	2	6	4	8	13	8	18	9	12	21	12	10	48	7	27	44
7								1	.	.	.	3	4	.	5	.	6	.	4	15	.	10	14	.	10	26
8									1	1	1	2	1	4	6	5	8	6	7	9	8	4	28	5	15	24
9										1	.	1	1	2	2	2	6	3	2	6	4	4	16	3	12	13
10											1	.	.	2	3	3	3	3	4	3	6	1	14	4	7	12
11												1	1	.	2	.	3	.	2	6	.	4	8	.	6	13
12													1	.	.	.	.	.	.	3	.	4	.	.	.	5
13														1	1	1	2	2	1	2	3	1	8	2	5	6
14															1	.	1	.	1	2	.	1	4	.	3	6
15																1	.	1	1	.	2	.	4	2	2	3
16																	1	.	.	1	.	1	2	.	3	2
17																		1	.	.	1	.	2	1	1	1

NUMBER	26	27	28	29	30	31	32	33	34	35	36	37	38	39	40	41	42	43	44	45	46	47	48	49	50	51
S.P.	92	108	100	108	132	108	108	172	116	116	124	132	140	124	124	148	132	148	132	180	140	148	140	148	228	164
O.S.	192	192	192	64	192	192	192	64	384	96	64	192	192	192	192	192	192	64	192	192	384	64	192	96	64	96
LEVEL	53	55	55	55	57	57	57	59	59	59	59	61	61	61	61	63	63	63	63	65	65	65	65	65	67	67
DIM'N	67 200	78 624	76 032	56 320	68 640	104 832	99 792	27 456	219 648	112 320	59 136	150 528	149 760	160 160	228 800	206 976	269 568	144 144	219 648	152 064	465 696	128 128	359 424	144 144	54 912	181 440
WEIGHT	1 1 0 > 3	2 2 0 > 1	1 0 2 > 1	3 0 0 > 3	4 1 0 > 1	0 2 1 > 1	1 0 1 > 3	6 0 0 > 1	2 1 1 > 1	0 2 0 > 3	1 0 0 > 5	1 3 0 > 1	4 0 1 > 1	0 1 2 > 1	2 1 0 > 3	3 2 0 > 1	2 0 2 > 1	4 0 0 > 3	0 1 1 > 3	5 1 0 > 1	1 2 1 > 1	0 0 3 > 1	2 0 1 > 3	0 1 0 > 5	7 0 0 > 1	0 4 0 > 1
0	304	321	312	224	245	400	375	84	772	392	196	480	460	511	736	596	798	405	648	384	1 297	336	984	380	120	450
1	201	213	210	153	165	271	258	56	535	272	138	333	325	361	523	422	573	294	466	273	934	246	720	278	84	324
2	132	135	140	102	105	182	178	35	362	190	98	226	222	254	367	288	407	209	336	186	666	180	524	206	56	233
3	76	102	87	67	90	118	108	35	254	119	61	169	172	169	247	234	285	153	220	165	478	125	360	137	56	177
4	87	81	93	67	62	121	123	20	239	131	71	149	146	178	255	188	285	146	244	120	466	132	381	153	35	162
5	56	45	60	44	33	76	87	10	152	92	52	92	92	124	177	116	198	101	176	72	322	96	276	117	20	111
6	49	60	57	41	52	79	74	20	164	83	44	112	109	118	167	150	197	102	160	105	334	91	258	103	35	126
7	13	33	15	22	42	27	18	20	80	27	10	58	77	40	76	102	87	67	52	90	162	35	108	33	35	69
8	31	33	37	25	27	50	52	10	102	57	32	69	66	83	113	91	135	67	116	62	228	67	185	78	20	85
9	16	24	22	12	21	34	30	10	66	36	20	54	44	55	68	70	90	41	76	52	163	46	120	53	20	69
10	21	15	24	16	12	31	36	4	62	38	25	40	38	55	75	51	90	44	86	33	148	49	135	60	10	54
11	8	18	9	12	21	18	12	10	48	19	7	37	44	28	49	60	57	41	38	52	111	25	74	25	20	48
12	.	6	.	4	15	.	.	10	14	.	.	10	26	.	13	33	15	22	.	42	27	.	18	.	20	15
13	10	11	15	7	9	20	21	4	39	23	15	30	24	38	44	39	60	25	56	27	104	35	87	40	10	42
14	5	8	6	7	9	11	8	4	28	13	5	21	24	19	31	33	37	25	28	27	72	19	52	19	10	31
15	6	4	8	4	3	10	16	1	20	16	12	14	12	25	30	18	40	16	40	12	65	24	62	33	4	25
16	2	6	3	2	6	8	4	4	16	8	3	18	13	13	16	24	22	12	18	21	52	13	30	13	10	27
17	2	3	6	1	2	6	9	1	12	8	7	10	6	18	15	14	27	7	26	9	43	19	39	21	4	18

1	.	.	.	2	.	1	3	3	3	3	4	3	6	6	1	14	9	4	10	12	13	21	15	24	16	20	12	45	13	36	15	4	18	18
	1	.	1	.	.	.	2	.	3	.	2	6	.	.	4	8	.	.	6	13	.	8	18	9	12	.	21	18	.	12	.	10	10	19
		1	.	.	1	.	.	2	.	3	1	.	3	6	.	6	5	7	4	3	9	9	5	15	4	20	3	22	12	30	18	1	9	20
			1	.	.	.	.	.	.	.	.	3	.	.	4	.	.	.	.	5	.	.	6	.	4	.	15	.	.	.	.	10	.	21
				1	.	1	1	1	2	2	1	2	4	3	1	8	5	2	8	6	9	10	11	15	7	13	9	31	10	21	10	4	15	22
					1	.	.	.	.	.	.	.	.	3	.	.	2	4	.	.	3	3	.	6	1	8	.	7	4	12	12	.	3	23
						1	.	.	1	.	.	1	2	.	1	2	2	.	6	2	3	2	6	3	2	4	6	16	4	4	3	4	12	24
							1	.	1	.	1	2	.	.	1	4	.	.	3	6	.	5	8	6	7	.	9	11	.	8	.	4	6	25
								1	.	1	1	.	2	2	.	4	3	2	3	3	5	6	4	8	4	10	3	16	7	16	8	1	7	26
									1	.	.	1	.	.	1	2	.	.	3	2	.	2	6	3	2	.	6	8	.	4	.	4	6	27
										1	.	.	1	1	.	2	1	1	2	1	4	2	3	6	1	6	2	10	6	9	5	1	5	28
											1	.	.	.	.	2	.	.	1	3	.	3	3	3	4	.	3	6	.	6	.	1	3	29
												1	.	.	1	.	.	.	.	2	.	.	3	.	2	.	6	.	.	.	.	4	.	30
													1	.	.	1	1	.	2	1	2	1	2	2	1	3	2	8	3	3	2	1	5	31
														1	.	.	1	1	.	.	2	2	.	3	1	4	.	5	3	6	5	.	2	32
															1	.	.	.	.	.	.	.	.	.	.	.	3	.	.	.	.	4	.	33
																1	.	.	1	1	.	1	2	2	1	.	2	4	.	3	.	1	3	34
																	1	.	.	.	1	1	.	1	1	2	.	4	2	2	2	.	2	35
																		1	.	.	.	.	.	.	.	2	.	.	1	3	3	.	.	36
																			1	.	.	.	1	.	.	.	1	2	.	.	.	1	3	37
																				1	.	.	1	.	1	.	2	.	.	.	.	1	.	38
																					1	.	.	1	.	1	.	2	2	1	1	.	1	39
																						1	.	1	1	.	.	2	.	2	.	.	1	40
																							1	.	.	.	1	.	.	.	.	1	.	41
																								1	.	.	.	1	.	1	.	.	1	42
																									1	.	.	.	.	.	.	.	.	43
																										1	.	.	1	1	1	.	.	44
																											1	.	.	.	.	1	.	45
																												1	.	.	.	.	1	46
																													1	.	.	.	.	47
																														1	.	.	.	48
																															1	.	.	49
																																1	.	50
																																	1	51

Table 3. *(continued)*

ALGEBRA B5, REPRESENTATIONS OF CLASS 0

NUMBER	0	1	2	3	4	5	6	7	8	9	10	11	12	13	14	15	16	17	18	19	20	21	22	23	24	25
S.P.	0	4	8	16	12	20	16	36	20	24	32	40	28	64	32	36	44	52	40	68	48	48	44	100	52	56
O.S.	1	10	40	10	80	80	80	10	32	240	40	80	320	10	160	240	240	80	480	80	80	320	320	10	160	480
LEVEL	1	11	19	21	25	29	29	31	31	35	37	39	39	41	41	43	45	47	47	49	49	49	49	51	51	53
DIM'N	1	11	55	65	165	429	330	275	462	1430	1144	2025	3003	935	4290	5005	7128	7150	11583	7293	7865	15400	17160	2717	22275	33033
	0	1	0	2	0	1	0	3	0	1	0	2	1	4	1	0	2	1	0	3	0	2	0	5	2	1
	0	0	1	0	0	1	0	0	0	0	2	1	0	0	0	1	0	2	1	1	0	0	1	0	0	1
	0	0	0	0	1	0	0	0	0	1	0	0	0	0	0	1	1	0	0	0	2	0	0	0	0	1
	0	0	0	0	0	0	1	0	0	0	0	0	1	0	0	0	0	0	1	0	0	1	0	0	0	0
	>	>	>	>	>	>	>	>	>	>	>	>	>	>	>	>	>	>	>	>	>	>	>	>	>	>
WEIGHT	0	0	0	0	0	0	0	0	2	0	0	0	0	0	2	0	0	0	0	0	0	0	2	0	2	0
0	1	1	5	5	5	9	10	5	10	30	24	35	35	15	50	45	56	50	105	45	65	120	120	15	135	231
1		1	1	1	4	9	4	5	6	12	8	13	28	5	30	41	52	50	57	45	33	68	88	15	102	119
2			1	1	1	2	3	1	3	11	8	13	13	5	21	17	22	19	48	17	30	58	56	5	66	112
3				1	.	1	.	1	.	4	4	9	4	5	6	7	12	12	18	13	15	28	20	5	30	63
4					1	2	1	1	2	3	2	3	11	1	12	16	21	19	24	17	14	29	42	5	50	50
5						1	.	1	.	1	1	2	3	1	3	7	11	12	9	13	6	13	15	5	21	27
6							1	.	1	3	2	3	4	1	9	5	6	5	21	4	13	26	26	1	31	48
7								1	.	.	.	1	.	1	.	.	4	4	.	9	.	4	.	5	6	7
8									1	.	.	.	4	.	5	5	6	5	9	4	5	10	20	1	25	16
9										1	1	2	1	1	2	2	3	3	8	3	6	11	9	1	12	26
10											1	1	.	1	.	1	1	2	3	2	3	3	3	1	3	14
11												1	.	1	.	.	1	1	.	2	.	3	.	1	3	7
12													1	.	1	2	3	3	3	3	2	4	7	1	9	8
13														1	.	.	.	.	.	1	.	.	.	1	.	.
14															1	.	.	.	3	.	2	4	4	.	5	8
15																1	1	2	1	2	1	1	2	1	2	4
16																	1	1	.	2	.	1	.	1	2	2
17																		1	.	1	.	.	.	1	.	1

NUMBER	26	27	28	29	30	31	32	33	34	35	36	37	38	39	40	41	42	43	44	45	46	47	48	49	50	51
S.P.	52	72	72	56	80	60	64	104	68	76	64	68	144	76	80	80	84	72	100	108	80	76	116	88	88	84
O.S.	320	40	240	320	80	960	80	80	240	320	640	160	10	240	160	32	480	960	80	240	480	960	80	240	960	320
LEVEL	53	55	55	55	57	57	57	59	59	59	59	59	61	61	61	61	63	63	65	65	65	65	67	67	67	67
DIM'N	23595	13650	26520	37752	28798	78650	23595	21945	57915	58344	117975	47190	7007	70070	85085	28314	137445	178750	72930	81510	175175	289575	91960	162162	333234	188760
	0	0	3	0	2	1	0	4	1	3	1	0	6	0	3	0	2	1	1	4	0	1	3	0	2	1
	0	3	0	0	2	1	0	1	0	0	1	0	0	2	0	0	1	0	3	0	2	0	2	1	1	0
	1	0	1	1	0	0	0	0	2	0	0	0	0	1	0	0	1	1	0	1	0	1	0	2	0	0
	1	0	0	0	0	1	2	0	0	1	0	1	0	0	0	0	0	1	0	0	1	0	0	0	1	2
	>	>	>	>	>	>	>	>	>	>	>	>	>	>	>	>	>	>	>	>	>	>	>	>	>	>
WEIGHT	0	0	0	2	0	0	0	0	0	0	2	2	0	0	2	4	0	0	0	0	0	2	0	0	0	0
0	125	90	160	200	170	350	115	115	225	224	525	190	35	250	325	102	445	690	210	230	625	925	240	534	1074	520
1	105	42	84	136	82	302	72	55	207	196	365	146	15	234	230	74	423	446	210	220	401	737	240	330	702	444
2	61	42	80	104	82	177	60	55	114	116	285	103	15	126	182	57	229	387	105	118	353	530	120	306	624	302
3	26	26	52	44	60	90	28	45	63	68	140	46	15	74	102	28	147	210	74	84	197	276	92	186	378	164
4	52	17	35	70	33	155	38	21	106	103	193	80	5	120	124	42	219	241	105	114	217	423	120	178	384	259
5	23	11	22	29	23	78	17	17	60	58	92	37	5	71	66	20	140	126	74	80	120	220	92	105	225	143
6	29	17	34	54	33	83	32	21	53	54	152	57	5	57	99	33	103	211	45	51	194	300	50	167	345	174
7	.	4	12	.	12	18	.	13	15	28	20	.	5	22	30	.	63	26	36	52	33	44	60	30	90	28
8	25	5	10	36	9	75	20	5	50	50	100	45	1	55	65	25	100	125	45	50	110	240	50	89	195	150
9	12	11	21	23	23	40	15	17	28	29	71	26	5	33	50	16	63	112	31	35	107	153	37	100	200	93
10	5	8	11	9	16	18	8	13	15	13	30	12	5	20	21	8	37	58	22	22	60	74	27	60	108	49
11	.	4	11	.	12	9	.	13	6	13	15	.	5	10	21	.	27	23	15	22	30	29	23	30	78	17
12	11	3	6	15	6	36	9	4	27	26	45	21	1	32	31	12	61	63	31	34	59	123	37	51	108	82
13	.	.	4	.	4	.	.	9	.	4	.	.	5	.	6	.	7	.	4	12	.	.	12	.	18	.
14	5	3	6	12	6	15	8	4	10	10	36	15	1	11	25	10	20	57	9	10	54	84	10	49	99	52
15	5	2	3	6	4	16	4	3	15	11	18	10	1	19	12	6	36	31	22	21	31	60	27	30	56	45
16	.	1	3	.	3	8	.	3	6	11	9	.	1	10	12	.	26	12	15	21	16	23	23	14	40	15
17	.	1	1	.	2	3	.	2	3	3	3	.	1	7	3	.	14	5	12	11	9	9	16	9	18	8

1	.	1	1	1	.	1	4	2	2	3	5	4	6	4	3	5	4	14	7	1	6	9	5	11	29	6	6	28	40	7	29	51	27	18
	1	.	.	.	1	.	.	.	.	1	.	1	.	.	2	.	3	.	.	1	.	3	.	7	.	4	11	.	.	12	.	9	.	19
		1	.	.	.	.	2	1	1	1	2	3	2	2	2	3	1	4	3	1	3	2	3	6	15	4	3	13	18	5	18	24	14	20
			1	.	.	1	2	.	1	3	.	3	3	.	3	2	4	7	.	1	3	9	.	8	11	4	6	15	15	6	14	36	9	21
				1	.	1	.	2	.	.	3	.	6	2	.	5	4	8	6	.	6	5	4	11	13	6	6	12	33	7	11	21	25	22
					1	.	.	.	.	.	.	.	.	.	1	.	.	.	.	1	.	.	.	.	.	.	4	.	.	4	.	.	.	23
						1	.	.	.	.	.	.	3	.	.	2	4	4	.	.	3	5	.	8	5	4	6	6	12	6	5	15	8	24
							1	.	1	1	.	2	1	.	2	1	1	2	.	1	2	2	.	4	5	3	3	8	6	4	9	16	4	25
								1	.	.	1	.	2	1	.	3	1	2	3	.	3	1	2	6	6	4	3	5	15	5	6	9	14	26
									1	.	.	1	.	.	1	.	.	.	.	1	1	.	.	1	.	2	1	3	.	2	3	3	.	27
										1	.	1	.	.	2	.	1	.	.	1	.	2	.	2	.	1	3	.	.	3	.	8	.	28
											1	.	.	1	.	.	.	2	2	.	.	1	2	.	6	.	.	5	9	.	6	9	7	29
												1	.	.	1	.	.	.	.	1	.	.	.	1	.	1	1	.	.	2	.	3	.	30
													1	.	.	1	1	1	.	.	2	1	.	4	2	3	3	3	5	4	3	6	4	31
														1	.	.	.	.	1	.	.	.	1	.	3	.	.	2	3	.	3	3	4	32
															1	.	.	.	.	1	.	.	.	.	.	.	1	.	.	1	.	.	.	33
																1	.	.	.	.	1	.	.	2	1	2	1	1	2	3	2	2	2	34
																	1	.	.	.	.	1	.	2	.	1	3	.	.	3	.	3	.	35
																		1	.	.	.	1	.	.	2	.	.	3	3	.	3	6	2	36
																			1	.	.	.	1	.	.	.	.	.	3	.	.	.	4	37
																				1	.	.	.	.	.	.	.	.	.	.	.	.	.	38
																					1	.	.	1	.	2	1	1	.	2	1	1	.	39
																						1	.	.	.	.	.	.	.	.	.	3	.	40
																							1	.	.	.	.	.	.	.	.	.	.	41
																								1	.	1	1	.	.	2	.	1	.	42
																									1	.	.	1	1	.	2	2	1	43
																										1	.	.	.	1	.	.	.	44
																											1	.	.	1	.	.	.	45
																												1	.	.	1	1	.	46
																													1	.	.	.	1	47
																														1	.	.	.	48
																															1	.	.	49
																																1	.	50
																																	1	51

Table 3. *(continued)*

ALGEBRA B5, REPRESENTATIONS OF CLASS 1

NUMBER	0	1	2	3	4	5	6	7	8	9	10	11	12
S.P.	5	13	21	29	29	37	37	53	45	45	53	61	53
O.S.	32	160	320	160	320	640	160	160	32	960	320	640	640
LEVEL	16	26	34	36	40	44	44	46	46	50	52	54	54
DIM'N	32	320	1 408	1 760	3 520	10 240	5 280	7 040	4 224	28 512	24 960	45 760	45 056
WEIGHT	0 0 0 0 > 1	1 0 0 0 > 1	0 1 0 0 > 1	2 0 0 0 > 1	0 0 1 0 > 1	1 1 0 0 > 1	0 0 0 1 > 1	3 0 0 0 > 1	0 0 0 0 > 3	1 0 1 0 > 1	0 2 0 0 > 1	2 1 0 0 > 1	1 0 0 1 > 1
0	1	5	14	15	25	60	30	35	21	126	100	165	164
1		1	4	5	9	24	12	15	9	57	46	80	80
2			1	1	3	8	5	5	4	24	20	34	38
3				1	.	4	.	5	.	9	10	24	12
4					1	2	2	1	2	9	7	12	18
5						1	.	1	.	3	4	8	5
6							1	.	1	3	2	3	8
7								1	.	.	.	4	.
8									1	.	.	.	4
9										1	1	2	2
10											1	1	.
11												1	.
12													1
13													
14													
15													

NUMBER	13	14	15	16	17	18	19	20	21	22	23	24	25
S.P.	85	61	61	69	77	69	93	77	77	77	125	85	85
O.S.	160	160	960	960	640	960	640	320	640	320	160	160	1 920
LEVEL	56	56	58	60	62	62	64	64	64	64	66	66	68
DIM'N	22 880	36 960	91 520	134 784	147 840	160 160	157 696	128 128	219 648	137 280	64 064	183 040	573 440
WEIGHT	4 0 0 0 > 1	1 0 0 0 > 3	0 1 1 0 > 1	2 0 1 0 > 1	1 2 0 0 > 1	0 1 0 1 > 1	3 1 0 0 > 1	0 0 2 0 > 1	2 0 0 1 > 1	0 1 0 0 > 3	5 0 0 0 > 1	2 0 0 0 > 3	1 1 1 0 > 1
0	70	120	285	381	385	420	364	304	524	325	126	395	1 200
1	35	61	145	201	205	226	200	168	292	181	70	227	684
2	15	31	71	99	102	121	100	91	157	101	35	128	376
3	15	9	32	57	66	50	80	44	80	40	35	61	220
4	5	16	33	45	46	63	44	48	82	57	15	72	196
5	5	4	15	24	30	26	34	23	38	22	15	31	113
6	1	9	13	18	17	33	16	24	42	32	5	41	96
7	5	.	.	9	10	.	24	.	12	.	15	9	32
8	.	5	5	6	5	15	4	10	20	20	1	25	40
9	1	2	6	9	11	13	12	12	18	12	5	16	54
10	1	.	3	3	8	5	8	6	5	4	5	4	30
11	1	.	.	3	4	.	8	.	5	.	5	4	15
12	.	1	2	3	3	6	3	5	8	7	1	9	22
13	1	.	.	.	.	.	4	.	.	.	5	.	.
14		1	.	.	.	3	.	2	4	4	.	5	8
15			1	1	2	2	2	3	2	2	1	2	12

NUMBER	26	27	28	29	30	31	32	33	34	35	36	37	38
S.P.	85	101	101	93	109	93	101	133	101	109	101	109	173
O.S.	640	320	960	320	640	1 920	160	640	960	640	640	160	160
LEVEL	68	70	70	70	72	72	72	74	74	74	74	74	76
DIM'N	292 864	264 000	480 480	274 560	570 240	1 034 880	251 680	457 600	896 896	798 720	901 120	302 016	160 160
WEIGHT	0 0 1 1 > 1	0 3 0 0 > 1	3 0 1 0 > 1	0 0 1 0 > 3	2 2 0 0 > 1	1 1 0 1 > 1	0 0 0 2 > 1	4 1 0 0 > 1	1 0 2 0 > 1	3 0 0 1 > 1	1 1 0 0 > 3	0 0 0 1 > 3	6 0 0 0 > 1
0	596	495	895	510	990	1 860	425	700	1 479	1 280	1 480	469	210
1	344	285	525	302	584	1 108	256	420	895	784	904	289	126
2	198	159	291	180	325	648	156	235	530	464	548	180	70
3	92	105	201	82	240	354	76	200	312	292	288	88	70
4	114	82	151	108	168	372	95	120	306	266	330	114	35
5	53	56	99	49	122	198	47	100	178	157	168	56	35
6	64	39	72	66	78	207	59	54	169	148	200	73	15
7	.	20	57	.	66	50	.	80	44	80	40	.	35
8	36	15	30	40	30	115	35	20	89	80	120	49	5
9	30	26	45	30	56	107	29	44	98	82	98	36	15
10	14	20	24	13	40	52	15	34	56	38	44	18	15
11	.	10	24	.	30	26	.	34	23	38	22	.	15
12	17	10	18	18	21	57	18	16	51	42	57	24	5
13	.	.	9	.	10	.	.	24	.	12	.	.	15
14	8	3	6	12	6	27	11	4	22	20	36	16	1
15	8	7	9	8	15	26	9	12	30	18	24	12	5

NUMBER	39	40	41	42	43	44	45	46	47	48	49	50	51
S.P.	109	117	125	117	109	133	141	117	117	149	125	125	125
O.S.	960	160	32	1 920	1 920	640	960	960	960	640	960	1 920	640
LEVEL	76	76	76	78	78	80	80	80	80	82	82	82	82
DIM'N	1 140 480	672 672	151 008	2 288 000	2 114 112	1 351 680	1 425 600	2 168 320	2 013 440	1 757 184	2 358 720	4 212 000	1 921 920
WEIGHT	0 2 1 0 > 1	3 0 0 0 > 3	0 0 0 0 > 5	2 1 1 0 > 1	1 0 1 1 > 1	1 3 0 0 > 1	4 0 1 0 > 1	0 2 0 1 > 1	1 0 1 0 > 3	3 2 0 0 > 1	0 1 2 0 > 1	2 1 0 1 > 1	1 0 0 2 > 1
0	1 755	986	207	3 275	3 041	1 760	1 800	2 910	2 670	2 086	2 925	5 265	2 340
1	1 071	619	131	2 035	1 904	1 100	1 140	1 840	1 706	1 326	1 875	3 388	1 516
2	639	380	84	1 221	1 182	664	690	1 153	1 087	804	1 185	2 135	981
3	399	227	43	828	672	484	525	688	606	640	747	1 368	568
4	369	230	55	703	726	382	396	709	693	460	735	1 317	635
5	231	128	28	466	408	279	291	422	381	360	462	820	366
6	201	139	37	383	442	204	213	430	441	244	444	795	409
7	75	61	.	220	92	140	201	128	82	240	144	354	76
8	105	85	26	195	260	100	105	250	285	116	255	465	265
9	125	72	19	248	246	148	151	251	240	188	279	481	235
10	81	31	10	155	133	112	99	148	125	142	177	275	134
11	42	31	.	113	53	76	99	77	49	122	87	198	47
12	63	41	13	123	144	71	72	144	153	88	159	273	152
13	.	9	.	32	.	20	57	.	.	66	.	50	.
14	27	25	10	52	84	28	30	82	96	34	87	153	95
15	39	16	7	75	78	52	45	81	79	66	102	151	86

1	1	.	2	.	2	.	1	2	6	.	4	9	.	11	13	.	12	12	18	12	.	5	21	16	.	54	30	36	45	44	30	56	50	107	29	16
	1	.	1	.	.	.	1	.	3	.	4	3	.	8	5	.	8	6	5	4	.	5	15	4	.	30	14	30	24	26	13	40	33	52	15	17
		1	.	1	1	1	.	1	4	4	2	3	5	4	12	6	3	13	8	14	8	1	16	9	5	31	45	20	18	44	49	25	54	81	55	18
			1	.	.	.	1	.	.	.	.	3	.	4	.	.	8	.	5	.	.	5	.	4	.	15	.	10	24	.	.	30	.	26	.	19
				1	.	.	.	.	2	2	1	1	2	3	4	3	2	9	2	4	4	1	9	2	3	18	24	14	9	21	24	19	36	39	31	20
					1	.	.	1	2	.	1	3	.	3	6	.	3	5	8	7	.	1	9	9	.	22	17	14	18	24	18	21	27	57	18	21
						1	.	1	.	2	.	.	3	.	6	3	.	5	4	8	6	.	6	5	4	11	22	6	6	21	33	7	24	39	35	22
							1	.	.	.	.	.	.	.	.	.	4	.	.	.	.	5	.	.	.	.	.	.	9	.	.	10	.	.	.	23
								1	.	.	.	.	.	.	3	.	.	2	4	4	.	.	3	5	.	8	8	4	6	12	12	6	12	27	11	24
									1	.	1	1	.	2	2	.	2	3	2	2	.	1	6	2	.	12	8	11	9	13	8	15	18	26	9	25
										1	.	.	1	.	2	2	.	3	1	2	3	.	3	1	2	6	12	4	3	10	15	5	15	18	21	26
											1	.	.	1	.	.	1	.	.	.	.	1	3	.	.	3	.	8	3	5	.	8	6	5	.	27
												1	.	1	.	.	2	.	2	.	.	1	.	2	.	6	.	4	9	.	.	11	.	13	.	28
													1	.	.	1	.	.	.	2	2	.	.	1	2	.	6	.	.	5	9	.	6	9	11	29
														1	.	.	1	.	.	.	.	1	.	.	.	3	.	4	3	.	.	8	.	5	.	30
															1	.	.	1	1	1	.	.	2	1	.	4	4	3	3	6	5	4	8	12	6	31
																1	.	.	.	.	1	.	.	.	1	.	3	.	.	2	3	.	3	3	8	32
																	1	.	.	.	.	1	.	.	.	.	.	.	3	.	.	4	.	.	.	33
																		1	.	.	.	.	1	.	.	2	2	2	1	2	2	3	6	4	3	34
																			1	.	.	.	.	1	.	2	.	1	3	.	.	3	.	6	.	35
																				1	.	.	.	1	.	.	2	.	.	3	3	.	3	6	3	36
																					1	.	.	.	1	.	.	.	.	.	3	.	.	.	4	37
																						1	.	.	.	.	.	.	.	.	.	.	.	.	.	38
																							1	.	.	1	.	2	1	2	.	2	3	2	.	39
																								1	.	.	.	.	.	.	.	.	.	3	.	40
																									1	.	.	.	.	.	.	.	.	.	.	41
																										1	.	1	1	.	.	2	.	2	.	42
																											1	.	.	1	1	.	2	2	2	43
																												1	.	.	.	1	.	.	.	44
																													1	.	.	1	.	.	.	45
																														1	.	.	1	1	.	46
																															1	.	.	.	1	47
																																1	.	.	.	48
																																	1	.	.	49
																																		1	.	50
																																			1	51

Table 3. *(continued)*

ALGEBRA B6, REPRESENTATIONS OF CLASS 0

0	4	8	16	12	20	36	16	20	24	24	32	40	64	28	36	32	44	36	52	68	40	100	48	48	44	S.P.
1	12	60	12	160	120	12	240	192	480	64	60	120	12	960	480	960	480	384	120	120	1440	12	160	960	1920	O.S.
1	13	23	25	31	35	37	37	41	43	43	45	47	49	49	53	53	55	55	57	59	59	61	61	61	63	LEVEL
1	13	78	90	286	715	442	715	1287	2925	1716	2275	3927	1729	7722	12285	14300	17017	19305	16575	16302	36036	5733	23660	46410	70070	DIM'N
0	1	0	2	0	1	3	0	0	1	0	0	2	4	1	0	1	2	1	1	3	0	5	0	2	0	
0	0	1	0	0	1	0	0	0	0	0	2	1	0	0	1	0	0	0	2	1	1	0	0	0	1	
0	0	0	0	1	0	0	0	0	1	0	0	0	0	0	1	0	1	0	0	0	0	0	2	0	0	
0	0	0	0	0	0	0	1	0	0	0	0	0	0	1	0	0	0	0	0	0	1	0	0	1	0	
0	0	0	0	0	0	0	0	1	0	0	0	0	0	0	0	1	0	0	0	0	0	0	0	0	1	
>	>	>	>	>	>	>	>	>	>	>	>	>	>	>	>	>	>	>	>	>	>	>	>	>	>	
0	0	0	0	0	0	0	0	0	0	2	0	0	0	0	0	0	0	2	0	0	0	0	0	0	0	WEIGHT
0	1	2	3	4	5	6	7	8	9	10	11	12	13	14	15	16	17	18	19	20	21	22	23	24	25	NUMBER
1	1	6	6	6	11	6	15	15	45	20	35	51	21	54	69	100	85	105	75	66	204	21	124	230	250	0
	1	1	1	5	11	6	5	10	15	10	10	16	6	45	64	50	80	70	75	66	94	21	54	110	195	1
		1	1	1	2	1	4	4	14	6	10	16	6	17	22	38	28	40	24	21	82	6	50	97	103	2
			1	.	1	1	.	.	5	.	5	11	6	5	9	10	15	10	15	16	30	6	24	45	35	3
				1	2	1	1	3	3	3	2	3	1	15	21	17	27	27	24	21	33	6	19	39	83	4
					1	1	.	.	1	.	1	2	1	4	9	4	14	6	15	16	12	6	8	17	28	5
						1	.	.	.	.	.	1	1	.	.	.	5	.	5	11	.	6	.	5	.	6
							1	1	3	2	2	3	1	4	5	14	6	15	5	4	30	1	18	36	40	7
								1	.	1	.	.	.	4	5	5	6	11	5	4	9	1	5	10	34	8
									1	.	1	2	1	1	2	3	3	3	3	3	11	1	8	15	13	9
										1	.	.	.	.	.	5	.	6	.	.	9	.	5	10	14	10
											1	1	1	.	1	.	1	.	2	2	4	1	4	4	4	11
												1	1	.	.	.	1	.	1	2	.	1	.	4	.	12
													1	.	.	.	.	.	.	1	.	1	.	.	.	13
														1	2	1	3	2	3	3	3	1	2	4	11	14
															1	.	1	.	2	2	1	1	1	1	3	15
																1	.	1	.	.	3	.	2	4	4	16

56	52	48	72	72	52	56	80	104	60	56	144	68	76	64	60	76	64	84	80	68	68	100	108	72	84	S.P.
960	960	960	60	480	960	384	120	120	2880	1920	12	480	960	240	1280	480	3840	960	960	960	1920	120	480	2880	384	O.S.
65	65	65	67	67	67	67	69	71	71	71	73	73	73	73	73	75	75	77	77	77	77	79	79	79	79	LEVEL
94809	87516	96525	37400	73150	91091	119119	77064	56056	287300	195195	16744	204204	203775	117117	278850	238680	568854	456456	389025	325325	790075	230945	257400	812175	532532	DIM'N
1	2	0	0	3	0	2	2	4	1	0	6	1	3	0	0	0	1	2	3	0	1	1	4	1	3	
1	0	1	3	0	0	0	2	1	1	0	0	0	0	0	0	2	1	1	0	0	1	3	0	0	0	
1	0	0	0	1	1	0	0	0	0	1	0	2	0	0	1	1	0	1	0	0	0	0	1	1	0	
0	0	0	0	0	1	0	0	0	1	0	0	0	1	2	0	0	0	0	0	1	0	0	0	1	0	
0	1	0	0	0	0	0	0	0	0	1	0	0	0	0	0	0	1	0	1	1	0	0	0	0	0	
>	>	>	>	>	>	>	>	>	>	>	>	>	>	>	>	>	>	>	>	>	>	>	>	>	>	
0	0	2	0	0	0	2	0	0	0	0	0	0	0	0	2	0	0	0	0	0	2	0	0	0	2	WEIGHT
26	27	28	29	30	31	32	33	34	35	36	37	38	39	40	41	42	43	44	45	46	47	48	49	50	51	NUMBER
429	276	345	160	290	259	375	300	196	700	555	56	444	435	309	650	480	1386	844	825	605	1575	385	420	1755	924	0
189	220	215	65	130	224	240	125	81	620	325	21	414	390	170	475	455	826	809	500	485	1175	385	405	985	700	1
180	118	152	65	125	110	172	125	81	306	262	21	194	194	146	311	209	679	373	416	298	780	166	185	876	470	2
99	50	50	40	80	45	70	90	66	150	105	21	104	110	65	120	120	315	234	220	125	350	115	130	455	240	3
66	96	93	22	45	97	106	42	26	276	147	6	183	177	78	229	201	381	360	234	242	582	166	180	459	354	4
35	38	30	14	28	41	40	29	21	134	57	6	100	97	33	88	116	171	225	118	105	254	115	125	229	172	5
9	10	.	5	15	.	10	15	16	30	.	6	24	45	.	.	35	35	99	50	.	50	55	80	45	70	6
64	46	67	22	44	43	77	42	26	118	121	6	74	74	68	148	78	320	138	199	146	377	59	66	414	230	7
16	40	40	5	10	39	46	9	5	110	64	1	71	70	34	110	76	164	135	100	120	285	59	65	195	176	8
34	17	21	14	27	17	27	29	21	55	48	6	38	39	30	56	44	141	83	96	61	159	40	45	209	106	9
16	15	30	5	10	14	36	9	5	35	54	1	21	20	30	70	21	144	35	90	70	180	14	15	180	111	10
18	4	6	10	14	7	6	20	16	24	18	6	20	17	15	20	26	56	48	38	26	60	28	28	104	40	11
9	4	.	5	14	.	6	15	16	12	.	6	8	17	.	.	13	28	35	38	.	30	19	28	41	40	12
.	.	.	.	5	.	.	5	11	.	.	6	.	5	.	.	.	.	9	10	.	.	5	15	.	10	13
8	14	12	3	6	16	15	6	4	51	24	1	37	36	14	42	43	69	81	46	52	117	40	44	93	77	14
4	3	3	2	3	7	3	4	3	22	9	1	20	15	6	15	25	26	47	17	23	42	28	27	44	27	15
8	5	9	3	6	5	11	6	4	15	21	1	10	10	13	26	11	60	20	40	29	71	9	10	87	46	16

1	.	1	2	.	1	.	1	.	2	3	.	1	3	.	3	3	3	11	.	1	8	15	.	.	13	13	34	17	.	21	19	27	17	27	17
	1	.	.	.	.	.	.	4	.	5	5	.	.	5	6	.	.	15	9	.	10	10	5	20	11	24	20	15	25	55	9	10	30	36	18
		1	1	.	1	.	.	.	1	.	.	1	1	.	.	2	2	4	.	1	4	4	.	.	9	4	18	4	.	6	15	14	7	6	19
			1	.	1	.	.	.	.	.	.	.	1	.	.	1	2	.	.	1	.	4	.	.	.	.	9	4	.	.	5	14	.	6	20
				1	.	1	1	1	4	1	2	2	3	2	2	4	3	6	8	1	5	4	6	9	6	22	11	14	12	24	6	6	42	15	21
					1	.	.	.	.	.	.	.	.	.	.	.	1	.	.	1	.	.	.	.	.	.	.	.	.	.	.	5	.	.	22
						1	.	.	2	.	.	1	1	1	.	3	2	2	3	1	3	1	3	3	3	6	6	3	5	6	4	3	21	3	23
							1	.	2	1	.	1	3	.	2	3	3	3	.	1	2	4	.	.	3	11	8	14	.	12	4	6	16	15	24
								1	.	1	1	.	.	2	1	.	.	6	3	.	5	4	2	7	6	8	11	5	11	18	6	6	13	11	25
									1	.	.	1	1	.	.	2	2	1	.	1	1	1	.	.	2	3	4	3	.	3	3	3	7	3	26
										1	.	.	.	.	1	.	.	3	.	.	2	4	.	.	3	4	8	5	.	9	4	6	5	11	27
											1	.	.	.	1	.	.	.	3	.	.	.	2	4	.	8	.	5	5	10	.	.	13	6	28
												1	.	.	.	1	1	.	.	1	.	.	.	.	1	.	1	.	.	.	2	1	.	.	29
													1	.	.	1	2	.	.	1	.	1	.	.	.	.	2	3	.	.	1	3	.	3	30
														1	.	.	.	2	1	.	3	1	1	2	3	2	6	1	5	4	4	3	6	2	31
															1	.	.	.	.	.	.	.	.	.	.	4	.	5	.	5	.	.	5	6	32
																1	1	.	.	1	.	.	.	.	.	.	1	.	.	.	1	1	.	.	33
																	1	.	.	1	.	.	.	.	.	.	.	.	.	.	.	1	.	.	34
																		1	.	.	1	1	.	.	2	1	4	1	.	2	3	3	2	2	35
																			1	.	.	.	1	1	.	2	.	1	2	2	.	.	6	1	36
																				1	.	.	.	.	.	.	.	.	.	.	.	.	.	.	37
																					1	.	.	.	1	.	2	.	.	.	2	1	1	.	38
																						1	.	.	.	.	2	1	.	.	1	3	.	2	39
																							1	.	.	.	.	.	1	.	.	.	3	.	40
																								1	.	.	.	.	2	2	.	.	.	1	41
																									1	.	1	.	.	.	2	1	.	.	42
																										1	.	1	.	1	.	.	2	1	43
																											1	.	.	.	1	1	.	.	44
																												1	.	.	.	.	.	1	45
																													1	.	.	.	.	.	46
																														1	.	.	.	1	47
																															1	.	.	.	48
																																1	.	.	49
																																	1	.	50
																																		1	51

Table 3. *(continued)*

ALGEBRA B6, REPRESENTATIONS OF CLASS 1

NUMBER	0	1	2	3	4	5	6	7	8	9	10	11	12	13	14	15	16	17	18	19	20	21	22	23	24	25
S.P.	6	14	22	30	30	38	54	38	46	46	54	54	62	86	54	62	62	70	70	78	94	70	126	78	78	78
O.S.	64	384	960	384	1280	1920	384	960	384	3840	64	960	1920	384	3840	3840	1920	3840	384	1920	1920	5760	384	1280	3840	3840
LEVEL	22	34	44	46	52	56	58	58	62	64	64	66	68	70	70	74	74	76	76	78	80	80	82	82	82	84
DIM'N	64	768	4160	4992	13312	35840	23296	27456	36608	128128	27456	105600	187200	87360	279552	499200	384384	709632	292864	732160	745472	1213056	279552	873600	1601600	1757184
WEIGHT	0 0 0 0 0 > 1	1 0 0 0 0 > 1	0 1 0 0 0 > 1	2 0 0 0 0 > 1	0 0 1 0 0 > 1	1 1 0 0 0 > 1	3 0 0 0 0 > 1	0 0 0 1 0 > 1	0 0 0 0 1 > 1	1 0 1 0 0 > 1	0 0 0 0 0 > 3	0 2 0 0 0 > 1	2 1 0 0 0 > 1	4 0 0 0 0 > 1	1 0 0 1 0 > 1	0 1 1 0 0 > 1	1 0 0 0 1 > 1	2 0 1 0 0 > 1	1 0 0 0 0 > 3	1 2 0 0 0 > 1	3 1 0 0 0 > 1	0 1 0 1 0 > 1	5 0 0 0 0 > 1	0 0 2 0 0 > 1	2 0 0 1 0 > 1	0 1 0 0 1 > 1
0	1	6	20	21	44	100	56	69	76	259	51	195	315	126	444	690	516	900	356	860	784	1350	252	890	1640	1680
1		1	5	6	14	35	21	25	30	104	21	80	135	56	195	315	240	424	171	410	385	660	126	445	825	860
2			1	1	4	10	6	9	12	38	9	30	50	21	82	136	110	184	82	180	170	315	56	215	396	436
3				1	.	5	6	.	.	14	.	15	35	21	25	60	30	104	21	115	135	126	56	100	195	164
4					1	2	1	3	5	12	4	9	15	6	33	54	50	72	40	70	65	144	21	99	181	219
5						1	1	.	.	4	.	5	10	6	9	24	12	38	9	45	50	57	21	46	82	80
6							1	.	.	.	.	.	5	6	.	.	.	14	.	15	35	.	21	.	25	.
7								1	2	3	2	2	3	1	12	18	23	24	20	22	20	63	6	42	78	108
8									1	.	1	.	.	.	4	5	10	6	11	5	4	24	1	15	30	54
9										1	.	1	2	1	3	8	5	12	4	14	15	24	6	20	33	38
10											1	.	.	.	.	.	5	.	6	.	.	9	.	5	10	24
11												1	1	1	.	4	.	4	.	10	10	9	6	10	9	12
12													1	1	.	.	.	4	.	5	10	.	6	.	9	.
13														1	.	.	.	.	.	.	5	.	6	.	.	.
14															1	2	2	3	2	3	3	9	1	7	12	18

NUMBER	26	27	28	29	30	31	32	33	34	35	36	37	38	39	40	41	42	43	44	45	46	47	48	49	50	51
S.P.	86	86	86	102	102	86	94	110	134	94	94	174	102	110	102	102	110	102	118	118	110	110	134	142	110	126
O.S.	7680	1920	960	960	3840	3840	384	1920	1920	11520	3840	384	3840	3840	960	1280	3840	7680	7680	1920	1920	1920	1920	3840	11520	384
LEVEL	86	86	86	88	88	88	88	90	92	92	92	94	94	94	94	94	96	96	98	98	98	98	100	100	100	100
DIM'N	3670016	2246400	1372800	1555840	2928640	2795520	1729728	3267264	2475200	9225216	4484480	792064	7188480	6758400	3179904	3660800	8712704	13631488	16972800	9609600	6662656	10762752	9225216	9957376	23795200	7454720
WEIGHT	1 1 1 0 0 > 1	2 0 0 0 1 > 1	0 1 0 0 0 > 3	0 3 0 0 0 > 1	3 0 1 0 0 > 1	0 0 1 1 0 > 1	2 0 0 0 0 > 3	2 2 0 0 0 > 1	4 1 0 0 0 > 1	1 1 0 1 0 > 1	0 0 1 0 1 > 1	6 0 0 0 0 > 1	1 0 2 0 0 > 1	3 0 0 1 0 > 1	0 0 0 2 0 > 1	0 0 1 0 0 > 3	0 2 1 0 0 > 1	1 1 0 0 1 > 1	2 1 1 0 0 > 1	3 0 0 0 1 > 1	0 0 0 1 1 > 1	1 1 0 0 0 > 3	1 3 0 0 0 > 1	4 0 1 0 0 > 1	1 0 1 1 0 > 1	3 0 0 0 0 > 3
0	3336	1980	1200	1275	2394	2340	1395	2499	1680	6906	3270	462	4980	4560	2130	2460	5656	8904	10320	5665	3930	6484	5124	5384	13810	4056
1	1716	1040	630	665	1264	1235	751	1329	910	3746	1790	252	2740	2535	1185	1375	3141	5004	5820	3260	2255	3719	2914	3094	7930	2381
2	840	532	331	330	624	642	398	660	455	1974	976	126	1460	1350	657	770	1686	2768	3140	1824	1296	2114	1504	1684	4473	1370
3	480	240	120	215	424	285	171	480	385	1041	420	126	830	825	304	325	1026	1384	2080	1040	596	1025	1134	1264	2435	751
4	384	267	174	150	284	327	210	300	205	1005	529	56	747	687	360	433	863	1506	1605	996	746	1195	810	859	2468	778
5	216	110	61	100	184	145	82	215	170	516	226	56	420	396	168	181	525	728	1040	532	344	558	580	624	1326	398
6	60	30	.	35	104	.	21	115	135	126	.	56	100	195	.	.	165	164	480	240	.	120	285	424	285	171
7	160	132	93	61	116	160	112	120	80	488	285	21	360	332	195	244	412	808	764	533	428	674	380	404	1325	440
8	56	65	50	20	40	75	61	39	25	225	150	6	160	150	101	140	181	424	330	280	246	385	159	170	680	251
9	88	50	31	40	72	71	40	85	65	243	121	21	201	181	91	101	248	376	484	267	198	302	270	284	703	210
10	16	30	30	5	10	30	36	9	5	90	80	1	60	60	51	80	66	224	120	145	136	220	54	60	335	146
11	48	12	9	30	38	32	9	60	50	114	50	21	112	82	44	40	156	160	296	110	92	122	200	184	365	82
12	24	12	.	15	38	.	9	45	50	57	.	21	46	82	.	.	80	80	216	110	.	61	135	184	145	82
13	.	.	.	.	14	.	.	15	35	.	.	21	.	25	.	.	.	.	60	30	.	.	35	104	.	21
14	30	23	16	13	24	33	20	27	20	108	63	6	88	78	48	57	106	190	204	132	114	165	111	116	356	112

1	.	1	.	2	2	3	1	4	3	5	16	5	4	9	12	15	4	19	15	48	26	6	50	33	23	22	64	76	122	50	53	62	80	72	184	40	15
	1	.	1	.	.	3	.	2	4	8	8	10	9	3	6	13	11	6	4	42	33	1	32	30	24	32	38	96	72	65	64	91	37	40	171	61	16
		1	.	1	2	.	1	.	3	.	8	5	.	5	12	.	4	14	15	24	.	6	20	33	.	.	34	38	88	50	.	31	55	72	71	40	17
			1	.	.	.	.	.	.	4	.	5	5	.	.	5	6	.	.	15	15	.	10	10	10	20	11	44	20	30	36	55	9	10	70	36	18
				1	1	.	1	.	.	.	4	.	.	5	4	.	.	10	10	9	.	6	10	9	.	.	24	12	48	12	.	9	45	38	32	9	19
					1	.	1	.	.	.	.	.	.	.	4	.	.	5	10	.	.	6	.	9	.	.	.	.	24	12	.	.	15	38	.	9	20
						1	.	1	1	2	4	2	2	2	3	6	2	4	3	18	13	1	18	12	12	12	22	36	42	23	30	32	26	24	87	20	21
							1	.	.	.	.	.	.	.	.	.	.	.	5	.	.	6	.	.	.	.	.	.	.	.	.	.	.	14	.	.	22
								1	.	.	2	.	.	1	1	3	.	3	2	6	5	1	12	3	6	4	12	10	24	5	14	8	18	12	45	4	23
									1	.	2	2	.	1	3	.	2	3	3	9	.	1	7	12	.	.	12	18	30	23	.	16	18	24	33	20	24
										1	.	1	1	.	.	2	1	.	.	6	6	.	5	4	5	7	6	16	11	10	17	18	6	6	35	11	25
											1	.	.	1	1	.	.	2	2	3	.	1	4	3	.	.	8	5	16	5	.	4	14	12	15	4	26
												1	.	.	.	.	1	.	.	3	.	.	2	4	.	.	3	8	8	10	.	9	4	6	13	11	27
													1	.	.	.	1	.	.	.	3	.	.	.	2	4	.	8	.	5	8	10	.	.	13	6	28
														1	.	.	.	1	1	.	.	1	.	.	.	.	4	.	4	.	.	.	10	4	.	.	29
															1	.	.	1	2	.	.	1	.	3	.	.	.	.	8	5	.	.	5	12	.	4	30
																1	.	.	.	2	2	.	3	1	3	2	3	4	6	2	8	4	4	3	18	2	31
																	1	.	.	.	.	.	.	.	.	.	.	4	.	5	.	5	.	.	5	6	32
																		1	1	.	.	1	.	.	.	.	.	.	4	.	.	.	5	4	.	.	33
																			1	.	.	1	.	.	.	.	.	.	.	.	.	.	.	4	.	.	34
																				1	.	.	1	1	.	.	2	2	4	2	.	2	3	3	6	2	35
																					1	.	.	.	1	1	.	2	.	1	4	2	.	.	6	1	36
																						1	.	.	.	.	.	.	.	.	.	.	.	.	.	.	37
																							1	.	.	.	1	.	2	.	.	.	2	1	3	.	38
																								1	.	.	.	.	2	2	.	.	1	3	.	2	39
																									1	.	.	.	.	.	2	.	.	.	3	.	40
																										1	.	.	.	.	2	2	.	.	.	1	41
																											1	.	1	.	.	.	2	1	.	.	42
																												1	.	1	.	1	.	.	2	1	43
																													1	.	.	.	1	1	.	.	44
																														1	.	.	.	.	.	1	45
																															1	.	.	.	.	.	46
																																1	.	.	.	1	47
																																	1	.	.	.	48
																																		1	.	.	49
																																			1	.	50
																																				1	51

Table 3. *(continued)*

ALGEBRA B7, REPRESENTATIONS OF CLASS 0

0	4	8	16	12	20	36	16	24	20	32	40	24	64	28	28	36	44	32	52	68	36	100	40	40	48	S.P.
1	14	84	14	280	168	14	560	840	672	84	168	448	14	128	2240	840	840	3360	168	168	2688	14	3360	896	280	O.S.
1	15	27	29	37	41	43	45	51	51	53	55	55	57	57	59	63	65	65	67	69	69	71	71	71	73	LEVEL
1	15	105	119	455	1105	665	1365	5355	3003	4080	6916	5005	2940	6435	17017	26180	35568	38675	33915	32487	65637	10948	92820	85085	59500	DIM'N
0	1	0	2	0	1	3	0	1	0	0	2	0	4	0	1	0	2	1	1	3	1	5	0	1	0	
0	0	1	0	0	1	0	0	0	0	2	1	0	0	0	0	1	0	0	2	1	0	0	1	0	0	
0	0	0	0	1	0	0	0	1	0	0	0	0	0	0	0	1	1	0	0	0	0	0	0	0	2	
0	0	0	0	0	0	0	1	0	0	0	0	0	0	0	1	0	0	0	0	0	0	0	1	0	0	
0	0	0	0	0	0	0	0	0	1	0	0	0	0	0	0	0	0	1	0	0	0	0	0	0	0	
0	0	0	0	0	0	0	0	0	0	0	0	1	0	0	0	0	0	0	0	0	1	0	0	0	0	
>	>	>	>	>	>	>	>	>	>	>	>	>	>	>	>	>	>	>	>	>	>	>	>	>	>	
0	0	0	0	0	0	0	0	0	0	0	0	0	0	2	0	0	0	0	0	0	0	0	0	2	0	WEIGHT
0	1	2	3	4	5	6	7	8	9	10	11	12	13	14	15	16	17	18	19	20	21	22	23	24	25	NUMBER
1	1	7	7	7	13	7	21	63	21	48	70	35	28	35	77	98	120	175	105	91	189	28	350	245	210	0
	1	1	1	6	13	7	6	18	15	12	19	15	7	20	66	92	114	75	105	91	135	28	140	140	80	1
		1	1	1	2	1	5	17	5	12	19	10	7	10	21	27	34	60	29	25	65	7	125	90	75	2
			1	.	1	1	.	6	.	6	13	.	7	.	6	11	18	15	18	19	15	7	45	20	35	3
				1	2	1	1	3	4	2	3	4	1	6	19	26	33	22	29	25	48	7	42	50	24	4
					1	1	.	1	.	1	2	.	1	.	5	11	17	5	18	19	10	7	15	10	10	5
						1	.	.	.	.	1	.	1	.	.	.	6	.	6	13	.	7	.	.	.	6
							1	3	1	2	3	3	1	3	4	5	6	19	5	4	21	1	39	33	23	7
								1	.	1	2	.	1	.	1	2	3	4	3	3	4	1	14	6	10	8
									1	.	.	1	.	2	4	5	6	5	5	4	17	1	9	18	5	9
										1	1	.	1	.	.	1	1	.	2	2	.	1	5	.	5	10
											1	.	1	.	.	.	1	.	1	2	.	1	.	.	.	11
												1	.	1	.	.	.	5	.	.	6	.	9	13	5	12
													1	.	.	.	.	.	.	1	.	1	.	.	.	13
														1	.	.	.	.	.	.	6	.	.	7	.	14
															1	2	3	1	3	3	3	1	3	3	2	15
																1	1	.	2	2	.	1	1	.	1	16

48	56	44	72	72	52	80	52	48	104	56	52	144	60	60	68	76	56	76	64	84	64	60	100	108	80	S.P.
2240	1680	6720	84	840	3360	168	2240	6720	168	2688	2688	14	6720	896	840	2240	6720	840	560	1680	13440	8960	160	040	3360	O.S.
73	77	77	79	79	79	81	81	81	83	83	83	85	85	85	87	87	87	89	89	91	91	91	93	93	93	LEVEL
117040	230945	222768	88179	173264	271700	178605	278460	388960	125580	466830	510510	35700	847875	608608	587860	583440	742560	671517	433160	1261260	2078505	1361360	617253	686205	1374450	DIM'N
2	1	0	0	3	2	2	0	0	4	2	0	6	1	2	1	3	0	0	0	2	1	0	1	4	3	
0	1	1	3	0	0	2	0	1	1	0	1	0	1	0	0	0	0	2	0	1	1	0	3	0	0	
0	1	0	0	1	0	0	1	0	0	0	0	0	0	0	2	0	1	1	0	1	0	1	0	1	0	
1	0	0	0	0	0	0	1	0	0	0	0	0	1	0	0	1	0	0	2	0	0	0	0	0	0	
0	0	1	0	0	1	0	0	0	0	0	0	0	0	0	0	0	1	0	0	0	1	0	0	0	1	
0	0	0	0	0	0	0	0	1	0	1	0	0	0	0	0	0	0	0	0	0	0	1	0	0	0	
>	>	>	>	>	>	>	>	>	>	>	>	>	>	>	>	>	>	>	>	>	>	>	>	>	>	
0	0	0	0	0	0	0	0	0	0	0	2	0	0	2	0	0	0	0	0	0	0	0	0	0	0	WEIGHT
26	27	28	29	30	31	32	33	34	35	36	37	38	39	40	41	42	43	44	45	46	47	48	49	50	51	NUMBER
392	715	448	259	476	490	483	462	784	308	840	840	84	1225	896	770	748	1232	819	672	1428	3003	1568	637	693	1750	0
162	275	364	93	186	405	177	408	424	112	465	570	28	1105	620	725	682	632	783	328	1377	1563	1204	637	672	925	1
146	264	164	93	180	185	177	173	320	112	355	345	28	470	380	295	292	528	313	288	552	1329	684	241	267	795	2
66	143	54	57	114	75	126	69	100	91	135	105	28	225	140	155	162	204	177	124	341	594	250	165	186	405	3
49	82	138	27	55	157	51	156	164	31	183	236	7	432	262	281	271	252	303	132	536	633	534	241	261	379	4
21	43	45	17	34	60	35	64	50	25	65	70	7	205	90	150	146	94	172	54	330	274	195	165	180	185	5
6	11	.	6	18	15	18	.	.	19	15	.	7	45	20	35	66	.	51	.	143	54	.	78	114	75	6
46	80	54	27	54	61	51	57	128	31	144	140	7	153	156	95	94	216	99	118	173	552	290	73	81	334	7
19	42	17	17	33	22	35	22	38	25	48	40	7	70	50	48	49	82	55	50	103	235	103	49	55	157	8
10	16	48	5	10	55	9	53	60	5	67	98	1	145	110	92	90	92	97	48	170	228	232	73	80	135	9
5	22	5	12	17	5	24	9	10	19	10	10	7	30	10	25	21	30	32	24	59	90	35	34	34	60	10
5	11	.	6	17	5	18	.	.	19	10	.	7	15	10	10	21	.	16	.	43	45	.	23	34	60	11
10	16	14	5	10	15	9	14	50	5	57	57	1	35	64	21	20	82	21	44	35	208	118	14	15	125	12
.	.	.	.	6	.	6	.	.	13	.	.	7	.	.	.	6	.	.	.	11	.	.	6	18	15	13
.	.	14	.	.	15	.	14	20	.	21	42	.	35	49	21	20	28	21	14	35	64	98	14	15	35	14
4	8	15	3	6	19	6	21	17	4	21	27	1	66	33	47	46	33	54	19	101	93	83	49	54	61	15
1	4	4	2	3	4	4	9	4	3	4	6	1	28	6	25	19	12	31	8	58	34	28	34	33	22	16

1	.	1	2	.	1	.	.	.	1	2	.	1	3	4	3	.	.	3	4	.	1	14	6	10	19	.	16	.	42	17	.	23	33	22	17
	1	.	.	1	.	3	2	2	4	8	4	3	6	5	6	5	14	4	17	15	1	15	18	10	10	30	11	18	20	84	40	9	10	55	18
		1	1	.	1	.	.	.	.	1	.	1	1	.	2	.	.	2	.	.	1	5	.	5	5	.	11	.	22	5	.	18	17	5	19
			1	.	1	.	.	.	.	.	.	.	1	.	1	.	.	2	.	.	1	.	.	.	5	.	.	.	11	.	.	6	17	5	20
				1	.	.	1	.	.	.	4	.	.	5	.	5	5	.	6	11	.	15	13	10	10	9	11	5	20	24	34	9	10	15	21
					1	.	.	.	.	.	.	.	.	.	.	.	.	1	.	.	1	.	.	.	.	.	.	.	.	.	.	.	6	.	22
						1	.	1	1	4	1	2	3	1	4	2	3	3	3	3	1	6	3	5	4	11	6	8	11	30	13	6	6	19	23
							1	.	.	.	.	.	.	.	.	.	5	.	6	6	.	.	7	.	.	9	.	5	.	24	14	.	.	15	24
								1	.	2	.	1	1	.	3	1	.	2	.	.	1	2	.	3	1	4	3	4	6	8	4	4	3	4	25
									1	2	.	1	3	1	3	.	.	3	3	.	1	3	3	2	4	.	3	.	8	15	.	4	6	19	26
										1	.	1	1	.	2	.	.	2	.	.	1	1	.	1	1	.	2	.	4	4	.	3	3	4	27
											1	.	.	1	.	2	1	.	1	2	.	6	2	5	4	3	6	2	11	8	11	6	6	5	28
												1	.	.	1	.	.	1	.	.	1	.	.	.	.	.	1	.	1	.	.	2	1	.	29
													1	.	1	.	.	2	.	.	1	.	.	.	1	.	.	.	2	.	.	1	3	4	30
														1	.	.	.	.	1	.	.	3	2	2	4	.	3	.	8	4	.	4	6	5	31
															1	.	.	1	.	.	1	.	.	.	.	.	.	.	1	.	.	1	1	.	32
																1	.	.	.	.	.	2	.	3	1	1	3	1	6	2	3	4	3	1	33
																	1	.	1	1	.	.	1	.	.	3	.	2	.	8	4	.	.	5	34
																		1	.	.	1	.	.	.	.	.	.	.	.	.	.	.	1	.	35
																			1	.	.	.	1	.	.	.	.	.	.	4	.	.	.	5	36
																				1	.	.	1	.	.	.	.	.	.	.	4	.	.	.	37
																					1	.	.	.	.	.	.	.	.	.	.	.	.	.	38
																						1	.	1	1	.	2	.	4	1	.	3	3	1	39
																							1	.	.	.	.	.	.	.	.	.	.	.	40
																								1	.	.	1	.	2	.	.	2	1	.	41
																									1	.	.	.	2	.	.	1	3	1	42
																										1	.	1	.	2	1	.	.	1	43
																											1	.	1	.	.	2	1	.	44
																												1	.	.	.	.	.	.	45
																													1	.	.	1	1	.	46
																														1	.	.	.	1	47
																															1	.	.	.	48
																																1	.	.	49
																																	1	.	50
																																		1	51

Table 3. *(continued)*

ALGEBRA B7, REPRESENTATIONS OF CLASS 1

0	1	2	3	4	5	6	7	8	9	10	11	12	13	14	15	16	17	18	19	20	21	22	23	24	25	NUMBER
7	15	23	31	31	39	55	39	47	47	55	63	55	87	63	55	63	71	63	79	95	71	127	71	79	79	S.P.
128	896	2 688	896	4 480	5 376	896	4 480	13 440	2 688	2 688	5 376	896	896	128	17 920	13 440	13 440	13 440	5 376	5 376	5 376	896	26 880	896	4 480	O.S.
29	43	55	57	65	69	71	73	79	79	81	83	83	85	85	87	91	93	93	95	97	97	99	99	99	101	LEVEL
128	1 792	11 520	13 440	44 800	114 688	71 680	116 480	499 200	209 664	396 032	685 440	256 256	304 640	183 040	1 376 256	2 284 800	3 168 256	2 562 560	3 144 960	3 096 576	3 194 880	1 096 704	7 037 184	2 306 304	4 787 200	DIM'N
0	1	0	2	0	1	3	0	1	0	0	2	0	4	0	1	0	2	1	1	3	1	5	0	1	0	
0	0	1	0	0	1	0	0	0	0	2	1	0	0	0	0	1	0	0	2	1	0	0	1	0	0	
0	0	0	0	1	0	0	0	1	0	0	0	0	0	0	0	1	1	0	0	0	0	0	0	0	2	
0	0	0	0	0	0	0	1	0	0	0	0	0	0	0	1	0	0	0	0	0	0	0	1	0	0	
0	0	0	0	0	0	0	0	0	1	0	0	0	0	0	0	0	0	1	0	0	0	0	0	0	0	
0	0	0	0	0	0	0	0	0	0	0	0	1	0	0	0	0	0	0	0	0	1	0	0	0	0	
>	>	>	>	>	>	>	>	>	>	>	>	>	>	>	>	>	>	>	>	>	>	>	>	>	>	
1	1	1	1	1	1	1	1	1	1	1	1	1	1	3	1	1	1	1	1	1	1	1	1	3	1	WEIGHT
1	7	27	28	70	154	84	133	470	189	343	546	196	210	127	980	1 435	1 841	1 477	1 701	1 512	1 588	462	3 430	1 050	2 155	0
	1	6	7	20	48	28	44	170	69	127	210	76	84	51	392	595	785	633	735	672	712	210	1 537	483	985	1
		1	1	5	12	7	14	55	25	42	69	30	28	21	148	230	305	264	289	266	316	84	663	222	430	2
			1	.	6	7	.	20	.	21	48	.	28	.	44	100	170	69	183	210	76	84	259	51	195	3
				1	2	1	4	15	9	11	18	12	7	9	52	80	105	107	99	90	140	28	270	103	176	4
					1	1	.	5	.	6	12	.	7	.	14	35	55	25	63	69	30	28	104	21	80	5
						1	.	.	.	.	6	.	7	.	.	.	20	.	21	48	.	28	.	.	.	6
							1	3	3	2	3	5	1	4	16	23	30	42	27	24	62	7	102	49	65	7
								1	.	1	2	.	1	.	4	10	15	9	17	18	12	7	38	9	30	8
									1	.	.	2	.	2	4	5	6	15	5	4	28	1	33	24	20	9
										1	1	.	1	.	.	5	5	.	12	12	.	7	14	.	15	10
											1	.	1	.	.	.	5	.	6	12	.	7	.	.	.	11
												1	.	1	.	.	.	5	.	.	12	.	9	13	5	12
													1	.	.	.	.	.	.	6	.	7	.	.	.	13

26	27	28	29	30	31	32	33	34	35	36	37	38	39	40	41	42	43	44	45	46	47	48	49	50	51	NUMBER
79	87	79	103	103	87	111	87	87	135	95	95	175	95	103	103	111	95	111	103	119	103	103	135	143	119	S.P.
17 920	26 880	26 880	2 688	13 440	13 440	5 376	17 920	13 440	5 376	5 376	2 688	896	53 760	896	13 440	17 920	26 880	13 440	4 480	26 880	53 760	17 920	5 376	13 440	13 440	O.S.
101	105	105	107	107	107	109	109	109	111	111	111	113	113	113	115	115	115	117	117	119	119	119	121	121	121	LEVEL
9 052 160	19 267 584	13 861 120	7 745 920	14 851 200	17 233 920	15 934 464	19 496 960	17 821 440	11 576 320	21 781 760	13 069 056	3 472 896	61 501 440	15 841 280	45 239 040	43 450 368	42 771 456	53 114 880	27 417 600	101 292 800	123 863 040	57 933 824	52 183 040	56 908 800	84 015 360	DIM'N
2	1	0	0	3	2	2	0	0	4	2	0	6	1	2	1	3	0	0	0	2	1	0	1	4	3	
0	1	1	3	0	0	2	0	1	1	0	1	0	1	0	0	0	0	2	0	1	1	0	3	0	0	
0	1	0	0	1	0	0	1	0	0	0	0	0	0	0	2	0	1	1	0	1	0	1	0	1	0	
1	0	0	0	0	0	0	1	0	0	0	0	0	1	0	0	1	0	0	2	0	0	0	0	0	0	
0	0	1	0	0	1	0	0	0	0	0	0	0	0	0	0	0	1	0	0	0	1	0	0	0	1	
0	0	0	0	0	0	0	0	1	0	1	0	0	0	0	0	0	0	0	0	0	0	1	0	0	0	
>	>	>	>	>	>	>	>	>	>	>	>	>	>	>	>	>	>	>	>	>	>	>	>	>	>	
1	1	1	1	1	1	1	1	1	1	1	3	1	1	3	1	1	1	1	1	1	1	1	1	1	1	WEIGHT
4 094	7 812	5 572	2 849	5 460	6 441	5 502	6 958	6 258	3 570	7 119	4 235	924	19 824	4 781	13 587	12 732	12 789	15 008	7 595	26 950	33 516	15 296	12 740	13 560	20 700	0
1 880	3 666	2 638	1 358	2 625	3 117	2 667	3 384	3 072	1 764	3 568	2 123	462	9 880	2 445	6 861	6 488	6 504	7 652	3 920	13 930	17 442	8 016	6 650	7 140	11 001	1
816	1 620	1 224	608	1 170	1 458	1 197	1 604	1 496	798	1 752	1 062	210	4 741	1 234	3 326	3 144	3 265	3 736	1 995	6 835	8 868	4 180	3 290	3 535	5 652	2
392	910	444	392	785	633	861	690	516	672	712	356	210	2 437	483	1 845	1 880	1 350	2 231	890	4 450	4 278	1 680	2 324	2 625	3 117	3
332	660	554	247	475	660	486	734	722	322	848	532	84	2 171	620	1 530	1 440	1 611	1 720	993	3 145	4 392	2 168	1 516	1 625	2 808	4
148	365	195	162	305	264	345	315	240	266	316	171	84	1 087	222	840	816	660	1 025	445	2 005	2 049	860	1 070	1 170	1 458	5
44	100	.	56	170	69	183	.	.	210	76	.	84	259	51	195	392	.	315	.	910	444	.	518	785	633	6
124	240	242	88	170	288	171	318	347	111	407	268	28	934	313	654	616	777	730	480	1 330	2 112	1 116	634	680	1 347	7
52	130	82	57	105	107	120	136	110	90	140	82	28	454	103	356	332	315	430	215	830	950	436	444	475	660	8
40	72	102	25	50	120	48	128	165	30	194	138	7	370	161	252	240	361	278	220	500	976	572	232	250	620	9
14	70	25	42	55	25	84	60	30	69	30	21	28	208	21	195	148	126	265	100	500	390	164	324	305	264	10
14	35	.	21	55	25	63	.	.	69	30	.	28	104	21	80	148	.	135	.	365	195	.	218	305	264	11
10	16	38	5	10	45	9	44	80	5	93	72	1	125	85	81	80	162	87	95	155	432	288	68	75	270	12
.	.	.	.	20	.	21	.	.	48	.	.	28	.	.	.	44	.	.	.	100	.	.	56	170	69	13

1	.	.	.	.	.	.	6	.	.	7	.	.	.	14	.	.	15	.	14	35	.	42	42	.	35	49	21	20	63	21	35	35	168	148	14	15	105	14
	1	2	3	3	3	3	5	1	12	4	9	16	38	33	16	30	42	33	54	50	24	62	40	7	174	49	135	124	144	160	99	305	423	219	160	170	288	15
		1	1	.	2	2	.	1	4	.	5	4	20	9	11	15	9	23	24	12	18	12	9	7	76	9	75	52	57	95	46	180	164	80	114	105	107	16
			1	.	1	2	.	1	.	.	.	4	10	.	6	15	9	17	.	.	18	12	.	7	38	9	30	52	.	50	.	130	82	.	78	105	107	17
				1	.	.	2	.	3	2	2	4	8	12	3	6	15	6	18	23	4	28	20	1	57	24	42	40	63	49	42	92	180	108	46	50	120	18
					1	1	.	1	.	.	.	.	5	.	6	5	.	12	.	.	12	.	.	7	14	.	15	14	.	35	.	70	25	.	63	55	25	19
						1	.	1	.	.	.	.	.	.	.	5	.	6	.	.	12	.	.	7	.	.	.	14	.	.	.	35	.	.	21	55	25	20
							1	.	.	1	.	.	.	4	.	.	5	.	5	10	.	12	11	.	15	13	10	10	24	11	15	20	68	54	9	10	45	21
								1	.	.	.	.	.	.	.	.	.	.	.	.	6	.	.	7	.	.	.	.	.	.	.	.	.	.	.	20	.	22
									1	.	1	1	4	3	2	3	3	4	8	5	3	5	4	1	24	4	23	16	24	28	20	53	66	38	32	30	42	23
										1	.	.	.	.	.	.	.	.	.	5	.	6	6	.	.	7	.	.	9	.	5	.	24	24	.	.	15	24
											1	.	2	.	1	1	.	3	4	.	2	.	.	1	8	.	15	4	9	15	10	30	18	12	22	15	9	25
												1	2	.	1	3	3	3	.	.	3	5	.	1	12	4	9	16	.	15	.	38	33	.	22	30	42	26
													1	.	1	1	.	2	.	.	2	.	.	1	4	.	5	4	.	10	.	20	9	.	17	15	9	27
														1	.	.	1	.	2	2	.	2	2	.	6	2	5	4	9	6	7	11	24	18	6	6	15	28
															1	.	.	1	.	.	1	.	.	1	.	.	.	.	.	5	.	5	.	.	12	5	.	29
																1	.	1	.	.	2	.	.	1	.	.	.	4	.	.	.	10	.	.	6	15	9	30
																	1	.	.	.	.	2	.	.	3	2	2	4	.	3	.	8	12	.	4	6	15	31
																		1	.	.	1	.	.	1	.	.	.	.	.	.	.	5	.	.	6	5	.	32
																			1	.	.	.	.	.	2	.	3	1	3	3	4	6	6	5	4	3	3	33
																				1	.	1	1	.	.	1	.	.	3	.	2	.	8	8	.	.	5	34
																					1	.	.	1	.	.	.	.	.	.	.	.	.	.	.	5	.	35
																						1	.	.	.	1	.	.	.	.	.	.	4	.	.	.	5	36
																							1	.	.	1	.	.	.	.	.	.	.	4	.	.	.	37
																								1	.	.	.	.	.	.	.	.	.	.	.	.	.	38
																									1	.	1	1	.	2	.	4	3	.	3	3	3	39
																										1	.	.	.	.	.	.	.	.	.	.	.	40
																											1	.	.	1	.	2	.	.	2	1	.	41
																												1	.	.	.	2	.	.	1	3	3	42
																													1	.	1	.	2	2	.	.	1	43
																														1	.	1	.	.	2	1	.	44
																															1	.	.	.	.	.	.	45
																																1	.	.	1	1	.	46
																																	1	.	.	.	1	47
																																		1	.	.	.	48
																																			1	.	.	49
																																				1	.	50
																																					1	51

Table 3. *(continued)*

ALGEBRA B8, REPRESENTATIONS OF CLASS 0

NUMBER	0	1	2	3	4	5	6	7	8	9	10	11	12	13	14	15	16	17	18	19	20	21	22	23	24	25
S.P.	0	4	8	16	12	20	36	16	24	32	20	40	64	24	28	28	36	32	44	52	32	68	100	40	36	48
O.S.	1	16	112	16	448	224	16	1 120	1 344	112	1 792	224	16	1 792	4 480	1 024	1 344	256	1 344	224	8 960	224	16	6 720	10 752	448
LEVEL	1	17	31	33	43	47	49	53	59	61	61	63	65	67	69	71	73	73	75	77	77	79	81	83	83	85
DIM'N	1	17	136	152	680	1 615	952	2 380	9 044	6 783	6 188	11 340	4 692	12 376	33 592	19 448	50 388	24 310	67 473	63 308	90 440	59 432	19 380	209 304	184 756	131 784
WEIGHT	0 0 0 0 0 0 0 > 0	1 0 0 0 0 0 0 > 0	0 1 0 0 0 0 0 > 0	2 0 0 0 0 0 0 > 0	0 0 1 0 0 0 0 > 0	1 1 0 0 0 0 0 > 0	3 0 0 0 0 0 0 > 0	0 0 0 1 0 0 0 > 0	1 0 1 0 0 0 0 > 0	0 2 0 0 0 0 0 > 0	0 0 0 0 1 0 0 > 0	2 1 0 0 0 0 0 > 0	4 0 0 0 0 0 0 > 0	0 0 0 0 0 1 0 > 0	1 0 0 1 0 0 0 > 0	0 0 0 0 0 0 1 > 0	0 1 1 0 0 0 0 > 0	0 0 0 0 0 0 0 > 2	2 0 1 0 0 0 0 > 0	1 2 0 0 0 0 0 > 0	1 0 0 0 1 0 0 > 0	3 1 0 0 0 0 0 > 0	5 0 0 0 0 0 0 > 0	0 1 0 1 0 0 0 > 0	1 0 0 0 0 1 0 > 0	0 0 2 0 0 0 0 > 0
0	1	1	8	8	8	15	8	28	84	63	28	92	36	56	104	56	132	70	161	140	280	120	36	552	308	328
1		1	1	1	7	15	8	7	21	14	21	22	8	21	91	35	125	35	154	140	105	120	36	195	231	111
2			1	1	1	2	1	6	20	14	6	22	8	15	25	15	32	20	40	34	87	29	8	177	96	105
3				1	.	1	1	.	7	7	.	15	8	.	7	.	13	.	21	21	21	22	8	63	21	48
4					1	2	1	1	3	2	5	3	1	5	23	10	31	10	39	34	27	29	8	51	75	29
5						1	1	.	1	1	.	2	1	.	6	.	13	.	20	21	6	22	8	18	15	12
6							1	.	.	.	.	1	1	.	.	.	.	.	7	7	.	15	8	.	.	.
7								1	3	2	1	3	1	4	4	4	5	6	6	5	24	4	1	48	27	28
8									1	1	.	2	1	.	1	.	2	.	3	3	5	3	1	17	5	12
9										1	.	1	1	.	.	.	1	.	1	2	.	2	1	6	.	6
10											1	.	.	1	4	3	5	3	6	5	5	4	1	9	23	5
11												1	1	.	.	.	.	.	1	1	.	2	1	.	.	.
12													1	.	.	.	.	.	.	.	.	1	1	.	.	.
13														1	.	1	.	2	.	.	5	.	.	9	6	5
14															1	.	2	.	3	3	1	3	1	3	4	2
15																1	.	1	.	.	.	.	.	.	6	.
16																	1	.	1	2	.	2	1	1	.	1

NUMBER	26	27	28	29	30	31	32	33	34	35	36	37	38	39	40	41	42	43	44	45	46	47	48	49	50	51
S.P.	48	40	56	44	72	72	44	80	52	104	52	144	48	60	56	68	76	52	76	56	60	56	84	64	64	100
O.S.	4 480	7 168	2 688	2 048	112	1 344	17 920	224	8 960	224	4 480	16	26 880	13 440	10 752	1 344	4 480	21 504	1 344	17 920	7 160	7 168	2 688	1 120	2 048	224
LEVEL	85	87	89	89	91	91	91	93	93	95	95	97	97	99	99	101	101	101	103	103	103	103	105	105	105	107
DIM'N	259 896	293 930	499 681	369 512	186 200	367 080	596 904	372 232	716 040	255 000	724 812	69 768	1 259 700	2 152 472	1 484 406	1 465 128	1 448 655	2 040 714	1 646 008	2 306 220	2 382 380	2 586 584	3 049 800	1 317 840	3 007 368	1 456 084
WEIGHT	2 0 0 1 0 0 0 > 0	1 0 0 0 0 0 1 > 0	1 1 1 0 0 0 0 > 0	1 0 0 0 0 0 0 > 2	0 3 0 0 0 0 0 > 0	3 0 1 0 0 0 0 > 0	0 1 0 0 1 0 0 > 0	2 2 0 0 0 0 0 > 0	2 0 0 0 1 0 0 > 0	4 1 0 0 0 0 0 > 0	0 0 1 1 0 0 0 > 0	6 0 0 0 0 0 0 > 0	0 1 0 0 0 1 0 > 0	1 1 0 1 0 0 0 > 0	2 0 0 0 0 1 0 > 0	1 0 2 0 0 0 0 > 0	3 0 0 1 0 0 0 > 0	0 1 0 0 0 0 1 > 0	0 2 1 0 0 0 0 > 0	0 0 1 0 1 0 0 > 0	2 0 0 0 0 0 1 > 0	0 1 0 0 0 0 0 > 2	2 1 1 0 0 0 0 > 0	0 0 0 2 0 0 0 > 0	2 0 0 0 0 0 0 > 2	1 3 0 0 0 0 0 > 0
0	616	490	1 105	504	392	728	728	728	792	456	748	120	1 540	1 960	1 638	1 224	1 183	1 722	1 288	2 380	1 820	2 184	2 232	1 280	2 296	980
1	224	245	377	315	126	252	609	238	672	148	671	36	735	1 792	798	1 161	1 092	1 239	1 239	1 085	1 330	1 309	2 162	560	1 400	980
2	205	170	364	175	126	245	239	238	267	148	250	36	580	669	636	417	410	654	438	930	710	865	766	500	935	330
3	91	35	195	35	77	154	77	168	105	120	98	36	175	315	231	216	224	189	245	350	245	245	468	210	315	224
4	59	80	98	110	32	65	207	60	233	36	229	8	255	623	281	400	385	480	426	385	525	510	747	200	555	330
5	25	15	51	20	20	40	66	41	87	29	92	8	75	291	96	210	205	135	239	140	170	140	455	80	175	224
6	7	.	13	.	7	21	.	21	21	22	.	8	.	63	21	48	91	.	70	.	35	.	195	.	35	105
7	56	58	96	60	32	64	68	60	76	36	71	8	209	188	232	116	114	240	120	339	263	340	208	182	372	87
8	23	10	50	10	20	39	21	41	27	29	27	8	60	85	75	58	59	65	66	125	80	90	123	75	110	58
9	6	.	26	.	14	20	6	28	6	22	11	8	15	36	15	30	25	15	38	45	15	20	70	35	20	40
10	10	25	16	39	5	10	62	9	70	5	67	1	80	180	88	113	110	183	118	120	202	197	205	62	216	87
11	6	.	13	.	7	20	.	21	6	22	.	8	.	18	15	12	25	.	19	.	15	.	51	.	20	27
12	.	.	.	.	.	7	.	7	.	15	.	8	.	.	.	.	7	.	.	.	.	.	13	.	.	7
13	10	20	16	21	5	10	14	9	15	5	14	1	70	35	78	21	20	84	21	110	92	135	35	58	149	14
14	4	4	8	6	3	6	19	6	24	4	26	1	22	81	27	57	56	48	65	42	58	50	121	24	60	58
15	.	7	.	15	.	.	14	.	15	.	14	.	20	35	21	21	20	69	21	28	77	77	35	14	85	14
16	1	.	4	.	2	3	5	4	5	3	11	1	5	34	5	30	23	10	37	15	10	10	69	10	10	40

1	.	.	.	.	.	.	.	.	.	7	.	8	.	.	.	.	.	.	.	.	20	.	21	.	.	27	.	28	28	56	.	14	64	.	17
	1	1	.	2	1	.	.	.	1	.	2	.	1	3	.	3	5	3	.	1	.	17	5	12	23	.	19	.	10	.	50	.	10	27	18
		1	.	1	1	.	.	.	.	.	1	.	1	1	.	2	.	2	.	1	.	6	.	6	6	.	13	.	.	.	26	.	.	21	19
			1	.	.	3	1	2	4	3	8	3	3	6	4	6	5	4	5	1	19	15	23	10	10	21	11	39	25	33	20	23	39	9	20
				1	1	.	.	.	.	.	.	.	.	1	.	1	.	2	.	1	.	.	.	.	6	.	.	.	.	.	13	.	.	7	21
					1	.	.	.	.	.	.	.	.	.	.	.	.	1	.	1	.	.	.	.	.	.	.	.	.	.	.	.	.	.	22
						1	.	1	1	.	4	.	2	3	1	4	1	3	2	1	4	6	4	5	4	4	6	14	4	6	11	10	6	6	23
							1	.	.	1	.	2	.	.	4	.	5	.	5	.	5	15	6	10	10	17	11	9	20	18	20	5	21	9	24
								1	.	.	2	.	1	1	.	3	.	2	1	1	.	2	.	3	1	.	3	5	.	.	6	5	.	4	25
									1	.	2	.	1	3	.	3	1	3	.	1	.	3	4	2	4	.	3	.	4	.	8	.	6	4	26
										1	.	1	.	.	.	.	.	.	.	.	5	.	6	.	.	6	.	9	7	13	.	5	15	.	27
											1	.	1	1	.	2	.	2	.	1	.	1	.	1	1	.	2	.	.	.	4	.	.	3	28
												1	.	.	.	.	.	.	.	.	.	.	.	.	.	6	.	.	7	7	.	.	8	.	29
													1	.	.	1	.	1	.	1	.	.	.	.	.	.	1	.	.	.	1	.	.	2	30
														1	.	1	.	2	.	1	.	.	.	.	1	.	.	.	.	.	2	.	.	1	31
															1	.	1	.	2	.	1	6	1	5	4	3	6	3	3	3	11	2	3	6	32
																1	.	1	.	1	.	.	.	.	.	.	.	.	.	.	1	.	.	1	33
																	1	.	.	.	.	3	1	2	4	.	3	.	3	.	8	.	3	4	34
																		1	.	1	.	.	.	.	.	.	.	.	.	.	.	.	.	.	35
																			1	.	.	2	.	3	1	.	3	1	.	.	6	1	.	4	36
																				1	.	.	.	.	.	.	.	.	.	.	.	.	.	.	37
																					1	.	1	.	.	1	.	3	1	2	.	2	2	.	38
																						1	.	1	1	.	2	.	.	.	4	.	.	3	39
																							1	.	.	.	.	.	1	.	.	.	2	.	40
																								1	.	.	1	.	.	.	2	.	.	2	41
																									1	.	.	.	.	.	2	.	.	1	42
																										1	.	.	1	1	.	.	1	.	43
																											1	.	.	.	1	.	.	2	44
																												1	.	.	.	1	.	.	45
																													1	.	.	.	1	.	46
																														1	.	.	1	.	47
																															1	.	.	1	48
																																1	.	.	49
																																	1	.	50
																																		1	51

Table 3. *(continued)*

ALGEBRA B8, REPRESENTATIONS OF CLASS 1

NUMBER	0	1	2	3	4	5	6	7	8	9	10	11	12
S.P.	8	16	24	32	32	40	56	40	48	56	48	64	88
O.S.	256	2 048	7 168	2 048	14 336	14 336	2 048	17 920	43 008	7 168	14 336	14 336	2 048
LEVEL	37	53	67	69	79	83	85	89	95	97	97	99	101
DIM'N	256	4 096	30 464	34 816	139 264	344 064	208 896	435 200	1 762 560	1 361 920	974 848	2 315 264	992 256
	0	1	0	2	0	1	3	0	1	0	0	2	4
	0	0	1	0	0	1	0	0	0	2	0	1	0
	0	0	0	0	1	0	0	0	1	0	0	0	0
	0	0	0	0	0	0	0	1	0	0	0	0	0
	0	0	0	0	0	0	0	0	0	0	1	0	0
	0	0	0	0	0	0	0	0	0	0	0	0	0
	0	0	0	0	0	0	0	0	0	0	0	0	0
WEIGHT	> 1	> 1	> 1	> 1	> 1	> 1	> 1	> 1	> 1	> 1	> 1	> 1	> 1
0	1	8	35	36	104	224	120	230	783	560	392	882	330
1		1	7	8	27	63	36	70	258	189	133	308	120
2			1	1	6	14	8	20	75	56	44	91	36
3				1	.	7	8	.	27	28	.	63	36
4					1	2	1	5	18	13	14	21	8
5						1	1	.	6	7	.	14	8
6							1	.	.	.	.	7	8
7								1	3	2	4	3	1
8									1	1	.	2	1
9										1	.	1	1
10											1	.	.
11												1	1
12													1
13													
14													

NUMBER	13	14	15	16	17	18	19	20	21	22	23	24	25
S.P.	56	56	64	64	72	72	80	64	96	128	72	72	80
O.S.	7 168	71 680	2 048	43 008	256	43 008	14 336	71 680	14 336	2 048	107 520	43 008	14 336
LEVEL	103	105	107	109	109	111	113	113	115	117	119	119	121
DIM'N	1 584 128	5 849 088	1 810 432	9 261 056	1 244 672	12 607 488	12 185 600	13 578 240	11 698 176	3 969 024	34 398 208	22 560 768	22 573 824
	0	1	0	0	0	2	1	1	3	5	0	1	0
	0	0	0	1	0	0	2	0	1	0	1	0	0
	0	0	0	1	0	1	0	0	0	0	0	0	2
	0	1	0	0	0	0	0	0	0	0	1	0	0
	0	0	0	0	0	0	0	1	0	0	0	0	0
	1	0	0	0	0	0	0	0	0	0	0	1	0
	0	0	1	0	0	0	0	0	0	0	0	0	0
WEIGHT	> 1	> 1	> 1	> 1	> 3	> 1	> 1	> 1	> 1	> 1	> 1	> 1	> 1
0	518	1 904	512	2 688	323	3 408	3 080	3 452	2 688	792	7 512	4 752	4 585
1	189	700	196	1 022	127	1 329	1 218	1 372	1 092	330	3 109	1 995	1 932
2	69	240	76	358	51	468	434	525	392	120	1 225	822	770
3	.	70	.	154	.	258	273	133	308	120	470	189	343
4	25	75	30	111	21	144	133	192	119	36	450	333	284
5	.	20	.	48	.	75	84	44	91	36	170	69	127
6	.	.	.	.	.	27	28	.	63	36	.	.	.
7	9	20	12	28	9	36	32	66	28	8	150	132	93
8	.	5	.	12	.	18	20	14	21	8	55	25	42
9	.	.	.	6	.	6	14	.	14	8	20	.	21
10	3	4	5	5	4	6	5	20	4	1	42	51	25
11	.	.	.	.	.	6	7	.	14	8	.	.	.
12	.	.	.	.	.	.	.	.	7	8	.	.	.
13	1	.	2	.	2	.	.	5	.	.	9	18	5
14		1	.	2	.	3	3	4	3	1	15	9	11

NUMBER	26	27	28	29	30	31	32	33	34	35	36	37	38
S.P.	80	80	88	88	104	104	80	112	88	136	88	176	88
O.S.	71 680	14 336	86 016	2 048	7 168	43 008	143 360	14 336	71 680	14 336	71 680	2 048	107 520
LEVEL	121	123	125	125	127	127	127	129	129	131	131	133	133
DIM'N	43 411 200	26 138 112	88 080 384	18 104 320	34 119 680	66 150 400	84 672 512	69 092 352	103 194 624	48 620 544	111 132 672	13 891 584	145 530 880
	2	1	1	1	0	3	0	2	2	4	0	6	0
	0	0	1	0	3	0	1	2	0	1	0	0	1
	0	0	1	0	0	1	0	0	0	0	1	0	0
	1	0	0	0	0	0	0	0	0	0	1	0	0
	0	0	0	0	0	0	1	0	1	0	0	0	0
	0	0	0	0	0	0	0	0	0	0	0	0	1
	0	1	0	0	0	0	0	0	0	0	0	0	0
WEIGHT	> 1	> 1	> 1	> 3	> 1	> 1	> 1	> 1	> 1	> 1	> 1	> 1	> 1
0	8 855	4 823	16 240	3 088	5 740	11 136	14 736	10 962	16 808	6 930	17 360	1 716	21 300
1	3 745	2 100	7 014	1 373	2 520	4 920	6 490	4 886	7 546	3 150	7 833	792	9 806
2	1 485	908	2 828	610	1 029	2 001	2 776	2 002	3 252	1 302	3 414	330	4 444
3	700	196	1 568	127	658	1 329	980	1 428	1 372	1 092	1 435	330	1 477
4	545	392	1 040	273	378	734	1 144	735	1 341	476	1 420	120	1 977
5	240	76	568	51	245	468	392	518	525	392	595	120	633
6	70	.	154	.	84	258	.	273	133	308	.	120	.
7	180	170	336	124	120	234	448	231	524	147	552	36	862
8	75	30	180	21	77	144	148	161	192	119	230	36	264
9	20	.	96	.	56	75	44	112	44	91	100	36	69
10	50	74	88	58	30	60	164	57	190	35	195	8	365
11	20	.	48	.	28	75	.	84	44	91	.	36	.
12	.	.	.	.	.	27	.	28	.	63	.	36	.
13	10	33	16	28	5	10	52	9	60	5	58	1	150
14	20	12	46	9	19	36	52	39	66	28	80	8	107

NUMBER	39	40	41	42	43	44	45	46	47	48	49	50	51
S.P.	96	96	104	112	96	112	96	104	104	120	104	112	136
O.S.	215 040	43 008	43 008	71 680	43 008	43 008	143 360	14 336	7 168	86 016	17 920	2 048	14 336
LEVEL	135	135	137	137	137	139	139	139	139	141	141	141	143
DIM'N	339 456 000	174 140 928	240 787 456	233 472 000	171 991 040	276 369 408	305 614 848	203 743 232	120 393 728	518 619 136	185 551 872	141 892 608	257 359 872
	1	2	1	3	0	0	0	2	0	2	0	2	1
	1	0	0	0	1	2	0	0	1	1	0	0	3
	0	0	2	0	0	1	1	0	0	1	0	0	0
	1	0	0	1	0	0	0	0	0	0	2	0	0
	0	0	0	0	0	0	1	0	0	0	0	0	0
	0	1	0	0	0	0	0	0	0	0	0	0	0
	0	0	0	0	1	0	0	1	0	0	0	0	0
WEIGHT	> 1	> 1	> 1	> 1	> 1	> 1	> 1	> 1	> 3	> 1	> 1	> 3	> 1
0	48 280	23 883	32 088	30 400	22 280	34 776	38 696	24 752	14 504	61 712	21 916	16 038	28 224
1	22 275	11 193	14 987	14 315	10 569	16 401	18 375	11 942	7 000	29 470	10 548	7 869	13 622
2	9 820	5 112	6 670	6 370	4 976	7 350	8 546	5 668	3 369	13 272	4 964	3 818	6 188
3	4 950	1 995	3 633	3 745	1 588	4 326	3 430	2 100	1 050	8 526	2 155	1 373	4 326
4	4 090	2 280	2 790	2 655	2 328	3 078	3 869	2 660	1 621	5 558	2 264	1 844	2 597
5	2 010	822	1 505	1 485	712	1 806	1 537	908	483	3 500	985	610	1 813
6	470	189	343	700	.	546	.	196	.	1 568	.	127	868
7	1 580	993	1 072	1 020	1 084	1 176	1 690	1 240	784	2 116	990	893	980
8	755	333	573	545	316	682	663	392	222	1 306	430	273	679
9	340	69	310	240	76	414	259	76	51	778	195	51	490
10	550	420	365	350	505	396	701	577	383	705	406	437	319
11	170	69	127	240	.	210	.	76	.	568	.	51	329
12	.	.	.	70	.	.	.	.	.	154	.	.	84
13	160	171	102	100	234	108	272	268	192	190	153	219	82
14	255	132	192	180	140	225	270	170	103	426	176	124	218

1	.	1	.	.	.	.	.	.	6	.	.	14	.	15	.	.	14	.	15	.	14	.	55	35	63	21	20	111	21	91	126	98	35	49	113	14	15
	1	.	1	2	.	2	1	5	.	6	5	.	24	.	13	18	14	27	14	21	35	8	25	110	25	105	75	30	132	104	30	21	249	80	21	154	16
		1	.	.	.	.	.	.	.	.	.	7	.	8	.	.	.	.	.	.	.	.	20	.	21	.	.	48	.	28	56	56	.	14	64	.	17
			1	1	.	2	1	.	.	.	5	.	12	.	7	18	.	20	14	21	.	8	.	55	25	42	75	.	69	.	30	.	180	.	21	105	18
				1	.	1	1	.	.	.	.	.	6	.	7	6	.	14	.	14	.	8	.	20	.	21	20	.	48	.	.	.	96	.	.	84	19
					1	.	.	3	3	2	4	5	8	4	3	6	16	6	20	4	23	1	42	72	51	52	50	62	60	102	74	49	112	65	58	55	20
						1	1	.	.	.	.	.	.	.	.	6	.	7	.	14	.	8	.	.	.	.	20	.	.	.	.	.	48	.	.	28	21
							1	.	.	.	.	.	.	.	.	.	.	.	.	7	.	8	.	.	.	.	.	.	.	.	.	.	.	.	.	.	22
								1	.	1	1	.	4	.	2	3	4	4	4	3	10	1	9	30	9	28	20	12	34	38	12	9	64	30	9	38	23
									1	.	.	2	.	2	.	.	4	.	5	.	5	.	15	15	18	10	10	28	11	33	33	24	20	20	28	9	24
										1	.	.	2	.	1	1	.	3	.	2	5	1	.	10	.	18	5	.	18	14	.	.	36	15	.	26	25
											1	.	2	.	1	3	.	3	4	3	.	1	.	15	9	11	20	.	18	.	12	.	46	.	9	26	26
												1	.	1	.	.	.	.	.	.	.	.	5	.	6	.	.	12	.	9	14	13	.	5	15	.	27
													1	.	1	1	.	2	.	2	.	1	.	5	.	6	5	.	12	.	.	.	24	.	.	20	28
														1	.	.	.	.	.	.	.	.	.	.	.	.	.	6	.	.	7	7	.	.	8	.	29
															1	.	.	1	.	1	.	1	.	.	.	.	.	.	6	.	.	.	6	.	.	14	30
																1	.	1	.	2	.	1	.	.	.	.	5	.	.	.	.	.	12	.	.	7	31
																	1	.	1	.	2	.	3	6	3	5	4	5	6	12	5	4	11	9	4	6	32
																		1	.	1	.	1	.	.	.	.	.	.	.	.	.	.	6	.	.	7	33
																			1	.	.	.	.	3	3	2	4	.	3	.	5	.	8	.	4	4	34
																				1	.	1	.	.	.	.	.	.	.	.	.	.	.	.	.	.	35
																					1	.	.	2	.	3	1	.	3	4	.	.	6	5	.	4	36
																						1	.	.	.	.	.	.	.	.	.	.	.	.	.	.	37
																							1	.	1	.	.	2	.	3	2	2	.	2	2	.	38
																								1	.	1	1	.	2	.	.	.	4	.	.	3	39
																									1	.	.	.	.	.	2	.	.	.	2	.	40
																										1	.	.	1	.	.	.	2	.	.	2	41
																											1	.	.	.	.	.	2	.	.	1	42
																												1	.	.	1	1	.	.	1	.	43
																													1	.	.	.	1	.	.	2	44
																														1	.	.	.	1	.	.	45
																															1	.	.	.	1	.	46
																																1	.	.	1	.	47
																																	1	.	.	1	48
																																		1	.	.	49
																																			1	.	50
																																				1	51

Table 3. *(continued)*

ALGEBRA B9, REPRESENTATIONS OF CLASS 0

S.P.	0	4	8	16	12	20	36	16	24	32	20	40	64	24	28	36	28	44	52	32	32	68	36	100	40	48
O.S.	1	18	144	18	672	288	18	2 016	2 016	144	4 032	288	18	5 376	8 064	2 016	4 608	2 016	288	2 304	20 160	288	512	18	12 096	672
LEVEL	1	19	35	37	49	53	55	61	67	69	71	71	73	79	79	83	85	85	87	89	89	89	91	91	95	97
DIM'N	1	19	171	189	969	2 261	1 311	3 876	14 364	10 640	11 628	17 595	7 125	27 132	61 047	89 775	50 388	118 864	110 124	75 582	189 924	101 745	92 378	32 319	427 329	265 335
	0	1	0	2	0	1	3	0	1	0	0	2	4	0	1	0	0	2	1	0	1	3	0	5	0	0
	0	0	1	0	0	1	0	0	0	2	0	1	0	0	0	1	0	0	2	0	0	1	0	0	1	0
	0	0	0	0	1	0	0	0	1	0	0	0	0	0	0	1	0	1	0	0	0	0	0	0	0	2
	0	0	0	0	0	0	0	1	0	0	0	0	0	0	1	0	0	0	0	0	0	0	0	0	1	0
	0	0	0	0	0	0	0	0	0	0	1	0	0	0	0	0	0	0	0	0	1	0	0	0	0	0
	0	0	0	0	0	0	0	0	0	0	0	0	0	1	0	0	0	0	0	0	0	0	0	0	0	0
	0	0	0	0	0	0	0	0	0	0	0	0	0	0	0	0	1	0	0	0	0	0	0	0	0	0
	0	0	0	0	0	0	0	0	0	0	0	0	0	0	0	0	0	0	0	1	0	0	0	0	0	0
	>	>	>	>	>	>	>	>	>	>	>	>	>	>	>	>	>	>	>	>	>	>	>	>	>	>
WEIGHT	0	0	0	0	0	0	0	0	0	0	0	0	0	0	0	0	0	0	0	0	0	0	2	0	0	0
NUMBER	0	1	2	3	4	5	6	7	8	9	10	11	12	13	14	15	16	17	18	19	20	21	22	23	24	25
0	1	1	9	9	9	17	9	36	108	80	36	117	45	84	135	171	84	208	180	126	420	153	126	45	819	483
1		1	1	1	8	17	9	8	24	16	28	25	9	28	120	163	56	200	180	56	140	153	70	45	259	147
2			1	1	1	2	1	7	23	16	7	25	9	21	29	37	21	46	39	35	119	33	35	9	238	140
3				1	.	1	1	.	8	8	.	17	9	.	8	15	.	24	24	.	28	25	.	9	84	63
4					1	2	1	1	3	2	6	3	1	6	27	36	15	45	39	15	32	33	20	9	60	34
5						1	1	.	1	1	.	2	1	.	7	15	.	23	24	.	7	25	.	9	21	14
6							1	.	.	.	.	1	1	.	.	.	.	8	8	.	.	17	.	9	.	.
7								1	3	2	1	3	1	5	4	5	5	6	5	10	29	4	10	1	57	33
8									1	1	.	2	1	.	1	2	.	3	3	.	6	3	.	1	20	14
9										1	.	1	1	.	.	1	.	1	2	.	.	2	.	1	7	7
10											1	.	.	1	4	5	4	6	5	4	5	4	6	1	9	5
11												1	1	.	.	.	.	1	1	.	.	2	.	1	.	.
12													1	.	.	.	.	.	.	.	.	1	.	1	.	.
13														1	.	.	1	.	.	3	5	.	3	.	9	5
14															1	2	.	3	3	.	1	3	.	1	3	2
15																1	.	1	2	.	.	2	.	1	1	1
16																	1	.	.	1	.	.	2	.	.	.

S.P.	36	48	56	72	40	72	44	80	44	52	104	52	48	144	48	60	68	56	76	76	56	52	84	64	100	60
O.S.	32 256	8 064	4 032	144	32 256	2 016	40 320	288	18 432	20 160	288	8 064	4 608	18	80 640	24 192	2 016	32 256	8 064	2 016	40 320	96 768	4 032	2 016	288	32 256
LEVEL	97	97	101	103	103	103	105	105	107	107	107	109	109	109	113	113	115	115	115	117	119	119	119	121	121	121
DIM'N	453 492	524 400	988 057	361 284	854 658	714 000	1 410 864	714 780	1 293 292	1 671 525	479 655	1 675 401	1 587 222	127 281	3 488 400	4 882 164	3 276 189	4 056 234	3 230 000	3 634 092	6 186 096	6 715 170	6 661 875	3 479 679	3 120 180	7 726 160
	1	2	1	0	1	3	0	2	1	2	4	0	1	6	0	1	1	2	3	0	0	0	2	0	1	2
	0	0	1	3	0	0	1	2	0	0	1	0	0	0	1	1	0	0	0	2	0	1	1	0	3	0
	0	0	1	0	0	1	0	0	0	0	0	1	0	0	0	0	2	0	0	1	1	0	1	0	0	0
	0	1	0	0	0	0	0	0	0	0	0	1	0	0	0	1	0	0	1	0	0	0	0	2	0	0
	0	0	0	0	0	0	1	0	0	1	0	0	0	0	0	0	0	0	0	0	1	0	0	0	0	0
	1	0	0	0	0	0	0	0	0	0	0	0	0	0	1	0	0	1	0	0	0	0	0	0	0	0
	0	0	0	0	1	0	0	0	0	0	0	0	0	0	0	0	0	0	0	0	0	1	0	0	0	1
	0	0	0	0	0	0	0	0	1	0	0	0	0	0	0	0	0	0	0	0	0	0	0	0	0	0
	>	>	>	>	>	>	>	>	>	>	>	>	>	>	>	>	>	>	>	>	>	>	>	>	>	>
WEIGHT	0	0	0	0	0	0	0	0	0	0	0	0	2	0	0	0	0	0	0	0	0	0	0	0	0	0
NUMBER	26	27	28	29	30	31	32	33	34	35	36	37	38	39	40	41	42	43	44	45	46	47	48	49	50	51
0	468	912	1 615	564	882	1 056	1 104	1 044	924	1 197	645	1 131	1 134	165	2 736	2 940	1 827	2 898	1 760	1 908	4 176	3 150	3 291	2 223	1 428	3 312
1	364	296	495	164	392	328	944	308	616	1 036	189	1 027	630	45	1 168	2 716	1 743	1 260	1 640	1 844	1 712	2 366	3 199	880	1 428	2 520
2	133	274	480	164	287	320	328	308	301	364	189	341	385	45	952	903	560	1 036	548	584	1 496	1 106	1 015	796	433	1 190
3	28	120	255	100	56	200	104	216	56	140	153	132	70	45	280	420	287	364	296	324	552	308	615	328	292	392
4	108	69	114	37	117	75	290	69	205	324	41	316	210	9	366	849	540	400	519	570	546	852	993	282	433	923
5	21	29	59	23	21	46	91	47	35	119	33	125	35	9	105	392	280	133	274	317	195	231	600	111	292	287
6	.	8	15	8	.	24	.	24	.	28	25	.	.	9	.	84	63	28	120	92	.	.	255	.	136	56
7	33	66	112	37	90	74	82	69	95	91	41	85	130	9	310	223	137	341	134	141	490	367	243	260	101	398
8	6	27	58	23	15	45	25	47	15	32	33	32	20	9	87	100	68	108	69	77	177	96	143	105	67	117
9	.	7	30	16	.	23	7	32	.	7	25	13	.	9	21	42	35	21	29	44	63	21	81	48	46	21
10	29	10	16	5	32	10	76	9	68	85	5	81	70	1	100	215	134	109	130	139	148	295	240	76	101	322
11	.	7	15	8	.	23	.	24	.	7	25	.	.	9	.	21	14	21	29	22	.	.	59	.	31	21
12	.	.	.	.	.	8	.	8	.	.	17	.	.	9	.	.	.	.	8	.	.	.	15	.	8	.
13	6	10	16	5	27	10	14	9	29	15	5	14	45	1	90	35	21	99	20	21	138	111	35	72	14	120
14	5	4	8	3	5	6	23	6	10	29	4	31	10	1	27	96	67	33	66	76	51	75	141	29	67	90
15	.	1	4	2	.	3	6	4	.	6	3	13	.	1	6	40	35	6	27	43	18	15	80	12	46	15
16	6	.	.	.	7	.	14	.	23	15	.	14	24	.	20	35	21	21	20	21	28	96	35	14	14	105

1	1	.	.	2	.	1	.	.	.	1	2	1	.	3	.	3	.	6	3	.	.	1	.	20	14	6	27	22	.	.	58	.	31	15	17
	1	.	.	1	.	1	.	.	.	.	1	1	.	1	.	2	.	.	2	.	.	1	.	7	7	.	7	15	.	.	30	.	24	.	18
		1	.	.	1	.	.	.	.	.	.	.	7	.	.	.	8	.	.	.	17	.	20	.	.	21	.	.	28	27	.	14	.	28	19
			1	.	.	.	3	2	1	4	8	3	4	6	4	6	4	5	4	5	6	1	24	15	10	29	10	11	48	27	20	28	9	32	20
				1	.	1	.	.	.	.	.	.	.	1	.	1	.	.	2	.	.	1	.	.	.	.	7	.	.	.	15	.	8	.	21
					1	.	.	.	.	.	.	.	.	.	.	.	8	.	.	.	9	.	.	.	.	.	.	.	.	27	.	.	.	28	22
						1	.	.	.	.	.	.	.	.	.	.	.	.	1	.	.	1	.	.	.	.	.	.	.	.	.	.	.	.	23
							1	1	.	1	4	2	.	3	1	4	.	1	3	2	.	1	5	6	5	5	4	6	17	5	11	12	6	5	24
								1	.	.	2	1	.	1	.	3	.	.	2	1	.	1	.	2	3	.	1	3	6	.	6	6	4	.	25
									1	.	.	.	1	.	4	.	3	5	.	5	3	.	5	15	10	6	10	11	9	23	20	5	9	27	26
										1	2	1	.	3	.	3	.	1	3	.	.	1	.	3	2	5	4	3	.	.	8	.	4	5	27
											1	1	.	1	.	2	.	.	2	.	.	1	.	1	1	.	1	2	.	.	4	.	3	.	28
												1	.	.	.	1	.	.	1	.	.	1	.	.	.	.	.	1	.	.	1	.	2	.	29
													1	.	.	.	1	.	.	.	2	.	5	.	.	6	.	.	9	6	.	5	.	7	30
														1	.	1	.	.	2	.	.	1	.	.	.	.	1	.	.	.	2	.	1	.	31
															1	.	.	1	.	2	.	.	1	6	5	1	4	6	3	4	11	2	6	4	32
																1	.	.	1	.	.	1	.	.	.	.	.	.	.	.	1	.	1	.	33
																	1	.	.	.	1	.	.	.	.	.	.	.	.	6	.	.	.	7	34
																		1	.	.	.	.	.	3	2	1	4	3	.	.	8	.	4	4	35
																			1	.	.	1	.	.	.	.	.	.	.	.	.	.	.	.	36
																				1	.	.	.	2	3	.	1	3	1	.	6	1	4	.	37
																					1	.	.	.	.	.	.	.	.	.	.	.	.	.	38
																						1	.	.	.	.	.	.	.	.	.	.	.	.	39
																							1	.	.	1	.	.	3	1	.	2	.	1	40
																								1	1	.	1	2	.	.	4	.	3	.	41
																									1	.	.	1	.	.	2	.	2	.	42
																										1	.	.	.	.	.	.	.	1	43
																											1	.	.	.	2	.	1	.	44
																												1	.	.	1	.	2	.	45
																													1	.	.	1	.	.	46
																														1	.	.	.	1	47
																															1	.	1	.	48
																																1	.	.	49
																																	1	.	50
																																		1	51

Table 3. *(continued)*

ALGEBRA B9, REPRESENTATIONS OF CLASS 1

9	17	25	33	33	41	57	41	49	57	49	65	89	57	57	65	65	73	S.P.
512	4 608	18 432	4 608	43 008	36 864	4 608	64 512	129 024	18 432	64 512	36 864	4 608	43 008	258 048	129 024	18 432	129 024	O.S.
46	64	80	82	94	98	100	106	112	114	116	116	118	124	124	128	130	130	LEVEL
512	9 216	77 824	87 552	408 576	983 040	583 680	1 488 384	5 788 160	4 386 816	3 969 024	7 354 368	3 064 320	7 938 048	22 413 312	34 320 384	11 907 072	46 061 568	DIM'N
0	1	0	2	0	1	3	0	1	0	0	2	4	0	1	0	0	2	
0	0	1	0	0	1	0	0	0	2	0	1	0	0	0	1	0	0	
0	0	0	0	1	0	0	0	1	0	0	0	0	0	0	1	0	1	
0	0	0	0	0	0	0	1	0	0	0	0	0	0	1	0	0	0	
0	0	0	0	0	0	0	0	0	0	1	0	0	0	0	0	0	0	
0	0	0	0	0	0	0	0	0	0	0	0	0	1	0	0	0	0	
0	0	0	0	0	0	0	0	0	0	0	0	0	0	0	0	1	0	
0	0	0	0	0	0	0	0	0	0	0	0	0	0	0	0	0	0	
>	>	>	>	>	>	>	>	>	>	>	>	>	>	>	>	>	>	
1	1	1	1	1	1	1	1	1	1	1	1	1	1	1	1	1	1	WEIGHT
0	1	2	3	4	5	6	7	8	9	10	11	12	13	14	15	16	17	NUMBER
1	9	44	45	147	312	165	369	1 225	864	726	1 350	495	1 140	3 384	4 662	1 422	5 859	0
	1	8	9	35	80	45	104	371	268	230	432	165	392	1 152	1 638	518	2 107	1
		1	1	7	16	9	27	98	72	70	116	45	133	362	525	189	679	2
			1	.	8	9	.	35	36	.	80	45	.	104	224	.	371	3
				1	2	1	6	21	15	20	24	9	44	102	147	69	189	4
					1	1	.	7	8	.	16	9	.	27	63	.	98	5
						1	.	.	.	.	8	9	.	.	.	.	35	6
							1	3	2	5	3	1	14	24	33	25	42	7
								1	1	.	2	1	.	6	14	.	21	8
									1	.	1	1	.	.	7	.	7	9
										1	.	.	4	4	5	9	6	10
											1	1	.	.	.	.	7	11
												1	.	.	.	.	.	12
													1	.	.	3	.	13
														1	2	.	3	14
															1	.	1	15

81	73	65	97	81	129	73	81	73	81	89	105	81	105	81	113	89	S.P.
36 864	4 608	322 560	36 864	512	4 608	387 072	43 008	258 048	258 048	258 048	18 432	129 024	129 024	645 120	36 864	36 864	O.S.
132	134	134	134	136	136	140	142	142	142	146	148	148	148	150	150	152	LEVEL
43 659 264	12 899 328	62 016 000	41 091 072	8 599 552	13 483 008	148 925 952	95 334 400	127 008 768	185 221 120	363 331 584	136 931 328	193 489 920	267 412 992	437 584 896	273 862 656	211 681 280	DIM'N
1	0	1	3	0	5	0	0	1	2	1	0	1	3	0	2	1	
2	0	0	1	0	0	1	0	0	0	1	3	0	0	1	2	0	
0	0	0	0	0	0	0	2	0	0	1	0	0	1	0	0	0	
0	0	0	0	0	0	1	0	0	1	0	0	0	0	0	0	0	
0	0	1	0	0	0	0	0	0	0	0	0	0	0	1	0	0	
0	0	0	0	0	0	0	0	1	0	0	0	0	0	0	0	0	
0	0	0	0	0	0	0	0	0	0	0	0	1	0	0	0	0	
0	1	0	0	0	0	0	0	0	0	0	0	0	0	0	0	1	
>	>	>	>	>	>	>	>	>	>	>	>	>	>	>	>	>	
1	1	1	1	3	1	1	1	1	1	1	1	1	1	1	1	1	WEIGHT
18	19	20	21	22	23	24	25	26	27	28	29	30	31	32	33	34	NUMBER
5 214	1 353	7 095	4 488	835	1 287	14 823	8 876	11 640	17 312	30 912	10 686	14 967	20 895	33 705	20 232	14 528	0
1 902	512	2 630	1 680	323	495	5 703	3 472	4 592	6 792	12 376	4 350	6 174	8 547	13 881	8 352	6 176	1
620	196	930	552	127	165	2 070	1 274	1 764	2 481	4 592	1 634	2 513	3 199	5 509	3 150	2 612	2
388	.	230	432	.	165	783	560	392	1 152	2 520	1 038	518	2 107	1 904	2 232	512	3
172	76	310	152	51	45	693	428	658	830	1 540	548	1 011	1 071	2 085	1 056	1 104	4
108	.	70	116	.	45	258	189	133	362	833	352	189	679	700	740	196	5
36	.	.	80	.	45	.	.	.	104	224	120	.	371	.	388	.	6
37	30	95	32	21	9	207	126	236	246	448	157	402	308	740	300	468	7
23	.	20	24	.	9	75	56	44	102	238	100	69	189	240	208	76	8
16	.	.	16	.	9	27	28	.	27	126	72	.	98	70	144	.	9
5	12	25	4	9	1	51	30	80	60	104	35	157	70	240	66	200	10
8	.	.	16	.	9	.	.	.	27	63	36	.	98	.	108	.	11
.	.	.	8	.	9	.	.	.	.	.	.	.	35	.	36	.	12
.	5	5	.	4	.	9	5	24	10	16	5	60	10	66	9	86	13
3	.	5	3	.	1	18	13	14	24	54	22	25	42	75	45	30	14
2	.	.	2	.	1	6	7	.	6	28	15	.	21	20	31	.	15

89	137	89	97	177	89	97	105	97	113	113	97	97	121	105	137	105	S.P.
322 560	36 864	258 048	4 608	4 608	645 120	774 144	129 024	258 048	258 048	129 024	645 120	387 072	258 048	64 512	36 864	129 024	O.S.
152	152	154	154	154	158	158	160	160	160	162	164	164	164	166	166	166	LEVEL
525 312 000	187 944 960	549 126 144	141 892 608	51 684 864	928 751 616	1 637 222 400	1 132 572 672	1 093 962 240	1 102 970 880	1 278 025 728	1 805 905 920	1 446 709 248	2 368 106 496	1 059 641 856	1 143 078 912	1 685 512 192	DIM'N
2	4	0	1	6	0	1	1	2	3	0	0	0	2	0	1	2	
0	1	0	0	0	1	1	0	0	0	2	0	1	1	0	3	0	
0	0	1	0	0	0	0	2	0	0	1	1	0	1	0	0	0	
0	0	1	0	0	0	1	0	0	1	0	0	0	0	2	0	0	
1	0	0	0	0	0	0	0	0	0	0	1	0	0	0	0	0	
0	0	0	0	0	1	0	0	1	0	0	0	0	0	0	0	0	
0	0	0	0	0	0	0	0	0	0	0	0	1	0	0	0	1	
0	0	0	0	0	0	0	0	0	0	0	0	0	0	0	0	0	
>	>	>	>	>	>	>	>	>	>	>	>	>	>	>	>	>	
1	1	1	3	1	1	1	1	1	1	1	1	1	1	1	1	1	WEIGHT
35	36	37	38	39	40	41	42	43	44	45	46	47	48	49	50	51	NUMBER
38 070	12 540	38 304	9 069	3 003	58 248	104 679	68 166	64 638	64 944	72 870	99 228	77 517	128 142	54 621	57 240	85 162	0
15 950	5 280	16 128	3 923	1 287	25 236	44 991	29 638	28 448	28 448	31 990	44 200	34 845	56 910	24 633	25 704	38 850	1
6 375	2 010	6 510	1 696	495	10 690	18 357	12 201	12 138	11 712	13 258	19 154	15 462	23 702	10 787	10 800	17 367	2
2 630	1 680	2 688	323	495	3 452	9 111	6 552	4 592	6 792	7 714	7 512	4 752	15 064	4 585	7 488	6 174	3
2 415	672	2 484	737	165	4 407	7 017	4 683	5 016	4 482	5 096	8 013	6 768	9 114	4 542	4 164	7 625	4
930	552	1 022	127	165	1 372	3 399	2 492	1 764	2 481	2 954	3 109	1 995	5 684	1 932	2 882	2 513	5
230	432	.	.	165	.	783	560	392	1 152	882	.	.	2 520	.	1 368	518	6
855	188	876	324	45	1 758	2 461	1 635	1 999	1 564	1 771	3 200	2 919	3 157	1 815	1 432	3 291	7
310	152	358	51	45	525	1 161	862	658	830	1 015	1 225	822	1 932	770	984	1 011	8
70	116	154	.	45	133	516	462	133	362	609	470	189	1 141	343	704	189	9
275	40	276	145	9	670	765	499	760	480	535	1 198	1 239	945	673	420	1 395	10
70	116	.	.	45	.	258	189	133	362	308	.	.	833	.	472	189	11
.	80	.	.	45	.	.	.	.	104	.	.	.	224	.	120	.	12
75	5	72	67	1	240	195	123	270	120	129	410	513	225	225	96	577	13
95	32	111	21	9	192	351	259	236	246	301	450	333	567	284	285	402	14
20	24	48	.	9	44	150	140	44	102	175	170	69	329	127	200	69	15

16	1	.	.	2	.	.	2	.	.	.	6	.	.	.	21	.	14	.	38	15	.	14	32	.	75	35	21	84	20	21	119	207	35	63	14	231
17		1	1	.	.	2	.	1	.	.	.	6	14	8	.	21	.	23	.	20	24	.	.	9	.	75	56	44	102	91	.	.	238	.	136	69
18			1	.	.	1	.	1	.	.	.	.	7	8	.	7	.	16	.	.	16	.	.	9	.	27	28	.	27	63	.	.	126	.	108	.
19				1	.	.	1	.	.	.	.	.	.	.	7	.	.	.	16	.	.	.	17	.	20	.	.	21	.	.	28	75	.	14	.	84
20					1	.	.	.	3	2	4	4	8	3	9	6	20	6	12	25	4	28	9	1	66	87	62	80	60	71	150	132	132	93	64	157
21						1	.	1	.	.	.	.	.	.	.	7	.	8	.	.	16	.	.	9	.	.	.	.	27	.	.	.	63	.	36	.
22							1	.	.	.	.	.	.	.	.	.	.	.	8	.	.	.	9	.	.	.	.	.	.	.	.	27	.	.	.	28
23								1	.	.	.	.	.	.	.	.	.	.	.	8	.	.	9	.	.	.	.	.	.	.	.	.	.	.	.
24									1	1	.	1	4	2	.	3	5	4	.	5	3	12	.	1	14	36	33	14	24	40	55	25	75	42	44	25
25										1	.	.	2	1	.	1	.	3	.	.	2	6	.	1	.	12	21	.	6	21	20	.	42	21	30	.
26											1	.	.	.	3	.	4	.	5	5	.	5	4	.	20	15	10	24	10	11	42	51	20	25	9	60
27												1	2	1	.	3	.	3	.	5	3	.	.	1	.	18	13	14	24	21	.	.	54	.	30	25
28													1	1	.	1	.	2	.	.	2	.	.	1	.	6	7	.	6	14	.	.	28	.	23	.
29														1	.	.	.	1	.	.	1	.	.	1	.	.	.	.	.	7	.	.	7	.	16	.
30															1	.	.	.	2	.	.	.	2	.	5	.	.	6	.	.	9	18	.	5	.	21
31																1	.	1	.	.	2	.	.	1	.	.	.	.	6	.	.	.	14	.	8	.
32																	1	.	.	1	.	2	.	.	4	6	5	4	4	6	15	9	11	11	6	9
33																		1	.	.	1	.	.	1	.	.	.	.	.	.	.	.	7	.	8	.
34																			1	.	.	.	1	.	.	.	.	.	.	.	.	6	.	.	.	7
35																				1	.	.	.	.	.	3	2	4	4	3	.	.	8	.	4	9
36																					1	.	.	1	.	.	.	.	.	.	.	.	.	.	.	.
37																						1	.	.	.	2	3	.	1	3	5	.	6	6	4	.
38																							1	.	.	.	.	.	.	.	.	.	.	.	.	.
39																								1	.	.	.	.	.	.	.	.	.	.	.	.
40																									1	.	.	1	.	.	3	3	.	2	.	3
41																										1	1	.	1	2	.	.	4	.	3	.
42																											1	.	.	1	.	.	2	.	2	.
43																												1	.	.	.	.	.	.	.	3
44																													1	.	.	.	2	.	1	.
45																														1	.	.	1	.	2	.
46																															1	.	.	1	.	.
47																																1	.	.	.	1
48																																	1	.	1	.
49																																		1	.	.
50																																			1	.
51																																				1

Table 3. *(continued)*

ALGEBRA B10, REPRESENTATIONS OF CLASS 0

S.P.	0	4	8	16	12	20	36	16	24	32	40	20	64	28	24	36	44	52	28	68	32	100	32	40	36	48
O.S.	1	20	180	20	960	360	20	3 360	2 880	180	360	8 064	20	13 440	13 440	2 880	2 880	360	15 360	360	40 320	20	11 520	20 160	5 120	960
LEVEL	1	21	39	41	55	59	61	69	75	77	79	81	81	89	91	93	95	97	99	99	101	101	105	107	109	109
DIM'N	1	21	210	230	1 330	3 059	1 750	5 985	21 735	15 939	26 125	20 349	10 395	104 006	54 264	150 535	197 505	181 125	116 280	165 186	367 080	51 359	203 490	807 576	293 930	495 880
	0	1	0	2	0	1	3	0	1	0	2	0	4	1	0	0	2	1	0	3	1	5	0	0	0	0
	0	0	1	0	0	1	0	0	0	2	1	0	0	0	0	1	0	2	0	1	0	0	0	1	0	0
	0	0	0	0	1	0	0	0	1	0	0	0	0	0	0	1	1	0	0	0	0	0	0	0	0	2
	0	0	0	0	0	0	0	1	0	0	0	0	0	1	0	0	0	0	0	0	0	0	0	1	0	0
	0	0	0	0	0	0	0	0	0	0	0	1	0	0	0	0	0	0	0	0	1	0	0	0	0	0
	0	0	0	0	0	0	0	0	0	0	0	0	0	0	1	0	0	0	0	0	0	0	0	0	0	0
	0	0	0	0	0	0	0	0	0	0	0	0	0	0	0	0	0	0	1	0	0	0	0	0	0	0
	0	0	0	0	0	0	0	0	0	0	0	0	0	0	0	0	0	0	0	0	0	0	1	0	0	0
	0	0	0	0	0	0	0	0	0	0	0	0	0	0	0	0	0	0	0	0	0	0	0	0	1	0
	>	>	>	>	>	>	>	>	>	>	>	>	>	>	>	>	>	>	>	>	>	>	>	>	>	>
WEIGHT	0	0	0	0	0	0	0	0	0	0	0	0	0	0	0	0	0	0	0	0	0	0	0	0	0	0
NUMBER	0	1	2	3	4	5	6	7	8	9	10	11	12	13	14	15	16	17	18	19	20	21	22	23	24	25
0	1	1	10	10	10	19	10	45	135	99	145	45	55	170	120	215	261	225	120	190	600	55	210	1 160	210	680
1		1	1	1	9	19	10	9	27	18	28	36	10	153	36	206	252	225	84	190	180	55	84	332	126	188
2			1	1	1	2	1	8	26	18	28	8	10	33	28	42	52	44	28	37	156	10	56	308	56	180
3				1	.	1	1	.	9	9	19	.	10	9	.	17	27	27	.	28	36	10	.	108	.	80
4					1	2	1	1	3	2	3	7	1	31	7	41	51	44	21	37	37	10	21	69	35	39
5						1	1	.	1	1	2	.	1	8	.	17	26	27	.	28	8	10	.	24	.	16
6							1	.	.	.	1	.	1	.	.	.	9	9	.	19	.	10	.	.	.	.
7								1	3	2	3	1	1	4	6	5	6	5	6	4	34	1	15	66	15	38
8									1	1	2	.	1	1	.	2	3	3	.	3	7	1	.	23	.	16
9										1	1	.	1	.	.	1	1	2	.	2	.	1	.	8	.	8
10											1	.	1	.	.	.	1	1	.	2	.	1	.	.	.	.
11												1	.	4	1	5	6	5	5	4	5	1	5	9	10	5
12													1	.	.	.	.	.	.	1	.	1	.	.	.	.
13														1	.	2	3	3	.	3	1	1	.	3	.	2
14															1	.	.	.	1	.	5	.	4	9	4	5
15																1	1	2	.	2	.	1	.	1	.	1
16																	1	1	.	2	.	1	.	.	.	.

S.P.	48	40	36	56	72	72	80	44	40	104	52	144	52	44	60	48	48	68	76	76	52	56	84	56	100	108
O.S.	13 440	1 024	80 640	5 760	180	2 880	360	80 640	107 520	360	40 320	20	13 440	92 160	40 320	201 600	46 080	2 880	13 440	2 880	10 240	80 640	5 760	80 640	360	2 880
LEVEL	109	111	111	113	115	115	117	119	119	119	121	121	123	125	127	129	129	129	129	131	131	131	133	135	135	135
DIM'N	981 750	352 716	1 002 915	1 820 105	655 500	1 297 890	1 285 956	3 028 410	2 184 126	848 540	3 553 000	219 604	3 533 145	3 863 080	10 143 000	8 580 495	5 616 324	6 729 800	6 619 239	7 389 900	6 760 390	9 875 250	13 430 340	14 839 209	6 194 475	6 843 420
	2	0	1	1	0	3	2	0	1	4	2	6	0	1	1	0	1	1	3	0	1	2	2	0	1	4
	0	0	0	1	3	0	2	1	0	1	0	0	0	0	1	1	0	0	0	2	0	0	1	0	3	0
	0	0	0	1	0	1	0	0	0	0	0	0	1	0	0	0	0	2	0	1	0	0	1	1	0	1
	1	0	0	0	0	0	0	0	0	0	0	0	1	0	1	0	0	0	1	0	0	0	0	0	0	0
	0	0	0	0	0	0	0	1	0	0	1	0	0	0	0	0	0	0	0	0	0	0	0	1	0	0
	0	0	1	0	0	0	0	0	0	0	0	0	0	0	0	1	0	0	0	0	0	1	0	0	0	0
	0	0	0	0	0	0	0	0	1	0	0	0	0	0	0	0	0	0	0	0	0	0	0	0	0	0
	0	0	0	0	0	0	0	0	0	0	0	0	0	1	0	0	0	0	0	0	0	0	0	0	0	0
	0	0	0	0	0	0	0	0	0	0	0	0	0	0	0	0	1	0	0	0	0	0	0	0	0	0
	>	>	>	>	>	>	>	>	>	>	>	>	>	>	>	>	>	>	>	>	>	>	>	>	>	>
WEIGHT	0	2	0	0	0	0	0	0	0	0	0	0	0	0	0	0	0	0	0	0	2	0	0	0	0	0
NUMBER	26	27	28	29	30	31	32	33	34	35	36	37	38	39	40	41	42	43	44	45	46	47	48	49	50	51
0	1 290	252	675	2 261	780	1 470	1 440	1 590	1 470	880	1 720	220	1 625	1 560	4 200	4 515	2 268	2 600	2 499	2 700	2 310	4 770	4 640	6 825	1 995	2 160
1	378	126	540	629	207	414	387	1 383	588	235	1 512	55	1 490	1 092	3 912	1 743	1 134	2 492	2 346	2 619	1 386	1 872	4 523	2 541	1 995	2 115
2	353	70	176	612	207	405	387	431	448	235	476	55	446	476	1 172	1 456	742	724	706	751	756	1 576	1 299	2 254	550	603
3	153	.	36	323	126	252	270	135	84	190	180	55	171	84	540	420	126	368	378	414	126	540	782	819	369	414
4	79	35	147	130	42	85	78	387	161	46	430	10	417	343	1 110	497	357	701	673	735	455	540	1 274	735	550	594
5	33	.	28	67	26	52	53	120	28	37	156	10	163	56	508	140	56	360	353	406	70	176	765	259	369	405
6	9	.	.	17	9	27	27	.	.	28	36	10	.	.	108	.	.	80	153	117	.	36	323	.	171	252
7	76	20	39	128	42	84	78	96	129	46	106	10	99	138	258	431	240	158	154	162	245	471	278	669	115	126
8	31	.	7	66	26	51	53	29	21	37	37	10	37	21	115	119	35	78	79	88	35	147	163	238	76	85
9	8	.	.	34	18	26	36	8	.	28	8	10	15	.	48	28	.	40	33	50	.	28	92	84	52	52
10	8	.	.	17	9	26	27	.	.	28	8	10	.	.	24	.	.	16	33	25	.	28	67	.	35	52
11	10	10	35	16	5	10	9	90	39	5	100	1	95	105	250	120	110	155	150	160	150	130	275	176	115	125
12	.	.	.	.	.	9	9	.	.	19	.	10	.	.	.	.	.	.	9	.	.	.	17	.	9	27
13	4	.	6	8	3	6	6	27	6	4	34	1	36	15	111	32	15	77	76	87	20	39	161	60	76	84
14	10	6	6	16	5	10	9	14	34	5	15	1	14	37	35	110	78	21	20	21	80	120	35	166	14	15
15	1	.	.	4	2	3	4	7	.	3	7	1	15	.	46	7	.	40	31	49	.	7	91	21	52	51
16	1	.	.	2	1	3	3	.	.	3	7	1	.	.	23	.	.	16	31	25	.	7	66	.	35	51

1	.	1	.	1	.	.	.	.	.	.	.	1	1	1	2	.	.	2	.	1	.	.	8	.	.	8	8	17	.	.	34	.	27	26	17
	1	.	.	.	1	.	3	.	.	3	6	.	.	.	.	14	7	.	15	.	14	31	35	20	33	21	20	21	51	21	35	28	14	15	18
		1	.	1	.	.	.	.	.	.	.	.	.	1	1	.	.	2	.	1	.	.	.	.	.	.	8	.	.	.	17	.	9	26	19
			1	.	.	3	.	2	4	.	1	8	3	6	6	4	5	4	5	1	5	5	15	29	10	10	10	11	10	35	20	57	9	10	20
				1	.	.	.	.	.	.	.	.	.	.	.	.	.	1	.	1	.	.	.	.	.	.	.	.	.	.	.	.	.	9	21
					1	.	1	.	.	2	.	.	.	.	.	.	7	.	.	.	.	8	.	20	26	.	.	.	27	21	.	28	.	.	22
						1	.	1	1	.	.	4	2	3	4	1	.	3	1	1	2	.	6	6	.	5	4	6	.	6	11	20	6	6	23
							1	.	.	1	.	.	.	.	.	.	.	.	.	.	.	8	.	.	9	.	.	.	19	.	.	.	.	.	24
								1	.	.	.	2	1	1	3	.	.	2	.	1	1	.	2	.	.	3	1	3	.	.	6	7	4	3	25
									1	.	.	2	1	3	3	.	.	3	1	1	.	.	3	.	.	2	4	3	.	6	8	.	4	6	26
										1	.	.	.	.	.	.	.	.	.	.	.	.	.	.	9	.	.	.	10	.	.	.	.	.	27
											1	.	.	.	.	4	1	.	5	.	5	4	15	5	4	10	10	11	6	6	20	9	9	10	28
												1	1	1	2	.	.	2	.	1	.	.	1	.	.	1	1	2	.	.	4	.	3	3	29
													1	.	1	.	.	1	.	1	.	.	.	.	.	.	.	1	.	.	1	.	2	1	30
														1	1	.	.	2	.	1	.	.	.	.	.	.	1	.	.	.	2	.	1	3	31
															1	.	.	1	.	1	.	.	.	.	.	.	.	.	.	.	1	.	1	1	32
																1	.	.	1	.	2	.	6	1	.	5	4	6	.	1	11	3	6	6	33
																	1	.	.	.	.	1	.	5	3	.	.	.	3	6	.	9	.	.	34
																		1	.	1	.	.	.	.	.	.	.	.	.	.	.	.	.	1	35
																			1	.	.	.	3	.	.	2	4	3	.	1	8	.	4	6	36
																				1	.	.	.	.	.	.	.	.	.	.	.	.	.	.	37
																					1	.	2	.	.	3	1	3	.	.	6	1	4	3	38
																						1	.	.	1	.	.	.	2	.	.	.	.	.	39
																							1	.	.	1	1	2	.	.	4	.	3	3	40
																								1	.	.	.	.	.	1	.	3	.	.	41
																									1	.	.	.	1	.	.	.	.	.	42
																										1	.	1	.	.	2	.	2	1	43
																											1	.	.	.	2	.	1	3	44
																												1	.	.	1	.	2	1	45
																													1	.	.	.	.	.	46
																														1	.	.	.	.	47
																															1	.	1	1	48
																																1	.	.	49
																																	1	.	50
																																		1	51

Table 3. *(continued)*

ALGEBRA B10, REPRESENTATIONS OF CLASS 1

10	18	26	34	34	42	58	42	50	58	66	50	90	58	58	66	74	82	66	98	66	130	74	74	82	82	S.P.
1 024	10 240	46 080	10 240	122 880	92 160	10 240	215 040	368 640	46 080	92 160	258 048	10 240	860 160	215 040	368 640	368 640	92 160	122 880	92 160	1 290 240	10 240	46 080	1 290 240	10 240	122 880	O.S.
56	76	94	96	110	114	116	124	130	132	134	136	136	144	146	148	150	152	154	154	156	156	160	162	164	164	LEVEL
1 024	20 480	193 536	215 040	1 146 880	2 703 360	1 576 960	4 766 720	17 977 344	13 424 640	22 256 640	14 708 736	9 067 520	79 478 784	34 728 960	118 702 080	157 515 776	147 087 360	63 504 384	136 249 344	254 679 040	43 524 096	89 303 040	588 564 480	92 610 560	369 954 816	DIM'N
0	1	0	2	0	1	3	0	1	0	2	0	4	1	0	0	2	1	0	3	1	5	0	0	0	0	
0	0	1	0	0	1	0	0	0	2	1	0	0	0	0	1	0	2	0	1	0	0	0	1	0	0	
0	0	0	0	1	0	0	0	1	0	0	0	0	0	0	1	1	0	0	0	0	0	0	0	0	2	
0	0	0	0	0	0	0	1	0	0	0	0	0	1	0	0	0	0	0	0	0	0	0	1	0	0	
0	0	0	0	0	0	0	0	0	0	0	1	0	0	0	0	0	0	0	0	1	0	0	0	0	0	
0	0	0	0	0	0	0	0	0	0	0	0	0	0	1	0	0	0	0	0	0	0	0	0	0	0	
0	0	0	0	0	0	0	0	0	0	0	0	0	0	0	0	0	0	1	0	0	0	0	0	0	0	
0	0	0	0	0	0	0	0	0	0	0	0	0	0	0	0	0	0	0	0	0	0	1	0	0	0	
0	0	0	0	0	0	0	0	0	0	0	0	0	0	0	0	0	0	0	0	0	0	0	0	1	0	
>	>	>	>	>	>	>	>	>	>	>	>	>	>	>	>	>	>	>	>	>	>	>	>	>	>	
1	1	1	1	1	1	1	1	1	1	1	1	1	1	1	1	1	1	1	1	1	1	1	1	1	1	WEIGHT
0	1	2	3	4	5	6	7	8	9	10	11	12	13	14	15	16	17	18	19	20	21	22	23	24	25	NUMBER
1	10	54	55	200	420	220	560	1 826	1 275	1 980	1 242	715	5 628	2 235	7 620	9 512	8 370	3 288	7 128	13 315	2 002	3 915	27 055	3 610	15 984	0
	1	9	10	44	99	55	147	512	366	585	369	220	1 785	726	2 490	3 176	2 835	1 140	2 475	4 626	715	1 422	9 730	1 353	5 838	1
		1	1	8	18	10	35	124	90	144	104	55	518	230	736	944	852	392	750	1 521	220	518	3 276	512	1 986	2
			1	.	9	10	.	44	45	99	.	55	147	.	312	512	531	.	585	369	220	.	1 225	.	864	3
				1	2	1	7	24	17	27	27	10	133	70	188	240	216	133	189	466	55	189	1 008	196	613	4
					1	1	.	8	9	18	.	10	35	.	80	124	135	.	144	104	55	.	371	.	268	5
						1	.	.	.	9	.	10	.	.	.	44	45	.	99	.	55	.	.	.	.	6
							1	3	2	3	6	1	28	20	38	48	42	44	36	129	10	69	273	76	164	7
								1	1	2	.	1	7	.	16	24	26	.	27	27	10	.	98	.	72	8
									1	1	.	1	.	.	8	8	18	.	18	.	10	.	35	.	36	9
										1	.	1	.	.	.	8	9	.	18	.	10	.	.	.	.	10
											1	.	4	5	5	6	5	14	4	30	1	25	60	30	35	11
												1	.	.	.	.	.	.	9	.	10	.	.	.	.	12

82	90	74	90	106	106	114	82	82	138	90	178	90	90	98	90	98	106	114	114	106	98	122	98	138	146	S.P.
860 160	1 024	1 290 240	737 280	46 080	368 640	92 160	2 580 480	860 160	92 160	1 290 240	10 240	860 160	368 640	2 580 480	3 225 600	92 160	368 640	860 160	368 640	10 240	1 290 240	737 280	2 580 480	92 160	368 640	O.S.
164	166	166	168	170	170	172	174	174	174	176	176	178	180	182	184	184	184	184	186	186	186	188	190	190	190	LEVEL
723 588 096	60 196 864	616 366 080	1 384 120 320	510 720 000	1 002 373 120	1 011 225 600	2 004 746 240	1 146 055 680	680 064 000	2 378 280 960	181 350 400	2 437 349 376	1 629 945 856	7 131 013 120	5 034 240 000	1 702 711 296	4 843 044 864	4 725 473 280	5 393 203 200	1 111 326 720	5 857 617 920	9 891 840 000	9 303 275 520	4 674 109 440	5 136 384 000	DIM'N
2	0	1	1	0	3	2	0	1	4	2	6	0	1	1	0	1	1	3	0	1	2	2	0	1	4	
0	0	0	1	3	0	2	1	0	1	0	0	0	0	1	1	0	0	0	2	0	0	1	0	3	0	
0	0	0	1	0	1	0	0	0	0	0	0	1	0	0	0	0	2	0	1	0	0	1	1	0	1	
1	0	0	0	0	0	0	0	0	0	0	0	1	0	1	0	0	0	1	0	0	0	0	0	0	0	
0	0	0	0	0	0	0	1	0	0	1	0	0	0	0	0	0	0	0	0	0	0	0	1	0	0	
0	0	1	0	0	0	0	0	0	0	0	0	0	0	0	1	0	0	0	0	0	1	0	0	0	0	
0	0	0	0	0	0	0	0	1	0	0	0	0	0	0	0	0	0	0	0	0	0	0	0	0	0	
0	0	0	0	0	0	0	0	0	0	0	0	0	1	0	0	0	0	0	0	0	0	0	0	0	0	
0	0	0	0	0	0	0	0	0	0	0	0	0	0	0	0	1	0	0	0	0	0	0	0	0	0	
>	>	>	>	>	>	>	>	>	>	>	>	>	>	>	>	>	>	>	>	>	>	>	>	>	>	
1	3	1	1	1	1	1	1	1	1	1	1	1	1	1	1	1	1	1	1	3	1	1	1	1	1	WEIGHT
26	27	28	29	30	31	32	33	34	35	36	37	38	39	40	41	42	43	44	45	46	47	48	49	50	51	NUMBER
31 374	2 188	25 080	54 960	18 690	36 740	35 145	69 520	38 130	21 450	77 940	5 005	77 112	46 472	207 970	138 225	43 527	133 488	127 560	141 240	26 620	152 195	246 600	226 125	108 240	116 600	0
11 487	835	9 330	20 520	7 095	14 000	13 518	26 896	14 928	8 415	30 627	2 002	30 450	18 882	83 713	56 550	18 138	54 324	52 269	58 026	11 257	63 150	102 420	94 878	45 465	49 280	1
3 892	323	3 356	7 056	2 469	4 856	4 725	9 960	5 732	2 970	11 418	715	11 452	7 596	31 808	22 485	7 529	20 816	20 034	22 380	4 758	25 280	39 696	38 502	17 778	19 280	2
1 785	.	726	3 840	1 560	3 176	3 330	3 384	1 140	2 475	4 626	715	4 662	1 422	15 589	7 095	1 353	11 052	11 487	12 900	835	9 330	25 008	14 823	12 243	14 000	3
1 197	127	1 160	2 176	762	1 496	1 458	3 487	2 156	915	4 002	220	4 039	3 031	11 242	8 635	3 124	7 386	7 091	7 954	2 019	9 731	14 118	14 976	6 342	6 866	4
518	.	230	1 168	486	944	1 017	1 152	392	750	1 521	220	1 638	518	5 383	2 630	512	3 888	3 892	4 566	323	3 356	8 736	5 703	4 359	4 856	5
147	.	.	312	165	512	531	.	.	585	369	220	.	.	1 225	.	.	864	1 785	1 350	.	726	3 840	.	2 055	3 176	6
322	51	380	576	199	392	378	1 132	791	234	1 296	55	1 304	1 200	3 612	3 175	1 300	2 364	2 268	2 536	864	3 575	4 488	5 511	2 004	2 168	7
133	.	70	304	126	240	261	362	133	189	466	55	525	189	1 687	930	196	1 233	1 197	1 440	127	1 160	2 728	2 070	1 368	1 496	8
35	.	.	160	90	124	180	104	.	144	104	55	224	.	742	230	.	656	518	856	.	230	1 600	783	972	944	9
35	.	.	80	45	124	135	.	.	144	104	55	.	.	371	.	.	268	518	432	.	230	1 168	.	651	944	10
70	21	115	120	40	80	75	330	280	45	375	10	371	471	1 015	1 100	544	654	630	695	375	1 235	1 220	1 880	535	580	11
.	.	.	.	.	44	45	.	.	99	.	55	.	.	.	.	.	.	147	.	.	.	312	.	165	512	12

1	.	2	3	3	.	3	6	1	.	21	.	15	28	.	20	62	25	48	51	102	44	36	129	10	147	69	462	310	76	336	322	388	51	380	728	693	361	392	13
	1	.	.	.	4	.	5	.	9	9	12	5	10	9	30	16	5	10	9	80	94	5	90	1	86	182	230	350	230	144	140	150	166	390	260	576	110	120	14
		1	1	2	.	2	.	1	.	7	.	8	7	.	.	32	17	24	35	27	.	27	27	10	63	.	196	70	.	180	133	224	.	70	420	258	252	240	15
			1	1	.	2	.	1	.	.	.	.	7	.	.	16	9	24	26	.	.	27	27	10	.	.	98	.	.	72	133	116	.	70	304	.	171	240	16
				1	.	1	.	1	.	.	.	.	.	.	.	8	9	8	18	.	.	18	.	10	.	.	35	.	.	36	35	80	.	.	160	.	135	124	17
					1	.	.	.	3	.	5	.	.	4	6	.	.	.	.	14	28	.	15	.	14	69	35	95	98	21	20	21	76	105	35	147	14	15	18
						1	.	1	.	.	.	.	.	.	.	.	.	8	9	.	.	18	.	10	.	.	.	.	.	.	35	.	.	.	80	.	45	124	19
							1	.	.	3	.	2	4	.	5	8	3	6	6	24	14	4	30	1	33	25	102	95	30	72	70	82	21	115	152	207	73	80	20
								1	.	.	.	.	.	.	.	.	.	.	.	.	.	9	.	10	.	.	.	.	.	.	.	.	.	.	.	.	.	44	21
									1	.	2	.	.	2	.	.	.	.	.	.	7	.	.	.	.	24	.	20	43	.	.	.	36	21	.	28	.	.	22
										1	.	1	1	.	.	4	2	3	4	6	.	3	6	1	14	.	42	20	.	38	28	46	.	20	86	75	50	48	23
											1	.	.	1	.	.	.	.	.	.	.	.	.	.	.	8	.	.	18	.	.	.	19	.	.	.	.	.	24
												1	.	.	.	2	1	1	3	.	.	2	.	1	7	.	14	.	.	24	7	24	.	.	48	27	34	24	25
													1	.	.	2	1	3	3	.	.	3	6	1	.	.	21	.	.	15	28	24	.	20	62	.	34	48	26
														1	.	.	.	.	.	.	.	.	.	.	.	.	.	.	9	.	.	.	10	.	.	.	.	.	27
															1	.	.	.	.	4	4	.	5	.	5	9	15	25	12	10	10	11	9	30	20	51	9	10	28
																1	1	1	2	.	.	2	.	1	.	.	7	.	.	8	7	16	.	.	32	.	26	24	29
																	1	.	1	.	.	1	.	1	.	.	.	.	.	.	.	8	.	.	8	.	18	8	30
																		1	1	.	.	2	.	1	.	.	.	.	.	.	7	.	.	.	16	.	9	24	31
																			1	.	.	1	.	1	.	.	.	.	.	.	.	.	.	.	8	.	9	8	32
																				1	.	.	1	.	2	.	6	5	.	5	4	6	.	5	11	18	6	6	33
																					1	.	.	.	.	3	.	5	5	.	.	.	4	6	.	9	.	.	34
																						1	.	1	.	.	.	.	.	.	.	.	.	.	.	.	.	8	35
																							1	.	.	.	3	.	.	2	4	3	.	5	8	.	4	6	36
																								1	.	.	.	.	.	.	.	.	.	.	.	.	.	.	37
																									1	.	2	.	.	3	1	3	.	.	6	6	4	3	38
																										1	.	.	2	.	.	.	2	.	.	.	.	.	39
																											1	.	.	1	1	2	.	.	4	.	3	3	40
																												1	.	.	.	.	.	1	.	3	.	.	41
																													1	.	.	.	1	.	.	.	.	.	42
																														1	.	1	.	.	2	.	2	1	43
																															1	.	.	.	2	.	1	3	44
																																1	.	.	1	.	2	1	45
																																	1	.	.	.	.	.	46
																																		1	.	.	.	.	47
																																			1	.	1	1	48
																																				1	.	.	49
																																					1	.	50
																																						1	51

Table 3. *(continued)*

ALGEBRA B11, REPRESENTATIONS OF CLASS 0

NUMBER	0	1	2	3	4	5	6	7	8	9	10	11	12
S.P.	0	4	8	16	12	20	36	16	24	32	40	64	20
O.S.	1	22	220	22	1 320	440	22	5 280	3 960	220	440	22	14 784
LEVEL	1	23	43	45	61	65	67	77	83	85	87	89	91
DIM'N	1	23	253	275	1 771	4 025	2 277	8 855	31 625	23 000	37 422	14 674	33 649
WEIGHT	0 0 0 0 0 0 0 0 0 0 > 0	1 0 0 0 0 0 0 0 0 0 > 0	0 1 0 0 0 0 0 0 0 0 > 0	2 0 0 0 0 0 0 0 0 0 > 0	0 0 1 0 0 0 0 0 0 0 > 0	1 1 0 0 0 0 0 0 0 0 > 0	3 0 0 0 0 0 0 0 0 0 > 0	0 0 0 1 0 0 0 0 0 0 > 0	1 0 1 0 0 0 0 0 0 0 > 0	0 2 0 0 0 0 0 0 0 0 > 0	2 1 0 0 0 0 0 0 0 0 > 0	4 0 0 0 0 0 0 0 0 0 > 0	0 0 0 0 1 0 0 0 0 0 > 0
0	1	1	11	11	11	21	11	55	165	120	176	66	55
1		1	1	1	10	21	11	10	30	20	31	11	45
2			1	1	1	2	1	9	29	20	31	11	9
3				1	.	1	1	.	10	10	21	11	.
4					1	2	1	1	3	2	3	1	8
5						1	1	.	1	1	2	1	.
6							1	.	.	.	1	1	.
7								1	3	2	3	1	1
8									1	1	2	1	.
9										1	1	1	.
10											1	1	.
11												1	.
12													1

NUMBER	13	14	15	16	17	18	19	20	21	22	23	24	25
S.P.	28	24	36	44	52	68	100	28	32	40	32	48	48
O.S.	21 120	29 568	3 960	3 960	440	440	22	42 240	73 920	31 680	42 240	1 320	21 120
LEVEL	99	103	103	105	107	109	111	113	113	119	121	121	121
DIM'N	168 245	100 947	240 350	312 984	284 625	256 795	78 430	245 157	664 125	1 434 510	490 314	872 850	1 730 520
WEIGHT	1 0 0 1 0 0 0 0 0 0 > 0	0 0 0 0 0 1 0 0 0 0 > 0	0 1 1 0 0 0 0 0 0 0 > 0	2 0 1 0 0 0 0 0 0 0 > 0	1 2 0 0 0 0 0 0 0 0 > 0	3 1 0 0 0 0 0 0 0 0 > 0	5 0 0 0 0 0 0 0 0 0 > 0	0 0 0 0 0 0 1 0 0 0 > 0	1 0 0 0 1 0 0 0 0 0 > 0	0 1 0 1 0 0 0 0 0 0 > 0	0 0 0 0 0 0 0 1 0 0 > 0	0 0 2 0 0 0 0 0 0 0 > 0	2 0 0 1 0 0 0 0 0 0 > 0
0	209	165	264	320	275	231	66	165	825	1 584	330	924	1 760
1	190	45	254	310	275	231	66	120	225	414	120	234	470
2	37	36	47	58	49	41	11	36	198	387	84	225	442
3	10	.	19	30	30	31	11	.	45	135	.	99	190
4	35	8	46	57	49	41	11	28	42	78	28	44	89
5	9	.	19	29	30	31	11	.	9	27	.	18	37
6	.	.	.	10	10	21	11	.	.	.	.	.	10
7	4	7	5	6	5	4	1	7	39	75	21	43	86
8	1	.	2	3	3	3	1	.	8	26	.	18	35
9	.	.	1	1	2	2	1	.	.	9	.	9	9
10	.	.	.	1	1	2	1	.	.	.	.	.	9
11	.	.	.	.	.	1	1	.	.	.	.	.	.
12	4	1	5	6	5	4	1	6	5	9	6	5	10
13	1	.	2	3	3	3	1	.	1	3	.	2	4
14		1	.	.	.	.	.	1	5	9	5	5	10
15			1	1	2	2	1	.	.	1	.	1	1

NUMBER	26	27	28	29	30	31	32	33	34	35	36	37	38
S.P.	36	56	36	72	72	80	40	104	44	44	144	40	52
O.S.	177 408	7 920	28 160	220	3 960	440	11 264	440	2 048	147 840	22	295 680	73 920
LEVEL	125	125	127	127	127	129	131	131	133	133	133	135	135
DIM'N	2 042 975	3 166 065	817 190	1 126 125	2 233 000	2 193 763	1 144 066	1 427 426	1 352 078	6 021 400	361 790	5 047 350	7 008 606
WEIGHT	1 0 0 0 0 1 0 0 0 0 > 0	1 1 1 0 0 0 0 0 0 0 > 0	0 0 0 0 0 0 0 0 1 0 > 0	0 3 0 0 0 0 0 0 0 0 > 0	3 0 1 0 0 0 0 0 0 0 > 0	2 2 0 0 0 0 0 0 0 0 > 0	0 0 0 0 0 0 0 0 0 1 > 0	4 1 0 0 0 0 0 0 0 0 > 0	0 0 0 0 0 0 0 0 0 0 > 2	0 1 0 0 1 0 0 0 0 0 > 0	6 0 0 0 0 0 0 0 0 0 > 0	1 0 0 0 0 0 1 0 0 0 > 0	2 0 0 0 1 0 0 0 0 0 > 0
0	935	3 059	330	1 045	1 980	1 925	462	1 166	462	2 200	286	2 310	2 376
1	765	779	210	255	510	475	210	286	252	1 940	66	840	2 115
2	225	760	84	255	500	475	126	286	126	548	66	660	603
3	45	399	.	155	310	330	.	231	.	170	66	120	225
4	192	146	56	47	95	87	56	51	70	498	11	212	551
5	36	75	.	29	58	59	.	41	.	153	11	36	198
6	.	19	.	10	30	30	.	31	.	.	11	.	45
7	45	144	21	47	94	87	35	51	35	110	11	175	121
8	8	74	.	29	57	59	.	41	.	33	11	28	42
9	.	38	.	20	29	40	.	31	.	9	11	.	9
10	.	19	.	10	29	30	.	31	.	.	11	.	9
11	.	.	.	.	10	10	.	21	.	.	11	.	.
12	41	16	15	5	10	9	15	5	20	104	1	46	115
13	7	8	.	3	6	6	.	4	.	31	1	7	39
14	6	16	5	5	10	9	10	5	10	14	1	41	15
15	.	4	.	2	3	4	.	3	.	8	1	.	8

NUMBER	39	40	41	42	43	44	45	46	47	48	49	50	51
S.P.	52	60	44	68	76	48	76	56	84	48	100	108	56
O.S.	21 120	63 360	337 920	3 960	21 120	443 520	3 960	177 408	7 920	253 440	440	3 960	147 840
LEVEL	137	141	143	143	143	145	145	147	147	149	149	149	151
DIM'N	6 924 610	19 639 125	10 214 875	12 910 590	12 673 000	19 228 000	14 060 475	21 948 003	25 371 346	17 160 990	11 555 775	12 744 875	32 586 400
WEIGHT	0 0 1 1 0 0 0 0 0 0 > 0	1 1 0 1 0 0 0 0 0 0 > 0	1 0 0 0 0 0 0 1 0 0 > 0	1 0 2 0 0 0 0 0 0 0 > 0	3 0 0 1 0 0 0 0 0 0 > 0	0 1 0 0 0 1 0 0 0 0 > 0	0 2 1 0 0 0 0 0 0 0 > 0	2 0 0 0 0 1 0 0 0 0 > 0	2 1 1 0 0 0 0 0 0 0 > 0	1 0 0 0 0 0 0 0 1 0 > 0	1 3 0 0 0 0 0 0 0 0 > 0	4 0 1 0 0 0 0 0 0 0 > 0	0 0 1 0 1 0 0 0 0 0 > 0
0	2 244	5 775	2 475	3 564	3 420	7 040	3 685	7 425	6 314	4 158	2 695	2 915	10 560
1	2 074	5 415	1 800	3 429	3 230	2 480	3 585	2 655	6 169	1 890	2 695	2 860	3 600
2	565	1 476	708	909	884	2 112	939	2 277	1 618	1 302	681	745	3 232
3	215	675	120	459	470	600	515	765	969	210	455	510	1 160
4	532	1 406	532	883	847	648	921	701	1 590	560	681	735	952
5	206	639	84	450	442	180	506	225	950	84	455	500	332
6	.	135	.	99	190	.	145	45	399	.	210	310	.
7	113	293	189	179	174	572	183	622	313	399	129	141	876
8	42	130	28	88	89	156	99	192	183	56	85	95	308
9	17	54	.	45	37	36	56	36	103	.	58	58	108
10	.	27	.	18	37	.	28	36	75	.	39	58	.
11	.	.	.	.	10	.	.	.	19	.	10	30	.
12	109	285	150	176	170	140	181	151	310	159	129	140	204
13	41	126	21	87	86	37	98	45	181	21	85	94	69
14	14	35	45	21	20	130	21	141	35	120	14	15	194
15	17	52	.	45	35	8	55	8	102	.	58	57	24

1	1	2	1	.	.	.	.	.	1	.	2	.	1	3	3	.	3	.	.	1	.	8	.	26	.	18	35	.	28	8	74	.	39	57	.	16
	1	1	1	.	.	.	.	.	.	.	1	.	1	1	2	.	2	.	.	1	.	.	.	9	.	9	9	.	19	.	38	.	30	29	.	17
		1	1	.	.	.	.	.	.	.	.	.	.	1	1	.	2	.	.	1	.	.	.	.	.	.	9	.	.	.	19	.	10	29	.	18
			1	.	.	.	.	.	.	.	.	.	.	.	.	.	1	.	.	1	.	.	.	.	.	.	.	.	.	.	.	.	.	10	.	19
				1	.	.	1	.	.	6	.	4	.	.	.	4	.	6	14	.	7	15	14	35	39	21	20	20	21	21	35	42	14	15	28	20
					1	3	.	2	4	1	8	.	3	6	6	.	4	.	4	1	6	5	5	15	6	10	10	34	11	41	20	15	9	10	66	21
						1	.	1	1	.	4	.	2	3	4	.	3	.	1	1	.	1	2	6	.	5	4	7	6	7	11	.	6	6	23	22
							1	.	.	.	.	1	.	.	.	3	.	3	.	.	7	.	.	.	8	.	.	20	.	21	.	35	.	.	28	23
								1	.	.	2	.	1	1	3	.	2	.	.	1	.	.	1	2	.	3	1	.	3	.	6	.	4	3	8	24
									1	.	2	.	1	3	3	.	3	.	.	1	.	1	.	3	.	2	4	.	3	7	8	.	4	6	.	25
										1	.	.	.	.	.	.	.	.	4	.	1	5	5	15	5	10	10	5	11	6	20	5	9	10	9	26
											1	.	1	1	2	.	2	.	.	1	.	.	.	1	.	1	1	.	2	.	4	.	3	3	.	27
												1	.	.	.	1	.	2	.	.	.	.	.	.	8	.	.	.	.	.	.	9	.	.	.	28
													1	.	1	.	1	.	.	1	.	.	.	.	.	.	.	.	1	.	1	.	2	1	.	29
														1	1	.	2	.	.	1	.	.	.	.	.	.	1	.	.	.	2	.	1	3	.	30
															1	.	1	.	.	1	.	.	.	.	.	.	.	.	.	.	1	.	1	1	.	31
																1	.	1	.	.	.	.	.	.	.	.	.	.	.	.	.	9	.	.	.	32
																	1	.	.	1	.	.	.	.	.	.	.	.	.	.	.	.	.	1	.	33
																		1	.	.	.	.	.	.	.	.	.	.	.	.	.	.	.	.	.	34
																			1	.	.	1	2	6	.	5	4	1	6	1	11	.	6	6	3	35
																				1	.	.	.	.	.	.	.	.	.	.	.	.	.	.	.	36
																					1	.	.	.	1	.	.	5	.	6	.	4	.	.	9	37
																						1	.	3	.	2	4	.	3	1	8	.	4	6	.	38
																							1	2	.	3	1	.	3	.	6	.	4	3	1	39
																								1	.	1	1	.	2	.	4	.	3	3	.	40
																									1	.	.	.	.	.	.	1	.	.	.	41
																										1	.	.	1	.	2	.	2	1	.	42
																											1	.	.	.	2	.	1	3	.	43
																												1	.	1	.	.	.	.	3	44
																													1	.	1	.	2	1	.	45
																														1	.	.	.	.	.	46
																															1	.	1	1	.	47
																																1	.	.	.	48
																																	1	.	.	49
																																		1	.	50
																																			1	51

Table 3. *(continued)*

ALGEBRA B11, REPRESENTATIONS OF CLASS 1

11	19	27	35	35	43	59	43	51	59	67	91	51	59	59	67	75	83	99	131	67	67	75	75	83	83	75	91	83	107	107	115	91	139	99	83	179	83	91	91	99	91	107	115	91	115	99	123	99	139	147	99	S.P.
2 048	22 528	112 640	22 528	337 920	225 280	22 528	675 840	1 013 760	112 640	225 280	22 528	946 176	2 703 360	946 176	1 013 760	1 013 760	225 280	225 280	22 528	675 840	4 730 880	4 055 040	337 920	337 920	2 703 360	5 677 056	2 027 520	112 640	112 640	1 013 760	225 280	22 528	225 280	2 048	9 461 760	22 528	4 730 880	4 730 880	2 703 360	8 110 080	2 703 360	1 013 760	2 703 360	14 192 640	1 013 760	5 677 056	2 027 520	1 013 760	225 280	1 013 760	9 461 760	O.S.
67	89	109	111	127	131	133	143	149	151	153	155	157	165	169	169	171	173	175	177	179	179	185	187	187	187	191	191	193	193	193	195	197	197	199	199	199	201	201	203	207	209	209	209	211	211	213	213	215	215	215	217	LEVEL
2 048	45 056	471 040	518 144	3 108 864	7 208 960	4 145 152	14 508 032	53 415 936	39 424 000	64 768 000	25 907 200	50 778 112	265 289 728	137 826 304	388 608 000	510 935 040	471 511 040	431 095 808	134 717 440	295 342 080	964 784 128	2 165 841 920	502 081 536	1 342 464 000	2 637 824 000	2 686 058 496	4 948 230 144	669 442 048	1 795 368 960	3 536 332 800	3 525 451 776	669 442 048	2 331 648 000	426 008 576	8 378 388 480	606 228 480	5 857 617 920	9 844 736 000	9 948 364 800	28 667 871 232	10 081 009 664	19 197 235 200	18 743 296 000	24 223 232 000	21 152 710 656	27 909 826 560	38 472 192 000	13 556 201 472	17 873 895 424	19 641 802 752	43 161 395 200	DIM'N
0	1	0	2	0	1	3	0	1	0	2	4	0	1	0	0	2	1	3	5	0	1	0	0	0	2	1	1	0	0	3	2	0	4	0	0	6	1	2	0	1	1	1	3	0	0	2	2	1	1	4	0	
0	0	1	0	0	1	0	0	0	2	1	0	0	0	0	1	0	2	1	0	0	0	1	0	0	0	0	1	0	3	0	2	0	1	0	1	0	0	0	0	1	0	0	0	1	2	0	1	0	3	0	0	
0	0	0	0	1	0	0	0	1	0	0	0	0	0	0	1	1	0	0	0	0	0	0	0	2	0	0	1	0	0	1	0	0	0	0	0	0	0	0	1	0	0	2	0	0	1	0	1	0	0	1	1	
0	0	0	0	0	0	0	1	0	0	0	0	0	1	0	0	0	0	0	0	0	0	1	0	0	1	0	0	0	0	0	0	0	0	0	0	0	0	0	1	1	0	0	1	0	0	0	0	0	0	0	0	
0	0	0	0	0	0	0	0	0	0	0	0	1	0	0	0	0	0	0	0	0	1	0	0	0	0	0	0	0	0	0	0	0	0	0	1	0	0	1	0	0	0	0	0	0	0	0	0	0	0	0	1	
0	0	0	0	0	0	0	0	0	0	0	0	0	0	1	0	0	0	0	0	0	0	0	0	0	0	1	0	0	0	0	0	0	0	0	0	0	0	0	0	0	0	0	0	1	0	1	0	0	0	0	0	
0	0	0	0	0	0	0	0	0	0	0	0	0	0	0	0	0	0	0	0	1	0	0	0	0	0	0	0	0	0	0	0	0	0	0	0	0	1	0	0	0	0	0	0	0	0	0	0	0	0	0	0	
0	0	0	0	0	0	0	0	0	0	0	0	0	0	0	0	0	0	0	0	0	0	0	1	0	0	0	0	0	0	0	0	0	0	0	0	0	0	0	0	0	1	0	0	0	0	0	0	0	0	0	0	
0	0	0	0	0	0	0	0	0	0	0	0	0	0	0	0	0	0	0	0	0	0	0	0	0	0	0	0	1	0	0	0	0	0	0	0	0	0	0	0	0	0	0	0	0	0	0	0	1	0	0	0	
0	0	0	0	0	0	0	0	0	0	0	0	0	0	0	0	0	0	0	0	0	0	0	0	0	0	0	0	0	0	0	0	1	0	0	0	0	0	0	0	0	0	0	0	0	0	0	0	0	0	0	0	
>	>	>	>	>	>	>	>	>	>	>	>	>	>	>	>	>	>	>	>	>	>	>	>	>	>	>	>	>	>	>	>	>	>	>	>	>	>	>	>	>	>	>	>	>	>	>	>	>	>	>	>	
1	1	1	1	1	1	1	1	1	1	1	1	1	1	1	1	1	1	1	1	1	1	1	1	1	1	1	1	1	1	1	1	1	1	3	1	1	1	1	1	1	1	1	1	1	1	1	1	1	1	1	1	WEIGHT
0	1	2	3	4	5	6	7	8	9	10	11	12	13	14	15	16	17	18	19	20	21	22	23	24	25	26	27	28	29	30	31	32	33	34	35	36	37	38	39	40	41	42	43	44	45	46	47	48	49	50	51	NUMBER
1	11	65	66	264	550	286	814	2 619	1 815	2 805	1 001	2 002	8 888	4 037	11 880	14 751	12 870	10 868	3 003	6 765	23 331	46 464	9 438	27 180	53 580	49 302	92 532	10 813	31 075	61 314	58 113	9 713	35 035	5 798	132 594	8 008	85 525	147 785	144 540	385 836	122 408	245 025	234 520	296 175	257 202	324 126	446 490	142 857	193 358	208 494	471 240	0
	1	10	11	54	120	66	200	684	485	770	286	560	2 640	1 242	3 630	4 599	4 070	3 520	1 001	2 235	7 630	15 694	3 288	9 315	18 400	17 352	32 382	3 915	11 055	21 879	20 933	3 610	12 870	2 188	48 384	3 003	31 845	54 705	53 760	146 116	47 568	93 765	90 320	114 785	99 342	127 242	174 240	57 285	76 318	82 764	186 935	1
		1	1	9	20	11	44	153	110	175	66	147	712	369	996	1 269	1 135	990	286	726	2 345	4 930	1 140	2 955	5 816	5 868	10 380	1 422	3 585	7 074	6 820	1 353	4 235	835	16 800	1 001	11 565	19 117	18 936	51 974	18 216	33 621	32 400	43 006	35 847	47 979	63 180	22 797	27 918	30 294	71 351	2
			1	.	10	11	.	54	55	120	66	.	200	.	420	684	705	770	286	.	560	1 826	.	1 275	2 640	1 242	5 610	.	2 255	4 599	4 785	.	3 520	.	5 628	1 001	2 235	7 630	7 620	25 206	3 288	17 685	18 400	13 315	20 502	17 352	39 510	3 915	19 118	21 879	27 055	3
				1	2	1	8	27	19	30	11	35	168	104	234	297	265	230	66	230	665	1 404	392	844	1 656	1 890	2 964	518	1 025	2 019	1 950	512	1 210	323	5 474	286	4 082	6 237	6 212	17 090	6 872	11 097	10 672	15 459	11 853	17 286	20 910	9 018	9 270	10 044	25 942	4
					1	1	.	9	10	20	11	.	44	.	99	153	165	175	66	.	147	512	.	366	712	369	1 581	.	650	1 269	1 355	.	990	.	1 785	286	726	2 345	2 490	8 106	1 140	5 790	5 816	4 626	6 750	5 868	12 855	1 422	6 335	7 074	9 730	5
						1	.	.	.	10	11	.	.	.	.	54	55	120	66	.	.	.	.	.	200	.	420	.	220	684	705	.	770	.	.	286	.	560	.	1 826	.	1 275	2 640	.	1 980	1 242	5 610	.	2 970	4 599	.	6
							1	3	2	3	1	7	32	27	43	54	47	40	11	70	168	348	133	207	408	570	720	189	246	486	465	196	285	127	1 638	66	1 390	1 862	1 850	5 068	2 548	3 280	3 152	5 273	3 492	5 892	6 144	3 549	2 710	2 934	8 855	7
								1	1	2	1	.	8	.	18	27	29	30	11	.	35	124	.	90	168	104	378	.	155	297	320	.	230	.	518	66	230	665	736	2 348	392	1 696	1 656	1 521	1 968	1 890	3 714	518	1 840	2 019	3 276	8
									1	1	1	.	.	.	9	9	20	20	11	.	.	44	.	45	44	.	198	.	110	153	220	.	175	.	147	66	.	147	312	1 024	.	897	712	369	1 161	369	2 166	.	1 300	1 269	1 225	9
										1	1	.	.	.	.	9	10	20	11	.	.	.	.	.	44	.	99	.	55	153	165	.	175	.	.	66	.	147	.	512	.	366	712	.	585	369	1 581	.	870	1 269	.	10
											1	.	.	.	.	.	.	10	11	.	.	.	.	.	.	.	.	.	.	54	55	.	120	.	.	66	.	.	.	.	.	.	200	.	.	.	420	.	220	684	.	11
												1	4	6	5	6	5	4	1	20	35	69	44	40	80	156	136	69	45	90	84	76	50	51	434	11	450	490	480	1 300	924	830	800	1 675	876	1 866	1 530	1 389	664	720	2 775	12

13	[illegible]	[illegible]	[illegible]	[illegible]	[illegible]	[illegible]	[illegible]	[illegible]	[illegible]	[illegible]	[illegible]	[illegible]	[illegible]	[illegible]	[illegible]	[illegible]	[illegible]	[illegible]	[illegible]	[illegible]	[illegible]	[illegible]	[illegible]	[illegible]	[illegible]	[illegible]	[illegible]	[illegible]	[illegible]	[illegible]	[illegible]	[illegible]	[illegible]	[illegible]	[illegible]	[illegible]	[illegible]	[illegible]
14	1	.	.	.	.	.	5	5	9	14	5	10	36	16	25	5	10	9	30	5	21	94	1	135	105	100	265	324	165	160	480	171	531	295	540	124	135	770
15		1	1	2	2	1	.	.	8	.	9	8	.	36	.	19	27	39	.	30	.	35	11	.	35	80	248	.	225	168	104	279	104	522	.	310	297	371
16			1	1	2	1	.	.	.	.	.	8	.	18	.	10	27	29	.	30	.	.	11	.	35	.	124	.	90	168	.	144	104	378	.	210	297	.
17				1	1	1	.	.	.	.	.	.	.	9	.	10	9	20	.	20	.	.	11	.	.	.	44	.	45	44	.	99	.	198	.	165	153	.
18					1	1	.	.	.	.	.	.	.	.	.	.	9	10	.	20	.	.	11	.	.	.	.	.	.	44	.	.	.	99	.	55	153	.
19						1	.	.	.	.	.	.	.	.	.	.	.	.	.	10	.	.	11	.	.	.	.	.	.	.	.	.	.	.	.	.	54	.
20							1	.	.	4	.	.	6	.	9	.	.	.	12	.	9	14	.	35	15	14	35	108	21	20	115	21	126	35	207	14	15	175
21								1	3	.	2	4	6	8	.	3	6	6	.	4	.	28	1	20	35	38	117	44	82	80	129	93	156	172	69	82	90	273
22									1	.	1	1	.	4	.	2	3	4	.	3	.	7	1	.	7	16	48	.	43	32	27	52	27	97	.	56	54	98
23										1	.	.	.	.	3	.	.	.	5	.	4	.	.	7	.	.	.	32	.	.	20	.	21	.	78	.	.	28
24											1	.	.	2	.	1	1	3	.	2	.	.	1	.	.	8	16	.	27	8	.	27	.	54	.	38	27	35
25												1	.	2	.	1	3	3	.	3	.	.	1	.	7	.	24	.	17	32	.	27	27	70	.	38	54	.
26													1	.	.	.	.	.	.	.	.	4	.	5	5	5	15	14	10	10	30	11	36	20	25	9	10	60
27														1	.	1	1	2	.	2	.	.	1	.	.	.	8	.	9	8	.	18	.	36	.	29	27	.
28															1	.	.	.	2	.	2	.	.	.	.	.	.	8	.	.	.	.	.	.	27	.	.	.
29																1	.	1	.	1	.	.	1	.	.	.	.	.	.	.	.	9	.	9	.	20	9	.
30																	1	1	.	2	.	.	1	.	.	.	.	.	.	8	.	.	.	18	.	10	27	.
31																		1	.	1	.	.	1	.	.	.	.	.	.	.	.	.	.	9	.	10	9	.
32																			1	.	1	.	.	.	.	.	.	.	.	.	.	.	.	.	9	.	.	.
33																				1	.	.	1	.	.	.	.	.	.	.	.	.	.	.	.	.	9	.
34																					1	.	.	.	.	.	.	.	.	.	.	.	.	.	.	.	.	.
35																						1	.	.	1	2	6	.	5	4	6	6	6	11	.	6	6	21
36																							1	.	.	.	.	.	.	.	.	.	.	.	.	.	.	.
37																								1	.	.	.	4	.	.	5	.	6	.	9	.	.	9
38																									1	.	3	.	2	4	.	3	6	8	.	4	6	.
39																										1	2	.	3	1	.	3	.	6	.	4	3	7
40																											1	.	1	1	.	2	.	4	.	3	3	.
41																												1	.	.	.	.	.	.	3	.	.	.
42																													1	.	.	1	.	2	.	2	1	.
43																														1	.	.	.	2	.	1	3	.
44																															1	.	1	.	.	.	.	3
45																																1	.	1	.	2	1	.
46																																	1	.	.	.	.	.
47																																		1	.	1	1	.
48																																			1	.	.	.
49																																				1	.	.
50																																					1	.
51																																						1

Table 3. *(continued)*

ALGEBRA B12, REPRESENTATIONS OF CLASS 0

0	4	8	16	12	20	36	16	24	32	40	64	20	28	36	24	44	52	68	100	32	28	40	48	48	32	S.P.
1	24	264	24	1760	528	24	7920	5280	264	528	24	25344	31680	5280	59136	5280	528	528	24	126720	101376	47520	1760	31680	126720	O.S.
1	25	47	49	67	71	73	85	91	93	95	97	101	109	113	115	115	117	119	121	125	127	131	133	133	137	LEVEL
1	25	300	324	2300	5175	2900	12650	44550	32175	52026	20150	53130	260820	368550	177100	476905	430650	385020	115830	1138500	480700	2421900	1462500	2902900	1081575	DIM'N
0	1	0	2	0	1	3	0	1	0	2	4	0	1	0	0	2	1	3	5	1	0	0	0	2	0	
0	0	1	0	0	1	0	0	0	2	1	0	0	0	1	0	0	2	1	0	0	0	1	0	0	0	
0	0	0	0	1	0	0	0	1	0	0	0	0	0	1	0	1	0	0	0	0	0	0	2	0	0	
0	0	0	0	0	0	0	1	0	0	0	0	0	1	0	0	0	0	0	0	0	0	1	0	1	0	
0	0	0	0	0	0	0	0	0	0	0	0	1	0	0	0	0	0	0	0	1	0	0	0	0	0	
0	0	0	0	0	0	0	0	0	0	0	0	0	0	0	1	0	0	0	0	0	0	0	0	0	0	
0	0	0	0	0	0	0	0	0	0	0	0	0	0	0	0	0	0	0	0	0	1	0	0	0	0	
0	0	0	0	0	0	0	0	0	0	0	0	0	0	0	0	0	0	0	0	0	0	0	0	0	1	
0	0	0	0	0	0	0	0	0	0	0	0	0	0	0	0	0	0	0	0	0	0	0	0	0	0	
0	0	0	0	0	0	0	0	0	0	0	0	0	0	0	0	0	0	0	0	0	0	0	0	0	0	
0	0	0	0	0	0	0	0	0	0	0	0	0	0	0	0	0	0	0	0	0	0	0	0	0	0	
0	0	0	0	0	0	0	0	0	0	0	0	0	0	0	0	0	0	0	0	0	0	0	0	0	0	
>	>	>	>	>	>	>	>	>	>	>	>	>	>	>	>	>	>	>	>	>	>	>	>	>	>	
0	0	0	0	0	0	0	0	0	0	0	0	0	0	0	0	0	0	0	0	0	0	0	0	0	0	WEIGHT
0	1	2	3	4	5	6	7	8	9	10	11	12	13	14	15	16	17	18	19	20	21	22	23	24	25	NUMBER
1	1	12	12	12	23	12	66	198	143	210	78	66	252	318	220	385	330	276	78	1100	220	2100	1220	2332	495	0
	1	1	1	11	23	12	11	33	22	34	12	55	231	307	55	374	330	276	78	275	165	505	285	572	165	1
		1	1	1	2	1	10	32	22	34	12	10	41	52	45	64	54	45	12	245	45	475	275	541	120	2
			1	.	1	1	.	11	11	23	12	.	11	21	.	33	33	34	12	55	.	165	120	231	.	3
				1	2	1	1	3	2	3	1	9	39	51	9	63	54	45	12	47	36	87	49	99	36	4
					1	1	.	1	1	2	1	.	10	21	.	32	33	34	12	10	.	30	20	41	.	5
						1	.	.	.	1	1	.	.	.	.	11	11	23	12	.	.	.	.	11	.	6
							1	3	2	3	1	1	4	5	8	6	5	4	1	44	8	84	48	96	28	7
								1	1	2	1	.	1	2	.	3	3	3	1	9	.	29	20	39	.	8
									1	1	1	.	.	1	.	1	2	2	1	.	.	10	10	10	.	9
										1	1	.	.	.	.	1	1	2	1	.	.	.	.	10	.	10
											1	.	.	.	.	.	.	1	1	.	.	.	.	.	.	11
												1	4	5	1	6	5	4	1	5	7	9	5	10	7	12
													1	2	.	3	3	3	1	1	.	3	2	4	.	13
														1	.	1	2	2	1	.	.	1	1	1	.	14

56	72	36	72	80	104	36	144	44	52	40	52	40	44	60	48	68	76	76	48	44	84	100	56	108	116	S.P.
10560	264	354816	5280	528	528	112640	24	253440	126720	67584	31680	709632	24576	95040	4096	5280	31680	5280	887040	1013760	10560	528	354816	5280	528	O.S.
137	139	139	139	141	143	145	145	147	149	151	151	151	155	155	157	157	157	159	161	161	161	163	163	163	165	LEVEL
5252625	1848924	3893670	3670524	3580500	2302300	2042975	573300	11248380	13006500	3268760	12782250	10758825	4457400	35887500	5200300	23411700	22940775	25322220	39961350	24515700	45418100	20470230	45305975	22542300	22808500	DIM'N
1	0	1	3	2	4	0	6	0	2	0	0	1	0	1	0	1	3	0	0	1	2	1	2	4	3	
1	3	0	0	2	1	0	0	1	0	0	0	0	0	1	0	0	0	2	1	0	1	3	0	0	2	
1	0	0	1	0	0	0	0	0	0	0	1	0	0	0	0	2	0	1	0	0	1	0	0	1	0	
0	0	0	0	0	0	0	0	0	0	0	1	0	0	1	0	0	1	0	0	0	0	0	0	0	0	
0	0	0	0	0	0	0	0	1	1	0	0	0	0	0	0	0	0	0	0	0	0	0	0	0	0	
0	0	1	0	0	0	0	0	0	0	0	0	0	0	0	0	0	0	0	1	0	0	0	1	0	0	
0	0	0	0	0	0	0	0	0	0	0	0	1	0	0	0	0	0	0	0	0	0	0	0	0	0	
0	0	0	0	0	0	0	0	0	0	0	0	0	0	0	0	0	0	0	0	1	0	0	0	0	0	
0	0	0	0	0	0	1	0	0	0	0	0	0	0	0	0	0	0	0	0	0	0	0	0	0	0	
0	0	0	0	0	0	0	0	0	0	1	0	0	0	0	0	0	0	0	0	0	0	0	0	0	0	
0	0	0	0	0	0	0	0	0	0	0	0	0	1	0	0	0	0	0	0	0	0	0	0	0	0	
0	0	0	0	0	0	0	0	0	0	0	0	0	0	0	0	0	0	0	0	0	0	0	0	0	0	
>	>	>	>	>	>	>	>	>	>	>	>	>	>	>	>	>	>	>	>	>	>	>	>	>	>	
0	0	0	0	0	0	0	0	0	0	0	0	0	0	0	2	0	0	0	0	0	0	0	0	0	0	WEIGHT
26	27	28	29	30	31	32	33	34	35	36	37	38	39	40	41	42	43	44	45	46	47	48	49	50	51	NUMBER
4025	1364	1254	2596	2508	1508	495	364	2948	3180	792	3002	3465	792	7700	924	4740	4543	4884	10494	3740	8348	3542	11055	3828	3740	0
945	308	1045	616	572	342	330	78	2629	2860	330	2793	1155	462	7260	462	4575	4312	4763	3399	2805	8172	3542	3630	3762	3740	1
924	308	280	605	572	342	120	78	679	745	210	698	930	210	1815	252	1115	1082	1148	2940	1005	1972	826	3160	902	869	2
483	187	55	374	396	276	.	78	209	275	.	264	165	.	825	.	560	572	627	825	165	1176	550	1045	616	638	3
162	52	243	105	96	56	84	12	623	687	84	661	270	126	1737	126	1086	1041	1128	819	780	1941	826	883	891	869	4
83	32	45	64	65	45	.	12	190	245	.	254	45	.	785	.	550	541	617	225	120	1155	550	280	605	638	5
21	11	.	33	33	34	.	12	.	55	.	.	.	.	165	.	120	231	176	.	.	483	253	55	374	396	6
160	52	51	104	96	56	28	12	124	136	56	127	228	56	328	70	200	194	204	733	248	348	143	794	156	148	7
82	32	9	63	65	45	.	12	37	47	.	47	36	.	145	.	98	99	110	198	36	203	94	243	105	107	8
42	22	.	32	44	34	.	12	10	10	.	19	.	.	60	.	50	41	62	45	.	114	64	45	64	76	9
21	11	.	32	33	34	.	12	.	10	.	.	.	.	30	.	20	41	31	.	.	83	43	45	64	65	10
.	.	.	11	11	23	.	12	.	.	.	.	.	.	.	.	.	11	.	.	.	21	11	.	33	33	11
16	5	47	10	9	5	21	1	118	130	21	123	53	35	320	35	197	190	202	160	203	345	143	172	155	148	12
8	3	8	6	6	4	.	1	35	44	.	46	8	.	141	.	97	96	109	42	28	201	94	51	104	107	13
4	2	.	3	4	3	.	1	9	9	.	19	.	.	58	.	50	39	61	9	.	113	64	9	63	76	14

1	.	.	.	.	5	1	9	5	10	6	16	5	6	10	9	5	6	1	14	15	15	14	48	15	35	20	21	20	21	150	53	35	14	162	15	14	15
	1	1	2	1	.	.	.	.	1	.	2	1	.	3	3	3	.	1	.	9	.	.	.	.	29	.	20	39	31	.	.	82	43	9	63	65	16
		1	1	1	.	.	.	.	.	.	1	1	.	1	2	2	.	1	.	.	.	.	.	.	10	.	10	10	21	.	.	42	33	.	32	44	17
			1	1	.	.	.	.	.	.	.	.	.	1	1	2	.	1	.	.	.	.	.	.	.	.	.	10	.	.	.	21	11	.	32	33	18
				1	.	.	.	.	.	.	.	.	.	.	.	1	.	1	.	.	.	.	.	.	.	.	.	.	.	.	.	.	.	.	11	11	19
					1	.	3	2	4	.	8	3	1	6	6	4	.	1	4	5	.	5	7	.	15	.	10	10	11	39	7	20	9	47	10	10	20
						1	.	.	.	1	.	.	6	.	.	.	5	.	14	15	5	14	7	10	35	10	21	20	21	20	47	35	14	21	15	14	21
							1	1	1	.	4	2	.	3	4	3	.	1	1	1	.	2	.	.	6	.	5	4	6	8	.	11	6	8	6	7	22
								1	.	.	2	1	.	1	3	2	.	1	.	.	.	1	.	.	2	.	3	1	3	.	.	6	4	.	3	5	23
									1	.	2	1	.	3	3	3	.	1	.	1	.	.	.	.	3	.	2	4	3	.	.	8	4	8	6	6	24
										1	.	.	.	.	.	.	1	.	.	.	4	.	7	4	.	6	.	.	.	20	8	.	.	21	.	.	25
											1	1	.	1	2	2	.	1	.	.	.	.	.	.	1	.	1	1	2	.	.	4	3	.	3	4	26
												1	.	.	1	1	.	1	.	.	.	.	.	.	.	.	.	.	1	.	.	1	2	.	1	2	27
													1	.	.	.	.	.	4	5	.	5	1	.	15	.	10	10	11	5	6	20	9	6	10	10	28
														1	1	2	.	1	.	.	.	.	.	.	.	.	.	1	.	.	.	2	1	.	3	3	29
															1	1	.	1	.	.	.	.	.	.	.	.	.	.	.	.	.	1	1	.	1	2	30
																1	.	1	.	.	.	.	.	.	.	.	.	.	.	.	.	.	.	.	1	1	31
																	1	.	.	.	1	.	.	3	.	3	.	.	.	.	8	.	.	.	.	.	32
																		1	.	.	.	.	.	.	.	.	.	.	.	.	.	.	.	.	.	.	33
																			1	1	.	2	.	.	6	.	5	4	6	1	.	11	6	1	6	7	34
																				1	.	.	.	.	3	.	2	4	3	.	.	8	4	1	6	6	35
																					1	.	.	1	.	2	.	.	.	.	.	.	.	.	.	.	36
																						1	.	.	2	.	3	1	3	.	.	6	4	.	3	5	37
																							1	.	.	.	.	.	.	5	1	.	.	6	.	.	38
																								1	.	1	.	.	.	.	.	.	.	.	.	.	39
																									1	.	1	1	2	.	.	4	3	.	3	4	40
																										1	.	.	.	.	.	.	.	.	.	.	41
																											1	.	1	.	.	2	2	.	1	3	42
																												1	.	.	.	2	1	.	3	3	43
																													1	.	.	1	2	.	1	2	44
																														1	.	.	.	1	.	.	45
																															1	.	.	.	.	.	46
																																1	1	.	1	2	47
																																	1	.	.	1	48
																																		1	.	.	49
																																			1	1	50
																																				1	51

Table 3. *(continued)*

ALGEBRA B12, REPRESENTATIONS OF CLASS 1

NUMBER	0	1	2	3	4	5	6	7	8	9	10	11	12
S.P.	12	20	28	36	36	44	60	44	52	60	68	92	52
O.S.	4 096	49 152	270 336	49 152	901 120	540 672	49 152	2 027 520	2 703 360	270 336	540 672	49 152	3 244 032
LEVEL	79	103	125	127	145	149	151	163	169	171	173	175	179
DIM'N	4 096	98 304	1 126 400	1 228 800	8 192 000	18 743 296	10 649 600	42 393 600	153 088 000	111 919 104	182 476 800	71 884 800	165 806 080
WEIGHT	00000000000>1	10000000000>1	01000000000>1	20000000000>1	00100000000>1	11000000000>1	30000000000>1	00010000000>1	10100000000>1	02000000000>1	21000000000>1	40000000000>1	00001000000>1
0	1	12	77	78	340	704	364	1 143	3 640	2 508	3 861	1 365	3 080
1		1	11	12	65	143	78	264	890	627	990	364	814
2			1	1	10	22	12	54	185	132	209	78	200
3				1	.	11	12	.	65	66	143	78	.
4					1	2	1	9	30	21	33	12	44
5						1	1	.	10	11	22	12	.
6							1	.	.	.	11	12	.
7								1	3	2	3	1	8
8									1	1	2	1	.
9										1	1	1	.
10											1	1	.
11												1	.
12													1

NUMBER	13	14	15	16	17	18	19	20	21	22	23	24	25
S.P.	60	68	60	76	84	100	132	68	68	76	84	84	76
O.S.	8 110 080	2 703 360	3 784 704	2 703 360	540 672	540 672	49 152	16 220 160	3 244 032	12 165 120	901 120	8 110 080	2 027 520
LEVEL	187	191	193	193	195	197	199	203	205	209	211	211	215
DIM'N	843 448 320	1 216 512 000	507 781 120	1 587 216 384	1 450 803 200	1 312 030 720	402 554 880	3 429 171 200	1 243 545 600	7 524 679 680	4 612 608 000	9 093 427 200	2 461 184 000
WEIGHT	10010000000>1	01100000000>1	00000100000>1	20100000000>1	12000000000>1	31000000000>1	50000000000>1	10001000000>1	00000010000>1	01010000000>1	00200000000>1	20010000000>1	00000001000>1
0	13 464	17 820	6 853	22 032	19 096	16 016	4 368	38 732	12 804	75 978	44 110	87 219	20 240
1	3 762	5 115	2 002	6 445	5 665	4 862	1 365	11 968	4 037	24 201	14 245	28 215	6 765
2	948	1 310	560	1 660	1 474	1 276	364	3 464	1 242	7 128	4 235	8 361	2 235
3	264	550	.	890	913	990	364	814	.	2 619	1 815	3 762	.
4	207	285	147	360	319	275	78	912	369	1 890	1 126	2 217	726
5	54	120	.	185	198	209	78	200	.	684	485	948	.
6	.	.	.	65	66	143	78	.	.	.	.	264	.
7	36	48	35	60	52	44	12	212	104	432	255	504	230
8	9	20	.	30	32	33	12	44	.	153	110	207	.
9	.	10	.	10	22	22	12	.	.	54	55	54	.
10	.	.	.	10	11	22	12	.	.	.	.	54	.
11	.	.	.	.	.	11	12	.	.	.	.	.	.
12	4	5	7	6	5	4	1	40	27	78	45	90	70

NUMBER	26	27	28	29	30	31	32	33	34	35	36	37	38
S.P.	92	108	76	108	116	140	84	180	84	92	92	92	84
O.S.	5 406 720	270 336	22 708 224	2 703 360	540 672	540 672	901 120	49 152	32 440 320	16 220 160	270 336	8 110 080	22 708 224
LEVEL	215	217	217	217	219	221	223	223	225	227	229	229	229
DIM'N	16 793 993 216	6 010 470 400	10 777 395 200	11 870 208 000	11 720 417 280	7 643 750 400	3 937 894 400	1 945 681 920	32 558 284 800	37 955 174 400	5 020 815 360	37 955 174 400	26 876 129 280
WEIGHT	11100000000>1	03000000000>1	10000100000>1	30100000000>1	22000000000>1	41000000000>1	00000000100>1	60000000000>1	01001000000>1	20001000000>1	00000000010>1	00110000000>1	10000010000>1
0	148 984	49 544	90 376	98 020	92 235	55 055	27 016	12 376	237 744	263 736	29 964	255 816	174 999
1	49 082	16 588	30 184	32 900	31 251	19 019	9 438	4 368	82 116	92 312	10 813	89 958	62 106
2	14 740	5 038	9 632	9 965	9 537	5 863	3 288	1 365	26 848	30 368	3 915	29 808	21 389
3	7 920	3 157	2 002	6 445	6 666	4 862	.	1 365	8 888	11 968	.	11 880	4 037
4	3 920	1 342	2 905	2 650	2 541	1 562	1 140	364	8 184	9 270	1 422	9 144	7 110
5	2 080	847	560	1 660	1 760	1 276	.	364	2 640	3 454	.	3 630	1 242
6	550	286	.	890	913	990	.	364	.	814	.	.	.
7	880	298	812	590	561	341	392	78	2 272	2 568	518	2 528	2 259
8	460	187	147	360	385	275	.	78	712	912	.	996	369
9	240	132	.	185	264	209	.	78	200	200	.	420	.
10	120	66	.	185	198	209	.	78	.	200	.	.	.
11	.	.	.	65	66	143	.	78	.	.	.	.	.
12	152	50	203	100	93	55	133	12	552	620	189	603	674

NUMBER	39	40	41	42	43	44	45	46	47	48	49	50	51
S.P.	100	100	108	108	116	116	92	92	124	140	100	148	156
O.S.	49 152	24 330 240	4 096	2 703 360	8 110 080	2 703 360	56 770 560	16 220 160	5 406 720	540 672	22 708 224	2 703 360	540 672
LEVEL	233	233	235	235	235	237	239	239	239	241	241	241	243
DIM'N	4 868 669 440	108 018 892 800	3 042 918 400	71 530 905 600	69 837 520 896	78 136 115 200	106 323 148 800	53 886 976 000	141 115 392 000	64 664 371 200	121 521 254 400	71 032 832 000	72 747 417 600
WEIGHT	00000000001>1	11010000000>1	00000000000>3	10200000000>1	30010000000>1	02100000000>1	01000100000>1	10000001000>1	21100000000>1	13000000000>1	20000100000>1	40100000000>1	32000000000>1
0	26 324	677 304	15 511	426 624	408 672	445 016	586 938	284 240	768 900	329 472	639 198	355 420	348 348
1	9 713	242 265	5 798	154 132	148 533	162 239	216 117	105 765	283 085	122 683	238 161	133 055	131 274
2	3 610	81 036	2 188	51 952	50 094	55 022	76 566	38 610	96 470	42 174	84 889	45 770	45 408
3	.	38 952	.	27 115	28 215	31 262	23 331	6 765	59 950	28 743	30 184	32 900	34 254
4	1 353	24 921	835	16 035	15 435	17 015	25 872	13 800	29 865	13 101	28 749	14 200	14 124
5	.	11 727	.	8 305	8 361	9 625	7 630	2 235	18 260	8 910	9 632	9 965	10 538
6	.	2 619	.	1 815	3 762	2 805	.	.	7 920	4 158	2 002	6 445	6 666
7	512	6 864	323	4 404	4 236	4 660	8 232	4 808	8 160	3 564	9 142	3 860	3 828
8	.	3 159	.	2 261	2 217	2 610	2 345	726	4 910	2 409	2 905	2 650	2 827
9	.	1 368	.	1 190	948	1 530	560	.	2 850	1 694	560	1 660	2 046
10	.	684	.	485	948	770	.	.	2 080	1 133	560	1 660	1 760
11	.	.	.	.	264	.	.	.	550	286	.	890	913
12	196	1 620	127	1 027	990	1 078	2 415	1 620	1 875	807	2 674	875	858

1	2	.	3	3	3	1	8	.	27	19	36	.	78	31	35	60	63	44	.	12	168	212	.	234	104	.	729	.	520	504	595	665	230	110	540	812	590	627	13
	1	.	1	2	2	1	.	.	9	10	9	.	40	21	.	30	43	33	.	12	44	44	.	99	.	.	306	.	275	207	340	147	.	635	374	147	360	451	14
		1	.	.	.	.	5	6	9	5	10	20	16	5	42	10	9	5	44	1	108	120	69	114	183	76	300	51	186	180	192	630	520	330	138	693	150	144	15
			1	1	2	1	.	.	.	.	9	.	20	11	.	30	32	33	.	12	.	44	.	.	.	.	153	.	110	207	175	.	.	460	253	147	360	385	16
				1	1	1	.	.	.	.	.	.	10	11	.	10	22	22	.	12	.	.	.	.	.	.	54	.	55	54	120	.	.	240	198	.	185	264	17
					1	1	.	.	.	.	.	.	.	.	.	10	11	22	.	12	.	.	.	.	.	.	.	.	.	54	.	.	.	120	66	.	185	198	18
						1	.	.	.	.	.	.	.	.	.	.	.	11	.	12	.	.	.	.	.	.	.	.	.	.	.	.	.	.	.	.	65	66	19
							1	.	3	2	4	.	8	3	7	6	6	4	.	1	32	40	.	43	27	.	132	.	92	90	104	168	70	192	91	203	100	104	20
								1	.	.	.	5	.	.	6	.	.	.	14	.	14	15	25	14	42	30	35	21	21	20	21	135	155	35	14	147	15	14	21
									1	1	1	.	4	2	.	3	4	3	.	1	8	8	.	18	.	.	54	.	48	36	58	35	.	108	62	35	60	74	22
										1	.	.	2	1	.	1	3	2	.	1	.	.	.	9	.	.	18	.	30	9	30	.	.	60	42	.	30	54	23
											1	.	2	1	.	3	3	3	.	1	.	8	.	.	.	.	27	.	19	36	30	.	.	78	42	35	60	63	24
												1	.	.	.	.	.	.	4	.	.	.	9	.	7	12	.	9	.	.	.	20	40	.	.	21	.	.	25
													1	1	.	1	2	2	.	1	.	.	.	.	.	.	9	.	10	9	20	.	.	40	32	.	30	43	26
														1	.	.	1	1	.	1	.	.	.	.	.	.	.	.	.	.	10	.	.	10	22	.	10	22	27
															1	.	.	.	.	.	4	5	.	5	6	.	15	.	19	10	11	35	20	20	9	42	10	10	28
																1	1	2	.	1	.	.	.	.	.	.	.	.	.	9	.	.	.	20	11	.	30	32	29
																	1	1	.	1	.	.	.	.	.	.	.	.	.	.	.	.	.	10	11	.	10	22	30
																		1	.	1	.	.	.	.	.	.	.	.	.	.	.	.	.	.	.	.	10	11	31
																			1	.	.	.	3	.	.	5	.	4	.	.	.	.	8	.	.	.	.	.	32
																				1	.	.	.	.	.	.	.	.	.	.	.	.	.	.	.	.	.	.	33
																					1	1	.	2	.	.	6	.	5	4	6	7	.	11	6	7	6	7	34
																						1	.	.	.	.	3	.	2	4	3	.	.	8	4	7	6	6	35
																							1	.	.	2	.	2	.	.	.	.	.	.	.	.	.	.	36
																								1	.	.	2	.	3	1	3	.	.	6	4	.	3	5	37
																									1	.	.	.	.	.	.	5	5	.	.	6	.	.	38
																										1	.	1	.	.	.	.	.	.	.	.	.	.	39
																											1	.	1	1	2	.	.	4	3	.	3	4	40
																												1	.	.	.	.	.	.	.	.	.	.	41
																													1	.	1	.	.	2	2	.	1	3	42
																														1	.	.	.	2	1	.	3	3	43
																															1	.	.	1	2	.	1	2	44
																																1	.	.	.	1	.	.	45
																																	1	.	.	.	.	.	46
																																		1	1	.	1	2	47
																																			1	.	.	1	48
																																				1	.	.	49
																																					1	1	50
																																						1	51

Table 3. *(continued)*

ALGEBRA C2, REPRESENTATIONS OF CLASS 0

	0	1	2	3	4	5	6	7	8	9	10	11	12	13	14	15	16	17	18	19	20	21	22	23	24	25
S.P.	0	4	8	16	20	32	36	40	52	64	72	68	80	100	100	104	128	116	144	136	148	164	160	196	200	180
O.S.	1	4	4	4	8	4	4	8	8	4	4	8	8	4	8	8	4	8	4	8	8	8	8	4	4	8
LEVEL	1	5	7	9	11	13	13	15	17	17	19	19	21	21	23	23	25	25	25	27	27	29	29	29	31	31
DIM'N	1	5	10	14	35	35	30	81	105	55	84	154	220	91	231	260	165	390	140	455	405	429	625	204	286	770
WEIGHT	0<0	0<1	2<0	0<2	2<1	4<0	0<3	2<2	4<1	0<4	6<0	2<3	4<2	0<5	6<1	2<4	8<0	4<3	0<6	6<2	2<5	8<1	4<4	0<7	10<0	6<3
NUMBER	0	1	2	3	4	5	6	7	8	9	10	11	12	13	14	15	16	17	18	19	20	21	22	23	24	25
0	1	1	2	2	3	3	2	5	5	3	4	6	8	3	7	8	5	10	4	11	9	9	13	4	6	14
1		1	1	1	3	2	2	4	5	2	3	6	7	3	7	7	4	10	3	10	9	9	12	4	5	14
2			1	1	2	2	1	4	4	2	3	5	7	2	6	7	4	9	3	10	8	8	12	3	5	13
3				1	1	1	1	3	3	2	2	4	6	2	5	6	3	8	3	9	7	7	11	3	4	12
4					1	1	1	2	3	1	2	4	5	2	5	5	3	8	2	8	7	7	10	3	4	12
5						1	.	1	2	1	2	2	4	1	4	4	3	6	2	7	5	6	9	2	4	10
6							1	1	1	1	1	3	3	2	3	4	2	6	2	6	6	5	8	3	3	10
7								1	1	1	1	2	3	1	3	4	2	5	2	6	5	5	8	2	3	9
8									1	.	1	1	2	1	3	2	2	4	1	5	4	5	6	2	3	8
9										1	.	1	1	1	1	3	1	3	2	3	4	3	6	2	2	6
10											1	.	1	.	2	1	2	2	1	4	2	4	4	1	3	6
11												1	1	1	1	2	1	3	1	3	4	3	5	2	2	6
12													1	.	1	1	1	2	1	3	2	3	4	1	2	5
13														1	.	1	.	1	1	1	3	1	3	2	1	3
14															1	.	1	1	.	2	1	3	2	1	2	4
15																1	.	1	1	1	2	1	3	1	1	3
16																	1	.	.	1	.	2	1	.	2	2
17																		1	.	1	1	1	2	1	1	3
18																			1	.	1	.	1	1	.	1
19																				1	.	1	1	.	1	2
20																					1	.	1	1	.	1

	26	27	28	29	30	31	32	33	34	35	36	37	38	39	40	41	42	43	44	45	46	47	48	49	50	51
S.P.	200	208	212	256	244	232	260	288	260	272	324	296	292	328	340	320	340	400	392	356	360	404	400	388	416	484
O.S.	8	8	8	4	8	8	8	4	8	8	4	8	8	8	8	8	8	4	4	8	8	8	8	8	8	4
LEVEL	31	33	33	33	35	35	35	37	37	37	37	39	39	39	41	41	41	41	43	43	43	43	45	45	45	45
DIM'N	595	810	935	285	715	1190	836	455	1326	1330	385	1309	1729	1134	1105	1995	1820	506	680	2090	2401	1495	1976	2835	2415	650
WEIGHT	2<6	8<2	4<5	0<8	10<1	6<4	2<7	12<0	8<3	4<6	0<9	10<2	6<5	2<8	12<1	8<4	4<7	0<10	14<0	10<3	6<6	2<9	12<2	8<5	4<8	0<11
NUMBER	26	27	28	29	30	31	32	33	34	35	36	37	38	39	40	41	42	43	44	45	46	47	48	49	50	51
0	11	14	15	5	11	18	12	7	18	18	5	17	21	14	13	23	20	6	8	22	25	15	20	27	23	6
1	10	13	15	4	11	17	12	6	18	17	5	16	21	13	13	22	20	5	7	22	24	15	19	27	22	6
2	10	13	14	4	10	17	11	6	17	17	4	16	20	13	12	22	19	5	7	21	24	14	19	26	22	5
3	9	12	13	4	9	16	10	5	16	16	4	15	19	12	11	21	18	5	6	20	23	13	18	25	21	5
4	8	11	13	3	9	15	10	5	16	15	4	14	19	11	11	20	18	4	6	20	22	13	17	25	20	5
5	7	10	11	3	8	14	8	5	14	14	3	13	17	10	10	19	16	4	6	18	21	11	16	23	19	4
6	7	9	11	3	7	13	9	4	14	13	4	12	17	10	9	18	16	4	5	18	20	12	15	23	18	5
7	7	9	10	3	7	13	8	4	13	13	3	12	16	10	9	18	15	4	5	17	20	11	15	22	18	4
8	5	8	9	2	7	11	7	4	12	11	3	11	15	8	9	16	14	3	5	16	18	10	14	21	16	4
9	6	6	8	3	5	10	7	3	10	11	3	9	13	9	7	15	13	4	4	14	17	10	12	19	16	4
10	4	7	6	2	6	9	5	4	10	9	2	10	12	7	8	14	11	3	5	14	16	8	13	18	14	3
11	5	6	8	2	5	9	7	3	10	10	3	9	13	8	7	14	13	3	4	14	16	10	12	19	15	4
12	4	6	6	2	5	8	5	3	9	9	2	9	11	7	7	13	11	3	4	13	15	8	12	17	14	3
13	4	3	6	2	3	6	6	2	6	8	3	6	10	7	5	10	11	3	3	10	13	9	9	15	13	4
14	2	5	4	1	5	6	4	3	8	6	2	8	9	5	7	11	9	2	4	12	12	7	11	15	11	3
15	4	3	5	2	3	6	5	2	6	8	2	6	9	7	5	10	10	3	3	10	13	8	9	14	13	3
16	1	4	2	1	4	4	2	3	6	4	1	7	6	4	6	9	6	2	4	10	9	5	10	12	9	2
17	2	3	4	1	3	5	4	2	6	6	2	6	8	5	5	9	9	2	3	10	11	7	9	13	11	3
18	3	1	3	2	1	3	4	1	3	6	2	3	6	6	3	6	8	3	2	6	10	7	6	10	11	3
19	1	3	2	1	3	4	2	2	5	4	1	6	6	4	5	8	6	2	3	9	9	5	9	11	9	2
20	2	1	3	1	1	3	4	1	3	5	2	3	6	5	3	6	8	2	2	6	9	7	6	10	10	3

1	.	.	1	1	.	2	1	.	3	2	1	2	4	2	1	5	4	2	5	6	4	1	3	8	6	4	8	9	6	2	21
	1	.	.	1	1	1	2	1	1	3	2	1	3	4	1	3	5	4	3	6	6	2	2	6	8	5	6	9	9	2	22
		1	.	.	1	.	1	1	.	1	3	.	1	3	2	1	3	4	1	3	6	2	1	3	6	6	3	6	8	3	23
			1	.	.	1	.	.	2	1	.	2	2	1	.	4	2	1	4	4	2	1	3	6	4	2	7	6	4	1	24
				1	.	1	1	.	1	2	1	1	3	2	1	3	4	2	3	5	4	1	2	6	6	4	6	8	6	2	25
					1	.	1	1	.	1	2	.	1	3	1	1	3	4	1	3	5	2	1	3	6	5	3	6	8	2	26
						1	.	.	1	1	.	1	2	1	.	3	2	1	3	4	2	1	2	5	4	2	6	6	4	1	27
							1	.	.	1	1	.	1	2	1	1	3	2	1	3	4	1	1	3	5	4	3	6	6	2	28
								1	.	.	1	.	.	1	1	.	1	3	.	1	3	2	.	1	3	4	1	3	6	2	29
									1	.	.	1	1	.	.	2	1	.	3	2	1	.	2	4	2	1	5	4	2	1	30
										1	.	.	1	1	.	1	2	1	1	3	2	1	1	3	4	2	3	5	4	1	31
											1	.	.	1	1	.	1	2	.	1	3	1	.	1	3	4	1	3	5	2	32
												1	.	.	.	1	.	.	2	1	.	.	2	2	1	.	4	2	1	.	33
													1	.	.	1	1	.	1	2	1	.	1	3	2	1	3	4	2	1	34
														1	.	.	1	1	.	1	2	1	.	1	3	2	1	3	4	1	35
															1	.	.	1	.	.	1	1	.	.	1	3	.	1	3	2	36
																1	.	.	1	1	.	.	1	2	1	.	3	2	1	.	37
																	1	.	.	1	1	.	.	1	2	1	1	3	2	1	38
																		1	.	.	1	1	.	.	1	2	.	1	3	1	39
																			1	.	.	.	1	1	.	.	2	1	.	.	40
																				1	.	.	.	1	1	.	1	2	1	.	41
																					1	.	.	.	1	1	.	1	2	1	42
																						1	.	.	.	1	.	.	1	1	43
																							1	.	.	.	1	.	.	.	44
																								1	.	.	1	1	.	.	45
																									1	.	.	1	1	.	46
																										1	.	.	1	1	47
																											1	.	.	.	48
																												1	.	.	49
																													1	.	50
																														1	51

Table 3. *(continued)*

ALGEBRA C2, REPRESENTATIONS OF CLASS 1

2	10	18	26	34	50	50	58	74	82	98	90	106	122	130	130	162	146	170	170	178	202	194	226	242	218	S.P.
4	8	4	8	8	4	8	8	8	8	4	8	8	8	8	8	4	8	8	8	8	8	8	8	4	8	O.S.
4	8	10	12	14	16	16	18	20	20	22	22	24	24	26	26	28	28	28	30	30	32	32	32	34	34	LEVEL
4	16	20	40	64	56	80	140	160	140	120	256	324	224	320	420	220	560	336	616	640	560	880	480	364	1 024	DIM'N
1<0	1<1	3<0	1<2	3<1	5<0	1<3	3<2	5<1	1<4	7<0	3<3	5<2	1<5	7<1	3<4	9<0	5<3	1<6	7<2	3<5	9<1	5<4	1<7	11<0	7<3	WEIGHT
0	1	2	3	4	5	6	7	8	9	10	11	12	13	14	15	16	17	18	19	20	21	22	23	24	25	NUMBER
1	2	2	3	4	3	4	6	6	5	4	8	9	6	8	10	5	12	7	12	12	10	15	8	6	16	0
	1	1	2	3	2	3	5	5	4	3	7	8	5	7	9	4	11	6	11	11	9	14	7	5	15	1
		1	1	2	2	2	4	4	3	3	6	7	4	6	8	4	10	5	10	10	8	13	6	5	14	2
			1	1	1	2	3	3	3	2	5	6	4	5	7	3	9	5	9	9	7	12	6	4	13	3
				1	1	1	2	3	2	2	4	5	3	5	6	3	8	4	8	8	7	11	5	4	12	4
					1	.	1	2	1	2	2	4	2	4	4	3	6	3	7	6	6	9	4	4	10	5
						1	1	1	2	1	3	3	3	3	5	2	6	4	6	7	5	9	5	3	10	6
							1	1	1	1	2	3	2	3	4	2	5	3	6	6	5	8	4	3	9	7
								1	.	1	1	2	1	3	2	2	4	2	5	4	5	6	3	3	8	8
									1	.	1	1	2	1	3	1	3	3	3	5	3	6	4	2	6	9
										1	.	1	.	2	1	2	2	1	4	2	4	4	2	3	6	10
											1	1	1	1	2	1	3	2	3	4	3	5	3	2	6	11
												1	.	1	1	1	2	1	3	2	3	4	2	2	5	12
													1	.	1	.	1	2	1	3	1	3	3	1	3	13
														1	.	1	1	.	2	1	3	2	1	2	4	14
															1	.	1	1	1	2	1	3	2	1	3	15
																1	.	.	1	.	2	1	.	2	2	16
																	1	.	1	1	1	2	1	1	3	17
																		1	.	1	.	1	2	.	1	18

234	250	250	290	290	274	298	338	306	314	362	346	338	370	394	370	386	442	450	410	410	450	458	442	466	530	S.P.
8	8	8	8	8	8	8	4	8	8	8	8	8	8	8	8	8	8	4	8	8	8	8	8	8	8	O.S.
34	36	36	36	38	38	38	40	40	40	40	42	42	42	44	44	44	44	46	46	46	46	48	48	48	48	LEVEL
924	1 040	1 296	660	896	1 560	1 280	560	1 680	1 820	880	1 620	2 240	1 716	1 344	2 500	2 464	1 144	816	2 560	3 080	2 240	2 380	3 520	3 240	1 456	DIM'N
3<6	9<2	5<5	1<8	11<1	7<4	3<7	13<0	9<3	5<6	1<9	11<2	7<5	3<8	13<1	9<4	5<7	1<10	15<0	11<3	7<6	3<9	13<2	9<5	5<8	1<11	WEIGHT
26	27	28	29	30	31	32	33	34	35	36	37	38	39	40	41	42	43	44	45	46	47	48	49	50	51	NUMBER
14	15	18	9	12	20	16	7	20	21	10	18	24	18	14	25	24	11	8	24	28	20	21	30	27	12	0
13	14	17	8	11	19	15	6	19	20	9	17	23	17	13	24	23	10	7	23	27	19	20	29	26	11	1
12	13	16	7	10	18	14	6	18	19	8	16	22	16	12	23	22	9	7	22	26	18	19	28	25	10	2
11	12	15	7	9	17	13	5	17	18	8	15	21	15	11	22	21	9	6	21	25	17	18	27	24	10	3
10	11	14	6	9	16	12	5	16	17	7	14	20	14	11	21	20	8	6	20	24	16	17	26	23	9	4
8	10	12	5	8	14	10	5	14	15	6	13	18	12	10	19	18	7	6	18	22	14	16	24	21	8	5
9	9	12	6	7	14	11	4	14	15	7	12	18	13	9	19	18	8	5	18	22	15	15	24	21	9	6
8	9	11	5	7	13	10	4	13	14	6	12	17	12	9	18	17	7	5	17	21	14	15	23	20	8	7
6	8	9	4	7	11	8	4	12	12	5	11	15	10	9	16	15	6	5	16	19	12	14	21	18	7	8
7	6	9	5	5	10	9	3	10	12	6	9	14	11	7	15	15	7	4	14	18	13	12	20	18	8	9
4	7	6	3	6	9	6	4	10	9	4	10	12	8	8	14	12	5	5	14	16	10	13	18	15	6	10
6	6	8	4	5	9	8	3	10	11	5	9	13	10	7	14	14	6	4	14	17	12	12	19	17	7	11
4	6	6	3	5	8	6	3	9	9	4	9	11	8	7	13	12	5	4	13	15	10	12	17	15	6	12
5	3	6	4	3	6	7	2	6	9	5	6	10	9	5	10	12	6	3	10	14	11	9	15	15	7	13
2	5	4	2	5	6	4	3	8	6	3	8	9	6	7	11	9	4	4	12	12	8	11	15	12	5	14
4	3	5	3	3	6	6	2	6	8	4	6	9	8	5	10	11	5	3	10	13	10	9	14	14	6	15
1	4	2	1	4	4	2	3	6	4	2	7	6	4	6	9	6	3	4	10	9	6	10	12	9	4	16
2	3	4	2	3	5	4	2	6	6	3	6	8	6	5	9	9	4	3	10	11	8	9	13	12	5	17
3	1	3	3	1	3	5	1	3	6	4	3	6	7	3	6	9	5	2	6	10	9	6	10	12	6	18

1	.	1	1	.	1	2	1	3	2	1	3	4	2	2	5	4	2	6	6	4	5	8	6	3	3	9	9	6	9	11	9	4	19
	1	.	1	1	.	1	2	1	3	2	1	3	4	1	3	5	3	3	6	6	3	6	8	4	2	6	9	8	6	10	11	5	20
		1	.	.	1	1	.	2	1	.	3	2	1	2	4	2	1	5	4	2	5	6	4	2	3	8	6	4	8	9	6	3	21
			1	.	.	1	1	1	2	1	1	3	2	1	3	4	2	3	5	4	3	6	6	3	2	6	8	6	6	9	9	4	22
				1	.	.	1	.	1	2	.	1	3	.	1	3	3	1	3	5	1	3	6	4	1	3	6	7	3	6	9	5	23
					1	.	.	1	.	.	2	1	.	2	2	1	.	4	2	1	4	4	2	1	3	6	4	2	7	6	4	2	24
						1	.	1	1	.	1	2	1	1	3	2	1	3	4	2	3	5	4	2	2	6	6	4	6	8	6	3	25
							1	.	1	1	.	1	2	.	1	3	2	1	3	4	1	3	5	3	1	3	6	6	3	6	8	4	26
								1	.	.	1	1	.	1	2	1	.	3	2	1	3	4	2	1	2	5	4	2	6	6	4	2	27
									1	.	.	1	1	.	1	2	1	1	3	2	1	3	4	2	1	3	5	4	3	6	6	3	28
										1	.	.	1	.	.	1	2	.	1	3	.	1	3	3	.	1	3	5	1	3	6	4	29
											1	.	.	1	1	.	.	2	1	.	3	2	1	.	2	4	2	1	5	4	2	1	30
												1	.	.	1	1	.	1	2	1	1	3	2	1	1	3	4	2	3	5	4	2	31
													1	.	.	1	1	.	1	2	.	1	3	2	.	1	3	4	1	3	5	3	32
														1	.	.	.	1	.	.	2	1	.	.	2	2	1	.	4	2	1	.	33
															1	.	.	1	1	.	1	2	1	.	1	3	2	1	3	4	2	1	34
																1	.	.	1	1	.	1	2	1	.	1	3	2	1	3	4	2	35
																	1	.	.	1	.	.	1	2	.	.	1	3	.	1	3	3	36
																		1	.	.	1	1	.	.	1	2	1	.	3	2	1	.	37
																			1	.	.	1	1	.	.	1	2	1	1	3	2	1	38
																				1	.	.	1	1	.	.	1	2	.	1	3	2	39
																					1	.	.	.	1	1	.	.	2	1	.	.	40
																						1	.	.	.	1	1	.	1	2	1	.	41
																							1	.	.	.	1	1	.	1	2	1	42
																								1	.	.	.	1	.	.	1	2	43
																									1	.	.	.	1	.	.	.	44
																										1	.	.	1	1	.	.	45
																											1	.	.	1	1	.	46
																												1	.	.	1	1	47
																													1	.	.	.	48
																														1	.	.	49
																															1	.	50
																																1	51

Table 3. *(continued)*

ALGEBRA C3, REPRESENTATIONS OF CLASS 0

0	4	8	12	16	20	24	32	28	36	36	40	44	52	48	72	52	64	60	68	68	76	72	80	76	96	100	84	88	128	100	92	100	104	108	108	108	116	132	112	132	136	124	124	136	164	144	132	144	140	200	148	S.P.
1	12	6	24	12	24	8	6	48	12	24	24	24	24	24	6	48	12	48	24	24	24	24	24	48	8	24	48	24	6	12	48	48	24	48	24	24	24	24	48	24	24	48	48	24	24	12	48	24	48	6	24	O.S.
1	9	11	15	17	19	19	21	23	25	25	27	27	29	29	31	31	33	33	33	35	35	35	37	37	37	39	39	39	41	41	41	41	43	43	43	43	45	45	45	45	47	47	47	47	49	49	49	49	49	51	51	LEVEL
1	14	21	70	90	189	84	126	512	385	525	924	594	924	1078	462	2205	1274	3072	1386	3276	2079	2457	3900	5720	1001	3003	7168	4914	1287	3528	11319	7168	9450	10752	7700	8190	12375	6006	19683	5460	11550	19404	22750	10395	7722	8568	32768	15092	24024	3003	23562	DIM'N
0 0 < 0	0 1 < 0	2 0 < 0	1 0 < 1	0 2 < 0	2 1 < 0	0 0 < 2	4 0 < 0	1 1 < 1	0 3 < 0	3 0 < 1	2 2 < 0	0 1 < 2	4 1 < 0	2 0 < 2	6 0 < 0	1 2 < 1	0 4 < 0	3 1 < 1	1 0 < 3	2 3 < 0	5 0 < 1	0 2 < 2	4 2 < 0	2 1 < 2	0 0 < 4	6 1 < 0	1 3 < 1	4 0 < 2	8 0 < 0	0 5 < 0	3 2 < 1	1 1 < 3	2 4 < 0	5 1 < 1	0 3 < 2	3 0 < 3	4 3 < 0	7 0 < 1	2 2 < 2	0 1 < 4	6 2 < 0	1 4 < 1	4 1 < 2	2 0 < 4	8 1 < 0	0 6 < 0	3 3 < 1	6 0 < 2	1 2 < 3	10 0 < 0	2 5 < 0	WEIGHT
0	1	2	3	4	5	6	7	8	9	10	11	12	13	14	15	16	17	18	19	20	21	22	23	24	25	26	27	28	29	30	31	32	33	34	35	36	37	38	39	40	41	42	43	44	45	46	47	48	49	50	51	NUMBER
1	2	3	4	6	9	4	6	16	11	13	24	12	20	22	10	39	22	48	18	48	27	33	52	68	11	35	80	54	15	36	113	64	90	96	64	66	105	46	159	36	90	144	164	69	54	60	224	100	148	21	150	0
	1	1	3	3	6	2	3	12	8	11	17	10	15	16	6	32	16	40	16	39	24	26	41	58	8	28	68	44	10	30	99	56	75	84	57	60	90	42	137	33	75	127	146	59	45	50	200	86	134	15	132	1
		1	1	2	4	2	3	8	5	7	14	7	12	14	6	24	13	32	12	32	18	23	36	48	8	24	56	40	10	24	83	48	66	72	48	50	79	34	122	28	68	109	129	55	40	44	176	80	118	15	117	2
			1	1	2	1	1	6	4	6	9	6	8	10	3	20	10	26	11	25	16	18	27	41	6	18	48	32	6	21	72	42	54	62	43	46	66	31	105	26	55	97	114	47	32	38	156	68	107	10	102	3
				1	1	1	1	4	3	3	7	4	6	9	3	15	9	20	8	20	11	16	23	34	6	15	40	29	6	18	59	36	47	52	36	38	57	24	94	22	49	84	100	44	28	35	136	63	94	10	90	4
					1	.	1	2	2	3	5	3	6	5	3	11	6	16	7	17	11	11	20	27	4	15	32	22	6	15	51	30	40	46	31	34	52	24	77	20	45	72	87	36	28	29	120	53	83	10	81	5
						1	.	2	1	1	3	3	2	6	1	9	5	12	6	12	6	12	13	24	5	8	28	21	3	12	41	28	33	36	28	29	39	16	72	18	33	64	77	36	18	26	104	49	76	6	69	6
							1	.	.	1	2	.	4	2	3	3	3	8	2	8	7	5	14	12	3	12	16	14	6	8	29	16	25	32	16	18	36	18	47	12	36	44	56	25	24	20	80	40	52	10	55	7
								1	1	1	2	2	2	3	1	7	4	9	5	10	6	8	11	19	3	8	23	15	3	11	35	23	28	31	24	26	35	16	59	16	30	56	66	29	18	23	91	40	67	6	62	8
									1	.	1	1	1	1	1	4	3	4	3	7	3	5	7	12	2	6	16	9	3	9	23	16	21	20	18	19	26	11	43	12	23	43	48	22	15	19	68	29	52	6	50	9
										1	1	.	2	1	1	3	2	6	2	6	6	4	9	10	2	8	14	10	3	7	25	14	20	26	15	17	28	16	40	11	27	40	48	21	18	17	70	32	48	6	48	10
											1	.	1	1	1	2	2	4	1	5	3	4	7	8	2	6	12	9	3	6	19	12	18	20	13	13	23	11	36	9	23	35	41	20	15	16	60	29	42	6	43	11
												1	.	1	.	2	1	2	3	3	1	4	3	9	2	2	10	6	1	5	15	14	13	12	14	15	16	6	33	11	13	31	36	19	8	13	50	21	44	3	36	12
													1	.	1	.	.	2	.	2	3	.	5	3	.	6	4	5	3	3	11	4	9	16	6	7	17	11	17	5	20	19	27	9	15	9	38	22	23	6	28	13
														1	.	1	1	2	1	2	1	3	3	6	2	2	8	6	1	4	12	10	11	12	10	10	14	6	28	8	13	26	30	17	8	12	44	21	35	3	31	14
															1	.	.	.	.	.	1	.	2	.	.	4	.	2	3	.	3	.	3	8	.	2	8	7	5	.	14	5	12	3	12	4	16	14	6	6	12	15
																1	1	1	1	2	1	2	2	4	1	2	7	3	1	4	10	8	10	9	9	9	12	6	22	7	11	24	24	13	8	11	38	15	31	3	29	16
																	1	.	.	1	.	1	1	1	1	1	4	1	1	3	4	4	7	4	5	3	7	3	13	4	7	16	12	9	6	9	24	9	19	3	21	17

1	.	1	1	.	2	2	.	2	3	3	1	2	7	3	6	9	5	5	10	6	13	4	11	15	19	7	8	7	28	15	19	3	21	18
	1	.	.	1	.	2	1	.	2	1	.	1	3	6	3	2	5	6	4	1	12	6	3	11	12	10	2	5	18	6	22	1	14	19
		1	.	.	1	1	.	1	2	1	1	2	4	2	5	4	4	3	7	3	9	3	7	12	12	5	6	6	20	9	15	3	18	20
			1	.	1	.	.	2	.	1	1	.	3	.	2	6	.	2	6	6	4	.	9	5	10	2	8	3	14	10	6	3	10	21
				1	.	1	1	.	2	1	.	1	2	4	3	2	4	3	3	1	10	4	3	10	9	9	2	5	16	6	17	1	13	22
					1	.	.	1	.	1	1	.	2	.	2	4	.	1	5	3	4	.	7	4	8	2	6	3	12	9	5	3	9	23
						1	.	.	1	1	.	1	2	2	2	2	3	3	3	1	7	3	3	8	9	5	2	4	13	6	13	1	11	24
							1	.	.	.	.	.	.	2	.	.	1	1	.	.	3	3	.	2	2	6	.	1	4	1	9	.	3	25
								1	.	.	1	.	.	.	.	2	.	.	2	3	.	.	5	.	3	.	6	.	4	5	.	3	3	26
									1	.	.	1	1	1	2	1	2	1	2	1	4	2	2	7	4	3	2	4	10	3	9	1	10	27
										1	.	.	1	.	1	2	.	1	2	1	3	.	3	3	6	2	2	2	8	6	4	1	6	28
											1	.	.	.	.	.	.	.	.	1	.	.	2	.	.	.	4	.	.	2	.	3	.	29
												1	.	.	1	.	1	.	1	.	1	1	1	4	1	1	1	3	4	1	4	1	7	30
													1	.	1	1	.	1	2	1	2	.	2	3	4	1	2	2	7	3	4	1	6	31
														1	.	.	1	1	.	.	2	2	.	2	2	3	.	1	3	1	7	.	3	32
															1	.	.	.	1	.	1	.	1	2	1	1	1	2	4	1	2	1	5	33
																1	.	.	1	1	.	.	2	.	2	.	2	.	3	3	.	1	2	34
																	1	.	.	.	1	1	.	2	1	1	.	1	2	1	4	.	3	35
																		1	.	.	1	.	.	1	2	1	.	1	2	1	3	.	2	36
																			1	.	.	.	1	.	1	.	1	.	2	1	.	1	2	37
																				1	.	.	1	.	.	.	2	.	.	1	.	1	.	38
																					1	.	.	1	1	1	.	1	2	1	2	.	2	39
																						1	.	.	.	1	.	.	.	.	2	.	.	40
																							1	.	.	.	1	.	.	1	.	1	.	41
																								1	.	.	.	1	1	.	1	.	2	42
																									1	.	.	.	1	1	.	.	1	43
																										1	.	.	.	.	1	.	.	44
																											1	.	.	.	.	1	.	45
																												1	.	.	.	.	1	46
																													1	.	.	.	1	47
																														1	.	.	.	48
																															1	.	.	49
																																1	.	50
																																	1	51

Table 3. *(continued)*

ALGEBRA C3, REPRESENTATIONS OF CLASS 1

NUMBER	0	1	2	3	4	5	6	7	8	9	10	11	12	13	14	15	16	17	18	19	20	21	22	23	24	25
S.P.	2	6	10	18	18	22	26	34	34	50	38	42	54	50	54	58	58	74	66	66	98	70	82	82	82	86
O.S.	6	8	24	6	24	24	24	24	24	6	24	48	8	24	24	24	48	24	24	24	6	48	24	24	48	24
LEVEL	6	10	14	16	18	20	22	24	24	26	26	28	28	30	30	32	32	34	34	34	36	36	36	38	38	38
DIM'N	6	14	64	56	126	216	350	448	378	252	616	1 386	330	1 344	1 100	2 016	2 240	1 728	2 184	2 464	792	5 460	2 002	4 116	6 006	3 744
WEIGHT	1 0 < 0	0 0 < 1	1 1 < 0	3 0 < 0	0 1 < 1	2 0 < 1	1 2 < 0	3 1 < 0	1 0 < 2	5 0 < 0	0 2 < 1	2 1 < 1	0 0 < 3	1 3 < 0	4 0 < 1	3 2 < 0	1 1 < 2	5 1 < 0	0 3 < 1	3 0 < 2	7 0 < 0	2 2 < 1	0 1 < 3	1 4 < 0	4 1 < 1	2 0 < 3
0	1	1	4	3	5	7	11	12	9	6	13	27	5	24	18	32	32	24	29	32	10	66	21	46	65	36
1		1	2	1	3	6	7	8	6	3	11	21	5	18	16	24	26	18	23	25	6	57	17	37	55	33
2			1	1	2	3	5	6	5	3	7	16	3	14	11	20	21	15	19	22	6	45	15	31	46	26
3				1	.	1	2	4	2	3	3	8	2	8	7	14	12	12	11	14	6	29	10	21	32	18
4					1	1	2	2	3	1	4	9	2	8	6	11	14	8	13	15	3	30	11	21	31	19
5						1	1	2	1	1	3	6	2	6	6	9	10	8	9	10	3	25	8	17	26	17
6							1	1	1	1	2	4	1	5	3	7	8	6	8	9	3	19	7	15	20	13
7								1	.	1	.	2	.	2	3	5	3	6	4	5	3	11	4	9	16	7
8									1	.	1	2	1	2	1	3	6	2	5	6	1	12	6	9	12	10
9										1	.	.	.	.	1	2	.	4	.	2	3	3	.	3	8	2
10											1	1	1	2	1	2	4	2	4	3	1	10	4	8	9	9
11												1	.	1	1	2	2	2	3	3	1	7	3	6	9	5
12													1	.	.	.	2	.	1	1	.	3	3	2	2	6
13														1	.	1	1	1	2	1	1	4	2	5	4	3
14															1	1	.	2	.	1	1	3	.	2	6	2
15																1	.	1	.	1	1	2	.	2	4	1
16																	1	.	1	1	.	2	2	2	2	3
17																		1	.	.	1	.	.	.	2	.
18																			1	.	.	1	1	2	1	1
19																				1	.	1	.	1	2	1

NUMBER	26	27	28	29	30	31	32	33	34	35	36	37	38	39	40	41	42	43	44	45	46	47	48	49	50	51
S.P.	90	102	90	106	102	98	114	130	106	114	118	162	122	118	122	150	130	138	130	134	146	166	146	138	154	170
O.S.	24	24	48	24	24	48	24	24	48	24	24	6	24	48	48	8	24	48	48	24	24	24	24	48	48	24
LEVEL	40	40	40	42	42	42	42	44	44	44	44	46	46	46	46	46	48	48	48	48	50	50	50	50	50	52
DIM'N	6 720	3 640	8 316	6 930	6 300	12 096	4 004	4 928	16 632	8 918	7 392	2 002	10 752	21 000	17 550	2 548	18 480	18 018	24 192	15 750	21 120	9 450	15 708	39 312	18 816	18 304
WEIGHT	3 3 < 0	6 0 < 1	1 2 < 2	5 2 < 0	0 4 < 1	3 1 < 2	1 0 < 4	7 1 < 0	2 3 < 1	5 0 < 2	0 2 < 3	9 0 < 0	1 5 < 0	4 2 < 1	2 1 < 3	0 0 < 5	3 4 < 0	6 1 < 1	1 3 < 2	4 0 < 3	5 3 < 0	8 0 < 1	0 5 < 1	3 2 < 2	1 1 < 4	7 2 < 0
0	68	34	78	63	54	104	30	40	135	69	52	15	80	154	119	14	127	119	156	97	132	55	94	237	100	104
1	56	31	66	51	48	90	25	32	117	58	48	10	68	138	105	14	109	105	138	90	114	51	82	210	90	88
2	48	24	57	45	39	78	23	28	101	53	40	10	59	118	93	11	97	92	122	77	102	42	73	189	81	80
3	36	18	39	36	27	56	17	24	73	40	30	10	44	90	70	9	77	72	92	60	84	34	55	147	64	68
4	32	16	42	30	29	58	18	18	75	40	31	6	44	88	73	9	72	69	96	61	76	31	58	150	66	59
5	28	16	33	27	25	48	14	18	63	32	28	6	38	79	62	9	64	62	82	56	70	31	49	129	58	55
6	23	11	29	23	21	41	13	15	54	29	23	6	34	65	55	7	56	52	73	47	61	24	45	116	52	49
7	17	11	17	20	13	27	9	15	38	22	15	6	24	51	38	5	44	46	51	34	52	24	33	86	39	45
8	14	6	22	13	15	30	11	8	39	21	19	3	24	46	44	6	40	36	58	37	42	16	35	91	44	33
9	8	7	5	14	5	12	3	12	16	14	6	6	12	29	16	3	25	32	24	18	36	18	17	47	20	36
10	12	6	17	11	14	24	8	8	33	15	17	3	22	40	37	6	35	31	50	34	38	16	31	77	38	30
11	10	6	13	11	10	19	7	8	28	15	12	3	18	35	30	4	31	31	41	26	35	16	27	67	32	30
12	4	1	9	3	6	12	6	2	15	6	12	1	10	18	24	5	17	12	30	21	18	6	17	45	28	13
13	7	3	9	7	8	12	5	6	20	9	9	3	15	23	22	3	24	20	32	19	26	11	22	50	25	23
14	6	6	4	9	5	10	2	8	14	10	6	3	10	25	14	3	20	26	22	17	28	16	15	40	18	27
15	5	3	4	7	4	8	2	6	12	9	5	3	9	19	12	2	18	20	20	13	23	11	14	36	16	23
16	3	1	7	3	5	9	5	2	13	6	8	1	9	15	19	3	15	12	25	15	16	6	16	39	23	13
17	2	3	.	5	.	3	.	6	4	5	.	3	3	11	4	.	9	16	6	7	17	11	6	17	5	20
18	2	1	4	2	4	4	3	2	10	3	5	1	8	10	12	2	12	9	18	9	12	6	14	27	16	11
19	2	1	3	3	3	6	2	2	8	6	4	1	6	12	10	2	11	12	16	10	14	6	11	28	14	13

1	.	.	.	.	.	.	1	.	2	.	.	.	4	.	2	.	3	.	3	.	.	3	8	.	2	8	7	.	5	.	14	20
	1	.	1	1	1	2	1	2	2	3	4	1	2	7	3	4	1	6	10	8	2	10	9	14	9	12	6	10	22	12	11	21
		1	.	.	1	.	.	2	.	1	2	3	.	3	1	4	.	2	3	9	2	4	2	10	6	4	1	6	15	14	3	22
			1	.	.	1	.	1	1	2	1	1	1	4	1	2	1	5	4	4	1	7	4	9	3	7	3	8	13	8	7	23
				1	.	1	1	.	2	.	2	.	2	3	3	.	1	2	7	3	.	6	9	5	5	10	6	5	13	4	11	24
					1	.	.	1	.	1	2	1	.	2	1	3	.	2	3	6	2	3	2	8	6	4	1	5	12	10	3	25
						1	.	.	1	.	1	.	1	2	1	.	1	2	4	2	.	5	4	4	3	7	3	4	9	3	7	26
							1	.	1	.	.	.	2	.	1	.	1	.	3	.	.	2	6	.	2	6	6	.	4	.	9	27
								1	.	1	1	1	.	2	1	2	.	2	2	4	1	3	2	7	3	3	1	5	10	8	3	28
									1	.	.	.	1	.	1	.	1	.	2	.	.	2	4	.	1	5	3	.	4	.	7	29
										1	.	.	.	1	.	1	.	2	1	1	1	2	1	4	1	2	1	4	4	4	2	30
											1	.	.	1	1	.	.	1	2	2	.	2	2	3	3	3	1	3	7	3	3	31
												1	.	.	.	1	.	.	.	2	1	.	.	2	1	.	.	1	3	6	.	32
													1	.	.	.	1	.	.	.	.	.	2	.	.	2	3	.	.	.	5	33
														1	.	.	.	1	1	1	.	2	1	2	1	2	1	3	4	2	2	34
															1	.	.	.	1	.	.	1	2	.	1	2	1	.	3	.	3	35
																1	.	.	.	1	1	.	.	2	1	.	.	1	2	4	.	36
																	1	.	.	.	.	.	.	.	.	.	1	.	.	.	2	37
																		1	.	.	.	1	.	1	.	1	.	2	1	1	1	38
																			1	.	.	1	1	.	1	2	1	.	2	.	2	39
																				1	.	.	.	1	1	.	.	1	2	2	.	40
																					1	.	.	.	.	.	.	.	.	2	.	41
																						1	.	.	.	1	.	.	1	.	1	42
																							1	.	.	1	1	.	.	.	2	43
																								1	.	.	.	1	1	1	.	44
																									1	.	.	.	1	.	.	45
																										1	.	.	.	.	1	46
																											1	.	.	.	1	47
																												1	.	.	.	48
																													1	.	.	49
																														1	.	50
																															1	51

Table 3. *(continued)*

ALGEBRA C4, REPRESENTATIONS OF CLASS 0

NUMBER	0	1	2	3	4	5	6	7	8	9	10	11	12
S.P.	0	4	8	8	12	16	20	32	20	24	24	32	28
O.S.	1	24	8	16	96	24	48	8	96	32	64	16	192
LEVEL	1	13	15	17	23	25	27	29	29	31	31	33	35
DIM'N	1	27	36	42	315	308	594	330	792	825	1 155	594	4 096
WEIGHT	0 0 0 < 0	0 1 0 < 0	2 0 0 < 0	0 0 0 < 1	1 0 1 < 0	0 2 0 < 0	2 1 0 < 0	4 0 0 < 0	0 1 0 < 1	0 0 2 < 0	2 0 0 < 1	0 0 0 < 2	1 1 1 < 0
0	1	3	4	2	11	12	18	10	16	17	19	10	64
1		1	1	1	6	5	9	4	10	9	13	5	40
2			1	.	2	3	6	4	4	6	6	4	24
3				1	3	2	3	1	6	5	10	3	24
4					1	1	2	1	3	3	4	2	14
5						1	1	1	1	2	1	2	8
6							1	1	.	.	1	.	4
7								1	.	.	.	.	.
8									1	1	1	1	4
9										1	.	1	2
10											1	.	2
11												1	.
12													1
13													
14													
15													
16													
17													

NUMBER	13	14	15	16	17	18	19	20	21	22	23	24	25
S.P.	36	36	40	36	52	40	72	44	44	48	56	52	52
O.S.	24	96	48	192	48	96	8	96	192	96	64	96	192
LEVEL	37	37	39	39	41	41	43	43	43	45	45	45	47
DIM'N	2 184	3 696	4 914	6 237	4 290	7 020	1 716	10 010	14 300	16 848	9 009	9 009	29 106
WEIGHT	0 3 0 < 0	3 0 1 < 0	2 2 0 < 0	1 0 1 < 1	4 1 0 < 0	0 2 0 < 1	6 0 0 < 0	0 1 2 < 0	2 1 0 < 1	2 0 2 < 0	4 0 0 < 1	0 1 0 < 2	1 2 1 < 0
0	32	48	66	69	50	68	20	90	124	144	65	65	226
1	19	32	39	46	30	48	10	63	91	98	50	47	160
2	12	20	30	30	24	29	10	42	58	72	32	32	112
3	10	20	21	30	15	36	4	43	66	66	40	33	111
4	7	12	15	20	12	21	4	29	42	47	24	24	77
5	5	6	11	13	9	13	4	19	24	34	13	17	53
6	3	6	8	5	9	7	4	11	17	20	13	9	37
7	.	2	3	.	6	.	4	.	4	6	6	.	9
8	2	3	4	9	3	9	1	13	18	21	10	13	34
9	1	1	3	6	2	5	1	9	9	15	4	9	22
10	1	3	3	3	3	6	1	7	12	12	10	6	24
11	.	.	.	3	.	2	.	3	3	6	1	6	7
12	1	1	2	2	2	3	1	5	6	8	4	5	16
13	1	.	1	.	1	1	1	2	1	2	1	2	8
14		1	1	.	2	.	1	.	3	3	4	.	6
15			1	.	1	.	1	.	1	2	1	.	4
16				1	.	1	.	2	2	3	1	3	5
17					1	.	1	.	.	.	1	.	.

NUMBER	26	27	28	29	30	31	32	33	34	35	36	37	38
S.P.	56	56	64	60	72	68	76	60	80	60	68	68	100
O.S.	64	64	24	192	16	48	96	384	48	96	96	192	48
LEVEL	47	47	49	49	49	51	51	51	53	53	53	53	55
DIM'N	12 012	13 728	11 340	36 864	4 719	27 720	21 021	65 536	29 700	42 042	41 250	63 063	19 305
WEIGHT	0 0 2 < 1	2 0 0 < 2	0 4 0 < 0	3 1 1 < 0	0 0 0 < 3	2 3 0 < 0	5 0 1 < 0	1 1 1 < 1	4 2 0 < 0	1 0 3 < 0	0 3 0 < 1	3 0 1 < 1	6 1 0 < 0
0	76	96	84	256	23	184	125	384	180	218	218	335	105
1	58	66	55	184	18	127	92	288	122	167	167	253	70
2	39	52	42	136	12	100	68	208	102	122	118	192	60
3	46	46	36	128	16	82	65	216	78	127	127	188	42
4	30	35	27	92	10	64	47	156	63	93	92	142	36
5	20	28	22	64	7	49	32	112	51	67	65	105	30
6	13	14	15	52	5	40	32	72	45	48	46	73	30
7	.	4	6	24	.	18	20	16	30	11	13	30	24
8	16	19	12	40	6	28	20	84	27	51	50	77	15
9	11	16	9	26	4	21	12	60	21	37	33	56	12
10	11	9	9	32	5	22	20	54	24	36	35	51	15
11	6	10	3	8	3	6	3	32	6	19	16	30	3
12	7	7	7	20	3	17	12	38	18	26	25	36	12
13	3	2	5	8	2	11	6	16	11	13	13	13	9
14	.	2	3	14	.	10	12	12	15	8	11	20	12
15	.	2	3	8	.	8	6	8	11	5	7	13	9
16	4	4	2	6	2	5	3	20	5	14	12	18	3
17	.	.	.	4	.	3	6	.	8	.	.	5	9

NUMBER	39	40	41	42	43	44	45	46	47	48	49	50	51
S.P.	72	72	76	128	76	84	80	84	88	104	84	84	100
O.S.	96	192	192	8	192	192	96	192	96	64	192	192	24
LEVEL	55	55	55	57	57	57	57	59	59	59	59	59	61
DIM'N	66 528	96 228	56 628	6 435	146 250	91 000	67 760	147 456	111 375	42 042	122 850	142 155	47 124
WEIGHT	0 2 2 < 0	2 2 0 < 1	1 0 1 < 2	8 0 0 < 0	2 1 2 < 0	4 1 0 < 1	0 2 0 < 2	1 3 1 < 0	4 0 2 < 0	6 0 0 < 1	0 1 2 < 1	2 1 0 < 2	0 5 0 < 0
0	336	468	244	35	682	392	288	640	471	154	474	555	188
1	250	366	193	20	520	312	218	488	357	126	374	434	139
2	192	267	142	20	400	232	172	376	288	92	288	342	110
3	186	288	153	10	391	246	168	368	267	105	294	336	100
4	141	206	114	10	299	181	131	282	211	74	228	266	81
5	108	145	84	10	227	127	106	216	166	50	175	209	66
6	74	112	59	10	169	109	68	168	130	50	125	147	53
7	24	42	12	10	66	58	22	72	72	32	39	54	26
8	78	111	69	4	165	98	80	156	117	40	140	163	45
9	60	75	51	4	123	64	65	116	90	24	108	129	35
10	54	87	48	4	121	82	53	122	90	40	99	111	36
11	30	36	33	1	63	30	42	56	45	10	66	81	17
12	41	57	36	4	90	53	42	92	67	24	77	88	30
13	22	27	19	4	47	24	22	56	34	13	42	45	22
14	17	32	10	4	46	42	16	52	47	24	32	40	18
15	14	20	7	4	33	24	14	40	34	13	24	31	15
16	21	27	23	1	45	24	26	44	32	10	48	54	14
17	.	7	.	4	11	17	.	12	20	13	.	9	6

1	.	1	1	1	1	1	4	3	1	2	4	2	4	3	12	4	10	9	9	3	15	21	16	1	31	18	17	36	21	10	34	33	12	18
	1	.	.	.	.	.	.	.	.	.	.	.	.	2	.	3	.	.	.	6	.	.	.	4	.	4	.	.	6	6	.	.	.	19
		1	.	1	.	1	2	2	1	1	2	1	3	1	8	3	8	5	6	2	11	11	12	1	23	9	13	24	15	4	27	27	9	20
			1	1	1	.	2	.	1	1	4	.	3	3	6	4	4	6	9	3	9	15	6	1	22	18	10	26	21	10	20	23	9	21
				1	.	.	1	.	1	1	2	.	2	1	4	3	3	3	6	2	7	8	4	1	16	9	9	18	15	4	15	18	7	22
					1	.	.	.	.	.	2	.	1	3	.	3	.	.	3	3	.	6	.	1	7	12	.	8	12	10	.	6	3	23
						1	.	1	1	.	.	1	.	.	4	.	3	2	3	.	4	4	9	.	9	3	9	8	6	1	16	18	3	24
							1	.	.	1	1	.	2	1	2	2	2	3	2	2	5	6	3	1	10	6	5	16	8	4	11	10	7	25
								1	.	.	.	1	.	.	2	.	3	1	1	.	3	3	6	.	6	2	5	6	3	1	12	9	2	26
									1	.	.	.	.	.	2	.	1	1	3	.	3	3	3	.	6	3	6	6	6	1	9	12	2	27
										1	.	.	1	.	.	1	.	1	.	1	2	1	.	1	2	1	2	8	2	1	3	2	5	28
											1	.	1	1	.	2	.	.	2	2	.	3	.	1	5	6	.	6	8	4	.	5	3	29
												1	.	.	.	.	.	.	.	.	.	.	3	.	.	.	2	.	.	.	3	3	.	30
													1	.	.	1	.	.	.	1	.	1	.	1	2	1	.	4	2	1	.	2	3	31
														1	.	1	.	.	.	2	.	.	.	1	.	3	.	.	3	4	.	.	.	32
															1	.	1	1	1	.	2	2	2	.	4	2	3	5	3	1	7	6	2	33
																1	.	.	.	1	.	.	.	1	.	1	.	.	2	1	.	.	.	34
																	1	.	.	.	1	.	1	.	2	.	1	2	1	.	4	2	1	35
																		1	.	.	1	1	.	.	1	1	1	4	1	1	3	1	2	36
																			1	.	.	1	.	.	2	2	.	2	3	1	.	3	1	37
																				1	.	.	.	1	.	.	.	.	.	1	.	.	.	38
																					1	.	.	.	1	.	1	2	1	.	2	1	1	39
																						1	.	.	1	1	.	2	1	1	.	1	1	40
																							1	.	.	.	1	.	.	.	2	2	.	41
																								1	.	.	.	.	.	.	.	.	.	42
																									1	.	.	1	1	.	.	1	1	43
																										1	.	.	1	1	.	.	.	44
																											1	.	.	.	1	1	.	45
																												1	.	.	.	.	1	46
																													1	.	.	.	.	47
																														1	.	.	.	48
																															1	.	.	49
																																1	.	50
																																	1	51

Table 3. *(continued)*

ALGEBRA C4, REPRESENTATIONS OF CLASS 1

NUMBER	0	1	2	3	4	5	6	7	8	9	10	11	12	13	14	15	16	17	18	19	20	21	22	23	24	25
S.P.	2	6	10	18	14	18	22	26	26	34	50	30	34	38	38	42	42	50	54	46	58	54	50	74	54	62
O.S.	8	32	48	8	64	96	96	48	64	48	8	192	96	64	96	64	192	48	96	192	48	32	192	48	192	64
LEVEL	8	16	20	22	24	28	30	32	32	34	36	36	38	38	40	40	42	44	44	44	46	46	46	48	48	48
DIM'N	8	48	160	120	288	792	1 232	1 512	1 056	1 728	792	4 200	4 752	3 520	6 552	3 696	13 728	9 408	9 360	13 728	12 936	8 008	22 464	9 504	31 680	10 296
WEIGHT	1 0 0 < 0	0 0 1 < 0	1 1 0 < 0	3 0 0 < 0	1 0 0 < 1	0 1 1 < 0	2 0 1 < 0	1 2 0 < 0	0 0 1 < 1	3 1 0 < 0	5 0 0 < 0	1 1 0 < 1	1 0 2 < 0	3 0 0 < 1	0 2 1 < 0	1 0 0 < 2	2 1 1 < 0	1 3 0 < 0	4 0 1 < 0	0 1 1 < 1	3 2 0 < 0	0 0 3 < 0	2 0 1 < 1	5 1 0 < 0	1 2 0 < 1	0 0 1 < 2
0	1	2	6	4	6	15	20	24	14	24	10	45	48	32	60	30	116	74	68	94	92	48	144	60	186	51
1		1	2	1	4	8	12	12	8	12	4	31	30	24	40	19	77	46	47	68	57	36	102	36	140	40
2			1	1	1	4	6	8	5	9	4	17	20	13	25	14	52	34	32	45	45	24	73	30	94	27
3				1	.	.	2	3	.	6	4	4	6	6	9	4	24	18	20	16	30	11	30	24	42	12
4					1	2	3	3	3	3	1	12	12	10	16	9	32	19	20	33	24	18	51	15	71	22
5						1	1	2	2	2	1	6	8	4	10	7	20	14	12	22	18	12	36	12	46	15
6							1	1	.	2	1	3	3	4	6	2	14	10	12	12	15	8	20	12	32	10
7								1	.	1	1	1	2	1	4	2	8	8	6	8	11	5	13	9	20	7
8									1	.	.	2	3	1	3	4	6	4	3	11	5	6	18	3	21	9
9										1	1	.	.	1	.	.	4	3	6	.	8	.	5	9	7	.
10											1	.	.	.	.	.	.	.	2	.	3	.	.	6	.	.
11												1	1	1	2	1	4	3	3	6	4	4	9	3	15	6
12													1	.	1	1	2	2	1	4	3	3	6	2	8	4
13														1	.	.	2	1	3	.	3	.	3	3	6	.
14															1	.	1	2	1	2	2	2	2	2	6	3
15																1	.	.	.	2	.	1	3	.	3	3
16																	1	1	1	.	2	.	2	2	3	.
17																		1	.	.	1	.	.	1	1	.
18																			1	.	1	.	.	2	.	.
19																				1	.	1	1	.	2	2

NUMBER	26	27	28	29	30	31	32	33	34	35	36	37	38	39	40	41	42	43	44	45	46	47	48	49	50	51
S.P.	98	58	62	66	66	78	66	70	70	74	82	82	74	86	90	102	82	78	106	86	86	90	90	130	90	94
O.S.	8	192	192	96	96	64	192	192	192	64	48	192	192	64	48	96	96	384	48	96	192	192	192	48	192	192
LEVEL	50	50	50	52	52	52	52	54	54	54	56	56	56	56	58	58	58	58	60	60	60	60	60	62	62	62
DIM'N	3 432	47 040	38 808	36 960	46 800	20 384	44 352	87 480	64 064	39 312	44 352	85 800	95 040	27 456	67 584	43 120	78 624	205 920	61 776	138 600	166 320	151 200	114 400	36 608	274 560	239 616
WEIGHT	7 0 0 < 0	1 1 2 < 0	3 1 0 < 1	0 3 1 < 0	3 0 2 < 0	5 0 0 < 1	1 1 0 < 2	2 2 1 < 0	1 0 2 < 1	3 0 0 < 2	1 4 0 < 0	4 1 1 < 0	0 2 1 < 1	1 0 0 < 3	3 3 0 < 0	6 0 1 < 0	0 1 3 < 0	2 1 1 < 1	5 2 0 < 0	2 0 3 < 0	1 3 0 < 1	4 0 1 < 1	0 1 1 < 2	7 1 0 < 0	1 2 2 < 0	3 2 0 < 1
0	20	260	205	188	236	92	204	414	270	166	194	365	384	92	272	160	288	778	225	476	580	516	360	120	908	768
1	10	188	157	136	170	74	152	306	212	122	136	270	294	76	192	120	222	608	156	376	461	402	285	80	705	620
2	10	137	109	100	130	50	115	228	154	98	108	208	223	55	157	92	172	463	132	289	347	318	224	70	556	473
3	10	66	58	52	72	32	54	132	78	52	68	136	114	30	110	68	96	256	102	170	199	192	124	60	332	294
4	4	97	82	71	90	40	86	162	122	71	72	146	171	47	103	65	132	359	84	226	275	243	179	42	426	377
5	4	70	53	52	67	24	66	117	88	57	57	107	130	34	82	47	102	270	69	172	204	188	142	36	334	278
6	4	46	42	37	47	24	40	92	62	35	45	92	88	26	70	47	73	196	63	132	158	142	100	36	252	230
7	4	33	24	28	34	13	31	65	44	28	37	64	67	19	55	32	56	144	51	99	116	105	80	30	198	163
8	1	34	24	24	32	10	39	54	51	34	26	48	75	22	37	20	60	155	30	100	115	108	92	15	194	156
9	4	11	17	12	20	13	9	37	13	14	21	52	27	5	40	32	24	72	45	48	64	73	32	30	106	112
10	4	.	4	.	6	6	.	9	.	4	6	24	.	.	18	20	.	16	30	11	13	30	.	24	24	42
11	1	22	18	18	21	10	23	42	36	19	21	40	51	17	31	20	43	110	27	76	90	77	65	15	146	127
12	1	16	9	12	15	4	18	28	26	16	16	26	38	12	24	12	34	80	21	57	62	56	52	12	113	86
13	1	7	12	8	12	10	6	24	11	9	12	32	21	5	22	20	18	54	24	36	51	51	26	15	78	87
14	1	10	6	10	8	4	10	22	18	7	14	20	26	10	20	12	24	54	18	43	50	36	37	12	85	68
15	.	6	3	4	6	1	12	9	15	10	5	8	21	9	7	3	18	44	6	31	31	30	36	3	59	42
16	1	5	6	6	8	4	5	16	7	7	10	20	16	3	17	12	14	38	18	26	36	36	21	12	61	57
17	1	2	1	4	2	1	2	8	3	2	8	8	8	2	11	6	8	16	11	13	20	13	12	9	36	27
18	1	.	3	.	3	4	.	6	.	2	3	14	.	.	10	12	.	12	15	8	11	20	.	12	17	32
19	.	4	2	3	3	1	6	7	11	4	4	6	14	7	6	3	14	29	5	24	24	18	25	3	44	33

1	.	.	1	.	.	1	.	1	.	2	1	.	4	.	2	3	8	.	.	8	6	.	8	11	5	7	13	.	9	14	20	20
	1	.	.	.	1	.	2	.	1	1	.	2	3	6	1	2	2	6	4	4	1	9	12	3	15	10	6	14	2	24	14	21
		1	.	1	.	.	2	2	2	3	1	3	5	4	4	3	6	9	2	5	3	8	20	5	14	18	18	14	3	32	27	22
			1	.	.	1	.	.	.	.	1	.	.	.	.	.	4	.	.	3	6	.	.	8	.	.	5	.	9	.	7	23
				1	.	.	1	1	2	1	1	1	4	3	1	3	4	6	2	4	3	6	12	4	10	15	9	10	3	24	21	24
					1	.	.	.	.	.	.	2	.	3	1	.	.	3	4	.	.	3	6	.	6	4	3	11	.	9	5	25
						1	.	.	.	.	.	.	.	.	.	.	.	.	.	.	2	.	.	3	.	.	.	.	6	.	.	26
							1	.	1	1	.	1	2	2	1	2	2	4	1	3	1	5	8	3	8	8	6	8	2	18	11	27
								1	.	1	1	.	2	.	1	1	4	.	.	3	3	.	6	4	4	6	9	.	3	9	15	28
									1	.	.	.	1	.	.	2	1	2	.	2	1	2	2	2	2	6	2	3	2	10	6	29
										1	.	.	1	.	1	1	2	.	.	2	1	.	4	3	3	3	6	.	2	7	8	30
											1	.	.	.	.	.	2	.	.	1	3	.	.	3	.	.	3	.	3	.	6	31
												1	.	1	1	.	.	2	1	.	.	2	4	.	3	3	3	6	.	7	4	32
													1	.	.	1	1	.	.	2	1	.	2	2	2	3	2	.	2	5	6	33
														1	.	.	.	1	1	.	.	2	2	.	3	2	1	4	.	5	3	34
															1	.	.	.	.	.	.	.	2	.	1	1	3	.	.	3	3	35
																1	.	.	.	1	.	.	.	1	.	1	.	.	1	2	1	36
																	1	.	.	1	1	.	.	2	.	.	2	.	2	.	3	37
																		1	.	.	.	1	1	.	1	2	1	2	.	4	2	38
																			1	.	.	.	.	.	.	.	.	2	.	.	.	39
																				1	.	.	.	1	.	.	.	.	1	.	1	40
																					1	.	.	1	.	.	.	.	2	.	.	41
																						1	.	.	1	.	.	1	.	2	.	42
																							1	.	1	1	1	.	.	2	2	43
																								1	.	.	.	.	1	.	.	44
																									1	.	.	.	.	1	.	45
																										1	.	.	.	1	1	46
																											1	.	.	.	1	47
																												1	.	.	.	48
																													1	.	.	49
																														1	.	50
																															1	51

Table 3. *(continued)*

ALGEBRA C5, REPRESENTATIONS OF CLASS 0

	0	1	2	3	4	5	6	7	8	9	10	11	12	13	14	15	16	17	18	19	20	21	22	23	24	25
S.P.	0	4	8	8	12	16	20	16	32	20	24	24	28	28	36	36	32	40	32	40	52	40	72	36	40	44
O.S.	1	40	10	80	240	40	80	160	10	480	80	320	480	320	40	240	80	80	640	32	80	160	10	960	480	240
LEVEL	1	17	19	25	31	33	35	35	37	41	43	43	47	47	49	49	49	51	51	51	53	53	55	55	57	59
DIM'N	1	44	55	165	891	780	1430	1155	715	5005	4004	6864	17920	8580	8250	15015	7865	17820	28160	4719	14300	21021	5005	66066	67760	73710
WEIGHT	0 0 0 0 < 0	0 1 0 0 < 0	2 0 0 0 < 0	0 0 0 1 < 0	1 0 1 0 < 0	0 2 0 0 < 0	2 1 0 0 < 0	1 0 0 0 < 1	4 0 0 0 < 0	0 1 0 1 < 0	0 0 2 0 < 0	2 0 0 1 < 0	1 1 1 0 < 0	0 0 1 0 < 1	0 3 0 0 < 0	3 0 1 0 < 0	0 0 0 2 < 0	2 2 0 0 < 0	1 1 0 0 < 1	0 0 0 0 < 2	4 1 0 0 < 0	3 0 0 0 < 1	6 0 0 0 < 0	1 0 1 1 < 0	0 2 0 1 < 0	0 1 2 0 < 0
NUMBER	0	1	2	3	4	5	6	7	8	9	10	11	12	13	14	15	16	17	18	19	20	21	22	23	24	25
0	1	4	5	5	21	20	30	15	15	55	44	64	160	60	70	115	55	140	160	27	100	101	35	346	320	330
1		1	1	2	9	7	12	8	5	29	21	36	84	35	35	63	29	70	100	14	50	66	15	208	198	201
2			1	.	3	4	8	2	5	11	12	16	48	16	22	39	17	52	48	10	40	34	15	120	112	126
3				1	3	2	3	5	1	15	10	20	40	20	15	30	16	30	64	8	20	45	5	123	122	117
4					1	1	2	1	1	6	5	8	22	10	10	18	9	21	30	5	16	21	5	70	68	73
5						1	1	.	1	2	3	2	12	4	7	9	6	15	12	4	12	8	5	39	38	45
6							1	.	1	.	.	2	6	.	4	9	.	11	6	.	12	8	5	15	20	24
7								1	.	3	2	4	8	6	3	6	5	6	20	3	4	15	1	40	40	39
8									1	.	.	.	.	.	.	3	.	4	.	.	8	2	5	.	.	.
9										1	1	1	4	3	2	3	3	4	8	2	3	5	1	22	21	24
10											1	.	2	1	1	1	2	3	2	2	2	1	1	12	10	15
11												1	2	.	1	3	.	3	4	.	3	5	1	8	12	12
12													1	.	1	1	.	2	1	.	2	1	1	4	6	8
13														1	.	.	1	.	2	1	.	1	.	6	5	6
14															1	.	.	1	.	.	1	.	1	.	2	3
15																1	.	1	.	.	2	1	1	.	.	.

	26	27	28	29	30	31	32	33	34	35	36	37	38	39	40	41	42	43	44	45	46	47	48	49	50	51
S.P.	44	44	48	56	52	48	64	60	52	52	68	76	56	56	56	60	80	64	64	100	60	80	64	128	68	68
O.S.	960	640	240	320	480	960	40	480	480	960	80	240	320	320	640	320	80	640	160	80	1920	160	640	10	240	480
LEVEL	59	59	61	61	63	63	65	65	65	65	67	67	67	67	67	67	69	69	69	71	71	71	71	73	73	73
DIM'N	131625	73216	118580	75075	188760	178750	61710	225280	185328	274428	145860	115830	210210	268125	330330	125840	145860	372736	173745	85085	1048576	168168	366080	24310	480480	583440
WEIGHT	2 1 0 1 < 0	1 0 0 1 < 1	2 0 2 0 < 0	4 0 0 1 < 0	1 2 1 0 < 0	0 1 1 0 < 1	0 4 0 0 < 0	3 1 1 0 < 0	0 1 0 2 < 0	2 0 1 0 < 1	2 3 0 0 < 0	5 0 1 0 < 0	0 0 2 1 < 0	2 0 0 2 < 0	1 2 0 0 < 1	0 1 0 0 < 2	4 2 0 0 < 0	3 1 0 0 < 1	2 0 0 0 < 2	6 1 0 0 < 0	1 1 1 1 < 0	5 0 0 0 < 1	0 0 1 1 < 1	8 0 0 0 < 0	1 0 3 0 < 0	0 3 0 1 < 0
NUMBER	26	27	28	29	30	31	32	33	34	35	36	37	38	39	40	41	42	43	44	45	46	47	48	49	50	51
0	565	288	500	275	740	590	230	800	576	836	500	360	590	785	920	320	460	928	425	245	2688	356	800	70	1120	1360
1	359	180	302	180	454	392	133	496	378	565	298	225	399	513	636	216	270	656	280	140	1824	258	564	35	762	934
2	215	112	204	112	304	236	98	352	240	352	228	165	254	349	390	144	220	416	202	120	1216	166	368	35	528	624
3	222	112	176	115	264	263	73	288	246	378	162	129	269	333	443	144	144	464	186	70	1224	190	400	15	506	634
4	129	70	114	68	174	157	53	198	158	235	122	93	170	222	267	98	114	286	131	60	808	116	262	15	349	422
5	69	44	75	36	114	93	41	132	102	138	91	63	107	149	155	66	90	164	96	50	528	66	172	15	238	278
6	49	14	45	36	78	45	28	108	48	80	73	63	60	77	103	30	80	132	44	50	328	66	94	15	158	192
7	75	44	60	40	90	106	25	100	102	156	56	45	113	141	186	66	50	200	87	24	528	85	188	5	222	278
8	11	.	12	16	18	.	10	48	.	16	32	39	.	17	19	.	52	48	10	40	64	34	.	15	32	48
9	39	28	37	20	57	62	18	64	66	93	41	30	71	93	105	46	38	112	62	20	340	45	124	5	151	179
10	18	18	24	8	35	36	13	40	42	51	30	18	45	63	55	32	29	56	47	16	216	21	82	5	102	110
11	27	8	22	20	39	31	13	52	30	51	32	30	40	46	71	19	34	88	27	20	210	45	68	5	98	125
12	12	5	13	8	25	17	10	31	20	27	24	18	25	29	37	14	25	43	19	16	132	21	45	5	67	80
13	9	12	9	4	13	24	4	14	27	36	9	6	28	38	38	23	8	40	31	4	132	15	60	1	59	67
14	2	.	3	2	12	4	7	12	6	4	15	9	9	6	12	4	15	12	4	12	48	8	16	5	30	38
15	6	.	5	8	9	.	4	22	.	10	14	18	.	9	13	.	21	30	5	16	40	21	.	5	19	32

1	.	.	1	.	.	.	3	2	3	3	8	6	1	7	13	3	8	18	18	6	3	17	26	17	16	6	16	25	3	80	5	40	1	38	38	16
	1	.	.	1	.	1	.	.	.	2	.	3	2	6	.	4	12	.	4	11	9	.	6	6	.	15	12	4	12	24	8	.	5	12	20	17
		1	.	.	1	.	2	3	3	6	3	5	4	9	12	3	11	12	18	7	6	16	17	26	9	7	30	12	4	80	15	33	1	38	48	18
			1	.	.	.	.	.	.	.	4	.	.	.	5	.	.	6	6	.	.	5	10	5	10	.	4	15	.	24	1	20	.	10	10	19
				1	.	1	.	.	.	.	.	.	2	.	.	.	6	.	.	4	9	.	.	.	.	11	6	.	12	.	8	.	5	.	.	20
					1	.	.	.	.	3	.	2	4	3	.	1	8	.	6	4	6	.	5	10	.	6	20	3	4	24	15	.	1	10	20	21
						1	.	.	.	.	.	.	.	.	.	.	.	.	.	.	3	.	.	.	.	4	.	.	8	.	2	.	5	.	.	22
							1	1	2	2	2	3	1	5	6	2	6	8	9	5	3	10	11	11	7	5	12	9	3	48	5	22	1	26	27	23
								1	1	1	.	1	1	4	3	2	4	3	3	4	3	6	3	8	2	4	8	2	3	28	5	12	1	17	21	24
									1	.	.	1	.	2	1	1	2	2	1	3	1	4	2	2	2	3	2	2	2	16	1	8	1	13	10	25
										1	.	1	1	2	.	1	4	.	3	3	3	.	3	4	.	4	8	2	3	14	5	.	1	7	12	26
											1	.	.	.	2	.	.	3	3	.	.	3	4	3	4	.	3	5	.	14	1	11	.	7	7	27
												1	.	1	.	1	2	.	1	2	1	.	2	1	.	3	2	2	2	8	1	.	1	5	6	28
													1	.	.	.	2	.	.	1	3	.	.	.	.	3	4	.	3	.	5	.	1	.	.	29
														1	.	1	1	.	.	2	1	.	.	1	.	2	1	.	2	4	1	.	1	3	6	30
															1	.	.	1	1	.	.	2	1	2	1	.	2	1	.	8	1	6	.	5	5	31
																1	.	.	.	1	.	.	.	.	.	1	.	.	1	.	.	.	1	.	2	32
																	1	.	.	1	1	.	.	.	.	2	1	.	2	.	1	.	1	.	.	33
																		1	.	.	.	1	1	.	1	.	.	1	.	4	.	4	.	3	2	34
																			1	.	.	.	1	1	.	.	2	1	.	4	1	.	.	2	3	35
																				1	.	.	.	.	.	1	.	.	1	.	.	.	1	.	.	36
																					1	.	.	.	.	1	.	.	2	.	1	.	1	.	.	37
																						1	.	.	.	.	.	.	.	2	.	2	.	3	1	38
																							1	.	.	.	.	1	.	2	.	.	.	1	1	39
																								1	.	.	1	.	.	2	1	.	.	1	3	40
																									1	.	.	1	.	.	.	2	.	.	.	41
																										1	.	.	1	.	.	.	1	.	.	42
																											1	.	.	.	1	.	.	.	.	43
																												1	.	.	.	.	.	.	.	44
																													1	.	.	.	1	.	.	45
																														1	.	.	.	1	1	46
																															1	.	.	.	.	47
																																1	.	.	.	48
																																	1	.	.	49
																																		1	.	50
																																			1	51

Table 3. *(continued)*

ALGEBRA C5, REPRESENTATIONS OF CLASS 1

S.P.	2	6	10	10	18	14	18	22	26	22	34	26	50	26	30	34	34	38
O.S.	10	80	80	32	10	320	240	240	80	320	80	160	10	320	960	160	240	320
LEVEL	10	22	26	26	28	34	38	40	42	42	44	44	46	46	50	50	52	52
DIM'N	10	110	320	132	220	1 408	2 860	4 212	4 620	4 290	4 928	5 720	2 002	9 152	32 340	9 438	28 028	24 960
	1	0	1	0	3	1	0	2	1	0	3	2	5	0	1	0	1	3
	0	0	1	0	0	0	1	0	2	1	1	0	0	0	1	0	0	0
	0	1	0	0	0	0	1	1	0	0	0	0	0	1	0	0	2	0
	0	0	0	0	0	1	0	0	0	0	0	0	0	1	1	1	0	1
	<	<	<	<	<	<	<	<	<	<	<	<	<	<	<	<	<	<
WEIGHT	0	0	0	1	0	0	0	0	0	1	0	1	0	0	0	1	0	0
NUMBER	0	1	2	3	4	5	6	7	8	9	10	11	12	13	14	15	16	17
0	1	3	8	2	5	16	30	39	42	29	40	34	15	56	167	43	138	112
1		1	2	1	1	8	13	18	17	17	16	21	5	30	97	24	75	68
2			1	.	1	2	6	9	11	6	12	8	5	15	49	14	45	36
3				1	.	4	5	6	5	10	4	15	1	16	55	14	39	40
4					1	.	.	3	4	.	8	2	5	.	11	.	12	16
5						1	2	3	3	4	3	5	1	8	27	8	22	20
6							1	1	2	1	2	1	1	4	12	5	13	8
7								1	1	.	2	1	1	.	6	.	5	8
8									1	.	1	.	1	.	2	.	3	2
9										1	.	1	.	2	6	3	5	4
10											1	.	1	.	.	.	.	2
11												1	.	.	3	.	2	4
12													1	.	.	.	.	.
13														1	2	2	3	1
14															1	.	1	1
15																1	.	.
16																	1	.

S.P.	38	42	38	50	54	42	42	58	46	50	74	46	58	98	54	50	54
O.S.	240	480	960	80	240	320	320	80	640	160	80	960	160	10	80	960	960
LEVEL	54	56	56	58	58	58	58	60	60	60	62	62	62	64	64	64	66
DIM'N	35 640	71 500	62 920	42 240	44 550	60 060	60 500	54 912	115 830	37 752	36 608	183 040	63 700	11 440	76 440	285 120	366 080
	0	2	1	1	4	1	0	3	2	1	5	0	4	7	0	2	1
	2	1	0	3	0	0	2	2	1	0	1	1	0	0	0	0	2
	1	1	1	0	1	0	0	0	0	0	0	1	0	0	3	1	0
	0	0	0	0	0	2	0	0	0	0	0	1	0	0	0	1	1
	<	<	<	<	<	<	<	<	<	<	<	<	<	<	<	<	<
WEIGHT	0	0	1	0	0	0	1	0	1	2	0	0	1	0	0	0	0
NUMBER	18	19	20	21	22	23	24	25	26	27	28	29	30	31	32	33	34
0	162	302	228	168	165	205	199	200	351	106	120	544	166	35	204	792	944
1	90	168	144	88	93	124	131	104	236	65	60	348	116	15	131	508	620
2	54	108	80	63	63	77	71	80	132	44	50	212	66	15	84	324	390
3	45	85	90	40	45	75	90	46	160	41	24	220	85	5	80	320	400
4	18	48	16	32	39	17	19	52	48	10	40	64	34	15	32	120	160
5	27	52	51	28	30	46	47	34	88	27	20	133	45	5	52	200	247
6	16	31	27	20	18	29	24	25	43	19	16	80	21	5	33	122	148
7	9	22	10	14	18	9	13	21	30	5	16	40	21	5	19	70	100
8	6	12	4	11	9	6	6	15	12	4	12	24	8	5	12	39	58
9	6	11	18	6	6	17	16	7	30	12	4	48	15	1	19	72	88
10	.	6	.	4	9	.	.	11	6	.	12	.	8	5	.	15	20
11	3	8	6	4	6	5	10	6	20	3	4	24	15	1	10	40	60
12	.	.	.	.	3	.	.	4	.	.	8	.	2	5	.	.	.
13	3	6	9	4	3	11	7	5	12	9	3	28	5	1	12	42	48
14	2	4	3	3	3	3	4	4	8	2	3	14	5	1	7	22	33
15	.	.	3	.	.	4	2	.	3	5	.	8	1	.	3	12	12
16	1	2	1	2	1	2	1	3	2	2	2	8	1	1	5	12	16

S.P.	54	58	62	58	58	66	66	78	62	70	62	70	82	82	66	70	70
O.S.	960	480	960	320	640	240	240	320	320	480	1 920	320	80	480	960	320	960
LEVEL	66	68	68	68	68	70	70	70	70	72	72	72	74	74	74	74	74
DIM'N	231 660	413 952	420 420	214 500	330 330	291 720	387 200	197 120	280 280	666 468	1 034 880	220 220	283 140	607 750	1 098 240	534 820	915 200
	0	1	3	0	2	0	3	5	0	2	1	0	1	4	1	0	3
	1	1	1	0	0	3	0	0	0	2	1	0	4	1	1	3	0
	0	2	0	2	0	1	2	0	1	1	1	1	0	1	0	0	1
	1	0	1	0	1	0	0	1	2	0	0	0	0	0	2	0	0
	<	<	<	<	<	<	<	<	<	<	<	<	<	<	<	<	<
WEIGHT	1	0	0	1	1	0	0	0	0	0	1	2	0	0	0	1	1
NUMBER	35	36	37	38	39	40	41	42	43	44	45	46	47	48	49	50	51
0	537	1 016	984	454	721	662	872	400	554	1 422	2 000	365	562	1 175	2 000	939	1 616
1	360	646	655	316	481	419	555	272	384	909	1 400	261	343	750	1 376	672	1 143
2	231	439	427	199	324	292	394	180	250	648	915	175	263	556	946	437	770
3	240	396	425	225	320	255	340	180	265	549	980	185	195	450	940	480	800
4	69	192	215	68	112	142	204	112	90	354	360	65	158	352	384	191	352
5	156	264	269	139	214	174	233	115	174	384	637	127	147	324	643	312	535
6	101	176	161	87	143	120	159	68	114	264	406	86	112	228	440	196	344
7	46	112	129	48	70	84	114	68	63	204	248	47	87	198	256	136	235
8	30	75	69	29	44	60	75	36	42	138	151	32	69	132	174	84	138
9	69	96	95	61	95	63	84	40	79	138	282	64	53	115	294	139	238
10	.	24	49	.	14	24	45	36	.	78	45	.	38	108	48	32	80
11	30	60	75	36	44	45	60	40	43	105	172	33	41	100	168	96	156
12	.	.	11	.	.	.	12	16	.	18	.	.	10	48	.	.	16
13	45	62	51	38	64	41	55	20	52	90	174	44	39	76	200	81	146
14	20	39	39	21	28	30	37	20	29	69	104	24	31	64	113	58	93
15	21	16	12	16	30	10	14	4	23	21	72	24	9	17	88	31	60
16	13	26	18	13	18	19	24	8	19	44	60	17	23	40	76	31	51

1	.	2	.	1	3	.	.	3	4	.	3	.	5	1	.	8	12	.	12	27	.	8	12	22	20	.	39	31	.	17	52	30	24	51	17
	1	1	.	2	1	.	1	2	1	.	2	4	1	1	3	4	12	5	16	12	7	5	16	13	8	11	34	34	9	20	31	40	24	27	18
		1	.	1	1	.	.	2	1	.	2	.	1	1	.	4	6	.	8	12	.	5	9	13	8	.	25	17	.	14	31	20	13	27	19
			1	.	.	1	1	.	2	1	.	4	1	.	2	6	8	9	10	9	9	12	7	9	4	13	16	42	13	7	14	48	22	36	20
				1	.	.	.	1	.	.	1	.	.	1	.	.	2	.	3	2	.	.	6	3	2	.	12	4	.	11	12	6	6	4	21
					1	.	.	1	.	.	2	.	1	1	.	.	.	.	.	6	.	.	.	5	8	.	9	.	.	4	22	.	.	10	22
						1	.	.	.	1	.	2	.	.	1	3	3	6	6	3	5	8	4	6	1	9	9	22	9	5	8	33	10	18	23
							1	.	1	.	.	2	1	.	1	2	6	3	6	6	6	3	6	5	4	7	12	24	6	6	11	24	16	18	24
								1	.	.	1	.	.	1	.	.	.	.	.	2	.	.	.	3	2	.	6	.	.	4	12	.	.	4	25
									1	.	.	.	1	.	.	2	3	.	3	6	.	3	3	5	4	.	9	12	.	4	11	12	10	18	26
										1	.	.	.	.	.	.	.	3	.	.	2	4	.	.	.	3	.	8	6	.	.	12	3	6	27
											1	.	.	1	.	.	.	.	.	.	.	.	.	2	.	.	.	.	.	6	.	.	.	.	28
												1	.	.	1	1	2	2	4	2	3	2	3	3	1	5	7	12	5	4	6	16	7	9	29
													1	.	.	.	.	.	.	3	.	.	.	2	4	.	3	.	.	1	8	.	.	6	30
														1	.	.	.	.	.	.	.	.	.	.	.	.	.	.	.	.	.	.	.	.	31
															1	.	.	.	2	.	1	.	1	1	.	2	3	2	2	2	2	4	1	1	32
																1	1	.	2	2	.	2	2	3	1	.	5	6	.	3	6	8	4	9	33
																	1	.	1	1	.	.	2	1	1	.	4	3	.	3	4	3	4	3	34
																		1	.	.	1	1	.	.	.	2	.	4	3	.	.	6	2	3	35
																			1	.	.	.	1	1	.	.	2	1	.	2	2	2	1	1	36
																				1	.	.	.	1	1	.	2	.	.	1	4	.	.	3	37
																					1	.	.	.	.	1	.	2	1	.	.	2	1	1	38
																						1	.	.	.	.	.	2	.	.	.	3	1	3	39
																							1	.	.	.	1	.	.	2	1	.	1	.	40
																								1	.	.	1	.	.	1	2	.	.	1	41
																									1	.	.	.	.	.	2	.	.	.	42
																										1	.	.	1	.	.	2	.	.	43
																											1	.	.	1	1	.	.	.	44
																												1	.	.	.	1	1	1	45
																													1	.	.	.	.	.	46
																														1	.	.	.	.	47
																															1	.	.	.	48
																																1	.	.	49
																																	1	.	50
																																		1	51

Table 3. *(continued)*

ALGEBRA C6, REPRESENTATIONS OF CLASS 0

0	4	8	8	12	12	16	20	32	16	20	24	24	24	28	28	36	36	40	28	52	32	72	32	40	36	S.P.
1	60	12	240	64	480	60	120	12	960	1440	160	960	960	960	384	60	480	120	1920	120	240	12	3840	960	960	O.S.
1	21	23	33	37	39	41	43	45	47	53	55	55	57	59	59	61	61	63	63	65	65	67	67	69	69	LEVEL
1	65	78	429	429	2002	1650	2925	1365	6006	18954	13650	25025	21450	57344	27027	24310	45760	51051	70070	38675	49686	12376	212992	150150	92950	DIM'N
0	0	2	0	0	1	0	2	4	1	0	0	2	0	1	2	0	3	2	0	4	0	6	1	3	0	
0	1	0	0	0	0	2	1	0	0	1	0	0	1	1	0	3	0	2	0	1	0	0	1	0	0	
0	0	0	0	0	1	0	0	0	0	0	2	0	0	1	0	0	1	0	1	0	0	0	0	0	0	
0	0	0	1	0	0	0	0	0	0	1	0	1	0	0	0	0	0	0	0	0	2	0	0	0	1	
0	0	0	0	0	0	0	0	0	1	0	0	0	0	0	0	0	0	0	1	0	0	0	1	1	0	
<	<	<	<	<	<	<	<	<	<	<	<	<	<	<	<	<	<	<	<	<	<	<	<	<	<	
0	0	0	0	1	0	0	0	0	0	0	0	0	1	0	1	0	0	0	0	0	0	0	0	0	1	WEIGHT
0	1	2	3	4	5	6	7	8	9	10	11	12	13	14	15	16	17	18	19	20	21	22	23	24	25	NUMBER
1	5	6	9	5	34	30	45	21	46	126	90	145	90	320	99	130	224	255	250	175	174	56	640	390	230	0
	1	1	3	2	12	9	15	6	21	57	38	69	47	144	54	56	104	110	131	75	86	21	352	222	128	1
		1	.	.	4	5	10	6	5	21	20	30	15	80	20	35	64	80	55	60	44	21	160	110	60	2
			1	1	3	2	3	1	10	24	15	30	26	56	31	20	40	39	68	25	43	6	192	125	71	3
				1	.	.	.	.	5	9	5	10	15	16	21	5	10	9	35	5	21	1	104	70	40	4
					1	1	2	1	2	9	7	12	8	30	10	13	24	27	29	20	21	6	84	57	35	5
						1	1	1	.	3	4	3	2	16	2	9	12	19	11	15	12	6	32	21	16	6
							1	1	.	.	.	3	.	8	2	5	12	14	.	15	.	6	16	21	.	7
								1	.	.	.	.	.	.	.	.	4	5	.	10	.	6	.	5	.	8
									1	3	2	4	5	8	6	3	6	6	15	4	10	1	44	30	20	9
										1	1	1	1	4	1	2	3	4	6	3	5	1	16	10	10	10
											1	.	.	2	.	1	1	3	2	2	3	1	4	2	4	11
												1	.	2	1	1	3	3	.	3	.	1	8	10	.	12
													1	.	1	.	.	.	3	.	2	.	8	5	6	13
														1	.	1	1	2	.	2	.	1	2	2	.	14
															1	.	.	.	.	.	.	.	4	5	.	15

36	40	40	48	44	44	40	48	56	44	52	44	48	64	60	60	68	76	48	52	80	52	52	100	56	56	S.P.
2880	192	1440	64	480	2880	3840	480	960	960	960	3840	1920	60	960	384	120	480	5760	1920	120	1440	5760	120	960	960	O.S.
71	71	73	73	75	75	75	77	77	77	79	79	79	81	81	81	83	83	83	83	85	85	85	87	87	87	LEVEL
371800	78078	361998	40898	353430	681615	501930	551616	364650	442442	836550	958464	812175	247247	958464	408408	570570	459459	2088450	858858	543400	1667666	3097094	293930	1750320	2333760	DIM'N
1	0	0	0	0	2	1	2	4	0	1	1	2	0	3	4	2	5	0	1	4	0	2	6	0	2	
0	0	2	0	1	1	0	0	0	2	2	0	1	4	1	0	3	0	1	0	2	1	0	1	0	0	
1	0	0	0	2	0	1	2	0	0	1	0	0	0	1	0	0	1	1	0	0	0	1	0	2	0	
1	0	1	0	0	1	0	0	1	0	0	1	0	0	0	0	0	0	0	0	0	2	0	0	1	2	
0	2	0	0	0	0	0	0	0	0	0	1	0	0	0	0	0	0	1	1	0	0	1	0	0	0	
<	<	<	<	<	<	<	<	<	<	<	<	<	<	<	<	<	<	<	<	<	<	<	<	<	<	
0	0	0	2	0	0	1	0	0	1	0	0	1	0	0	1	0	0	0	1	0	0	0	0	0	0	WEIGHT
26	27	28	29	30	31	32	33	34	35	36	37	38	39	40	41	42	43	44	45	46	47	48	49	50	51	NUMBER
1040	186	930	82	870	1611	990	1280	750	794	1830	1728	1335	515	1920	576	1110	819	3270	1206	980	2482	4502	490	2400	3264	0
556	99	506	43	466	894	588	684	425	481	984	1008	826	264	1040	366	580	444	1976	724	505	1482	2759	245	1458	1956	1
300	55	275	27	280	516	290	440	260	239	640	560	430	190	720	200	435	324	1105	430	405	880	1600	210	880	1224	2
286	54	264	24	233	468	351	342	225	296	487	584	516	124	512	236	270	214	1186	435	230	869	1665	105	866	1150	3
140	30	126	14	105	225	210	154	105	190	210	336	330	49	216	161	105	84	705	264	85	499	990	35	495	660	4
150	29	141	14	138	261	172	211	132	144	311	322	257	88	344	120	199	154	655	258	180	514	947	90	518	701	5
78	17	75	10	82	135	77	132	69	66	198	176	115	66	224	54	146	104	355	155	140	306	513	75	308	428	6
30	.	39	.	42	96	29	80	69	34	134	56	81	45	184	54	116	104	167	43	125	138	298	75	162	224	7
.	.	.	.	.	21	.	20	30	.	30	.	15	15	80	20	50	64	.	.	80	.	55	60	.	44	8
70	16	66	8	60	120	104	90	60	89	130	184	160	34	140	76	75	60	384	156	65	290	552	30	294	392	9
35	9	33	5	35	60	47	53	30	39	80	100	68	24	88	31	54	40	204	94	49	170	294	25	173	232	10
18	6	15	4	21	27	18	33	12	14	48	54	24	17	54	10	39	24	105	58	37	99	147	20	102	141	11
13	.	18	.	17	42	17	32	30	21	54	30	47	17	72	31	42	40	97	24	44	75	165	25	90	118	12
13	5	12	3	11	21	30	16	10	25	23	56	45	6	24	21	13	10	115	58	11	89	165	5	90	120	13
6	.	9	.	11	18	6	18	12	8	34	15	16	13	42	10	31	24	49	14	32	45	78	20	54	66	14
5	.	6	.	5	15	10	10	10	15	15	16	30	4	20	21	10	10	55	14	10	39	90	5	45	60	15

1	.	1	.	1	.	1	.	.	.	.	.	3	.	4	3	.	4	3	2	16	.	2	9	16	2	19	12	11	.	19	12	11	15	18	12	16
	1	1	.	2	.	1	.	2	.	.	.	.	.	.	9	.	7	12	.	12	.	8	5	30	10	18	24	.	.	27	.	29	20	.	21	17
		1	.	1	.	1	.	.	.	.	.	.	.	.	3	.	4	3	.	8	.	2	5	16	2	14	12	.	.	19	.	11	15	.	12	18
			1	.	1	.	2	1	3	6	3	5	2	6	9	12	9	4	9	13	30	15	4	14	6	9	6	57	36	8	51	81	4	51	69	19
				1	.	1	.	.	.	.	.	.	.	.	.	.	.	3	.	.	.	.	.	8	2	5	12	.	.	14	.	.	15	.	.	20
					1	.	.	.	1	3	2	2	2	3	3	3	6	1	2	7	16	3	3	8	1	6	3	26	23	6	30	36	3	28	42	21
						1	.	.	.	.	.	.	.	.	.	.	.	.	.	.	.	.	.	.	.	.	4	.	.	5	.	.	10	.	.	22
							1	1	.	2	.	3	.	3	6	4	5	4	5	9	8	10	3	11	6	7	6	27	8	7	22	42	4	27	32	23
								1	.	.	.	.	.	.	3	.	2	4	.	3	.	5	1	8	6	4	6	.	.	6	.	15	4	.	10	24
									1	.	1	.	1	.	.	3	.	.	2	.	8	3	.	.	1	.	.	13	15	.	12	18	.	11	16	25
										1	.	1	.	2	2	1	3	1	1	5	4	2	2	6	1	5	3	12	5	5	13	18	3	16	18	26
											1	.	1	.	.	.	.	.	.	.	4	.	.	.	.	.	.	5	10	.	6	6	.	5	10	27
												1	.	1	1	.	1	1	1	4	.	1	2	4	1	4	3	6	.	4	5	6	3	9	5	28
													1	.	.	.	.	.	.	.	.	.	.	.	.	.	.	.	5	.	.	.	.	.	.	29
														1	.	.	1	.	.	2	.	.	1	2	.	3	1	2	.	3	3	2	2	6	3	30
															1	.	1	1	.	2	.	1	1	4	1	3	3	.	.	4	.	6	3	.	5	31
																1	.	.	1	.	2	2	.	.	1	.	.	6	3	.	5	9	.	6	7	32
																	1	.	.	1	.	.	1	2	.	2	1	.	.	3	.	2	2	.	3	33
																		1	.	.	.	.	.	2	1	1	3	.	.	3	.	.	3	.	.	34
																			1	.	.	1	.	.	1	.	.	3	.	.	2	3	.	3	2	35
																				1	.	.	1	1	.	2	1	.	.	2	.	.	2	.	.	36
																					1	.	.	.	.	.	.	2	2	.	3	3	.	3	4	37
																						1	.	.	1	.	.	.	.	.	.	3	.	.	2	38
																							1	.	.	1	.	.	.	1	.	.	1	.	.	39
																								1	.	1	1	.	.	2	.	.	2	.	.	40
																									1	.	.	.	.	.	.	.	.	.	.	41
																										1	.	.	.	1	.	.	1	.	.	42
																											1	.	.	1	.	.	2	.	.	43
																												1	.	.	1	1	.	2	1	44
																													1	.	.	.	.	.	.	45
																														1	.	.	1	.	.	46
																															1	.	.	1	1	47
																																1	.	.	1	48
																																	1	.	.	49
																																		1	.	50
																																			1	51

Table 3. *(continued)*

ALGEBRA C6, REPRESENTATIONS OF CLASS 1

NUMBER	0	1	2	3	4	5	6	7	8	9	10	11	12	13	14	15	16	17	18	19	20	21	22	23	24	25
S.P.	2	6	10	18	10	14	18	18	22	26	34	50	22	26	26	30	30	34	38	38	34	34	42	42	50	54
O.S.	12	160	120	12	192	960	480	384	480	120	120	12	1920	960	960	2880	1280	480	960	480	960	1920	960	384	120	480
LEVEL	12	28	32	34	36	44	48	48	50	52	54	56	56	58	60	64	64	66	66	68	68	68	70	70	72	72
DIM'N	12	208	560	364	572	4368	7800	4576	11088	11440	11648	4368	27456	35100	43680	144144	57200	112320	105600	136136	104104	168168	265200	116480	144144	155584
	1	0	1	3	0	1	0	1	2	1	3	5	0	2	0	1	0	1	3	0	0	1	2	3	1	4
	0	0	1	0	0	0	1	0	0	2	1	0	1	0	0	1	0	0	0	2	0	1	1	0	3	0
	0	1	0	0	0	0	1	0	1	0	0	0	0	0	1	0	1	2	0	1	0	0	1	0	0	1
	0	0	0	0	0	1	0	0	0	0	0	0	0	0	1	1	0	0	1	0	1	0	0	0	0	0
	0	0	0	0	1	0	0	0	0	0	0	0	1	1	0	0	0	0	0	0	1	0	0	0	0	0
	<	<	<	<	<	<	<	<	<	<	<	<	<	<	<	<	<	<	<	<	<	<	<	<	<	<
WEIGHT	0	0	0	0	0	0	0	1	0	0	0	0	0	0	0	0	1	0	0	0	0	1	0	1	0	0
0	1	4	10	6	5	30	50	20	64	65	60	21	96	110	140	401	130	300	260	340	214	329	620	200	320	324
1		1	2	1	2	12	18	10	24	22	20	6	48	57	66	198	72	141	132	161	114	191	294	120	144	154
2			1	1	.	3	8	2	12	14	15	6	16	21	30	96	29	80	69	94	56	81	184	54	101	104
3				1	.	.	.	.	4	5	10	6	.	5	.	21	.	20	30	30	.	15	80	20	50	64
4					1	4	5	6	6	5	4	1	24	30	30	90	40	60	60	66	61	115	120	76	54	60
5						1	2	1	3	3	3	1	8	10	13	42	17	32	30	38	30	47	72	31	37	40
6							1	.	1	2	2	1	2	2	6	18	6	18	12	22	15	16	42	10	26	24
7								1	.	.	.	.	4	5	5	15	10	10	10	11	16	30	20	21	9	10
8									1	1	2	1	.	2	.	9	.	7	12	12	.	8	30	10	18	24
9										1	1	1	.	.	.	3	.	4	3	8	.	2	16	2	14	12
10											1	1	.	.	.	.	.	.	3	.	.	.	8	2	5	12
11												1	.	.	.	.	.	.	.	.	.	.	.	.	.	4
12													1	1	2	6	4	5	4	6	8	10	11	6	6	6
13														1	.	3	.	2	4	3	.	5	8	6	4	6
14															1	2	1	3	1	3	4	2	6	1	4	3
15																1	.	1	1	2	.	1	4	1	3	3

NUMBER	26	27	28	29	30	31	32	33	34	35	36	37	38	39	40	41	42	43	44	45	46	47	48	49	50	51
S.P.	42	58	38	74	42	42	98	46	46	58	46	54	50	50	54	50	58	58	62	54	66	66	78	54	58	70
O.S.	384	120	5760	120	960	1920	12	3840	2880	960	3840	160	2880	960	2880	3840	384	960	2880	3840	480	480	960	5760	1920	960
LEVEL	72	74	74	76	76	76	78	78	80	80	80	82	82	82	84	84	84	86	86	86	88	88	88	88	88	90
DIM'N	89232	180180	608400	112320	448448	551616	31824	1021020	1237600	523600	864864	448800	1867008	743600	2274480	1750320	401544	2306304	2501928	2574208	1521520	2068560	1089088	3675672	2839200	3390816
	0	3	1	5	1	0	7	2	0	4	1	0	2	1	1	0	1	1	3	2	0	3	5	0	1	2
	0	2	0	1	0	2	0	1	1	0	0	0	0	0	2	1	0	1	1	0	3	0	0	1	2	2
	0	0	1	0	0	0	0	0	1	0	0	3	1	0	0	1	0	2	0	1	1	2	0	0	0	1
	0	0	0	0	2	0	0	0	1	0	1	0	1	0	1	0	0	0	1	0	0	0	1	1	0	0
	1	0	1	0	0	1	0	1	0	1	0	0	0	2	0	0	0	0	0	0	0	0	0	1	0	0
	<	<	<	<	<	<	<	<	<	<	<	<	<	<	<	<	<	<	<	<	<	<	<	<	<	<
WEIGHT	1	0	0	0	0	0	0	0	0	0	1	0	0	0	0	1	2	0	0	1	0	0	0	0	1	0
0	142	370	1035	210	724	864	56	1485	1780	665	1084	590	2520	875	2880	1930	397	2800	2901	2652	1720	2320	1110	3676	2720	3630
1	78	165	588	90	399	502	21	876	1008	402	651	332	1434	516	1656	1206	237	1568	1683	1678	956	1300	650	2271	1752	2028
2	43	125	298	75	224	256	21	464	578	222	356	204	858	304	999	657	149	1018	1059	939	644	884	425	1338	978	1404
3	.	80	55	60	44	64	21	160	160	110	60	70	300	55	390	160	27	420	516	290	300	440	260	346	340	740
4	44	60	330	30	216	290	6	509	550	240	390	175	784	305	906	760	144	824	921	1060	494	680	354	1394	1140	1041
5	24	44	165	25	118	144	6	260	310	125	215	107	454	178	534	410	89	522	561	587	326	446	225	818	621	705
6	14	32	78	20	66	68	6	120	174	57	115	65	258	106	306	214	58	332	324	304	217	291	132	478	320	471
7	14	10	90	5	60	80	1	144	155	70	130	50	225	105	261	260	55	239	270	370	144	200	105	495	405	306
8	.	27	29	20	21	36	6	84	84	57	35	35	150	29	204	100	14	208	261	172	150	211	132	208	210	361
9	.	19	11	15	12	16	6	32	48	21	16	22	78	17	114	48	10	134	135	77	104	132	69	120	100	238
10	.	14	.	15	.	.	6	16	.	21	.	.	30	.	39	.	.	42	96	29	40	80	69	.	34	134
11	.	5	.	10	.	.	6	.	.	5	.	.	.	.	.	.	.	.	21	.	.	20	30	.	.	30
12	8	7	42	4	32	36	1	64	84	30	71	30	122	62	144	134	35	146	150	192	92	124	60	287	202	198
13	.	6	15	4	10	20	1	44	40	30	20	15	70	16	96	64	8	90	120	104	64	90	60	123	134	150
14	5	5	18	3	18	14	1	24	45	10	37	18	66	38	75	65	24	90	78	90	58	78	30	165	91	126
15	.	4	6	3	5	8	1	16	22	10	10	10	35	9	51	30	5	56	60	47	42	53	30	70	60	96

1	.	.	.	2	2	.	1	.	.	3	.	9	.	7	7	.	12	17	5	24	6	24	22	27	42	15	28	27	60	17	23	10	96	60	36	16
	1	.	1	.	.	2	.	2	1	.	3	2	2	3	2	1	4	12	2	4	7	18	6	24	12	4	36	27	18	26	33	12	39	22	60	17
		1	.	.	.	2	1	1	3	.	3	.	3	.	.	1	8	.	10	.	.	13	.	18	.	.	17	42	17	16	32	30	.	21	54	18
			1	.	.	1	.	2	1	.	2	.	2	.	2	1	2	6	2	.	4	6	.	18	6	.	22	18	6	22	18	12	15	16	46	19
				1	.	.	.	.	.	2	.	3	.	4	2	.	3	8	1	12	3	12	14	12	18	11	16	12	24	10	14	4	54	22	21	20
					1	.	1	.	.	.	.	3	.	2	4	.	8	8	5	6	3	13	5	18	20	3	17	21	30	12	16	10	40	40	27	21
						1	.	1	1	.	2	.	2	.	.	1	2	.	2	.	.	6	.	9	.	.	11	18	6	12	18	12	.	8	34	22
							1	.	.	.	.	.	.	.	.	.	4	.	5	.	.	5	.	6	.	.	5	15	10	4	10	10	.	15	15	23
								1	.	.	1	.	1	.	.	1	.	.	.	.	.	.	.	3	.	.	4	3	.	8	4	3	.	2	16	24
									1	.	1	.	2	.	.	1	.	.	2	.	.	.	.	.	.	.	.	9	.	.	7	12	.	.	12	25
										1	.	.	.	.	.	.	.	.	.	4	.	.	5	.	5	6	.	.	6	.	.	.	15	5	.	26
											1	.	1	.	.	1	.	.	.	.	.	.	.	.	.	.	.	3	.	.	4	3	.	.	8	27
												1	.	1	1	.	2	4	1	3	2	6	3	8	8	2	10	9	12	7	9	4	22	14	16	28
													1	.	.	1	.	.	.	.	.	.	.	.	.	.	.	.	.	.	.	3	.	.	.	29
														1	.	.	.	2	.	1	1	3	2	3	2	2	6	3	3	4	6	1	12	3	9	30
															1	.	1	2	1	.	1	2	.	6	4	.	6	6	4	6	5	4	8	10	12	31
																1	.	.	.	.	.	.	.	.	.	.	.	.	.	.	.	.	.	.	.	32
																	1	.	1	.	.	2	.	3	.	.	3	6	4	3	5	4	.	5	9	33
																		1	.	.	1	1	.	2	1	.	4	2	1	3	3	1	4	2	7	34
																			1	.	.	.	.	.	.	.	.	3	.	.	2	4	.	.	3	35
																				1	.	.	1	.	2	1	.	.	3	.	.	.	6	3	.	36
																					1	.	.	.	.	.	2	.	.	1	1	.	.	.	3	37
																						1	.	1	.	.	2	2	1	2	3	1	.	1	5	38
																							1	.	.	1	.	.	.	.	.	.	3	.	.	39
																								1	.	.	1	1	.	2	1	1	.	1	4	40
																									1	.	.	.	1	.	.	.	2	2	.	41
																										1	.	.	.	.	.	.	.	.	.	42
																											1	.	.	1	1	.	.	.	2	43
																												1	.	.	1	1	.	.	2	44
																													1	.	.	.	.	1	.	45
																														1	.	.	.	.	1	46
																															1	.	.	.	1	47
																																1	.	.	.	48
																																	1	.	.	49
																																		1	.	50
																																			1	51

Table 3. *(continued)*

ALGEBRA C7, REPRESENTATIONS OF CLASS 0

NUMBER	0	1	2	3	4	5	6	7	8	9	10	11	12
S.P.	0	4	8	8	12	16	12	20	32	16	20	20	24
O.S.	1	84	14	560	840	84	448	168	14	3 360	896	3 360	280
LEVEL	1	25	27	41	47	49	49	51	53	59	63	65	67
DIM'N	1	90	105	910	3 900	3 094	2 002	5 355	2 380	20 020	18 018	54 978	37 400
WEIGHT	000000<0	010000<0	200000<0	000100<0	101000<0	020000<0	000001<0	210000<0	400000<0	100010<0	100000<1	010100<0	002000<0
0	1	6	7	14	50	42	14	63	28	98	56	238	160
1		1	1	4	15	11	5	18	7	39	25	94	60
2			1	.	5	6	.	12	7	9	5	34	30
3				1	3	2	2	3	1	15	12	33	20
4					1	1	.	2	1	3	2	12	9
5						1	.	1	1	.	.	4	5
6							1	.	.	5	7	9	5
7								1	1	.	.	.	.
8									1	.	.	.	.
9										1	1	3	2
10											1	.	.
11												1	1
12													1

NUMBER	13	14	15	16	17	18	19	20	21	22	23	24	25
S.P.	24	28	36	36	24	40	28	52	72	28	32	32	32
O.S.	2 240	1 680	84	840	6 720	168	2 688	168	14	6 720	560	13 440	4 480
LEVEL	67	71	73	73	73	75	75	77	79	79	81	83	83
DIM'N	70 720	150 528	60 515	116 025	139 230	124 488	170 170	90 440	27 132	334 152	214 200	967 680	340 340
WEIGHT	200100<0	111000<0	030000<0	301000<0	010001<0	220000<0	200001<0	410000<0	600000<0	001010<0	000200<0	110010<0	001000<1
0	272	560	217	385	322	420	350	280	84	672	420	1 680	476
1	112	220	82	155	151	159	170	105	28	316	190	820	246
2	48	120	51	95	46	114	60	84	28	126	90	360	90
3	40	72	25	50	72	48	83	30	7	144	84	384	130
4	16	38	16	30	21	33	26	24	7	57	38	162	47
5	4	20	11	15	5	23	5	18	7	21	20	60	15
6	10	16	5	10	35	9	42	5	1	63	35	168	70
7	4	10	6	15	.	17	5	18	7	.	.	30	.
8	.	.	.	5	.	6	.	12	7	.	.	.	.
9	4	8	3	6	10	6	12	4	1	24	15	68	26
10	.	.	.	.	5	.	6	.	.	9	5	24	15
11	1	4	2	3	2	4	2	3	1	9	7	24	8
12	.	2	1	1	.	3	.	2	1	3	4	6	2
13	1	2	1	3	.	3	2	3	1	.	.	12	.
14		1	1	1	.	2	.	2	1	.	.	3	.
15			1	.	.	1	.	1	1	.	.	.	.

NUMBER	26	27	28	29	30	31	32	33	34	35	36	37	38
S.P.	40	36	36	40	36	44	44	44	40	48	56	52	40
O.S.	3 360	6 720	5 376	3 360	6 720	896	840	6 720	672	840	2 240	1 680	26 880
LEVEL	85	87	87	89	89	89	91	91	91	93	93	95	95
DIM'N	656 370	1 491 750	931 840	1 404 480	928 200	612 612	1 293 292	2 586 584	606 424	1 975 050	1 322 685	2 900 625	4 594 590
WEIGHT	300010<0	101100<0	110000<1	020100<0	000101<0	300000<1	012000<0	210100<0	000020<0	202000<0	400100<0	121000<0	101001<0
0	1 002	2 430	1 120	2 128	1 120	644	1 890	3 640	700	2 730	1 645	3 815	4 578
1	503	1 160	600	1 028	580	355	900	1 786	354	1 300	820	1 820	2 483
2	243	600	240	544	250	155	525	1 006	174	810	496	1 160	1 152
3	240	519	328	462	300	199	391	804	180	564	370	780	1 338
4	108	260	124	240	131	79	224	438	86	338	216	488	608
5	39	130	40	124	55	25	130	222	44	205	112	305	256
6	105	210	188	182	155	119	147	315	91	210	140	280	717
7	39	50	20	64	.	25	65	158	.	125	112	205	96
8	9	.	.	.	.	5	.	34	.	30	48	45	.
9	45	100	68	92	68	43	81	165	43	120	80	170	320
10	15	30	42	26	35	28	21	45	21	30	20	40	168
11	15	48	20	45	29	12	46	81	21	69	40	103	132
12	3	24	4	20	11	2	27	36	12	42	16	61	48
13	15	18	10	24	.	12	22	57	.	42	40	69	48
14	3	8	2	12	.	2	14	24	.	23	16	43	16
15	.	.	.	4	.	.	5	4	.	5	4	20	.

NUMBER	39	40	41	42	43	44	45	46	47	48	49	50	51
S.P.	44	64	60	44	48	68	76	44	48	56	80	60	100
O.S.	2 688	84	1 680	6 720	448	168	840	13 440	13 440	128	168	2 688	168
LEVEL	95	97	97	97	97	99	99	99	99	99	101	101	103
DIM'N	1 021 020	802 620	3 225 600	3 890 250	804 440	1 819 272	1 469 650	6 580 224	6 963 880	379 236	1 670 760	3 325 608	854 658
WEIGHT	000010<1	040000<0	311000<0	020001<0	000002<0	230000<0	501000<0	100110<0	210001<0	000000<2	420000<0	400001<0	610000<0
0	896	1 008	3 920	3 570	672	2 156	1 610	6 048	5 908	268	1 848	2 456	882
1	486	462	1 880	1 965	356	1 001	770	3 240	3 307	142	847	1 405	392
2	230	327	1 280	930	186	740	560	1 680	1 642	82	672	740	336
3	264	189	800	1 077	192	406	320	1 704	1 832	78	336	794	147
4	128	132	530	496	99	295	230	868	869	43	261	392	126
5	60	97	340	215	55	214	155	440	371	27	201	170	105
6	145	63	280	590	105	133	105	876	1 010	44	105	448	42
7	.	66	280	110	.	169	155	140	261	.	180	170	105
8	.	21	120	.	.	72	95	.	46	.	114	60	84
9	71	43	180	263	54	94	75	436	460	24	80	208	36
10	40	9	40	140	30	19	15	210	245	14	15	112	6
11	35	30	112	108	29	67	50	216	186	14	60	83	30
12	16	21	68	37	17	48	30	108	63	10	45	26	24
13	.	21	92	57	.	52	50	66	129	.	54	83	30
14	.	16	53	21	.	38	30	30	42	.	39	26	24
15	.	11	20	5	.	23	15	.	5	.	23	5	18

16	1	.	1	.	2	1	.	.	.	.	3	.	.	.	.	2	.	12	.	9	16	15	.	.	6	38	.	.	22	30	.	21	.	33	26	24
17		1	.	1	.	.	3	2	8	5	5	13	12	12	15	7	11	21	10	16	10	23	68	20	6	24	55	16	13	10	100	95	8	11	42	5
18			1	.	1	1	.	.	.	.	.	.	.	.	.	.	.	4	.	5	4	10	.	.	6	20	.	.	17	15	.	5	.	23	5	18
19				1	.	.	.	.	4	.	5	5	6	6	.	7	5	15	.	10	10	15	24	.	4	20	30	.	10	10	30	65	.	10	42	5
20					1	1	.	.	.	.	.	.	.	.	.	.	.	.	.	.	4	.	.	.	.	10	.	.	6	15	.	.	.	17	5	18
21						1	.	.	.	.	.	.	.	.	.	.	.	.	.	.	.	.	.	.	.	.	.	.	.	5	.	.	.	6	.	12
22							1	1	2	1	1	6	2	5	6	1	6	9	5	9	4	13	24	10	4	14	18	9	9	6	48	30	5	8	12	4
23								1	.	.	.	3	.	2	2	.	3	3	3	6	1	7	6	4	3	8	4	6	6	3	24	6	4	6	2	3
24									1	.	1	2	1	3	.	1	3	6	.	5	4	9	8	.	3	11	10	.	7	6	13	20	.	7	12	4
25										1	.	.	2	.	3	1	.	.	2	.	.	.	12	6	.	.	9	5	.	.	18	15	3	.	6	.
26											1	.	.	.	.	1	.	3	.	2	4	3	.	.	1	8	.	.	4	6	.	10	.	6	12	4
27												1	.	1	.	.	2	2	.	3	1	5	2	.	2	6	2	.	5	3	6	4	.	5	2	3
28													1	.	.	1	.	.	.	.	.	.	4	.	.	.	5	.	.	.	5	10	.	.	6	.
29														1	.	.	1	1	.	1	1	4	.	.	2	4	2	.	4	3	.	2	.	4	2	3
30															1	.	.	.	1	.	.	.	3	3	.	.	2	3	.	.	8	3	2	.	1	.
31																1	.	.	.	.	.	.	.	.	.	.	.	.	.	.	.	5	.	.	6	.
32																	1	.	.	1	.	2	.	.	1	2	.	.	3	1	.	.	.	3	.	2
33																		1	.	1	1	2	.	.	1	4	.	.	3	3	.	2	.	4	2	3
34																			1	.	.	.	.	1	.	.	.	2	.	.	4	.	2	.	.	.
35																				1	.	1	.	.	1	2	.	.	2	1	.	.	.	3	.	2
36																					1	.	.	.	.	2	.	.	1	3	.	.	.	3	2	3
37																						1	.	.	1	1	.	.	2	1	.	.	.	2	.	2
38																							1	.	.	.	1	.	.	.	2	2	.	.	1	.
39																								1	.	.	.	1	.	.	.	.	1	.	.	.
40																									1	.	.	.	1	.	.	.	.	1	.	1
41																										1	.	.	1	1	.	.	.	2	.	2
42																											1	.	.	.	.	1	.	.	1	.
43																												1	.	.	.	.	1	.	.	.
44																													1	.	.	.	.	1	.	1
45																														1	.	.	.	1	.	2
46																															1	.	.	.	.	.
47																																1	.	.	1	.
48																																	1	.	.	.
49																																		1	.	1
50																																			1	.
51																																				1

Table 3. *(continued)*

ALGEBRA C7, REPRESENTATIONS OF CLASS 1

2	6	10	18	10	14	14	18	22	26	18	34	50	S.P.
14	280	168	14	672	128	2 240	840	840	168	2 688	168	14	O.S.
14	34	38	40	46	50	54	58	60	62	62	64	66	LEVEL
14	350	896	560	1 638	1 430	10 752	17 850	24 752	24 570	24 960	24 192	8 568	DIM'N
1	0	1	3	0	0	1	0	2	1	1	3	5	
0	0	1	0	0	0	0	1	0	2	0	1	0	
0	1	0	0	0	0	0	1	1	0	0	0	0	
0	0	0	0	0	0	1	0	0	0	0	0	0	
0	0	0	0	1	0	0	0	0	0	0	0	0	
0	0	0	0	0	0	0	0	0	0	1	0	0	
<	<	<	<	<	<	<	<	<	<	<	<	<	
0	0	0	0	0	1	0	0	0	0	0	0	0	WEIGHT
0	1	2	3	4	5	6	7	8	9	10	11	12	NUMBER
1	5	12	7	9	5	48	75	95	93	60	84	28	0
	1	2	1	3	2	16	23	30	27	26	24	7	1
		1	1	.	.	4	10	15	17	5	18	7	2
			1	.	.	.	.	5	6	.	12	7	3
				1	1	4	5	6	5	12	4	1	4
					1	.	.	.	.	6	.	.	5
						1	2	3	3	2	3	1	6
							1	1	2	.	2	1	7
								1	1	.	2	1	8
									1	.	1	1	9
										1	.	.	10
											1	1	11
												1	12
													13
													14
													15

22	26	26	26	30	30	34	38	38	30	42	50	54	S.P.
6 720	3 360	2 240	2 688	896	6 720	840	2 240	840	8 960	1 680	168	840	O.S.
70	72	74	74	76	78	80	80	82	82	84	86	86	LEVEL
108 290	134 640	152 320	102 102	123 760	480 480	353 430	339 456	414 960	452 608	791 350	407 680	444 600	DIM'N
0	2	0	0	2	1	1	3	0	0	2	1	4	
1	0	0	1	0	1	0	0	2	0	1	3	0	
0	0	1	0	0	0	2	0	1	1	1	0	1	
0	0	1	0	0	1	0	1	0	0	0	0	0	
1	1	0	0	0	0	0	0	0	0	0	0	0	
0	0	0	0	0	0	0	0	0	1	0	0	0	
<	<	<	<	<	<	<	<	<	<	<	<	<	
0	0	0	1	1	0	0	0	0	0	0	0	0	WEIGHT
13	14	15	16	17	18	19	20	21	22	23	24	25	NUMBER
215	243	280	141	155	782	555	496	615	512	1 105	544	560	0
93	108	116	70	79	334	228	216	253	256	455	214	230	1
30	39	50	20	25	158	125	112	145	96	280	148	155	2
.	9	.	.	5	34	30	48	45	.	120	72	95	3
38	45	44	37	43	125	81	80	87	128	155	68	75	4
14	15	14	21	28	35	21	20	21	64	35	14	15	5
12	15	18	10	12	57	42	40	49	48	92	46	50	6
3	3	8	2	2	24	23	16	28	16	53	32	30	7
.	3	.	.	2	12	9	16	15	.	38	22	30	8
.	.	.	.	.	4	5	4	10	.	20	17	15	9
4	5	5	6	7	15	10	10	11	24	20	9	10	10
.	.	.	.	.	.	.	4	.	.	10	6	15	11
.	.	.	.	.	.	.	.	.	.	.	.	5	12
1	1	2	1	1	6	5	4	6	8	11	6	6	13
	1	.	.	1	3	2	4	3	.	8	4	6	14
		1	.	.	2	3	1	3	2	6	4	3	15

34	34	58	42	74	38	98	38	42	42	42	46	42	S.P.
3 360	13 440	168	2 688	168	4 480	14	20 160	2 240	6 720	2 688	13 440	13 440	O.S.
86	86	88	88	90	90	92	92	94	94	94	96	96	LEVEL
618 800	1 261 260	495 040	837 760	293 760	729 300	77 520	3 346 110	2 227 680	2 926 000	1 188 096	5 290 740	3 480 750	DIM'N
0	1	3	3	5	0	7	1	1	0	0	2	1	
0	1	2	0	1	0	0	0	0	2	0	1	0	
0	0	0	0	0	0	0	1	0	0	0	0	1	
1	0	0	0	0	1	0	0	2	0	0	0	0	
1	0	0	0	0	0	0	1	0	1	1	1	0	
0	1	0	1	0	0	0	0	0	0	1	0	0	
<	<	<	<	<	<	<	<	<	<	<	<	<	
0	0	0	0	0	1	0	0	0	0	0	0	1	WEIGHT
26	27	28	29	30	31	32	33	34	35	36	37	38	NUMBER
640	1 252	616	740	336	570	84	2 943	1 870	2 395	840	4 050	2 295	0
312	647	240	392	126	306	28	1 494	928	1 233	444	2 109	1 287	1
140	261	180	170	105	130	28	717	490	605	214	1 080	575	2
.	46	114	60	84	.	28	126	90	145	.	360	90	3
150	335	74	208	36	165	7	732	441	610	236	1 047	730	4
70	175	14	112	6	90	1	342	196	280	126	483	420	5
66	129	54	83	30	72	7	342	225	291	114	516	321	6
30	42	39	26	24	29	7	153	118	132	56	231	128	7
.	21	33	26	24	.	7	57	38	69	.	162	47	8
.	5	23	5	18	.	7	21	20	30	.	60	15	9
30	65	10	42	5	40	1	153	95	130	61	228	185	10
.	.	17	5	18	.	7	.	.	.	.	30	.	11
.	.	6	.	12	.	7	.	.	.	.	.	.	12
13	20	7	12	4	17	1	66	47	56	30	98	73	13
.	10	6	12	4	.	1	24	15	30	.	68	26	14
6	4	5	2	3	6	1	27	25	21	15	36	24	15

46	58	46	50	54	50	50	54	46	62	58	62	50	S.P.
6 720	3 360	2 688	896	280	6 720	5 376	6 720	26 880	896	1 680	6 720	3 360	O.S.
98	98	98	98	100	100	100	102	102	102	104	104	104	LEVEL
5 788 160	2 586 584	2 904 720	884 884	1 939 938	8 536 320	5 173 168	10 053 120	10 005 996	2 441 880	9 609 600	10 720 710	6 683 040	DIM'N
0	4	0	0	0	2	2	1	1	4	1	3	1	
1	0	2	0	0	0	1	2	0	0	1	1	0	
1	0	0	0	3	1	0	0	0	0	2	0	0	
1	0	0	0	0	1	0	1	1	0	0	1	0	
0	1	0	0	0	0	0	0	0	0	0	0	2	
0	0	0	1	0	0	0	0	1	0	0	0	0	
<	<	<	<	<	<	<	<	<	<	<	<	<	
0	0	1	1	0	0	1	0	0	1	0	0	0	WEIGHT
39	40	41	42	43	44	45	46	47	48	49	50	51	NUMBER
4 400	1 752	1 773	498	1 365	6 120	2 944	6 816	5 768	1 220	6 280	6 709	3 642	0
2 216	928	1 011	270	681	3 096	1 702	3 464	3 215	721	3 122	3 430	2 004	1
1 220	503	459	142	405	1 780	800	2 032	1 615	355	1 965	2 110	1 078	2
320	243	99	.	130	600	240	768	250	155	780	1 006	174	3
1 048	465	587	149	312	1 464	1 000	1 628	1 781	435	1 420	1 606	1 093	4
448	210	357	84	126	624	609	672	980	280	560	651	588	5
564	240	258	78	183	813	448	932	888	199	871	958	579	6
304	108	104	43	108	444	178	520	429	79	538	542	310	7
144	108	54	.	56	260	124	344	131	79	334	438	86	8
80	39	20	.	35	130	40	188	55	25	210	222	44	9
232	105	151	44	71	330	264	372	485	119	330	375	307	10
.	39	.	.	.	50	20	64	.	25	65	158	.	11
.	9	.	.	.	.	.	.	.	5	.	34	.	12
120	45	58	24	41	172	99	200	233	43	196	205	161	13
56	45	31	.	20	100	68	132	68	43	120	165	43	14
62	15	18	14	24	90	30	102	107	12	118	105	87	15

1	1	.	.	.	.	4	.	.	.	5	10	.	6	.	10	.	24	15	20	16	34	45	36	15	36	14	11	50	63	56	125	28	50	55	81	16
	1	.	.	.	.	.	.	.	.	.	5	.	6	.	.	.	9	5	10	.	24	15	16	15	21	.	5	30	42	36	35	28	30	45	21	17
		1	1	1	2	.	4	3	3	.	2	4	2	3	.	1	9	7	12	.	24	8	30	15	10	.	13	48	20	69	29	12	73	81	21	18
			1	.	1	.	2	2	1	.	.	3	.	2	.	1	3	4	3	.	6	2	16	3	2	.	9	24	4	32	11	2	46	36	12	19
				1	.	.	2	1	3	.	.	3	2	3	.	1	.	.	.	.	12	.	.	15	.	.	.	18	10	24	.	12	22	57	.	20
					1	.	1	2	1	.	.	2	.	2	.	1	.	.	3	.	3	.	8	3	2	.	5	8	2	24	.	2	28	24	.	21
						1	.	.	.	2	2	.	1	.	4	.	9	7	7	8	12	15	17	5	11	8	6	24	18	27	57	7	28	27	42	22
							1	1	1	.	.	2	.	2	.	1	.	.	.	.	3	.	.	3	.	.	.	8	2	12	.	2	14	24	.	23
								1	.	.	.	1	.	1	.	1	.	.	.	.	.	.	.	.	.	.	.	.	.	4	.	.	5	4	.	24
									1	.	.	1	.	2	.	1	.	.	.	.	.	.	.	3	.	.	.	.	.	.	.	2	.	12	.	25
										1	.	.	.	.	1	.	3	4	2	4	3	3	8	1	2	5	3	12	3	12	24	1	16	12	23	26
											1	.	1	.	.	.	3	2	4	.	8	5	8	5	6	.	3	13	12	18	15	7	17	21	10	27
												1	.	1	.	1	.	.	.	.	.	.	.	.	.	.	.	.	.	.	.	.	.	4	.	28
													1	.	.	.	.	.	.	.	4	.	.	5	.	.	.	5	6	6	.	7	5	15	.	29
														1	.	1	.	.	.	.	.	.	.	.	.	.	.	.	.	.	.	.	.	.	.	30
															1	.	.	.	.	2	.	3	.	.	2	3	.	.	3	.	12	1	.	.	9	31
																1	.	.	.	.	.	.	.	.	.	.	.	.	.	.	.	.	.	.	.	32
																	1	1	1	.	2	1	4	1	1	.	2	6	2	8	6	1	10	9	5	33
																		1	.	.	.	.	2	.	.	.	1	3	.	3	2	.	6	3	3	34
																			1	.	1	.	2	1	1	.	1	2	1	6	.	1	6	6	.	35
																				1	.	.	.	.	.	2	.	.	.	.	4	.	.	.	5	36
																					1	.	.	1	.	.	.	2	1	3	.	1	3	6	.	37
																						1	.	.	1	.	.	.	2	.	3	1	.	.	2	38
																							1	.	.	.	1	1	.	2	.	.	4	2	.	39
																								1	.	.	.	.	.	.	.	1	.	3	.	40
																									1	.	.	.	1	.	.	1	.	.	.	41
																										1	.	.	.	.	.	.	.	.	.	42
																											1	.	.	.	.	.	2	.	.	43
																												1	.	1	.	.	2	2	.	44
																													1	.	.	1	.	.	.	45
																														1	.	.	1	1	.	46
																															1	.	.	.	1	47
																																1	.	.	.	48
																																	1	.	.	49
																																		1	.	50
																																			1	51

Table 3. *(continued)*

ALGEBRA C8, REPRESENTATIONS OF CLASS 0

0	4	8	8	12	16	20	32	12	16	16	20	24	24	20	28	36	36	40	52	24	72	28	28	28	32	S.P.
1	112	16	1 120	1 344	112	224	16	1 792	256	8 960	6 720	448	4 480	7 168	2 688	112	1 344	224	224	26 880	16	10 752	7 168	17 920	2 048	O.S.
1	29	31	49	55	57	59	61	61	65	71	77	79	79	79	83	85	85	87	89	89	91	91	93	95	95	LEVEL
1	119	136	1 700	6 885	5 320	9 044	3 876	6 188	4 862	53 040	134 368	88 179	169 575	102 102	344 064	133 280	258 400	269 892	189 924	568 480	54 264	680 238	470 288	1 193 808	554 268	DIM'N
0	0	2	0	1	0	2	4	0	0	1	0	0	2	1	1	0	3	2	4	0	6	2	0	0	2	
0	1	0	0	0	2	1	0	0	0	0	1	0	0	0	1	3	0	2	1	1	0	0	1	0	0	
0	0	0	0	1	0	0	0	0	0	0	0	2	0	0	1	0	1	0	0	0	0	0	0	1	0	
0	0	0	1	0	0	0	0	0	0	0	1	0	1	0	0	0	0	0	0	0	0	0	0	0	0	
0	0	0	0	0	0	0	0	0	0	1	0	0	0	0	0	0	0	0	0	0	0	0	0	1	0	
0	0	0	0	0	0	0	0	1	0	0	0	0	0	0	0	0	0	0	0	1	0	1	0	0	0	
0	0	0	0	0	0	0	0	0	0	0	0	0	0	1	0	0	0	0	0	0	0	0	0	0	0	
<	<	<	<	<	<	<	<	<	<	<	<	<	<	<	<	<	<	<	<	<	<	<	<	<	<	
0	0	0	0	0	0	0	0	0	1	0	0	0	0	0	0	0	0	0	0	0	0	0	1	0	1	WEIGHT
0	1	2	3	4	5	6	7	8	9	10	11	12	13	14	15	16	17	18	19	20	21	22	23	24	25	NUMBER
1	7	8	20	69	56	84	36	28	14	176	400	259	455	182	896	336	608	644	420	784	120	846	448	1 456	476	0
	1	1	5	18	13	21	8	9	5	62	140	87	165	74	312	113	216	217	140	332	36	369	206	618	225	1
		1	.	6	7	14	8	.	.	14	50	42	70	14	168	70	132	154	112	98	36	126	56	238	70	2
			1	3	2	3	1	3	2	20	42	25	50	31	88	30	60	57	35	138	8	156	98	248	109	3
				1	1	2	1	.	.	4	15	11	20	5	46	19	36	39	28	39	8	48	25	94	30	4
					1	1	1	.	.	.	5	6	5	.	24	13	18	27	21	9	8	9	5	34	5	5
						1	1	.	.	.	.	.	5	.	12	7	18	20	21	.	8	9	.	.	5	6
							1	.	.	.	.	.	.	.	.	.	6	7	14	.	8	.	.	.	.	7
								1	1	5	9	5	10	14	16	5	10	9	5	55	1	63	50	91	57	8
									1	.	.	.	.	7	.	.	.	.	.	20	.	21	28	28	36	9
										1	3	2	4	2	8	3	6	6	4	15	1	18	12	33	14	10
											1	1	1	.	4	2	3	4	3	3	1	3	2	12	2	11
												1	.	.	2	1	1	3	2	.	1	.	.	4	.	12
													1	.	2	1	3	3	3	.	1	3	.	.	2	13

32	32	40	36	32	40	44	44	36	48	56	36	44	52	40	64	60	40	68	76	40	80	44	100	44	48	S.P.
1 120	35 840	8 960	13 440	35 840	6 720	1 344	13 440	43 008	1 344	4 480	26 880	7 168	2 688	1 792	112	2 688	17 920	224	1 344	107 520	224	26 880	224	35 840	53 760	O.S.
97	99	101	103	103	105	107	107	107	109	109	109	109	111	111	113	113	113	115	115	115	117	117	119	119	119	LEVEL
724 812	3 342 336	2 204 475	4 804 625	2 645 370	4 418 232	3 921 372	8 002 800	6 963 200	5 890 976	3 956 750	5 337 252	4 434 144	8 456 140	3 162 816	2 233 392	9 191 424	4 996 810	4 992 288	4 029 102	24 984 050	4 457 400	20 599 920	2 184 126	31 752 192	36 212 176	DIM'N
0	1	3	1	0	0	0	2	1	2	4	0	3	1	0	0	3	0	2	5	1	4	0	6	1	2	
0	1	0	0	0	2	1	1	1	0	0	0	0	2	0	4	1	0	3	0	0	2	2	1	0	1	
0	0	0	1	1	0	2	0	0	2	0	0	0	1	0	0	1	0	0	1	1	0	0	0	0	0	
2	0	0	1	0	1	0	1	0	0	1	1	0	0	0	0	0	1	0	0	0	0	0	0	1	0	
0	1	1	0	0	0	0	0	0	0	0	0	0	0	2	0	0	0	0	0	0	0	0	0	1	0	
0	0	0	0	0	0	0	0	0	0	0	1	0	0	0	0	0	0	0	0	1	0	1	0	0	1	
0	0	0	0	1	0	0	0	1	0	0	0	1	0	0	0	0	0	0	0	0	0	0	0	0	0	
<	<	<	<	<	<	<	<	<	<	<	<	<	<	<	<	<	<	<	<	<	<	<	<	<	<	
0	0	0	0	0	0	0	0	0	0	0	0	0	0	0	0	0	1	0	0	0	0	0	0	0	0	WEIGHT
26	27	28	29	30	31	32	33	34	35	36	37	38	39	40	41	42	43	44	45	46	47	48	49	50	51	NUMBER
860	3 584	2 107	4 865	1 946	4 200	3 612	7 120	4 480	5 152	3 150	3 444	2 528	7 084	1 952	1 792	7 168	2 394	3 808	2 862	13 650	3 192	10 464	1 470	16 128	17 136	0
355	1 568	944	2 090	928	1 820	1 545	3 125	2 200	2 206	1 400	1 647	1 272	3 032	920	741	3 080	1 223	1 589	1 224	6 765	1 316	5 247	588	7 920	8 713	1
160	672	448	1 050	322	945	882	1 730	840	1 344	840	672	532	1 904	420	518	2 072	476	1 162	888	3 010	1 036	2 400	504	3 920	4 186	2
139	640	390	822	445	716	591	1 230	1 080	842	550	777	636	1 143	428	268	1 152	632	570	447	3 276	462	2 568	196	3 760	4 298	3
60	264	174	400	151	365	331	660	390	495	320	316	244	705	190	185	756	246	410	321	1 420	357	1 141	168	1 820	1 976	4
30	96	62	195	46	185	189	330	120	294	165	126	74	435	90	134	480	90	295	216	575	273	477	140	880	818	5
.	48	62	75	.	95	93	235	60	180	165	.	74	291	.	91	396	.	232	216	215	245	243	140	280	575	6
.	.	14	.	.	.	.	50	.	42	70	.	14	63	.	28	168	.	98	132	.	154	.	112	.	98	7
49	232	140	280	215	238	189	405	536	266	175	360	322	350	194	77	344	330	161	126	1 539	125	1 215	49	1 704	2 040	8
14	64	35	70	105	56	42	90	272	56	35	162	168	70	84	14	64	175	28	21	693	20	540	7	720	910	9
20	92	60	130	72	118	102	210	184	150	100	144	114	210	84	52	220	130	113	90	648	95	522	42	800	900	10
9	32	20	61	21	57	57	102	52	85	50	57	31	126	38	36	136	47	80	60	255	71	207	35	376	354	11
5	8	4	30	5	25	33	45	10	51	20	21	5	74	20	25	82	15	57	36	90	53	69	28	180	117	12
.	16	20	23	.	30	27	72	26	52	50	.	31	84	.	25	112	.	62	60	93	64	108	35	116	246	13

1	.	.	.	.	.	5	.	6	7	9	8	5	24	15	30	35	26	21	45	90	30	20	63	56	40	35	9	40	70	19	15	280	15	225	6	322	385	14
	1	1	1	2	2	.	1	.	.	.	.	.	4	4	10	.	15	17	30	5	28	20	.	5	52	.	19	64	.	45	36	30	46	39	28	50	78	15
		1	.	1	1	.	1	.	.	.	.	.	.	.	.	.	5	6	5	.	6	5	.	.	24	.	13	24	.	27	18	.	27	9	21	.	9	16
			1	1	2	.	1	.	.	.	.	.	.	4	.	.	.	.	15	.	11	20	.	5	18	.	7	46	.	26	36	.	39	.	28	.	39	17
				1	1	.	1	.	.	.	.	.	.	.	.	.	.	.	5	.	6	5	.	.	12	.	7	24	.	20	18	.	27	.	21	.	9	18
					1	.	1	.	.	.	.	.	.	.	.	.	.	.	.	.	.	5	.	.	.	.	.	12	.	7	18	.	20	.	21	.	.	19
						1	.	1	1	3	1	2	8	5	13	10	12	11	21	24	16	10	24	14	23	15	6	24	26	13	10	106	11	85	5	144	145	20
							1	.	.	.	.	.	.	.	.	.	.	.	.	.	.	.	.	.	.	.	.	.	.	.	6	.	7	.	14	.	.	21
								1	.	.	1	.	4	5	5	.	6	5	15	12	10	10	.	14	15	.	4	20	.	10	10	38	10	45	5	44	100	22
									1	.	1	.	.	.	.	5	.	.	.	12	.	.	9	7	.	5	.	.	15	.	.	38	.	30	.	44	50	23
										1	.	1	2	1	6	2	5	6	9	4	9	4	9	2	13	7	4	14	8	9	6	36	8	27	4	66	45	24
											1	.	.	.	.	.	.	.	.	6	.	.	.	7	.	.	.	.	.	.	.	14	.	15	.	14	35	25
												1	.	.	3	.	2	3	3	.	6	1	3	.	7	4	3	8	2	6	3	9	6	6	3	32	9	26
													1	1	2	.	3	3	6	2	5	4	.	2	9	.	3	11	.	7	6	12	7	15	4	18	30	27
														1	.	.	.	.	3	.	2	4	.	2	3	.	1	8	.	4	6	.	6	.	4	.	15	28
															1	.	1	2	2	.	3	1	.	.	5	.	2	6	.	5	3	3	5	3	3	8	6	29
																1	.	.	.	2	.	.	3	1	.	2	.	.	5	.	.	12	.	9	.	18	15	30
																	1	1	1	.	1	1	.	.	4	.	2	4	.	4	3	.	4	3	3	.	3	31
																		1	.	.	1	.	.	.	2	.	1	2	.	3	1	.	3	.	2	.	.	32
																			1	.	1	1	.	.	2	.	1	4	.	3	3	.	4	.	3	.	3	33
																				1	.	.	.	1	.	.	.	.	.	.	.	4	.	5	.	5	10	34
																					1	.	.	.	1	.	1	2	.	2	1	.	3	.	2	.	.	35
																						1	.	.	.	.	.	2	.	1	3	.	3	.	3	.	.	36
																							1	.	.	1	.	.	1	.	.	3	.	2	.	8	3	37
																								1	.	.	.	.	.	.	.	.	.	.	.	.	5	38
																									1	.	1	1	.	2	1	.	2	.	2	.	.	39
																										1	.	.	.	.	.	.	.	.	.	4	.	40
																											1	.	.	1	.	.	1	.	1	.	.	41
																												1	.	1	1	.	2	.	2	.	.	42
																													1	.	.	.	.	.	.	.	.	43
																														1	.	.	1	.	1	.	.	44
																															1	.	1	.	2	.	.	45
																																1	.	1	.	2	2	46
																																	1	.	1	.	.	47
																																		1	.	.	1	48
																																			1	.	.	49
																																				1	.	50
																																					1	51

Table 3. *(continued)*

ALGEBRA C8, REPRESENTATIONS OF CLASS 1

2	6	10	18	10	14	14	18	22	26	34	50	18	22	22	26	26	30	26	34	38	30	38	42	50	54	30	58	74	34	34	34	98	42	38	38	46	42	42	38	46	46	58	42	54	50	42	54	46	46	58	62	S.P.
16	448	224	16	1 792	4 480	1 024	1 344	1 344	224	224	16	10 752	2 048	17 920	8 960	4 480	13 440	21 504	1 344	4 480	7 168	1 344	2 688	224	1 344	35 840	224	224	8 960	53 760	14 336	16	10 752	14 336	53 760	2 048	4 480	17 920	35 840	35 840	13 440	8 960	10 752	448	13 440	107 520	13 440	21 504	14 336	2 688	13 440	O.S.
16	40	44	46	56	64	64	68	70	72	74	76	76	80	84	86	88	92	92	94	94	94	96	98	100	100	100	102	104	104	104	104	106	106	108	110	110	112	112	112	114	116	116	116	118	118	118	120	120	120	122	122	LEVEL
16	544	1 344	816	3 808	22 848	7 072	36 176	49 248	47 600	45 696	15 504	88 128	70 720	330 752	403 104	434 112	1 326 000	671 840	940 576	912 000	795 872	1 079 568	2 025 856	1 005 312	1 101 600	2 170 560	1 193 808	682 176	2 635 680	5 834 400	1 881 152	170 544	3 762 304	4 900 896	13 529 824	3 100 800	8 527 200	11 541 504	6 852 768	20 511 792	21 271 488	9 682 400	8 434 176	6 777 680	30 844 800	30 749 600	35 473 152	24 923 360	9 137 024	32 672 640	36 951 200	DIM'N
1 0 0 0 0 0 0 < 0	0 0 1 0 0 0 0 < 0	1 1 0 0 0 0 0 < 0	3 0 0 0 0 0 0 < 0	0 0 0 0 1 0 0 < 0	1 0 0 1 0 0 0 < 0	0 0 0 0 0 0 1 < 0	0 1 1 0 0 0 0 < 0	2 0 1 0 0 0 0 < 0	1 2 0 0 0 0 0 < 0	3 1 0 0 0 0 0 < 0	5 0 0 0 0 0 0 < 0	1 0 0 0 0 1 0 < 0	1 0 0 0 0 0 0 < 1	0 1 0 0 1 0 0 < 0	2 0 0 0 1 0 0 < 0	0 0 1 1 0 0 0 < 0	1 1 0 1 0 0 0 < 0	0 1 0 0 0 0 1 < 0	1 0 2 0 0 0 0 < 0	3 0 0 1 0 0 0 < 0	2 0 0 0 0 0 1 < 0	0 2 1 0 0 0 0 < 0	2 1 1 0 0 0 0 < 0	1 3 0 0 0 0 0 < 0	4 0 1 0 0 0 0 < 0	0 0 1 0 0 1 0 < 0	3 2 0 0 0 0 0 < 0	5 1 0 0 0 0 0 < 0	0 0 0 1 1 0 0 < 0	1 1 0 0 0 1 0 < 0	0 0 1 0 0 0 0 < 1	7 0 0 0 0 0 0 < 0	3 0 0 0 0 1 0 < 0	1 1 0 0 0 0 0 < 1	1 0 1 0 1 0 0 < 0	3 0 0 0 0 0 0 < 1	1 0 0 2 0 0 0 < 0	0 2 0 0 1 0 0 < 0	0 0 0 1 0 0 1 < 0	2 1 0 0 1 0 0 < 0	0 1 1 1 0 0 0 < 0	4 0 0 0 1 0 0 < 0	0 0 0 0 1 1 0 < 0	0 0 3 0 0 0 0 < 0	2 0 1 1 0 0 0 < 0	1 0 1 0 0 0 1 < 0	1 2 0 1 0 0 0 < 0	0 2 0 0 0 0 1 < 0	0 0 0 0 1 0 0 < 1	1 1 2 0 0 0 0 < 0	3 1 0 1 0 0 0 < 0	WEIGHT
0	1	2	3	4	5	6	7	8	9	10	11	12	13	14	15	16	17	18	19	20	21	22	23	24	25	26	27	28	29	30	31	32	33	34	35	36	37	38	39	40	41	42	43	44	45	46	47	48	49	50	51	NUMBER
1	6	14	8	14	70	14	105	132	126	112	36	126	70	400	448	490	1 345	490	924	840	532	1 008	1 792	854	888	1 330	952	504	1 490	3 184	854	120	1 848	1 974	6 636	1 120	4 010	5 312	2 590	8 883	9 170	3 752	2 912	2 730	12 600	9 996	13 790	7 560	2 492	12 292	13 350	0
	1	2	1	4	20	5	28	36	32	28	8	48	30	152	174	180	505	220	336	320	244	366	651	298	321	600	329	168	660	1 473	422	36	872	1 001	3 024	580	1 795	2 444	1 293	4 124	4 140	1 764	1 440	1 220	5 710	5 152	6 260	3 945	1 312	5 476	6 085	1
		1	1	.	5	.	12	18	20	21	8	9	5	48	62	75	235	60	180	165	74	207	396	204	216	215	245	140	280	575	141	36	369	361	1 400	225	910	1 168	512	2 064	2 215	944	640	708	3 195	2 180	3 600	1 710	570	3 371	3 685	2
			1	.	.	.	.	6	7	14	8	.	.	.	14	.	50	.	42	70	14	63	168	98	132	.	154	112	.	98	.	36	126	56	238	70	160	272	.	672	560	448	.	217	1 050	322	1 330	350	.	1 302	1 730	3
				1	4	2	5	6	5	4	1	18	14	52	60	58	160	101	102	100	114	108	190	82	90	264	88	42	281	660	214	8	396	521	1 296	308	750	1 050	646	1 774	1 714	760	708	491	2 360	2 650	2 566	2 056	694	2 184	2 480	4
					1	.	2	3	3	3	1	3	2	16	20	23	72	26	52	50	31	60	112	55	60	93	64	35	116	246	70	8	156	177	582	109	367	488	256	856	895	390	312	280	1 277	1 092	1 441	860	306	1 311	1 460	5
						1	.	.	.	.	.	6	8	14	15	14	35	48	21	20	56	21	35	14	15	112	14	6	112	280	112	1	168	287	504	176	280	400	323	672	616	280	344	168	840	1 365	888	1 078	370	728	840	6
							1	1	2	2	1	.	.	4	4	10	30	5	28	20	5	34	64	38	36	30	46	28	50	78	20	8	48	50	252	30	185	216	96	376	470	174	140	162	680	412	790	330	130	794	815	7
								1	1	2	1	.	.	.	4	.	15	.	11	20	5	18	46	26	36	.	39	28	.	39	.	8	48	25	94	30	60	112	.	264	220	174	.	82	400	151	520	170	.	490	660	8
									1	1	1	.	.	.	.	.	5	.	6	5	.	12	24	20	18	.	27	21	.	9	.	8	9	5	34	5	30	48	.	96	120	62	.	51	195	46	280	60	.	303	330	9
										1	1	.	.	.	.	.	.	.	.	5	.	.	12	7	18	.	20	21	.	.	.	8	9	.	.	5	.	.	.	48	.	62	.	.	75	.	95	.	.	93	235	10
											1	.	.	.	.	.	.	.	.	.	.	.	.	.	6	.	7	14	.	.	.	8	.	.	.	.	.	.	.	.	.	14	.	.	.	.	.	.	.	.	50	11
												1	1	4	5	5	15	12	10	10	14	11	20	9	10	38	10	5	44	100	37	1	63	93	216	57	130	180	128	312	309	140	150	92	435	550	483	435	165	421	480	12
													1	.	.	.	.	6	.	.	7	.	.	.	.	14	.	.	14	35	21	.	21	56	63	36	35	50	64	84	77	35	70	21	105	280	111	224	90	91	105	13
														1	1	2	6	2	5	4	2	6	11	6	6	12	7	4	18	30	10	1	18	24	90	14	62	76	48	132	156	60	66	52	222	201	256	158	72	246	260	14
															1	.	3	.	2	4	2	3	8	4	6	.	6	4	.	15	.	1	18	12	33	14	20	40	.	92	72	60	.	25	130	72	168	83	.	150	210	15

1	2	.	3	1	.	3	6	4	3	3	5	3	8	6	2	1	3	4	36	2	32	28	16	48	79	20	30	30	114	63	129	47	29	146	132	16
	1	.	1	1	.	2	4	3	3	.	4	3	.	3	.	1	3	2	12	2	9	16	.	32	38	20	.	16	61	21	87	26	.	90	102	17
		1	.	.	1	.	.	.	.	4	.	.	5	10	6	.	6	14	24	8	15	20	24	34	36	15	30	11	50	100	56	78	40	50	55	18
			1	.	.	1	2	2	1	.	3	2	.	.	.	1	.	.	4	.	5	4	.	8	20	4	.	11	30	5	40	5	.	56	45	19
				1	.	.	2	1	3	.	3	3	.	.	.	1	3	.	.	2	.	.	.	16	.	20	.	.	23	.	30	.	.	27	72	20
					1	.	.	.	.	.	.	.	.	5	.	.	6	7	9	8	5	10	.	24	16	15	.	5	30	35	36	42	.	30	45	21
						1	1	2	1	.	2	2	.	.	.	1	.	.	.	.	.	.	4	.	4	10	4	.	6	10	.	30	5	.	34	30	22
							1	1	1	.	2	2	.	.	.	1	.	.	.	.	.	.	.	4	.	4	.	.	10	.	15	.	.	17	30	23
								1	.	.	1	1	.	.	.	1	.	.	.	.	.	.	.	.	.	.	.	.	.	.	5	.	.	6	5	24
									1	.	1	2	.	.	.	1	.	.	.	.	.	.	.	.	.	4	.	.	.	.	.	.	.	.	15	25
										1	.	.	2	2	1	.	1	2	9	1	7	7	8	12	17	5	13	6	24	30	27	22	17	28	27	26
											1	1	.	.	.	1	.	.	.	.	.	.	.	.	.	.	.	.	.	.	.	.	.	.	5	27
												1	.	.	.	1	.	.	.	.	.	.	.	.	.	.	.	.	.	.	.	.	.	.	.	28
													1	.	.	.	.	.	3	.	4	2	2	3	8	1	6	3	12	6	12	4	6	16	12	29
														1	.	.	1	1	3	1	2	4	.	8	8	5	.	3	13	10	18	12	.	17	21	30
															1	.	.	2	.	1	.	.	4	.	.	.	5	.	.	15	.	11	10	.	.	31
																1	.	.	.	.	.	.	.	.	.	.	.	.	.	.	.	.	.	.	.	32
																	1	.	.	1	.	.	.	4	.	5	.	.	5	.	6	.	.	5	15	33
																		1	.	1	.	.	.	.	.	.	.	.	.	5	.	6	.	.	.	34
																			1	.	1	1	.	2	4	1	.	2	6	2	8	2	.	10	9	35
																				1	.	.	.	.	.	.	.	.	.	.	.	.	.	.	.	36
																					1	.	.	.	2	.	.	1	3	.	3	.	.	6	3	37
																						1	.	1	2	1	.	1	2	.	6	2	.	6	6	38
																							1	.	.	.	2	.	.	3	.	2	4	.	.	39
																								1	.	1	.	.	2	.	3	.	.	3	6	40
																									1	.	.	1	1	.	2	.	.	4	2	41
																										1	.	.	.	.	.	.	.	.	3	42
																											1	.	.	.	.	.	1	.	.	43
																												1	.	.	.	.	.	2	.	44
																													1	.	1	.	.	2	2	45
																														1	.	1	.	.	.	46
																															1	.	.	1	1	47
																																1	.	.	.	48
																																	1	.	.	49
																																		1	.	50
																																			1	51

Table 3. *(continued)*

ALGEBRA C9, REPRESENTATIONS OF CLASS 0

S.P.	0	4	8	8	12	16	20	32	12	16	16	20	24	24	20	28	36	36	24	40	52	72	24	28	28	32
O.S.	1	144	18	2 016	2 016	144	288	18	5 376	2 304	20 160	12 096	672	8 064	32 256	4 032	144	2 016	4 608	288	288	18	80 640	32 256	40 320	2 016
LEVEL	1	33	35	57	63	65	67	69	73	81	83	89	91	91	95	95	97	97	99	99	101	103	105	107	111	113
DIM'N	1	152	171	2 907	11 305	8 568	14 364	5 985	15 504	25 194	121 125	290 871	186 200	361 760	377 910	709 632	267 444	522 291	277 134	534 888	367 080	100 947	1 813 968	2 136 645	3 527 160	2 069 613
	0	0	2	0	1	0	2	4	0	0	1	0	0	2	1	1	0	3	1	2	4	6	0	2	0	0
	0	1	0	0	0	2	1	0	0	0	0	1	0	0	0	1	3	0	0	2	1	0	1	0	0	0
	0	0	0	0	1	0	0	0	0	0	0	0	2	0	0	1	0	1	0	0	0	0	0	0	1	0
	0	0	0	1	0	0	0	0	0	0	0	1	0	1	0	0	0	0	0	0	0	0	0	0	0	2
	0	0	0	0	0	0	0	0	0	0	1	0	0	0	0	0	0	0	0	0	0	0	0	0	1	0
	0	0	0	0	0	0	0	0	1	0	0	0	0	0	0	0	0	0	0	0	0	0	1	1	0	0
	0	0	0	0	0	0	0	0	0	0	0	0	0	0	1	0	0	0	0	0	0	0	0	0	0	0
	0	0	0	0	0	0	0	0	0	1	0	0	0	0	0	0	0	0	0	0	0	0	0	0	0	0
	<	<	<	<	<	<	<	<	<	<	<	<	<	<	<	<	<	<	<	<	<	<	<	<	<	<
WEIGHT	0	0	0	0	0	0	0	0	0	0	0	0	0	0	0	0	0	0	1	0	0	0	0	0	0	0
NUMBER	0	1	2	3	4	5	6	7	8	9	10	11	12	13	14	15	16	17	18	19	20	21	22	23	24	25
0	1	8	9	27	91	72	108	45	48	42	285	621	392	704	414	1 344	492	903	210	936	600	165	1 584	1 701	2 760	1 575
1		1	1	6	21	15	24	9	14	14	90	195	119	228	154	420	149	287	84	284	180	45	610	672	1 065	595
2			1	.	7	8	16	9	.	.	20	69	56	96	28	224	92	175	14	200	144	45	176	224	400	259
3				1	3	2	3	1	4	5	25	51	30	60	57	104	35	70	35	66	40	9	224	250	380	208
4					1	1	2	1	.	.	5	18	13	24	9	54	22	42	5	45	32	9	62	76	140	87
5						1	1	1	.	.	.	6	7	6	.	28	15	21	.	31	24	9	14	14	50	42
6							1	1	.	.	.	.	.	6	.	14	8	21	.	23	24	9	.	14	.	.
7								1	.	.	.	.	.	.	.	.	.	7	.	8	16	9	.	.	.	.
8									1	2	5	9	5	10	21	16	5	10	16	9	5	1	75	84	119	63
9										1	.	.	.	.	7	.	.	.	9	.	.	.	20	21	28	14
10											1	3	2	4	3	8	3	6	2	6	4	1	20	24	42	25
11												1	1	1	.	4	2	3	.	4	3	1	4	4	15	11
12													1	.	.	2	1	1	.	3	2	1	.	.	5	6

S.P.	28	32	32	40	36	40	32	44	44	48	56	36	36	52	36	64	44	60	40	40	68	76	48	80	40	100
O.S.	64 512	80 640	18 432	20 160	24 192	12 096	161 280	2 016	24 192	2 016	8 064	43 008	193 536	4 032	80 640	144	32 256	4 032	4 032	36 864	288	2 016	4 608	288	322 560	288
LEVEL	113	115	115	117	119	121	123	123	123	125	125	127	127	127	129	129	129	129	131	131	131	131	133	133	135	135
DIM'N	3 133 746	9 630 720	3 627 936	6 217 750	13 231 372	11 956 320	12 790 800	10 345 860	21 377 925	15 343 944	10 296 594	9 876 048	32 744 448	21 651 336	22 682 352	5 523 300	20 369 349	23 113 728	12 758 823	24 806 400	12 209 400	9 828 567	15 242 370	10 659 000	102 070 584	5 047 350
	0	1	2	3	1	0	0	0	2	2	4	0	1	1	0	0	3	3	0	1	2	5	3	4	1	6
	1	1	0	0	0	2	0	1	1	0	0	0	1	2	0	4	0	1	0	1	3	0	0	2	0	1
	0	0	0	0	1	0	1	2	0	2	0	1	0	1	0	0	0	1	0	0	0	1	0	0	1	0
	0	0	0	0	1	1	0	0	1	0	1	0	0	0	1	0	0	0	0	0	0	0	0	0	0	0
	0	1	0	1	0	0	0	0	0	0	0	0	0	0	0	0	0	0	2	0	0	0	0	0	0	0
	0	0	0	0	0	0	0	0	0	0	0	0	0	0	1	0	0	0	0	0	0	0	0	0	1	0
	0	0	0	0	0	0	1	0	0	0	0	0	1	0	0	0	1	0	0	0	0	0	0	0	0	0
	1	0	1	0	0	0	0	0	0	0	0	0	0	0	0	0	0	0	0	0	0	0	0	0	0	0
	<	<	<	<	<	<	<	<	<	<	<	<	<	<	<	<	<	<	<	<	<	<	<	<	<	<
WEIGHT	0	0	0	0	0	0	0	0	0	0	0	1	0	0	0	0	0	0	0	1	0	0	1	0	0	0
NUMBER	26	27	28	29	30	31	32	33	34	35	36	37	38	39	40	41	42	43	44	45	46	47	48	49	50	51
0	1 638	6 720	1 728	3 910	8 764	7 488	5 328	6 300	12 609	8 904	5 490	3 024	12 096	12 096	8 400	2 964	6 741	12 096	4 527	6 720	6 264	4 725	3 690	5 160	32 616	2 310
1	700	2 660	756	1 580	3 416	2 938	2 346	2 443	5 001	3 458	2 200	1 414	5 460	4 690	3 710	1 115	3 108	4 704	1 980	3 220	2 372	1 827	1 806	1 932	14 829	840
2	182	1 120	224	740	1 680	1 504	784	1 372	2 733	2 072	1 312	448	2 016	2 912	1 456	772	1 260	3 136	860	1 120	1 720	1 323	686	1 512	6 400	720
3	303	960	332	575	1 195	1 026	1 020	833	1 746	1 176	765	672	2 416	1 576	1 592	361	1 395	1 568	840	1 560	762	595	889	608	6 480	252
4	74	390	88	255	570	516	332	459	927	682	444	206	846	962	618	247	523	1 022	355	510	544	427	309	468	2 720	216
5	14	140	14	90	273	258	98	259	459	399	228	56	252	588	238	177	154	644	160	140	389	287	84	356	1 072	180
6	.	70	14	90	105	132	.	126	327	245	228	.	126	392	.	120	154	532	.	70	305	287	84	320	400	180
7	.	.	.	20	.	.	.	.	69	56	96	.	.	84	.	36	28	224	.	.	128	175	14	200	.	144
8	135	296	150	175	350	294	435	231	495	322	210	329	1 044	420	655	91	609	408	339	780	189	147	452	145	2 676	56
9	63	64	72	35	70	56	180	42	90	56	35	168	432	70	252	14	252	64	126	416	28	21	249	20	1 008	7
10	31	116	36	75	160	144	138	123	255	180	120	98	348	250	248	61	213	260	139	240	132	105	144	110	1 088	48
11	5	40	5	25	74	69	39	68	123	101	60	25	96	149	94	42	57	160	60	60	93	70	35	82	416	40
12	.	10	.	5	36	30	9	39	54	60	24	5	18	87	34	29	9	96	30	10	66	42	5	61	144	32

1	.	2	1	3	.	3	3	1	.	4	.	.	.	20	5	25	28	36	.	32	81	62	60	.	48	99	.	25	51	132	.	30	12	10	35	14	132	40	13
	1	.	.	.	1	.	.	.	5	6	9	5	14	24	16	15	30	26	55	21	45	30	20	50	138	40	91	9	84	40	49	122	19	15	73	15	392	6	14
		1	1	1	.	2	2	1	.	.	.	.	.	5	.	5	12	18	.	20	36	33	24	.	9	61	.	22	9	75	.	5	52	42	5	53	48	32	15
			1	.	.	1	1	1	.	.	.	.	.	.	.	.	.	6	.	7	6	7	6	.	.	28	.	15	.	28	.	.	31	21	.	31	.	24	16
				1	.	1	2	1	.	.	.	.	.	.	.	5	.	.	.	.	18	13	24	.	.	21	.	8	9	54	.	.	30	42	5	45	.	32	17
					1	.	.	.	.	.	.	.	7	.	8	.	.	.	20	.	.	.	.	28	48	.	28	.	28	.	14	72	.	.	45	.	112	.	18
						1	1	1	.	.	.	.	.	.	.	.	.	.	.	.	6	7	6	.	.	14	.	8	.	28	.	.	23	21	.	31	.	24	19
							1	1	.	.	.	.	.	.	.	.	.	.	.	.	.	.	6	.	.	.	.	.	.	14	.	.	8	21	.	23	.	24	20
								1	.	.	.	.	.	.	.	.	.	.	.	.	.	.	.	.	.	.	.	.	.	.	.	.	.	7	.	8	.	16	21
									1	1	3	2	2	8	2	5	13	12	15	11	21	16	10	12	36	23	33	6	21	24	20	28	13	10	16	11	144	5	22
										1	.	.	.	4	2	5	5	6	.	5	15	10	10	.	18	15	.	4	21	20	.	14	10	10	16	10	52	5	23
											1	1	.	2	.	1	6	5	3	6	9	9	4	2	6	13	12	4	3	14	9	4	9	6	2	8	48	4	24
												1	.	.	.	.	3	2	.	3	3	6	1	.	.	7	4	3	.	8	5	.	6	3	.	6	12	3	25
													1	.	1	.	.	.	5	.	.	.	.	7	12	.	9	.	7	.	5	16	.	.	9	.	38	.	26
														1	.	1	2	3	.	3	6	5	4	.	3	9	.	3	3	11	.	2	7	6	2	7	16	4	27
															1	.	.	.	.	.	.	.	.	.	6	.	.	.	7	.	.	8	.	.	9	.	14	.	28
																1	.	.	.	.	3	2	4	.	.	3	.	1	3	8	.	.	4	6	2	6	.	4	29
																	1	1	.	2	2	3	1	.	.	5	.	2	.	6	.	.	5	3	.	5	4	3	30
																		1	.	1	1	1	1	.	.	4	.	2	.	4	.	.	4	3	.	4	.	3	31
																			1	.	.	.	.	1	2	.	3	.	1	.	2	2	.	.	1	.	12	.	32
																				1	.	1	.	.	.	2	.	1	.	2	.	.	3	1	.	3	.	2	33
																					1	1	1	.	.	2	.	1	.	4	.	.	3	3	.	4	.	3	34
																						1	.	.	.	1	.	1	.	2	.	.	2	1	.	3	.	2	35
																							1	.	.	.	.	.	.	2	.	.	1	3	.	3	.	3	36
																								1	.	.	.	.	.	.	.	2	.	.	1	.	.	.	37
																									1	.	.	.	1	.	.	1	.	.	1	.	4	.	38
																										1	.	1	.	1	.	.	2	1	.	2	.	2	39
																											1	.	.	.	1	.	.	.	.	.	3	.	40
																												1	.	.	.	.	1	.	.	1	.	1	41
																													1	.	.	.	.	.	1	.	.	.	42
																														1	.	.	1	1	.	2	.	2	43
																															1	.	.	.	.	.	.	.	44
																																1	.	.	1	.	.	.	45
																																	1	.	.	1	.	1	46
																																		1	.	1	.	2	47
																																			1	.	.	.	48
																																				1	.	1	49
																																					1	.	50
																																						1	51

Table 3. *(continued)*

ALGEBRA C9, REPRESENTATIONS OF CLASS 1

NUMBER	0	1	2	3	4	5	6	7	8	9	10	11	12
S.P.	2	6	10	18	10	14	14	18	22	18	26	34	50
O.S.	18	672	288	18	4 032	8 064	4 608	2 016	2 016	512	288	288	18
LEVEL	18	46	50	52	66	74	78	78	80	82	82	84	86
DIM'N	18	798	1 920	1 140	7 752	43 776	23 256	67 032	89 964	16 796	85 272	80 256	26 334
WEIGHT	10000000<0	00100000<0	11000000<0	30000000<0	00001000<0	10010000<0	00000010<0	01100000<0	20100000<0	00000000<1	12000000<0	31000000<0	50000000<0
0	1	7	16	9	20	96	28	140	175	14	164	144	45
1		1	2	1	5	24	9	33	42	5	37	32	9
2			1	1	.	6	.	14	21	.	23	24	9
3				1	.	.	.	.	7	.	8	16	9
4					1	4	3	5	6	2	5	4	1
5						1	.	2	3	.	3	3	1
6							1	.	.	1	.	.	.
7								1	1	.	2	2	1
8									1	.	1	2	1
9										1	.	.	.
10											1	1	1
11												1	1
12													1
13													

NUMBER	13	14	15	16	17	18	19	20	21	22	23	24	25
S.P.	18	22	22	26	26	30	34	38	26	38	30	42	30
O.S.	32 256	40 320	18 432	20 160	8 064	24 192	2 016	8 064	96 768	2 016	32 256	4 032	18 432
LEVEL	90	98	98	100	102	106	108	108	110	110	112	112	114
DIM'N	248 064	854 658	413 440	1 026 000	1 072 512	3 197 700	2 212 056	2 154 240	2 825 604	2 496 144	3 292 016	4 625 208	2 116 296
WEIGHT	10000100<0	01001000<0	10000001<0	20001000<0	00110000<0	11010000<0	10200000<0	30010000<0	01000010<0	02100000<0	20000010<0	21100000<0	01000000<1
0	224	665	224	740	784	2 125	1 428	1 312	1 170	1 540	1 260	2 716	644
1	76	225	88	255	258	711	465	444	477	500	523	882	285
2	14	70	14	90	105	327	245	228	126	280	154	532	70
3	.	.	.	20	.	69	56	96	.	84	28	224	.
4	24	66	36	75	72	195	123	120	192	129	213	225	131
5	4	20	5	25	28	87	62	60	48	71	57	132	30
6	6	14	16	15	14	35	21	20	75	21	84	35	65
7	.	5	.	5	12	36	33	24	9	40	9	75	5
8	.	.	.	5	.	18	13	24	.	21	9	54	.
9	.	.	8	.	.	.	.	.	27	.	28	.	36
10	.	.	.	.	.	6	7	6	.	14	.	28	.
11	.	.	.	.	.	.	.	6	.	.	.	14	.
12	.	.	.	.	.	.	.	.	.	.	.	.	.
13	1	4	2	5	5	15	10	10	18	11	21	20	14

NUMBER	26	27	28	29	30	31	32	33	34	35	36	37	38
S.P.	50	54	34	58	30	74	98	34	34	42	34	38	42
O.S.	288	2 016	4 608	288	107 520	288	18	20 160	161 280	32 256	129 024	120 960	40 320
LEVEL	114	114	116	116	118	118	120	122	122	124	126	128	130
DIM'N	2 232 576	2 451 570	2 441 880	2 604 672	7 938 048	1 444 608	346 104	9 021 390	20 785 050	13 108 480	14 470 400	44 602 272	37 363 500
WEIGHT	13000000<0	40100000<0	20000000<1	32000000<0	00100100<0	51000000<0	70000000<0	00011000<0	11000100<0	30000100<0	00100001<0	10101000<0	02001000<0
0	1 264	1 323	686	1 392	2 816	720	165	2 975	6 645	3 808	3 360	12 960	10 255
1	396	427	309	432	1 152	216	45	1 200	2 774	1 616	1 540	5 340	4 255
2	269	287	84	320	400	180	45	490	1 058	672	490	2 410	1 995
3	128	175	14	200	.	144	45	.	176	224	.	400	455
4	96	105	144	102	448	48	9	454	1 090	640	712	2 022	1 610
5	64	70	35	74	152	40	9	180	398	250	220	885	735
6	14	15	73	14	160	6	1	154	385	224	334	666	520
7	44	42	5	53	48	32	9	75	124	76	60	375	320
8	30	42	5	45	.	32	9	.	62	76	.	140	165
9	.	.	45	.	48	.	.	42	105	56	160	162	120
10	23	21	.	31	.	24	9	.	14	14	.	50	70
11	8	21	.	23	.	24	9	.	.	14	.	.	.
12	.	7	.	8	.	16	9	.	.	.	.	.	.
13	9	10	16	10	52	5	1	58	135	84	101	279	230

NUMBER	39	40	41	42	43	44	45	46	47	48	49	50	51
S.P.	42	38	46	46	38	46	58	54	50	42	42	54	42
O.S.	8 064	129 024	80 640	18 432	161 280	24 192	20 160	672	24 192	32 256	64 512	24 192	483 840
LEVEL	130	130	132	132	134	134	134	136	136	138	138	138	140
DIM'N	27 159 132	36 581 688	65 523 780	22 573 824	38 663 100	65 780 736	30 107 000	20 231 904	94 140 288	42 418 944	31 274 152	106 375 680	166 337 248
WEIGHT	10020000<0	11000001<0	21001000<0	30000001<0	00010010<0	01110000<0	40001000<0	00300000<0	20110000<0	00001100<0	00010000<1	12010000<0	10100010<0
0	7 595	7 622	17 010	4 256	7 630	17 024	7 060	4 928	23 184	7 712	4 788	25 056	28 672
1	3 084	3 568	7 110	2 028	3 525	6 948	2 980	1 991	9 492	3 540	2 357	10 260	13 580
2	1 519	1 232	3 500	756	1 330	3 640	1 580	1 134	5 208	1 490	854	5 814	5 530
3	259	182	1 120	224	.	896	740	336	1 680	.	.	2 112	784
4	1 143	1 679	2 690	968	1 608	2 548	1 125	712	3 472	1 592	1 175	3 720	6 310
5	544	547	1 280	332	600	1 303	575	398	1 846	660	422	2 061	2 493
6	364	805	861	472	721	784	350	210	1 056	694	595	1 104	2 860
7	267	148	555	88	215	672	255	227	966	280	141	1 116	907
8	87	74	390	88	.	312	255	113	570	.	.	732	332
9	84	399	189	240	315	168	70	42	216	288	308	216	1 250
10	42	14	140	14	.	168	90	70	273	.	.	390	98
11	.	.	70	14	.	.	90	.	105	.	.	132	.
12	.	.	.	.	.	.	20	.	.	.	.	.	.
13	165	249	396	150	264	386	175	113	540	281	214	594	1 095

1	.	1	2	6	5	4	3	6	3	11	2	6	6	2	7	16	4	1	23	40	24	26	114	96	77	62	166	36	93	192	75	63	272	116	70	312	384	14
	1	.	.	.	.	.	6	.	7	.	8	.	.	9	.	14	.	.	14	35	21	48	63	50	35	119	84	72	112	77	35	21	105	112	112	111	460	15
		1	.	3	2	4	.	3	3	8	.	4	6	2	6	.	4	1	.	20	24	.	42	50	25	31	116	36	.	88	75	30	160	.	.	204	138	16
			1	2	3	1	.	3	.	6	.	4	3	.	5	4	3	1	10	8	4	5	45	35	39	10	60	5	30	96	25	36	138	50	20	156	117	17
				1	1	1	.	2	.	4	.	3	3	.	4	.	3	1	.	4	4	.	15	20	11	5	40	5	.	46	25	19	74	.	.	105	39	18
					1	.	.	1	.	2	.	2	1	.	3	.	2	1	.	.	.	.	5	5	6	.	10	.	.	24	5	13	36	.	.	48	9	19
						1	.	.	.	2	.	1	3	.	3	.	3	1	.	.	4	.	.	.	.	.	20	5	.	.	25	.	28	.	.	36	.	20
							1	.	1	.	1	.	.	1	.	4	.	.	5	10	6	12	24	20	15	28	34	16	38	36	15	11	50	44	37	56	155	21
								1	.	1	.	2	1	.	2	.	2	1	.	.	.	.	.	5	.	.	5	.	.	12	5	7	12	.	.	36	.	22
									1	.	.	.	.	1	.	.	.	.	.	5	6	.	9	10	5	14	24	16	.	16	15	5	30	.	.	36	55	23
										1	.	1	1	.	2	.	2	1	.	.	.	.	.	.	.	.	5	.	.	.	5	.	12	.	.	18	.	24
											1	.	.	1	.	.	.	.	.	.	.	6	.	.	.	14	.	8	14	.	.	.	.	14	21	.	55	25
												1	.	.	1	.	1	1	.	.	.	.	.	.	.	.	.	.	.	.	.	.	.	.	.	6	.	26
													1	.	1	.	2	1	.	.	.	.	.	.	.	.	.	.	.	.	5	.	.	.	.	.	.	27
														1	.	.	.	.	.	.	.	.	.	.	.	7	.	8	.	.	.	.	.	.	.	.	20	28
															1	.	1	1	.	.	.	.	.	.	.	.	.	.	.	.	.	.	.	.	.	.	.	29
																1	.	.	2	2	1	2	9	7	7	4	12	2	12	17	5	6	24	18	10	27	45	30
																	1	1	.	.	.	.	.	.	.	.	.	.	.	.	.	.	.	.	.	.	.	31
																		1	.	.	.	.	.	.	.	.	.	.	.	.	.	.	.	.	.	.	.	32
																			1	.	.	.	3	2	4	.	3	.	3	8	1	3	12	8	2	12	9	33
																				1	1	.	3	4	2	2	8	2	.	8	5	3	13	.	.	18	15	34
																					1	.	.	.	.	.	4	2	.	.	5	.	5	.	.	6	.	35
																						1	.	.	.	2	.	1	4	.	.	.	.	5	6	.	15	36
																							1	1	1	.	2	.	.	4	1	2	6	.	.	8	3	37
																								1	.	.	1	.	.	2	1	1	2	.	.	6	.	38
																									1	.	.	.	.	2	.	1	3	.	.	3	.	39
																										1	.	1	.	.	.	.	.	.	.	.	5	40
																											1	.	.	.	1	.	2	.	.	3	.	41
																												1	.	.	.	.	.	.	.	.	.	42
																													1	.	.	.	.	2	1	.	3	43
																														1	.	1	1	.	.	2	.	44
																															1	.	.	.	.	.	.	45
																																1	.	.	.	.	.	46
																																	1	.	.	1	.	47
																																		1	.	.	.	48
																																			1	.	.	49
																																				1	.	50
																																					1	51

Table 3. *(continued)*

ALGEBRA C10, REPRESENTATIONS OF CLASS 0

NUMBER	0	1	2	3	4	5	6	7	8	9	10	11	12	13	14	15	16	17	18	19	20	21	22	23	24	25
S.P.	0	4	8	8	12	16	20	32	12	16	16	20	20	24	24	28	36	36	20	40	52	72	24	24	28	28
O.S.	1	180	20	3 360	2 880	180	360	20	13 440	40 320	11 520	1 024	20 160	960	13 440	5 760	180	2 880	107 520	360	360	20	46 080	201 600	80 640	80 640
LEVEL	1	37	39	65	71	73	75	77	85	95	97	101	101	103	103	107	109	109	111	111	113	115	119	121	123	127
DIM'N	1	189	210	4 655	17 556	13 110	21 735	8 855	33 915	248 710	87 210	58 786	574 770	361 284	706 629	1 351 680	498 750	978 880	1 119 195	987 525	664 125	177 100	1 662 804	4 916 250	5 720 330	9 085 230
WEIGHT	000000000<0	010000000<0	200000000<0	000100000<0	101000000<0	020000000<0	210000000<0	400000000<0	000001000<0	100010000<0	000000010<0	000000000<1	010100000<0	002000000<0	200100000<0	111000000<0	030000000<0	301000000<0	100000100<0	220000000<0	410000000<0	600000000<0	100000001<0	010001000<0	200001000<0	001010000<0
0	1	9	10	35	116	90	135	55	75	430	90	42	910	564	1 029	1 920	690	1 280	795	1 305	825	220	708	2 850	3 050	4 770
1		1	1	7	24	17	27	10	20	123	28	14	259	156	301	544	190	368	272	360	225	55	266	1 005	1 100	1 685
2			1	.	8	9	18	10	.	27	.	.	91	72	126	288	117	224	48	252	180	55	42	285	360	621
3				1	3	2	3	1	5	30	9	5	60	35	70	120	40	80	90	75	45	10	102	330	365	540
4					1	1	2	1	.	6	.	.	21	15	28	62	25	48	14	51	36	10	14	90	110	195
5						1	1	1	.	.	.	.	7	8	7	32	17	24	.	35	27	10	.	20	20	69
6							1	1	.	.	.	.	.	.	7	16	9	24	.	26	27	10	.	.	20	.
7								1	.	.	.	.	.	.	.	.	.	8	.	9	18	10	.	.	.	.
8									1	5	3	2	9	5	10	16	5	10	28	9	5	1	41	95	105	147
9										1	.	.	3	2	4	8	3	6	4	6	4	1	5	25	30	51
10											1	1	.	.	.	.	.	.	7	.	.	.	18	20	21	28
11												1	.	.	.	.	.	.	.	.	.	.	9	.	.	.
12													1	1	1	4	2	3	.	4	3	1	.	5	5	18
13														1	.	2	1	1	.	3	2	1	.	.	.	6
14															1	2	1	3	.	3	3	1	.	.	5	.
15																1	1	1	.	2	2	1	.	.	.	.

NUMBER	26	27	28	29	30	31	32	33	34	35	36	37	38	39	40	41	42	43	44	45	46	47	48	49	50	51
S.P.	32	32	28	40	36	32	32	40	44	36	44	48	56	32	52	64	60	36	68	76	36	44	80	40	36	100
O.S.	3 360	161 280	322 560	40 320	40 320	92 160	46 080	20 160	2 880	10 240	40 320	2 880	13 440	537 600	5 760	180	5 760	645 120	360	2 880	201 600	107 520	360	8 064	430 080	360
LEVEL	129	131	133	133	135	135	137	137	139	139	139	141	141	143	143	145	145	147	147	147	149	149	149	151	151	151
DIM'N	5 206 740	24 330 240	13 520 780	15 444 891	32 376 456	15 444 891	9 360 540	28 863 450	24 515 700	10 623 470	51 074 375	35 985 600	24 089 625	47 805 615	50 106 420	12 442 815	52 715 520	119 848 960	27 255 690	21 871 850	78 738 660	73 220 224	23 362 020	42 868 826	75 508 356	10 758 825
WEIGHT	000200000<0	110010000<0	010000010<0	300010000<0	101100000<0	200000010<0	010000000<1	020100000<0	012000000<0	200000000<1	210100000<0	202000000<0	400100000<0	001000100<0	121000000<0	040000000<0	311000000<0	110000100<0	230000000<0	501000000<0	000101000<0	300000100<0	420000000<0	000020000<0	001000001<0	610000000<0
0	2 660	11 520	4 140	6 651	14 616	4 347	2 100	12 390	10 260	2 190	20 755	14 400	8 925	11 955	19 380	4 635	19 200	26 880	9 750	7 370	17 700	14 848	7 920	9 254	12 252	3 465
1	924	4 160	1 646	2 446	5 208	1 764	882	4 438	3 636	938	7 504	5 112	3 255	4 875	6 864	1 598	6 816	11 200	3 378	2 600	7 240	6 304	2 715	3 759	5 398	1 155
2	392	1 728	414	1 134	2 520	504	210	2 247	2 016	252	4 060	3 024	1 932	1 584	4 224	1 098	4 512	4 032	2 433	1 880	2 760	2 496	2 115	1 575	1 638	990
3	291	1 344	650	795	1 638	705	378	1 392	1 117	407	2 352	1 566	1 015	1 930	2 079	468	2 048	4 496	982	764	2 835	2 560	774	1 458	2 390	315
4	119	540	154	351	770	182	84	693	608	98	1 239	899	588	610	1 259	318	1 328	1 540	697	548	1 065	944	594	595	700	270
5	56	192	28	123	364	28	14	343	340	14	609	520	301	176	764	226	832	448	496	368	400	272	450	259	182	225
6	.	96	.	123	140	28	.	175	164	14	434	320	301	.	508	153	688	224	388	368	.	272	405	.	.	225
7	.	.	.	27	.	.	.	.	.	.	91	72	126	.	108	45	288	.	162	224	.	48	252	.	.	180
8	77	360	255	210	420	279	169	350	273	184	585	378	245	730	490	105	472	1 712	217	168	1 040	980	165	526	1 071	63
9	30	140	57	90	190	66	35	170	144	40	300	210	140	224	290	70	300	560	151	120	380	340	125	208	303	54
10	14	64	98	35	70	108	82	56	42	91	90	56	35	255	70	14	64	592	28	21	342	336	20	168	490	7
11	.	.	35	.	.	36	45	.	.	55	.	.	.	75	.	.	.	160	.	.	90	84	.	42	231	.
12	13	48	9	30	87	9	5	81	79	5	144	117	70	62	172	48	184	152	106	80	140	90	93	87	74	45
13	7	12	.	6	42	.	.	35	45	.	63	69	28	14	100	33	110	28	75	48	50	14	69	42	14	36
14	.	24	.	30	33	9	.	42	37	5	102	72	70	.	114	33	152	76	82	80	.	90	84	.	.	45
15	.	6	.	6	14	.	.	21	23	.	42	38	28	.	70	25	86	14	59	48	.	14	60	.	.	36

1	.	.	1	1	1	.	.	.	.	.	.	.	.	.	.	.	7	8	.	7	8	7	.	32	17	32	.	35	24	.	.	35	.	.	27	16
	1	.	1	2	1	.	.	.	.	.	.	.	6	.	.	.	.	.	.	21	15	28	.	24	9	62	.	34	48	.	14	51	.	.	36	17
		1	.	.	.	2	5	6	9	5	24	21	15	30	24	16	26	21	18	45	30	20	75	40	9	40	186	19	15	119	112	15	63	135	6	18
			1	1	1	.	.	.	.	.	.	.	.	.	.	.	.	.	.	7	8	7	.	16	9	32	.	26	24	.	.	35	.	.	27	19
				1	1	.	.	.	.	.	.	.	.	.	.	.	.	.	.	.	.	7	.	.	.	16	.	9	24	.	.	26	.	.	27	20
					1	.	.	.	.	.	.	.	.	.	.	.	.	.	.	.	.	.	.	.	.	.	.	.	8	.	.	9	.	.	18	21
						1	.	.	.	.	.	7	.	.	8	9	.	.	10	.	.	.	20	.	.	.	48	.	.	28	28	.	14	63	.	22
							1	1	3	2	8	3	5	13	3	2	12	11	2	21	16	10	20	23	6	24	48	13	10	42	28	11	25	31	5	23
								1	.	.	4	.	5	5	3	.	6	5	2	15	10	10	.	15	4	20	24	10	10	.	28	10	.	.	5	24
									1	1	2	.	1	6	.	.	5	6	.	9	9	4	4	13	4	14	8	9	6	15	4	8	11	5	4	25
										1	.	.	.	3	.	.	2	3	.	3	6	1	.	7	3	8	.	6	3	5	.	6	6	.	3	26
											1	.	1	2	.	.	3	3	.	6	5	4	.	9	3	11	4	7	6	.	4	7	.	.	4	27
												1	.	.	1	1	.	.	1	.	.	.	5	.	.	.	12	.	.	9	7	.	5	14	.	28
													1	.	.	.	.	.	.	3	2	4	.	3	1	8	.	4	6	.	4	6	.	.	4	29
														1	.	.	1	2	.	2	3	1	.	5	2	6	.	5	3	.	.	5	.	.	3	30
															1	.	.	.	1	.	.	.	.	.	.	.	6	.	.	.	7	.	.	.	.	31
																1	.	.	1	.	.	.	.	.	.	.	.	.	.	.	.	.	.	7	.	32
																	1	1	.	1	1	1	.	4	2	4	.	4	3	.	.	4	.	.	3	33
																		1	.	.	1	.	.	2	1	2	.	3	1	.	.	3	.	.	2	34
																			1	.	.	.	.	.	.	.	.	.	.	.	.	.	.	.	.	35
																				1	1	1	.	2	1	4	.	3	3	.	.	4	.	.	3	36
																					1	.	.	1	1	2	.	2	1	.	.	3	.	.	2	37
																						1	.	.	.	2	.	1	3	.	.	3	.	.	3	38
																							1	.	.	.	2	.	.	3	1	.	2	2	.	39
																								1	1	1	.	2	1	.	.	2	.	.	2	40
																									1	.	.	1	.	.	.	1	.	.	1	41
																										1	.	1	1	.	.	2	.	.	2	42
																											1	.	.	.	1	.	.	.	.	43
																												1	.	.	.	1	.	.	1	44
																													1	.	.	1	.	.	2	45
																														1	.	.	1	.	.	46
																															1	.	.	.	.	47
																																1	.	.	1	48
																																	1	.	.	49
																																		1	.	50
																																			1	51

Table 3. *(continued)*

ALGEBRA C10, REPRESENTATIONS OF CLASS 1

NUMBER	0	1	2	3	4	5	6	7	8	9	10	11	12
S.P.	2	6	10	18	10	14	18	22	14	26	34	50	18
O.S.	20	960	360	20	8 064	13 440	2 880	2 880	15 360	360	360	20	5 120
LEVEL	20	52	56	58	76	84	88	90	92	92	94	96	100
DIM'N	20	1 120	2 640	1 540	14 364	77 616	115 920	153 824	62 016	143 640	133 056	42 504	90 440
	1	0	1	3	0	1	0	2	0	1	3	5	0
	0	0	1	0	0	0	1	0	0	2	1	0	0
	0	1	0	0	0	0	1	1	0	0	0	0	0
	0	0	0	0	0	1	0	0	0	0	0	0	0
	0	0	0	0	1	0	0	0	0	0	0	0	0
	0	0	0	0	0	0	0	0	0	0	0	0	0
	0	0	0	0	0	0	0	0	1	0	0	0	0
	0	0	0	0	0	0	0	0	0	0	0	0	0
	0	0	0	0	0	0	0	0	0	0	0	0	1
	<	<	<	<	<	<	<	<	<	<	<	<	<
WEIGHT	0	0	0	0	0	0	0	0	0	0	0	0	0
0	1	8	18	10	27	126	180	224	48	207	180	55	42
1		1	2	1	6	28	38	48	14	42	36	10	14
2			1	1	.	7	16	24	.	26	27	10	.
3				1	.	.	.	8	.	9	18	10	.
4					1	4	5	6	4	5	4	1	5
5						1	2	3	.	3	3	1	.
6							1	1	.	2	2	1	.
7								1	.	1	2	1	.
8									1	.	.	.	2
9										1	1	1	.
10											1	1	.
11												1	.
12													1
13													

NUMBER	13	14	15	16	17	18	19	20	21	22	23	24	25
S.P.	18	22	26	26	22	26	30	34	38	38	42	26	50
O.S.	80 640	80 640	40 320	13 440	92 160	10 240	40 320	2 880	13 440	2 880	5 760	322 560	360
LEVEL	104	112	114	116	116	120	120	122	122	124	126	128	128
DIM'N	601 920	1 957 760	2 322 540	2 380 224	1 591 744	1 085 280	6 963 880	4 729 536	4 614 720	5 266 800	9 660 000	9 360 540	4 564 560
	1	0	2	0	1	1	1	1	3	0	2	0	1
	0	1	0	0	0	0	1	0	0	2	1	1	3
	0	0	0	1	0	0	0	2	0	1	1	0	0
	0	0	0	1	0	0	1	0	1	0	0	0	0
	0	1	1	0	0	0	0	0	0	0	0	0	0
	1	0	0	0	0	0	0	0	0	0	0	0	0
	0	0	0	0	0	0	0	0	0	0	0	1	0
	0	0	0	0	1	0	0	0	0	0	0	0	0
	0	0	0	0	0	0	0	0	0	0	0	0	0
	<	<	<	<	<	<	<	<	<	<	<	<	<
WEIGHT	0	0	0	0	0	1	0	0	0	0	0	0	0
0	360	1 024	1 134	1 176	504	252	3 157	2 088	1 932	2 232	3 912	2 331	1 788
1	110	312	351	350	182	98	952	615	588	655	1 148	868	508
2	20	96	123	140	28	14	434	320	301	364	688	224	343
3	.	.	27	.	.	.	91	72	126	108	288	.	162
4	30	80	90	86	66	40	230	144	140	150	260	310	110
5	5	24	30	33	9	5	102	72	70	82	152	76	73
6	.	6	6	14	.	.	42	38	28	46	86	14	50
7	.	.	6	.	.	.	21	15	28	24	62	.	34
8	6	14	15	14	24	18	35	21	20	21	35	102	14
9	.	.	.	.	.	.	7	8	7	16	32	.	26
10	.	.	.	.	.	.	.	.	7	.	16	.	9
11	.	.	.	.	.	.	.	.	.	.	.	.	.
12	.	.	.	.	8	10	.	.	.	.	.	27	.
13	1	4	5	5	3	2	15	10	10	11	20	24	9

NUMBER	26	27	28	29	30	31	32	33	34	35	36	37	38
S.P.	54	30	58	74	98	30	30	34	34	34	42	38	34
O.S.	2 880	107 520	360	360	20	268 800	184 320	46 080	40 320	403 200	80 640	241 920	645 120
LEVEL	128	130	130	132	134	136	136	138	140	140	142	146	148
DIM'N	5 016 000	10 767 680	5 249 244	2 841 696	657 800	24 227 280	14 263 680	16 224 936	26 380 816	62 177 500	38 536 960	126 758 520	70 605 216
	4	2	3	5	7	0	0	2	0	1	3	1	0
	0	0	2	1	0	0	1	0	0	1	0	0	0
	1	0	0	0	0	1	0	0	0	0	0	1	1
	0	0	0	0	0	0	0	0	1	0	0	0	0
	0	0	0	0	0	0	0	0	1	0	0	1	0
	0	0	0	0	0	1	0	0	0	1	1	0	0
	0	1	0	0	0	0	0	0	0	0	0	0	0
	0	0	0	0	0	0	0	0	0	0	0	0	1
	0	0	0	0	0	0	1	1	0	0	0	0	0
	<	<	<	<	<	<	<	<	<	<	<	<	<
WEIGHT	0	0	0	0	0	0	0	0	0	0	0	0	0
0	1 880	2 496	1 950	990	220	5 250	2 304	2 436	5 348	12 260	6 960	22 923	8 964
1	548	944	549	270	55	1 960	952	1 022	1 974	4 655	2 680	8 604	3 810
2	368	272	405	225	55	665	224	266	784	1 745	1 100	3 810	1 170
3	224	48	252	180	55	.	.	42	.	285	360	621	.
4	120	340	116	54	10	680	400	434	669	1 625	940	2 910	1 604
5	80	90	84	45	10	225	88	102	258	585	365	1 251	477
6	48	14	60	36	10	70	14	14	105	180	110	522	126
7	48	14	51	36	10	.	.	14	.	90	110	195	.
8	15	112	14	6	1	208	174	191	196	490	280	828	666
9	24	.	35	27	10	.	.	.	.	20	20	69	.
10	24	.	26	27	10	.	.	.	.	.	20	.	.
11	8	.	9	18	10	.	.	.	.	.	.	.	.
12	.	28	.	.	.	48	80	90	42	105	56	162	270
13	10	28	10	5	1	66	36	41	72	170	105	342	192

NUMBER	39	40	41	42	43	44	45	46	47	48	49	50	51
S.P.	42	42	46	38	46	38	58	54	50	46	38	42	54
O.S.	80 640	13 440	161 280	122 880	40 320	645 120	40 320	960	40 320	92 160	537 600	92 160	40 320
LEVEL	148	148	150	152	152	152	152	154	154	154	156	156	156
DIM'N	104 720 000	75 373 760	181 704 600	49 922 880	178 516 800	174 594 420	81 719 000	53 488 800	252 806 400	105 719 040	164 745 504	121 938 960	281 801 520
	0	1	2	0	0	1	4	0	2	3	0	1	1
	2	0	1	0	1	1	0	0	0	0	0	1	2
	0	0	0	1	1	0	0	3	1	0	0	0	0
	0	2	0	0	1	0	0	0	1	0	1	0	1
	1	0	1	0	0	0	1	0	0	0	0	0	0
	0	0	0	0	0	0	0	0	0	0	0	0	0
	0	0	0	0	0	0	0	0	0	0	1	0	0
	0	0	0	0	0	1	0	0	0	1	0	0	0
	0	0	0	0	0	0	0	0	0	0	0	0	0
	<	<	<	<	<	<	<	<	<	<	<	<	<
WEIGHT	0	0	0	1	0	0	0	0	0	0	0	1	0
0	17 984	13 160	29 646	4 872	29 064	20 079	12 141	8 244	39 312	11 088	18 072	10 758	42 084
1	6 786	4 879	11 256	2 198	10 808	8 688	4 646	3 036	14 658	4 872	7 731	4 942	15 680
2	3 136	2 352	5 472	644	5 572	2 906	2 446	1 704	7 924	1 764	2 816	1 568	8 785
3	704	392	1 728	.	1 344	414	1 134	492	2 520	504	.	210	3 150
4	2 290	1 620	3 795	1 010	3 550	3 705	1 560	975	4 800	2 100	3 216	2 305	5 090
5	1 032	756	1 788	285	1 788	1 173	795	537	2 520	705	1 152	687	2 792
6	444	364	768	70	910	308	351	303	1 302	182	400	168	1 498
7	228	119	540	.	420	154	351	149	770	182	.	84	980
8	640	448	1 050	480	952	1 554	420	252	1 272	888	1 284	1 115	1 320
9	96	56	192	.	224	28	123	92	364	28	.	14	518
10	.	.	96	.	.	.	123	.	140	28	.	.	175
11	.	.	.	.	.	.	27	.	.	.	.	.	.
12	120	84	189	240	168	630	70	42	216	360	480	579	216
13	280	200	480	131	463	468	210	134	645	279	448	313	705

1	1	2	.	.	6	5	4	6	11	4	6	6	4	7	4	1	20	5	5	28	50	30	138	48	116	92	200	30	228	114	90	74	322	66	152	70	368	14
	1	.	.	.	3	2	4	3	8	.	4	6	4	6	4	1	.	.	5	.	25	30	51	.	60	30	140	.	104	57	90	35	190	66	.	35	240	15
		1	.	.	2	3	1	3	6	.	4	3	.	5	3	1	5	.	.	12	10	5	54	9	42	46	72	5	113	18	30	42	162	9	48	10	183	16
			1	1	.	.	.	.	.	6	.	.	7	.	.	.	14	16	18	14	35	21	63	75	50	35	84	65	77	182	35	21	105	108	160	155	111	17
				1	.	.	.	.	.	.	.	.	.	.	.	.	.	8	9	.	.	.	.	27	.	.	.	36	.	63	.	.	.	36	48	90	.	18
					1	1	1	2	4	.	3	3	.	4	3	1	.	.	.	.	5	5	18	.	24	13	48	.	54	9	30	22	87	9	.	5	123	19
						1	.	1	2	.	2	1	.	3	2	1	.	.	.	.	.	.	6	.	6	7	12	.	28	.	6	15	42	.	.	.	56	20
							1	.	2	.	1	3	.	3	3	1	.	.	.	.	.	5	.	.	.	.	24	.	.	.	30	.	33	9	.	.	42	21
								1	1	.	2	1	.	2	2	1	.	.	.	.	.	.	.	.	6	.	6	.	14	.	6	8	14	.	.	.	42	22
									1	.	1	1	.	2	2	1	.	.	.	.	.	.	.	.	.	.	6	.	.	.	6	.	14	.	.	.	21	23
										1	.	.	1	.	.	.	4	2	2	5	10	6	24	18	20	15	34	14	36	42	15	11	50	24	52	32	56	24
											1	.	.	1	1	1	.	.	.	.	.	.	.	.	.	.	.	.	.	.	.	.	.	.	.	.	7	25
												1	.	1	2	1	.	.	.	.	.	.	.	.	.	.	.	.	.	.	6	.	.	.	.	.	.	26
													1	.	.	.	.	.	2	.	5	6	9	.	10	5	24	.	16	21	15	5	30	24	.	16	36	27
														1	1	1	.	.	.	.	.	.	.	.	.	.	.	.	.	.	.	.	.	.	.	.	.	28
															1	1	.	.	.	.	.	.	.	.	.	.	.	.	.	.	.	.	.	.	.	.	.	29
																1	.	.	.	.	.	.	.	.	.	.	.	.	.	.	.	.	.	.	.	.	.	30
																	1	.	.	2	2	1	9	3	7	7	12	2	17	6	5	6	24	3	16	4	27	31
																		1	1	.	.	.	.	6	.	.	.	8	.	14	.	.	.	8	14	18	.	32
																			1	.	.	.	.	.	.	.	.	.	.	7	.	.	.	8	.	9	.	33
																				1	.	.	3	.	2	4	3	.	8	.	1	3	12	.	4	.	12	34
																					1	1	3	.	4	2	8	.	8	3	5	3	13	3	.	2	18	35
																						1	.	.	.	.	4	.	.	.	5	.	5	3	.	.	6	36
																							1	.	1	1	2	.	4	.	1	2	6	.	.	.	8	37
																								1	.	.	.	1	.	2	.	.	.	1	4	2	.	38
																									1	.	1	.	2	.	1	1	2	.	.	.	6	39
																										1	.	.	2	.	.	1	3	.	.	.	3	40
																											1	.	.	.	1	.	2	.	.	.	3	41
																												1	.	.	.	.	.	.	.	2	.	42
																													1	.	.	1	1	.	.	.	2	43
																														1	.	.	.	1	.	1	.	44
																															1	.	.	.	.	.	.	45
																																1	.	.	.	.	.	46
																																	1	.	.	.	1	47
																																		1	.	.	.	48
																																			1	.	.	49
																																				1	.	50
																																					1	51

Table 3. *(continued)*

ALGEBRA C11, REPRESENTATIONS OF CLASS 0

0	4	8	8	12	16	20	32	12	16	16	20	24	24	28	20	36	36	40	52	20	72	24	24	28	28	S.P.
1	220	22	5 280	3 960	220	440	22	29 568	73 920	42 240	31 680	1 320	21 120	7 920	11 264	220	3 960	440	440	295 680	22	443 520	253 440	177 408	147 840	O.S.
1	41	43	73	79	81	83	85	97	107	113	113	115	115	119	121	121	121	123	125	127	127	137	139	139	143	LEVEL
1	230	253	7 084	26 082	19 250	31 625	12 650	67 298	471 086	245 157	1 057 540	655 500	1 288 000	2 416 128	326 876	876 645	1 726 725	1 721 412	1 138 500	2 860 165	296 010	11 827 750	6 619 239	13 627 845	21 074 900	DIM'N
0	0	2	0	1	0	2	4	0	1	0	0	0	2	1	0	0	3	2	4	1	6	0	1	2	0	
0	1	0	0	0	2	1	0	0	0	0	1	0	0	1	0	3	0	2	1	0	0	1	0	0	0	
0	0	0	0	1	0	0	0	0	0	0	0	2	0	1	0	0	1	0	0	0	0	0	0	0	1	
0	0	0	1	0	0	0	0	0	0	0	1	0	1	0	0	0	0	0	0	0	0	0	0	0	0	
0	0	0	0	0	0	0	0	0	1	0	0	0	0	0	0	0	0	0	0	0	0	0	0	0	1	
0	0	0	0	0	0	0	0	1	0	0	0	0	0	0	0	0	0	0	0	0	0	1	0	1	0	
0	0	0	0	0	0	0	0	0	0	0	0	0	0	0	0	0	0	0	0	1	0	0	0	0	0	
0	0	0	0	0	0	0	0	0	0	1	0	0	0	0	0	0	0	0	0	0	0	0	0	0	0	
0	0	0	0	0	0	0	0	0	0	0	0	0	0	0	0	0	0	0	0	0	0	0	1	0	0	
0	0	0	0	0	0	0	0	0	0	0	0	0	0	0	1	0	0	0	0	0	0	0	0	0	0	
<	<	<	<	<	<	<	<	<	<	<	<	<	<	<	<	<	<	<	<	<	<	<	<	<	<	
0	0	0	0	0	0	0	0	0	0	0	0	0	0	0	0	0	0	0	0	0	0	0	0	0	0	WEIGHT
0	1	2	3	4	5	6	7	8	9	10	11	12	13	14	15	16	17	18	19	20	21	22	23	24	25	NUMBER
1	10	11	44	144	110	165	66	110	616	165	1 276	780	1 440	2 640	132	935	1 749	1 760	1 100	1 375	286	4 730	1 683	5 049	7 700	0
	1	1	8	27	19	30	11	27	161	48	332	198	384	684	42	236	459	445	275	435	66	1 537	594	1 674	2 506	1
		1	.	9	10	20	11	.	35	.	116	90	160	360	.	145	279	310	220	75	66	430	90	540	910	2
			1	3	2	3	1	6	35	14	69	40	80	136	14	45	90	84	50	130	11	456	210	501	728	3
				1	1	2	1	.	7	.	24	17	32	70	.	28	54	57	40	20	11	123	28	150	259	4
					1	1	1	.	.	.	8	9	8	36	.	19	27	39	30	.	11	27	.	27	91	5
						1	1	.	.	.	.	.	8	18	.	10	27	29	30	.	11	.	.	27	.	6
							1	.	.	.	.	.	.	.	.	9	10	20	.	11	.	.	.	.	.	7
								1	5	4	9	5	10	16	5	5	10	9	5	35	1	115	75	126	175	8
									1	.	3	2	4	8	.	3	6	6	4	5	1	30	9	36	60	9
										1	.	.	.	.	2	.	.	.	.	7	.	20	27	21	28	10
											1	1	1	4	.	2	3	4	3	.	1	6	.	6	21	11
												1	.	2	.	1	1	3	2	.	1	.	.	.	7	12
													1	2	.	1	3	3	3	.	1	.	.	6	.	13

28	32	32	40	36	28	40	44	32	44	48	56	52	32	64	60	32	36	68	76	80	36	100	36	44	128	S.P.
22 528	5 280	295 680	73 920	63 360	1 182 720	31 680	3 960	337 920	63 360	3 960	21 120	7 920	506 880	220	7 920	1 478 400	112 640	440	3 960	440	1 774 080	440	443 520	295 680	22	O.S.
143	145	147	149	151	153	153	155	155	155	157	157	159	161	161	161	163	163	163	163	165	167	167	169	169	169	LEVEL
4 249 388	11 870 760	55 570 944	34 796 125	72 208 500	46 267 375	63 683 136	53 326 350	52 300 160	111 731 400	77 613 822	51 801 750	106 924 125	63 740 820	25 987 654	111 141 888	150 218 750	71 504 125	56 494 900	45 187 065	47 690 500	370 472 960	21 460 725	235 669 500	223 092 870	4 292 145	DIM'N
1	0	1	3	1	0	0	0	2	2	2	4	1	0	0	3	0	2	2	5	4	1	6	0	3	8	
0	0	1	0	0	1	2	1	0	1	0	0	2	1	4	1	0	0	3	0	2	1	1	0	0	0	
0	0	0	0	1	0	0	2	0	0	2	0	1	0	0	1	1	0	0	1	0	0	0	0	0	0	
0	2	0	0	1	0	1	0	0	1	0	1	0	0	0	0	0	0	0	0	0	0	0	1	0	0	
0	0	1	1	0	0	0	0	0	0	0	0	0	0	0	0	0	0	0	0	0	0	0	0	0	0	
0	0	0	0	0	0	0	0	0	0	0	0	0	0	0	0	0	0	0	0	0	0	0	1	0	0	
0	0	0	0	0	0	0	0	0	0	0	0	0	0	0	0	1	0	0	0	0	1	0	0	1	0	
0	0	0	0	0	1	0	0	1	0	0	0	0	0	0	0	0	0	0	0	0	0	0	0	0	0	
0	0	0	0	0	0	0	0	0	0	0	0	0	0	0	0	0	0	0	0	0	0	0	0	0	0	
0	0	0	0	0	0	0	0	0	0	0	0	0	1	0	0	0	1	0	0	0	0	0	0	0	0	
<	<	<	<	<	<	<	<	<	<	<	<	<	<	<	<	<	<	<	<	<	<	<	<	<	<	
1	0	0	0	0	0	0	0	0	0	0	0	0	0	0	0	0	0	0	0	0	0	0	0	0	0	WEIGHT
26	27	28	29	30	31	32	33	34	35	36	37	38	39	40	41	42	43	44	45	46	47	48	49	50	51	NUMBER
792	4 224	18 480	10 605	22 980	8 745	19 360	15 840	9 152	32 296	22 110	13 750	29 535	7 788	6 930	29 040	23 650	8 085	14 520	10 989	11 660	52 800	5 005	33 660	28 974	1 001	0
294	1 356	6 132	3 577	7 536	3 243	6 376	5 166	3 456	10 724	7 224	4 600	9 624	3 102	2 204	9 480	8 965	3 276	4 635	3 564	3 685	20 400	1 540	12 798	11 385	286	1
42	564	2 520	1 645	3 600	795	3 200	2 835	960	5 756	4 230	2 720	5 880	708	1 505	6 240	2 850	840	3 320	2 574	2 860	7 200	1 320	4 770	4 425	286	2
112	388	1 792	1 050	2 151	1 174	1 814	1 443	1 264	3 048	2 012	1 300	2 652	1 246	589	2 592	3 250	1 330	1 230	954	960	7 480	385	4 596	4 215	66	3
14	156	714	462	1 000	272	896	778	320	1 596	1 146	752	1 596	266	398	1 674	1 005	308	869	684	735	2 520	330	1 685	1 535	66	4
.	72	252	161	468	48	440	432	48	780	657	384	963	42	281	1 044	285	42	616	459	555	720	275	621	435	66	5
.	.	126	161	180	.	224	207	48	556	405	384	639	.	190	864	.	42	481	459	500	360	275	.	435	66	6
.	.	.	35	.	.	.	.	.	116	90	160	135	.	55	360	.	.	200	279	310	.	220	.	75	66	7
45	91	424	245	490	410	406	315	444	675	434	280	560	511	119	536	1 100	550	245	189	185	2 540	70	1 515	1 435	11	8
5	35	164	105	220	90	196	165	104	345	240	160	330	102	79	340	330	116	170	135	140	820	60	540	495	11	9
20	14	64	35	70	133	56	42	144	90	56	35	70	218	14	64	330	237	28	21	20	752	7	432	420	1	10
.	15	56	35	100	14	93	90	14	165	133	80	195	14	54	208	90	14	119	90	104	220	50	195	130	11	11
.	8	14	7	48	.	40	51	.	72	78	32	113	.	37	124	20	.	84	54	77	40	40	69	20	11	12
.	.	28	35	38	.	48	42	14	117	82	80	129	.	37	172	.	14	92	90	94	110	50	.	130	11	13

1	.	1	1	2	2	.	1	.	.	.	.	.	.	7	7	16	.	24	26	.	48	43	32	79	.	28	97	.	.	66	54	67	20	40	.	20	11	14
	1	.	.	.	.	.	.	.	9	.	.	11	.	.	.	.	35	.	.	36	.	.	.	.	99	.	.	75	110	.	.	.	160	.	90	84	.	15
		1	.	1	1	.	1	.	.	.	.	.	.	.	.	.	.	8	9	.	8	9	8	36	.	19	36	.	.	39	27	39	.	30	.	.	11	16
			1	1	2	.	1	.	.	.	.	.	.	.	7	.	.	.	.	.	24	17	32	27	.	10	70	.	.	38	54	57	.	40	.	20	11	17
				1	1	.	1	.	.	.	.	.	.	.	.	.	.	.	.	.	8	9	8	18	.	10	36	.	.	29	27	39	.	30	.	.	11	18
					1	.	1	.	.	.	.	.	.	.	.	.	.	.	.	.	.	.	8	.	.	.	18	.	.	10	27	29	.	30	.	.	11	19
						1	.	5	3	6	9	2	5	24	15	30	28	26	21	32	45	30	20	40	41	9	40	95	46	19	15	15	234	6	147	140	1	20
							1	.	.	.	.	.	.	.	.	.	.	.	.	.	.	.	.	.	.	.	.	.	.	.	9	10	.	20	.	.	11	21
								1	.	1	3	.	2	8	5	13	4	12	11	4	21	16	10	23	5	6	24	25	5	13	10	11	60	5	51	35	1	22
									1	.	.	1	.	.	.	.	7	.	.	8	.	.	.	.	18	.	.	20	20	.	.	.	48	.	28	28	.	23
										1	.	.	.	4	5	5	.	6	5	4	15	10	10	15	.	4	20	.	5	10	10	10	30	5	.	35	1	24
											1	.	1	2	1	6	.	5	6	.	9	9	4	13	.	4	14	5	.	9	6	8	10	4	18	5	1	25
												1	.	.	.	.	.	.	.	.	.	.	.	.	9	.	.	.	10	.	.	.	.	.	.	.	.	26
													1	.	.	3	.	2	3	.	3	6	1	7	.	3	8	.	.	6	3	6	.	3	6	.	1	27
														1	1	2	.	3	3	.	6	5	4	9	.	3	11	.	.	7	6	7	5	4	.	5	1	28
															1	.	.	.	.	.	3	2	4	3	.	1	8	.	.	4	6	6	.	4	.	5	1	29
																1	.	1	2	.	2	3	1	5	.	2	6	.	.	5	3	5	.	3	.	.	1	30
																	1	.	.	1	.	.	.	.	2	.	.	5	2	.	.	.	12	.	9	7	.	31
																		1	1	.	1	1	1	4	.	2	4	.	.	4	3	4	.	3	.	.	1	32
																			1	.	.	1	.	2	.	1	2	.	.	3	1	3	.	2	.	.	1	33
																				1	.	.	.	.	.	.	.	.	2	.	.	.	6	.	.	7	.	34
																					1	1	1	2	.	1	4	.	.	3	3	4	.	3	.	.	1	35
																						1	.	1	.	1	2	.	.	2	1	3	.	2	.	.	1	36
																							1	.	.	.	2	.	.	1	3	3	.	3	.	.	1	37
																								1	.	1	1	.	.	2	1	2	.	2	.	.	1	38
																									1	.	.	.	1	.	.	.	.	.	.	.	.	39
																										1	.	.	.	1	.	1	.	1	.	.	1	40
																											1	.	.	1	1	2	.	2	.	.	1	41
																												1	.	.	.	.	2	.	3	1	.	42
																													1	.	.	.	.	.	.	.	.	43
																														1	.	1	.	1	.	.	1	44
																															1	1	.	2	.	.	1	45
																																1	.	1	.	.	1	46
																																	1	.	.	1	.	47
																																		1	.	.	1	48
																																			1	.	.	49
																																				1	.	50
																																					1	51

Table 3. *(continued)*

ALGEBRA C11, REPRESENTATIONS OF CLASS 1

S.P.	2	6	10	18	10	14	18	22	26	34	14	50	18	18	22	22	26	26	22	30	34	38	38	42	26	50
O.S.	22	1 320	440	22	14 784	21 120	3 960	3 960	440	440	42 240	22	28 160	177 408	2 048	147 840	73 920	21 120	337 920	63 360	3 960	21 120	3 960	7 920	112 640	440
LEVEL	22	58	62	64	86	94	98	100	102	104	106	106	118	118	122	126	128	130	134	134	136	136	138	140	142	142
DIM'N	22	1 518	3 520	2 024	24 794	129 536	189 750	249 480	230 230	210 496	144 210	65 780	326 876	1 311 552	208 012	4 091 010	4 807 000	4 857 600	4 922 368	13 997 984	9 373 650	9 152 000	10 328 472	18 785 250	6 656 384	8 727 488
	1	0	1	3	0	1	0	2	1	3	0	5	0	1	0	0	2	0	1	1	1	3	0	2	1	1
	0	0	1	0	0	0	1	0	2	1	0	0	0	0	0	1	0	0	0	1	0	0	2	1	0	3
	0	1	0	0	0	0	1	1	0	0	0	0	0	0	0	0	0	1	0	0	2	0	1	1	0	0
	0	0	0	0	0	1	0	0	0	0	0	0	0	0	0	0	0	1	0	1	0	1	0	0	0	0
	0	0	0	0	1	0	0	0	0	0	0	0	0	0	0	1	1	0	0	0	0	0	0	0	0	0
	0	0	0	0	0	0	0	0	0	0	0	0	0	1	0	0	0	0	0	0	0	0	0	0	0	0
	0	0	0	0	0	0	0	0	0	0	1	0	0	0	0	0	0	0	0	0	0	0	0	0	0	0
	0	0	0	0	0	0	0	0	0	0	0	0	0	0	0	0	0	0	1	0	0	0	0	0	0	0
	0	0	0	0	0	0	0	0	0	0	0	0	1	0	0	0	0	0	0	0	0	0	0	0	0	0
	0	0	0	0	0	0	0	0	0	0	0	0	0	0	0	0	0	0	0	0	0	0	0	0	1	0
	<	<	<	<	<	<	<	<	<	<	<	<	<	<	<	<	<	<	<	<	<	<	<	<	<	<
WEIGHT	0	0	0	0	0	0	0	0	0	0	0	0	0	0	1	0	0	0	0	0	0	0	0	0	0	0
NUMBER	0	1	2	3	4	5	6	7	8	9	10	11	12	13	14	15	16	17	18	19	20	21	22	23	24	25
0	1	9	20	11	35	160	225	279	255	220	75	66	90	540	42	1 491	1 645	1 680	960	4 476	2 925	2 720	3 105	5 415	840	2 440
1		1	2	1	7	32	43	54	47	40	20	11	28	150	14	413	462	456	320	1 228	786	752	831	1 449	308	634
2			1	1	.	8	18	27	29	30	.	11	.	27	.	126	161	180	48	556	405	384	459	864	42	426
3				1	.	.	.	9	10	20	.	11	.	.	.	.	35	.	.	116	90	160	135	360	.	200
4					1	4	5	6	5	4	5	1	9	36	5	94	105	100	104	265	165	160	171	295	116	124
5						1	2	3	3	3	.	1	.	6	.	28	35	38	14	117	82	80	93	172	14	82
6							1	1	2	2	.	1	.	.	.	7	7	16	.	48	43	32	52	97	.	56
7								1	1	2	.	1	.	.	.	.	7	.	.	24	17	32	27	70	.	38
8									1	1	.	1	.	.	.	.	.	.	.	8	9	8	18	36	.	29
9										1	.	1	.	.	.	.	.	.	.	.	.	8	.	18	.	10
10											1	.	3	6	2	14	15	14	32	35	21	20	21	35	46	14
11												1	.	.	.	.	.	.	.	.	.	.	.	.	.	.
12													1	.	1	.	.	.	8	.	.	.	.	.	20	.
13														1	.	4	5	5	4	15	10	10	11	20	5	9

S.P.	54	58	26	74	30	98	30	34	30	34	34	42	34	38	38	42	42	46	34	46	58	54	50	38	54	58
O.S.	3 960	440	887 040	440	295 680	22	591 360	73 920	1 013 760	887 040	253 440	177 408	112 640	443 520	22 528	147 840	21 120	295 680	2 365 440	63 360	73 920	1 320	63 360	2 365 440	63 360	7 920
LEVEL	142	144	146	146	148	148	154	158	158	158	160	160	162	164	164	166	166	168	170	170	170	172	172	174	174	176
DIM'N	9 590 724	9 917 600	26 438 500	5 262 400	30 107 000	1 184 040	64 768 000	68 431 440	62 923 630	163 601 438	70 794 000	99 985 600	40 859 500	321 499 750	45 762 640	262 692 936	187 867 680	451 870 650	269 192 000	437 184 000	199 699 500	128 378 250	613 806 336	654 405 752	676 825 600	587 915 328
	4	3	0	5	2	7	0	0	0	1	2	3	0	1	2	0	1	2	0	0	4	0	2	1	1	1
	0	2	1	1	0	0	0	0	1	1	0	0	1	0	0	2	0	1	0	1	0	0	0	1	2	1
	1	0	0	0	0	0	1	0	0	0	0	0	0	1	0	0	0	0	1	1	0	3	1	0	0	2
	0	0	0	0	0	0	0	1	0	0	0	0	0	0	0	0	2	0	0	1	0	0	1	0	1	0
	0	0	0	0	0	0	0	1	0	0	0	0	0	1	0	1	0	1	0	0	1	0	0	0	0	0
	0	0	0	0	0	0	1	0	0	1	0	1	0	0	0	0	0	0	0	0	0	0	0	0	0	0
	0	0	1	0	1	0	0	0	0	0	0	0	0	0	0	0	0	0	0	0	0	0	0	0	0	0
	0	0	0	0	0	0	0	0	0	0	0	0	0	0	0	0	0	0	1	0	0	0	0	1	0	0
	0	0	0	0	0	0	0	0	1	0	1	0	0	0	0	0	0	0	0	0	0	0	0	0	0	0
	0	0	0	0	0	0	0	0	0	0	0	0	0	0	0	0	0	0	0	0	0	0	0	0	0	0
	<	<	<	<	<	<	<	<	<	<	<	<	<	<	<	<	<	<	<	<	<	<	<	<	<	<
WEIGHT	0	0	0	0	0	0	0	0	0	0	0	0	1	0	1	0	0	0	0	0	0	0	0	0	0	0
NUMBER	26	27	28	29	30	31	32	33	34	35	36	37	38	39	40	41	42	43	44	45	46	47	48	49	50	51
0	2 574	2 640	4 150	1 320	4 425	286	8 960	8 904	5 733	20 759	6 030	11 700	2 850	37 695	2 982	29 379	21 324	48 195	19 740	46 560	19 530	13 005	62 640	43 817	66 560	56 400
1	684	680	1 420	330	1 535	66	3 072	3 024	2 212	7 221	2 358	4 120	1 162	12 978	1 232	10 157	7 264	16 751	7 798	15 888	6 832	4 397	21 424	17 579	22 736	19 018
2	459	500	360	275	435	66	1 024	1 176	504	2 671	594	1 674	252	5 663	294	4 641	3 444	8 064	2 331	8 088	3 577	2 439	11 448	5 739	12 624	11 253
3	279	310	.	220	75	66	.	.	.	430	90	540	.	910	42	1 029	564	2 520	.	1 920	1 645	690	3 600	795	4 480	4 140
4	135	130	455	60	495	11	960	926	851	2 265	915	1 296	485	3 960	519	3 090	2 181	5 089	3 000	4 720	2 065	1 280	6 344	6 830	6 676	5 484
5	90	94	110	50	130	11	312	350	182	807	210	501	98	1 680	112	1 379	1 003	2 380	868	2 350	1 050	697	3 299	2 118	3 634	3 177
6	54	67	20	40	20	11	96	140	28	246	28	150	14	693	14	588	476	1 015	224	1 184	462	390	1 688	544	1 936	1 862
7	54	57	.	40	20	11	.	.	.	123	28	150	.	259	14	301	156	714	.	544	462	190	1 000	272	1 264	1 138
8	27	39	.	30	.	11	.	.	.	27	.	27	.	91	.	126	72	252	.	288	161	117	468	48	664	684
9	27	29	.	30	.	11	.	.	.	.	.	27	.	.	.	.	.	126	.	.	161	.	180	.	224	207
10	15	14	129	6	140	1	256	238	327	595	354	336	212	990	229	760	532	1 239	1 108	1 120	490	294	1 488	2 534	1 536	1 232
11	9	10	.	20	.	11	.	.	.	.	.	.	.	.	.	.	.	.	.	.	35	.	.	.	.	.
12	.	.	27	.	28	.	48	42	124	105	135	56	101	162	111	120	84	189	380	168	70	42	216	861	216	168
13	10	10	30	5	35	1	80	86	66	205	75	126	40	405	45	330	235	564	310	540	245	155	750	750	816	694

.	.	.	.	.	.	.	.	.	.	10	.	.	.	.	.	.	.	.	.	44	.	45	.	55	.	66	.	.	.	110	.	.	.	.	231	.	.	14
	1	1	2	.	6	5	4	6	11	.	6	6	7	5	4	5	1	24	33	9	60	9	36	5	162	5	136	107	234	76	264	105	85	372	180	424	396	15
		1	.	.	3	2	4	3	8	.	4	6	6	.	4	5	1	.	.	.	30	9	36	.	60	5	70	35	164	.	120	105	40	220	90	276	240	16
			1	.	2	3	1	3	6	.	4	3	5	.	3	.	1	6	14	.	12	.	6	.	63	.	49	53	84	14	130	35	48	186	28	210	230	17
				1	.	.	.	.	.	2	.	.	.	6	.	7	.	14	14	24	35	27	21	18	63	20	50	35	84	102	77	35	21	105	245	111	91	18
					1	1	1	2	4	.	3	3	4	.	3	.	1	.	.	.	6	.	6	.	21	.	28	15	56	.	62	35	25	100	14	141	141	19
						1	.	1	2	.	2	1	3	.	2	.	1	.	.	.	.	.	.	.	7	.	7	8	14	.	32	7	17	48	.	64	86	20
							1	.	2	.	1	3	3	.	3	.	1	.	.	.	.	.	6	.	.	.	.	.	28	.	.	35	.	38	.	48	42	21
								1	1	.	2	1	2	.	2	.	1	.	.	.	.	.	.	.	.	.	7	.	7	.	16	7	9	16	.	48	52	22
									1	.	1	1	2	.	2	.	1	.	.	.	.	.	.	.	.	.	.	.	7	.	.	7	.	16	.	24	26	23
										1	.	.	.	.	.	.	.	.	.	8	.	9	.	10	.	11	.	.	.	27	.	.	.	.	63	.	.	24
											1	.	1	.	1	.	1	.	.	.	.	.	.	.	.	.	.	.	.	.	.	.	.	.	.	8	9	25
												1	1	.	2	.	1	.	.	.	.	.	.	.	.	.	.	.	.	.	.	7	.	.	.	.	.	26
													1	.	1	.	1	.	.	.	.	.	.	.	.	.	.	.	.	.	.	.	.	.	.	.	.	27
														1	.	1	.	4	5	3	10	3	6	2	24	2	20	15	34	24	36	15	11	50	56	56	50	28
															1	.	1	.	.	.	.	.	.	.	.	.	.	.	.	.	.	.	.	.	.	.	.	29
																1	.	.	.	.	5	3	6	.	9	2	10	5	24	.	16	15	5	30	28	36	30	30
																	1	.	.	.	.	.	.	.	.	.	.	.	.	.	.	.	.	.	.	.	.	31
																		1	2	.	2	.	1	.	9	.	7	7	12	4	17	5	6	24	8	27	28	32
																			1	.	.	.	.	.	3	.	2	4	3	.	8	1	3	12	.	12	16	33
																				1	.	1	.	1	.	1	.	.	.	6	.	.	.	.	14	.	.	34
																					1	.	1	.	3	.	4	2	8	.	8	5	3	13	4	18	17	35
																						1	.	.	.	1	.	.	.	.	.	.	.	.	7	.	.	36
																							1	.	.	.	.	.	4	.	.	5	.	5	.	6	5	37
																								1	.	1	.	.	.	.	.	.	.	.	.	.	.	38
																									1	.	1	1	2	.	4	1	2	6	.	8	10	39
																										1	.	.	.	.	.	.	.	.	.	.	.	40
																											1	.	1	.	2	1	1	2	.	6	6	41
																												1	.	.	2	.	1	3	.	3	6	42
																													1	.	.	1	.	2	.	3	3	43
																														1	.	.	.	.	2	.	.	44
																															1	.	1	1	.	2	4	45
																																1	.	.	.	.	.	46
																																	1	.	.	.	2	47
																																		1	.	1	2	48
																																			1	.	.	49
																																				1	1	50
																																					1	51

Table 3. *(continued)*

ALGEBRA C12, REPRESENTATIONS OF CLASS 0

S.P.	0	4	8	8	12	16	20	32	12	16	20	24	24
O.S.	1	264	24	7 920	5 280	264	528	24	59 136	126 720	47 520	1 760	31 680
LEVEL	1	45	47	81	87	89	91	93	109	119	125	127	127
DIM'N	1	275	300	10 350	37 375	27 324	44 550	17 550	123 970	837 200	1 837 080	1 126 125	2 220 075
	0	0	2	0	1	0	2	4	0	1	0	0	2
	0	1	0	0	0	2	1	0	0	0	1	0	0
	0	0	0	0	1	0	0	0	0	0	0	2	0
	0	0	0	1	0	0	0	0	0	0	1	0	1
	0	0	0	0	0	0	0	0	0	1	0	0	0
	0	0	0	0	0	0	0	0	1	0	0	0	0
	0	0	0	0	0	0	0	0	0	0	0	0	0
	0	0	0	0	0	0	0	0	0	0	0	0	0
	0	0	0	0	0	0	0	0	0	0	0	0	0
	0	0	0	0	0	0	0	0	0	0	0	0	0
	0	0	0	0	0	0	0	0	0	0	0	0	0
	<	<	<	<	<	<	<	<	<	<	<	<	<
WEIGHT	0	0	0	0	0	0	0	0	0	0	0	0	0
NUMBER	0	1	2	3	4	5	6	7	8	9	10	11	12
0	1	11	12	54	175	132	198	78	154	848	1 728	1 045	1 947
1		1	1	9	30	21	33	12	35	204	414	245	477
2			1	.	10	11	22	12	.	44	144	110	198
3				1	3	2	3	1	7	40	78	45	90
4					1	1	2	1	.	8	27	19	36
5						1	1	1	.	.	9	10	9
6							1	1	.	.	.	.	9
7								1	.	.	.	.	.
8									1	5	9	5	10
9										1	3	2	4
10											1	1	1
11												1	.
12													1

S.P.	16	28	36	36	40	52	72	20	20	24	24	28	28
O.S.	126 720	10 560	264	5 280	528	528	24	67 584	709 632	4 096	887 040	354 816	253 440
LEVEL	129	131	133	133	135	137	139	141	143	145	153	155	159
DIM'N	600 875	4 100 096	1 467 400	2 898 000	2 861 430	1 866 150	475 020	1 225 785	6 561 555	742 900	25 957 800	29 668 275	44 993 520
	0	1	0	3	2	4	6	0	1	0	0	2	0
	0	1	3	0	2	1	0	0	0	0	1	0	0
	0	1	0	1	0	0	0	0	0	0	0	0	1
	0	0	0	0	0	0	0	0	0	0	0	0	0
	0	0	0	0	0	0	0	0	0	0	0	0	1
	0	0	0	0	0	0	0	0	0	0	1	1	0
	0	0	0	0	0	0	0	0	1	0	0	0	0
	1	0	0	0	0	0	0	0	0	0	0	0	0
	0	0	0	0	0	0	0	0	0	0	0	0	0
	0	0	0	0	0	0	0	1	0	0	0	0	0
	0	0	0	0	0	0	0	0	0	0	0	0	0
	<	<	<	<	<	<	<	<	<	<	<	<	<
WEIGHT	0	0	0	0	0	0	0	0	0	1	0	0	0
NUMBER	13	14	15	16	17	18	19	20	21	22	23	24	25
0	275	3 520	1 232	2 320	2 310	1 430	364	297	2 211	132	7 392	7 875	11 792
1	75	840	287	560	539	330	78	90	650	42	2 226	2 415	3 556
2	.	440	176	340	374	264	78	.	110	.	616	770	1 276
3	20	152	50	100	93	55	12	28	177	14	602	658	944
4	.	78	31	60	63	44	12	.	27	.	161	196	332
5	.	40	21	30	43	33	12	.	.	.	35	35	116
6	.	20	11	30	32	33	12	.	.	.	.	35	.
7	.	.	.	10	11	22	12	.	.	.	.	.	.
8	5	16	5	10	9	5	1	9	42	5	135	147	203
9	.	8	3	6	6	4	1	.	6	.	35	42	69
10	.	4	2	3	4	3	1	.	.	.	7	7	24
11	.	2	1	1	3	2	1	.	.	.	.	.	8
12	.	2	1	3	3	3	1	.	.	.	.	7	.

S.P.	24	32	32	40	28	36	40	44	44	28	48	56	52
O.S.	1 013 760	7 920	506 880	126 720	270 336	95 040	47 520	5 280	95 040	3 548 160	5 280	31 680	10 560
LEVEL	159	161	163	165	167	167	169	171	171	173	173	173	175
DIM'N	21 246 940	25 008 750	117 145 600	72 522 450	26 558 675	149 428 125	130 630 500	108 162 054	227 561 724	135 481 500	156 310 000	104 001 975	213 468 750
	1	0	1	3	1	1	0	0	2	0	2	4	1
	0	0	1	0	0	0	2	1	1	1	0	0	2
	0	0	0	0	0	1	0	2	0	0	2	0	1
	0	2	0	0	0	1	1	0	1	0	0	1	0
	0	0	1	1	0	0	0	0	0	0	0	0	0
	0	0	0	0	0	0	0	0	0	0	0	0	0
	0	0	0	0	0	0	0	0	0	0	0	0	0
	0	0	0	0	0	0	0	0	0	1	0	0	0
	1	0	0	0	0	0	0	0	0	0	0	0	0
	0	0	0	0	0	0	0	0	0	0	0	0	0
	0	0	0	0	1	0	0	0	0	0	0	0	0
	<	<	<	<	<	<	<	<	<	<	<	<	<
WEIGHT	0	0	0	0	0	0	0	0	0	0	0	0	0
NUMBER	26	27	28	29	30	31	32	33	34	35	36	37	38
0	3 388	6 390	28 160	16 082	2 739	34 485	28 908	23 430	48 060	16 500	32 560	20 295	43 230
1	1 125	1 905	8 640	5 008	972	10 470	8 808	7 075	14 751	5 725	9 850	6 270	13 040
2	165	780	3 520	2 288	132	4 950	4 389	3 850	7 866	1 375	5 720	3 696	7 920
3	368	499	2 304	1 340	350	2 734	2 292	1 811	3 834	1 910	2 514	1 620	3 295
4	48	198	912	588	42	1 260	1 125	969	1 998	435	1 423	936	1 973
5	.	90	320	204	.	585	549	535	972	75	810	477	1 185
6	.	.	160	204	.	225	279	255	693	.	500	477	785
7	.	.	.	44	.	.	.	.	144	.	110	198	165
8	118	105	488	280	130	560	462	357	765	600	490	315	630
9	14	40	188	120	14	250	222	186	390	130	270	180	370
10	.	17	64	40	.	113	105	101	186	20	149	90	218
11	.	9	16	8	.	54	45	57	81	.	87	36	126
12	.	.	32	40	.	43	54	47	132	.	92	90	144

S.P.	32	64	60	68	76	80	32	100	32	128	36	36	36
O.S.	1 013 760	264	10 560	528	5 280	528	3 548 160	528	3 041 280	24	4 257 792	675 840	887 040
LEVEL	175	177	177	179	179	181	183	183	185	185	187	187	189
DIM'N	151 853 130	50 979 500	219 648 000	110 115 720	87 793 875	91 763 100	415 154 025	40 495 455	286 715 700	7 888 725	1 010 380 800	318 704 100	629 021 250
	2	0	3	2	5	4	0	6	0	8	1	2	0
	0	4	1	3	0	2	0	1	1	0	1	0	0
	0	0	1	0	1	0	1	0	0	0	0	0	0
	0	0	0	0	0	0	0	0	0	0	0	0	1
	0	0	0	0	0	0	0	0	0	0	0	0	0
	0	0	0	0	0	0	0	0	0	0	0	0	1
	0	0	0	0	0	0	1	0	0	0	1	0	0
	1	0	0	0	0	0	0	0	0	0	0	0	0
	0	0	0	0	0	0	0	0	0	0	0	0	0
	0	0	0	0	0	0	0	0	1	0	0	1	0
	0	0	0	0	0	0	0	0	0	0	0	0	0
	<	<	<	<	<	<	<	<	<	<	<	<	<
WEIGHT	0	0	0	0	0	0	0	0	0	0	0	0	0
NUMBER	39	40	41	42	43	44	45	46	47	48	49	50	51
0	17 226	9 988	42 240	20 856	15 795	16 588	42 801	7 007	20 196	1 365	95 040	20 900	59 290
1	6 075	2 947	12 760	6 171	4 740	4 862	15 141	2 002	7 623	364	34 200	8 010	21 035
2	1 650	2 002	8 360	4 400	3 420	3 762	4 730	1 716	1 683	364	11 880	1 980	7 700
3	2 045	724	3 200	1 506	1 165	1 166	5 055	462	2 868	78	11 528	3 042	6 965
4	510	487	2 060	1 060	835	891	1 537	396	594	78	3 834	684	2 506
5	75	342	1 280	749	560	671	430	330	90	78	1 080	90	910
6	75	231	1 060	584	560	605	.	330	.	78	540	90	.
7	.	66	440	242	340	374	.	264	.	78	.	.	.
8	645	133	600	273	210	205	1 545	77	1 080	12	3 528	1 153	2 080
9	150	88	380	189	150	155	456	66	210	12	1 128	238	728
10	20	60	232	132	100	115	123	55	28	12	300	28	259
11	.	41	138	93	60	85	27	44	.	12	54	.	91
12	20	41	192	102	100	104	.	55	.	12	150	28	.

1	.	.	.	.	.	.	3	7	2	20	21	28	36	14	64	35	51	70	56	42	90	168	56	35	70	180	14	64	28	21	20	405	7	408	1	912	438	522	13
	1	1	1	2	2	1	.	.	.	.	.	.	.	.	8	8	.	18	27	29	54	.	48	36	88	.	31	108	73	60	74	.	44	.	12	27	.	.	14
		1	.	1	1	1	.	.	.	.	.	.	.	.	.	.	.	.	9	10	9	.	10	9	40	.	21	40	43	30	43	.	33	.	12	.	.	.	15
			1	1	2	1	.	.	.	.	.	.	.	.	.	8	.	.	.	.	27	.	19	36	30	.	11	78	42	60	63	.	44	.	12	.	.	.	16
				1	1	1	.	.	.	.	.	.	.	.	.	.	.	.	.	.	9	.	10	9	20	.	11	40	32	30	43	.	33	.	12	.	.	.	17
					1	1	.	.	.	.	.	.	.	.	.	.	.	.	.	.	.	.	.	9	.	.	.	20	11	30	32	.	33	.	12	.	.	.	18
						1	.	.	.	.	.	.	.	.	.	.	.	.	.	.	.	.	.	.	.	.	.	.	.	10	11	.	22	.	12	.	.	.	19
							1	.	1	.	.	.	9	.	.	.	22	.	.	.	.	35	.	.	.	36	.	.	.	.	.	75	.	153	.	160	165	90	20
								1	.	5	6	9	4	5	24	15	5	30	26	21	45	35	30	20	40	40	9	40	19	15	15	115	6	75	1	282	84	175	21
									1	.	.	.	.	.	.	.	11	.	.	.	.	.	.	.	.	.	.	.	.	.	.	.	.	54	.	.	55	.	22
										1	1	3	.	2	8	5	.	13	12	11	21	5	16	10	23	5	6	24	13	10	11	30	5	9	1	72	9	60	23
											1	.	.	.	4	5	.	5	6	5	15	.	10	10	15	5	4	20	10	10	10	.	5	.	1	36	9	.	24
												1	.	1	2	1	.	6	5	6	9	.	9	4	13	.	4	14	9	6	8	6	4	.	1	12	.	21	25
													1	.	.	.	2	.	.	.	.	7	.	.	.	8	.	.	.	.	.	20	.	27	.	48	30	28	26
														1	.	.	.	3	2	3	3	.	6	1	7	.	3	8	6	3	6	.	3	.	1	.	.	7	27
															1	1	.	2	3	3	6	.	5	4	9	.	3	11	7	6	7	.	4	.	1	6	.	.	28
																1	.	.	.	.	3	.	2	4	3	.	1	8	4	6	6	.	4	.	1	.	.	.	29
																	1	.	.	.	.	.	.	.	.	.	.	.	.	.	.	.	.	9	.	.	10	.	30
																		1	1	2	2	.	3	1	5	.	2	6	5	3	5	.	3	.	1	.	.	.	31
																			1	1	1	.	1	1	4	.	2	4	4	3	4	.	3	.	1	.	.	.	32
																				1	.	.	1	.	2	.	1	2	3	1	3	.	2	.	1	.	.	.	33
																					1	.	1	1	2	.	1	4	3	3	4	.	3	.	1	.	.	.	34
																						1	.	.	.	1	.	.	.	.	.	5	.	3	.	12	3	9	35
																							1	.	1	.	1	2	2	1	3	.	2	.	1	.	.	.	36
																								1	.	.	.	2	1	3	3	.	3	.	1	.	.	.	37
																									1	.	1	1	2	1	2	.	2	.	1	.	.	.	38
																										1	.	.	.	.	.	.	.	.	.	6	3	.	39
																											1	.	1	.	1	.	1	.	1	.	.	.	40
																												1	1	1	2	.	2	.	1	.	.	.	41
																													1	.	1	.	1	.	1	.	.	.	42
																														1	1	.	2	.	1	.	.	.	43
																															1	.	1	.	1	.	.	.	44
																																1	.	.	.	2	.	3	45
																																	1	.	1	.	.	.	46
																																		1	.	.	1	.	47
																																			1	.	.	.	48
																																				1	.	.	49
																																					1	.	50
																																						1	51

Table 3. *(continued)*

ALGEBRA C12, REPRESENTATIONS OF CLASS 1

S.P.	2	6	10	18	10	14	18	22	26	34	50	14	18	18	22	26	22	26	30	34	38	38	22	42	50	54
O.S.	24	1 760	528	24	25 344	31 680	5 280	5 280	528	528	24	101 376	354 816	112 640	253 440	126 720	24 576	31 680	95 040	5 280	31 680	5 280	1 013 760	10 560	528	5 280
LEVEL	24	64	68	70	96	104	108	110	112	114	116	120	132	136	140	142	144	144	148	150	150	152	152	154	156	156
DIM'N	24	2 000	4 576	2 600	40 480	205 920	297 000	387 504	354 200	320 320	98 280	303 600	2 631 200	961 400	7 948 800	9 266 400	1 188 640	9 266 400	26 371 800	17 463 600	17 050 176	19 076 200	13 156 000	34 452 000	15 787 200	17 342 000
	1	0	1	3	0	1	0	2	1	3	5	0	1	0	0	2	0	0	1	1	3	0	1	2	1	4
	0	0	1	0	0	0	1	0	2	1	0	0	0	0	1	0	0	0	1	0	0	2	0	1	3	0
	0	1	0	0	0	0	1	1	0	0	0	0	0	0	0	0	0	1	0	2	0	1	0	1	0	1
	0	0	0	0	0	1	0	0	0	0	0	0	0	0	0	0	0	1	1	0	1	0	0	0	0	0
	0	0	0	0	1	0	0	0	0	0	0	0	0	0	1	1	0	0	0	0	0	0	0	0	0	0
	0	0	0	0	0	0	0	0	0	0	0	0	1	0	0	0	0	0	0	0	0	0	0	0	0	0
	0	0	0	0	0	0	0	0	0	0	0	1	0	0	0	0	0	0	0	0	0	0	0	0	0	0
	0	0	0	0	0	0	0	0	0	0	0	0	0	0	0	0	0	0	0	0	0	0	1	0	0	0
	0	0	0	0	0	0	0	0	0	0	0	0	0	1	0	0	0	0	0	0	0	0	0	0	0	0
	0	0	0	0	0	0	0	0	0	0	0	0	0	0	0	0	0	0	0	0	0	0	0	0	0	0
	0	0	0	0	0	0	0	0	0	0	0	0	0	0	0	0	1	0	0	0	0	0	0	0	0	0
	<	<	<	<	<	<	<	<	<	<	<	<	<	<	<	<	<	<	<	<	<	<	<	<	<	<
WEIGHT	0	0	0	0	0	0	0	0	0	0	0	0	0	0	0	0	0	0	0	0	0	0	0	0	0	0
NUMBER	0	1	2	3	4	5	6	7	8	9	10	11	12	13	14	15	16	17	18	19	20	21	22	23	24	25
0	1	10	22	12	44	198	275	340	308	264	78	110	770	165	2 080	2 288	132	2 310	6 117	3 960	3 696	3 180	4 650	1 260	7 234	3 420
1		1	2	1	8	36	48	60	52	44	12	27	196	48	528	588	42	576	1 539	978	936	1 028	510	1 785	774	835
2			1	1	.	9	20	30	32	33	12	.	35	.	160	204	.	225	693	500	477	565	75	1 060	518	560
3				1	.	.	.	10	11	22	12	.	.	.	.	44	.	.	144	110	198	165	.	440	242	340
4					1	4	5	6	5	4	1	6	42	14	108	120	14	114	300	186	180	192	150	330	138	150
5						1	2	3	3	3	1	.	7	.	32	40	.	43	132	92	90	104	20	192	91	100
6							1	1	2	2	1	.	.	.	8	8	.	18	54	48	36	58	.	108	62	60
7								1	1	2	1	.	.	.	.	8	.	.	27	19	36	30	.	78	42	60
8									1	1	1	.	.	.	.	.	.	.	9	10	9	20	.	40	32	30
9										1	1	.	.	.	.	.	.	.	.	.	9	.	.	20	11	30
10											1	.	.	.	.	.	.	.	.	.	.	.	.	.	.	10
11												1	6	4	14	15	5	14	35	21	20	21	40	35	14	15
12													1	.	4	5	.	5	15	10	10	11	5	20	9	10
13														1	.	.	2	.	.	.	.	.	8	.	.	.

S.P.	58	74	98	26	26	30	30	30	34	34	42	30	38	34	42	42	46	34	46	58	54	38	50	34	54	58
O.S.	528	528	24	2 128 896	675 840	709 632	49 152	1 182 720	126 720	1 774 080	354 816	4 055 040	760 320	1 013 760	253 440	31 680	506 880	1 351 680	95 040	126 720	1 760	270 336	95 040	7 096 320	95 040	10 560
LEVEL	158	160	162	164	164	166	168	172	176	176	178	180	182	182	184	184	186	188	188	188	190	190	190	192	192	194
DIM'N	17 760 600	9 266 400	2 035 800	66 397 320	27 268 800	74 989 200	16 640 960	156 293 280	161 460 000	389 716 600	235 435 200	221 506 560	744 876 000	247 033 800	603 200 000	429 483 600	1 030 114 800	280 816 200	985 327 200	448 672 224	284 734 296	311 493 600	1 373 486 400	868 296 000	1 501 156 800	1 289 288 000
	3	5	7	0	1	2	1	0	0	1	3	0	1	2	0	1	2	0	0	4	0	2	2	0	1	1
	2	1	0	1	0	0	0	0	0	1	0	1	0	0	2	0	1	1	1	0	0	0	0	0	2	1
	0	0	0	0	0	0	0	1	0	0	0	0	1	0	0	0	0	0	1	0	3	0	1	1	0	2
	0	0	0	0	0	0	0	0	1	0	0	0	0	0	0	2	0	0	1	0	0	0	1	0	1	0
	0	0	0	0	0	0	0	0	1	0	0	0	1	0	1	0	1	0	0	1	0	0	0	0	0	0
	0	0	0	0	0	0	0	1	0	1	1	0	0	0	0	0	0	0	0	0	0	0	0	0	0	0
	0	0	0	1	0	1	0	0	0	0	0	0	0	0	0	0	0	0	0	0	0	0	0	0	0	0
	0	0	0	0	0	0	0	0	0	0	0	0	0	0	0	0	0	0	0	0	0	0	0	1	0	0
	0	0	0	0	0	0	0	0	0	0	0	1	0	1	0	0	0	0	0	0	0	0	0	0	0	0
	0	0	0	0	1	0	0	0	0	0	0	0	0	0	0	0	0	0	0	0	0	0	0	0	0	0
	0	0	0	0	0	0	0	0	0	0	0	0	0	0	0	0	0	1	0	0	0	1	0	0	0	0
	<	<	<	<	<	<	<	<	<	<	<	<	<	<	<	<	<	<	<	<	<	<	<	<	<	<
WEIGHT	0	0	0	0	0	0	1	0	0	0	0	0	0	0	0	0	0	0	0	0	0	0	0	0	0	0
NUMBER	26	27	28	29	30	31	32	33	34	35	36	37	38	39	40	41	42	43	44	45	46	47	48	49	50	51
0	3 476	1 716	364	6 831	1 980	7 260	924	14 322	13 980	32 977	18 480	11 968	58 608	12 540	45 440	32 790	74 250	10 395	70 950	29 832	19 580	10 824	95 040	38 500	100 386	84 260
1	825	396	78	2 160	684	2 324	336	4 536	4 392	10 577	5 992	4 320	18 624	4 581	14 488	10 323	23 784	4 032	22 356	9 608	6 116	4 248	30 006	14 180	31 644	26 252
2	605	330	78	540	90	650	42	1 491	1 680	3 871	2 415	960	8 032	1 125	6 560	4 830	11 360	840	11 265	5 008	3 360	972	15 885	4 150	17 442	15 405
3	374	264	78	.	.	110	.	.	.	616	770	.	1 276	165	1 440	780	3 520	.	2 640	2 288	935	132	4 950	.	6 138	5 610
4	144	66	12	627	238	678	126	1 288	1 225	3 010	1 708	1 528	5 172	1 632	4 010	2 826	6 572	1 582	6 058	2 640	1 627	1 680	8 104	5 010	8 478	6 920
5	104	55	12	150	28	177	14	413	456	1 064	658	320	2 172	368	1 776	1 285	3 056	308	2 989	1 340	878	350	4 183	1 420	4 587	3 981
6	74	44	12	27	.	27	.	126	180	322	196	48	888	48	752	603	1 296	42	1 494	588	488	42	2 124	360	2 430	2 318
7	63	44	12	.	.	27	.	.	.	161	196	.	332	48	384	198	912	.	684	588	236	42	1 260	.	1 584	1 414
8	43	33	12	.	.	.	.	.	.	35	35	.	116	.	160	90	320	.	360	204	145	.	585	.	828	845
9	32	33	12	.	.	.	.	.	.	.	35	.	.	.	.	.	160	.	.	204	.	.	225	.	279	255
10	11	22	12	.	.	.	.	.	.	.	.	.	.	.	.	.	.	.	.	44	.	.	.	.	.	.
11	14	6	1	156	84	168	50	304	280	700	392	524	1 152	562	880	616	1 428	636	1 288	560	336	680	1 704	1 660	1 752	1 400
12	10	5	1	36	9	42	5	94	100	240	147	104	468	118	380	270	648	116	617	280	176	130	855	455	927	785
13	.	.	.	27	30	28	22	48	42	105	56	168	162	180	120	84	189	267	168	70	42	288	216	490	216	168

1	1	.	2	6	5	4	6	.	11	6	6	7	4	1	6	.	6	.	28	38	70	42	14	186	14	156	122	268	14	300	120	96	14	422	110	480	446	14
	1	.	.	3	2	4	3	.	8	4	6	6	4	1	.	.	6	.	.	.	35	42	.	69	14	80	40	188	.	136	120	45	14	250	.	312	270	15
		1	.	.	.	.	.	.	.	.	.	.	.	.	.	10	.	12	.	.	.	.	44	.	45	.	.	.	120	.	.	.	132	.	110	.	.	16
			1	2	3	1	3	.	6	4	3	5	3	1	.	.	.	.	7	16	14	7	.	72	.	56	60	96	.	147	40	54	.	210	20	237	258	17
				1	1	1	2	.	4	3	3	4	3	1	.	.	.	.	.	.	7	7	.	24	.	32	17	64	.	70	40	28	.	113	.	159	158	18
					1	.	1	.	2	2	1	3	2	1	.	.	.	.	.	.	.	.	.	8	.	8	9	16	.	36	8	19	.	54	.	72	96	19
						1	.	.	2	1	3	3	3	1	.	.	.	.	.	.	.	7	.	.	.	.	.	32	.	.	40	.	.	43	.	54	47	20
							1	.	1	2	1	2	2	1	.	.	.	.	.	.	.	.	.	.	.	8	.	8	.	18	8	10	.	18	.	54	58	21
								1	.	.	.	.	.	.	6	3	7	2	14	14	35	21	32	63	36	50	35	84	46	77	35	21	51	105	129	111	91	22
									1	1	1	2	2	1	.	.	.	.	.	.	.	.	.	.	.	.	.	8	.	.	8	.	.	18	.	27	29	23
										1	.	1	1	1	.	.	.	.	.	.	.	.	.	.	.	.	.	.	.	.	.	.	.	.	.	9	10	24
											1	1	2	1	.	.	.	.	.	.	.	.	.	.	.	.	.	.	.	.	8	.	.	.	.	.	.	25
												1	1	1	.	.	.	.	.	.	.	.	.	.	.	.	.	.	.	.	.	.	.	.	.	.	.	26
													1	1	.	.	.	.	.	.	.	.	.	.	.	.	.	.	.	.	.	.	.	.	.	.	.	27
														1	.	.	.	.	.	.	.	.	.	.	.	.	.	.	.	.	.	.	.	.	.	.	.	28
															1	.	1	.	4	5	10	6	4	24	4	20	15	34	5	36	15	11	5	50	30	56	50	29
																1	.	1	.	.	.	.	8	.	9	.	.	.	20	.	.	.	22	.	27	.	.	30
																	1	.	.	.	5	6	.	9	4	10	5	24	.	16	15	5	5	30	.	36	30	31
																		1	.	.	.	.	.	.	.	.	.	.	10	.	.	.	11	.	.	.	.	32
																			1	2	2	1	.	9	.	7	7	12	.	17	5	6	.	24	5	27	28	33
																				1	.	.	.	3	.	2	4	3	.	8	1	3	.	12	.	12	16	34
																					1	1	.	3	.	4	2	8	.	8	5	3	.	13	.	18	17	35
																						1	.	.	.	.	.	4	.	.	5	.	.	5	.	6	5	36
																							1	.	1	.	.	.	2	.	.	.	2	.	6	.	.	37
																								1	.	1	1	2	.	4	1	2	.	6	.	8	10	38
																									1	.	.	.	.	.	.	.	2	.	.	.	.	39
																										1	.	1	.	2	1	1	.	2	.	6	6	40
																											1	.	.	2	.	1	.	3	.	3	6	41
																												1	.	.	1	.	.	2	.	3	3	42
																													1	.	.	.	1	.	.	.	.	43
																														1	.	1	.	1	.	2	4	44
																															1	.	.	.	.	.	.	45
																																1	.	.	.	.	2	46
																																	1	.	.	.	.	47
																																		1	.	1	2	48
																																			1	.	.	49
																																				1	1	50
																																					1	51

Table 3. *(continued)*

ALGEBRA D4, REPRESENTATIONS OF CLASS 0

NUMBER	0	1	2	3	4	5	6	7	8	9	10	11	12
S.P.	0	8	16	16	16	24	32	40	40	40	64	48	64
O.S.	1	24	8	8	8	96	24	48	48	48	8	32	8
LEVEL	1	11	13	13	13	19	21	23	23	23	25	25	25
DIM'N	1	28	35	35	35	350	300	567	567	567	294	840	294
WEIGHT	0 0 0 0	0 0 1 0	2 0 0 0	0 0 0 2	0 2 0 0	1 1 0 1	0 0 2 0	2 0 1 0	0 0 1 2	0 2 1 0	4 0 0 0	2 0 0 2	0 0 0 4
0	1	4	3	3	3	14	12	15	15	15	6	16	6
1		1	1	1	1	7	5	8	8	8	3	10	3
2			1	.	.	3	2	6	3	3	3	6	1
3				1	.	3	2	3	6	3	1	6	3
4					1	3	2	3	3	6	1	6	1
5						1	1	2	2	2	1	3	1
6							1	1	1	1	1	1	1
7								1	.	.	1	1	.
8									1	.	.	1	1
9										1	.	.	.
10											1	.	.
11												1	.
12													1

NUMBER	13	14	15	16	17	18	19	20	21	22	23	24	25
S.P.	48	48	64	56	72	72	72	72	80	80	80	104	88
O.S.	32	32	8	192	24	96	96	96	48	48	48	48	96
LEVEL	25	25	25	29	31	31	31	31	33	33	33	35	35
DIM'N	840	840	294	4096	1925	3675	3675	3675	4312	4312	4312	3696	8910
WEIGHT	2 2 0 0	0 2 0 2	0 4 0 0	1 1 1 1	0 0 3 0	3 1 0 1	1 1 0 3	1 3 0 1	2 0 2 0	0 0 2 2	0 2 2 0	4 0 1 0	2 0 1 2
0	16	16	6	64	29	43	43	43	48	48	48	32	78
1	10	10	3	40	16	29	29	29	31	31	31	21	56
2	6	6	1	24	9	21	18	18	24	18	18	18	39
3	6	6	1	24	9	18	21	18	18	24	18	12	39
4	6	6	3	24	9	18	18	21	18	18	24	12	39
5	3	3	1	14	6	12	12	12	13	13	13	10	26
6	1	1	1	8	5	7	7	7	9	9	9	8	16
7	1	.	.	4	2	7	3	3	7	4	4	8	12
8	.	1	.	4	2	3	7	3	4	7	4	3	12
9	1	1	1	4	2	3	3	7	4	4	7	3	9
10	.	.	.	.	.	3	.	.	2	.	.	6	3
11	.	.	.	2	1	3	3	1	3	3	3	3	9
12	.	.	.	.	.	.	3	.	.	2	.	.	3
13	1	.	.	2	1	3	1	3	3	3	3	3	6
14		1	.	2	1	1	3	3	3	3	3	2	6
15			1	.	.	.	.	3	.	.	2	.	.
16				1	1	1	1	1	2	2	2	2	4
17					1	.	.	.	1	1	1	1	1
18						1	.	.	1	.	.	2	2

NUMBER	26	27	28	29	30	31	32	33	34	35	36	37	38
S.P.	104	88	88	104	144	112	112	144	112	96	112	112	112
O.S.	48	96	96	48	8	32	32	8	32	96	32	32	32
LEVEL	35	35	35	35	37	37	37	37	37	37	37	37	37
DIM'N	3696	8910	8910	3696	1386	5775	5775	1386	5775	15092	5775	5775	5775
WEIGHT	0 0 1 4	2 2 1 0	0 2 1 2	0 4 1 0	6 0 0 0	4 0 0 2	2 0 0 4	0 0 0 6	4 2 0 0	2 2 0 2	0 2 0 4	2 4 0 0	0 4 0 2
0	32	78	78	32	10	39	39	10	39	116	39	39	39
1	21	56	56	21	6	29	29	6	29	84	29	29	29
2	12	39	39	12	6	22	21	3	22	60	21	21	21
3	18	39	39	12	3	21	22	6	21	60	22	21	21
4	12	39	39	18	3	21	21	3	21	60	21	22	22
5	10	26	26	10	3	15	15	3	15	42	15	15	15
6	8	16	16	8	3	10	10	3	10	29	10	10	10
7	3	12	9	3	3	10	6	1	10	19	6	6	6
8	8	9	12	3	1	6	10	3	6	19	10	6	6
9	3	12	12	8	1	6	6	1	6	19	6	10	10
10	.	3	.	.	3	6	1	.	6	6	.	1	.
11	3	6	6	2	1	6	6	1	6	12	6	3	3
12	6	.	3	.	.	1	6	3	.	6	6	.	1
13	2	9	6	3	1	6	3	1	6	12	3	6	6
14	3	6	9	3	1	3	6	1	3	12	6	6	6
15	.	3	3	6	.	.	.	.	1	6	1	6	6
16	2	4	4	2	1	3	3	1	3	7	3	3	3
17	1	1	1	1	1	1	1	1	1	1	1	1	1
18	.	2	.	.	1	3	1	.	3	3	.	1	.

NUMBER	39	40	41	42	43	44	45	46	47	48	49	50	51
S.P.	144	104	128	120	120	120	136	136	152	136	152	136	136
O.S.	8	192	24	192	192	192	48	48	96	96	96	48	96
LEVEL	37	39	41	41	41	41	43	43	43	43	43	43	43
DIM'N	1386	25725	8918	32768	32768	32768	21840	21840	18711	32340	18711	21840	32340
WEIGHT	0 6 0 0	1 1 2 1	0 0 4 0	3 1 1 1	1 1 1 3	1 3 1 1	2 0 3 0	0 0 3 2	5 1 0 1	3 1 0 3	1 1 0 5	0 2 3 0	3 3 0 1
0	10	189	62	192	192	192	120	120	87	156	87	120	156
1	6	134	40	144	144	144	87	87	66	122	66	87	122
2	3	94	27	112	104	104	70	60	54	94	48	60	94
3	3	94	27	104	112	104	60	70	48	94	54	60	94
4	6	94	27	104	104	112	60	60	48	94	48	70	94
5	3	66	20	78	78	78	47	47	38	71	38	47	71
6	3	47	17	56	56	56	36	36	29	52	29	36	52
7	1	31	10	48	36	36	29	23	29	40	18	23	40
8	1	31	10	36	48	36	23	29	18	40	29	23	37
9	3	31	10	36	36	48	23	23	18	37	18	29	40
10	.	7	3	24	8	8	12	6	21	18	3	6	18
11	1	21	7	30	30	24	18	18	18	30	18	18	27
12	.	7	3	8	24	8	6	12	3	18	21	6	10
13	1	21	7	30	24	30	18	18	18	27	12	18	30
14	1	21	7	24	30	30	18	18	12	27	18	18	27
15	3	7	3	8	8	24	6	6	3	10	3	12	18
16	1	15	6	19	19	19	14	14	12	19	12	14	19
17	1	8	5	8	8	8	9	9	7	7	7	9	7
18	.	5	2	14	6	6	8	5	12	12	3	5	12

1	.	.	1	.	.	2	2	.	2	.	.	1	3	1	.	3	3	.	1	.	5	2	6	14	6	5	8	3	12	12	5	6	19
	1	.	.	1	.	.	.	2	2	2	.	.	.	.	1	3	1	3	3	1	5	2	6	6	14	5	5	3	6	3	8	12	20
		1	.	.	1	1	.	1	.	.	1	1	1	.	1	1	.	1	.	.	4	2	8	4	4	7	4	7	7	3	4	7	21
			1	.	.	1	1	.	1	.	.	1	1	1	.	1	1	.	1	.	4	2	4	8	4	4	7	3	7	7	4	3	22
				1	.	.	.	1	1	1	.	.	.	.	1	1	1	1	1	1	4	2	4	4	8	4	4	3	3	3	7	7	23
					1	.	.	.	.	.	1	1	.	.	1	.	.	.	.	.	.	.	4	.	.	2	.	7	3	.	.	3	24
						1	.	.	.	.	.	1	1	.	.	1	.	.	.	.	2	1	4	4	2	3	3	3	7	3	3	3	25
							1	.	.	.	.	.	1	1	.	.	1	.	.	.	.	.	.	4	.	.	2	.	3	7	.	.	26
								1	.	.	.	.	.	.	1	1	.	1	.	.	2	1	4	2	4	3	3	3	3	1	3	7	27
									1	.	.	.	.	.	.	1	1	.	1	.	2	1	2	4	4	3	3	1	3	3	3	3	28
										1	.	.	.	.	.	.	.	1	1	1	.	.	.	.	4	.	.	.	.	.	2	3	29
											1	.	.	.	.	.	.	.	.	.	.	.	.	.	.	.	.	3	.	.	.	.	30
												1	.	.	.	.	.	.	.	.	.	.	2	.	.	1	.	3	3	.	.	1	31
													1	.	.	.	.	.	.	.	.	.	.	2	.	.	1	.	3	3	.	.	32
														1	.	.	.	.	.	.	.	.	.	.	.	.	.	.	.	3	.	.	33
															1	.	.	.	.	.	.	.	2	.	.	1	.	3	1	.	.	3	34
																1	.	.	.	.	1	1	2	2	2	2	2	1	3	1	2	3	35
																	1	.	.	.	.	.	.	2	.	.	1	.	1	3	.	.	36
																		1	.	.	.	.	.	.	2	.	.	.	.	.	1	3	37
																			1	.	.	.	.	.	2	.	.	.	.	.	1	1	38
																				1	.	.	.	.	.	.	.	.	.	.	.	.	39
																					1	1	1	1	1	2	2	1	1	1	2	1	40
																						1	.	.	.	1	1	.	.	.	1	.	41
																							1	.	.	1	.	1	1	.	.	1	42
																								1	.	.	1	.	1	1	.	.	43
																									1	.	.	.	.	.	1	1	44
																										1	.	.	.	.	.	.	45
																											1	.	.	.	.	.	46
																												1	.	.	.	.	47
																													1	.	.	.	48
																														1	.	.	49
																															1	.	50
																																1	51

Table 3. *(continued)*

ALGEBRA D4, REPRESENTATIONS OF CLASS 1

4	12	20	36	28	28	36	44	52	52	52	68	60	60	100	76	84	76	68	84	76	84	92	92	100	108	100	124	100	108	124	116	108	108	148	124	132	124	116	132	196	156	148	172	132	156	132	140	148	140	172	140	S.P.
0	32	48	0	32	32	96	96	32	32	48	48	96	96	0	32	32	32	96	32	96	192	96	96	48	96	96	32	96	32	32	48	96	96	48	96	96	96	192	96	0	32	32	32	96	32	96	96	32	96	32	192	O.S.
7	13	17	19	19	19	23	25	25	25	27	29	29	29	31	31	31	31	31	31	33	35	35	35	37	37	37	37	37	37	37	39	39	39	41	41	41	41	41	41	43	43	43	43	43	43	43	43	43	43	43	45	LEVEL
8	56	160	112	224	224	840	1296	672	672	1400	1568	2800	2800	672	2400	1680	2400	4536	1680	6160	12936	7392	7392	7840	8800	12320	3696	12320	6600	3696	10752	18144	18144	7776	22680	16800	22680	39200	16800	2640	12320	15400	7392	30576	12320	39424	28512	15400	28512	7392	72800	DIM'N
1	0	1	3	1	1	0	2	0	0	1	3	1	1	5	3	1	3	1	1	0	2	0	0	1	4	2	0	2	0	0	3	1	1	5	3	1	3	1	1	7	5	3	1	0	5	3	1	3	1	1	2	
0 0	1 0	0 1	0 0	0 0	2 0	1 1	1 0	1 0	3 0	0 2	0 1	0 1	2 1	0 0	0 0	0 0	2 0	2 0	4 0	1 2	1 1	1 1	3 1	0 3	1 0	1 0	1 0	3 0	3 0	5 0	0 2	0 2	2 2	0 1	0 1	0 1	2 1	2 1	4 1	0 0	0 0	0 0	0 0	1 3	2 0	2 0	2 0	4 0	4 0	6 0	1 2	
0	1	0	0	2	0	1	1	3	1	0	0	2	0	0	2	4	0	2	0	1	1	3	1	0	1	3	5	1	3	1	0	2	0	0	2	4	0	2	0	0	2	4	6	1	0	2	4	0	2	0	1	WEIGHT
0	1	2	3	4	5	6	7	8	9	10	11	12	13	14	15	16	17	18	19	20	21	22	23	24	25	26	27	28	29	30	31	32	33	34	35	36	37	38	39	40	41	42	43	44	45	46	47	48	49	50	51	NUMBER
1	3	6	3	6	6	17	21	10	10	21	18	33	33	6	22	15	22	45	15	57	101	54	54	56	54	78	21	78	39	21	62	108	108	36	111	80	111	204	80	10	48	61	28	147	48	174	120	61	120	28	306	0
	1	2	1	3	3	9	12	6	6	11	10	21	21	3	15	10	15	30	10	37	69	38	38	36	38	57	15	57	29	15	42	78	78	24	84	60	84	152	60	6	37	48	21	107	37	133	93	48	93	21	231	1
		1	1	1	1	4	7	3	3	7	8	12	12	3	10	6	10	19	6	23	48	25	25	26	29	40	10	40	21	10	34	54	54	21	63	43	63	111	43	6	29	37	15	77	29	102	70	37	70	15	177	2
			1	.	.	.	3	.	.	2	6	3	3	3	6	1	6	6	1	7	24	8	8	12	21	18	3	18	10	3	24	24	24	18	39	18	39	54	18	6	22	21	6	37	22	60	36	21	36	6	105	3
				1	.	2	3	3	1	3	3	9	6	1	6	6	6	12	3	15	30	21	15	16	18	30	10	27	15	6	20	42	36	12	45	38	45	81	29	3	21	29	15	57	21	75	57	28	51	10	129	4
					1	2	3	1	3	3	3	6	9	1	6	3	6	12	6	15	30	15	21	16	18	27	6	30	15	10	20	36	42	12	45	29	45	81	38	3	21	28	10	57	21	75	51	29	57	15	129	5
						1	1	1	1	2	2	4	4	1	3	3	3	7	3	10	19	12	12	12	12	19	6	19	10	6	15	27	27	10	31	25	31	58	25	3	15	21	10	43	15	55	40	21	40	10	96	6
							1	.	.	1	2	2	2	1	3	1	3	3	1	5	14	6	6	8	12	12	3	12	6	3	13	18	18	10	26	15	26	38	15	3	15	15	6	29	15	42	27	15	27	6	75	7
								1	.	.	.	2	.	.	1	3	.	3	.	3	6	9	3	4	3	12	6	6	6	1	5	15	9	3	15	21	12	30	8	1	6	15	10	21	6	27	30	10	18	3	48	8
									1	.	.	.	2	.	.	.	1	3	3	3	6	3	9	4	3	6	1	12	6	6	5	9	15	3	12	8	15	30	21	1	6	10	3	21	6	27	18	15	30	10	48	9
										1	1	1	1	1	1	1	1	1	1	4	8	4	4	7	7	7	3	7	3	3	9	13	13	8	16	12	16	26	12	3	10	10	6	24	10	29	19	10	19	6	55	10
											1	.	.	1	1	.	1	.	.	.	4	.	.	2	7	3	.	3	.	.	7	4	4	8	12	3	12	9	3	3	10	6	1	8	10	19	6	6	6	1	31	11
												1	.	.	1	1	.	1	.	2	4	4	2	3	3	7	3	3	3	1	4	10	7	3	12	12	9	19	6	1	6	10	6	16	6	19	19	6	12	3	37	12
													1	.	.	.	1	1	1	2	4	2	4	3	3	3	1	7	3	3	4	7	10	3	9	6	12	19	12	1	6	6	3	16	6	19	12	10	19	6	37	13
														1	.	.	.	.	.	.	.	.	.	.	3	.	.	.	.	.	2	.	.	6	3	.	3	.	.	3	6	1	.	.	6	6	.	1	.	.	7	14
															1	.	.	.	.	.	2	.	.	1	3	3	.	1	.	.	3	3	3	3	9	3	6	6	2	1	6	6	1	6	6	12	6	3	3	1	21	15
																1	.	.	.	.	.	2	.	.	.	3	3	.	1	.	.	3	.	.	3	9	.	6	.	.	1	6	6	4	.	6	12	.	3	.	9	16
																	1	.	.	.	2	.	.	1	3	1	.	3	.	.	3	3	3	3	6	2	9	6	3	1	6	3	1	6	6	12	3	6	6	1	21	17
																		1	.	1	2	2	2	2	1	3	1	3	3	1	3	5	5	2	6	6	6	14	6	1	3	6	3	11	3	12	12	6	12	3	26	18
																			1	.	.	.	2	.	.	.	.	3	1	3	.	.	3	.	.	.	3	6	9	.	.	.	.	4	1	6	3	6	12	6	9	19

1	1	1	1	2	1	1	1	1	1	1	2	4	4	2	4	4	4	8	4	1	3	3	3	10	3	7	7	3	7	3	20	20
	1	.	.	1	1	1	.	1	.	.	2	2	2	2	4	2	4	4	2	1	3	3	1	5	3	7	3	3	3	1	15	21
		1	.	.	.	1	1	.	1	.	.	2	.	.	2	4	.	4	.	.	1	3	3	3	.	3	7	.	3	.	7	22
			1	.	.	.	.	1	1	1	.	.	2	.	.	.	2	4	4	.	.	.	.	3	1	3	3	3	7	3	7	23
				1	.	.	.	.	.	.	1	1	1	1	1	1	1	1	1	1	1	1	1	4	1	1	1	1	1	1	8	24
					1	.	.	.	.	.	1	.	.	2	2	.	2	.	.	1	3	1	.	.	3	3	.	1	.	.	5	25
						1	.	.	.	.	.	1	.	.	2	2	.	2	.	.	1	3	1	2	.	3	3	.	1	.	5	26
							1	.	.	.	.	.	.	.	.	2	.	.	.	.	.	1	3	.	.	.	3	.	.	.	.	27
								1	.	.	.	.	1	.	.	.	2	2	2	.	.	.	.	2	1	3	1	3	3	1	5	28
									1	.	.	.	.	.	.	.	.	2	.	.	.	.	.	1	.	1	3	.	3	.	3	29
										1	.	.	.	.	.	.	.	.	2	.	.	.	.	.	.	.	.	1	3	3	.	30
											1	.	.	1	1	.	1	.	.	1	1	1	.	.	1	1	.	1	.	.	4	31
												1	.	.	1	1	.	1	.	.	1	1	1	2	.	1	1	.	1	.	4	32
													1	.	.	.	1	1	1	.	.	.	.	2	1	1	1	1	1	1	4	33
														1	.	.	.	.	.	1	1	.	.	.	1	.	.	.	.	.	.	34
															1	.	.	.	.	.	1	1	.	.	.	1	.	.	.	.	2	35
																1	.	.	.	.	.	1	1	.	.	.	1	.	.	.	.	36
																	1	.	.	.	.	.	.	.	1	1	.	1	.	.	2	37
																		1	.	.	.	.	.	1	.	1	1	.	1	.	2	38
																			1	.	.	.	.	.	.	.	.	1	1	1	.	39
																				1	.	.	.	.	.	.	.	.	.	.	.	40
																					1	.	.	.	.	.	.	.	.	.	.	41
																						1	.	.	.	.	.	.	.	.	.	42
																							1	.	.	.	.	.	.	.	.	43
																								1	.	.	.	.	.	.	1	44
																									1	.	.	.	.	.	.	45
																										1	.	.	.	.	1	46
																											1	.	.	.	.	47
																												1	.	.	.	48
																													1	.	.	49
																														1	.	50
																															1	51

Table 3. *(continued)*

ALGEBRA D5, REPRESENTATIONS OF CLASS 0

0	8	16	16	24	32	32	32	40	64	40	48	48	S.P.
1	40	10	80	240	40	80	80	80	10	480	80	320	O.S.
1	15	17	21	27	29	29	29	31	33	35	37	37	LEVEL
1	45	54	210	945	770	1 050	1 050	1 386	660	5 940	4 125	8 085	DIM'N
0 0 0 0 0	0 1 0 0 0	2 0 0 0 0	0 0 1 0 1	1 0 0 1 0	0 2 0 0 0	1 0 0 0 2	1 0 2 0 0	2 1 0 0 0	4 0 0 0 0	0 1 1 0 1	0 0 0 2 0	2 0 1 0 1	WEIGHT
0	1	2	3	4	5	6	7	8	9	10	11	12	NUMBER
1	5	4	10	25	20	20	20	26	10	80	45	85	0
	1	1	3	10	7	9	9	11	4	38	23	45	1
		1	.	4	3	3	3	8	4	14	12	24	2
			1	3	2	4	4	3	1	18	11	22	3
				1	1	1	1	2	1	7	5	10	4
					1	.	.	1	1	3	2	3	5
						1	.	.	.	3	2	4	6
							1	.	.	3	2	4	7
								1	1	.	.	3	8
									1	.	.	.	9
										1	1	1	10
											1	.	11
												1	12
													13
													14
													15
													16
													17

56	56	56	80	64	80	72	72	64	64	80	80	80	S.P.
160	160	480	16	80	16	40	240	320	320	80	80	80	O.S.
39	39	41	41	41	41	43	43	43	43	45	45	45	LEVEL
6 930	6 930	17 920	2 772	8 910	2 772	7 644	14 784	23 040	23 040	16 380	17 325	17 325	DIM'N
0 0 0 1 2	0 0 2 1 0	1 1 0 1 0	0 0 0 0 4	0 0 2 0 2	0 0 4 0 0	0 3 0 0 0	3 0 0 1 0	1 1 0 0 2	1 1 2 0 0	2 2 0 0 0	3 0 0 0 2	3 0 2 0 0	WEIGHT
13	14	15	16	17	18	19	20	21	22	23	24	25	NUMBER
60	60	160	16	70	16	64	104	160	160	110	95	95	0
33	33	84	10	37	10	31	58	92	92	59	58	58	1
15	15	48	6	16	6	18	40	48	48	44	36	36	2
18	18	40	6	21	6	14	28	52	52	27	34	34	3
8	8	22	3	10	3	9	18	26	26	19	19	19	4
3	3	12	1	6	1	7	10	12	12	13	9	9	5
6	3	8	3	6	1	3	6	16	12	6	10	10	6
3	6	8	1	6	3	3	6	12	16	6	10	10	7
.	.	6	.	.	.	3	10	6	6	10	9	9	8
.	.	.	.	.	.	.	4	.	.	3	3	3	9
2	2	4	1	3	1	2	3	6	6	4	4	4	10
1	1	2	1	1	1	1	1	2	2	3	1	1	11
.	.	2	.	.	.	1	3	3	3	3	4	4	12
1	.	.	1	1	.	.	.	2	.	.	1	.	13
	1	.	.	1	1	.	.	.	2	.	.	1	14
		1	.	.	.	1	1	1	1	2	1	1	15
			1	.	.	.	.	.	.	.	.	.	16
				1	.	.	.	.	.	.	.	.	17

104	72	144	80	88	88	88	88	96	96	112	96	104	S.P.
80	960	10	480	320	320	240	960	240	480	320	480	480	O.S.
47	47	49	49	49	49	51	51	53	53	53	53	55	LEVEL
12 870	72 765	4 290	73 710	50 688	50 688	70 070	143 000	112 320	128 700	81 081	128 700	174 636	DIM'N
4 1 0 0 0	1 0 1 1 1	6 0 0 0 0	0 2 1 0 1	1 0 1 0 3	1 0 3 0 1	0 1 0 2 0	2 1 1 0 1	2 0 0 2 0	0 1 0 1 2	4 0 1 0 1	0 1 2 1 0	1 2 0 1 0	WEIGHT
26	27	28	29	30	31	32	33	34	35	36	37	38	NUMBER
70	405	20	380	224	224	310	620	440	480	281	480	646	0
38	242	10	224	140	140	191	389	276	309	182	309	400	1
32	136	10	125	80	80	120	248	184	186	128	186	268	2
17	142	4	132	88	88	113	234	166	198	112	198	237	3
14	78	4	75	50	50	69	142	107	118	74	118	157	4
11	41	4	45	28	28	41	80	68	69	45	69	105	5
4	44	1	42	36	28	38	78	58	81	40	69	84	6
4	44	1	42	28	36	38	78	58	69	40	81	84	7
11	17	4	21	10	10	21	59	44	33	45	33	69	8
8	.	4	.	.	.	.	14	12	.	24	.	15	9
3	23	1	23	18	18	23	42	35	44	22	44	54	10
2	12	1	11	10	10	15	21	21	26	10	26	34	11
3	9	1	12	6	6	11	30	22	21	22	21	36	12
.	6	.	5	10	4	6	9	9	20	4	11	13	13
.	6	.	5	4	10	6	9	9	11	4	20	13	14
2	4	1	7	3	3	7	14	12	12	10	12	24	15
.	.	.	.	4	.	.	.	.	5	.	.	.	16
.	3	.	2	4	4	3	3	6	8	1	8	7	17

104	120	104	104	120	128	120	112	128	112	112	112	128	S.P.
480	160	480	480	160	40	480	320	80	320	320	320	80	O.S.
55	55	55	55	55	57	57	57	57	57	57	57	57	LEVEL
199 017	64 350	199 017	189 189	64 350	52 920	207 360	242 550	90 090	210 210	242 550	274 560	90 090	DIM'N
2 0 0 1 2	0 1 0 0 4	2 0 2 1 0	0 1 2 0 2	0 1 4 0 0	0 4 0 0 0	3 1 0 1 0	1 2 0 0 2	2 0 0 0 4	0 0 1 2 1	1 2 2 0 0	2 0 2 0 2	2 0 4 0 0	WEIGHT
39	40	41	42	43	44	45	46	47	48	49	50	51	NUMBER
651	180	651	637	180	180	640	740	230	620	740	800	230	0
432	124	432	409	124	105	412	489	158	414	489	538	158	1
276	81	276	244	81	72	304	313	104	267	313	352	104	2
282	84	282	267	84	60	248	320	109	274	320	358	109	3
176	53	176	164	53	43	176	202	71	174	202	229	71	4
105	31	105	105	31	34	120	126	46	108	126	142	46	5
114	42	108	108	31	22	92	133	52	114	123	150	46	6
108	31	114	108	42	22	92	123	46	114	133	150	52	7
66	15	66	45	15	21	102	82	25	57	82	82	25	8
15	.	15	.	.	6	48	16	6	.	16	16	6	9
65	23	65	67	23	16	60	77	31	72	77	93	31	10
36	16	36	39	16	11	38	44	19	47	44	58	19	11
39	10	39	30	10	11	50	51	16	37	51	51	16	12
27	16	18	27	6	4	14	29	19	29	21	37	10	13
18	6	27	27	16	4	14	21	10	29	29	37	19	14
20	6	20	19	6	9	30	30	9	23	30	29	9	15
6	10	.	6	.	.	.	5	10	5	.	10	.	16
12	6	12	18	6	3	8	12	10	18	12	22	10	17

1	.	.	.	.	.	.	.	.	.	.	.	.	4	.	.	.	.	.	5	.	.	.	6	6	10	.	.	.	.	5	5	10	10	18
	1	.	.	.	1	.	.	1	.	1	3	.	.	2	3	2	3	3	3	12	3	1	3	6	1	7	12	12	1	7	12	6	1	19
		1	.	.	1	1	1	2	.	1	.	.	.	.	7	5	.	10	.	8	8	.	8	.	.	3	22	10	3	.	10	10	3	20
			1	.	.	1	.	.	2	.	3	3	1	3	6	5	9	4	6	9	15	6	9	12	3	3	11	19	9	15	15	18	4	21
				1	.	.	1	.	2	.	3	1	3	3	6	5	6	4	9	9	9	3	15	12	6	3	11	15	4	15	19	18	9	22
					1	.	.	1	.	1	.	.	.	.	3	2	.	3	.	6	3	.	3	.	.	3	12	6	1	.	6	6	1	23
						1	.	.	.	.	.	.	.	.	3	2	.	4	.	3	6	.	3	.	.	1	8	6	3	.	6	6	1	24
							1	.	.	.	.	.	.	.	3	2	.	4	.	3	3	.	6	.	.	1	8	6	1	.	6	6	3	25
								1	.	1	.	.	.	.	.	.	.	3	.	.	.	.	.	.	.	.	6	.	.	.	.	.	.	26
									1	.	1	1	1	2	2	3	4	1	4	5	6	3	6	7	3	2	6	8	4	10	8	10	4	27
										1	.	.	.	.	.	.	.	.	.	.	.	.	.	.	.	.	.	.	.	.	.	.	.	28
											1	.	.	1	1	1	2	1	2	4	2	1	2	3	1	2	4	6	1	5	6	3	1	29
												1	.	.	.	.	2	.	.	.	3	3	.	3	.	.	.	3	4	3	.	4	.	30
													1	.	.	.	.	.	2	.	.	.	3	3	3	.	.	.	.	3	3	4	4	31
														1	.	1	1	.	1	2	1	1	1	1	1	1	2	2	1	4	2	1	1	32
															1	1	.	1	.	2	2	.	2	.	.	1	4	3	1	.	3	3	1	33
																1	.	.	.	1	1	.	1	.	.	1	2	1	1	.	1	1	1	34
																	1	.	.	.	1	1	.	1	.	.	.	2	1	2	.	1	.	35
																		1	.	.	.	.	.	.	.	.	2	.	.	.	.	.	.	36
																			1	.	.	.	1	1	1	.	.	.	.	2	2	1	1	37
																				1	.	.	.	.	.	1	1	1	.	.	1	.	.	38
																					1	.	.	.	.	.	.	1	1	.	.	1	.	39
																						1	.	.	.	.	.	.	1	.	.	.	.	40
																							1	.	.	.	.	.	.	.	1	1	1	41
																								1	.	.	.	.	.	1	.	1	.	42
																									1	.	.	.	.	.	.	.	1	43
																										1	.	.	.	.	.	.	.	44
																											1	.	.	.	.	.	.	45
																												1	.	.	.	.	.	46
																													1	.	.	.	.	47
																														1	.	.	.	48
																															1	.	.	49
																																1	.	50
																																	1	51

Table 3. *(continued)*

ALGEBRA D5, REPRESENTATIONS OF CLASS 1

4	12	20	20	20	36	28	36	44	44	44	52	52	52	68	52	100	68	68	60	68	76	76	76	76	84	84	100	84	84	100	100	108	92	92	116	116	92	116	148	108	100	108	108	196	116	108	116	116	116	116	140	S.P.
10	80	16	16	80	10	320	240	240	160	160	80	80	80	80	320	10	80	80	960	240	320	240	480	480	480	160	80	160	320	80	80	240	320	320	80	80	960	80	80	80	960	480	480	10	160	960	320	160	320	480	160	O.S.
9	19	21	21	23	25	29	33	35	35	35	37	37	37	39	39	41	41	41	43	45	45	47	47	47	49	49	49	49	49	49	51	51	51	51	53	53	53	53	55	55	55	55	55	57	57	57	57	57	57	59	59	LEVEL
10	120	126	126	320	210	1728	2970	4312	3696	3696	4410	4950	4950	4608	10560	1782	6930	6930	36750	27720	28160	34398	48114	48114	68640	46800	20790	46800	64680	20790	37632	42120	90090	90090	48510	50050	192192	50050	31680	70070	299520	150150	150150	9438	144144	380160	216216	144144	216216	376320	102960	DIM'N
1 0 0 0 0	0 0 0 1 0	0 0 0 0 2	0 0 2 0 0	1 1 0 0 0	3 0 0 0 0	1 0 1 0 1	0 1 0 1 0	2 0 0 1 0	0 1 0 0 2	0 1 2 0 0	1 2 0 0 0	2 0 0 0 2	2 0 2 0 0	3 1 0 0 0	0 0 1 1 1	5 0 0 0 0	0 0 1 0 3	0 0 3 0 1	1 1 1 0 1	1 0 0 2 0	3 0 1 0 1	0 2 0 1 0	1 0 0 1 2	1 0 2 1 0	2 1 0 1 0	0 2 0 0 2	1 0 0 0 4	0 2 2 0 0	1 0 2 0 2	1 0 4 0 0	1 3 0 0 0	4 0 0 1 0	2 1 0 0 2	2 1 2 0 0	3 2 0 0 0	4 0 0 0 2	0 1 1 1 1	4 0 2 0 0	5 1 0 0 0	0 0 0 3 0	2 0 1 1 1	0 1 1 0 3	0 1 3 0 1	7 0 0 0 0	0 0 0 2 2	1 2 1 0 1	2 0 1 0 3	0 0 2 2 0	2 0 3 0 1	1 1 0 2 0	0 0 0 1 4	WEIGHT
0	1	2	3	4	5	6	7	8	9	10	11	12	13	14	15	16	17	18	19	20	21	22	23	24	25	26	27	28	29	30	31	32	33	34	35	36	37	38	39	40	41	42	43	44	45	46	47	48	49	50	51	NUMBER
1	4	3	3	8	4	24	33	40	32	32	38	36	36	32	72	10	37	37	205	136	128	156	198	198	272	176	65	176	236	65	136	136	297	297	148	134	600	134	80	186	832	388	388	20	336	984	500	336	500	880	192	0
	1	1	1	2	1	10	14	18	16	16	16	19	19	14	38	4	21	21	113	76	74	86	120	120	155	107	42	107	145	42	74	79	187	187	83	87	378	87	44	120	538	256	256	10	226	634	338	226	338	572	135	1
		1	.	.	.	4	5	6	10	6	5	10	10	4	20	1	15	10	60	40	40	44	75	69	80	70	30	60	90	25	36	40	115	115	40	55	236	55	20	75	340	180	160	4	154	400	230	150	226	360	100	2
			1	.	.	4	5	6	6	10	5	10	10	4	20	1	10	15	60	40	40	44	69	75	80	60	25	70	90	30	36	40	115	115	40	55	236	55	20	75	340	160	180	4	150	400	226	154	230	360	90	3
				1	1	3	6	10	6	6	10	9	9	11	17	4	10	10	59	44	45	51	66	66	102	60	25	60	82	25	52	58	114	114	65	58	227	58	38	78	343	157	157	10	144	408	218	144	218	387	90	4
					1	.	.	4	.	.	3	3	3	8	.	4	.	.	14	12	24	15	15	15	48	16	6	16	16	6	24	40	48	48	44	36	64	36	32	29	136	43	43	10	48	168	80	48	80	168	32	5
						1	2	3	3	3	3	4	4	3	9	1	6	6	30	22	22	26	39	39	50	36	16	36	51	16	25	28	67	67	30	34	141	34	17	48	211	105	105	4	96	253	145	96	145	239	63	6
							1	1	1	1	2	1	1	2	4	1	3	3	14	12	10	16	20	20	30	20	9	20	29	9	18	18	36	36	22	19	84	19	14	30	126	65	65	4	60	158	90	60	90	158	41	7
								1	.	.	1	1	1	2	.	1	.	.	7	5	10	8	8	8	22	10	3	10	10	3	12	18	26	26	19	19	40	19	14	16	78	29	29	4	31	104	50	31	50	100	21	8
									1	.	.	1	.	.	2	.	3	1	6	5	4	6	15	9	11	13	9	9	18	4	6	6	22	18	7	10	50	10	4	18	74	51	37	1	45	90	66	36	55	90	37	9
										1	.	.	1	.	2	.	1	3	6	5	4	6	9	15	11	9	4	13	18	9	6	6	18	22	7	10	50	10	4	18	74	37	51	1	36	90	55	45	66	90	24	10
											1	.	.	1	.	1	.	.	3	2	3	6	3	3	12	6	1	6	6	1	10	10	12	12	13	9	24	9	11	9	41	18	18	4	18	66	28	18	28	65	12	11
												1	.	.	.	.	.	.	3	2	4	3	6	3	8	6	3	6	6	1	4	6	16	12	6	10	24	10	4	9	44	21	18	1	24	60	36	18	28	56	18	12
													1	.	.	.	.	.	3	2	4	3	3	6	8	6	1	6	6	3	4	6	12	16	6	10	24	10	4	9	44	18	21	1	18	60	28	24	36	56	12	13
														1	.	1	.	.	.	.	3	.	.	.	6	.	.	.	.	.	3	10	6	6	10	9	.	9	11	.	17	.	.	4	.	21	10	.	10	21	.	14
															1	.	1	1	2	3	1	3	6	6	6	5	4	5	10	4	4	3	9	9	5	4	29	4	3	12	42	27	27	1	26	51	37	26	37	58	21	15
																1	.	.	.	.	.	.	.	.	.	.	.	.	.	.	.	4	.	.	3	3	.	3	8	.	.	.	.	4	.	.	.	.	.	.	.	16
																	1	.	.	.	.	.	3	.	.	2	4	.	4	.	.	.	3	.	.	1	8	.	.	3	12	18	6	.	14	12	22	6	10	16	18	17
																		1	.	.	.	.	.	3	.	.	.	2	4	4	.	.	.	3	.	.	8	1	.	3	12	6	18	.	6	12	10	14	22	16	4	18

1	1	1	2	2	2	4	3	1	3	3	1	3	3	6	6	4	4	14	4	3	6	23	12	12	1	13	35	18	13	18	36	10	19
	1	.	1	1	1	2	1	1	1	1	1	2	1	2	2	3	1	8	1	2	5	12	7	7	1	9	18	10	9	10	24	8	20
		1	.	.	.	2	.	.	.	.	.	1	3	3	3	3	4	.	4	3	.	9	.	.	1	.	12	6	.	6	11	.	21
			1	.	.	1	1	.	1	.	.	2	1	1	1	2	1	4	1	2	2	4	3	3	1	4	14	3	4	3	14	3	22
				1	.	.	1	1	.	1	.	.	.	2	.	.	1	4	.	.	2	6	7	3	.	7	8	10	4	4	10	8	23
					1	.	.	.	1	1	1	.	.	.	2	.	.	4	1	.	2	6	3	7	.	4	8	4	7	10	10	3	24
						1	.	.	.	.	.	1	1	1	1	2	1	.	1	2	.	4	.	.	1	.	7	3	.	3	7	.	25
							1	.	.	.	.	.	.	1	.	.	1	2	.	.	1	2	3	1	.	3	6	3	3	1	6	3	26
								1	.	.	.	.	.	.	.	.	.	.	.	.	.	.	3	.	.	2	.	4	.	.	.	6	27
									1	.	.	.	.	.	1	.	.	2	1	.	1	2	1	3	.	3	6	1	3	3	6	2	28
										1	.	.	.	.	.	.	.	2	.	.	1	3	3	3	.	3	3	4	3	4	6	3	29
											1	.	.	.	.	.	.	.	.	.	.	.	.	3	.	.	.	.	2	4	.	.	30
												1	.	.	.	1	.	.	.	1	.	.	.	.	1	.	3	.	.	.	2	.	31
													1	.	.	1	1	.	1	2	.	.	.	.	1	.	.	.	.	.	.	.	32
														1	.	.	1	.	.	.	.	2	.	.	.	.	3	3	.	1	3	.	33
															1	.	.	.	1	.	.	2	.	.	.	.	3	1	.	3	3	.	34
																1	.	.	.	1	.	.	.	.	1	,	.	.	.	.	.	.	35
																	1	.	.	.	.	.	.	.	.	.	.	.	.	.	.	.	36
																		1	.	.	1	1	1	1	.	2	2	1	2	1	4	2	37
																			1	.	.	.	.	.	.	.	.	.	.	.	.	.	38
																				1	.	.	.	.	1	.	.	.	.	.	.	.	39
																					1	.	.	.	.	1	.	.	1	.	2	1	40
																						1	.	.	.	.	1	1	.	1	2	.	41
																							1	.	.	1	.	1	.	.	.	2	42
																								1	.	.	.	.	1	1	.	.	43
																									1	.	.	.	.	.	.	.	44
																										1	.	.	.	.	.	1	45
																											1	.	.	.	1	.	46
																												1	.	.	.	.	47
																													1	.	.	.	48
																														1	.	.	49
																															1	.	50
																																1	51

Table 3. *(continued)*

ALGEBRA D5, REPRESENTATIONS OF CLASS 2

NUMBER	0	1	2	3	4	5	6	7	8	9	10	11	12	13	14	15	16	17	18	19	20	21	22	23	24	25
S.P.	5	13	21	29	29	45	37	37	53	45	53	53	61	61	61	85	69	77	69	77	85	77	77	93	85	93
O.S.	16	80	160	80	160	16	80	320	80	480	160	320	80	320	480	80	480	160	480	320	80	160	320	320	320	160
LEVEL	11	19	25	27	29	31	31	33	35	37	39	39	39	41	43	43	45	45	45	47	47	47	47	49	49	49
DIM'N	16	144	560	720	1200	672	1440	3696	2640	8800	8064	11088	5280	15120	25200	7920	38016	17280	34992	43680	23760	30800	49280	48048	55440	29568
WEIGHT	0	1	0	2	0	0	0	1	3	1	0	1	1	2	0	4	2	0	0	1	2	0	2	3	0	0
	0	0	1	0	0	0	0	1	0	0	2	0	0	1	1	0	0	1	1	2	0	0	0	1	0	0
	1 0	0 0	1 0	1 0	0 1	0 0	2 0	0 0	0 0	1 1	1 0	1 0	3 0	1 0	0 1	1 0	0 1	0 0	2 0	0 0	0 0	1 2	2 0	0 0	1 1	3 1
	0	1	0	0	1	3	1	1	1	0	0	2	0	0	1	0	1	3	1	1	3	0	1	1	2	0
0	1	4	10	10	16	6	16	36	20	66	64	69	30	86	130	36	170	70	166	180	86	120	190	164	190	94
1		1	3	4	6	3	6	16	10	32	26	36	16	46	71	20	98	42	87	103	52	71	114	100	115	58
2			1	1	2	1	3	6	4	15	13	18	9	22	37	10	53	23	51	56	31	41	66	56	69	37
3				1	.	.	.	3	4	6	6	6	3	16	17	10	32	10	21	36	16	21	36	46	33	18
4					1	1	1	2	1	6	5	10	4	9	20	4	27	16	27	29	19	23	37	28	44	21
5						1	.	.	.	.	.	4	.	.	5	.	6	10	6	5	10	5	10	4	20	4
6							1	.	.	3	2	4	4	3	8	1	12	6	18	12	10	14	22	12	24	18
7								1	1	2	3	3	1	6	9	4	15	6	12	19	9	11	18	22	21	10
8									1	.	.	.	.	3	.	4	6	.	.	6	3	.	6	16	.	.
9										1	1	1	1	2	4	1	6	3	7	8	4	7	10	9	12	8
10											1	.	.	1	2	1	2	1	3	6	1	3	3	6	6	3
11												1	.	.	2	.	3	3	3	3	4	3	4	3	9	3
12													1	.	.	.	.	.	3	.	.	2	4	.	3	6
13														1	.	1	2	.	.	3	1	.	3	6	.	.
14															1	.	1	1	1	2	1	2	1	2	4	2
15																1	.	.	.	.	.	.	.	3	.	.
16																	1	.	.	1	1	.	1	2	.	.
17																		1	.	.	1	.	.	.	2	.
18																			1	.	.	1	1	.	2	2

NUMBER	26	27	28	29	30	31	32	33	34	35	36	37	38	39	40	41	42	43	44	45	46	47	48	49	50	51
S.P.	125	85	109	101	125	101	101	93	101	101	109	109	117	109	117	133	109	117	125	149	173	117	125	133	133	141
O.S.	80	960	80	80	16	160	480	960	320	480	320	320	80	480	480	320	960	960	160	80	80	480	320	320	320	480
LEVEL	51	51	51	51	51	53	53	53	53	55	55	55	55	57	57	57	57	59	59	59	59	59	59	59	61	61
DIM'N	20592	144144	26400	39600	9504	70560	124800	205920	102960	196560	155232	164736	80080	258720	196560	129360	364000	529200	198000	68640	48048	388080	274560	185328	332640	343200
WEIGHT	5	1	0	0	0	0	3	1	1	1	2	3	3	0	1	4	1	2	0	1	6	0	1	1	1	4
	0	1	0	0	0	3	0	1	1	0	2	0	0	2	0	1	0	1	2	0	0	2	0	0	3	0
	0 0	1 1	1 0	3 0	5 0	1 0	1 1	1 0	3 0	0 2	1 0	1 0	3 0	0 1	0 1	1 0	2 1	0 1	0 0	0 0	1 0	2 0	2 0	4 0	0 0	0 1
	1	0	4	2	0	0	0	2	0	1	0	2	0	1	3	0	1	1	3	5	0	1	3	1	1	1
0	56	400	70	116	22	210	360	600	280	616	406	406	186	630	430	280	860	1166	400	116	84	840	560	349	660	626
1	35	292	46	70	15	126	224	378	179	332	259	268	124	406	289	185	558	768	272	82	56	552	370	238	436	430
2	20	174	28	46	10	77	136	234	116	209	157	171	81	257	190	115	367	496	179	56	35	369	249	166	282	284
3	20	104	15	22	6	49	98	132	64	128	118	114	52	162	112	100	216	346	109	36	35	222	148	100	205	224
4	10	98	21	28	6	43	78	147	70	131	90	106	52	163	132	66	236	311	127	42	20	237	173	109	178	179
5	1	24	15	10	1	10	20	60	20	50	21	40	20	60	75	15	90	110	70	30	4	90	90	40	60	60
6	4	56	10	21	6	24	42	84	52	77	48	64	34	94	78	34	161	185	74	25	10	162	112	88	104	107
7	10	58	10	15	3	30	53	81	38	78	66	66	31	103	75	56	138	212	74	25	20	147	103	66	131	136
8	10	17	.	.	.	10	32	21	10	21	36	36	16	33	18	46	33	104	22	6	20	45	22	15	64	98
9	4	32	6	10	3	16	27	46	26	46	35	37	19	61	46	28	92	123	47	16	10	96	68	50	77	78
10	4	18	3	6	1	13	15	24	12	26	25	18	9	41	25	22	51	79	28	9	10	60	42	28	60	53
11	1	14	6	6	1	7	12	30	12	29	15	22	10	36	39	12	54	71	36	16	4	57	51	28	43	42
12	.	8	1	6	3	3	6	12	16	12	6	10	10	16	12	4	44	32	14	4	1	42	28	36	19	20
13	4	9	.	.	.	6	15	12	6	11	19	18	9	21	10	22	21	58	15	3	10	30	15	10	40	53
14	1	8	3	3	1	5	6	14	6	17	11	10	4	23	20	9	31	44	20	9	4	34	29	18	32	27
15	4	.	.	.	.	.	6	.	.	.	6	6	3	.	.	16	.	17	.	.	10	.	.	.	10	32
16	1	4	.	.	.	3	6	7	3	7	8	10	4	12	8	9	12	32	10	3	4	19	10	6	22	27
17	.	.	3	1	.	.	.	6	.	5	.	4	.	6	15	.	9	11	13	9	.	9	18	4	6	6
18	.	4	1	3	1	2	3	6	6	8	4	4	4	10	9	3	23	20	9	4	1	23	18	18	14	12

1	.	.	.	1	.	.	1	2	.	.	.	3	2	3	1	3	6	3	1	9	3	6	6	18	6	1	4	12	6	3	19	15	19
	1	.	.	.	.	.	.	.	.	.	.	.	.	3	.	2	.	4	.	3	6	.	3	8	6	3	.	6	6	1	4	6	20
		1	.	.	1	1	.	2	1	1	1	1	1	2	2	6	3	1	1	6	6	2	12	12	5	4	1	11	10	10	10	6	21
			1	.	.	.	.	2	.	.	.	1	3	3	3	3	3	4	4	6	3	3	9	14	6	1	1	12	6	6	10	12	22
				1	.	.	1	.	.	.	.	.	.	2	.	.	.	3	3	1	.	.	6	.	9	.	.	4	.	.	.	6	15	23
					1	.	.	.	1	1	.	.	.	2	.	3	.	1	.	3	6	.	6	6	5	4	.	5	10	4	4	3	24
						1	.	.	.	1	1	.	.	.	2	.	.	.	1	.	.	.	6	.	.	.	.	5	4	10	.	.	25
							1	.	.	.	.	.	.	.	.	.	.	.	.	.	.	3	.	.	.	.	4	.	.	.	.	6	26
								1	.	.	.	1	1	1	1	2	2	1	1	4	2	2	4	8	3	1	1	7	3	3	8	6	27
									1	.	.	.	.	.	.	.	.	.	.	.	3	.	.	.	2	4	.	.	4	.	.	.	28
										1	.	.	.	.	.	.	.	.	.	.	.	.	3	.	.	.	.	2	4	4	.	.	29
											1	.	.	.	.	.	.	.	.	.	.	.	.	.	.	.	.	.	.	4	.	.	30
												1	.	.	.	.	1	.	.	2	.	1	.	2	1	.	1	3	.	.	6	2	31
													1	.	.	.	1	1	1	.	.	2	.	4	.	.	1	.	.	.	3	6	32
														1	.	1	.	1	.	2	2	.	2	4	3	1	.	3	3	1	3	3	33
															1	.	.	.	1	.	.	.	2	.	.	.	.	3	1	3	.	.	34
																1	.	.	.	1	1	.	1	2	1	1	.	1	1	1	2	1	35
																	1	.	.	.	.	1	.	2	.	.	1	.	.	.	3	2	36
																		1	.	.	.	.	.	2	.	.	.	.	.	.	1	3	37
																			1	.	.	.	.	.	.	.	.	.	.	.	.	.	38
																				1	.	.	.	1	1	.	.	1	.	.	2	1	39
																					1	.	.	.	1	1	.	.	1	.	.	.	40
																						1	.	.	.	.	1	.	.	.	.	2	41
																							1	.	.	.	.	1	1	1	.	.	42
																								1	.	.	.	.	.	.	1	1	43
																									1	.	.	.	.	.	.	.	44
																										1	.	.	.	.	.	.	45
																											1	.	.	.	.	.	46
																												1	.	.	.	.	47
																													1	.	.	.	48
																														1	.	.	49
																															1	.	50
																																1	51

Table 3. *(continued)*

ALGEBRA D6, REPRESENTATIONS OF CLASS 0

0	1	2	3	4	5	6	7	8	9	10	11	12	13	14	15	16	17	18	19	20	21	22	23	24	25	
0	8	16	16	24	24	24	32	40	64	32	40	48	48	48	48	56	56	56	72	72	56	80	64	104	72	S.P.
1	60	12	240	32	32	480	60	120	12	960	1440	160	960	480	480	192	192	960	60	480	1920	120	240	120	480	O.S.
1	19	21	29	31	31	35	37	39	41	41	47	49	49	49	49	51	51	53	55	55	55	57	57	59	59	LEVEL
1	66	77	495	462	462	2079	1638	2860	1287	8085	21021	14014	27456	21450	21450	27027	27027	57344	23100	45045	90090	48114	55055	35750	84942	DIM'N
0	0	2	0	0	0	1	0	2	4	1	0	0	2	0	0	2	2	1	0	3	0	2	0	4	0	
0	1	0	0	0	0	0	2	1	0	0	1	0	0	1	1	0	0	1	3	0	0	2	0	1	0	
0	0	0	0	0	0	1	0	0	0	0	0	2	0	0	0	0	0	1	0	1	1	0	0	0	0	
0 0	0 0	0 0	0 1	0 0	2 0	0 0	0 0	0 0	0 0	1 0	0 1	0 0	0 1	0 0	2 0	0 0	2 0	0 0	0 0	0 0	1 0	0 0	0 2	0 0	0 1	
0	0	0	0	2	0	0	0	0	0	1	0	0	0	2	0	2	0	0	0	0	1	0	0	0	2	WEIGHT
0	1	2	3	4	5	6	7	8	9	10	11	12	13	14	15	16	17	18	19	20	21	22	23	24	25	NUMBER
1	6	5	15	10	10	39	30	40	15	85	165	94	176	130	130	135	135	320	120	205	390	210	215	130	270	0
	1	1	4	3	3	13	9	14	5	34	69	41	80	59	59	66	66	144	51	97	193	96	105	60	141	1
		1	.	.	.	5	4	10	5	10	25	20	40	20	20	30	30	80	30	65	80	70	45	50	60	2
			1	1	1	3	2	3	1	13	27	16	32	27	27	31	31	56	19	38	94	36	52	22	75	3
				1	.	.	.	.	.	5	9	5	10	15	10	15	15	16	5	10	45	9	25	5	45	4
					1	.	.	.	.	5	9	5	10	10	15	15	15	16	5	10	45	9	25	5	36	5
						1	1	2	1	3	10	7	14	9	9	12	12	30	12	24	38	25	23	18	33	6
							1	1	1	.	4	3	4	3	3	3	3	16	9	13	14	17	12	14	15	7
								1	1	.	.	.	4	.	.	3	3	8	4	13	.	13	.	14	.	8
									1	.	.	.	.	.	.	.	.	.	.	5	.	4	.	10	.	9
										1	3	2	4	4	4	5	5	8	3	6	18	6	11	4	18	10
											1	1	1	1	1	1	1	4	2	3	7	4	5	3	8	11
												1	.	.	.	.	.	2	1	1	3	3	2	2	3	12
													1	.	.	1	1	2	1	3	.	3	.	3	.	13
														1	.	1	.	.	.	.	3	.	2	.	6	14
															1	.	1	.	.	.	3	.	2	.	3	15
																1	.	.	.	.	.	.	.	.	.	16

26	27	28	29	30	31	32	33	34	35	36	37	38	39	40	41	42	43	44	45	46	47	48	49	50	51	
64	72	144	96	80	80	96	72	80	80	80	88	88	88	88	96	112	96	88	96	104	120	104	120	104	128	S.P.
3840	480	12	32	960	192	32	2880	1440	1920	1920	480	2880	480	480	480	960	960	3840	960	960	192	960	192	960	60	O.S.
59	59	61	61	61	61	61	63	65	65	65	67	67	67	67	69	69	69	69	69	71	71	71	71	71	73	LEVEL
270336	84942	11011	28314	189189	99099	28314	396396	379456	462462	462462	343980	710775	408375	408375	533610	375375	750750	1171456	750750	792792	378378	693693	378378	693693	222156	DIM'N
1	0	6	0	3	0	0	1	0	1	1	0	2	0	0	2	4	2	1	2	1	4	1	4	1	0	
1	0	0	0	0	0	0	0	2	0	0	1	1	2	2	0	0	1	0	1	2	0	0	0	0	4	
0	0	0	0	0	0	0	1	0	1	1	2	0	0	0	2	0	0	0	0	1	0	0	0	0	0	
1 0	2 1	0 0	0 0	1 0	2 0	4 0	0 1	0 1	0 0	2 0	0 0	0 1	0 0	2 0	0 0	0 1	0 0	1 1	2 0	0 0	0 0	1 0	2 0	3 0	0 0	
1	0	0	4	1	2	0	0	0	2	0	0	0	2	0	0	0	2	1	0	0	2	3	0	1	0	WEIGHT
26	27	28	29	30	31	32	33	34	35	36	37	38	39	40	41	42	43	44	45	46	47	48	49	50	51	NUMBER
960	270	35	70	549	267	70	1176	1040	1130	1130	840	1695	915	915	1170	735	1470	2368	1470	1640	606	1145	606	1145	420	0
496	141	15	37	298	146	37	622	548	631	631	453	933	512	512	642	416	856	1360	856	892	364	682	364	682	219	1
240	60	15	16	170	70	16	330	296	320	320	270	565	265	265	410	280	490	720	490	580	230	370	230	370	150	2
248	75	5	21	153	80	21	315	278	348	348	229	480	285	285	328	217	485	776	485	450	211	408	211	408	107	3
120	36	1	15	75	45	10	150	130	195	185	105	225	165	150	150	100	270	440	270	200	120	255	120	239	45	4
120	45	1	10	75	45	15	150	130	185	195	105	225	150	165	150	100	270	440	270	200	120	239	120	255	45	5
114	33	5	10	79	37	10	162	150	172	172	133	277	146	146	202	138	263	406	263	288	124	220	124	220	75	6
48	15	5	6	34	16	6	81	84	80	80	77	149	75	75	123	80	131	208	131	186	66	117	66	117	57	7
24	.	5	.	34	.	.	33	40	32	32	38	109	36	36	79	80	95	72	95	122	66	37	66	37	36	8
.	.	5	.	10	.	.	.	.	.	.	.	25	.	.	20	40	20	.	20	26	30	.	30	.	10	9
52	18	1	6	35	21	6	74	68	92	92	59	123	78	78	88	60	139	228	139	124	65	133	65	133	31	10
20	8	1	3	13	10	3	36	35	42	42	34	63	37	37	51	32	65	116	65	77	31	71	31	71	22	11
6	3	1	1	3	6	1	18	16	18	18	21	30	15	15	30	14	27	58	27	47	12	38	12	38	15	12
10	.	1	.	13	.	.	14	18	16	16	16	45	19	19	32	32	46	38	46	51	31	21	31	21	15	13
8	3	.	3	5	6	1	13	12	26	18	11	21	21	16	16	10	35	64	30	23	15	51	15	38	6	14
8	6	.	1	5	6	3	13	12	18	26	11	21	16	21	16	10	30	64	35	23	15	38	15	51	6	15
4	.	.	.	5	.	.	5	6	10	6	5	15	10	10	10	10	25	20	20	15	15	15	15	10	4	16

1	.	.	.	.	.	.	.	.	4	.	.	.	5	.	.	5	6	6	10	5	15	10	10	10	10	20	20	25	15	15	10	15	15	4	17
	1	1	1	.	2	.	2	.	3	.	1	.	3	.	.	6	10	6	6	10	20	9	9	17	14	18	17	18	33	12	10	12	10	12	18
		1	.	.	1	.	1	.	.	.	1	.	.	.	.	.	4	.	.	3	4	3	3	3	4	3	.	3	16	3	.	3	.	9	19
			1	.	1	.	2	.	.	.	1	.	3	.	.	.	.	.	.	.	10	.	.	7	14	9	.	9	11	12	.	12	.	4	20
				1	.	1	.	2	2	2	.	1	1	3	1	6	5	9	9	6	9	7	7	9	4	12	32	12	13	5	24	5	24	4	21
					1	.	1	.	.	.	1	.	.	.	.	.	.	.	.	.	4	.	.	3	4	3	.	3	8	3	.	3	.	4	22
						1	.	1	.	1	.	1	.	1	1	3	2	3	3	3	3	2	2	6	1	3	16	3	7	1	13	1	13	3	23
							1	.	.	.	1	.	.	.	.	.	.	.	.	.	.	.	.	.	4	.	.	.	.	3	.	3	.	.	24
								1	.	.	.	1	.	1	.	.	.	3	.	.	.	2	.	.	.	3	8	.	.	1	13	.	5	.	25
									1	.	.	.	1	.	.	2	3	3	3	3	6	4	4	5	4	8	9	8	9	5	6	5	6	3	26
										1	.	.	.	1	1	.	.	.	3	.	.	.	2	.	.	.	8	3	.	.	5	1	13	.	27
											1	.	.	.	.	.	.	.	.	.	.	.	.	.	.	.	.	.	.	.	.	.	.	.	28
												1	.	.	.	.	.	.	.	.	.	.	.	.	.	.	.	.	.	.	5	.	.	.	29
													1	.	.	.	.	.	.	.	3	.	.	2	4	4	.	4	3	5	.	5	.	1	30
														1	.	.	.	.	.	.	.	.	.	.	.	.	4	.	.	.	5	.	5	.	31
															1	.	.	.	.	.	.	.	.	.	.	.	.	.	.	.	.	.	5	.	32
																1	1	1	1	2	2	1	1	3	1	2	4	2	5	1	3	1	3	2	33
																	1	.	.	1	1	1	1	1	1	1	.	1	4	1	.	1	.	2	34
																		1	.	.	.	1	.	.	.	2	2	.	.	1	3	.	1	.	35
																			1	.	.	.	1	.	.	.	2	2	.	.	1	1	3	.	36
																				1	.	.	.	1	.	.	.	.	2	.	.	.	.	1	37
																					1	.	.	1	1	1	.	1	2	1	.	1	.	1	38
																						1	.	.	.	1	.	.	.	1	.	.	.	.	39
																							1	.	.	.	.	1	.	.	.	1	.	.	40
																								1	.	.	.	.	1	.	.	.	.	1	41
																									1	.	.	.	.	1	.	1	.	.	42
																										1	.	.	.	1	.	.	.	.	43
																											1	.	.	.	1	.	1	.	44
																												1	.	.	.	1	.	.	45
																													1	.	.	.	.	1	46
																														1	.	.	.	.	47
																															1	.	.	.	48
																																1	.	.	49
																																	1	.	50
																																		1	51

Table 3. *(continued)*

ALGEBRA D6, REPRESENTATIONS OF CLASS 1

NUMBER	S.P.	O.S.	LEVEL	DIM'N	WEIGHT	0	1	2	3	4	5	6	7	8	9	10	11	12	13	14	15	16
0	4	12	11	12	1 0 0 0 0 0	1																
1	12	160	25	220	0 0 1 0 0 0	5	1															
2	20	120	29	560	1 1 0 0 0 0	10	2	1														
3	36	12	31	352	3 0 0 0 0 0	5	1	1	1													
4	20	192	31	792	0 0 0 1 0 1	10	3	.	.	1												
5	28	960	39	4 928	1 0 0 0 1 0	40	14	4	.	4	1											
6	36	192	41	4 752	1 0 0 0 0 2	30	12	3	.	5	1	1										
7	36	192	41	4 752	1 0 0 2 0 0	30	12	3	.	5	1	.	1									
8	36	480	43	8 008	0 1 1 0 0 0	54	19	8	.	5	2	.	.	1								
9	44	480	45	11 232	2 0 1 0 0 0	65	24	13	5	6	3	.	.	1	1							
10	52	120	47	11 088	1 2 0 0 0 0	60	21	13	4	5	3	.	.	2	1	1						
11	68	120	49	11 088	3 1 0 0 0 0	50	18	14	10	4	3	.	.	2	2	1	1					
12	44	1 920	49	36 036	0 1 0 1 0 1	155	69	24	.	30	10	4	4	3	.	.	.	1				
13	100	12	51	4 004	5 0 0 0 0 0	15	5	5	5	1	1	.	.	1	1	1	1	.	1			
14	52	960	51	45 760	2 0 0 1 0 1	170	79	34	10	35	13	5	5	3	3	.	.	1	.	1		
15	52	960	53	48 048	0 0 1 0 1 0	170	78	33	.	34	14	5	5	6	.	.	.	2	.	.	1	
16	60	640	55	54 912	0 0 1 0 0 2	160	80	32	.	40	16	10	6	6	.	.	.	3	.	.	1	1
17	60	640	55	54 912	0 0 1 2 0 0	160	80	32	.	40	16	6	10	6	.	.	.	3	.	.	1	.
18	60	2 880	57	155 232	1 1 0 0 1 0	461	219	109	25	95	45	15	15	20	10	4	.	6	.	3	2	.
19	68	480	59	112 112	1 0 2 0 0 0	300	143	79	20	61	32	10	10	17	7	3	.	5	.	2	3	.
20	76	960	59	112 320	3 0 0 0 1 0	280	138	80	40	60	32	10	10	14	14	4	4	4	.	4	1	.
21	68	960	59	161 700	1 1 0 0 0 2	395	205	95	20	105	46	25	20	18	9	3	.	8	.	4	2	2
22	68	960	59	132 132	0 0 0 1 1 1	315	164	72	.	86	38	20	20	17	.	.	.	9	.	.	4	2
23	68	960	59	161 700	1 1 0 2 0 0	395	205	95	20	105	46	20	25	18	9	3	.	8	.	4	2	.
24	76	480	61	133 056	0 2 1 0 0 0	330	156	90	26	65	37	11	11	22	11	8	.	6	.	3	3	.
25	84	192	61	112 112	3 0 0 0 0 2	230	124	66	30	65	31	15	15	12	12	3	3	5	.	5	1	1
26	84	192	61	75 504	0 0 0 1 0 3	146	80	37	.	45	21	15	10	10	.	.	.	6	.	.	3	3
27	84	192	61	112 112	3 0 0 2 0 0	230	124	66	30	65	31	15	15	12	12	3	3	5	.	5	1	.
28	84	192	61	75 504	0 0 0 3 0 1	146	80	37	.	45	21	10	15	10	.	.	.	6	.	.	3	1
29	84	960	63	257 400	2 1 1 0 0 0	570	276	176	80	115	70	20	20	41	30	16	8	11	.	8	6	.
30	100	120	65	133 056	1 3 0 0 0 0	270	126	87	40	50	34	9	9	24	16	13	4	6	.	4	4	.
31	108	480	65	148 500	4 0 1 0 0 0	275	135	97	65	55	38	10	10	24	24	13	13	6	5	6	3	.
32	76	5 760	65	755 040	1 0 1 1 0 1	1 440	786	398	80	420	207	105	105	97	38	14	.	48	.	18	21	9
33	116	120	67	164 736	3 2 0 0 0 0	290	138	106	70	54	40	10	10	29	25	17	13	7	4	6	5	.
34	84	960	67	480 480	1 0 0 0 2 0	825	454	241	45	245	128	65	65	67	23	12	.	33	.	11	17	7
35	84	1 920	67	677 600	0 2 0 1 0 1	1 185	651	341	85	350	179	90	90	90	45	24	.	42	.	22	17	7
36	148	120	69	100 672	5 1 0 0 0 0	150	70	60	50	26	22	5	5	18	18	14	14	4	10	4	3	.
37	92	1 920	69	756 756	1 0 0 0 1 2	1 119	648	339	60	375	195	120	105	99	33	15	.	57	.	18	28	21
38	92	3 840	69	1 250 964	2 1 0 1 0 1	1 959	1 104	624	240	600	322	160	160	159	114	48	24	74	.	52	30	12
39	92	1 920	69	756 756	1 0 0 2 1 0	1 119	648	339	60	375	195	105	120	99	33	15	.	57	.	18	28	12
40	196	12	71	27 456	7 0 0 0 0 0	35	15	15	15	5	5	1	1	5	5	5	5	1	5	1	1	.
41	92	2 880	71	1 281 280	0 1 1 0 1 0	1 920	1 074	608	160	580	323	160	160	180	84	48	.	86	.	40	46	17
42	116	192	71	264 264	1 0 0 0 0 4	310	187	102	16	115	63	45	35	35	10	6	.	22	.	6	12	12
43	116	960	71	637 000	4 0 0 1 0 1	830	480	298	170	265	153	75	75	79	79	34	34	35	10	35	13	5
44	100	960	71	906 048	1 0 0 2 0 2	1 190	705	382	70	419	225	135	135	119	37	16	.	72	.	21	39	24
45	116	192	71	264 264	1 0 0 4 0 0	310	187	102	16	115	63	35	45	35	10	6	.	22	.	6	12	5
46	108	160	73	428 064	0 0 3 0 0 0	560	317	194	64	170	102	49	49	61	31	18	.	29	.	14	18	6
47	100	2 880	73	1 924 560	2 0 1 0 1 0	2 610	1 488	894	330	810	470	230	230	264	162	81	33	124	.	74	66	24
48	100	1 920	73	1 544 400	0 1 1 0 0 2	1 990	1 176	647	160	690	375	225	205	200	92	48	.	112	.	52	56	36
49	100	1 920	73	1 544 400	0 1 1 2 0 0	1 990	1 176	647	160	690	375	205	225	200	92	48	.	112	.	52	56	24
50	108	2 880	75	2 306 304	1 2 0 0 1 0	2 920	1 662	1 020	400	900	539	260	260	318	208	124	40	146	.	96	78	27
51	108	1 920	75	2 276 736	2 0 1 0 0 2	2 656	1 602	928	320	950	535	310	300	286	172	80	32	159	.	92	78	48

1	.	.	.	.	2	2	.	.	1	1	3	.	.	.	9	.	7	7	.	12	12	21	.	17	5	5	24	12	6	24	24	36	27	36	17
	1	1	1	1	.	1	2	1	.	1	.	4	3	3	7	4	5	10	3	8	20	8	1	22	3	13	10	3	9	36	26	26	53	42	18
		1	.	.	.	.	1	.	.	.	.	2	2	1	3	3	2	3	2	3	6	3	1	12	1	3	6	1	7	18	12	12	26	18	19
			1	.	.	.	.	1	.	1	.	2	1	3	.	3	.	.	3	.	10	.	1	.	.	13	.	.	.	14	.	.	18	16	20
				1	.	.	.	1	.	.	.	.	.	.	3	.	2	4	.	6	8	3	.	8	3	5	6	1	3	13	16	12	18	26	21
					1	.	.	.	1	.	1	.	.	.	3	.	4	2	.	8	3	8	.	8	5	1	13	5	3	12	13	13	12	18	22
						1	.	.	.	1	.	.	.	.	3	.	2	4	.	3	8	6	.	8	1	5	6	3	3	13	12	16	18	18	23
							1	.	.	.	.	1	2	1	.	2	.	3	2	.	3	.	1	6	.	3	.	.	3	6	6	6	20	6	24
								1	.	.	.	.	.	.	.	.	.	.	.	.	4	.	.	.	.	5	.	.	.	5	.	.	6	10	25
									1	.	.	.	.	.	.	.	.	.	.	4	.	.	.	.	5	.	5	.	.	.	5	.	.	6	26
										1	.	.	.	.	.	.	.	.	.	.	4	.	.	.	.	5	.	.	.	5	.	.	6	6	27
											1	.	.	.	.	.	.	.	.	.	.	4	.	.	.	.	5	5	.	.	.	5	.	.	28
												1	1	1	.	2	.	.	2	.	3	.	1	.	.	3	.	.	.	6	.	.	10	6	29
													1	.	.	1	.	.	1	.	.	.	1	.	.	.	.	.	.	.	.	.	4	.	30
														1	.	1	.	.	2	.	.	.	1	.	.	3	.	.	.	.	.	.	.	.	31
															1	.	1	1	.	2	2	2	.	4	1	1	3	1	2	6	6	6	8	9	32
																1	.	.	1	.	.	.	1	.	.	.	.	.	.	.	.	.	.	.	33
																	1	.	.	1	.	1	.	2	1	.	1	1	1	3	2	2	3	3	34
																		1	.	.	1	.	.	2	.	1	.	.	1	2	3	3	6	3	35
																			1	.	.	.	1	.	.	.	.	.	.	.	.	.	.	.	36
																				1	.	.	.	.	1	.	1	.	.	.	2	.	.	3	37
																					1	.	.	.	.	1	.	.	.	2	.	.	3	3	38
																						1	.	.	.	.	1	1	.	.	.	2	.	.	39
																							1	.	.	.	.	.	.	.	.	.	.	.	40
																								1	.	.	.	.	1	1	1	1	2	1	41
																									1	.	.	.	.	.	.	.	.	.	42
																										1	.	.	.	.	.	.	.	.	43
																											1	.	.	.	.	.	.	.	44
																												1	.	.	.	.	.	.	45
																													1	.	.	.	.	.	46
																														1	.	.	1	1	47
																															1	.	.	1	48
																																1	.	.	49
																																	1	.	50
																																		1	51

Table 3. *(continued)*

ALGEBRA D6, REPRESENTATIONS OF CLASS 2

6	14	22	30	30	38	38	54	54	46	46	54	62	54	86	62	70	62	70	70	78	78	78	86	94	78	S.P.
32	192	480	192	640	480	960	32	192	192	1 920	480	960	1 920	192	960	192	1 920	1 920	2 880	960	640	1 920	480	960	1 920	O.S.
16	26	34	36	40	44	44	46	46	46	50	52	54	54	56	56	56	58	60	62	62	64	64	64	64	64	LEVEL
32	352	1 728	2 112	4 928	8 800	13 728	4 224	9 152	9 504	43 680	36 960	66 528	82 368	32 032	91 520	41 184	155 232	224 224	325 248	237 600	244 608	436 800	174 240	247 104	379 456	DIM'N
0 0 0 0 0 1	1 0 0 1 0 0	0 1 0 0 0 1	2 0 0 0 0 1	0 0 1 1 0 0	0 0 0 0 1 1	1 1 0 1 0 0	0 0 0 0 0 3	3 0 0 1 0 0	0 0 0 2 0 1	1 0 1 0 0 1	0 2 0 0 0 1	2 1 0 0 0 1	1 0 0 1 1 0	4 0 0 0 0 1	1 0 0 1 0 2	1 0 0 3 0 0	0 1 1 1 0 0	2 0 1 1 0 0	0 1 0 0 1 1	1 2 0 1 0 0	0 0 2 0 0 1	2 0 0 0 1 1	0 1 0 0 0 3	3 1 0 1 0 0	0 1 0 2 0 1	WEIGHT
0	1	2	3	4	5	6	7	8	9	10	11	12	13	14	15	16	17	18	19	20	21	22	23	24	25	NUMBER
1	5	15	15	29	40	65	15	35	36	150	115	180	225	70	215	90	360	461	615	450	415	735	270	399	625	0
	1	4	5	10	15	25	6	15	15	65	50	85	105	35	105	45	175	235	315	230	220	395	145	215	335	1
		1	1	3	6	8	3	5	6	26	21	35	47	15	51	22	81	111	161	110	113	203	81	105	177	2
			1	.	.	4	.	5	.	10	10	25	15	15	15	6	36	65	65	70	54	105	30	85	69	3
				1	2	2	1	1	3	9	7	12	21	5	24	12	36	48	78	48	56	100	42	45	96	4
					1	.	1	.	1	3	2	3	8	1	13	5	13	18	39	17	27	48	27	16	48	5
						1	.	1	.	3	4	8	6	5	6	3	16	26	32	31	26	47	16	35	36	6
							1	.	.	.	.	.	.	.	5	.	.	.	9	.	5	10	15	.	10	7
								1	.	.	.	4	.	5	.	.	.	10	.	10	.	15	.	25	.	8
									1	.	.	.	4	.	5	5	5	6	15	5	10	20	10	4	30	9
										1	1	2	2	1	3	1	6	9	15	11	13	21	9	12	18	10
											1	1	.	1	.	.	3	3	6	8	6	6	3	8	6	11
												1	.	1	.	.	.	3	.	4	.	6	.	8	.	12
													1	.	1	1	2	3	6	3	5	8	4	3	10	13
														1	.	.	.	.	.	.	.	.	.	4	.	14
															1	.	.	.	3	.	2	4	4	.	4	15

94	126	86	86	86	102	102	94	102	110	102	94	134	118	102	110	102	110	110	174	150	110	118	126	126	134	S.P.
192	192	960	3 840	1 920	480	1 920	1 920	640	960	480	5 760	960	480	1 920	1 920	3 840	960	960	192	32	1 920	960	192	192	192	O.S.
66	66	66	68	68	70	70	70	70	72	72	72	74	74	74	74	74	74	74	76	76	76	76	76	76	76	LEVEL
224 224	96 096	494 208	1 057 056	672 672	464 640	862 400	864 864	411 840	988 416	672 672	2 288 000	768 768	576 576	1 862 784	1 716 000	2 718 144	1 121 120	1 258 400	256 256	151 008	2 306 304	1 965 600	704 704	896 896	453 024	DIM'N
2 0 0 0 0 3	5 0 0 1 0 0	2 0 0 2 0 1	1 1 1 0 0 1	0 0 1 1 1 0	0 3 0 0 0 1	3 0 1 0 0 1	0 0 1 1 0 2	0 0 1 3 0 0	2 2 0 0 0 1	0 0 0 0 2 1	1 1 0 1 1 0	4 1 0 0 0 1	0 0 0 0 1 3	1 0 2 1 0 0	3 0 0 1 1 0	1 1 0 1 0 2	0 0 0 2 1 1	1 1 0 3 0 0	6 0 0 0 0 1	0 0 0 0 0 3	0 2 1 1 0 0	3 0 0 1 0 2	0 0 0 2 0 3	3 0 0 3 0 0	0 0 0 4 0 1	WEIGHT
26	27	28	29	30	31	32	33	34	35	36	37	38	39	40	41	42	43	44	45	46	47	48	49	50	51	NUMBER
310	126	729	1 555	950	610	1 095	1 080	485	1 170	765	2 805	770	555	2 085	1 825	2 955	1 135	1 290	210	115	2 400	1 860	611	801	375	0
175	70	405	851	525	335	625	615	280	669	440	1 605	455	330	1 214	1 085	1 739	675	770	126	70	1 405	1 135	375	495	231	1
97	35	219	448	287	180	335	351	161	360	257	897	250	203	687	619	1 013	401	454	70	46	802	673	231	299	143	2
45	35	105	260	130	115	235	156	70	265	120	485	215	94	400	395	525	190	230	70	22	491	405	115	175	70	3
53	15	117	222	157	89	167	198	97	180	147	491	125	120	378	339	579	244	272	35	28	444	390	143	178	93	4
31	5	61	104	81	41	76	117	52	81	87	255	55	81	197	178	336	143	153	15	21	229	223	93	105	55	5
22	15	51	127	71	60	111	87	42	130	71	259	105	58	216	203	291	115	133	35	15	273	219	70	97	46	6
15	.	15	16	16	5	10	45	10	9	30	55	5	45	39	40	120	45	40	1	15	45	75	45	35	15	7
6	15	15	36	.	20	65	.	.	70	.	65	85	.	54	105	69	.	30	35	.	85	105	.	45	.	8
15	1	35	40	45	15	30	60	40	30	46	135	20	40	100	90	180	95	105	5	10	115	125	55	65	45	9
12	5	24	58	37	27	48	51	23	58	41	131	45	37	112	100	162	69	73	15	10	141	117	46	53	28	10
3	5	6	32	17	21	26	21	10	41	21	64	35	18	62	47	72	33	32	15	6	91	51	22	22	15	11
3	5	6	16	.	10	26	.	.	31	.	32	35	.	26	47	36	.	16	15	.	46	51	.	22	.	12
5	1	13	22	20	10	18	27	16	21	23	66	16	21	56	48	85	44	46	5	6	68	61	28	31	21	13
.	5	.	.	.	.	10	.	.	10	.	.	25	.	.	15	.	.	.	15	.	.	15	.	6	.	14
5	.	5	8	8	3	6	18	6	6	14	27	4	18	22	20	52	24	20	1	6	27	35	21	15	10	15

1	.	.	.	.	.	.	.	.	4	.	.	5	.	5	.	.	6	10	.	5	15	.	4	10	10	20	20	25	.	1	11	15	10	15	15	16
	1	1	2	2	3	2	1	2	3	1	1	3	12	9	7	9	12	6	15	11	30	12	10	32	21	36	21	18	5	3	42	24	15	12	10	17
		1	.	1	.	2	.	2	.	1	1	3	6	.	4	9	.	.	11	.	15	12	.	13	21	18	.	9	5	.	22	24	.	12	.	18
			1	.	1	1	1	.	1	1	.	1	4	4	2	3	7	3	4	7	12	3	8	13	8	20	12	8	1	3	16	13	10	5	6	19
				1	.	.	.	1	.	.	1	.	3	.	4	3	.	.	8	.	6	8	.	6	6	6	.	3	5	.	16	6	.	3	.	20
					1	.	.	.	.	.	.	.	2	2	1	1	3	1	3	3	4	2	3	9	2	6	6	2	1	1	9	3	6	1	3	21
						1	.	.	.	1	.	1	2	.	1	3	.	.	3	.	6	3	.	5	8	10	.	4	1	.	9	13	.	5	.	22
							1	.	.	1	.	.	.	.	.	.	3	.	.	2	.	.	6	.	.	8	3	.	.	3	.	5	6	.	1	23
								1	.	.	1	.	.	.	.	3	.	.	4	.	.	8	.	.	6	.	.	.	5	.	.	6	.	3	.	24
									1	.	.	1	.	2	.	.	3	3	.	3	6	.	3	5	4	8	9	8	.	1	6	5	6	5	6	25
										1	.	.	.	.	.	.	.	.	.	.	.	.	.	.	.	4	.	.	.	.	.	5	.	.	.	26
											1	.	.	.	.	.	.	.	.	.	.	4	.	.	.	.	.	.	5	.	.	.	.	.	.	27
												1	.	.	.	.	.	.	.	.	3	.	.	2	4	4	.	4	.	.	3	5	.	5	.	28
													1	.	1	1	.	.	2	.	2	2	.	3	2	3	.	1	1	.	6	3	.	1	.	29
														1	.	.	1	1	.	2	2	.	2	3	1	2	4	2	.	1	3	1	3	1	3	30
															1	.	.	.	1	.	.	1	.	.	.	.	.	.	1	.	3	.	.	.	.	31
																1	.	.	1	.	.	2	.	.	2	.	.	.	1	.	.	3	.	1	.	32
																	1	.	.	1	.	.	2	.	.	2	2	.	.	1	.	1	3	.	1	33
																		1	.	.	.	.	.	.	.	.	2	2	.	.	.	.	1	1	3	34
																			1	.	.	1	.	.	.	.	.	.	1	.	.	.	.	.	.	35
																				1	.	.	1	.	.	.	1	.	.	1	.	.	1	.	1	36
																					1	.	.	1	1	1	.	1	.	.	2	1	.	1	.	37
																						1	.	.	.	.	.	.	1	.	.	.	.	.	.	38
																							1	.	.	.	.	.	.	1	.	.	1	.	.	39
																								1	.	.	.	.	.	.	1	.	.	.	.	40
																									1	.	.	.	.	.	.	1	.	1	.	41
																										1	.	.	.	.	.	1	.	.	.	42
																											1	.	.	.	.	.	1	.	1	43
																												1	.	.	.	.	.	1	.	44
																													1	.	.	.	.	.	.	45
																														1	.	.	.	.	.	46
																															1	.	.	.	.	47
																																1	.	.	.	48
																																	1	.	.	49
																																		1	.	50
																																			1	51

Table 3. *(continued)*

ALGEBRA D7, REPRESENTATIONS OF CLASS 0

0	8	16	16	24	24	32	40	64	32	40	40	40	48	48	56	48	72	72	56	80	104	56	144	64	64	64	64	80	72	72	72	88	72	88	80	80	88	88	88	88	96	112	112	80	96	112	104	88	128	120	88	S.P.
1	84	14	560	840	448	84	168	14	3 360	448	448	3 360	280	2 240	1 680	6 720	84	840	2 688	168	168	6 720	14	560	2 240	2 240	13 440	3 360	2 688	2 688	6 720	448	6 720	448	3 360	672	840	6 720	1 344	1 344	840	2 240	64	26 880	448	64	1 680	6 720	84	1 680	13 440	O.S.
1	23	25	37	43	43	45	47	49	53	55	55	59	61	61	65	65	67	67	67	69	71	71	73	73	73	73	75	77	77	77	79	79	79	79	81	81	83	83	83	83	85	85	85	85	85	85	87	87	89	89	89	LEVEL
1	91	104	1 001	4 004	3 003	3 080	5 265	2 275	24 024	21 021	21 021	58 968	38 220	75 075	150 528	200 200	58 344	114 400	243 243	119 119	85 085	392 392	24 752	231 868	375 375	375 375	1 118 208	750 750	1 025 024	1 025 024	1 561 560	672 672	1 301 300	672 672	1 447 992	728 728	1 272 348	2 650 725	1 054 053	1 054 053	1 930 656	1 337 050	306 735	6 324 318	1 145 144	306 735	2 788 500	5 309 304	741 741	3 075 072	7 547 904	DIM'N
0	0	2	0	1	0	0	2	4	1	1	1	0	0	2	1	0	0	3	2	2	4	0	6	0	0	0	1	3	1	1	1	3	0	3	0	0	0	2	0	0	2	4	0	1	0	0	1	0	0	3	1	
0	1	0	0	0	0	2	1	0	0	0	0	1	0	0	1	1	3	0	0	2	1	0	0	0	0	0	1	0	1	1	0	0	0	0	2	0	1	1	0	0	0	0	0	0	0	0	2	2	4	1	0	
0	0	0	0	1	0	0	0	0	0	0	0	0	2	0	1	0	0	1	0	0	0	1	0	0	1	1	0	0	0	0	1	0	0	0	0	0	2	0	0	0	2	0	0	1	0	0	1	0	0	1	0	
0	0	0	1	0	0	0	0	0	0	0	0	1	0	1	0	0	0	0	0	0	0	0	0	2	0	0	0	0	0	0	1	0	1	0	1	0	0	1	0	0	0	1	0	0	0	0	0	0	0	0	1	
0 0	0 0	0 0	0 0	0 0	1 0	0 0	0 0	0 0	0 1	0 0	2 0	0 0	0 0	0 0	0 0	1 0	0 0	0 0	1 0	0 0	0 0	0 1	0 0	0 0	0 0	2 0	0 1	0 1	0 0	2 0	0 0	0 0	1 0	2 0	0 0	0 2	0 0	0 0	0 1	2 1	0 0	0 0	0 0	1 0	2 0	4 0	0 0	1 0	0 0	0 0	0 1	
0	0	0	0	0	1	0	0	0	0	2	0	0	0	0	0	1	0	0	1	0	0	0	0	0	2	0	0	0	2	0	0	2	1	0	0	0	0	0	2	0	0	0	4	1	2	0	0	1	0	0	0	WEIGHT
0	1	2	3	4	5	6	7	8	9	10	11	12	13	14	15	16	17	18	19	20	21	22	23	24	25	26	27	28	29	30	31	32	33	34	35	36	37	38	39	40	41	42	43	44	45	46	47	48	49	50	51	NUMBER
1	7	6	21	56	35	42	57	21	154	105	105	294	168	315	560	630	202	356	651	357	217	938	56	504	735	735	2 240	1 260	1 680	1 680	2 688	924	1 960	924	2 310	966	1 848	3 759	1 197	1 197	2 548	1 590	267	7 602	1 162	267	3 480	5 866	847	3 360	7 728	0
	1	1	5	16	10	11	17	6	55	40	40	109	64	125	220	265	76	146	290	142	87	419	21	224	345	345	1 040	610	820	820	1 268	470	960	470	1 090	485	884	1 834	621	621	1 238	795	146	3 916	607	146	1 680	3 041	396	1 660	4 060	1
		1	.	6	.	5	12	6	15	10	10	39	30	60	120	85	45	96	120	102	72	165	21	94	130	130	480	330	360	360	648	240	390	240	575	215	510	1 074	270	270	768	520	70	1 860	267	70	1 070	1 491	270	1 160	2 040	2
			1	3	3	2	3	1	18	15	15	36	21	42	72	110	24	48	123	45	27	180	6	97	161	161	456	273	392	392	558	230	471	230	480	242	387	816	324	324	546	357	80	1 977	324	80	733	1 548	168	728	2 088	3
				1	.	1	2	1	4	3	3	13	9	18	38	34	15	30	44	31	22	69	6	41	59	59	200	135	162	162	276	106	193	106	251	105	218	457	141	141	327	222	37	910	146	37	459	742	116	492	1 030	4
					1	.	.	.	5	6	6	9	5	10	16	45	5	10	51	9	5	73	1	39	75	75	184	110	186	186	220	111	230	111	186	120	147	315	171	171	206	135	45	978	176	45	270	770	59	264	1 044	5
						1	1	1	.	.	.	5	4	5	20	10	11	16	10	21	17	25	6	20	20	20	80	55	60	60	134	40	80	40	135	45	124	239	60	60	194	125	16	395	70	16	290	350	86	320	500	6
							1	1	.	.	.	.	.	5	10	.	5	16	10	16	17	.	6	.	.	.	40	55	30	30	54	40	.	40	65	.	60	174	.	.	124	125	.	155	.	.	190	170	55	270	170	7
								1	.	.	.	.	.	.	.	.	.	6	.	5	12	.	6	.	.	.	.	15	.	.	.	10	.	10	.	.	.	39	.	.	30	60	.	.	.	.	40	.	15	120	.	8
									1	1	1	3	2	4	8	13	3	6	16	6	4	27	1	16	27	27	76	50	72	72	104	46	94	46	94	52	80	168	75	75	118	80	21	436	80	21	164	355	40	172	504	9
										1	.	.	.	.	.	5	.	.	6	.	.	9	.	5	15	10	24	15	36	30	30	21	45	21	26	25	21	45	45	36	30	20	15	204	45	10	40	165	9	40	234	10
											1	.	.	.	.	5	.	.	6	.	.	9	.	5	10	15	24	15	30	36	30	21	45	21	26	25	21	45	36	45	30	20	10	204	45	15	40	165	9	40	234	11
												1	1	1	4	3	2	3	3	4	3	10	1	7	9	9	28	18	24	24	49	15	38	15	47	23	45	84	33	33	67	42	10	183	37	10	100	155	28	108	240	12
													1	.	2	.	1	1	.	3	2	4	1	3	3	3	8	4	6	6	24	3	14	3	21	12	27	39	15	15	39	18	6	72	16	6	60	58	19	66	114	13
														1	2	.	1	3	3	3	3	.	1	.	.	.	14	18	12	12	19	15	.	15	24	.	21	60	.	.	42	42	.	69	.	.	66	79	19	90	78	14
															1	.	1	1	.	2	2	.	1	.	.	.	4	4	3	3	8	3	.	3	13	.	13	26	.	.	22	18	.	24	.	.	42	34	15	52	33	15

1	.	.	1	.	.	3	.	2	4	4	8	5	10	10	13	6	18	6	12	11	11	21	18	18	16	10	6	82	21	6	23	67	6	24	108	16
	1	.	.	1	1	.	1	.	.	.	.	.	.	.	.	.	.	.	5	.	4	5	.	.	4	5	.	.	.	.	20	10	11	20	.	17
		1	.	1	2	.	1	.	.	.	.	4	.	.	.	3	.	3	.	.	.	13	.	.	9	18	.	.	.	.	14	.	5	38	.	18
			1	.	.	.	.	.	.	.	4	5	5	5	5	6	.	6	6	.	5	15	.	.	10	10	.	30	.	.	15	35	4	20	34	19
				1	1	.	1	.	.	.	.	.	.	.	.	.	.	.	.	.	.	5	.	.	4	5	.	.	.	.	10	.	5	20	.	20
					1	.	1	.	.	.	.	.	.	.	.	.	.	.	.	.	.	.	.	.	.	5	.	.	.	.	.	.	.	10	.	21
						1	.	1	1	1	2	1	2	2	6	1	7	1	5	5	6	9	8	8	9	4	3	30	10	3	13	23	4	14	50	22
							1	.	.	.	.	.	.	.	.	.	.	.	.	.	.	.	.	.	.	.	.	.	.	.	.	.	.	.	.	23
								1	.	.	.	.	.	.	3	.	3	.	2	2	3	3	3	3	6	1	1	9	6	1	7	6	3	8	24	24
									1	.	.	.	2	.	.	1	3	.	.	2	.	.	6	3	.	.	3	12	6	1	.	9	.	.	18	25
										1	.	.	.	2	.	.	3	1	.	2	.	.	3	6	.	.	1	12	6	3	.	9	.	.	18	26
											1	1	1	1	2	1	.	1	3	.	3	6	.	.	5	4	.	10	.	.	9	13	3	11	14	27
												1	.	.	.	1	.	1	.	.	.	3	.	.	2	4	.	.	.	.	3	.	1	8	.	28
													1	.	.	1	.	.	.	.	.	.	.	.	.	.	.	4	.	.	.	5	.	.	5	29
														1	.	.	.	1	.	.	.	.	.	.	.	.	.	4	.	.	.	5	.	.	5	30
															1	.	.	.	1	.	2	2	.	.	3	1	.	3	.	.	5	3	2	6	6	31
																1	.	.	.	.	.	.	.	.	.	.	.	.	.	.	.	.	.	.	.	32
																	1	.	.	1	.	.	2	2	.	.	1	3	3	1	.	2	.	.	8	33
																		1	.	.	.	.	.	.	.	.	.	.	.	.	.	.	.	.	.	34
																			1	.	1	1	.	.	1	1	.	.	.	.	4	3	2	4	.	35
																				1	.	.	1	1	.	.	1	.	1	1	.	.	.	.	4	36
																					1	.	.	.	1	.	.	.	.	.	2	.	1	2	.	37
																						1	.	.	1	1	.	.	.	.	2	.	1	4	.	38
																							1	.	.	.	1	.	1	.	.	.	.	.	.	39
																								1	.	.	.	.	1	1	.	.	.	.	.	40
																									1	.	.	.	.	.	1	.	1	2	.	41
																										1	.	.	.	.	.	.	.	2	.	42
																											1	.	.	.	.	.	.	.	.	43
																												1	.	.	.	1	.	.	2	44
																													1	.	.	.	.	.	.	45
																														1	.	.	.	.	.	46
																															1	.	1	1	.	47
																																1	.	.	.	48
																																	1	.	.	49
																																		1	.	50
																																			1	51

Table 3. *(continued)*

ALGEBRA D7, REPRESENTATIONS OF CLASS 1

NUMBER	0	1	2	3	4	5	6	7	8	9	10	11	12	13	14	15	16	17	18	19	20	21	22	23	24	25
S.P.	4	12	20	36	20	28	28	28	36	44	36	52	68	100	44	52	52	52	52	60	60	60	68	76	60	76
O.S.	14	280	168	14	672	64	64	2 240	840	840	2 688	168	168	14	6 720	3 360	1 344	1 344	2 240	448	448	6 720	840	2 240	8 960	840
LEVEL	13	31	35	37	41	43	43	49	53	55	55	57	59	61	63	65	65	65	67	67	67	71	73	73	73	75
DIM'N	14	364	896	546	2 002	1 716	1 716	11 648	18 200	24 948	36 608	24 024	23 296	8 008	128 128	157 950	115 830	115 830	163 072	140 140	140 140	504 504	353 808	352 000	640 640	408 408
	1	0	1	3	0	0	0	1	0	2	1	1	3	5	0	2	0	0	0	2	2	1	1	3	0	0
	0	0	1	0	0	0	0	0	1	0	0	2	1	0	1	0	1	1	0	0	0	1	0	0	0	2
	0	1	0	0	0	0	0	0	1	1	0	0	0	0	0	0	0	0	1	0	0	0	2	0	1	1
	0	0	0	0	0	0	0	1	0	0	0	0	0	0	0	0	0	0	1	0	0	1	0	1	0	0
	0 0	0 0	0 0	0 0	0 1	0 0	2 0	0 0	0 0	0 0	1 0	0 0	0 0	0 0	0 1	0 1	0 0	2 0	0 0	0 0	2 0	0 0	0 0	0 0	1 0	0 0
WEIGHT	0	0	0	0	0	2	0	0	0	0	1	0	0	0	0	0	2	0	0	2	0	0	0	0	1	0
0	1	6	12	6	15	10	10	60	80	96	120	87	72	21	304	330	225	225	328	240	240	869	558	520	900	600
1		1	2	1	4	3	3	18	24	30	44	26	22	6	120	135	96	96	132	106	106	360	231	222	414	247
2			1	1	.	.	.	5	10	16	10	16	17	6	40	55	30	30	54	40	40	174	124	125	155	140
3				1	.	.	.	.	.	6	.	5	12	6	.	15	.	.	.	10	10	39	30	60	.	40
4					1	1	1	4	5	6	16	5	4	1	44	50	41	41	48	46	46	130	82	80	188	86
5						1	.	.	.	.	6	.	.	.	14	15	21	15	14	21	21	35	21	20	84	21
6							1	.	.	.	6	.	.	.	14	15	15	21	14	21	21	35	21	20	84	21
7								1	2	3	3	3	3	1	14	18	12	12	19	15	15	60	42	42	69	48
8									1	1	.	2	2	1	4	4	3	3	8	3	3	26	22	18	24	28
9										1	.	1	2	1	.	4	.	.	.	3	3	13	9	18	.	14
10											1	.	.	.	4	5	5	5	5	6	6	15	10	10	30	11
11												1	1	1	.	.	.	.	.	.	.	5	4	5	.	10
12													1	1	.	.	.	.	.	.	.	.	.	5	.	.
13														1	.	.	.	.	.	.	.	.	.	.	.	.
14															1	1	1	1	2	1	1	6	5	4	10	6
15																1	.	.	.	1	1	3	2	4	.	3
16																	1	.	.	1	.	.	.	.	4	.

NUMBER	26	27	28	29	30	31	32	33	34	35	36	37	38	39	40	41	42	43	44	45	46	47	48	49	50	51
S.P.	84	68	68	100	108	76	84	76	116	148	76	84	196	84	84	84	100	84	100	92	92	92	92	116	100	100
O.S.	1 680	3 360	13 440	168	840	2 240	2 688	2 240	168	168	20 160	2 688	14	2 240	6 720	6 720	448	6 720	448	13 440	1 344	1 344	6 720	3 360	2 688	2 688
LEVEL	77	77	77	79	79	79	79	79	81	83	83	83	85	85	85	85	85	85	85	87	87	87	89	89	89	89
DIM'N	773 500	728 728	1 765 764	384 384	427 856	780 780	1 164 800	780 780	462 462	269 568	3 832 920	1 665 664	68 952	2 354 352	3 309 696	3 718 000	858 858	3 718 000	858 858	5 955 950	3 097 094	3 097 094	5 940 480	2 879 800	5 513 508	5 513 508
	2	0	1	1	4	0	3	0	3	5	1	0	7	1	0	1	0	1	0	2	0	0	0	4	2	2
	1	0	1	3	0	0	0	0	2	1	0	0	0	0	2	0	0	0	0	1	2	2	1	0	1	1
	1	0	0	0	1	0	0	0	0	0	1	0	0	0	0	1	0	1	0	0	0	0	1	0	0	0
	0	1	0	0	0	1	0	1	0	0	0	0	0	2	0	0	0	0	0	0	0	0	1	0	0	0
	0 0	0 1	1 0	0 0	0 0	0 0	1 0	2 0	0 0	0 0	1 1	1 0	0 0	0 0	1 0	0 1	0 2	0 3	0 0	1 0	0 2	0 0	0 0	1 0	0 2	0 0
WEIGHT	0	0	1	0	0	2	1	0	0	0	0	1	0	0	0	2	3	0	1	0	2	0	0	0	2	0
0	1 030	864	2 121	472	486	790	1 200	790	502	252	3 735	1 412	56	2 094	2 976	3 090	607	3 090	607	4 865	2 370	2 370	4 680	1 995	3 834	3 834
1	432	408	1 007	192	206	393	588	393	207	102	1 830	726	21	1 031	1 468	1 597	324	1 597	324	2 448	1 236	1 236	2 332	1 026	2 040	2 040
2	270	170	435	131	146	160	290	160	157	87	880	315	21	524	736	740	146	740	146	1 320	595	595	1 270	610	1 050	1 050
3	120	.	85	60	96	.	120	.	102	72	165	.	21	94	176	130	.	130	.	480	135	135	320	330	360	360
4	150	189	465	64	70	196	276	196	68	32	858	376	6	484	690	815	176	815	176	1 159	636	636	1 092	490	1 061	1 061
5	35	84	210	14	15	105	126	91	14	6	378	196	1	210	300	420	105	405	91	504	336	315	462	210	546	546
6	35	84	210	14	15	91	126	105	14	6	378	196	1	210	300	405	91	420	105	504	315	336	462	210	546	546
7	90	78	189	43	48	80	123	80	50	27	399	164	6	240	336	366	80	366	80	594	296	296	582	273	516	516
8	52	33	68	30	30	32	44	32	36	22	178	72	6	120	160	154	37	154	37	280	132	132	312	135	228	228
9	38	.	34	20	30	.	44	.	31	22	69	.	6	41	80	59	.	59	.	200	66	66	144	135	162	162
10	20	34	80	9	10	40	51	40	10	5	168	86	1	100	140	180	45	180	45	244	145	145	237	110	251	251
11	20	.	10	16	16	.	10	.	21	17	25	.	6	20	40	20	.	20	.	80	30	30	80	55	60	60
12	10	.	.	5	16	.	10	.	16	17	.	.	6	.	.	.	.	.	.	40	.	.	.	55	30	30
13	.	.	.	.	6	.	.	.	5	12	.	.	6	.	.	.	.	.	.	.	.	.	.	15	.	.
14	11	14	26	6	6	16	16	16	7	4	72	38	1	48	62	73	21	73	21	108	60	60	122	50	103	103
15	8	.	13	4	6	.	16	.	6	4	27	.	1	16	32	27	.	27	.	76	31	31	56	50	72	72
16	.	5	10	.	.	10	6	6	.	.	24	20	.	15	20	40	15	30	10	34	31	25	36	15	51	45

1	.	.	1	.	.	.	4	.	.	5	10	.	.	6	6	10	.	.	24	20	.	15	20	30	10	40	15	34	25	31	36	15	45	51	17
	1	.	.	2	3	1	3	3	6	6	6	4	3	6	3	6	5	3	30	17	1	24	24	27	10	27	10	42	21	21	63	18	36	36	18
		1	.	.	.	.	.	.	.	.	5	.	.	.	6	.	.	.	9	.	.	5	10	15	.	10	.	24	15	15	16	15	36	30	19
			1	.	.	.	.	.	.	.	5	.	.	.	6	.	.	.	9	.	.	5	10	10	.	15	.	24	15	15	16	15	30	36	20
				1	1	1	.	2	4	.	3	3	3	.	3	.	4	3	10	.	1	7	14	9	.	9	.	28	12	12	30	18	24	24	21
					1	.	.	1	2	.	.	2	1	.	.	.	3	2	4	.	1	3	4	3	.	3	.	8	3	3	16	4	6	6	22
						1	.	.	2	.	.	1	3	.	3	.	3	3	.	.	1	.	.	.	.	.	.	14	.	.	.	18	12	12	23
							1	.	.	2	2	.	.	3	1	3	.	.	9	9	.	7	7	12	6	12	6	12	9	9	17	5	15	15	24
								1	1	.	.	2	1	.	.	.	2	2	.	.	1	.	4	.	.	.	.	4	3	3	8	4	3	3	25
									1	.	.	1	1	.	.	.	2	2	.	.	1	.	.	.	.	.	.	4	.	.	.	4	3	3	26
										1	.	.	.	1	.	1	.	.	3	4	.	4	2	3	3	3	3	3	2	2	8	1	3	3	27
											1	.	.	.	1	.	.	.	3	.	.	2	4	4	.	4	.	8	5	5	8	5	10	10	28
												1	.	.	.	.	1	1	.	.	1	.	.	.	.	.	.	.	.	.	.	.	.	.	29
													1	.	.	.	1	2	.	.	1	.	.	.	.	.	.	.	.	.	.	4	.	.	30
														1	.	.	.	.	.	2	.	.	.	3	3	.	1	.	2	.	.	.	3	.	31
															1	.	.	.	.	.	.	.	.	.	.	.	.	4	.	.	.	5	5	5	32
																1	.	.	.	2	.	.	.	.	1	3	3	.	.	2	.	.	.	3	33
																	1	1	.	.	1	.	.	.	.	.	.	.	.	.	.	.	.	.	34
																		1	.	.	1	.	.	.	.	.	.	.	.	.	.	.	.	.	35
																			1	.	.	1	1	1	.	1	.	2	1	1	4	1	2	2	36
																				1	.	.	.	.	1	.	1	.	.	.	.	.	.	.	37
																					1	.	.	.	.	.	.	.	.	.	.	.	.	.	38
																						1	.	.	.	.	.	.	.	.	2	.	.	.	39
																							1	.	.	.	.	1	1	1	2	1	1	1	40
																								1	.	.	.	.	1	.	.	.	2	.	41
																									1	.	.	.	.	.	.	.	.	.	42
																										1	.	.	.	1	.	.	.	2	43
																											1	.	.	.	.	.	.	.	44
																												1	.	.	.	1	1	1	45
																													1	.	.	.	1	.	46
																														1	.	.	.	1	47
																															1	.	.	.	48
																																1	.	.	49
																																	1	.	50
																																		1	51

Table 3. *(continued)*

ALGEBRA D7, REPRESENTATIONS OF CLASS 2

S.P.	7	15	23	31	31	39	39	55	47	47	55	63	55	63	87	55	63	63	79	71	71	79	71	95	79	79
O.S.	64	448	1 344	448	2 240	2 688	2 240	448	1 344	6 720	448	64	1 344	2 688	448	8 960	6 720	6 720	448	6 720	2 688	2 688	13 440	2 688	2 240	8 960
LEVEL	22	34	44	46	52	56	58	58	62	64	64	64	66	68	70	70	74	74	76	76	76	78	80	80	82	82
DIM'N	64	832	4 928	5 824	17 472	45 760	40 768	29 120	64 064	181 440	64 064	27 456	146 432	256 256	116 480	448 448	728 000	768 768	320 320	1 078 272	741 312	1 089 088	2 119 936	1 089 088	1 478 400	2 759 680
WEIGHT	0 0 0 0 0 0 1	1 0 0 0 1 0 0	0 1 0 0 0 0 1	2 0 0 0 0 0 1	0 0 1 0 1 0 0	1 1 0 0 1 0 0	0 0 0 1 0 0 1	3 0 0 0 1 0 0	0 0 0 0 1 1 0	1 0 1 0 0 0 1	0 0 0 0 1 0 2	0 0 0 0 3 0 0	0 2 0 0 0 0 1	2 1 0 0 0 0 1	4 0 0 0 0 0 1	1 0 0 1 1 0 0	1 0 0 0 0 1 1	0 1 1 0 1 0 0	1 0 0 0 0 0 3	2 0 1 0 1 0 0	1 0 0 0 2 0 1	1 2 0 0 1 0 0	0 1 0 1 0 0 1	3 1 0 0 1 0 0	0 0 2 0 0 0 1	2 0 0 1 0 0 1
NUMBER	0	1	2	3	4	5	6	7	8	9	10	11	12	13	14	15	16	17	18	19	20	21	22	23	24	25
0	1	6	21	21	49	106	84	56	105	294	91	36	216	336	126	553	735	819	266	1 035	657	966	1 764	840	1 120	2 079
1		1	5	6	15	36	29	21	40	114	36	15	85	141	56	234	330	360	126	474	306	446	829	406	540	1 014
2			1	1	4	10	10	6	15	40	15	6	31	51	21	94	145	149	60	199	141	190	380	176	250	469
3				1	.	5	.	6	.	15	.	.	15	36	21	29	40	65	15	114	36	120	150	141	115	234
4					1	2	3	1	6	12	6	3	9	15	6	36	62	57	28	75	66	72	165	66	110	205
5						1	.	1	.	4	.	.	5	10	6	10	15	25	6	40	15	46	65	51	50	94
6							1	.	2	3	3	1	2	3	1	12	27	18	15	24	30	22	69	20	45	84
7								1	.	.	.	.	.	5	6	.	.	.	.	15	.	15	.	36	.	29
8									1	.	1	1	.	.	.	4	10	5	6	6	16	5	24	4	15	30
9										1	.	.	1	2	1	3	6	8	3	12	6	14	26	15	21	36
10											1	.	.	.	.	.	5	.	6	.	6	.	9	.	5	10
11												1	.	.	.	.	.	.	.	.	6	.	.	.	.	.
12													1	1	1	.	.	4	.	4	.	10	10	10	10	10
13														1	1	.	.	.	.	4	.	5	.	10	.	10
14															1	.	.	.	.	.	.	.	.	5	.	.
15																1	2	2	1	3	3	3	9	3	7	12

S.P.	127	79	87	87	87	95	103	103	87	95	103	111	135	95	95	175	103	111	103	111	103	103	111	119	119	119
O.S.	448	13 440	13 440	6 720	6 720	1 344	1 344	6 720	8 960	2 688	448	2 688	2 688	13 440	26 880	448	2 240	2 240	6 720	8 960	8 960	26 880	6 720	6 720	2 688	13 440
LEVEL	82	84	86	86	86	86	88	88	88	88	88	90	92	92	92	94	94	94	94	94	94	96	96	98	98	98
DIM'N	396 032	3 634 176	6 067 776	4 576 000	3 807 232	1 665 664	2 498 496	4 752 384	5 381 376	4 717 440	2 050 048	5 189 184	3 843 840	10 250 240	17 326 400	1 188 096	6 849 024	5 033 600	13 069 056	12 446 720	11 275 264	30 338 880	15 567 552	20 944 000	14 158 144	29 981 952
WEIGHT	5 0 0 0 1 0 0	0 1 0 0 1 1 0	1 1 1 0 0 0 1	2 0 0 0 1 1 0	0 1 0 0 1 0 2	0 1 0 0 3 0 0	0 3 0 0 0 0 1	3 0 1 0 0 0 1	0 0 1 1 1 0 0	2 0 0 0 1 0 2	2 0 0 0 3 0 0	2 2 0 0 0 0 1	4 1 0 0 0 0 1	0 0 1 0 0 1 1	1 1 0 1 1 0 0	6 0 0 0 0 0 1	0 0 0 2 0 0 1	0 0 1 0 0 0 3	1 0 2 0 1 0 0	3 0 0 1 1 0 0	0 0 1 0 2 0 1	1 1 0 0 0 1 1	0 2 1 0 1 0 0	3 0 0 0 0 1 1	1 1 0 0 0 0 3	2 1 1 0 1 0 0
NUMBER	26	27	28	29	30	31	32	33	34	35	36	37	38	39	40	41	42	43	44	45	46	47	48	49	50	51
0	252	2 520	4 040	2 885	2 346	966	1 491	2 779	3 255	2 646	1 085	2 835	1 806	5 250	9 240	462	3 220	2 190	6 426	5 859	5 180	13 740	7 140	8 400	5 460	12 684
1	126	1 245	2 010	1 470	1 191	500	750	1 434	1 655	1 386	576	1 470	966	2 775	4 845	252	1 725	1 200	3 420	3 174	2 805	7 475	3 831	4 710	3 075	6 954
2	56	605	950	725	605	256	361	689	825	711	301	711	476	1 455	2 460	126	920	660	1 756	1 639	1 515	3 990	1 985	2 555	1 720	3 639
3	56	225	540	330	215	90	230	474	360	306	126	516	406	615	1 290	126	415	270	990	1 014	625	1 980	1 191	1 470	810	2 409
4	21	291	418	348	303	135	159	304	402	360	157	315	211	753	1 203	56	481	360	864	804	821	2 082	981	1 344	951	1 800
5	21	105	234	145	105	45	105	199	175	141	60	225	176	315	615	56	220	145	480	469	335	1 000	590	725	431	1 165
6	6	134	168	162	156	67	63	120	186	183	81	123	81	388	556	21	249	205	399	372	438	1 068	450	687	536	827
7	21	.	65	40	.	.	35	114	.	36	15	120	141	.	150	56	.	.	115	234	.	225	180	330	90	540
8	1	64	56	75	75	41	20	40	84	91	46	39	25	189	245	6	121	105	171	160	244	524	191	340	286	345
9	6	47	92	62	51	22	41	75	81	66	28	87	66	161	276	21	113	81	220	205	177	494	269	348	232	523
10	.	24	16	30	45	15	5	10	30	51	21	9	5	99	90	1	60	75	60	60	120	264	66	165	186	120
11	.	14	.	15	15	21	.	.	14	21	21	.	.	35	35	.	21	20	21	20	84	104	21	70	70	35
12	6	15	50	15	15	6	31	40	36	15	6	61	51	65	130	21	54	30	120	94	69	210	169	145	90	319
13	6	.	25	15	.	.	15	40	.	15	6	46	51	.	65	21	.	.	50	94	.	105	85	145	45	234
14	6	.	.	.	.	.	.	15	.	.	.	15	36	.	.	21	.	.	.	29	.	.	.	40	.	65
15	1	21	30	27	24	12	13	24	36	30	15	27	20	78	117	6	56	42	93	84	96	234	111	162	120	213

1	.	1	.	1	.	3	.	2	4	.	8	8	10	13	5	3	6	13	16	6	6	4	39	42	1	27	27	32	30	48	112	38	75	72	72	16
	1	.	1	.	2	3	2	4	3	1	6	16	6	6	3	9	12	16	6	3	19	15	32	52	6	26	16	52	36	36	94	67	62	44	127	17
		1	.	.	.	.	.	.	.	.	.	.	.	5	.	.	.	.	.	6	.	.	.	9	.	.	5	15	.	.	10	24	.	15	36	.	18
			1	.	1	.	2	.	3	1	.	8	6	.	.	5	12	.	6	3	14	15	.	26	6	.	.	21	36	.	47	35	62	22	92	19
				1	.	.	.	.	.	.	4	.	5	5	5	.	.	5	6	6	.	.	15	15	.	10	10	10	10	30	44	11	30	30	20	20
					1	.	1	.	.	1	.	4	.	.	.	5	4	.	.	.	10	10	.	10	6	.	.	10	10	.	15	25	15	6	50	21
						1	.	1	1	.	2	4	2	3	1	2	3	6	3	1	4	3	15	18	1	13	9	18	12	18	42	22	27	24	42	22
							1	.	.	1	.	.	.	.	.	.	4	.	.	.	5	10	.	.	6	.	.	.	10	.	.	.	15	.	25	23
								1	.	.	.	2	.	.	.	1	1	3	.	.	3	2	6	6	1	6	3	12	3	6	12	12	6	6	24	24
									1	.	.	2	2	.	.	1	3	.	3	1	3	3	.	9	1	.	.	7	12	.	21	12	27	12	30	25
										1	.	.	.	.	.	.	.	.	.	.	.	5	.	.	6	.	.	.	.	.	.	.	.	.	.	26
											1	.	1	1	1	.	.	2	1	1	.	.	6	6	.	5	4	5	4	10	16	6	10	10	11	27
												1	.	.	.	1	1	.	.	.	2	2	.	3	1	.	.	4	3	.	6	8	6	3	16	28
													1	.	.	.	.	.	1	1	.	.	.	3	.	.	.	2	4	.	8	3	10	5	8	29
														1	.	.	.	.	1	.	.	.	3	.	.	2	4	.	.	4	8	.	5	10	.	30
															1	.	.	.	.	1	.	.	.	.	.	.	.	.	.	4	.	.	.	.	.	31
																1	.	.	.	.	1	1	.	.	1	.	.	.	.	.	.	4	.	.	4	32
																	1	.	.	.	1	2	.	.	1	.	.	.	3	.	.	.	6	.	8	33
																		1	.	.	.	.	2	2	.	3	1	3	1	3	4	3	2	2	6	34
																			1	.	.	.	.	.	.	.	.	.	.	.	4	.	5	5	.	35
																				1	.	.	.	.	.	.	.	.	.	.	.	.	.	.	.	36
																					1	1	.	.	1	.	.	.	.	.	.	.	.	.	4	37
																						1	.	.	1	.	.	.	.	.	.	.	.	.	.	38
																							1	.	.	1	1	.	.	1	2	.	1	2	.	39
																								1	.	.	.	1	1	.	2	2	2	1	4	40
																									1	.	.	.	.	.	.	.	.	.	.	41
																										1	.	.	.	.	.	.	.	.	.	42
																											1	.	.	.	.	.	.	2	.	43
																												1	.	.	.	1	.	.	2	44
																													1	.	.	.	2	.	2	45
																														1	.	.	.	.	.	46
																															1	.	1	1	.	47
																																1	.	.	1	48
																																	1	.	.	49
																																		1	.	50
																																			1	51

Table 3. *(continued)*

ALGEBRA D8, REPRESENTATIONS OF CLASS 0

S.P.	0	8	16	16	24	32	40	24	64	32	32	32	40	40	48	48	56	72	72	80	48	104	56	56	56	144
O.S.	1	112	16	1 120	1 344	112	224	1 792	16	128	128	8 960	6 720	7 168	448	4 480	2 688	112	1 344	224	26 880	224	10 752	3 584	3 584	16
LEVEL	1	27	29	45	51	53	55	55	57	57	57	65	71	71	73	73	77	79	79	81	81	83	83	83	83	85
DIM'N	1	120	135	1 820	7 020	5 304	8 925	8 008	3 740	6 435	6 435	60 060	141 372	162 162	89 760	176 800	344 064	129 675	255 255	260 832	716 040	180 880	850 850	595 595	595 595	50 388
	0	0	2	0	1	0	2	0	4	0	0	1	0	1	0	2	1	0	3	2	0	4	2	0	0	6
	0	1	0	0	0	2	1	0	0	0	0	0	1	0	0	0	1	3	0	2	1	1	0	1	1	0
	0	0	0	0	1	0	0	0	0	0	0	0	0	0	2	0	1	0	1	0	0	0	0	0	0	0
	0	0	0	1	0	0	0	0	0	0	0	0	1	0	0	1	0	0	0	0	0	0	0	0	0	0
	0	0	0	0	0	0	0	0	0	0	0	1	0	0	0	0	0	0	0	0	0	0	0	0	0	0
	0 0	0 0	0 0	0 0	0 0	0 0	0 0	0 1	0 0	0 0	2 0	0 0	0 0	1 0	0 0	0 0	0 0	0 0	0 0	0 0	0 1	0 0	0 1	0 0	2 0	0 0
WEIGHT	0	0	0	0	0	0	0	0	0	2	0	0	0	1	0	0	0	0	0	0	0	0	0	2	0	0
NUMBER	0	1	2	3	4	5	6	7	8	9	10	11	12	13	14	15	16	17	18	19	20	21	22	23	24	25
0	1	8	7	28	76	56	77	56	28	35	35	252	476	434	272	512	896	315	567	560	1 288	336	1 330	875	875	84
1		1	1	6	19	13	20	15	7	10	10	81	158	155	92	180	312	106	205	197	499	119	540	355	355	28
2			1	.	7	6	14	.	7	.	.	21	56	35	42	84	168	63	133	140	154	98	210	105	105	28
3				1	3	2	3	4	1	3	3	23	45	54	26	52	88	29	58	54	186	32	205	143	143	7
4					1	1	2	.	1	.	.	5	16	10	11	22	46	18	36	37	55	26	70	40	40	7
5						1	1	.	1	.	.	.	6	.	5	6	24	13	19	25	15	20	15	10	10	7
6							1	.	1	.	.	.	.	.	.	6	12	6	19	19	.	20	15	.	.	7
7								1	.	1	1	5	9	19	5	10	16	5	10	9	65	5	72	58	58	1
8									1	.	.	.	.	.	.	.	.	.	7	6	.	14	.	.	.	7
9										1	.	.	.	7	.	.	.	.	.	.	20	.	21	28	21	.
10											1	.	.	7	.	.	.	.	.	.	20	.	21	21	28	.
11												1	3	3	2	4	8	3	6	6	18	4	22	15	15	1
12													1	.	1	1	4	2	3	4	4	3	4	3	3	1
13														1	.	.	.	.	.	.	5	.	6	6	6	.
14															1	.	2	1	1	3	.	2	.	.	.	1
15																1	2	1	3	3	.	3	4	.	.	1

S.P.	64	64	56	64	64	80	64	72	80	72	88	88	72	88	96	112	80	80	80	104	128	120	80	136	152	88
O.S.	1 024	1 024	17 920	1 120	35 840	8 960	35 840	13 440	6 720	43 008	1 344	13 440	26 880	7 168	1 344	4 480	1 792	8 960	8 960	2 688	112	2 688	107 520	224	1 344	26 880
LEVEL	85	85	87	89	91	93	93	95	97	97	99	99	99	99	101	101	101	101	101	103	105	105	105	107	107	107
DIM'N	700 128	700 128	1 336 608	771 120	3 686 400	2 406 690	4 084 080	4 972 500	4 514 400	10 649 600	3 879 876	8 129 264	6 683 040	6 738 732	5 793 480	3 968 055	3 638 544	6 077 500	6 077 500	8 204 625	2 102 100	8 847 360	30 630 600	4 678 128	3 821 090	25 008 750
	2	2	0	0	1	3	0	1	0	1	0	2	0	3	2	4	0	0	0	1	0	3	1	2	5	0
	0	0	0	0	1	0	0	0	2	1	1	1	0	0	0	0	0	0	0	2	4	1	0	3	0	2
	0	0	1	0	0	0	1	1	0	0	2	0	0	0	2	0	0	0	0	1	0	1	1	0	1	0
	0	0	0	2	0	0	0	1	1	0	0	1	1	0	0	1	0	1	1	0	0	0	0	0	0	0
	0	0	1	0	1	1	0	0	0	0	0	0	0	0	0	0	2	0	0	0	0	0	0	0	0	0
	0 0	2 0	0 0	0 0	0 0	0 0	1 0	0 0	0 0	1 0	0 0	0 0	0 1	1 0	0 0	0 0	0 0	0 0	2 0	0 0	0 0	0 0	0 1	0 0	0 0	0 1
WEIGHT	2	0	0	0	0	0	1	0	0	1	0	0	0	1	0	0	0	2	0	0	0	0	0	0	0	0
NUMBER	26	27	28	29	30	31	32	33	34	35	36	37	38	39	40	41	42	43	44	45	46	47	48	49	50	51
0	896	896	1 904	1 008	4 480	2 498	4 032	5 300	4 480	8 960	3 556	7 280	5 264	4 844	4 872	3 031	2 576	4 060	4 060	6 545	1 540	6 272	19 768	3 136	2 338	14 910
1	380	380	772	408	1 880	1 087	1 774	2 250	1 906	4 080	1 526	3 182	2 396	2 285	2 120	1 354	1 192	1 915	1 915	2 834	650	2 768	9 375	1 357	1 036	7 125
2	140	140	294	168	840	567	630	1 120	988	1 680	861	1 820	938	1 085	1 288	868	504	735	735	1 778	441	1 904	4 256	994	784	3 360
3	156	156	294	156	728	428	770	871	738	1 808	587	1 242	1 074	1 033	820	532	540	903	903	1 086	243	1 064	4 300	508	394	3 291
4	50	50	109	64	310	206	265	420	378	700	324	682	419	445	482	326	224	345	345	670	166	710	1 888	365	292	1 510
5	10	10	39	30	120	81	85	200	198	240	182	350	165	155	281	180	94	130	130	417	121	456	784	262	205	675
6	10	10	.	.	60	81	.	80	96	120	87	254	.	155	179	180	.	.	.	273	78	384	304	205	205	330
7	64	64	101	53	248	145	330	290	242	784	189	405	469	453	262	170	237	425	425	340	73	328	1 891	151	117	1 450
8	.	.	.	.	.	21	.	.	.	.	.	56	.	35	42	84	.	.	.	57	21	168	.	84	133	.
9	28	28	28	14	64	35	140	70	56	336	42	90	196	196	56	35	98	210	190	70	14	64	784	28	21	595
10	28	28	28	14	64	35	140	70	56	336	42	90	196	196	56	35	98	190	210	70	14	64	784	28	21	595
11	18	18	36	21	100	65	110	134	120	284	101	213	180	177	148	100	97	161	161	204	49	212	800	106	85	639
12	3	3	13	9	36	23	34	62	59	88	56	105	69	54	83	52	41	59	59	123	34	132	320	75	58	266
13	7	7	9	5	24	15	45	30	26	114	21	45	73	70	30	20	39	75	75	40	9	40	316	19	15	250
14	.	.	5	4	10	5	10	30	26	20	33	48	25	10	48	22	20	20	20	73	23	80	120	54	36	95
15	3	3	.	.	18	23	.	24	30	44	26	75	.	54	52	52	.	.	.	81	23	110	120	58	58	135

1	1	1	2	.	2	.	.	.	1	.	.	.	.	5	5	.	10	16	10	16	32	.	10	27	22	.	.	.	51	18	63	40	42	36	55	16
	1	.	1	.	1	.	.	.	1	.	.	.	.	.	.	.	.	6	.	5	6	.	.	5	6	.	.	.	24	13	24	.	25	19	15	17
		1	1	.	2	.	.	.	1	.	.	.	.	.	5	.	.	.	.	.	16	.	10	11	22	.	.	.	17	6	46	.	24	36	.	18
			1	.	1	.	.	.	1	.	.	.	.	.	.	.	.	.	.	.	6	.	.	5	6	.	.	.	12	6	24	.	19	19	.	19
				1	.	1	1	1	.	1	1	3	2	8	5	13	13	12	32	11	21	27	19	16	10	16	27	27	23	6	24	120	13	10	97	20
					1	.	.	.	1	.	.	.	.	.	.	.	.	.	.	.	.	.	.	.	6	.	.	.	.	.	12	.	6	19	.	21
						1	.	.	.	1	1	.	.	4	5	.	5	6	16	5	15	.	19	10	10	.	.	.	15	4	20	44	10	10	50	22
							1	.	.	1	.	.	.	.	.	5	.	.	12	.	.	9	7	.	.	5	15	10	.	.	.	38	.	.	30	23
								1	.	.	1	.	.	.	.	5	.	.	12	.	.	9	7	.	.	5	10	15	.	.	.	38	.	.	30	24
									1	.	.	.	.	.	.	.	.	.	.	.	.	.	.	.	.	.	.	.	.	.	.	.	.	7	.	25
										1	.	.	.	.	.	.	.	.	6	.	.	.	7	.	.	.	.	.	.	.	.	14	.	.	15	26
											1	.	.	.	.	.	.	.	6	.	.	.	7	.	.	.	.	.	.	.	.	14	.	.	15	27
												1	1	2	1	3	6	5	6	6	9	10	3	9	4	7	9	9	13	4	14	42	9	6	32	28
													1	.	.	.	3	2	.	3	3	4	.	6	1	3	3	3	7	3	8	12	6	3	8	29
														1	1	.	2	3	3	3	6	.	3	5	4	.	.	.	9	3	11	14	7	6	18	30
															1	.	.	.	.	.	3	.	3	2	4	.	.	.	3	1	8	.	4	6	.	31
																1	.	.	2	.	.	3	1	.	.	2	4	4	.	.	.	12	.	.	9	32
																	1	1	.	2	2	.	.	3	1	.	.	.	5	2	6	4	5	3	4	33
																		1	.	1	1	.	.	1	1	.	.	.	4	2	4	.	4	3	4	34
																			1	.	.	.	1	.	.	.	.	.	.	.	.	4	.	.	5	35
																				1	.	.	.	1	.	.	.	.	2	1	2	.	3	1	.	36
																					1	.	.	1	1	.	.	.	2	1	4	.	3	3	.	37
																						1	.	.	.	1	1	1	.	.	.	3	.	.	2	38
																							1	.	.	.	.	.	.	.	.	.	.	.	.	39
																								1	.	.	.	.	1	1	2	.	2	1	.	40
																									1	.	.	.	.	.	2	.	1	3	.	41
																										1	.	.	.	.	.	.	.	.	.	42
																											1	.	.	.	.	.	.	.	.	43
																												1	.	.	.	.	.	.	.	44
																													1	1	1	.	2	1	.	45
																														1	.	.	1	.	.	46
																															1	.	1	1	.	47
																																1	.	.	1	48
																																	1	.	.	49
																																		1	.	50
																																			1	51

Table 3. *(continued)*

ALGEBRA D8, REPRESENTATIONS OF CLASS 1

NUMBER	0	1	2	3	4	5	6	7	8	9	10	11	12	13	14	15	16	17	18	19	20	21	22	23	24	25
S.P.	4	12	20	36	20	28	28	36	44	52	68	36	100	44	44	44	52	52	52	60	60	68	76	76	84	60
O.S.	16	448	224	16	1 792	1 024	4 480	1 344	1 344	224	224	10 752	16	1 024	1 024	17 920	8 960	4 480	21 504	13 440	7 168	1 344	4 480	1 344	2 688	35 840
LEVEL	15	37	41	43	51	57	59	63	65	67	69	69	71	71	71	77	79	81	83	85	85	87	87	89	91	91
DIM'N	16	560	1 344	800	4 368	11 440	24 192	36 720	49 504	46 800	44 352	112 320	14 688	91 520	91 520	371 280	448 800	456 960	1 050 192	1 372 800	1 237 600	942 480	933 504	1 067 040	1 989 680	2 715 648
WEIGHT	1,0,0,0,0,0 0,0	0,0,1,0,0,0 0,0	1,1,0,0,0,0 0,0	3,0,0,0,0,0 0,0	0,0,0,0,1,0 0,0	0,0,0,0,0,1 0,1	1,0,0,1,0,0 0,0	0,1,1,0,0,0 0,0	2,0,1,0,0,0 0,0	1,2,0,0,0,0 0,0	3,1,0,0,0,0 0,0	1,0,0,0,0,0 1,0	5,0,0,0,0,0 0,0	1,0,0,0,0,0 0,2	1,0,0,0,0,2 0,0	0,1,0,0,1,0 0,0	2,0,0,0,1,0 0,0	0,0,1,1,0,0 0,0	0,1,0,0,0,1 0,1	1,1,0,1,0,0 0,0	2,0,0,0,0,1 0,1	1,0,2,0,0,0 0,0	3,0,0,1,0,0 0,0	0,2,1,0,0,0 0,0	2,1,1,0,0,0 0,0	0,0,1,0,0,0 1,0
0	1	7	14	7	21	35	84	111	133	119	98	210	28	140	140	525	567	560	1 029	1 464	1 085	931	868	987	1 687	2 016
1		1	2	1	5	10	22	29	36	31	26	70	7	50	50	185	206	200	410	536	445	340	326	359	623	848
2			1	1	.	.	6	12	19	19	20	15	7	10	10	60	81	80	120	254	155	179	180	201	384	304
3				1	.	.	.	.	7	6	14	.	7	.	.	.	21	.	.	56	35	42	84	57	168	.
4					1	3	4	5	6	5	4	22	1	18	18	58	65	62	161	165	177	103	100	107	185	344
5						1	.	.	.	.	.	6	.	7	7	14	15	14	63	35	70	21	20	21	35	132
6							1	2	3	3	3	4	1	3	3	18	23	24	44	75	54	52	52	59	110	120
7								1	1	2	2	.	1	.	.	5	5	10	10	32	10	27	22	34	63	40
8									1	1	2	.	1	.	.	.	5	.	.	16	10	11	22	17	46	.
9										1	1	.	1	.	.	.	.	.	.	6	.	5	6	12	24	.
10											1	.	1	.	.	.	.	.	.	.	.	.	6	.	12	.
11												1	.	1	1	4	5	5	16	15	19	10	10	11	20	44
12													1	.	.	.	.	.	.	.	.	.	.	.	.	.
13														1	.	.	.	.	6	.	7	.	.	.	.	14
14															1	.	.	.	6	.	7	.	.	.	.	14
15																1	1	2	3	6	3	5	4	6	11	14

NUMBER	26	27	28	29	30	31	32	33	34	35	36	37	38	39	40	41	42	43	44	45	46	47	48	49	50	51
S.P.	100	108	68	68	116	68	68	148	84	76	76	196	92	92	76	76	84	84	92	84	92	116	92	84	92	108
O.S.	224	1 344	7 168	7 168	224	8 960	53 760	224	10 752	7 168	7 168	16	1 024	1 024	53 760	35 840	4 480	17 920	35 840	10 752	13 440	8 960	7 168	107 520	7 168	448
LEVEL	93	93	93	93	95	95	95	97	97	97	97	99	99	99	101	101	103	103	105	105	107	107	107	107	107	109
DIM'N	960 960	1 067 040	2 333 760	2 333 760	1 131 520	2 970 240	7 207 200	636 480	4 607 680	6 054 048	6 054 048	155 040	3 818 880	3 818 880	14 871 600	10 501 920	8 910 720	12 540 000	22 170 720	10 692 864	21 705 600	10 346 336	10 890 880	46 410 000	10 890 880	6 651 216
WEIGHT	1,3,0,0,0,0 0,0	4,0,1,0,0,0 0,0	0,0,1,0,0,0 0,2	0,0,1,0,0,2 0,0	3,2,0,0,0,0 0,0	0,0,0,1,1,0 0,0	1,1,0,0,0,0 1,0	5,1,0,0,0,0 0,0	3,0,0,0,0,0 1,0	1,1,0,0,0,0 0,2	1,1,0,0,0,2 0,0	7,0,0,0,0,0 0,0	3,0,0,0,0,0 0,2	3,0,0,0,0,2 0,0	1,0,1,0,1,0 0,0	0,0,0,1,0,1 0,1	1,0,0,2,0,0 0,0	0,2,0,0,1,0 0,0	2,1,0,0,1,0 0,0	0,0,0,0,1,0 1,0	0,1,1,1,0,0 0,0	4,0,0,0,1,0 0,0	0,0,0,0,1,0 0,2	1,0,1,0,0,1 0,1	0,0,0,0,1,2 0,0	0,0,3,0,0,0 0,0
0	756	784	1 470	1 470	798	1 904	4 648	392	2 590	3 311	3 311	84	1 820	1 820	8 001	4 998	4 424	6 265	10 170	4 368	9 660	4 088	3 892	18 515	3 892	2 667
1	272	292	650	650	290	816	2 009	140	1 150	1 505	1 505	28	850	850	3 534	2 334	1 960	2 785	4 595	2 088	4 320	1 882	1 926	8 925	1 926	1 197
2	184	205	225	225	218	328	829	119	540	595	595	28	380	380	1 641	900	966	1 355	2 400	864	2 290	1 087	790	3 895	790	687
3	84	133	.	.	140	.	154	98	210	105	105	28	140	140	294	.	168	315	840	.	560	567	.	630	.	202
4	78	85	286	286	82	334	825	38	478	673	673	7	386	386	1 458	1 083	807	1 150	1 907	990	1 772	785	956	4 220	956	486
5	14	15	126	126	14	126	315	6	182	301	301	1	175	175	540	498	294	420	693	462	630	280	476	1 960	476	168
6	52	58	96	96	60	132	321	32	205	249	249	7	156	156	654	414	387	543	950	408	918	428	393	1 777	393	273
7	36	36	30	30	43	54	110	26	70	80	80	7	50	50	283	155	188	250	435	170	480	206	160	700	160	156
8	24	36	.	.	37	.	55	26	70	40	40	7	50	50	109	.	64	125	310	.	220	206	.	265	.	76
9	19	19	.	.	25	.	15	20	15	10	10	7	10	10	39	.	30	60	120	.	120	81	.	85	.	45
10	6	19	.	.	19	.	.	20	15	.	.	7	10	10	.	.	.	.	60	.	.	81	.	.	.	.
11	9	10	41	41	10	48	115	5	72	104	104	1	64	64	231	188	135	190	328	189	314	145	196	795	196	91
12	.	7	.	.	6	.	.	14	.	.	.	7	.	.	.	.	.	.	.	.	.	21	.	.	.	.
13	.	.	21	15	.	14	35	.	21	49	42	.	28	28	63	84	35	50	84	84	77	35	105	350	91	21
14	.	.	15	21	.	14	35	.	21	42	49	.	28	28	63	84	35	50	84	84	77	35	91	350	105	21
15	6	6	12	12	7	19	36	4	22	30	30	1	18	18	96	69	63	82	142	78	158	65	80	299	80	51

1	.	.	3	3	2	4	3	8	.	4	6	.	.	6	.	18	4	22	15	15	1	18	18	36	.	21	42	100	.	72	65	.	110	.	24	16
	1	.	2	.	3	1	3	6	4	4	3	3	3	5	8	8	3	4	6	6	1	3	3	39	24	31	31	54	33	80	23	32	102	32	30	17
		1	.	1	.	.	.	.	4	.	.	5	5	.	5	10	.	6	12	12	.	7	7	24	30	15	20	34	34	36	15	40	125	40	11	18
			1	.	1	1	2	4	.	3	3	.	.	4	.	4	3	4	3	3	1	3	3	13	.	9	18	36	.	38	23	.	34	.	15	19
				1	.	.	.	.	.	.	.	.	.	.	.	5	.	6	6	6	.	7	7	9	.	5	10	24	.	16	15	.	45	.	5	20
					1	.	1	2	.	2	1	.	.	3	.	.	2	.	.	.	1	.	.	5	.	4	5	10	.	20	5	.	10	.	11	21
						1	.	2	.	1	3	.	.	3	.	.	3	4	.	.	1	3	3	.	.	.	.	18	.	.	23	.	.	.	.	22
							1	1	.	2	1	.	.	2	.	.	2	.	.	.	1	.	.	.	.	.	5	5	.	10	5	.	.	.	5	23
								1	.	1	1	.	.	2	.	.	2	.	.	.	1	.	.	.	.	.	.	5	.	.	5	.	.	.	.	24
									1	.	.	1	1	.	2	2	.	1	2	2	.	1	1	9	10	7	7	12	14	17	5	16	39	16	6	25
										1	.	.	.	1	.	.	1	.	.	.	.	1	.	.	.	.	.	.	.	.	.	.	.	.	.	26
											1	.	.	1	.	.	2	.	.	.	1	.	.	.	.	.	.	.	.	.	5	.	.	.	.	27
												1	.	.	.	.	.	.	2	.	.	1	.	.	4	.	.	.	5	.	.	10	15	6	.	28
													1	.	.	.	.	.	.	2	.	.	1	.	4	.	.	.	5	.	.	6	15	10	.	29
														1	.	.	1	.	.	.	1	.	.	.	.	.	.	.	.	.	.	.	.	.	.	30
															1	.	.	.	.	.	.	.	.	3	3	4	2	3	6	8	1	6	9	6	3	31
																1	.	1	1	1	.	1	1	3	.	2	4	8	.	8	5	.	13	.	3	32
																	1	.	.	.	1	.	.	.	.	.	.	.	.	.	.	.	.	.	.	33
																		1	.	.	.	1	1	.	.	.	.	4	.	.	5	.	.	.	.	34
																			1	.	.	1	.	.	.	.	.	.	.	.	.	.	5	.	.	35
																				1	.	.	1	.	.	.	.	.	.	.	.	.	5	.	.	36
																					1	.	.	.	.	.	.	.	.	.	.	.	.	.	.	37
																						1	.	.	.	.	.	.	.	.	.	.	.	.	.	38
																							1	.	.	.	.	.	.	.	.	.	.	.	.	39
																								1	.	1	1	2	.	4	1	.	3	.	2	40
																									1	.	.	.	2	.	.	3	3	3	.	41
																										1	.	.	.	2	.	.	.	.	1	42
																											1	1	.	2	1	.	.	.	1	43
																												1	.	.	1	.	.	.	.	44
																													1	.	.	1	.	1	.	45
																														1	.	.	.	.	1	46
																															1	.	.	.	.	47
																																1	.	.	.	48
																																	1	.	.	49
																																		1	.	50
																																			1	51

Table 3. *(continued)*

ALGEBRA D8, REPRESENTATIONS OF CLASS 2

NUMBER	0	1	2	3	4	5	6	7	8	9	10	11	12
S.P.	8	16	24	32	32	40	56	40	48	48	56	64	56
O.S.	128	1 024	3 584	1 024	7 168	7 168	1 024	8 960	7 168	21 504	3 584	7 168	3 584
LEVEL	29	43	55	57	65	69	71	73	79	79	81	83	83
DIM'N	128	1 920	13 312	15 360	56 320	141 440	87 040	161 280	326 144	670 208	524 160	898 560	465 920
	0	1	0	2	0	1	3	0	0	1	0	2	0
	0	0	1	0	0	1	0	0	0	0	2	1	0
	0	0	0	0	1	0	0	0	0	1	0	0	0
	0	0	0	0	0	0	0	1	0	0	0	0	0
	0	0	0	0	0	0	0	0	1	0	0	0	0
	1	0	1	1	0	0	0	1	0	1	1	1	1
	0	0	0	0	0	0	0	0	0	0	0	0	1
WEIGHT	0	1	0	0	1	1	1	0	1	0	0	0	0
0	1	7	28	28	76	161	84	154	238	518	371	574	280
1		1	6	7	21	49	28	49	84	182	133	217	105
2			1	1	5	12	7	15	29	57	43	70	40
3				1	.	6	7	.	.	21	21	49	.
4					1	2	1	4	10	15	11	18	15
5						1	1	.	.	5	6	12	.
6							1	.	.	.	.	6	.
7								1	3	3	2	3	6
8									1	.	.	.	2
9										1	1	2	.
10											1	1	.
11												1	.
12													1
13													
14													

NUMBER	13	14	15	16	17	18	19	20	21	22	23	24	25
S.P.	88	64	72	56	64	72	64	80	96	72	128	88	72
O.S.	1 024	1 024	128	35 840	21 504	21 504	35 840	7 168	7 168	21 504	1 024	1 024	53 760
LEVEL	85	85	85	87	91	93	93	95	97	97	99	99	99
DIM'N	391 680	439 296	183 040	2 036 736	3 294 720	4 523 904	4 264 960	4 426 240	4 299 776	6 223 360	1 488 384	2 489 344	11 197 440
	4	0	0	1	0	2	1	1	3	1	5	1	0
	0	0	0	0	1	0	0	2	1	0	0	0	1
	0	0	0	0	1	1	0	0	0	0	0	0	0
	0	0	0	1	0	0	0	0	0	0	0	0	1
	0	0	0	0	0	0	1	0	0	0	0	0	0
	1	1	3	0	0	0	1	0	0	0	0	0	1
	0	0	0	0	0	0	0	0	0	1	0	0	0
WEIGHT	0	2	0	1	1	1	0	1	1	1	1	3	0
0	210	232	91	1 156	1 638	2 051	1 904	1 862	1 596	2 330	462	784	4 200
1	84	91	36	448	658	854	791	784	700	1 015	210	357	1 821
2	28	36	15	163	246	323	318	301	273	435	84	162	759
3	28	.	.	49	106	182	84	189	217	105	84	36	294
4	7	15	6	55	83	108	123	101	91	185	28	75	297
5	7	.	.	15	36	57	29	64	70	40	28	15	114
6	7	.	.	.	.	21	.	21	49	.	28	.	.
7	1	6	3	16	23	30	46	27	24	77	7	34	108
8	.	3	1	4	5	6	15	5	4	33	1	18	33
9	1	.	.	4	10	15	10	17	18	15	7	6	40
10	1	.	.	.	5	5	.	12	12	.	7	.	15
11	1	.	.	.	.	5	.	6	12	.	7	.	.
12	.	1	1	.	.	.	5	.	.	12	.	7	9
13	1	.	.	.	.	.	.	.	6	.	7	.	.
14		1	.	.	.	.	.	.	.	6	.	7	.

NUMBER	26	27	28	29	30	31	32	33	34	35	36	37	38
S.P.	80	80	80	80	88	88	104	104	88	112	88	136	96
O.S.	7 168	7 168	35 840	71 680	43 008	35 840	3 584	21 504	35 840	7 168	53 760	7 168	21 504
LEVEL	99	101	101	105	105	107	107	107	109	109	109	111	111
DIM'N	5 940 480	7 468 032	14 257 152	24 837 120	29 560 960	30 551 040	11 648 000	22 471 680	33 606 144	23 761 920	37 440 000	16 972 800	45 239 040
	1	0	2	0	1	2	0	3	0	2	0	4	2
	0	0	0	1	1	0	3	0	0	2	1	1	0
	0	2	0	0	1	0	0	1	1	0	0	0	0
	0	0	1	0	0	0	0	0	1	0	0	0	0
	0	0	0	1	0	1	0	0	0	0	0	0	0
	2	1	1	0	1	0	1	1	0	1	1	1	1
	0	0	0	0	0	0	0	0	0	0	1	0	1
WEIGHT	1	0	0	1	0	1	0	0	1	0	0	0	0
0	1 981	2 576	4 900	7 476	9 065	8 456	3 220	6 132	8 946	6 076	9 590	3 780	10 675
1	889	1 141	2 191	3 430	4 137	3 983	1 491	2 891	4 213	2 884	4 570	1 848	5 215
2	397	481	922	1 536	1 776	1 805	651	1 260	1 928	1 267	2 155	826	2 485
3	91	216	448	553	994	791	413	854	819	910	735	700	1 015
4	177	190	362	669	702	787	258	499	850	504	1 000	329	1 160
5	36	85	163	234	387	318	168	323	360	357	330	273	435
6	.	.	49	.	106	84	56	182	.	189	.	217	105
7	81	68	130	278	248	328	90	174	353	174	461	112	533
8	36	20	40	112	72	130	25	50	137	48	205	30	239
9	15	31	55	94	134	123	58	108	149	122	145	91	185
10	.	15	15	29	72	29	43	57	65	85	40	70	40
11	.	.	15	.	36	29	21	57	.	64	.	70	40
12	19	5	10	38	16	45	5	10	44	9	95	5	108
13	.	.	.	.	.	.	.	21	.	21	.	49	.
14	7	.	.	14	.	15	.	.	14	.	35	.	42

NUMBER	39	40	41	42	43	44	45	46	47	48	49	50	51
S.P.	96	104	96	176	104	112	104	112	96	112	104	120	104
O.S.	21 504	3 584	107 520	1 024	7 168	1 024	21 504	35 840	71 680	21 504	8 960	43 008	71 680
LEVEL	111	111	113	113	113	113	115	115	115	117	117	119	119
DIM'N	36 391 680	15 375 360	104 186 880	4 961 280	43 563 520	18 330 624	75 119 616	72 558 720	83 166 720	87 127 040	51 701 760	164 736 000	132 387 840
	0	0	1	6	2	2	1	3	0	0	0	2	0
	1	1	1	0	0	0	0	0	0	2	0	1	0
	0	0	0	0	0	0	2	0	1	1	0	1	1
	0	0	1	0	0	0	0	1	0	0	2	0	0
	0	0	0	0	0	0	0	0	1	0	0	0	0
	1	3	0	1	1	3	0	0	1	0	1	0	0
	0	0	0	0	0	0	0	0	0	0	0	0	1
WEIGHT	2	0	1	0	2	0	1	1	0	1	0	1	1
0	8 379	3 360	24 752	924	9 254	3 696	16 576	15 413	18 074	18 004	10 394	31 668	24 780
1	4 095	1 659	11 984	462	4 627	1 869	8 134	7 679	8 911	8 918	5 194	15 981	12 615
2	1 989	821	5 574	210	2 275	933	3 823	3 625	4 323	4 229	2 555	7 644	6 370
3	657	266	2 856	210	889	357	2 107	2 191	1 764	2 499	1 120	4 977	2 520
4	966	403	2 468	84	1 108	463	1 703	1 611	2 053	1 891	1 224	3 424	3 190
5	306	126	1 233	84	397	162	927	922	829	1 118	540	2 182	1 245
6	.	.	294	84	91	36	216	448	.	336	.	994	.
7	465	204	1 022	28	537	232	704	666	951	778	569	1 408	1 574
8	231	98	390	7	264	115	263	250	419	288	249	515	774
9	141	60	496	28	177	75	380	362	380	456	250	878	605
10	36	15	228	28	36	15	205	163	150	281	115	528	225
11	.	.	114	28	36	15	85	163	.	141	.	387	.
12	108	58	125	1	127	64	81	80	181	87	104	155	362
13	.	.	.	28	.	.	.	49	.	.	.	106	.
14	63	21	35	.	70	28	21	20	63	21	35	35	182

1	.	.	.	.	.	.	.	.	.	.	7	.	.	.	.	.	.	.	.	.	20	.	21	21	28	.	.	28	28	.	.	28	.	14	.	64	15
	1	2	3	3	3	3	6	1	3	12	6	9	16	36	38	46	16	30	57	33	62	24	77	66	28	183	7	81	34	140	130	165	165	110	314	291	16
		1	1	.	2	2	.	1	.	4	.	5	4	10	20	10	11	15	25	23	15	18	15	15	6	80	7	15	6	77	55	65	98	50	185	105	17
			1	.	1	2	.	1	.	.	.	.	4	.	10	10	6	15	.	17	.	18	15	.	.	40	7	15	6	31	55	.	51	.	134	.	18
				1	.	.	2	.	1	3	3	2	4	12	8	15	3	6	18	6	27	4	33	30	15	57	1	36	18	42	40	69	49	45	92	134	19
					1	1	.	1	.	.	.	.	.	.	5	.	6	5	.	12	.	12	.	.	.	15	7	.	.	15	15	.	36	.	72	.	20
						1	.	1	.	.	.	.	.	.	.	.	.	5	.	6	.	12	.	.	.	.	7	.	.	.	15	.	.	.	36	.	21
							1	.	1	.	1	.	.	4	.	5	.	.	5	.	10	.	12	16	6	15	.	19	7	10	10	24	11	15	20	64	22
								1	.	.	.	.	.	.	.	.	.	.	.	.	.	6	.	.	.	.	7	.	.	.	.	.	.	.	.	.	23
									1	.	.	.	.	.	.	.	.	.	.	.	.	.	.	6	.	.	.	7	.	.	.	.	.	.	.	14	24
										1	.	1	1	3	4	3	2	3	8	4	6	3	6	6	3	24	1	6	3	23	16	26	28	21	53	47	25
											1	.	.	.	.	.	.	.	.	.	5	.	6	6	6	.	.	7	7	.	.	9	.	5	.	24	26
												1	.	.	2	.	1	1	4	3	.	2	.	.	.	8	1	.	.	15	4	10	15	10	30	15	27
													1	.	2	3	1	3	.	3	.	3	6	.	.	12	1	6	3	9	16	.	15	.	38	.	28
														1	.	1	.	.	2	.	2	.	2	3	1	6	.	3	1	5	4	9	6	7	11	21	29
															1	.	1	1	.	2	.	2	.	.	.	4	1	.	.	5	4	.	10	.	20	.	30
																1	.	.	.	.	.	.	2	.	.	3	.	3	1	2	4	.	3	.	8	.	31
																	1	.	.	1	.	1	.	.	.	.	1	.	.	.	.	.	.	5	.	5	.	32
																		1	.	1	.	2	.	.	.	.	1	.	.	.	4	.	.	.	10	.	33
																			1	.	.	.	.	.	.	2	.	.	.	3	1	3	3	4	6	6	34
																				1	.	1	.	.	.	.	1	.	.	.	.	.	.	.	5	.	35
																					1	.	1	1	1	.	.	1	1	.	.	3	.	2	.	8	36
																						1	.	.	.	.	1	.	.	.	.	.	.	.	.	.	37
																							1	.	.	.	.	1	1	.	.	.	.	.	.	.	38
																								1	.	.	.	1	.	.	.	.	.	.	.	4	39
																									1	.	.	.	1	.	.	.	.	.	.	.	40
																										1	.	.	.	1	1	.	2	.	4	.	41
																											1	.	.	.	.	.	.	.	.	.	42
																												1	.	.	.	.	.	.	.	.	43
																													1	.	.	.	.	.	.	.	44
																														1	.	.	1	.	2	.	45
																															1	.	.	.	2	.	46
																																1	.	1	.	2	47
																																	1	.	1	.	48
																																		1	.	.	49
																																			1	.	50
																																				1	51

Table 3. *(continued)*

ALGEBRA D9, REPRESENTATIONS OF CLASS 0

NUMBER	0	1	2	3	4	5	6	7	8	9	10	11	12
S.P.	0	8	16	16	24	32	40	64	24	32	32	40	48
O.S.	1	144	18	2 016	2 016	144	288	18	5 376	2 304	20 160	12 096	672
LEVEL	1	31	33	53	59	61	63	65	67	73	77	83	85
DIM'N	1	153	170	3 060	11 475	8 550	14 212	5 814	18 564	43 758	132 600	302 328	188 955
WEIGHT	0 0 0 0 0 0 0 0 0	0 1 0 0 0 0 0 0 0	2 0 0 0 0 0 0 0 0	0 0 0 1 0 0 0 0 0	1 0 1 0 0 0 0 0 0	0 2 0 0 0 0 0 0 0	2 1 0 0 0 0 0 0 0	4 0 0 0 0 0 0 0 0	0 0 0 0 0 1 0 0 0	0 0 0 0 0 0 1 0 1	1 0 0 0 1 0 0 0 0	0 1 0 1 0 0 0 0 0	0 0 2 0 0 0 0 0 0
0	1	9	8	36	99	72	100	36	84	126	384	720	411
1		1	1	7	22	15	23	8	21	35	112	216	125
2			1	.	8	7	16	8	.	.	28	76	56
3				1	3	2	3	1	5	10	28	54	31
4					1	1	2	1	.	.	6	19	13
5						1	1	1	.	.	.	7	6
6							1	1	.	.	.	.	.
7								1	.	.	.	.	.
8									1	3	5	9	5
9										1	.	.	.
10											1	3	2
11												1	1
12													1
13													
14													
15													

NUMBER	13	14	15	16	17	18	19	20	21	22	23	24	25
S.P.	48	40	48	48	56	72	72	80	104	48	144	56	56
O.S.	8 064	32 256	2 304	2 304	4 032	144	2 016	288	288	80 640	18	32 256	40 320
LEVEL	85	87	89	89	89	91	91	93	95	97	97	99	103
DIM'N	373 065	510 510	393 822	393 822	709 632	261 800	516 800	520 506	352 716	2 131 800	94 962	2 494 206	3 837 240
WEIGHT	2 0 0 1 0 0 0 0 0	1 0 0 0 0 0 0 1 0	1 0 0 0 0 0 2 0 0	1 0 0 0 0 0 0 0 2	1 1 1 0 0 0 0 0 0	0 3 0 0 0 0 0 0 0	3 0 1 0 0 0 0 0 0	2 2 0 0 0 0 0 0 0	4 1 0 0 0 0 0 0 0	0 1 0 0 0 1 0 0 0	6 0 0 0 0 0 0 0 0	2 0 0 0 0 1 0 0 0	0 0 1 0 1 0 0 0 0
0	777	798	504	504	1 344	464	848	828	492	2 352	120	2 430	3 456
1	245	266	175	175	420	141	274	261	156	840	36	903	1 280
2	112	56	35	35	224	84	176	184	128	252	36	336	476
3	62	85	60	60	104	34	68	63	37	282	8	308	436
4	26	15	10	10	54	21	42	43	30	81	8	102	158
5	7	.	.	.	28	15	22	29	23	21	8	21	56
6	7	.	.	.	14	7	22	22	23	.	8	21	.
7	.	.	.	.	.	.	8	7	16	.	8	.	.
8	10	26	21	21	16	5	10	9	5	85	1	93	129
9	.	7	8	8	.	.	.	.	.	20	.	21	28
10	4	4	3	3	8	3	6	6	4	23	1	28	45
11	1	.	.	.	4	2	3	4	3	5	1	5	16
12	.	.	.	.	2	1	1	3	2	.	1	.	6
13	1	.	.	.	2	1	3	3	3	.	1	5	.
14		1	1	1	.	.	.	.	.	5	.	6	9
15			1	.	.	.	.	.	.	.	.	.	.

NUMBER	26	27	28	29	30	31	32	33	34	35	36	37	38
S.P.	56	64	64	64	80	72	64	80	72	72	88	88	72
O.S.	64 512	2 016	18 432	80 640	20 160	24 192	161 280	12 096	21 504	21 504	2 016	24 192	193 536
LEVEL	103	105	105	107	109	111	113	113	115	115	115	115	117
DIM'N	5 290 740	2 174 436	6 096 948	10 340 352	6 613 425	13 590 225	17 005 950	12 150 138	13 579 566	13 579 566	10 270 260	21 607 560	43 084 800
WEIGHT	0 1 0 0 0 0 1 0 1	0 0 0 2 0 0 0 0 0	2 0 0 0 0 0 1 0 1	1 1 0 0 1 0 0 0 0	3 0 0 0 1 0 0 0 0	1 0 1 1 0 0 0 0 0	0 0 1 0 0 0 0 1 0	0 2 0 1 0 0 0 0 0	0 0 1 0 0 0 2 0 0	0 0 1 0 0 0 0 0 2	0 1 2 0 0 0 0 0 0	2 1 0 1 0 0 0 0 0	1 1 0 0 0 0 0 1 0
0	4 032	1 812	4 116	8 064	4 473	9 441	9 198	7 896	6 216	6 216	6 228	12 816	20 160
1	1 554	671	1 645	3 080	1 764	3 638	3 780	3 052	2 639	2 639	2 421	5 067	8 540
2	434	272	560	1 344	896	1 776	1 288	1 561	875	875	1 344	2 848	3 360
3	590	229	635	1 064	618	1 254	1 513	1 052	1 110	1 110	829	1 758	3 480
4	155	92	190	444	292	594	499	531	355	355	451	952	1 290
5	35	42	35	168	112	279	154	273	105	105	251	482	420
6	.	.	35	84	112	111	.	133	.	.	119	349	210
7	.	.	.	.	28	.	.	.	.	.	.	76	.
8	222	67	241	312	180	360	585	298	465	465	231	495	1 356
9	84	14	92	64	35	70	215	56	196	196	42	90	496
10	54	26	64	124	80	164	186	146	143	143	122	258	472
11	10	11	10	44	28	75	55	71	40	40	67	126	140
12	.	5	.	12	6	36	15	31	10	10	39	57	30
13	.	.	10	22	28	29	.	36	.	.	31	90	70
14	19	5	22	24	15	30	65	26	58	58	21	45	162
15	7	.	8	.	.	.	20	.	28	21	.	.	48

NUMBER	39	40	41	42	43	44	45	46	47	48	49	50	51
S.P.	96	112	72	80	80	104	88	80	128	96	96	120	136
O.S.	2 016	8 064	80 640	18 432	18 432	4 032	32 256	4 032	144	2 304	2 304	4 032	288
LEVEL	117	117	119	119	119	119	119	121	121	121	121	121	123
DIM'N	15 148 224	10 287 550	26 686 260	33 945 600	33 945 600	21 140 350	26 604 864	14 232 672	5 264 424	20 785 050	20 785 050	22 404 096	11 589 240
WEIGHT	2 0 2 0 0 0 0 0 0	4 0 0 1 0 0 0 0 0	0 0 0 1 0 1 0 0 0	1 1 0 0 0 0 2 0 0	1 1 0 0 0 0 0 0 2	1 2 1 0 0 0 0 0 0	3 0 0 0 0 0 0 1 0	0 0 0 0 2 0 0 0 0	0 4 0 0 0 0 0 0 0	3 0 0 0 0 0 2 0 0	3 0 0 0 0 0 0 0 2	3 1 1 0 0 0 0 0 0	2 3 0 0 0 0 0 0 0
0	8 496	5 278	11 844	13 440	13 440	11 284	10 800	5 760	2 592	7 140	7 140	10 752	5 280
1	3 344	2 128	5 005	5 880	5 880	4 424	4 718	2 464	995	3 220	3 220	4 284	2 067
2	2 000	1 344	1 904	2 240	2 240	2 744	2 128	1 008	672	1 400	1 400	2 912	1 500
3	1 150	742	2 049	2 520	2 520	1 509	1 956	1 016	332	1 405	1 405	1 464	690
4	667	450	772	900	900	921	806	408	225	555	555	968	492
5	384	245	294	280	280	567	266	168	162	175	175	616	351
6	244	245	.	140	140	371	266	.	105	175	175	518	274
7	56	112	.	.	.	77	56	.	28	35	35	224	112
8	318	205	798	1 068	1 068	410	768	396	87	602	602	392	179
9	56	35	286	456	456	70	280	140	14	260	260	64	28
10	178	120	294	356	356	244	290	156	58	216	216	252	125
11	99	62	109	100	100	146	85	64	40	60	60	156	88
12	57	26	39	20	20	86	15	30	27	10	10	94	63
13	62	62	.	50	50	96	85	.	27	60	60	130	68
14	30	20	101	142	142	40	98	53	9	85	85	40	19
15	.	.	28	64	56	.	28	14	.	36	36	.	.

1	.	.	.	.	.	.	.	.	.	7	.	8	.	.	.	20	.	21	28	.	.	48	.	.	28	56	64	.	28	14	.	36	36	.	.	16
	1	1	1	2	2	.	1	.	.	.	.	.	6	6	12	.	19	.	.	19	38	15	32	26	.	10	10	60	15	.	21	10	10	74	49	17
		1	.	1	1	.	1	.	.	.	.	.	.	.	.	.	7	.	.	6	7	.	6	7	.	.	.	28	.	.	15	.	.	28	29	18
			1	1	2	.	1	.	.	.	.	.	.	6	.	.	.	.	.	.	19	.	13	26	.	.	.	20	15	.	7	10	10	54	28	19
				1	1	.	1	.	.	.	.	.	.	.	.	.	.	.	.	.	7	.	6	7	.	.	.	14	.	.	7	.	.	28	22	20
					1	.	1	.	.	.	.	.	.	.	.	.	.	.	.	.	.	.	.	7	.	.	.	.	.	.	.	.	.	14	7	21
						1	.	1	3	3	2	3	8	5	13	18	12	15	15	11	21	44	16	10	36	36	36	23	26	21	6	21	21	24	13	22
							1	.	.	.	.	.	.	.	.	.	.	.	.	.	.	.	.	.	.	.	.	.	.	.	.	.	.	.	.	23
								1	.	.	.	3	4	5	5	.	6	.	.	5	15	22	10	10	.	18	18	15	26	.	4	21	21	20	10	24
									1	.	1	.	2	1	6	4	5	3	3	6	9	8	9	4	13	6	6	13	4	9	4	3	3	14	9	25
										1	.	1	.	.	.	5	.	6	6	.	.	12	.	.	9	14	14	.	7	5	.	8	8	.	.	26
											1	.	.	.	3	.	2	.	.	3	3	.	6	1	5	.	.	7	.	4	3	.	.	8	6	27
												1	.	.	.	.	.	.	.	.	.	6	.	.	.	7	7	.	7	.	.	8	8	.	.	28
													1	1	2	.	3	.	.	3	6	4	5	4	.	3	3	9	4	.	3	3	3	11	7	29
														1	.	.	.	.	.	.	3	.	2	4	.	.	.	3	4	.	1	3	3	8	4	30
															1	.	1	.	.	2	2	.	3	1	.	.	.	5	.	.	2	.	.	6	5	31
																1	.	1	1	.	.	2	.	.	3	2	2	.	1	2	.	1	1	.	.	32
																	1	.	.	1	1	.	1	1	.	.	.	4	.	.	2	.	.	4	4	33
																		1	.	.	.	.	.	.	.	2	.	.	.	.	.	1	.	.	.	34
																			1	.	.	.	.	.	.	.	2	.	.	.	.	.	1	.	.	35
																				1	.	.	1	.	.	.	.	2	.	.	1	.	.	2	3	36
																					1	.	1	1	.	.	.	2	.	.	1	.	.	4	3	37
																						1	.	.	.	1	1	.	1	.	.	1	1	.	.	38
																							1	.	.	.	.	1	.	.	1	.	.	2	2	39
																								1	.	.	.	.	.	.	.	.	.	2	1	40
																									1	.	.	.	.	1	.	.	.	.	.	41
																										1	.	.	.	.	.	1	.	.	.	42
																											1	.	.	.	.	.	1	.	.	43
																												1	.	.	1	.	.	1	2	44
																													1	.	.	1	1	.	.	45
																														1	.	.	.	.	.	46
																															1	.	.	.	1	47
																																1	.	.	.	48
																																	1	.	.	49
																																		1	1	50
																																			1	51

Table 3. *(continued)*

ALGEBRA D9, REPRESENTATIONS OF CLASS 1

NUMBER	0	1	2	3	4	5	6	7	8	9	10	11	12	13	14	15	16
S.P.	4	12	20	36	20	28	28	36	36	36	44	52	68	100	36	44	44
O.S.	18	672	288	18	4 032	8 064	4 608	256	256	2 016	2 016	288	288	18	32 256	18 432	40 320
LEVEL	17	43	47	49	61	69	71	73	73	73	75	77	79	81	83	89	91
DIM'N	18	816	1 920	1 122	8 568	45 696	31 824	24 310	24 310	67 830	90 288	84 150	78 336	25 194	293 760	707 200	930 240
	1	0	1	3	0	1	0	0	0	0	2	1	3	5	1	1	0
	0	0	1	0	0	0	0	0	0	1	0	2	1	0	0	0	1
	0	1	0	0	0	0	0	0	0	1	1	0	0	0	0	0	0
	0	0	0	0	0	1	0	0	0	0	0	0	0	0	0	0	0
	0	0	0	0	1	0	0	0	0	0	0	0	0	0	0	0	1
	0	0	0	0	0	0	0	0	0	0	0	0	0	0	1	0	0
	0	0	0	0	0	0	0	2	0	0	0	0	0	0	0	1	0
	0	0	0	0	0	0	1	0	0	0	0	0	0	0	0	0	0
WEIGHT	0	0	0	0	0	0	0	0	2	0	0	0	0	0	0	1	0
0	1	8	16	8	28	112	56	35	35	147	176	156	128	36	336	560	832
1		1	2	1	6	26	15	10	10	34	42	36	30	8	102	190	264
2			1	1	.	7	.	.	.	14	22	22	23	8	21	35	84
3				1	.	.	.	.	.	.	8	7	16	8	.	.	.
4					1	4	4	3	3	5	6	5	4	1	28	64	72
5						1	.	.	.	2	3	3	3	1	5	10	22
6							1	1	1	.	.	.	.	.	6	22	14
7								1	.	.	.	.	.	.	.	8	.
8									1	.	.	.	.	.	.	8	.
9										1	1	2	2	1	.	.	6
10											1	1	2	1	.	.	.
11												1	1	1	.	.	.
12													1	1	.	.	.
13														1	.	.	.
14															1	3	4

NUMBER	17	18	19	20	21	22	23	24	25	26	27	28	29	30	31	32	33
S.P.	52	52	60	52	68	76	60	60	76	60	68	68	84	100	108	60	116
O.S.	20 160	8 064	24 192	96 768	2 016	8 064	9 216	9 216	2 016	32 256	2 304	2 304	4 032	288	2 016	107 520	288
LEVEL	93	95	99	101	101	101	103	103	103	103	105	105	105	107	107	109	109
DIM'N	1 108 536	1 116 288	3 281 850	3 779 100	2 217 072	2 188 800	2 956 096	2 956 096	2 474 010	4 377 296	3 401 190	3 401 190	4 558 176	2 154 240	2 386 800	9 302 400	2 494 206
	2	0	1	0	1	3	0	0	0	2	2	2	2	1	4	0	3
	0	0	1	1	0	0	1	1	2	0	0	0	1	3	0	0	2
	0	1	0	0	2	0	0	0	1	0	0	0	1	0	1	1	0
	0	1	1	0	0	1	0	0	0	0	0	0	0	0	0	0	0
	1	0	0	0	0	0	0	0	0	0	0	0	0	0	0	0	0
	0	0	0	0	0	0	0	0	0	0	0	0	0	0	0	1	0
	0	0	0	0	0	0	2	0	0	0	2	0	0	0	0	0	0
	0	0	0	1	0	0	0	0	0	1	0	0	0	0	0	0	0
WEIGHT	0	0	0	0	0	0	0	2	0	0	0	2	0	0	0	0	0
0	896	880	2 281	2 030	1 440	1 344	1 344	1 344	1 512	2 128	1 400	1 400	2 576	1 136	1 184	3 920	1 192
1	292	282	747	750	470	450	520	520	492	806	555	555	849	366	393	1 510	387
2	112	111	349	210	244	245	140	140	273	266	175	175	518	246	274	525	289
3	28	.	76	.	56	112	.	.	77	56	35	35	224	112	176	.	184
4	80	76	200	267	124	120	200	200	128	290	216	216	220	92	100	548	96
5	28	29	90	70	62	62	50	50	70	85	60	60	130	61	68	185	70
6	15	14	35	90	21	20	78	78	21	98	85	85	35	14	15	180	14
7	.	.	.	27	.	.	36	28	.	28	36	36	.	.	.	48	.
8	.	.	.	27	.	.	28	36	.	28	36	36	.	.	.	48	.
9	6	12	38	15	32	26	10	10	40	15	10	10	74	42	42	60	50
10	6	.	19	.	13	26	.	.	20	15	10	10	54	28	42	.	43
11	.	.	7	.	6	7	.	.	14	.	.	.	28	22	22	.	29
12	.	.	.	.	.	7	.	.	.	.	.	.	14	7	22	.	22
13	.	.	.	.	.	.	.	.	.	.	.	.	.	.	8	.	7
14	5	5	15	22	10	10	18	18	11	26	21	21	20	9	10	58	10

NUMBER	34	35	36	37	38	39	40	41	42	43	44	45	46	47	48	49	50	51
S.P.	148	68	68	196	68	84	76	76	84	84	92	76	92	84	84	92	116	84
O.S.	288	20 160	161 280	18	129 024	32 256	120 960	129 024	40 320	8 064	18 432	161 280	80 640	32 256	32 256	24 192	20 160	32 256
LEVEL	111	113	113	113	115	115	119	119	121	121	121	123	123	125	125	125	125	127
DIM'N	1 364 352	9 883 800	24 066 900	319 770	24 186 240	15 049 216	47 837 592	60 648 588	39 673 920	28 139 760	37 209 600	51 395 760	69 227 298	42 401 502	42 401 502	66 853 248	31 467 800	50 605 056
	5	0	1	7	0	3	1	1	0	1	3	0	2	0	0	0	4	0
	1	0	1	0	0	0	0	1	2	0	0	0	1	0	0	1	0	0
	0	0	0	0	1	0	1	0	0	0	0	0	0	0	0	1	0	0
	0	1	0	0	0	0	0	0	0	2	0	1	0	1	1	1	0	0
	0	1	0	0	0	0	1	0	1	0	0	0	1	0	0	0	1	1
	0	0	1	0	0	1	0	0	0	0	0	0	0	0	0	0	0	1
	0	0	0	0	1	0	0	1	0	0	1	0	0	2	0	0	0	0
	0	0	0	0	0	0	0	0	0	0	0	1	0	0	0	0	0	0
WEIGHT	0	0	0	0	1	0	0	1	0	0	1	0	0	0	2	0	0	0
0	576	3 660	8 926	120	7 504	4 928	15 120	16 548	11 712	8 280	8 960	12 488	18 921	8 967	8 967	17 808	7 504	10 752
1	184	1 430	3 521	36	3 138	1 990	6 060	7 089	4 722	3 325	3 930	5 441	7 734	4 044	4 044	7 206	3 118	4 768
2	156	560	1 407	36	1 029	903	2 744	2 639	2 248	1 602	1 645	2 016	3 948	1 470	1 470	3 745	1 764	1 904
3	128	.	252	36	.	336	476	434	512	272	560	.	1 344	.	.	896	896	.
4	44	521	1 290	8	1 296	736	2 220	2 969	1 730	1 214	1 668	2 324	2 844	1 816	1 816	2 620	1 150	2 064
5	37	200	488	8	410	308	972	1 035	800	569	635	848	1 390	650	650	1 331	618	816
6	6	168	420	1	530	238	702	1 225	540	378	694	965	882	813	813	798	350	860
7	.	42	105	.	216	56	162	504	120	84	288	384	189	378	351	168	70	336
8	.	42	105	.	216	56	162	504	120	84	288	384	189	351	378	168	70	336
9	30	80	162	8	120	102	412	310	360	271	190	304	624	225	225	684	292	328
10	30	.	81	8	.	102	158	155	180	92	190	.	444	.	.	312	292	.
11	23	.	21	8	.	21	56	35	84	42	35	.	168	.	.	168	112	.
12	23	.	.	8	.	21	.	.	.	.	35	.	84	.	.	.	112	.
13	16	.	.	8	.	.	.	.	.	.	.	.	.	.	.	.	28	.
14	5	62	150	1	161	93	294	399	240	170	241	344	412	286	286	391	180	334

1	.	.	.	.	6	.	.	7	7	.	7	8	8	.	.	.	14	.	.	14	35	.	63	21	63	154	50	35	92	132	84	126	126	77	35	126	15
	1	1	2	6	4	5	4	3	3	6	4	3	3	11	6	6	18	7	4	24	46	1	44	28	120	108	102	78	64	120	176	96	96	194	80	132	16
		1	.	3	.	2	4	.	.	3	4	3	3	8	4	6	.	6	4	.	23	1	.	28	45	54	52	26	64	.	124	.	.	88	80	.	17
			1	2	.	3	1	.	.	3	.	.	.	6	4	3	5	5	3	10	10	1	10	5	48	20	38	38	10	40	66	30	30	97	28	54	18
				1	.	1	1	.	.	2	.	.	.	4	3	3	.	4	3	.	5	1	.	5	16	10	22	11	10	.	44	.	.	46	28	.	19
					1	.	.	1	1	.	1	1	1	.	.	.	4	.	.	5	10	.	16	6	24	38	20	15	22	44	34	41	41	36	15	48	20
						1	.	.	.	1	.	.	.	2	2	1	.	3	2	.	.	1	.	.	6	.	6	5	.	.	12	.	.	24	6	.	21
							1	.	.	.	.	.	.	2	1	3	.	3	3	.	.	1	.	5	.	.	.	.	10	.	22	.	.	.	28	.	22
								1	.	.	.	1	.	.	.	.	.	.	.	.	.	.	6	.	.	14	.	.	8	14	.	21	15	.	.	14	23
									1	.	.	.	1	.	.	.	.	.	.	.	.	.	6	.	.	14	.	.	8	14	.	15	21	.	.	14	24
										1	.	.	.	1	2	1	.	2	2	.	.	1	.	.	.	.	6	.	.	.	6	.	.	12	6	.	25
											1	1	1	.	.	.	.	.	.	.	5	.	.	6	9	19	10	5	22	.	24	.	.	16	15	.	26
												1	.	.	.	.	.	.	.	.	.	.	.	.	.	7	.	.	8	.	.	.	.	.	.	.	27
													1	.	.	.	.	.	.	.	.	.	.	.	.	7	.	.	8	.	.	.	.	.	.	.	28
														1	1	1	.	2	2	.	.	1	.	.	.	.	.	.	.	.	6	.	.	.	6	.	29
															1	.	.	1	1	.	.	1	.	.	.	.	.	.	.	.	.	.	.	.	.	.	30
																1	.	1	2	.	.	1	.	.	.	.	.	.	.	.	.	.	.	.	6	.	31
																	1	.	.	2	2	.	3	1	9	6	7	7	3	14	12	12	12	17	5	19	32
																		1	1	.	.	1	.	.	.	.	.	.	.	.	.	.	.	.	.	.	33
																			1	.	.	1	.	.	.	.	.	.	.	.	.	.	.	.	.	.	34
																				1	.	.	.	.	3	.	2	4	.	4	3	3	3	8	1	8	35
																					1	.	.	1	3	3	4	2	3	.	8	.	.	8	5	.	36
																						1	.	.	.	.	.	.	.	.	.	.	.	.	.	.	37
																							1	.	.	2	.	.	1	4	.	5	5	.	.	5	38
																								1	.	.	.	.	3	.	4	.	.	.	5	.	39
																									1	.	1	1	.	.	2	.	.	4	1	.	40
																										1	.	.	1	.	.	.	.	.	.	.	41
																											1	.	.	.	1	.	.	2	1	.	42
																												1	.	.	.	.	.	2	.	.	43
																													1	.	.	.	.	.	.	.	44
																														1	.	1	1	.	.	2	45
																															1	.	.	.	1	.	46
																																1	.	.	.	.	47
																																	1	.	.	.	48
																																		1	.	.	49
																																			1	.	50
																																				1	51

Table 3. *(continued)*

ALGEBRA D9, REPRESENTATIONS OF CLASS 2

NUMBER	S.P.	O.S.	LEVEL	DIM'N	WEIGHT	0	1	2	3	4	5	6	7	8	9	10	11	12	13	14
0	9	256	37	256	0 0 0 0 0 0 1 0 0	1														
1	17	2 304	53	4 352	1 0 0 0 0 0 0 0 1	8	1													
2	25	9 216	67	34 560	0 1 0 0 0 0 1 0 0	36	7	1												
3	33	2 304	69	39 168	2 0 0 0 0 0 1 0 0	36	8	1	1											
4	33	21 504	79	169 728	0 0 1 0 0 0 0 0 1	111	28	6	.	1										
5	41	18 432	83	413 440	1 1 0 0 0 0 0 0 1	232	64	14	7	2	1									
6	57	2 304	85	248 064	3 0 0 0 0 0 0 0 1	120	36	8	8	1	1	1								
7	41	32 256	89	574 464	0 0 0 1 0 0 1 0 0	258	76	21	.	5	.	.	1							
8	49	64 512	95	2 276 352	1 0 1 0 0 0 1 0 0	846	272	77	28	18	6	.	3	1						
9	49	32 256	97	1 410 048	0 0 0 0 1 0 0 0 1	468	154	49	.	15	.	.	4	.	1					
10	57	9 216	97	1 740 800	0 2 0 0 0 0 1 0 0	596	196	57	28	13	7	.	2	1	.	1				
11	65	18 432	99	2 937 600	2 1 0 0 0 0 1 0 0	918	316	92	64	21	14	7	3	2	.	1	1			
12	89	2 304	101	1 240 320	4 0 0 0 0 0 1 0 0	330	120	36	36	8	8	8	1	1	.	1	1	1		
13	57	21 504	103	2 558 976	0 0 0 0 0 1 1 0 0	672	238	84	.	29	.	.	10	.	3	.	.	.	1	
14	57	129 024	105	8 186 112	1 0 0 1 0 0 0 0 1	2 169	776	258	76	78	21	.	20	5	4	.	.	.	.	1
15	65	9 216	107	3 394 560	0 0 0 0 0 0 0 1 1	750	280	105	.	40	.	.	15	.	6	.	.	.	2	.
16	73	2 304	109	3 055 104	0 0 0 0 0 0 2 0 1	603	232	91	.	36	.	.	15	.	6	.	.	.	3	.
17	81	256	109	1 244 672	0 0 0 0 0 0 0 0 3	232	91	36	.	15	.	.	6	.	3	.	.	.	1	.
18	65	64 512	109	12 729 600	0 1 1 0 0 0 0 0 1	2 988	1 106	377	161	114	49	.	28	12	5	6	.	.	.	2
19	73	64 512	111	17 199 104	2 0 1 0 0 0 0 0 1	3 716	1 420	489	272	147	77	28	36	18	6	6	6	.	.	3
20	65	161 280	113	20 837 376	1 0 0 0 1 0 1 0 0	4 200	1 624	602	154	212	49	.	70	15	20	.	.	.	5	4
21	81	18 432	113	16 450 560	1 2 0 0 0 0 0 0 1	3 312	1 282	448	280	135	85	28	32	20	5	14	7	.	.	3
22	97	18 432	115	15 628 032	3 1 0 0 0 0 0 0 1	2 808	1 128	400	316	120	92	64	28	21	4	14	14	7	.	3
23	129	2 304	117	5 209 344	5 0 0 0 0 0 0 0 1	792	330	120	120	36	36	36	8	8	1	8	8	8	.	1
24	73	193 536	119	51 270 912	0 1 0 1 0 0 1 0 0	8 820	3 549	1 358	518	483	182	.	156	57	42	21	.	.	9	15
25	73	129 024	119	38 697 984	1 0 0 0 0 1 0 0 1	6 300	2 576	1 029	238	402	84	.	152	29	56	.	.	.	18	10
26	81	21 504	121	33 196 800	0 0 2 0 0 0 1 0 0	5 292	2 170	840	371	301	133	.	96	43	25	21	.	.	5	11
27	81	129 024	121	64 204 800	2 0 0 1 0 0 1 0 0	10 212	4 220	1 630	776	581	258	76	186	78	50	21	21	.	10	20
28	81	64 512	123	52 093 440	1 0 0 0 0 0 1 1 0	7 245	3 080	1 295	280	540	105	.	225	40	92	.	.	.	39	15
29	97	2 304	125	19 348 992	1 0 0 0 0 0 3 0 0	2 304	1 016	448	91	198	36	.	90	15	40	.	.	.	21	6
30	89	18 432	125	47 297 536	1 0 0 0 0 0 1 0 2	5 930	2 584	1 121	232	488	91	.	213	36	96	.	.	.	42	15
31	89	129 024	125	127 918 336	1 1 1 0 0 0 1 0 0	18 304	7 720	3 038	1 680	1 090	594	161	344	184	88	98	49	.	16	46
32	81	322 560	127	138 498 048	0 1 0 0 1 0 0 0 1	18 516	7 932	3 291	1 156	1 312	448	.	494	163	174	49	.	.	52	55
33	105	9 216	127	48 829 440	0 3 0 0 0 0 1 0 0	6 336	2 716	1 086	686	391	252	84	122	78	30	57	28	.	5	19
34	105	64 512	127	95 047 680	3 0 1 0 0 0 1 0 0	12 228	5 312	2 120	1 420	762	489	272	238	147	60	77	77	28	10	36
35	89	161 280	129	167 443 200	2 0 0 0 1 0 0 0 1	20 763	9 100	3 815	1 624	1 524	602	154	574	212	200	49	49	.	60	70
36	113	18 432	129	98 233 344	2 2 0 0 0 0 1 0 0	11 880	5 202	2 094	1 492	756	532	280	234	163	57	113	85	28	9	39
37	89	129 024	131	177 365 760	0 0 1 1 0 0 0 0 1	21 204	9 303	3 935	1 638	1 586	658	.	596	246	204	106	.	.	58	83
38	137	18 432	131	68 217 600	4 1 0 0 0 0 1 0 0	7 260	3 270	1 338	1 128	484	400	316	148	120	35	92	92	64	5	28
39	89	322 560	133	266 342 400	0 1 0 0 0 1 1 0 0	29 232	13 104	5 768	1 904	2 481	791	.	1 042	318	420	84	.	.	165	123
40	177	2 304	133	19 100 928	6 0 0 0 0 0 1 0 0	1 716	792	330	330	120	120	120	36	36	8	36	36	36	1	8
41	97	387 072	135	534 825 984	1 1 0 1 0 0 0 0 1	57 624	25 928	11 130	5 600	4 506	2 212	518	1 688	806	570	364	182	.	160	264
42	97	129 024	135	316 033 536	2 0 0 0 0 1 1 0 0	32 235	14 756	6 559	2 576	2 834	1 029	238	1 189	402	480	84	84	.	186	152
43	97	193 536	137	366 220 800	0 1 0 0 0 0 0 1 1	34 740	16 085	7 370	2 330	3 345	1 015	.	1 500	435	671	105	.	.	291	185
44	105	64 512	137	374 250 240	1 0 2 0 0 0 0 0 1	37 674	17 164	7 441	4 032	3 030	1 624	371	1 133	602	376	322	133	.	102	197
45	113	129 024	137	363 827 200	3 0 0 1 0 0 0 0 1	35 404	16 344	7 111	4 220	2 890	1 630	776	1 080	581	360	258	258	76	100	186
46	97	322 560	139	535 818 240	0 0 1 0 1 0 1 0 0	50 688	23 416	10 570	4 200	4 632	1 821	.	1 954	759	778	294	.	.	291	297
47	105	64 512	139	336 627 200	0 1 0 0 0 0 2 0 1	29 008	13 706	6 447	1 981	3 019	889	.	1 418	397	663	91	.	.	321	177
48	113	9 216	139	138 498 048	0 1 0 0 0 0 0 0 3	11 376	5 432	2 584	784	1 233	357	.	586	162	288	36	.	.	135	75
49	113	64 512	139	425 677 824	0 2 1 0 0 0 0 0 1	40 284	18 536	8 105	4 732	3 312	1 932	574	1 234	713	406	433	217	.	108	230
50	105	64 512	139	429 977 600	2 0 0 0 0 0 0 1 1	37 960	17 920	8 295	3 080	3 780	1 295	280	1 700	540	758	105	105	.	331	225
51	105	32 256	141	319 258 368	0 0 0 2 0 0 1 0 0	28 125	13 160	6 012	2 576	2 655	1 141	.	1 123	481	444	216	.	.	162	190

1	1	1	.	.	.	.	.	.	.	6	.	.	14	8	22	.	14	.	.	15	.	14	.	55	.	35	63	132	21	20	91	147	78	21	147	49	15
	1	.	.	.	.	.	.	.	.	.	.	.	7	8	8	.	.	.	.	.	.	.	.	20	.	.	21	48	.	.	28	84	28	.	56	14	16
		1	.	.	.	.	.	.	.	.	.	.	.	.	8	.	.	.	.	.	.	.	.	.	.	.	.	27	.	.	.	28	36	.	28	.	17
			1	1	.	2	2	1	5	.	6	5	.	.	.	24	15	13	18	15	27	36	21	29	8	114	29	40	107	78	114	36	15	135	40	85	18
				1	.	1	2	1	.	.	.	5	.	.	.	12	.	7	18	15	20	.	21	.	8	57	29	.	43	78	.	.	.	70	40	.	19
					1	.	.	.	3	3	2	4	6	3	6	8	16	3	6	20	6	23	4	46	1	72	56	77	52	50	108	81	34	60	92	68	20
						1	1	1	.	.	.	.	.	.	.	6	.	7	6	.	14	.	14	.	8	21	.	.	21	21	.	.	.	49	.	.	21
							1	1	.	.	.	.	.	.	.	.	.	.	6	.	7	.	14	.	8	.	.	.	.	21	.	.	.	.	.	.	22
								1	.	.	.	.	.	.	.	.	.	.	.	.	.	.	7	.	8	.	.	.	.	.	.	.	.	.	.	.	23
									1	.	1	1	.	.	.	4	4	2	3	4	4	10	3	10	1	30	10	15	28	20	40	15	6	34	15	31	24
										1	.	.	2	1	3	.	4	.	.	5	.	5	.	15	.	15	18	33	10	10	33	36	18	11	39	20	25
											1	.	.	.	.	2	.	1	1	.	3	5	2	.	1	10	.	.	18	5	15	.	.	18	.	15	26
												1	.	.	.	2	.	1	3	4	3	.	3	.	1	15	10	.	11	20	.	.	.	18	15	.	27
													1	1	1	.	.	.	.	.	.	.	.	5	.	.	6	12	.	.	9	19	7	.	14	5	28
														1	.	.	.	.	.	.	.	.	.	.	.	.	.	.	.	.	.	7	.	.	.	.	29
															1	.	.	.	.	.	.	.	.	.	.	.	.	6	.	.	.	7	7	.	7	.	30
																1	.	1	1	.	2	.	2	.	1	5	.	.	6	5	.	.	.	12	.	.	31
																	1	.	.	1	.	2	.	3	.	6	3	6	5	4	12	6	3	6	6	9	32
																		1	.	.	1	.	1	.	1	.	.	.	.	.	.	.	.	6	.	.	33
																			1	.	1	.	2	.	1	.	.	.	.	5	.	.	.	.	.	.	34
																				1	.	.	.	.	.	3	3	.	2	4	.	.	.	3	6	.	35
																					1	.	1	.	1	.	.	.	.	.	.	.	.	.	.	.	36
																						1	.	.	.	2	.	.	3	1	4	.	.	3	.	5	37
																							1	.	1	.	.	.	.	.	.	.	.	.	.	.	38
																								1	.	.	1	2	.	.	3	3	1	.	2	2	39
																									1	.	.	.	.	.	.	.	.	.	.	.	40
																										1	.	.	1	1	.	.	.	2	.	.	41
																											1	.	.	.	.	.	.	.	2	.	42
																												1	.	.	.	1	1	.	1	.	43
																													1	.	.	.	.	1	.	.	44
																														1	.	.	.	.	.	.	45
																															1	.	.	.	.	1	46
																																1	.	.	.	.	47
																																	1	.	.	.	48
																																		1	.	.	49
																																			1	.	50
																																				1	51

Table 3. *(continued)*

ALGEBRA D10, REPRESENTATIONS OF CLASS 0

NUMBER	0	1	2	3	4	5	6	7	8	9	10	11	12
S.P.	0	8	16	16	24	32	40	64	24	32	32	40	40
O.S.	1	180	20	3360	2880	180	360	20	13440	11520	40320	512	512
LEVEL	1	35	37	61	67	69	71	73	79	89	89	91	91
DIM'N	1	190	209	4845	17765	13090	21546	8645	38760	125970	266475	92378	92378
WEIGHT	0000000000	0100000000	2000000000	0001000000	1010000000	0200000000	2100000000	4000000000	0000010000	0000000010	1000100000	0000000200	0000000002
0	1	10	9	45	125	90	126	45	120	210	555	126	126
1		1	1	8	25	17	26	9	28	56	148	35	35
2			1	.	9	8	18	9	.	.	36	.	.
3				1	3	2	3	1	6	15	33	10	10
4					1	1	2	1	.	.	7	.	.
5						1	1	1	.	.	.	.	.
6							1	1	.	.	.	.	.
7								1	.	.	.	.	.
8									1	4	5	3	3
9										1	.	1	1
10											1	.	.
11												1	.
12													1
13													
14													
15													

NUMBER	13	14	15	16	17	18	19	20	21	22	23	24	25
S.P.	40	48	48	56	72	40	72	80	104	48	144	48	56
O.S.	20160	960	13440	5760	180	107520	2880	360	360	46080	20	201600	80640
LEVEL	95	97	97	101	103	103	103	105	107	109	109	113	115
DIM'N	592515	365750	723520	1351680	490314	1385670	969969	965770	642390	3048474	168245	5542680	6409935
WEIGHT	0101000000	0020000000	2001000000	1110000000	0300000000	1000001000	3010000000	2200000000	4100000000	1000000101	6000000000	0100010000	2000010000
0	1035	590	1120	1920	654	1350	1209	1170	690	2058	165	3960	4095
1	283	163	320	544	181	420	353	334	198	686	45	1308	1400
2	99	72	144	288	108	84	225	234	162	126	45	384	504
3	63	36	72	120	39	123	78	72	42	225	9	398	432
4	22	15	30	62	24	21	48	49	34	35	9	112	140
5	8	7	8	32	17	.	25	33	26	.	9	28	28
6	.	.	8	16	8	.	25	25	26	.	9	.	28
7	.	.	.	.	.	.	9	8	18	.	9	.	.
8	9	5	10	16	5	33	10	9	5	74	1	105	114
9	.	.	.	.	.	7	.	.	.	25	.	20	21
10	3	2	4	8	3	5	6	6	4	10	1	28	34
11	.	.	.	.	.	.	.	.	.	9	.	.	.
12	.	.	.	.	.	.	.	.	.	9	.	.	.
13	1	1	1	4	2	.	3	4	3	.	1	6	6
14		1	.	2	1	.	1	3	2	.	1	.	.
15			1	2	1	.	3	3	3	.	1	.	6

NUMBER	26	27	28	29	30	31	32	33	34	35	36	37	38
S.P.	56	64	56	64	64	64	64	80	72	72	72	80	88
O.S.	80640	3360	322560	161280	23040	23040	92160	40320	40320	5120	5120	20160	2880
LEVEL	119	121	123	123	125	125	125	125	127	127	127	129	131
DIM'N	9699690	5420415	19150670	25681920	14284998	14284998	21767616	16166150	33078430	16166150	16166150	29226560	24386670
WEIGHT	0010100000	0002000000	0100000010	1100100000	0100000200	0100000002	2000000010	3000100000	1011000000	2000000200	2000000002	0201000000	0120000000
0	5790	3015	8190	13440	5166	5166	8352	7430	15610	5250	5250	12960	10170
1	1971	1027	2982	4704	1939	1939	3136	2676	5502	2030	2030	4584	3611
2	720	411	798	2016	504	504	1008	1332	2646	630	630	2320	1980
3	606	316	1057	1464	720	720	1128	843	1707	765	765	1422	1113
4	216	125	266	602	175	175	322	393	798	210	210	710	599
5	76	56	56	224	35	35	56	148	371	35	35	360	331
6	.	.	.	112	.	.	56	148	147	35	35	176	156
7	.	.	.	.	.	.	.	36	.	.	.	.	.
8	157	81	363	376	267	267	390	215	430	286	286	354	273
9	28	14	119	64	101	101	128	35	70	109	109	56	42
10	54	31	85	148	60	60	100	95	194	70	70	172	143
11	.	.	35	.	45	36	36	.	.	45	45	.	.
12	.	.	35	.	36	45	36	.	.	45	45	.	.
13	19	13	15	52	10	10	15	33	88	10	10	83	78
14	7	6	.	14	.	.	.	7	42	.	.	36	45
15	.	.	.	26	.	.	15	33	34	10	10	42	36

NUMBER	39	40	41	42	43	44	45	46	47	48	49	50	51
S.P.	88	64	96	112	104	128	72	120	72	72	136	88	152
O.S.	40320	537600	2880	13440	5760	180	645120	5760	201600	430080	360	107520	2880
LEVEL	131	133	133	133	135	137	137	137	139	139	139	139	139
DIM'N	51465375	58624500	35619870	24025386	49145096	11967150	145530880	51363840	89109240	135795660	26114550	88267179	21061215
WEIGHT	2101000000	0010001000	2020000000	4001000000	1210000000	0400000000	1100001000	3110000000	0001010000	0010000101	2300000000	3000001000	5010000000
0	21015	18540	13830	8586	18216	4110	40320	17280	23640	31500	8370	21483	6255
1	7579	7118	4966	3152	6520	1445	15904	6272	9292	12782	2990	8708	2277
2	4203	2352	2934	1968	4008	972	6048	4224	3456	4032	2154	3780	1701
3	2364	2626	1536	987	2002	435	5960	1928	3486	5118	900	3313	693
4	1267	840	882	594	1212	293	2142	1266	1280	1554	638	1323	509
5	635	252	503	320	740	209	672	800	476	434	453	420	353
6	459	.	319	320	484	136	336	672	.	.	353	420	353
7	99	.	72	144	100	36	.	288	.	.	144	84	225
8	585	915	374	240	480	101	2088	456	1217	2027	207	1167	159
9	90	290	56	35	70	14	656	64	376	798	28	364	21
10	303	282	208	140	284	67	708	292	436	590	144	431	115
11	.	75	.	.	.	.	160	.	90	315	.	84	.
12	.	75	.	.	.	.	160	.	90	315	.	84	.
13	147	81	115	72	169	46	204	180	158	155	101	123	78
14	66	21	66	30	99	31	42	108	56	35	72	21	48
15	105	.	72	72	111	31	102	150	.	.	78	123	78

1	1	.	1	2	2	.	1	.	.	.	.	.	7	.	.	.	7	14	.	.	22	22	44	.	37	30	69	24	21	85	.	.	56	21	48	16
	1	.	.	1	1	.	1	.	.	.	.	.	.	.	.	.	.	.	.	.	8	7	8	.	7	8	32	17	.	32	.	.	33	.	25	17
		1	.	.	.	3	.	5	6	9	5	26	24	21	21	30	15	30	24	24	26	21	45	85	30	20	40	9	210	40	129	222	19	126	15	18
			1	1	2	.	1	.	.	.	.	.	.	.	.	.	7	.	.	.	.	.	22	.	15	30	23	8	.	62	.	.	32	21	48	19
				1	1	.	1	.	.	.	.	.	.	.	.	.	.	.	.	.	.	.	8	.	7	8	16	8	.	32	.	.	25	.	25	20
					1	.	1	.	.	.	.	.	.	.	.	.	.	.	.	.	.	.	.	.	.	8	.	.	.	16	.	.	8	.	25	21
						1	.	.	.	.	.	7	.	8	8	8	.	.	9	9	.	.	.	20	.	.	.	.	48	.	28	84	.	28	.	22
							1	.	.	.	.	.	.	.	.	.	.	.	.	.	.	.	.	.	.	.	.	.	.	.	.	.	.	.	9	23
								1	1	3	2	4	8	3	3	4	5	13	3	3	12	11	21	23	16	10	23	6	56	24	45	54	13	33	10	24
									1	.	.	.	4	.	.	4	5	5	3	3	6	5	15	.	10	10	15	4	28	20	.	.	10	33	10	25
										1	1	.	2	.	.	.	1	6	.	.	5	6	9	5	9	4	13	4	10	14	16	10	9	5	6	26
											1	.	.	.	.	.	.	3	.	.	2	3	3	.	6	1	7	3	.	8	6	.	6	.	3	27
												1	.	1	1	1	.	.	1	1	.	.	.	5	.	.	.	.	12	.	9	19	.	7	.	28
													1	.	.	.	1	2	.	.	3	3	6	.	5	4	9	3	5	11	.	.	7	5	6	29
														1	.	.	.	.	1	.	.	.	.	.	.	.	.	.	.	.	.	7	.	.	.	30
															1	.	.	.	.	1	.	.	.	.	.	.	.	.	.	.	.	7	.	.	.	31
																1	.	.	1	1	.	.	.	.	.	.	.	.	6	.	.	.	.	7	.	32
																	1	.	.	.	.	.	3	.	2	4	3	1	.	8	.	.	4	5	6	33
																		1	.	.	1	2	2	.	3	1	5	2	.	6	.	.	5	.	3	34
																			1	.	.	.	.	.	.	.	.	.	.	.	.	.	.	.	.	35
																				1	.	.	.	.	.	.	.	.	.	.	.	.	.	.	.	36
																					1	1	1	.	1	1	4	2	.	4	.	.	4	.	3	37
																						1	.	.	1	.	2	1	.	2	.	.	3	.	1	38
																							1	.	1	1	2	1	.	4	.	.	3	.	3	39
																								1	.	.	.	.	2	.	3	3	.	1	.	40
																									1	.	1	1	.	2	.	.	2	.	1	41
																										1	.	.	.	2	.	.	1	.	3	42
																											1	1	.	1	.	.	2	.	1	43
																												1	.	.	.	.	1	.	.	44
																													1	.	.	.	.	1	.	45
																														1	.	.	1	.	1	46
																															1	.	.	.	.	47
																																1	.	.	.	48
																																	1	.	.	49
																																		1	.	50
																																			1	51

Table 3. *(continued)*

ALGEBRA D10, REPRESENTATIONS OF CLASS 1

NUMBER	0	1	2	3	4	5	6	7	8	9	10	11	12	13	14	15	16	17	18	19	20	21	22	23	24	25
S.P.	4	12	20	36	20	28	36	28	44	52	68	36	100	36	44	44	52	52	52	52	60	68	76	76	52	84
O.S.	20	960	360	20	8 064	13 440	2 880	15 360	2 880	360	360	5 120	20	80 640	80 640	92 160	40 320	13 440	5 120	5 120	40 320	2 880	13 440	2 880	322 560	5 760
LEVEL	19	49	53	55	71	79	83	85	85	87	89	91	91	97	105	107	107	109	109	109	113	115	115	117	119	119
DIM'N	20	1 140	2 640	1 520	15 504	80 256	117 040	77 520	154 224	142 120	130 416	167 960	40 964	682 176	2 089 164	2 273 920	2 462 400	2 457 840	1 679 600	1 679 600	7 106 000	4 740 120	4 667 520	5 230 016	11 511 720	9 544 080
	1	0	1	3	0	1	0	0	2	1	3	0	5	1	0	1	2	0	1	1	1	1	3	0	0	2
	0	0	1	0	0	0	1	0	0	2	1	0	0	0	1	0	0	0	0	0	1	0	0	2	1	1
	0	1	0	0	0	0	1	0	1	0	0	0	0	0	0	0	0	1	0	0	0	2	0	1	0	1
	0	0	0	0	0	1	0	0	0	0	0	0	0	0	0	0	0	1	0	0	1	0	1	0	0	0
	0	0	0	0	1	0	0	0	0	0	0	0	0	0	1	0	1	0	0	0	0	0	0	0	0	0
	0	0	0	0	0	0	0	0	0	0	0	0	0	1	0	0	0	0	0	0	0	0	0	0	0	0
	0	0	0	0	0	0	0	1	0	0	0	0	0	0	0	0	0	0	0	0	0	0	0	0	1	0
	0 0	0 0	0 0	0 0	0 0	0 0	0 0	0 0	0 0	0 0	0 0	1 0	0 0	0 0	0 0	0 1	0 0	0 0	2 0	0 0	0 0	0 0	0 0	0 0	0 0	0 0
WEIGHT	0	0	0	0	0	0	0	0	0	0	0	1	0	0	0	0	0	0	0	2	0	0	0	0	0	0
0	1	9	18	9	36	144	188	84	225	198	162	126	45	504	1 239	1 008	1 332	1 302	630	630	3 355	2 106	1 968	2 196	3 618	3 732
1		1	2	1	7	30	39	21	48	41	34	35	9	140	357	322	393	378	210	210	993	621	594	646	1 239	1 110
2			1	1	.	8	16	.	25	25	26	.	9	28	112	56	148	147	35	35	459	319	320	356	336	672
3				1	.	.	.	.	9	8	18	.	9	.	.	.	36	.	.	.	99	72	144	100	.	288
4					1	4	5	5	6	5	4	10	1	34	86	100	95	90	70	70	235	145	140	149	400	255
5						1	2	.	3	3	3	.	1	6	26	15	33	34	10	10	105	72	72	81	102	150
6							1	.	1	2	2	.	1	.	7	.	7	14	.	.	44	37	30	46	21	85
7								1	.	.	.	3	.	6	14	30	15	14	24	24	35	21	20	21	117	35
8									1	1	2	.	1	.	.	.	7	.	.	.	22	15	30	23	.	62
9										1	1	.	1	.	.	.	.	.	.	.	8	7	8	16	.	32
10											1	.	1	.	.	.	.	.	.	.	.	.	8	.	.	16
11												1	.	.	.	8	.	.	9	9	.	.	.	.	27	.
12													1	.	.	.	.	.	.	.	.	.	.	.	.	.
13														1	4	4	5	5	3	3	15	10	10	11	28	20
14															1	.	1	2	.	.	6	5	4	6	5	11

NUMBER	26	27	28	29	30	31	32	33	34	35	36	37	38	39	40	41	42	43	44	45	46	47	48	49	50	51
S.P.	100	60	108	116	60	148	60	68	196	68	68	84	68	76	76	76	84	84	76	92	92	84	84	92	116	76
O.S.	360	107 520	2 880	360	184 320	360	268 800	46 080	20	40 320	403 200	80 640	645 120	241 920	61 440	61 440	80 640	13 440	645 120	161 280	40 320	46 080	46 080	92 160	40 320	537 600
LEVEL	121	121	121	123	125	125	127	127	127	131	131	133	137	137	139	139	139	139	141	141	143	143	143	143	143	145
DIM'N	4 434 144	13 168 064	4 903 140	5 064 640	25 865 840	2 708 640	27 287 040	29 302 560	615 296	28 352 940	69 283 500	42 602 560	99 484 000	133 806 816	75 246 080	75 246 080	109 599 600	77 597 520	243 877 920	189 290 920	180 897 024	182 908 440	182 908 440	146 729 856	84 299 600	202 521 000
	1	2	4	3	0	5	0	2	7	0	1	3	0	1	0	0	0	1	1	2	0	1	1	3	4	0
	3	0	0	2	1	1	0	0	0	0	1	0	0	0	0	0	2	0	1	1	1	1	1	0	0	0
	0	0	1	0	0	0	1	0	0	0	0	0	1	1	1	1	0	0	0	0	1	0	0	0	0	0
	0	0	0	0	0	0	0	0	0	1	0	0	0	0	0	0	0	2	0	0	1	0	0	0	0	1
	0	0	0	0	0	0	0	0	0	1	0	0	0	1	0	0	1	0	0	1	0	0	0	0	1	0
	0	0	0	0	0	0	1	0	0	0	1	1	0	0	0	0	0	0	0	0	0	0	0	0	0	0
	0	1	0	0	0	0	0	0	0	0	0	0	0	0	0	0	0	0	0	0	0	0	0	0	0	1
	0 0	0 0	0 0	0 0	1 0	0 0	0 0	1 0	0 0	0 0	0 0	0 0	0 1	0 0	2 0	0 0	0 0	0 0	0 1	0 0	0 0	2 0	0 0	0 1	0 0	0 0
WEIGHT	0	0	0	0	1	0	0	1	0	0	0	0	0	0	0	2	0	0	0	0	0	0	2	0	0	0
0	1 626	3 780	1 701	1 698	5 964	810	6 912	6 174	165	6 399	15 615	8 568	16 380	26 136	10 752	10 752	20 097	14 211	35 694	32 354	30 240	23 226	23 226	19 152	12 708	26 850
1	474	1 323	509	498	2 205	234	2 448	2 331	45	2 292	5 648	3 164	6 440	9 570	4 368	4 368	7 399	5 210	14 336	12 054	11 158	9 639	9 639	7 854	4 804	10 905
2	317	420	353	370	560	198	832	686	45	880	2 204	1 400	2 030	4 252	1 344	1 344	3 465	2 467	5 110	6 048	5 712	3 339	3 339	3 136	2 676	3 920
3	144	84	225	234	.	162	.	126	45	.	384	504	.	720	.	.	777	411	798	2 016	1 344	504	504	1 008	1 332	.
4	106	431	115	110	805	50	800	860	9	750	1 860	1 050	2 470	3 144	1 760	1 760	2 430	1 705	5 565	3 970	3 636	3 935	3 935	3 084	1 585	4 280
5	70	123	78	80	190	42	264	225	9	282	690	432	750	1 353	520	520	1 107	786	1 863	1 914	1 821	1 275	1 275	1 128	843	1 510
6	48	21	48	57	35	34	84	35	9	111	224	140	210	565	140	140	490	369	532	847	924	350	350	322	393	525
7	14	126	15	14	293	6	228	315	1	210	525	294	918	864	708	708	660	462	2 079	1 071	966	1 597	1 597	1 158	420	1 597
8	32	21	48	49	.	34	.	35	9	.	112	140	.	216	.	.	245	125	266	602	420	175	175	322	393	.
9	25	.	25	33	.	26	.	.	9	.	28	28	.	76	.	.	112	56	56	224	224	35	35	56	148	.
10	8	.	25	25	.	26	.	.	9	.	.	28	.	.	.	.	.	.	.	112	.	.	.	56	148	.
11	.	28	.	.	108	.	48	117	.	42	105	56	326	162	288	288	120	84	735	189	168	657	657	408	70	549
12	.	.	9	8	.	18	.	.	9	.	.	.	.	.	.	.	.	.	.	.	.	.	.	.	36	.
13	9	33	10	10	64	5	72	74	1	76	185	114	267	357	200	200	290	205	653	496	468	483	483	390	215	548
14	6	5	6	7	10	4	22	10	1	29	56	34	70	144	50	50	122	93	170	210	230	120	120	100	95	185

1	.	.	1	1	.	.	.	.	6	.	.	7	.	.	22	.	14	25	.	14	35	21	90	63	78	78	50	35	217	84	77	186	186	128	35	180	15
	1	.	.	.	3	2	4	3	.	8	4	5	6	6	.	4	.	10	1	.	28	34	.	54	.	.	62	31	85	148	104	60	60	100	95	.	16
		1	.	.	2	3	1	3	.	6	4	.	3	5	.	3	6	.	1	12	12	6	15	57	10	10	45	45	30	78	114	20	20	15	33	60	17
			1	.	.	.	.	.	.	.	.	.	.	.	8	.	.	9	.	.	.	.	27	.	36	28	.	.	63	.	.	81	72	36	.	48	18
				1	.	.	.	.	.	.	.	.	.	.	8	.	.	9	.	.	.	.	27	.	28	36	.	.	63	.	.	72	81	36	.	48	19
					1	1	1	2	.	4	3	.	3	4	.	3	.	.	1	.	6	6	.	19	.	.	26	13	15	52	54	10	10	15	33	.	20
						1	.	1	.	2	2	.	1	3	.	2	.	.	1	.	.	.	.	7	.	.	7	6	.	14	28	.	.	.	7	.	21
							1	.	.	2	1	.	3	3	.	3	.	.	1	.	.	6	.	.	.	.	.	.	.	26	.	.	.	15	33	.	22
								1	.	1	2	.	1	2	.	2	.	.	1	.	.	.	.	.	.	.	7	.	.	7	14	.	.	.	7	.	23
									1	.	.	1	.	.	3	.	4	3	.	5	10	6	22	24	18	18	20	15	52	34	36	42	42	30	15	58	24
										1	1	.	1	2	.	2	.	.	1	.	.	.	.	.	.	.	.	.	.	7	.	.	.	.	7	.	25
											1	.	.	1	.	1	.	.	1	.	.	.	.	.	.	.	.	.	.	.	.	.	.	.	.	.	26
												1	.	.	.	.	.	3	.	.	5	6	.	9	.	.	10	5	26	24	16	21	21	30	15	.	27
													1	1	.	2	.	.	1	.	.	.	.	.	.	.	.	.	.	.	.	.	.	.	7	.	28
														1	.	1	.	.	1	.	.	.	.	.	.	.	.	.	.	.	.	.	.	.	.	.	29
															1	.	.	1	.	.	.	.	6	.	7	7	.	.	14	.	.	16	16	8	.	14	30
																1	.	.	1	.	.	.	.	.	.	.	.	.	.	.	.	.	.	.	.	.	31
																	1	.	.	2	2	1	4	9	3	3	7	7	8	12	17	6	6	4	5	18	32
																		1	.	.	.	.	.	.	.	.	.	.	7	.	.	8	8	8	.	.	33
																			1	.	.	.	.	.	.	.	.	.	.	.	.	.	.	.	.	.	34
																				1	.	.	.	3	.	.	2	4	.	3	8	.	.	.	1	5	35
																					1	1	.	3	.	.	4	2	4	8	8	3	3	4	5	.	36
																						1	.	.	.	.	.	.	.	4	.	.	.	4	5	.	37
																							1	.	1	1	.	.	2	.	.	2	2	1	.	4	38
																								1	.	.	1	1	.	2	4	.	.	.	1	.	39
																									1	.	.	.	.	.	.	2	.	.	.	.	40
																										1	.	.	.	.	.	.	2	.	.	.	41
																											1	.	.	1	2	.	.	.	1	.	42
																												1	.	.	2	.	.	.	.	.	43
																													1	.	.	1	1	1	.	.	44
																														1	.	.	.	.	1	.	45
																															1	.	.	.	.	.	46
																																1	.	.	.	.	47
																																	1	.	.	.	48
																																		1	.	.	49
																																			1	.	50
																																				1	51

Table 3. *(continued)*

ALGEBRA D10, REPRESENTATIONS OF CLASS 2

S.P.	10	18	26	34	34	42	58	42	50	58	50	66	90	58	58	66	66	74	82	74	66	98	82	90	130	74
O.S.	512	5 120	23 040	5 120	61 440	46 080	5 120	107 520	184 320	23 040	129 024	46 080	5 120	107 520	430 080	184 320	61 440	184 320	46 080	23 040	645 120	46 080	5 120	512	5 120	645 120
LEVEL	46	64	80	82	94	98	100	106	112	114	116	116	118	124	124	128	130	130	132	134	134	134	136	136	136	140
DIM'N	512	9 728	87 040	97 280	486 400	1 157 632	680 960	1 896 960	7 257 600	5 457 408	5 457 408	9 095 680	3 745 280	11 907 072	30 098 432	45 478 400	19 845 120	60 673 536	57 054 720	24 806 400	89 886 720	53 251 072	21 498 880	8 599 552	17 228 288	211 447 808
	0	1	0	2	0	1	3	0	1	0	0	2	4	0	1	0	0	2	1	0	1	3	0	0	5	0
	0	0	1	0	0	1	0	0	0	2	0	1	0	0	0	1	0	0	2	0	0	1	0	0	0	1
	0	0	0	0	1	0	0	0	1	0	0	0	0	0	0	1	0	1	0	0	0	0	0	0	0	0
	0	0	0	0	0	0	0	1	0	0	0	0	0	0	1	0	0	0	0	0	0	0	0	0	0	1
	0	0	0	0	0	0	0	0	0	0	1	0	0	0	0	0	0	0	0	0	1	0	0	0	0	0
	0	0	0	0	0	0	0	0	0	0	0	0	0	1	0	0	0	0	0	0	0	0	0	0	0	0
	0	0	0	0	0	0	0	0	0	0	0	0	0	0	0	0	1	0	0	0	0	0	0	0	0	0
	0 0	1 0	0 0	0 0	1 0	1 0	1 0	0 0	0 0	0 0	1 0	0 0	0 0	0 0	1 0	1 0	1 0	1 0	1 0	0 1	0 0	1 0	2 0	0 0	1 0	0 0
WEIGHT	1	0	1	1	0	0	0	1	1	1	0	1	1	1	0	0	0	0	0	1	1	0	1	3	0	1
NUMBER	0	1	2	3	4	5	6	7	8	9	10	11	12	13	14	15	16	17	18	19	20	21	22	23	24	25
0	1	9	45	45	155	321	165	405	1 305	909	837	1 395	495	1 398	3 763	5 085	1 890	6 291	5 535	2 025	8 310	4 653	1 585	603	1 287	16 902
1		1	8	9	36	81	45	111	387	276	258	441	165	468	1 251	1 746	672	2 223	1 983	750	3 006	1 725	603	232	495	6 346
2			1	1	7	16	9	28	100	73	76	117	45	154	383	547	238	703	636	280	1 034	561	232	91	165	2 246
3				1	.	8	9	.	36	36	.	81	45	.	111	232	.	387	396	.	258	441	.	.	165	846
4					1	2	1	6	21	15	21	24	9	49	105	150	84	192	174	105	334	153	91	36	45	732
5						1	1	.	7	8	.	16	9	.	28	64	.	100	109	.	76	117	.	.	45	272
6							1	.	.	.	.	8	9	.	.	.	.	36	36	.	.	81	.	.	45	.
7								1	3	2	5	3	1	15	24	33	29	42	37	40	99	32	36	15	9	213
8									1	1	.	2	1	.	6	14	.	21	23	.	21	24	.	.	9	77
9										1	.	1	1	.	.	7	.	7	16	.	.	16	.	.	9	28
10											1	.	.	4	4	5	10	6	5	15	25	4	15	6	1	51
11												1	1	.	.	.	.	7	8	.	.	16	.	.	9	.
12													1	.	.	.	.	.	.	.	.	8	.	.	9	.
13														1	.	.	3	.	.	6	5	.	6	3	.	9

S.P.	82	74	82	90	106	82	106	82	114	90	90	138	90	106	98	178	90	98	106	98	114	114	98	98	122	106
O.S.	61 440	645 120	430 080	368 640	23 040	430 080	184 320	1 290 240	46 080	184 320	645 120	46 080	430 080	5 120	46 080	5 120	1 612 800	1 290 240	184 320	645 120	430 080	184 320	1 290 240	1 290 240	368 640	107 520
LEVEL	142	142	142	146	148	148	148	150	150	152	152	152	154	154	154	154	158	158	160	160	160	162	164	164	164	166
DIM'N	134 041 600	200 931 840	261 381 120	505 885 184	188 528 640	340 343 808	369 177 600	670 940 160	375 072 768	429 977 600	800 419 840	254 679 040	827 706 880	150 492 160	375 072 768	68 913 152	1 555 361 280	2 443 284 480	1 673 605 120	1 820 206 080	1 631 764 992	1 875 363 840	2 941 542 912	2 695 959 552	3 456 358 400	1 704 102 400
	0	1	2	1	0	1	3	0	2	1	2	4	0	1	1	6	0	1	1	2	3	0	0	0	2	0
	0	0	0	1	3	0	0	1	2	0	0	1	0	0	0	0	1	1	0	0	0	2	0	1	1	0
	2	0	0	1	0	0	1	0	0	0	0	0	1	0	0	0	0	0	2	0	0	1	1	0	1	0
	0	0	1	0	0	0	0	0	0	0	0	0	1	0	0	0	0	1	0	0	1	0	0	0	0	2
	0	0	0	0	0	0	0	1	0	0	1	0	0	0	0	0	0	0	0	0	0	0	1	0	0	0
	0	1	0	0	0	0	0	0	0	0	0	0	0	0	0	0	1	0	0	1	0	0	0	0	0	0
	0	0	0	0	0	1	0	0	0	0	0	0	0	0	0	0	0	0	0	0	0	0	0	1	0	0
	0 0	1 0	0 0	0 0	0 0	0 0	0 0	1 0	0 0	1 1	1 0	0 0	1 0	3 0	1 0	0 0	0 0	1 0	1 0	0 0	1 0	1 0	0 0	1 0	1 0	0 0
WEIGHT	1	0	1	1	1	1	1	0	1	0	0	1	0	0	2	1	1	0	0	1	0	0	1	0	0	1
NUMBER	26	27	28	29	30	31	32	33	34	35	36	37	38	39	40	41	42	43	44	45	46	47	48	49	50	51
0	9 990	14 535	19 455	34 119	11 595	20 307	22 575	40 545	21 627	22 250	45 170	13 035	45 135	6 765	17 667	3 003	74 835	121 050	77 850	81 945	73 611	82 305	123 543	108 423	142 785	67 025
1	3 816	5 598	7 479	13 383	4 626	8 190	9 099	16 293	8 793	9 270	18 522	5 445	18 542	2 907	7 515	1 287	31 653	50 898	33 120	35 298	31 683	35 370	53 697	47 649	62 283	29 484
2	1 366	2 092	2 671	4 864	1 707	3 248	3 351	6 297	3 267	3 830	7 226	2 055	7 293	1 248	3 187	495	13 062	20 294	13 330	14 698	12 791	14 353	22 666	20 622	25 461	12 579
3	596	468	1 251	2 664	1 074	672	2 223	2 169	2 313	750	3 006	1 725	2 988	232	603	495	4 200	10 062	7 128	5 598	7 479	8 298	8 820	6 300	16 191	5 292
4	448	756	872	1 598	563	1 267	1 103	2 316	1 080	1 575	2 664	681	2 709	539	1 353	165	5 229	7 572	5 001	5 907	4 791	5 403	9 219	8 778	9 606	5 152
5	196	154	383	863	360	238	703	776	756	280	1 034	561	1 106	91	232	165	1 624	3 666	2 648	2 092	2 671	3 118	3 549	2 576	5 992	2 170
6	.	.	111	232	120	.	387	.	396	.	258	441	.	.	.	165	.	846	596	468	1 251	918	.	.	2 664	.
7	129	261	252	456	159	486	312	796	303	645	915	189	929	234	579	45	2 019	2 589	1 707	2 280	1 634	1 839	3 573	3 664	3 265	2 001
8	57	49	105	242	101	84	192	258	210	105	334	153	377	36	91	45	602	1 221	896	756	872	1 051	1 358	1 029	1 998	840
9	28	.	28	128	73	.	100	76	145	.	76	117	161	.	.	45	154	544	476	154	383	631	518	238	1 179	371
10	30	85	60	104	35	181	70	250	66	265	285	40	285	105	249	9	740	785	510	835	490	545	1 294	1 499	960	720
11	.	.	28	64	36	.	100	.	109	.	76	117	.	.	.	45	.	272	196	154	383	316	.	.	863	.
12	.	.	.	.	.	.	36	.	36	.	.	81	.	.	.	45	.	.	.	.	111	.	.	.	232	.
13	5	24	10	16	5	66	10	66	9	107	75	5	72	46	111	1	255	195	123	285	120	129	429	591	225	234

1	2	.	3	3	.	5	3	.	.	1	18	13	15	24	54	22	29	42	78	45	40	99	32	114	15	36	9	212	360	264	261	252	306	483	402	576	301	14
	1	.	1	2	.	.	2	.	.	1	6	7	.	6	28	15	.	21	21	31	.	21	24	49	.	.	9	49	154	142	49	105	178	182	84	334	133	15
		1	.	.	2	.	.	3	1	.	.	.	6	.	.	.	21	.	14	.	45	15	.	14	24	48	.	75	35	21	84	20	21	119	228	35	63	16
			1	1	.	.	2	.	.	1	.	.	.	6	14	8	.	21	.	23	.	21	24	.	.	.	9	.	77	57	49	105	92	.	.	242	.	17
				1	.	.	1	.	.	1	.	.	.	.	7	8	.	7	.	16	.	.	16	.	.	.	9	.	28	28	.	28	64	.	.	128	.	18
					1	.	.	1	1	.	.	.	.	.	.	.	7	.	.	.	16	.	.	.	9	25	.	20	.	.	21	.	.	28	75	.	14	19
						1	.	.	.	.	3	2	4	4	8	3	10	6	20	6	15	25	4	28	6	15	1	70	87	62	85	60	71	156	152	132	96	20
							1	.	.	1	.	.	.	.	.	.	.	7	.	8	.	.	16	.	.	.	9	.	.	.	.	28	.	.	.	64	.	21
								1	.	.	.	.	.	.	.	.	.	.	.	.	8	.	.	.	9	9	.	.	.	.	.	.	.	.	27	.	.	22
									1	.	.	.	.	.	.	.	.	.	.	.	.	.	.	.	.	9	.	.	.	.	.	.	.	.	.	.	.	23
										1	.	.	.	.	.	.	.	.	.	.	.	.	8	.	.	.	9	.	.	.	.	.	.	.	.	.	.	24
											1	1	.	1	4	2	.	3	5	4	.	5	3	12	.	.	1	15	36	33	15	24	40	57	29	75	43	25
												1	.	.	2	1	.	1	.	3	.	.	2	6	.	.	1	.	12	21	.	6	21	21	.	42	21	26
													1	.	.	.	3	.	4	.	6	5	.	5	3	6	.	20	15	10	24	10	11	42	56	20	25	27
														1	2	1	.	3	.	3	.	5	3	.	.	.	1	.	18	13	15	24	21	.	.	54	.	28
															1	1	.	1	.	2	.	.	2	.	.	.	1	.	6	7	.	6	14	.	.	28	.	29
																1	.	.	.	1	.	.	1	.	.	.	1	.	.	.	.	.	7	.	.	7	.	30
																	1	.	.	.	2	.	.	.	1	3	.	5	.	.	6	.	.	9	18	.	5	31
																		1	.	1	.	.	2	.	.	.	1	.	.	.	.	6	.	.	.	14	.	32
																			1	.	.	1	.	2	.	.	.	4	6	5	4	4	6	15	10	11	11	33
																				1	.	.	1	.	.	.	1	.	.	.	.	.	.	.	.	7	.	34
																					1	.	.	.	1	1	.	.	.	.	.	.	.	.	6	.	.	35
																						1	.	.	.	.	.	.	3	2	4	4	3	.	.	8	.	36
																							1	.	.	.	1	.	.	.	.	.	.	.	.	.	.	37
																								1	.	.	.	.	2	3	.	1	3	5	.	6	6	38
																									1	.	.	.	.	.	.	.	.	.	.	.	.	39
																										1	.	.	.	.	.	.	.	.	.	.	.	40
																											1	.	.	.	.	.	.	.	.	.	.	41
																												1	.	.	1	.	.	3	3	.	2	42
																													1	1	.	1	2	.	.	4	.	43
																														1	.	.	1	.	.	2	.	44
																															1	.	.	.	.	.	.	45
																																1	.	.	.	2	.	46
																																	1	.	.	1	.	47
																																		1	.	.	1	48
																																			1	.	.	49
																																				1	.	50
																																					1	51

Table 3. *(continued)*

ALGEBRA D11, REPRESENTATIONS OF CLASS 0

NUMBER	0	1	2	3	4	5	6	7	8	9	10	11	12	13	14	15	16	17	18	19	20	21	22	23	24	25
S.P.	0	8	16	16	24	32	40	64	24	32	32	40	48	48	40	56	72	72	80	40	104	144	48	48	56	56
O.S.	1	220	22	5 280	3 960	220	440	22	29 568	73 920	42 240	31 680	1 320	21 120	11 264	7 920	220	3 960	440	295 680	440	22	443 520	253 440	11 264	11 264
LEVEL	1	39	41	69	75	77	79	81	91	101	105	107	109	109	111	113	115	115	117	119	119	121	129	129	131	131
DIM'N	1	231	252	7 315	26 334	19 228	31 395	12 397	74 613	497 420	319 770	1 083 852	662 354	1 312 311	646 646	2 416 128	864 500	1 713 040	1 689 765	3 357 585	1 106 875	283 360	12 978 900	9 976 824	7 113 106	7 113 106
WEIGHT	0	0	2	0	1	0	2	4	0	1	0	0	0	2	0	1	0	3	2	1	4	6	0	1	1	1
	0	1	0	0	0	2	1	0	0	0	0	1	0	0	0	1	3	0	2	0	1	0	1	0	0	0
	0	0	0	0	1	0	0	0	0	0	0	0	2	0	0	1	0	1	0	0	0	0	0	0	0	0
	0	0	0	1	0	0	0	0	0	0	0	1	0	1	0	0	0	0	0	0	0	0	0	0	0	0
	0	0	0	0	0	0	0	0	0	1	0	0	0	0	0	0	0	0	0	0	0	0	0	0	0	0
	0	0	0	0	0	0	0	0	1	0	0	0	0	0	0	0	0	0	0	0	0	0	1	0	0	0
	0	0	0	0	0	0	0	0	0	0	0	0	0	0	0	0	0	0	0	1	0	0	0	0	0	0
	0	0	0	0	0	0	0	0	0	0	1	0	0	0	0	0	0	0	0	0	0	0	0	0	0	0
	0 0	0 0	0 0	0 0	0 0	0 0	0 0	0 0	0 0	0 0	0 0	0 0	0 0	0 0	1 0	0 0	0 0	0 0	0 0	0 0	0 0	0 0	0 0	0 1	2 0	0 0
	0	0	0	0	0	0	0	0	0	0	0	0	0	0	1	0	0	0	0	0	0	0	0	0	0	2
0	1	11	10	55	154	110	155	55	165	770	330	1 430	814	1 551	462	2 640	890	1 660	1 595	2 145	935	220	6 270	3 828	2 310	2 310
1		1	1	9	28	19	29	10	36	189	84	359	206	405	126	684	226	442	416	624	245	55	1 923	1 218	756	756
2			1	.	10	9	20	10	.	45	.	125	90	180	.	360	135	280	290	120	200	55	555	210	126	126
3				1	3	2	3	1	7	38	21	72	41	82	35	136	44	88	81	168	47	10	534	378	245	245
4					1	1	2	1	.	8	.	25	17	34	.	70	27	54	55	28	38	10	148	56	35	35
5						1	1	1	.	.	.	9	8	9	.	36	19	28	37	.	29	10	36	.	.	.
6							1	1	.	.	.	.	.	9	.	18	9	28	28	.	29	10	.	.	.	.
7								1	.	.	.	.	.	.	.	.	.	10	9	.	20	10	.	.	.	.
8									1	5	5	9	5	10	10	16	5	10	9	40	5	1	125	115	80	80
9										1	.	3	2	4	.	8	3	6	6	6	4	1	33	15	10	10
10											1	.	.	.	3	.	.	.	.	7	.	.	20	34	27	27
11												1	1	1	.	4	2	3	4	.	3	1	7	.	.	.
12													1	.	.	2	1	1	3	.	2	1	.	.	.	.
13														1	.	2	1	3	3	.	3	1	.	.	.	.
14															1	.	.	.	.	.	.	.	.	9	10	10

NUMBER	26	27	28	29	30	31	32	33	34	35	36	37	38	39	40	41	42	43	44	45	46	47	48	49	50	51
S.P.	56	56	64	64	80	56	72	80	64	88	88	64	96	112	104	72	64	128	120	136	152	72	160	72	88	200
O.S.	177 408	147 840	5 280	295 680	73 920	1 182 720	63 360	31 680	337 920	3 960	63 360	506 880	3 960	21 120	7 920	112 640	1 478 400	220	7 920	440	3 960	1 774 080	440	443 520	295 680	440
LEVEL	131	135	137	139	141	143	143	145	145	147	147	149	149	149	151	151	153	153	153	155	155	157	157	159	159	159
DIM'N	14 872 858	22 208 340	12 273 030	57 987 072	36 038 079	59 491 432	73 489 416	64 324 260	66 927 861	53 117 350	112 363 625	123 559 128	76 969 200	51 620 625	105 223 482	138 105 110	175 287 255	25 162 137	108 725 760	54 511 380	43 743 700	428 460 032	45 685 728	259 837 578	256 270 784	20 322 225
WEIGHT	2	0	0	1	3	0	1	0	2	0	2	0	2	4	1	2	0	0	3	2	5	1	4	0	3	6
	0	0	0	1	0	1	0	2	0	1	1	1	0	0	2	0	0	4	1	3	0	1	2	0	0	1
	0	1	0	0	0	0	1	0	0	2	0	0	2	0	1	0	1	0	1	0	1	0	0	0	0	0
	0	0	2	0	0	0	1	1	0	0	1	0	0	1	0	0	0	0	0	0	0	0	0	1	0	0
	0	1	0	1	1	0	0	0	0	0	0	0	0	0	0	0	0	0	0	0	0	0	0	0	0	0
	1	0	0	0	0	0	0	0	0	0	0	0	0	0	0	0	0	0	0	0	0	0	0	1	0	0
	0	0	0	0	0	0	0	0	0	0	0	0	0	0	0	0	1	0	0	0	0	1	0	0	1	0
	0	0	0	0	0	1	0	0	1	0	0	0	0	0	0	0	0	0	0	0	0	0	0	0	0	0
	0 0	0 0	0 0	0 0	0 0	0 0	0 0	0 0	0 0	0 0	0 0	1 0	0 0	0 0	0 0	1 0	0 0	0 0	0 0	0 0	0 0	0 0	0 0	0 0	0 0	0 0
	0	0	0	0	0	0	0	0	0	0	0	1	0	0	0	1	0	0	0	0	0	0	0	0	0	0
0	6 490	9 130	4 730	21 120	11 649	15 180	24 376	20 130	15 477	15 730	32 615	23 100	21 340	13 245	27 930	23 430	34 155	6 215	26 400	12 650	9 460	73 920	9 680	43 230	39 248	3 905
1	2 052	2 873	1 490	6 816	3 858	5 214	7 912	6 558	5 460	5 138	10 808	8 358	7 042	4 461	9 192	8 694	12 273	2 014	8 796	4 154	3 160	27 216	3 185	15 852	14 808	1 265
2	720	1 035	590	2 880	1 890	1 350	3 760	3 291	1 680	2 790	5 930	2 058	4 120	2 760	5 610	2 520	3 960	1 350	5 880	2 975	2 350	10 080	2 495	5 790	6 240	1 100
3	577	804	417	1 928	1 103	1 722	2 230	1 848	1 827	1 439	3 060	2 982	1 978	1 267	2 565	3 143	4 184	552	2 456	1 138	874	9 408	866	5 475	5 188	335
4	184	283	163	784	509	420	1 032	915	504	768	1 627	686	1 127	758	1 543	812	1 308	370	1 604	803	640	3 304	668	1 971	2 024	290
5	36	99	72	288	189	84	476	459	84	422	809	126	638	405	936	126	384	262	1 008	568	442	1 008	506	720	624	245
6	36	.	.	144	189	.	188	225	84	198	584	.	404	405	612	126	.	171	846	442	442	504	461	.	624	245
7	.	.	.	.	45	.	.	.	.	.	125	.	90	180	126	.	.	45	360	180	280	.	290	.	120	200
8	135	185	95	440	250	539	500	410	575	315	675	1 055	430	275	550	1 120	1 320	115	520	235	180	2 980	175	1 726	1 650	65
9	40	63	36	172	110	123	224	198	144	164	348	225	238	160	324	260	398	76	332	163	130	992	133	606	600	56
10	21	28	14	64	35	154	70	56	164	42	90	374	56	35	70	399	365	14	64	28	21	816	20	466	448	7
11	7	22	15	60	38	21	101	95	21	89	168	35	131	82	192	35	112	52	204	114	88	280	99	216	168	47
12	.	8	7	16	8	.	48	41	.	51	75	.	75	34	112	.	28	35	122	81	54	56	73	76	28	38
13	7	.	.	30	38	.	39	48	21	41	120	.	82	82	126	35	.	35	170	88	88	140	90	.	168	47
14	.	.	.	.	.	35	.	.	36	.	.	135	.	.	.	145	75	.	.	.	.	160	.	90	84	.

1	1	1	2	.	2	1	.	.	.	.	.	.	.	8	8	.	16	25	.	25	50	.	42	34	78	.	.	27	96	63	54	28	64	.	28	38	15
	1	.	1	.	1	1	.	.	.	.	.	.	.	.	.	.	.	9	.	8	9	.	8	9	36	.	.	19	36	37	28	.	37	.	.	29	16
		1	1	.	2	1	.	.	.	.	.	.	.	8	.	.	.	.	.	.	25	.	17	34	26	.	.	9	70	36	54	.	55	.	28	38	17
			1	.	1	1	.	.	.	.	.	.	.	.	.	.	.	.	.	.	9	.	8	9	18	.	.	9	36	28	28	.	37	.	.	29	18
				1	.	.	5	4	3	3	6	9	5	24	15	33	30	26	38	21	45	74	30	20	40	84	105	9	40	19	15	258	15	157	154	6	19
					1	1	.	.	.	.	.	.	.	.	.	.	.	.	.	.	.	.	.	9	.	.	.	.	18	9	28	.	28	.	.	29	20
						1	.	.	.	.	.	.	.	.	.	.	.	.	.	.	.	.	.	.	.	.	.	.	.	.	10	.	9	.	.	20	21
							1	.	.	.	1	3	2	8	5	5	13	12	5	11	21	10	16	10	23	10	28	6	24	13	10	68	11	54	40	5	22
								1	1	1	.	.	.	.	.	7	.	.	8	.	.	25	.	.	.	28	20	.	.	.	.	48	.	28	28	.	23
									1	.	.	.	.	.	.	.	.	.	.	.	.	9	.	.	.	10	.	.	.	.	.	.	.	.	.	.	24
										1	.	.	.	.	.	.	.	.	.	.	.	9	.	.	.	10	.	.	.	.	.	.	.	.	.	.	25
											1	.	.	4	5	.	5	6	5	5	15	.	10	10	15	10	.	4	20	10	10	34	10	.	40	5	26
												1	1	2	1	.	6	5	.	6	9	.	9	4	13	.	6	4	14	9	6	12	8	19	6	4	27
													1	.	.	.	3	2	.	3	3	.	6	1	7	.	.	3	8	6	3	.	6	7	.	3	28
														1	1	.	2	3	.	3	6	.	5	4	9	.	.	3	11	7	6	6	7	.	6	4	29
															1	.	.	.	.	.	3	.	2	4	3	.	.	1	8	4	6	.	6	.	6	4	30
																1	.	.	1	.	.	3	.	.	.	3	5	.	.	.	.	12	.	9	7	.	31
																	1	1	.	2	2	.	3	1	5	.	.	2	6	5	3	.	5	.	.	3	32
																		1	.	1	1	.	1	1	4	.	.	2	4	4	3	.	4	.	.	3	33
																			1	.	.	.	.	.	.	3	.	.	.	.	.	6	.	.	7	.	34
																				1	.	.	1	.	2	.	.	1	2	3	1	.	3	.	.	2	35
																					1	.	1	1	2	.	.	1	4	3	3	.	4	.	.	3	36
																						1	.	.	.	1	.	.	.	.	.	.	.	.	.	.	37
																							1	.	1	.	.	1	2	2	1	.	3	.	.	2	38
																								1	.	.	.	.	2	1	3	.	3	.	.	3	39
																									1	.	.	1	1	2	1	.	2	.	.	2	40
																										1	.	.	.	.	.	.	.	.	.	.	41
																											1	.	.	.	.	2	.	3	1	.	42
																												1	.	1	.	.	1	.	.	1	43
																													1	1	1	.	2	.	.	2	44
																														1	.	.	1	.	.	1	45
																															1	.	1	.	.	2	46
																																1	.	.	1	.	47
																																	1	.	.	1	48
																																		1	.	.	49
																																			1	.	50
																																				1	51

Table 3. *(continued)*

ALGEBRA D11, REPRESENTATIONS OF CLASS 1

NUMBER	0	1	2	3	4	5	6	7	8	9	10	11	12
S.P.	4	12	20	36	20	28	36	44	52	28	68	100	36
O.S.	22	1 320	440	22	14 784	21 120	3 960	3 960	440	42 240	440	22	28 160
LEVEL	21	55	59	61	81	89	93	95	97	99	99	101	109
DIM'N	22	1 540	3 520	2 002	26 334	133 056	191 268	249 964	228 228	170 544	206 976	63 756	497 420
	1	0	1	3	0	1	0	2	1	0	3	5	0
	0	0	1	0	0	0	1	0	2	0	1	0	0
	0	1	0	0	0	0	1	1	0	0	0	0	0
	0	0	0	0	0	1	0	0	0	0	0	0	0
	0	0	0	0	1	0	0	0	0	0	0	0	0
	0	0	0	0	0	0	0	0	0	0	0	0	0
	0	0	0	0	0	0	0	0	0	1	0	0	0
	0	0	0	0	0	0	0	0	0	0	0	0	0
	0 0	0 0	0 0	0 0	0 0	0 0	0 0	0 0	0 0	0 0	0 0	0 0	0 1
WEIGHT	0	0	0	0	0	0	0	0	0	0	0	0	0
0	1	10	20	10	45	180	234	280	245	120	200	55	210
1		1	2	1	8	34	44	54	46	28	38	10	56
2			1	1	.	9	18	28	28	.	29	10	.
3				1	.	.	.	10	9	.	20	10	.
4					1	4	5	6	5	6	4	1	15
5						1	2	3	3	.	3	1	.
6							1	1	2	.	2	1	.
7								1	1	.	2	1	.
8									1	.	1	1	.
9										1	.	.	4
10											1	1	.
11												1	.
12													1
13													
14													

NUMBER	13	14	15	16	17	18	19	20	21	22	23	24	25
S.P.	44	44	36	44	52	52	44	60	68	76	76	52	84
O.S.	1 024	1 024	177 408	147 840	73 920	21 120	337 920	63 360	3 960	21 120	3 960	112 640	7 920
LEVEL	111	111	111	119	121	123	125	127	129	129	131	131	133
DIM'N	352 716	352 716	1 444 608	4 307 072	5 032 170	4 987 136	6 366 976	14 226 212	9 393 384	9 229 440	10 270 260	13 023 360	18 595 500
	0	0	1	0	2	0	1	1	1	3	0	1	2
	0	0	0	1	0	0	0	1	0	0	2	0	1
	0	0	0	0	0	1	0	0	2	0	1	0	1
	0	0	0	0	0	1	0	1	0	1	0	0	0
	0	0	0	1	1	0	0	0	0	0	0	0	0
	0	0	1	0	0	0	0	0	0	0	0	0	0
	0	0	0	0	0	0	0	0	0	0	0	0	0
	0	0	0	0	0	0	1	0	0	0	0	0	0
	2 0	0 0	0 0	0 0	0 0	0 0	0 0	0 0	0 0	0 0	0 0	1 0	0 0
WEIGHT	0	2	0	0	0	0	0	0	0	0	0	1	0
0	126	126	720	1 760	1 890	1 840	1 680	4 721	2 950	2 760	3 060	2 520	5 190
1	35	35	184	464	509	488	504	1 274	793	758	821	812	1 406
2	.	.	36	144	189	188	84	584	404	405	450	126	846
3	.	.	.	.	45	.	.	125	90	180	126	.	360
4	10	10	40	100	110	104	144	270	166	160	170	260	290
5	.	.	7	30	38	39	21	120	82	82	92	35	170
6	.	.	.	8	8	16	.	50	42	34	52	.	96
7	.	.	.	.	8	.	.	25	17	34	26	.	70
8	.	.	.	.	.	.	.	9	8	9	18	.	36
9	3	3	6	14	15	14	38	35	21	20	21	84	35
10	.	.	.	.	.	.	.	.	.	9	.	.	18
11	.	.	.	.	.	.	.	.	.	.	.	.	.
12	1	1	.	.	.	.	8	.	.	.	.	28	.
13	1	.	.	.	.	.	.	.	.	.	.	10	.
14		1	.	.	.	.	.	.	.	.	.	10	.

NUMBER	26	27	28	29	30	31	32	33	34	35	36	37	38
S.P.	100	108	52	116	60	148	196	60	60	68	68	68	68
O.S.	440	3 960	887 040	440	295 680	440	22	591 360	1 013 760	73 920	56 320	56 320	887 040
LEVEL	135	135	137	137	139	139	141	145	147	149	149	149	149
DIM'N	8 520 512	9 405 000	30 889 782	9 623 614	34 994 960	5 051 904	1 118 260	71 066 688	94 140 288	72 547 244	67 603 900	67 603 900	177 827 650
	1	4	0	3	2	5	7	0	0	0	0	0	1
	3	0	1	2	0	1	0	0	1	0	1	1	1
	0	1	0	0	0	0	0	1	0	0	0	0	0
	0	0	0	0	0	0	0	0	0	1	0	0	0
	0	0	0	0	0	0	0	0	0	1	0	0	0
	0	0	0	0	0	0	0	1	0	0	0	0	1
	0	0	1	0	1	0	0	0	0	0	0	0	0
	0	0	0	0	0	0	0	0	0	0	0	0	0
	0 0	0 0	0 0	0 0	0 0	0 0	0 0	0 0	0 1	0 0	2 0	0 0	0 0
WEIGHT	0	0	0	0	0	0	0	0	0	0	0	2	0
0	2 240	2 350	5 985	2 330	6 240	1 100	220	11 340	11 808	10 430	7 350	7 350	25 480
1	596	640	1 904	623	2 024	290	55	3 710	4 144	3 444	2 660	2 660	8 495
2	397	442	504	461	624	245	55	1 239	1 008	1 302	630	630	3 255
3	180	280	.	290	120	200	55	.	.	.	.	.	555
4	120	130	560	124	600	56	10	1 100	1 420	1 021	955	955	2 535
5	79	88	140	90	168	47	10	357	322	378	210	210	927
6	54	54	28	64	28	38	10	112	56	147	35	35	296
7	36	54	.	55	28	38	10	.	.	.	.	.	148
8	28	28	.	37	.	29	10	.	.	.	.	.	36
9	14	15	144	14	154	6	1	276	474	252	344	344	630
10	9	28	.	28	.	29	10	.	.	.	.	.	.
11	.	10	.	9	.	20	10	.	.	.	.	.	.
12	.	.	27	.	28	.	.	48	152	42	127	127	105
13	.	.	.	.	.	.	.	.	44	.	55	45	.
14	.	.	.	.	.	.	.	.	44	.	45	55	.

NUMBER	39	40	41	42	43	44	45	46	47	48	49	50	51
S.P.	68	76	76	84	76	84	84	68	92	92	116	108	76
O.S.	253 440	11 264	11 264	177 408	443 520	147 840	21 120	2 365 440	295 680	63 360	73 920	1 320	2 365 440
LEVEL	149	151	151	151	155	157	157	159	159	161	161	163	163
DIM'N	105 462 084	75 508 356	75 508 356	107 903 488	335 675 340	272 272 000	192 472 280	345 181 056	466 375 140	442 041 600	204 297 500	127 481 640	832 233 402
	2	2	2	3	1	0	1	0	2	0	4	0	1
	0	0	0	0	0	2	0	0	1	1	0	0	1
	0	0	0	0	1	0	0	1	0	1	0	3	0
	0	0	0	0	0	0	2	0	0	1	0	0	0
	0	0	0	0	1	1	0	0	1	0	1	0	0
	0	0	0	1	0	0	0	0	0	0	0	0	0
	0	0	0	0	0	0	0	0	0	0	0	0	0
	0	0	0	0	0	0	0	1	0	0	0	0	1
	0 1	2 0	0 0	0 0	0 0	0 0	0 0	0 0	0 0	0 0	0 0	0 0	0 0
WEIGHT	0	0	2	0	0	0	0	0	0	0	0	0	0
0	12 180	7 560	7 560	13 920	42 255	32 320	22 850	32 040	61 894	48 240	20 235	12 870	69 297
1	4 354	2 786	2 786	4 728	14 226	10 936	7 699	11 828	17 744	16 344	7 010	4 362	26 074
2	1 218	756	756	2 052	6 228	5 056	3 596	3 618	8 784	8 268	3 858	2 394	8 994
3	210	126	126	720	1 035	1 120	590	.	2 880	1 920	1 890	654	1 350
4	1 506	1 010	1 010	1 420	4 230	3 250	2 280	4 204	5 285	4 820	2 090	1 275	9 365
5	378	245	245	577	1 797	1 464	1 038	1 239	2 522	2 388	1 103	687	3 045
6	56	35	35	184	742	640	482	336	1 104	1 200	509	381	840
7	56	35	35	184	283	320	163	.	784	544	509	181	420
8	.	.	.	36	99	144	72	.	288	288	189	108	84
9	505	366	366	350	1 026	780	546	1 416	1 260	1 134	490	294	3 164
10	.	.	.	36	.	.	.	.	144	.	189	.	.
11	.	.	.	.	.	.	.	.	.	.	45	.	.
12	162	136	136	56	162	120	84	436	189	168	70	42	966
13	45	55	55	.	.	.	.	110	.	.	.	.	231
14	45	55	55	.	.	.	.	110	.	.	.	.	231

1	4	5	5	5	15	10	10	11	10	20	9	10	34	10	40	5	1	86	100	90	70	70	220	115	80	80	135	420	340	240	400	580	545	250	154	970	15
	1	1	2	.	6	5	4	6	.	11	6	6	6	7	6	4	1	26	15	34	10	10	66	15	10	10	40	168	142	108	102	244	266	110	84	246	16
		1	.	.	3	2	4	3	.	8	4	6	.	6	6	4	1	.	.	.	.	.	33	15	10	10	40	63	72	36	.	172	120	110	39	123	17
			1	.	2	3	1	3	.	6	4	3	.	5	.	3	1	7	.	14	.	.	14	.	.	.	7	66	52	52	21	90	131	38	48	42	18
				1	.	.	.	.	3	.	.	.	6	.	7	.	.	14	30	14	24	24	35	34	27	27	21	63	50	35	117	84	77	35	21	280	19
					1	1	1	2	.	4	3	3	.	4	.	3	1	.	.	.	.	.	7	.	.	.	7	22	30	15	.	60	62	38	24	21	20
						1	.	1	.	2	2	1	.	3	.	2	1	.	.	.	.	.	.	.	.	.	.	8	8	7	.	16	32	8	17	.	21
							1	.	.	2	1	3	.	3	.	3	1	.	.	.	.	.	.	.	.	.	7	.	.	.	.	30	.	38	.	.	22
								1	.	1	2	1	.	2	.	2	1	.	.	.	.	.	.	.	.	.	.	.	8	.	.	8	16	8	8	.	23
									1	.	.	.	.	.	.	.	.	.	8	.	9	9	.	9	10	10	.	.	.	.	27	.	.	.	.	63	24
										1	1	1	.	2	.	2	1	.	.	.	.	.	.	.	.	.	.	.	.	.	.	8	.	8	.	.	25
											1	.	.	1	.	1	1	.	.	.	.	.	.	.	.	.	.	.	.	.	.	.	.	.	.	.	26
												1	.	1	.	2	1	.	.	.	.	.	.	.	.	.	.	.	.	.	.	.	.	8	.	.	27
													1	.	1	.	.	4	4	5	3	3	10	4	3	3	6	24	20	15	28	34	36	15	11	66	28
														1	.	1	1	.	.	.	.	.	.	.	.	.	.	.	.	.	.	.	.	.	.	.	29
															1	.	.	.	.	.	.	.	5	4	3	3	6	9	10	5	.	24	16	15	5	33	30
																1	1	.	.	.	.	.	.	.	.	.	.	.	.	.	.	.	.	.	.	.	31
																	1	.	.	.	.	.	.	.	.	.	.	.	.	.	.	.	.	.	.	.	32
																		1	.	2	.	.	2	.	.	.	1	9	7	7	5	12	17	5	6	10	33
																			1	.	1	1	.	1	1	1	.	.	.	.	6	.	.	.	.	14	34
																				1	.	.	.	.	.	.	.	3	2	4	.	3	8	1	3	.	35
																					1	.	.	.	1	.	.	.	.	.	.	.	.	.	.	.	36
																						1	.	.	.	1	.	.	.	.	.	.	.	.	.	.	37
																							1	.	.	.	1	3	4	2	.	8	8	5	3	5	38
																								1	1	1	.	.	.	.	.	.	.	.	.	7	39
																									1	.	.	.	.	.	.	.	.	.	.	.	40
																										1	.	.	.	.	.	.	.	.	.	.	41
																											1	.	.	.	.	4	.	5	.	.	42
																												1	1	1	.	2	4	1	2	.	43
																													1	.	.	1	2	1	1	.	44
																														1	.	.	2	.	1	.	45
																															1	.	.	.	.	2	46
																																1	.	1	.	.	47
																																	1	.	1	.	48
																																		1	.	.	49
																																			1	.	50
																																				1	51

Table 3. *(continued)*

ALGEBRA D11, REPRESENTATIONS OF CLASS 2

S.P.	11	19	27	35	35	43	59	43	51	59	67	51	91	59	59	67	75	83	67	99	67	131	75	75	83	83
O.S.	1 024	11 264	56 320	11 264	168 960	112 640	11 264	337 920	506 880	56 320	112 640	473 088	11 264	1 351 680	473 088	506 880	506 880	112 640	337 920	112 640	2 365 440	11 264	168 960	2 027 520	56 320	168 960
LEVEL	56	76	94	96	110	114	116	124	130	132	134	136	136	144	146	148	150	152	154	154	156	156	160	162	164	164
DIM'N	1 024	21 504	214 016	236 544	1 340 416	3 132 416	1 813 504	5 913 600	22 021 120	16 343 040	26 966 016	19 475 456	10 881 024	103 369 728	49 437 696	152 807 424	201 793 536	187 264 000	98 233 344	172 282 880	353 633 280	54 405 120	152 807 424	804 722 688	181 913 600	502 081 536
	0	1	0	2	0	1	3	0	1	0	2	0	4	1	0	0	2	1	0	3	1	5	0	0	0	0
	0	0	1	0	0	1	0	0	0	2	1	0	0	0	0	1	0	2	0	1	0	0	0	1	0	0
	0	0	0	0	1	0	0	0	1	0	0	0	0	0	0	1	1	0	0	0	0	0	0	0	0	2
	0	0	0	0	0	0	0	1	0	0	0	0	0	1	0	0	0	0	0	0	0	0	0	1	0	0
	0	0	0	0	0	0	0	0	0	0	0	1	0	0	0	0	0	0	0	0	1	0	0	0	0	0
	0	0	0	0	0	0	0	0	0	0	0	0	0	0	1	0	0	0	0	0	0	0	0	0	0	0
	0	0	0	0	0	0	0	0	0	0	0	0	0	0	0	0	0	0	1	0	0	0	0	0	0	0
	0	0	0	0	0	0	0	0	0	0	0	0	0	0	0	0	0	0	0	0	0	0	1	0	0	0
	0 0	1 0	0 0	0 0	1 0	1 0	1 0	0 0	0 0	0 0	0 0	1 0	0 0	1 0	0 0	1 0	1 0	1 0	1 0	1 0	0 0	1 0	0 0	0 0	1 1	0 0
WEIGHT	1	0	1	1	0	0	0	1	1	1	1	0	1	0	1	0	0	0	0	0	1	0	1	1	0	1
NUMBER	0	1	2	3	4	5	6	7	8	9	10	11	12	13	14	15	16	17	18	19	20	21	22	23	24	25
0	1	10	55	55	209	430	220	605	1 925	1 330	2 035	1 397	715	6 149	2 640	8 195	10 097	8 800	4 125	7 348	15 180	2 002	5 313	30 195	5 500	17 655
1		1	9	10	45	100	55	155	530	375	595	405	220	1 910	837	2 625	3 320	2 935	1 398	2 530	5 160	715	1 890	10 629	2 025	6 315
2			1	1	8	18	10	36	126	91	145	111	55	542	258	761	971	870	468	760	1 656	220	672	3 501	750	2 103
3				1	.	9	10	.	45	45	100	.	55	155	.	321	530	540	.	595	405	220	.	1 305	.	909
4					1	2	1	7	24	17	27	28	10	136	76	191	243	218	154	190	494	55	238	1 053	280	636
5						1	1	.	8	9	18	.	10	36	.	81	126	136	.	145	111	55	.	387	.	276
6							1	.	.	.	9	.	10	.	.	.	45	45	.	100	.	55	.	.	.	.
7								1	3	2	3	6	1	28	21	38	48	42	49	36	133	10	84	279	105	167
8									1	1	2	.	1	7	.	16	24	26	.	27	28	10	.	100	.	73
9										1	1	.	1	.	.	8	8	18	.	18	.	10	.	36	.	36
10											1	.	1	.	.	.	8	9	.	18	.	10	.	.	.	.
11												1	.	4	5	5	6	5	15	4	30	1	29	60	40	35
12													1	.	.	.	.	.	.	9	.	10	.	.	.	.
13														1	.	2	3	3	.	3	6	1	.	21	.	15

S.P.	83	99	91	75	91	107	107	115	83	83	139	91	179	91	91	99	91	99	107	115	115	107	115	99	123	99
O.S.	1 351 680	1 024	11 264	2 838 528	1 013 760	56 320	506 880	112 640	4 730 880	2 365 440	112 640	2 365 440	11 264	1 351 680	1 351 680	4 055 040	7 096 320	506 880	506 880	1 351 680	506 880	112 640	11 264	2 838 528	1 013 760	4 730 880
LEVEL	164	166	166	166	168	170	170	172	174	174	174	176	176	178	180	182	184	184	184	184	186	186	186	186	188	190
DIM'N	984 473 600	60 196 864	152 807 424	920 482 816	1 863 787 520	681 640 960	1 340 416 000	1 343 806 464	2 927 468 544	1 860 655 104	895 870 976	3 455 502 336	235 755 520	3 514 570 752	2 928 808 960	10 200 960 000	7 910 031 360	3 514 570 752	6 876 334 080	6 712 320 000	7 614 903 296	2 966 845 440	1 171 523 584	9 156 381 696	13 905 475 584	14 333 935 616
	2	0	0	1	1	0	3	2	0	1	4	2	6	0	1	1	0	1	1	3	0	1	1	2	2	0
	0	0	0	0	1	3	0	2	1	0	1	0	0	0	0	1	1	0	0	0	2	0	0	0	1	0
	0	0	0	0	1	0	1	0	0	0	0	0	0	1	0	0	0	0	2	0	1	0	0	0	1	1
	1	0	0	0	0	0	0	0	0	0	0	0	0	1	0	1	0	0	0	1	0	0	0	0	0	0
	0	0	0	0	0	0	0	0	1	0	0	1	0	0	0	0	0	0	0	0	0	0	0	0	0	1
	0	0	0	1	0	0	0	0	0	0	0	0	0	0	0	0	1	0	0	0	0	0	0	1	0	0
	0	0	0	0	0	0	0	0	0	1	0	0	0	0	0	0	0	0	0	0	0	0	0	0	0	0
	0	0	0	0	0	0	0	0	0	0	0	0	0	0	1	0	0	0	0	0	0	0	0	0	0	0
	0 0	3 0	1 0	1 0	0 0	0 0	0 0	0 0	1 0	0 0	0 0	1 0	0 0	1 0	1 0	1 0	0 0	0 1	1 0	1 0	1 0	2 0	0 0	0 0	1 0	0 0
WEIGHT	1	0	2	0	1	1	1	1	0	1	1	0	1	0	0	0	1	1	0	0	0	1	3	1	0	1
NUMBER	26	27	28	29	30	31	32	33	34	35	36	37	38	39	40	41	42	43	44	45	46	47	48	49	50	51
0	34 595	1 585	4 213	30 085	59 720	20 020	39 215	37 180	81 070	48 675	22 165	89 859	5 005	88 495	64 295	234 960	169 840	67 749	149 325	141 725	156 530	52 480	19 855	184 965	270 215	271 480
1	12 440	603	1 585	10 950	21 902	7 470	14 750	14 113	30 685	18 660	8 635	34 635	2 002	34 191	25 620	92 760	67 990	27 750	59 613	57 170	63 120	21 880	8 350	75 255	110 465	111 430
2	4 133	232	603	3 843	7 398	2 560	5 045	4 870	11 101	6 996	3 025	12 639	715	12 571	10 080	34 542	26 406	11 295	22 398	21 539	23 903	9 100	3 510	29 472	42 125	44 178
3	1 910	.	.	837	4 020	1 605	3 320	3 430	3 763	1 398	2 530	5 160	715	5 085	1 890	16 920	8 310	2 025	11 850	12 440	13 707	1 585	603	10 950	26 555	16 902
4	1 245	91	232	1 292	2 242	779	1 532	1 485	3 791	2 560	925	4 327	220	4 332	3 920	11 956	9 882	4 580	7 793	7 484	8 344	3 790	1 480	11 069	14 739	16 766
5	542	.	.	258	1 202	495	971	1 035	1 251	468	760	1 656	220	1 746	672	5 724	3 006	750	4 086	4 133	4 773	603	232	3 843	9 123	6 346
6	155	.	.	.	321	165	530	540	.	.	595	405	220	.	.	1 305	.	.	909	1 910	1 395	.	.	837	4 020	.
7	328	36	91	410	584	201	396	381	1 198	910	235	1 366	55	1 366	1 505	3 760	3 532	1 855	2 447	2 348	2 614	1 585	630	3 956	4 611	6 012
8	136	.	.	76	308	127	243	263	383	154	190	494	55	547	238	1 756	1 034	280	1 272	1 245	1 481	232	91	1 292	2 803	2 246
9	36	.	.	.	162	91	126	181	111	.	145	111	55	232	.	774	258	.	672	542	881	.	.	258	1 643	846
10	36	.	.	.	81	45	126	136	.	.	145	111	55	.	.	387	.	.	276	542	441	.	.	258	1 202	.
11	70	15	36	120	120	40	80	75	340	310	45	385	10	380	570	1 035	1 185	750	665	640	705	670	270	1 325	1 235	1 995
12	.	.	.	.	.	.	45	45	.	.	100	.	55	.	.	.	.	.	.	155	.	.	.	.	321	.
13	28	.	.	21	62	25	48	51	105	49	36	133	10	150	84	471	334	105	341	328	393	91	36	410	737	732

1	.	.	.	4	.	5	.	10	9	15	5	10	6	15	30	16	5	10	9	80	100	5	90	1	86	210	230	365	305	144	140	150	285	120	405	260	595	14
	1	1	2	.	2	.	1	.	7	.	8	7	.	.	.	32	17	24	35	28	.	27	28	10	64	.	200	76	.	182	136	227	.	.	76	425	272	15
		1	1	.	2	.	1	.	.	.	.	7	.	.	.	16	9	24	26	.	.	27	28	10	.	.	100	.	.	73	136	117	.	.	76	308	.	16
			1	.	1	.	1	.	.	.	.	.	.	.	.	8	9	8	18	.	.	18	.	10	.	.	36	.	.	36	36	81	.	.	.	162	.	17
				1	.	.	.	3	.	6	.	.	3	6	6	.	.	.	.	14	28	.	15	.	14	76	35	95	122	21	20	21	126	52	105	35	147	18
					1	.	1	.	.	.	.	.	.	.	.	.	.	8	9	.	.	18	.	10	.	.	.	.	.	.	36	.	.	.	.	81	.	19
						1	.	.	3	.	2	4	.	.	5	8	3	6	6	24	15	4	30	1	33	29	102	99	40	72	70	82	36	15	120	152	213	20
							1	.	.	.	.	.	.	.	.	.	.	.	.	.	.	9	.	10	.	.	.	.	.	.	.	.	.	.	.	.	.	21
								1	.	2	.	.	1	3	.	.	.	.	.	.	7	.	.	.	.	24	.	20	51	.	.	.	54	27	21	.	28	22
									1	.	1	1	.	.	.	4	2	3	4	6	.	3	6	1	14	.	42	21	.	38	28	46	.	.	21	86	77	23
										1	.	.	1	1	.	.	.	.	.	.	.	.	.	.	.	8	.	.	18	.	.	.	28	10	.	.	.	24
											1	.	.	.	.	2	1	1	3	.	.	2	.	1	7	.	14	.	.	24	7	24	.	.	.	48	28	25
												1	.	.	.	2	1	3	3	.	.	3	6	1	.	.	21	.	.	15	28	24	.	.	21	62	.	26
													1	.	.	.	.	.	.	.	.	.	.	.	.	.	.	.	.	.	.	.	10	.	.	.	.	27
														1	.	.	.	.	.	.	.	.	.	.	.	.	.	.	9	.	.	.	10	10	.	.	.	28
															1	.	.	.	.	4	4	.	5	.	5	10	15	25	15	10	10	11	15	6	30	20	51	29
																1	1	1	2	.	.	2	.	1	.	.	7	.	.	8	7	16	.	.	.	32	.	30
																	1	.	1	.	.	1	.	1	.	.	.	.	.	.	.	8	.	.	.	8	.	31
																		1	1	.	.	2	.	1	.	.	.	.	.	.	7	.	.	.	.	16	.	32
																			1	.	.	1	.	1	.	.	.	.	.	.	.	.	.	.	.	8	.	33
																				1	.	.	1	.	2	.	6	5	.	5	4	6	.	.	5	11	18	34
																					1	.	.	.	.	3	.	5	6	.	.	.	6	3	6	.	9	35
																						1	.	1	.	.	.	.	.	.	.	.	.	.	.	.	.	36
																							1	.	.	.	3	.	.	2	4	3	.	.	5	8	.	37
																								1	.	.	.	.	.	.	.	.	.	.	.	.	.	38
																									1	.	2	.	.	3	1	3	.	.	.	6	6	39
																										1	.	.	2	.	.	.	3	1	.	.	.	40
																											1	.	.	1	1	2	.	.	.	4	.	41
																												1	.	.	.	.	.	.	1	.	3	42
																													1	.	.	.	1	1	.	.	.	43
																														1	.	1	.	.	.	2	.	44
																															1	.	.	.	.	2	.	45
																																1	.	.	.	1	.	46
																																	1	.	.	.	.	47
																																		1	.	.	.	48
																																			1	.	.	49
																																				1	.	50
																																					1	51

Table 3. *(continued)*

ALGEBRA D12, REPRESENTATIONS OF CLASS 0

S.P.	0	8	16	16	24	32	40	64	24	32	40	32	48
O.S.	1	264	24	7 920	5 280	264	528	24	59 136	126 720	47 520	126 720	1 760
LEVEL	1	43	45	77	83	85	87	89	103	113	119	121	121
DIM'N	1	276	299	10 626	37 674	27 300	44 275	17 250	134 596	874 874	1 874 730	735 471	1 136 200
	0	0	2	0	1	0	2	4	0	1	0	0	0
	0	1	0	0	0	2	1	0	0	0	1	0	0
	0	0	0	0	1	0	0	0	0	0	0	0	2
	0	0	0	1	0	0	0	0	0	0	1	0	0
	0	0	0	0	0	0	0	0	0	1	0	0	0
	0	0	0	0	0	0	0	0	1	0	0	0	0
	0	0	0	0	0	0	0	0	0	0	0	0	0
	0	0	0	0	0	0	0	0	0	0	0	1	0
	0	0	0	0	0	0	0	0	0	0	0	0	0
	0 0	0 0	0 0	0 0	0 0	0 0	0 0	0 0	0 0	0 0	0 0	0 0	0 0
WEIGHT	0	0	0	0	0	0	0	0	0	0	0	0	0
NUMBER	0	1	2	3	4	5	6	7	8	9	10	11	12
0	1	12	11	66	186	132	187	66	220	1 034	1 914	495	1 088
1		1	1	10	31	21	32	11	45	235	444	120	254
2			1	.	11	10	22	11	.	55	154	.	110
3				1	3	2	3	1	8	43	81	28	46
4					1	1	2	1	.	9	28	.	19
5						1	1	1	.	.	10	.	9
6							1	1	.	.	.	.	.
7								1	.	.	.	.	.
8									1	5	9	6	5
9										1	3	.	2
10											1	.	1
11												1	.
12													1
13													
14													

S.P.	48	56	72	72	80	40	104	48	48	144	40	48	56
O.S.	31 680	10 560	264	5 280	528	67 584	528	2 048	2 048	24	709 632	887 040	354 816
LEVEL	121	125	127	127	129	131	131	133	133	133	135	145	147
DIM'N	2 254 000	4 100 096	1 450 449	2 877 875	2 816 856	1 961 256	1 821 600	1 352 078	1 352 078	457 470	7 436 429	27 956 500	31 798 305
	2	1	0	3	2	0	4	0	0	6	1	0	2
	0	1	3	0	2	0	1	0	0	0	0	1	0
	0	1	0	1	0	0	0	0	0	0	0	0	0
	1	0	0	0	0	0	0	0	0	0	0	0	0
	0	0	0	0	0	0	0	0	0	0	0	0	0
	0	0	0	0	0	0	0	0	0	0	0	1	1
	0	0	0	0	0	0	0	0	0	0	1	0	0
	0	0	0	0	0	0	0	0	0	0	0	0	0
	0	0	0	0	0	0	0	0	0	0	0	0	0
	0 0	0 0	0 0	0 0	0 0	0 1	0 0	2 0	0 0	0 0	0 0	0 0	0 0
WEIGHT	0	0	0	0	0	0	0	0	2	0	0	0	0
NUMBER	13	14	15	16	17	18	19	20	21	22	23	24	25
0	2 080	3 520	1 177	2 211	2 112	792	1 232	462	462	286	3 245	9 460	9 801
1	500	840	276	541	507	210	297	126	126	66	885	2 705	2 880
2	220	440	165	341	352	.	242	.	.	66	165	770	990
3	92	152	49	98	90	56	52	35	35	11	220	690	743
4	38	78	30	60	61	.	42	.	.	11	36	189	234
5	10	40	21	31	41	.	32	.	.	11	.	45	45
6	10	20	10	31	31	.	32	.	.	11	.	.	45
7	.	.	.	11	10	.	22	.	.	11	.	.	.
8	10	16	5	10	9	15	5	10	10	1	47	145	156
9	4	8	3	6	6	.	4	.	.	1	7	38	46
10	1	4	2	3	4	.	3	.	.	1	.	8	8
11	.	.	.	.	.	4	.	3	3	.	7	20	21
12	.	2	1	1	3	.	2	.	.	1	.	.	.
13	1	2	1	3	3	.	3	.	.	1	.	.	8
14		1	1	1	2	.	2	.	.	1	.	.	.

S.P.	48	56	64	56	64	80	72	80	56	88	88	64	96
O.S.	1 013 760	253 440	7 920	270 336	506 880	126 720	95 040	47 520	3 548 160	5 280	95 040	1 013 760	5 280
LEVEL	149	151	153	155	155	157	159	161	163	163	163	165	165
DIM'N	28 683 369	46 966 920	25 719 980	55 242 044	121 245 696	74 563 125	151 637 850	131 708 304	164 038 875	107 837 730	228 541 500	183 050 560	155 227 644
	1	0	0	1	1	3	1	0	0	0	2	2	2
	0	0	0	0	1	0	0	2	1	1	1	0	0
	0	1	0	0	0	0	1	0	0	2	0	0	2
	0	0	2	0	0	0	1	1	0	0	1	0	0
	0	1	0	0	1	1	0	0	0	0	0	0	0
	0	0	0	0	0	0	0	0	0	0	0	0	0
	0	0	0	0	0	0	0	0	0	0	0	0	0
	0	0	0	0	0	0	0	0	1	0	0	1	0
	1	0	0	0	0	0	0	0	0	0	0	0	0
	0 0	0 0	0 0	1 0	0 0	0 0	0 0	0 0	0 0	0 0	0 0	0 0	0 0
WEIGHT	0	0	0	1	0	0	0	0	0	0	0	0	0
NUMBER	26	27	28	29	30	31	32	33	34	35	36	37	38
0	6 633	13 728	7 084	9 372	31 680	17 445	36 378	29 920	26 235	23 298	48 444	26 752	31 548
1	2 010	4 014	2 074	2 982	9 480	5 345	10 938	9 030	8 505	7 044	14 844	8 880	9 628
2	330	1 430	814	462	3 960	2 585	5 148	4 500	2 145	3 795	8 074	2 640	5 588
3	588	1 030	532	938	2 456	1 398	2 823	2 330	2 620	1 807	3 846	2 768	2 476
4	84	359	206	126	990	640	1 296	1 146	624	958	2 032	744	1 402
5	.	125	90	.	360	235	594	570	120	524	1 004	120	789
6	.	.	.	.	180	235	234	280	.	245	724	120	499
7	.	.	.	.	.	55	.	.	.	.	154	.	110
8	165	213	109	295	504	285	570	466	750	357	765	796	486
9	21	72	41	35	196	125	254	224	168	185	393	196	268
10	.	25	17	.	68	43	114	107	28	100	189	28	147
11	43	28	14	94	64	35	70	56	189	42	90	200	56
12	.	9	8	.	18	9	54	46	.	57	84	.	84
13	.	.	.	.	34	43	44	54	.	46	135	28	92
14	.	.	.	.	9	9	18	28	.	28	56	.	47

S.P.	112	104	128	120	136	152	64	64	160	72	72	72	200
O.S.	31 680	10 560	264	10 560	528	5 280	3 548 160	3 041 280	528	135 168	135 168	675 840	528
LEVEL	165	167	169	169	171	171	173	173	173	175	175	175	175
DIM'N	103 603 500	210 608 125	49 612 794	215 547 904	106 826 720	85 353 345	468 682 500	451 980 360	88 444 200	314 618 150	314 618 150	500 528 875	38 629 305
	4	1	0	3	2	5	0	0	4	0	0	2	6
	0	2	4	1	3	0	0	1	2	1	1	0	1
	0	1	0	1	0	1	1	0	0	0	0	0	0
	1	0	0	0	0	0	0	0	0	0	0	0	0
	0	0	0	0	0	0	0	0	0	0	0	0	0
	0	0	0	0	0	0	0	0	0	0	0	0	0
	0	0	0	0	0	0	1	0	0	0	0	0	0
	0	0	0	0	0	0	0	0	0	0	0	0	0
	0	0	0	0	0	0	0	0	0	0	0	0	0
	0 0	0 0	0 0	0 0	0 0	0 0	0 0	0 1	0 0	2 0	0 0	0 1	0 0
WEIGHT	0	0	0	0	0	0	0	0	0	0	2	0	0
NUMBER	39	40	41	42	43	44	45	46	47	48	49	50	51
0	19 580	41 085	9 042	38 720	18 392	13 761	58 740	46 728	13 992	28 182	28 182	47 355	5 577
1	6 090	12 510	2 716	11 920	5 587	4 246	19 805	16 218	4 257	9 996	9 996	16 800	1 672
2	3 740	7 590	1 815	7 920	3 982	3 146	6 270	3 828	3 322	2 310	2 310	4 620	1 452
3	1 582	3 198	683	3 048	1 404	1 076	6 262	5 516	1 062	3 507	3 507	5 782	407
4	942	1 914	456	1 982	987	786	1 923	1 218	817	756	756	1 428	352
5	500	1 155	321	1 240	696	541	555	210	617	126	126	210	297
6	500	755	210	1 040	541	541	.	.	562	.	.	210	297
7	220	155	55	440	220	341	.	.	352	.	.	.	242
8	310	620	129	584	263	201	1 800	1 839	195	1 225	1 225	1 940	72
9	180	364	85	372	182	145	534	378	148	245	245	434	62
10	92	215	58	228	127	98	148	56	110	35	35	56	52
11	35	70	14	64	28	21	440	600	20	431	431	635	7
12	38	125	39	136	90	60	36	.	81	.	.	.	42
13	92	141	39	190	98	98	.	.	100	.	.	56	52
14	38	87	30	107	70	60	.	.	71	.	.	.	42

1	.	1	.	1	.	.	1	.	.	.	.	.	.	.	.	.	.	10	.	9	10	.	9	10	40	21	40	41	31	.	.	41	.	.	.	32	15
	1	1	.	2	.	.	1	.	.	.	.	.	.	.	.	9	.	.	.	.	28	.	19	38	29	10	78	40	60	.	.	61	.	.	.	42	16
		1	.	1	.	.	1	.	.	.	.	.	.	.	.	.	.	.	.	.	10	.	9	10	20	10	40	31	31	.	.	41	.	.	.	32	17
			1	.	1	1	.	.	.	.	9	.	.	31	.	.	.	.	35	.	.	36	.	.	.	.	.	.	.	75	189	.	156	156	200	.	18
				1	.	.	1	.	.	.	.	.	.	.	.	.	.	.	.	.	.	.	.	10	.	.	20	10	31	.	.	31	.	.	.	32	19
					1	.	.	.	.	.	.	.	.	11	.	.	.	.	.	.	.	.	.	.	.	.	.	.	.	.	54	.	66	55	55	.	20
						1	.	.	.	.	.	.	.	11	.	.	.	.	.	.	.	.	.	.	.	.	.	.	.	.	54	.	55	66	55	.	21
							1	.	.	.	.	.	.	.	.	.	.	.	.	.	.	.	.	.	.	.	.	.	11	.	.	10	.	.	.	22	22
								1	5	6	5	9	5	10	24	15	30	26	40	21	45	46	30	20	40	9	40	19	15	125	115	15	80	80	130	6	23
									1	1	.	3	2	.	8	5	13	12	6	11	21	6	16	10	23	6	24	13	10	33	15	11	10	10	15	5	24
										1	.	.	.	.	4	5	5	6	.	5	15	6	10	10	15	4	20	10	10	.	.	10	.	.	15	5	25
											1	.	.	3	.	.	.	.	7	.	.	8	.	.	.	.	.	.	.	20	34	.	27	27	38	.	26
												1	1	.	2	1	6	5	.	6	9	.	9	4	13	4	14	9	6	7	.	8	.	.	.	4	27
													1	.	.	.	3	2	.	3	3	.	6	1	7	3	8	6	3	.	.	6	.	.	.	3	28
														1	.	.	.	.	.	.	.	.	.	.	.	.	.	.	.	.	9	.	10	10	10	.	29
															1	1	2	3	.	3	6	.	5	4	9	3	11	7	6	.	.	7	.	.	.	4	30
																1	.	.	.	.	3	.	2	4	3	1	8	4	6	.	.	6	.	.	.	4	31
																	1	1	.	2	2	.	3	1	5	2	6	5	3	.	.	5	.	.	.	3	32
																		1	.	1	1	.	1	1	4	2	4	4	3	.	.	4	.	.	.	3	33
																			1	.	.	1	.	.	.	.	.	.	.	5	4	.	3	3	4	.	34
																				1	.	.	1	.	2	1	2	3	1	.	.	3	.	.	.	2	35
																					1	.	1	1	2	1	4	3	3	.	.	4	.	.	.	3	36
																						1	.	.	.	.	.	.	.	.	.	.	.	.	4	.	37
																							1	.	1	1	2	2	1	.	.	3	.	.	.	2	38
																								1	.	.	2	1	3	.	.	3	.	.	.	3	39
																									1	1	1	2	1	.	.	2	.	.	.	2	40
																										1	.	1	.	.	.	1	.	.	.	1	41
																											1	1	1	.	.	2	.	.	.	2	42
																												1	.	.	.	1	.	.	.	1	43
																													1	.	.	1	.	.	.	2	44
																														1	.	.	.	.	.	.	45
																															1	.	1	1	1	.	46
																																1	.	.	.	1	47
																																	1	.	.	.	48
																																		1	.	.	49
																																			1	.	50
																																				1	51

Table 3. *(continued)*

ALGEBRA D12, REPRESENTATIONS OF CLASS 1

4	12	20	36	20	28	36	44	52	68	100	28	36	S.P.
24	1 760	528	24	25 344	31 680	5 280	5 280	528	528	24	101 376	354 816	O.S.
23	61	65	67	91	99	103	105	107	109	111	113	125	LEVEL
24	2 024	4 576	2 576	42 504	210 496	299 000	388 080	351 624	315 744	95 680	346 104	2 841 696	DIM'N
1	0	1	3	0	1	0	2	1	3	5	0	1	
0	0	1	0	0	0	1	0	2	1	0	0	0	
0	1	0	0	0	0	1	1	0	0	0	0	0	
0	0	0	0	0	1	0	0	0	0	0	0	0	
0	0	0	0	1	0	0	0	0	0	0	0	0	
0	0	0	0	0	0	0	0	0	0	0	0	1	
0	0	0	0	0	0	0	0	0	0	0	1	0	
0	0	0	0	0	0	0	0	0	0	0	0	0	
0	0	0	0	0	0	0	0	0	0	0	0	0	
0 0	0 0	0 0	0 0	0 0	0 0	0 0	0 0	0 0	0 0	0 0	0 0	0 0	
0	0	0	0	0	0	0	0	0	0	0	0	0	WEIGHT
0	1	2	3	4	5	6	7	8	9	10	11	12	NUMBER
1	11	22	11	55	220	285	341	297	242	66	165	990	0
	1	2	1	9	38	49	60	51	42	11	36	234	1
		1	1	.	10	20	31	31	32	11	.	45	2
			1	.	.	.	11	10	22	11	.	.	3
				1	4	5	6	5	4	1	7	46	4
					1	2	3	3	3	1	.	8	5
						1	1	2	2	1	.	.	6
							1	1	2	1	.	.	7
								1	1	1	.	.	8
									1	1	.	.	9
										1	.	.	10
											1	6	11
												1	12
													13
													14

36	44	44	52	52	60	44	68	76	76	84	100	108	S.P.
112 640	24 576	253 440	126 720	31 680	95 040	1 013 760	5 280	31 680	5 280	10 560	528	5 280	O.S.
127	133	133	135	137	141	143	143	143	145	147	149	149	LEVEL
1 307 504	2 496 144	8 288 280	9 614 000	9 472 320	26 723 424	15 997 696	17 497 480	17 160 000	18 987 696	34 155 000	15 471 456	17 050 176	DIM'N
0	0	0	2	0	1	1	1	3	0	2	1	4	
0	0	1	0	0	1	0	0	0	2	1	3	0	
0	0	0	0	1	0	0	2	0	1	1	0	1	
0	0	0	0	1	1	0	0	1	0	0	0	0	
0	0	1	1	0	0	0	0	0	0	0	0	0	
0	0	0	0	0	0	0	0	0	0	0	0	0	
0	0	0	0	0	0	0	0	0	0	0	0	0	
0	0	0	0	0	0	1	0	0	0	0	0	0	
1	0	0	0	0	0	0	0	0	0	0	0	0	
0 0	1 0	0 0	0 0	0 0	0 0	0 0	0 0	0 0	0 0	0 0	0 0	0 0	
0	1	0	0	0	0	0	0	0	0	0	0	0	WEIGHT
13	14	15	16	17	18	19	20	21	22	23	24	25	NUMBER
330	462	2 409	2 585	2 508	6 414	2 640	3 993	3 740	4 125	6 985	2 992	3 146	0
84	126	585	640	612	1 590	744	986	942	1 017	1 737	732	786	1
.	.	180	235	234	724	120	499	500	555	1 040	486	541	2
.	.	.	55	.	154	.	110	220	155	440	220	341	3
21	35	114	125	118	305	196	187	180	191	325	134	145	4
.	.	34	43	44	135	28	92	92	103	190	88	98	5
.	.	9	9	18	56	.	47	38	58	107	60	60	6
.	.	.	9	.	28	.	19	38	29	78	40	60	7
.	.	.	.	.	10	.	9	10	20	40	31	31	8
.	.	.	.	.	.	.	.	10	.	20	10	31	9
.	.	.	.	.	.	.	.	.	.	.	.	11	10
5	10	14	15	14	35	46	21	20	21	35	14	15	11
.	.	4	5	5	15	6	10	10	11	20	9	10	12
1	3	.	.	.	.	8	.	.	.	.	.	.	13
	1	.	.	.	.	.	.	.	.	.	.	.	14

116	52	148	52	60	60	196	60	60	68	68	60	84	S.P.
528	675 840	528	2 128 896	24 576	24 576	24	709 632	1 182 720	126 720	1 774 080	4 055 040	354 816	O.S.
151	153	153	155	155	155	155	157	163	167	167	169	169	LEVEL
17 310 720	43 266 496	8 946 080	74 989 200	29 953 728	29 953 728	1 937 520	84 299 600	168 396 800	169 449 280	416 440 024	297 457 160	249 964 000	DIM'N
3	1	5	0	1	1	7	2	0	0	1	0	3	
2	0	1	1	0	0	0	0	0	0	1	1	0	
0	0	0	0	0	0	0	0	1	0	0	0	0	
0	0	0	0	0	0	0	0	0	1	0	0	0	
0	0	0	0	0	0	0	0	0	1	0	0	0	
0	0	0	0	0	0	0	0	1	0	1	0	1	
0	0	0	1	0	0	0	1	0	0	0	0	0	
0	0	0	0	0	0	0	0	0	0	0	0	0	
0	0	0	0	0	0	0	0	0	0	0	1	0	
0 0	0 1	0 0	0 0	2 0	0 0	0 0	0 0	0 0	0 0	0 0	0 0	0 0	
0	0	0	0	0	2	0	0	0	0	0	0	0	WEIGHT
26	27	28	29	30	31	32	33	34	35	36	37	38	NUMBER
3 102	4 620	1 452	9 350	2 772	2 772	286	9 735	17 600	16 104	39 391	21 483	21 450	0
762	1 428	352	2 772	882	882	66	2 937	5 344	4 928	12 167	7 140	6 738	1
562	210	297	720	126	126	66	885	1 760	1 840	4 595	1 680	2 880	2
352	.	242	.	.	.	66	165	.	.	770	.	990	3
138	434	62	747	280	280	11	797	1 448	1 334	3 315	2 289	1 846	4
100	56	52	184	35	35	11	220	464	488	1 199	504	743	5
71	.	42	36	.	.	11	36	144	188	378	84	234	6
61	.	42	.	.	.	11	36	.	.	189	.	234	7
41	.	32	.	.	.	11	.	.	.	45	.	45	8
31	.	32	.	.	.	11	.	.	.	.	.	45	9
10	.	22	.	.	.	11	.	.	.	.	.	.	10
14	130	6	171	90	90	1	182	324	294	735	699	406	11
10	15	5	40	10	10	1	47	100	104	255	144	156	12
.	38	.	27	30	30	.	28	48	42	105	196	56	13
.	10	.	.	11	11	.	.	.	.	.	44	.	14

68	76	68	84	84	76	92	92	116	68	108	100	108	S.P.
1 013 760	760 320	1 351 680	253 440	31 680	270 336	506 880	95 040	126 720	7 096 320	1 760	95 040	95 040	O.S.
171	173	175	175	175	177	177	179	179	181	181	181	183	LEVEL
330 372 000	771 599 400	579 462 000	620 910 576	438 357 920	640 676 960	1 056 321 000	994 593 600	456 456 000	1 049 848 800	283 210 200	1 381 064 256	1 499 784 000	DIM'N
2	1	0	0	1	2	2	0	4	0	0	2	1	
0	0	1	2	0	0	1	1	0	0	0	0	2	
0	1	0	0	0	0	0	1	0	1	3	1	0	
0	0	0	0	2	0	0	1	0	0	0	1	1	
0	1	0	1	0	0	1	0	1	0	0	0	0	
0	0	0	0	0	0	0	0	0	0	0	0	0	
0	0	0	0	0	0	0	0	0	0	0	0	0	
0	0	0	0	0	0	0	0	0	1	0	0	0	
1	0	0	0	0	0	0	0	0	0	0	0	0	
0 0	0 0	1 0	0 0	0 0	1 0	0 0	0 0	0 0	0 0	0 0	0 0	0 0	
0	0	1	0	0	1	0	0	0	0	0	0	0	WEIGHT
39	40	41	42	43	44	45	46	47	48	49	50	51	NUMBER
22 110	64 845	32 010	49 401	34 914	32 802	79 155	73 260	30 690	57 750	19 415	96 228	99 880	0
7 470	20 190	11 214	15 453	10 876	11 676	24 993	22 932	9 806	20 034	6 077	30 330	31 476	1
2 010	8 735	2 520	7 071	5 024	2 982	12 240	11 490	5 345	5 985	3 305	16 128	17 460	2
330	1 430	.	1 551	814	462	3 960	2 640	2 585	.	890	5 148	6 160	3
2 415	5 478	3 885	4 190	2 939	4 081	6 789	6 172	2 665	6 608	1 622	8 166	8 436	4
588	2 304	812	1 871	1 325	938	3 214	3 032	1 398	1 904	867	4 229	4 586	5
84	943	126	810	610	126	1 395	1 512	640	504	478	2 142	2 454	6
84	359	.	405	206	126	990	684	640	.	226	1 296	1 588	7
.	125	.	180	90	.	360	360	235	.	135	594	850	8
.	.	.	.	.	.	180	.	235	.	.	234	280	9
.	.	.	.	.	.	.	.	55	.	.	.	.	10
740	1 188	1 340	900	630	1 415	1 449	1 302	560	2 024	336	1 710	1 746	11
165	483	260	390	275	295	664	622	285	560	175	860	926	12
207	162	465	120	84	493	189	168	70	546	42	216	216	13
45	.	165	.	.	176	.	.	.	110	.	.	.	14

1	1	2	6	.	5	4	6	11	6	6	7	.	4	7	.	.	1	7	30	39	76	21	46	21	192	35	162	123	35	278	302	125	140	95	424	482	15
	1	.	3	.	2	4	3	8	4	6	6	.	4	.	.	.	1	7	.	.	38	.	46	21	72	.	82	41	35	196	136	125	.	44	254	312	16
		1	2	.	3	1	3	6	4	3	5	.	3	.	.	.	1	.	8	16	16	.	8	.	75	.	59	59	.	102	148	43	28	54	210	240	17
			1	.	1	1	2	4	3	3	4	.	3	.	.	.	1	.	.	.	8	.	8	.	25	.	34	17	.	68	70	43	.	27	114	161	18
				1	.	.	.	.	.	.	.	4	.	6	3	3	.	7	14	14	35	38	21	43	63	84	50	35	94	84	77	35	144	21	105	111	19
					1	.	1	2	2	1	3	.	2	.	.	.	1	.	.	.	.	.	.	.	9	.	9	8	.	18	36	9	.	19	54	74	20
						1	.	2	1	3	3	.	3	.	.	.	1	.	.	.	.	.	8	.	.	.	.	.	.	34	.	43	.	.	44	54	21
							1	1	2	1	2	.	2	.	.	.	1	.	.	.	.	.	.	.	.	.	9	.	.	9	18	9	.	9	18	56	22
								1	1	1	2	.	2	.	.	.	1	.	.	.	.	.	.	.	.	.	.	.	.	9	.	9	.	.	18	28	23
									1	.	1	.	1	.	.	.	1	.	.	.	.	.	.	.	.	.	.	.	.	.	.	.	.	.	.	10	24
										1	1	.	2	.	.	.	1	.	.	.	.	.	.	.	.	.	.	.	.	.	.	9	.	.	.	.	25
											1	.	1	.	.	.	1	.	.	.	.	.	.	.	.	.	.	.	.	.	.	.	.	.	.	.	26
												1	.	.	1	1	.	.	.	.	.	8	.	9	.	28	.	.	31	.	.	.	27	.	.	.	27
													1	.	.	.	1	.	.	.	.	.	.	.	.	.	.	.	.	.	.	.	.	.	.	.	28
														1	.	.	.	1	4	5	10	5	6	5	24	10	20	15	10	34	36	15	34	11	50	56	29
															1	.	.	.	.	.	.	.	.	.	.	10	.	.	11	.	.	.	.	.	.	.	30
																1	.	.	.	.	.	.	.	.	.	10	.	.	11	.	.	.	.	.	.	.	31
																	1	.	.	.	.	.	.	.	.	.	.	.	.	.	.	.	.	.	.	.	32
																		1	.	.	5	.	6	5	9	.	10	5	10	24	16	15	.	5	30	36	33
																			1	2	2	.	1	.	9	.	7	7	.	12	17	5	6	6	24	27	34
																				1	.	.	.	.	3	.	2	4	.	3	8	1	.	3	12	12	35
																					1	.	1	.	3	.	4	2	.	8	8	5	.	3	13	18	36
																						1	.	1	.	3	.	.	3	.	.	.	6	.	.	.	37
																							1	.	.	.	.	.	.	4	.	5	.	.	5	6	38
																								1	.	.	.	.	3	.	.	.	.	.	.	.	39
																									1	.	1	1	.	2	4	1	.	2	6	8	40
																										1	.	.	1	.	.	.	.	.	.	.	41
																											1	.	.	1	2	1	.	1	2	6	42
																												1	.	.	2	.	.	1	3	3	43
																													1	.	.	.	.	.	.	.	44
																														1	.	1	.	.	2	3	45
																															1	.	.	1	1	2	46
																																1	.	.	.	.	47
																																	1	.	.	.	48
																																		1	.	.	49
																																			1	1	50
																																				1	51

Table 3. *(continued)*

ALGEBRA D12, REPRESENTATIONS OF CLASS 2

NUMBER	0	1	2	3	4	5	6	7	8	9	10	11	12
S.P.	12	20	28	36	36	44	60	44	52	60	68	92	52
O.S.	2 048	24 576	135 168	24 576	450 560	270 336	24 576	1 013 760	1 351 680	135 168	270 336	24 576	1 622 016
LEVEL	67	89	109	111	127	131	133	143	149	151	153	155	157
DIM'N	2 048	47 104	516 096	565 248	3 579 904	8 243 200	4 710 400	17 616 896	64 204 800	47 151 104	77 156 352	30 617 600	65 286 144
	0	1	0	2	0	1	3	0	1	0	2	4	0
	0	0	1	0	0	1	0	0	0	2	1	0	0
	0	0	0	0	1	0	0	0	1	0	0	0	0
	0	0	0	0	0	0	0	1	0	0	0	0	0
	0	0	0	0	0	0	0	0	0	0	0	0	1
	0	0	0	0	0	0	0	0	0	0	0	0	0
	0	0	0	0	0	0	0	0	0	0	0	0	0
	0	0	0	0	0	0	0	0	0	0	0	0	0
	0	0	0	0	0	0	0	0	0	0	0	0	0
	1 0	0 0	1 0	1 0	0 0	0 0	0 0	1 0	1 0	1 0	1 0	1 0	0 0
WEIGHT	0	1	0	0	1	1	1	0	0	0	0	0	1
0	1	11	66	66	274	561	286	869	2 739	1 881	2 871	1 001	2 211
1		1	10	11	55	121	66	209	704	495	781	286	605
2			1	1	9	20	11	45	155	111	176	66	155
3				1	.	10	11	.	55	55	121	66	.
4					1	2	1	8	27	19	30	11	36
5						1	1	.	9	10	20	11	.
6							1	.	.	.	10	11	.
7								1	3	2	3	1	7
8									1	1	2	1	.
9										1	1	1	.
10											1	1	.
11												1	.
12													1
13													

NUMBER	13	14	15	16	17	18	19	20	21	22	23	24	25
S.P.	60	60	68	76	84	100	132	68	68	76	76	84	84
O.S.	4 055 040	1 892 352	1 351 680	1 351 680	270 336	270 336	24 576	1 622 016	8 110 080	6 082 560	1 013 760	450 560	4 055 040
LEVEL	165	169	169	171	173	175	177	179	179	185	187	187	187
DIM'N	336 322 560	188 604 416	488 656 896	639 907 840	587 575 296	534 159 360	165 335 040	433 168 384	1 295 360 000	2 873 155 584	797 423 616	1 770 496 000	3 485 081 600
	1	0	0	2	1	3	5	0	1	0	0	0	2
	0	0	1	0	2	1	0	0	0	1	0	0	0
	0	0	1	1	0	0	0	0	0	0	0	2	0
	1	0	0	0	0	0	0	0	0	1	0	0	1
	0	0	0	0	0	0	0	0	1	0	0	0	0
	0	1	0	0	0	0	0	0	0	0	0	0	0
	0	0	0	0	0	0	0	1	0	0	0	0	0
	0	0	0	0	0	0	0	0	0	0	1	0	0
	0	0	0	0	0	0	0	0	0	0	0	0	0
	0 0	1 0	0 0	0 0	0 0	0 0	0 0	0 0	1 0	1 0	1 0	1 0	1 0
WEIGHT	1	0	1	1	1	1	1	1	0	0	0	0	0
0	9 582	4 642	12 639	15 521	13 431	11 154	3 003	8 162	26 070	51 018	12 078	29 590	58 234
1	2 794	1 397	3 795	4 774	4 191	3 586	1 001	2 640	8 360	16 908	4 125	9 955	19 679
2	739	405	1 024	1 299	1 155	1 001	286	837	2 515	5 210	1 398	3 100	6 114
3	209	.	430	704	715	781	286	.	605	1 925	.	1 330	2 794
4	171	111	237	300	267	231	66	258	697	1 455	468	870	1 710
5	45	.	100	155	166	176	66	.	155	530	.	375	739
6	.	.	.	55	55	121	66	.	.	.	.	.	209
7	32	28	43	54	47	40	11	76	172	354	154	210	414
8	8	.	18	27	29	30	11	.	36	126	.	91	171
9	.	.	9	9	20	20	11	.	.	45	.	45	45
10	.	.	.	9	10	20	11	.	.	.	.	.	45
11	.	.	.	.	.	10	11	.	.	.	.	.	.
12	4	6	5	6	5	4	1	21	35	69	49	40	80
13	1	.	2	3	3	3	1	.	7	24	.	17	32

NUMBER	26	27	28	29	30	31	32	33	34	35	36	37	38
S.P.	76	92	84	108	108	116	92	140	108	100	84	180	84
O.S.	11 354 112	2 703 360	450 560	135 168	1 351 680	270 336	135 168	270 336	2 048	24 576	16 220 160	24 576	11 354 112
LEVEL	191	191	193	193	193	195	197	197	199	199	199	199	201
DIM'N	3 839 447 040	6 484 101 120	1 171 523 584	2 335 793 152	4 607 336 448	4 570 030 080	1 338 884 096	3 000 524 800	426 008 576	1 095 450 624	11 774 304 256	771 563 520	8 976 844 800
	1	1	0	0	3	2	0	4	0	0	0	6	1
	0	1	0	3	0	2	0	1	0	0	1	0	0
	0	1	0	0	1	0	0	0	0	0	0	0	0
	0	0	0	0	0	0	0	0	0	0	0	0	0
	0	0	0	0	0	0	0	0	0	0	1	0	0
	1	0	0	0	0	0	0	0	0	0	0	0	0
	0	0	0	0	0	0	0	0	0	0	0	0	1
	0	0	0	0	0	0	0	0	0	0	0	0	0
	0	0	1	0	0	0	0	0	0	0	0	0	0
	0 0	1 0	0 0	1 0	1 0	1 0	1 1	1 0	3 0	1 0	0 0	1 0	1 0
WEIGHT	1	0	1	0	0	0	0	0	0	2	1	0	0
0	57 453	99 341	14 938	32 956	64 834	60 984	15 026	36 036	4 213	11 298	151 074	8 008	104 742
1	19 822	34 221	5 313	11 550	22 869	21 714	5 500	13 156	1 585	4 213	54 054	3 003	38 247
2	6 557	10 800	1 890	3 696	7 304	6 996	2 025	4 301	603	1 585	18 384	1 001	13 590
3	1 397	5 830	.	2 310	4 774	4 906	.	3 586	.	.	6 149	1 001	2 640
4	2 061	3 038	672	1 044	2 059	1 980	750	1 221	232	603	5 861	286	4 680
5	405	1 619	.	660	1 299	1 375	.	1 001	.	.	1 910	286	837
6	.	430	.	220	704	715	.	781	.	.	.	286	.
7	605	728	238	248	490	468	280	286	91	232	1 714	66	1 550
8	111	382	.	156	300	322	.	231	.	.	542	66	258
9	.	200	.	111	155	221	.	176	.	.	155	66	.
10	.	100	.	55	155	166	.	176	.	.	.	66	.
11	.	.	.	.	55	55	.	121	.	.	.	66	.
12	161	136	84	45	90	84	105	50	36	91	444	11	486
13	28	70	.	28	54	57	.	40	.	.	136	11	76

NUMBER	39	40	41	42	43	44	45	46	47	48	49	50	51
S.P.	92	92	100	92	108	116	92	116	100	124	100	140	148
O.S.	8 110 080	4 055 040	12 165 120	8 110 080	1 351 680	4 055 040	28 385 280	1 351 680	11 354 112	2 703 360	4 055 040	270 336	1 351 680
LEVEL	201	203	207	209	209	209	211	211	213	213	215	215	215
DIM'N	13 777 920 000	13 845 278 720	39 638 016 000	16 736 051 200	26 387 472 384	25 764 710 400	36 252 463 104	28 943 523 840	41 594 009 600	52 449 173 504	24 808 734 720	24 192 143 360	26 580 787 200
	2	0	1	1	1	3	0	0	2	2	1	1	4
	0	0	1	0	0	0	1	2	0	1	0	3	0
	0	1	0	0	2	0	0	1	0	1	0	0	1
	0	1	1	0	0	1	0	0	0	0	0	0	0
	1	0	0	0	0	0	0	0	0	0	0	0	0
	0	0	0	0	0	0	1	0	1	0	0	0	0
	0	0	0	0	0	0	0	0	0	0	0	0	0
	0	0	0	1	0	0	0	0	0	0	0	0	0
	0	0	0	0	0	0	0	0	0	0	1	0	0
	0 0	0 0	0 0	0 0	0 0	0 0	1 0	0 0	1 0	0 0	1 0	0 0	0 0
WEIGHT	1	1	1	1	1	1	0	1	0	1	0	1	1
0	166 770	162 569	428 230	159 258	269 742	256 619	352 341	280 896	382 052	482 944	200 970	206 789	221 364
1	60 665	59 300	159 390	60 753	101 475	97 449	133 881	106 722	147 312	185 834	79 233	80 509	86 999
2	20 800	20 474	55 695	22 785	35 762	34 429	49 110	37 896	54 457	66 429	30 933	29 073	31 504
3	8 360	8 195	27 005	4 125	18 755	19 679	15 180	21 582	19 822	41 569	5 313	19 833	22 869
4	6 648	6 582	17 983	8 394	11 604	11 159	17 253	12 336	19 196	21 675	11 970	9 537	10 329
5	2 515	2 625	8 530	1 398	6 035	6 114	5 160	7 005	6 557	13 330	1 890	6 501	7 304
6	605	.	1 925	.	1 330	2 794	.	2 035	1 397	5 830	.	3 025	4 774
7	1 942	1 921	5 236	3 028	3 374	3 242	5 741	3 580	6 388	6 282	4 592	2 757	2 984
8	697	761	2 426	468	1 740	1 710	1 656	2 014	2 061	3 798	672	1 869	2 059
9	155	321	1 060	.	915	739	405	1 189	405	2 214	.	1 320	1 299
10	155	.	530	.	375	739	.	595	405	1 619	.	880	1 299
11	.	.	.	.	.	209	.	.	.	430	.	220	704
12	500	489	1 320	1 064	841	810	1 775	886	1 971	1 545	1 743	669	725
13	172	191	597	154	428	414	494	491	605	918	238	449	490

1	.	.	.	.	.	5	5	9	15	5	10	36	16	29	5	10	9	40	5	15	36	94	1	141	105	100	265	359	165	160	495	171	546	295	654	124	135	14
	1	1	2	2	1	.	.	8	.	9	8	.	36	.	19	27	39	.	30	.	.	36	11	.	36	81	252	.	227	171	111	282	111	527	.	312	300	15
		1	1	2	1	.	.	.	.	.	8	.	18	.	10	27	29	.	30	.	.	.	11	.	36	.	126	.	91	171	.	145	111	382	.	211	300	16
			1	1	1	.	.	.	.	.	.	.	9	.	10	9	20	.	20	.	.	.	11	.	.	.	45	.	45	45	.	100	.	200	.	166	155	17
				1	1	.	.	.	.	.	.	.	.	.	.	9	10	.	20	.	.	.	11	.	.	.	.	.	.	45	.	.	.	100	.	55	155	18
					1	.	.	.	.	.	.	.	.	.	.	.	.	.	10	.	.	.	11	.	.	.	.	.	.	.	.	.	.	.	.	.	55	19
						1	.	.	4	.	.	6	.	10	.	.	.	15	.	6	15	14	.	35	15	14	35	115	21	20	115	21	126	35	239	14	15	20
							1	3	.	2	4	6	8	.	3	6	6	.	4	.	.	28	1	21	35	38	117	49	82	80	133	93	161	172	84	82	90	21
								1	.	1	1	.	4	.	2	3	4	.	3	.	.	7	1	.	7	16	48	.	43	32	28	52	28	97	.	56	54	22
									1	.	.	.	.	3	.	.	.	6	.	3	6	.	.	7	.	.	.	32	.	.	20	.	21	.	86	.	.	23
										1	.	.	2	.	1	1	3	.	2	.	.	.	1	.	.	8	16	.	27	8	.	27	.	54	.	38	27	24
											1	.	2	.	1	3	3	.	3	.	.	.	1	.	7	.	24	.	17	32	.	27	28	70	.	38	54	25
												1	.	.	.	.	.	.	.	.	.	4	.	5	5	5	15	15	10	10	30	11	36	20	29	9	10	26
													1	.	1	1	2	.	2	.	.	.	1	.	.	.	8	.	9	8	.	18	.	36	.	29	27	27
														1	.	.	.	2	.	1	3	.	.	.	.	.	.	8	.	.	.	.	.	.	27	.	.	28
															1	.	1	.	1	.	.	.	1	.	.	.	.	.	.	.	.	9	.	9	.	20	9	29
																1	1	.	2	.	.	.	1	.	.	.	.	.	.	8	.	.	.	18	.	10	27	30
																	1	.	1	.	.	.	1	.	.	.	.	.	.	.	.	.	.	9	.	10	9	31
																		1	.	1	1	.	.	.	.	.	.	.	.	.	.	.	.	.	9	.	.	32
																			1	.	.	.	1	.	.	.	.	.	.	.	.	.	.	.	.	.	9	33
																				1	.	.	.	.	.	.	.	.	.	.	.	.	.	.	.	.	.	34
																					1	.	.	.	.	.	.	.	.	.	.	.	.	.	.	.	.	35
																						1	.	.	1	2	6	.	5	4	6	6	6	11	.	6	6	36
																							1	.	.	.	.	.	.	.	.	.	.	.	.	.	.	37
																								1	.	.	.	4	.	.	5	.	6	.	10	.	.	38
																									1	.	3	.	2	4	.	3	6	8	.	4	6	39
																										1	2	.	3	1	.	3	.	6	.	4	3	40
																											1	.	1	1	.	2	.	4	.	3	3	41
																												1	.	.	.	.	.	.	3	.	.	42
																													1	.	.	1	.	2	.	2	1	43
																														1	.	.	.	2	.	1	3	44
																															1	.	1	.	.	.	.	45
																																1	.	1	.	2	1	46
																																	1	.	.	.	.	47
																																		1	.	1	1	48
																																			1	.	.	49
																																				1	.	50
																																					1	51

Table 3. *(continued)*

ALGEBRA E6, REPRESENTATIONS OF CLASS 0

NUMBER	0	1	2	3	4	5	6	7	8	9	10	11	12	13	14	15	16	17
S.P.	0	6	12	18	24	24	24	36	36	30	36	42	42	42	48	54	48	48
O.S.	1	72	270	720	72	432	432	27	27	2 160	2 160	1 080	1 080	1 440	270	72	2 160	2 160
LEVEL	1	23	33	43	45	47	47	49	49	55	61	63	63	65	65	67	69	69
DIM'N	1	78	650	2 925	2 430	5 824	5 824	3 003	3 003	34 749	70 070	78 975	78 975	105 600	85 293	43 758	252 252	252 252
	0	0	1	0	0	1	0	3	0	1	0	2	0	0	2	0	1	0
	0	0	0	0	0	1	0	0	0	0	1	0	1	0	0	0	1	0
	0 0	1 0	0 0	0 1	2 0	0 0	0 0	0 0	0 0	1 0	0 0	0 0	0 0	1 1	0 0	3 0	1 0	1 0
	0	0	0	0	0	0	1	0	0	0	1	1	0	0	0	0	0	1
WEIGHT	0	0	1	0	0	0	1	0	3	1	0	0	2	0	2	0	0	1
0	1	6	20	45	36	64	64	24	24	261	374	360	360	480	351	180	900	900
1		1	5	15	11	24	24	10	10	106	169	170	170	220	165	80	445	445
2			1	4	3	8	8	4	4	40	72	76	76	96	76	35	211	211
3				1	1	2	2	1	1	14	28	31	31	42	33	17	95	95
4					1	.	.	.	.	5	9	10	10	20	15	11	40	40
5						1	.	1	.	4	10	15	10	15	14	6	45	36
6							1	.	1	4	10	10	15	15	14	6	36	45
7								1	.	.	.	5	.	.	4	.	10	.
8									1	.	.	.	5	.	4	.	.	10
9										1	3	4	4	6	5	3	16	16
10											1	1	1	2	1	1	6	6
11												1	.	.	1	.	3	.
12													1	.	1	.	.	3
13														1	.	1	2	2
14															1	.	.	.

NUMBER	18	19	20	21	22	23	24	25	26	27	28	29	30	31	32	33	34
S.P.	60	60	54	60	60	60	66	66	84	84	66	72	72	72	78	78	96
O.S.	432	432	6 480	1 080	2 160	1 080	2 160	2 160	270	270	6 480	720	4 320	4 320	1 440	2 160	72
LEVEL	71	71	75	77	77	77	79	79	81	81	83	85	85	85	87	87	89
DIM'N	146 432	146 432	852 930	600 600	812 175	600 600	972 972	972 972	371 800	371 800	2 453 814	1 337 050	2 977 975	2 977 975	1 911 195	3 490 695	537 966
	3	0	1	1	1	0	2	1	4	1	0	0	2	0	0	2	0
	0	0	0	0	0	2	1	0	0	0	1	0	0	1	0	0	0
	1 0	1 0	0 1	0 0	2 0	0 0	0 0	0 0	0 0	0 0	1 0	0 2	1 0	1 0	2 1	1 0	4 0
	0	0	0	2	0	0	0	1	0	0	1	0	1	0	0	0	0
WEIGHT	0	3	1	0	1	1	1	2	1	4	0	0	0	2	0	2	0
0	416	416	2 376	1 470	2 085	1 470	2 124	2 124	640	640	4 914	2 440	5 320	5 320	3 249	5 679	810
1	220	220	1 236	790	1 085	790	1 165	1 165	370	370	2 724	1 365	3 020	3 020	1 819	3 254	446
2	112	112	622	412	552	412	623	623	210	210	1 472	750	1 676	1 676	1 004	1 836	248
3	52	52	303	205	278	205	320	320	112	112	777	408	904	904	559	1 018	145
4	20	20	141	95	145	95	155	155	55	55	399	225	470	470	324	559	96
5	31	20	140	100	130	100	174	151	70	55	394	210	500	455	290	550	76
6	20	31	140	100	130	100	151	174	55	70	394	210	455	500	290	550	76
7	16	.	25	20	20	20	64	26	40	10	80	45	160	85	60	160	16
8	.	16	25	20	20	20	26	64	10	40	80	45	85	160	60	160	16
9	10	10	62	45	62	45	76	76	30	30	196	109	242	242	158	289	45
10	4	4	25	20	27	20	33	33	14	14	95	54	119	119	81	142	24
11	4	.	10	7	9	9	24	11	14	4	38	23	68	45	33	76	10
12	.	4	10	9	9	7	11	24	4	14	38	23	45	68	33	76	10
13	1	1	9	7	14	7	12	12	5	5	45	27	55	55	48	69	17
14	.	.	4	3	3	3	8	8	4	4	14	12	24	24	15	40	6

NUMBER	35	36	37	38	39	40	41	42	43	44	45	46	47	48	49	50	51
S.P.	78	78	84	90	90	90	84	90	96	96	84	96	96	114	96	96	114
O.S.	4 320	4 320	2 160	216	2 160	216	2 160	2 160	432	432	8 640	432	432	432	2 160	2 160	432
LEVEL	89	89	91	91	91	91	91	91	93	93	93	93	93	95	95	95	95
DIM'N	4 752 384	4 752 384	5 054 400	1 559 376	3 309 696	1 559 376	5 054 400	3 309 696	3 162 159	3 007 368	12 514 788	3 007 368	3 162 159	1 896 180	7 779 200	7 779 200	1 896 180
	1	0	1	0	3	0	0	0	3	2	1	0	0	4	3	1	0
	1	0	1	3	0	0	0	0	0	2	1	0	0	1	0	1	0
	0 1	0 1	2 0	0 0	0 1	0 0	2 0	0 1	2 0	0 0	0 0	0 0	2 0	0 0	0 0	0 0	0 0
	0	1	0	0	0	3	1	0	0	0	1	2	0	0	1	0	1
WEIGHT	0	1	0	0	0	0	1	3	0	0	1	2	3	0	1	3	4
0	7 296	7 296	7 200	2 052	4 320	2 052	7 200	4 320	3 816	3 600	16 596	3 600	3 816	1 860	8 960	8 960	1 860
1	4 256	4 256	4 220	1 235	2 610	1 235	4 220	2 610	2 326	2 195	9 925	2 195	2 326	1 180	5 480	5 480	1 180
2	2 432	2 432	2 432	728	1 544	728	2 432	1 544	1 392	1 318	5 832	1 318	1 392	737	3 304	3 304	737
3	1 358	1 358	1 382	418	885	418	1 382	885	808	772	3 352	772	808	445	1 950	1 950	445
4	736	736	780	230	480	230	780	480	450	439	1 870	439	450	255	1 120	1 120	255
5	761	720	785	240	540	234	740	474	501	468	1 896	429	439	290	1 180	1 105	255
6	720	761	740	234	474	240	785	540	439	429	1 896	468	501	255	1 105	1 180	290
7	240	160	240	80	240	64	160	104	216	200	576	110	95	160	480	320	70
8	160	240	160	64	104	80	240	240	95	110	576	200	216	70	320	480	160
9	392	392	418	129	274	129	418	274	260	249	1 038	249	260	155	644	644	155
10	202	202	224	69	145	69	224	145	144	134	554	134	144	86	352	352	86
11	116	84	120	42	108	31	92	58	98	98	296	59	58	76	236	176	38
12	84	116	92	31	58	42	120	108	58	59	296	98	98	38	176	236	76
13	98	98	123	34	69	34	123	69	75	66	278	66	75	42	182	182	42
14	48	48	48	18	40	18	48	40	36	44	152	44	36	32	112	112	32

1	.	.	.	.	.	.	5	.	.	.	.	.	9	5	10	10	20	15	11	16	16	40	5	10	5	40	10	20	9	55	9	20	5	40	40	5	15
	1	.	1	.	3	2	4	4	8	3	5	1	18	11	28	22	18	32	6	55	40	62	22	48	14	52	28	47	47	144	27	34	36	112	88	17	16
		1	.	1	3	4	4	2	3	8	1	5	18	11	22	28	18	32	6	40	55	52	14	28	22	62	48	34	27	144	47	47	17	88	112	36	17
			1	.	.	.	.	.	4	.	5	.	.	.	10	.	.	10	.	15	.	15	6	30	.	.	.	26	25	36	.	.	30	60	20	.	18
				1	.	.	.	.	.	4	.	5	.	.	.	10	.	10	.	.	15	.	.	.	6	15	30	.	.	36	25	26	.	20	60	30	19
					1	1	1	1	2	2	1	1	7	5	10	10	8	14	3	22	22	26	9	18	9	26	18	19	19	70	19	19	14	50	50	14	20
						1	.	.	.	1	.	1	3	2	4	3	3	4	1	8	12	12	3	6	7	12	10	10	6	32	13	9	4	20	24	11	21
							1	.	.	.	.	.	3	2	4	4	6	5	3	8	8	16	3	6	3	16	6	10	6	27	6	10	4	20	20	4	22
								1	1	.	1	.	3	2	3	4	3	4	1	12	8	12	7	10	3	12	6	9	13	32	6	10	11	24	20	4	23
									1	.	1	.	.	.	3	.	.	4	.	6	.	6	3	10	.	.	.	9	10	16	.	.	11	24	10	.	24
										1	.	1	.	.	.	3	.	4	.	.	6	.	.	.	3	6	10	.	.	16	10	9	.	10	24	11	25
											1	.	.	.	.	.	.	.	.	.	.	.	4	.	.	.	3	3	.	.	.	8	8	.	.	26
												1	.	.	.	.	.	.	.	.	.	.	.	.	.	4	.	.	.	3	3	.	.	8	8	27
													1	1	1	1	2	1	1	4	4	6	2	3	2	6	3	4	4	12	4	4	3	8	8	3	28
														1	.	.	1	.	1	2	2	2	1	1	1	2	1	1	3	4	3	1	2	2	2	2	29
															1	.	.	1	.	2	.	3	1	3	.	.	.	4	3	6	.	.	3	8	4	.	30
																1	.	1	.	.	2	.	.	.	1	3	3	.	.	6	3	4	.	4	8	3	31
																	1	.	1	.	.	2	.	.	.	2	.	1	.	.	.	1	.	.	.	.	32
																		1	.	.	.	.	.	.	.	.	.	.	.	3	.	.	.	4	4	.	33
																			1	.	.	.	.	.	.	.	.	.	.	.	.	.	.	.	.	.	34
																				1	.	1	1	1	.	.	.	1	2	2	.	.	2	2	1	.	35
																					1	.	.	.	1	1	1	.	.	2	2	1	.	1	2	2	36
																						1	.	.	.	.	.	1	.	.	.	.	.	.	.	.	37
																							1	.	.	.	.	.	1	.	.	.	1	.	.	.	38
																								1	.	.	.	1	1	.	.	.	2	2	.	.	39
																									1	.	.	.	.	.	1	.	.	.	.	1	40
																										1	.	.	.	.	.	1	.	.	.	.	41
																											1	.	.	.	1	1	.	.	2	2	42
																												1	.	.	.	.	.	.	.	.	43
																													1	.	.	.	1	.	.	.	44
																														1	.	.	.	1	1	.	45
																															1	.	.	.	.	1	46
																																1	.	.	.	.	47
																																	1	.	.	.	48
																																		1	.	.	49
																																			1	.	50
																																				1	51

Table 3. *(continued)*

ALGEBRA E6, REPRESENTATIONS OF CLASS 1

NUMBER	0	1	2	3	4	5	6	7	8	9	10	11	12	13	14	15	16	17	18	19	20	21	22	23	24	25
S.P.	4	10	16	16	22	28	28	34	34	40	40	46	40	64	52	46	52	52	58	58	58	64	64	70	64	76
O.S.	27	216	27	432	1080	270	1080	432	2160	432	216	432	2160	27	270	4320	2160	2160	1080	2160	4320	432	432	1080	2160	432
LEVEL	17	31	33	39	47	49	53	55	59	61	61	63	63	65	65	69	71	73	75	75	77	77	77	79	79	79
DIM'N	27	351	351	1728	7371	7722	17550	19305	51975	46332	34398	54054	112320	19305	61425	314496	359424	386100	393822	494208	967680	412776	459459	579150	1123200	359424
WEIGHT	1	0	0	1	0	2	0	0	1	1	0	2	1	4	1	0	2	0	0	0	1	0	0	3	1	0
	0	0	0	0	1	0	0	0	0	0	2	1	0	0	0	1	0	0	0	0	1	0	0	0	1	0
	0 0	0 0	0 0	1 0	0 0	0 0	1 0	1 0	0 1	2 0	0 0	0 0	0 0	0 0	0 0	1 0	1 0	0 1	2 0	0 1	0 0	0 0	2 0	0 0	0 0	0 0
	0	1	0	0	0	0	1	0	0	0	0	0	1	0	0	0	0	1	1	0	1	2	0	1	0	1
	0	0	2	0	1	1	0	2	0	0	0	0	1	0	3	1	1	0	0	2	0	1	2	0	2	3
0	1	5	4	16	44	40	80	75	180	146	104	144	320	40	145	720	736	750	704	856	1600	646	731	820	1680	480
1		1	1	4	15	14	30	31	75	61	46	66	144	20	70	344	360	375	355	446	836	344	386	445	900	268
2			1	.	5	4	10	16	25	20	20	26	64	10	40	160	160	180	170	240	416	184	216	225	480	160
3				1	4	5	10	10	30	26	19	30	60	10	30	155	175	180	174	216	425	175	190	240	465	140
4					1	1	3	4	10	9	7	11	24	4	14	68	76	83	83	108	204	88	98	118	236	76
5						1	.	.	4	3	3	8	8	4	4	24	40	33	32	40	104	38	36	76	112	32
6							1	1	3	4	2	3	8	1	5	28	32	37	41	48	92	41	47	54	112	36
7								1	.	.	.	.	4	.	5	10	10	15	15	30	36	25	26	20	60	30
8									1	1	1	2	2	1	1	10	14	14	16	18	44	16	19	31	50	14
9										1	.	.	.	.	.	4	5	5	10	6	15	5	10	10	20	4
10											1	1	.	1	.	4	4	5	6	6	20	5	10	15	20	4
11												1	.	1	.	.	4	.	.	.	10	.	.	15	10	.
12													1	.	1	3	4	6	6	10	16	10	9	10	24	11
13														1	.	.	.	.	.	.	.	.	.	5	.	.
14															1	.	.	.	.	4	.	3	3	.	8	8

NUMBER	26	27	28	29	30	31	32	33	34	35	36	37	38	39	40	41	42	43	44	45	46	47	48	49	50	51
S.P.	64	76	100	76	70	76	70	94	82	76	82	82	82	88	88	88	88	88	100	94	106	100	124	94	112	94
O.S.	4320	270	27	432	1080	2160	8640	432	2160	6480	2160	4320	6480	1080	2160	2160	4320	4320	2160	2160	1080	1080	270	4320	270	12960
LEVEL	81	81	81	83	83	85	85	87	87	89	91	91	91	93	93	93	93	95	95	95	95	97	97	97	97	99
DIM'N	1640925	638820	100386	741312	1253070	2088450	4200768	853281	2559843	5776056	4582656	6243237	7601958	5553900	7528950	6747300	7757100	10378368	6110208	9189180	3281850	5501925	1837836	14017536	3675672	30115800
WEIGHT	1	3	0	1	0	2	1	4	1	0	0	0	2	2	2	1	0	0	3	2	0	0	5	0	2	1
	0	0	0	0	2	1	0	0	0	1	1	1	0	0	0	2	1	0	1	0	1	0	0	0	0	1
	1 1	0 0	0 0	3 0	1 0	1 0	1 0	1 0	1 0	0 1	0 0	2 0	0 1	0 0	2 0	0 0	0 0	1 1	0 0	0 0	0 0	3 0	0 0	1 1	0 0	1 0
	0	0	0	0	0	0	1	0	0	0	2	0	0	2	0	0	1	1	0	1	0	1	0	0	0	1
	0	2	5	0	0	0	1	0	3	1	0	1	1	0	1	1	2	0	1	2	4	0	1	2	4	0
0	2380	830	105	960	1640	2480	5248	836	2769	6306	4576	6340	7596	5090	7000	6180	7090	9264	4928	7760	2525	4445	1206	11456	2636	23690
1	1270	455	62	516	904	1395	2920	494	1601	3615	2676	3663	4425	3025	4110	3675	4230	5500	3000	4680	1565	2660	766	6956	1644	14440
2	635	240	40	256	484	740	1600	279	960	2040	1536	2083	2485	1760	2320	2140	2540	3200	1760	2810	1000	1560	476	4256	1060	8600
3	670	250	35	281	485	780	1580	290	884	2021	1530	2074	2550	1770	2405	2150	2450	3214	1820	2770	935	1580	485	4106	996	8695
4	334	128	20	144	252	407	840	158	500	1111	858	1161	1414	1006	1352	1223	1420	1848	1052	1622	569	928	292	2440	613	5114
5	168	76	10	72	131	248	416	112	240	578	464	604	820	588	788	715	751	1008	688	920	306	508	210	1312	344	3028
6	168	64	10	80	126	206	432	80	263	593	466	645	768	555	762	674	787	1056	592	921	324	556	168	1404	359	2965
7	65	30	10	30	54	85	224	35	175	310	256	330	386	295	380	355	495	570	304	550	240	295	91	860	250	1615
8	80	33	4	38	65	116	208	52	126	304	245	337	423	312	423	381	409	570	362	507	171	306	112	762	199	1715
9	40	15	1	26	30	50	100	20	60	151	120	194	210	155	235	190	205	324	180	270	85	200	55	426	110	950
10	35	14	1	16	35	60	96	31	60	155	124	179	215	170	220	210	205	304	204	266	85	170	70	420	110	970
11	15	14	.	6	15	45	36	31	20	65	54	69	140	100	130	130	85	130	174	151	35	70	70	170	55	550
12	32	14	4	16	26	46	104	20	76	154	132	170	206	160	212	191	246	309	176	288	118	167	55	449	128	921
13	.	4	.	.	.	10	.	16	.	.	.	.	25	20	20	20	.	.	64	26	.	.	40	.	10	80
14	.	4	4	.	.	.	24	.	40	33	32	32	40	36	36	38	104	72	32	112	76	37	10	168	76	205

1	1	2	3	3	6	3	4	4	8	3	15	5	1	9	13	22	46	10	32	74	64	92	101	81	113	97	115	167	92	143	54	100	30	237	64	511	15
	1	.	.	.	3	.	.	4	4	.	6	5	.	3	6	16	16	10	10	32	26	36	62	45	62	61	46	71	76	76	20	42	30	98	30	283	16
		1	1	1	2	2	1	1	2	2	6	1	1	4	5	9	20	4	14	34	34	43	45	40	56	45	60	88	42	69	31	53	14	124	33	270	17
			1	.	.	.	1	.	.	.	3	.	.	4	2	3	8	1	5	13	12	28	18	16	32	17	21	48	16	30	10	41	5	64	15	140	18
				1	.	1	1	.	2	2	.	1	1	.	.	.	10	.	14	14	16	16	18	19	19	16	44	38	14	50	31	21	4	80	33	113	19
					1	.	.	1	1	.	2	1	.	1	3	6	6	4	4	15	13	18	25	20	27	29	22	37	33	33	10	23	14	53	14	146	20
						1	.	.	.	1	.	.	1	.	.	.	4	.	4	5	10	6	6	10	10	5	20	20	4	20	15	10	1	35	14	60	21
							1	.	.	.	.	.	.	.	.	.	4	.	5	5	6	10	6	10	10	5	15	20	4	20	10	15	1	40	15	60	22
								1	.	.	.	1	.	.	.	3	.	4	.	.	.	.	10	7	9	9	.	.	24	11	.	.	14	.	4	38	23
									1	.	.	1	.	.	.	.	3	.	4	6	6	6	10	9	9	10	16	17	11	24	10	10	4	32	14	59	24
										1	.	.	1	.	.	.	.	.	4	.	.	.	.	.	.	.	10	.	.	10	15	.	.	15	14	.	25
											1	.	.	1	1	2	2	1	1	6	5	10	9	7	14	11	9	20	12	12	4	16	5	27	5	72	26
												1	.	.	.	.	.	.	.	.	.	.	4	3	3	3	.	.	8	8	.	.	4	.	4	14	27
													1	.	.	.	.	.	.	.	.	.	.	.	.	.	.	.	.	.	5	.	.	.	4	.	28
														1	.	.	.	.	.	.	.	4	.	.	5	.	.	5	.	.	.	10	.	6	.	15	29
															1	1	.	1	.	3	2	4	3	2	4	8	3	8	8	3	1	6	5	12	1	36	30
																1	.	1	.	.	.	.	3	2	4	4	.	.	8	3	.	.	5	.	1	18	31
																	1	.	1	2	3	3	3	4	4	3	6	9	3	8	4	6	1	15	5	30	32
																		1	.	.	.	.	.	.	.	.	.	.	4	.	.	.	5	.	.	.	33
																			1	.	.	.	.	.	.	.	3	.	.	4	4	.	.	6	5	.	34
																				1	1	1	1	1	1	2	2	4	2	2	1	3	1	6	1	14	35
																					1	.	.	1	.	.	1	2	.	1	1	1	.	2	1	6	36
																						1	.	.	1	.	.	2	.	.	.	3	.	3	.	6	37
																							1	1	1	1	.	.	2	2	.	.	1	.	1	7	38
																								1	.	.	.	.	.	1	.	.	.	.	1	3	39
																									1	.	.	.	.	.	.	.	.	.	.	3	40
																										1	.	.	1	.	.	.	1	.	.	3	41
																											1	.	.	1	1	.	.	2	1	.	42
																												1	.	.	.	1	.	1	.	2	43
																													1	.	.	.	1	.	.	.	44
																														1	.	.	.	.	1	.	45
																															1	.	.	.	1	.	46
																																1	.	.	.	.	47
																																	1	.	.	.	48
																																		1	.	.	49
																																			1	.	50
																																				1	51

Table 3. *(continued)*

ALGEBRA E7, REPRESENTATIONS OF CLASS 0

S.P.	0	4	8	12	12	16	16	20	24	24	28	28	36
O.S.	1	126	756	56	2 016	126	4 032	7 560	1 512	10 080	576	4 032	126
LEVEL	1	35	53	55	67	69	77	87	89	97	99	101	103
DIM'N	1	133	1 539	1 463	8 645	7 371	40 755	152 152	150 822	365 750	253 935	573 440	238 602
	0	1	0	0	0	2	0	1	1	0	0	1	3
	0	0	0	0	1	0	0	0	0	0	0	1	0
	0 0	0 0	0 0	0 0	0 0	0 0	1 0	0 0	0 0	0 1	2 0	0 0	0 0
	0	0	0	0	0	0	0	0	0	0	0	0	0
	0	0	1	0	0	0	0	1	0	0	0	0	0
WEIGHT	0	0	0	2	0	0	1	0	2	0	0	0	0
NUMBER	0	1	2	3	4	5	6	7	8	9	10	11	12
0	1	7	27	21	77	63	225	588	504	980	603	1 344	504
1		1	6	5	22	17	75	212	190	390	248	544	203
2			1	1	5	4	22	70	67	145	96	208	79
3				1	.	.	6	20	27	45	36	64	24
4					1	1	5	21	20	50	34	78	32
5						1	.	6	5	15	10	32	17
6							1	5	6	15	11	24	10
7								1	1	4	3	8	4
8									1	.	.	.	.
9										1	1	2	1
10											1	.	.
11												1	1
12													1
13													
14													

S.P.	28	32	36	48	32	36	40	40	44	40	44	44	52
O.S.	12 096	756	1 512	56	24 192	30 240	7 560	12 096	1 512	20 160	24 192	30 240	4 032
LEVEL	103	105	107	109	111	119	121	121	123	125	129	131	131
DIM'N	980 343	617 253	915 705	293 930	3 424 256	7 142 499	6 619 239	7 482 618	6 760 390	11 316 305	23 969 792	24 386 670	12 609 597
	0	0	0	0	1	0	2	0	2	0	0	1	0
	0	0	0	0	0	1	0	1	0	0	0	0	0
	0 0	0 0	0 0	0 0	1 0	0 0	0 0	0 0	0 0	1 0	1 0	0 1	1 0
	1	0	0	0	0	0	0	0	0	1	0	0	0
	0	2	1	0	0	1	1	0	0	0	1	0	0
WEIGHT	1	0	2	4	1	0	0	2	2	0	1	0	3
NUMBER	13	14	15	16	17	18	19	20	21	22	23	24	25
0	2 079	1 197	1 575	420	5 696	9 387	8 073	8 856	7 420	12 341	22 848	22 680	10 521
1	879	510	695	190	2 528	4 366	3 783	4 230	3 575	5 976	11 360	11 265	5 371
2	351	209	295	85	1 072	1 957	1 719	1 951	1 673	2 800	5 488	5 439	2 674
3	135	84	135	45	432	828	729	891	783	1 260	2 608	2 484	1 350
4	129	80	115	35	430	845	762	855	750	1 262	2 550	2 565	1 276
5	39	30	40	15	160	350	342	345	325	538	1 120	1 185	571
6	45	29	45	15	162	343	315	366	327	545	1 154	1 149	605
7	13	9	14	5	56	133	129	142	133	223	488	507	262
8	5	4	10	5	16	44	40	66	57	80	224	180	150
9	3	2	3	1	18	47	48	51	51	88	194	213	105
10	.	.	.	.	6	14	15	15	21	35	70	84	35
11	.	.	.	.	5	15	21	15	20	29	65	90	35
12	.	.	.	.	.	.	6	.	5	.	.	15	.
13	1	1	2	1	4	15	14	20	18	30	82	75	53
14		1	1	1	.	5	4	5	4	10	32	25	22

S.P.	48	48	52	64	48	52	56	52	68	56	56	60	60
O.S.	4 032	2 016	4 032	126	60 480	7 560	15 120	60 480	1 512	24 192	60 480	4 032	12 096
LEVEL	133	133	135	137	137	139	141	143	143	145	149	151	151
DIM'N	18 372 354	13 728 792	19 046 664	5 248 750	72 847 026	48 316 905	73 853 000	132 793 375	25 273 458	130 945 815	209 868 813	109 120 648	163 601 438
	1	0	2	4	1	1	1	0	1	2	0	0	0
	0	2	1	0	0	0	0	1	0	0	0	0	0
	2 0	0 0	0 0	0 0	0 0	0 0	0 0	1 0	0 0	1 0	0 1	0 0	2 0
	0	0	0	0	1	0	0	0	0	0	0	2	0
	0	0	0	0	0	2	1	0	0	0	1	0	1
WEIGHT	0	0	0	0	1	0	2	1	4	1	0	0	0
NUMBER	26	27	28	29	30	31	32	33	34	35	36	37	38
0	15 624	11 844	15 372	3 750	56 952	34 965	49 020	88 845	14 364	81 963	120 771	58 324	87 332
1	7 916	5 938	7 756	1 899	29 297	18 195	25 980	46 965	7 809	43 673	65 475	32 022	48 000
2	3 903	2 908	3 834	955	14 701	9 267	13 490	24 280	4 181	22 810	34 807	17 256	25 879
3	1 857	1 338	1 764	440	7 191	4 617	7 010	12 230	2 295	11 574	18 063	9 126	13 689
4	1 868	1 402	1 885	488	7 168	4 605	6 790	12 250	2 169	11 684	18 114	9 103	13 650
5	863	682	954	273	3 373	2 235	3 290	6 005	1 089	5 895	9 189	4 678	6 995
6	862	638	869	230	3 395	2 223	3 375	6 010	1 125	5 809	9 198	4 700	7 050
7	385	291	411	116	1 546	1 037	1 586	2 862	545	2 838	4 549	2 362	3 539
8	146	104	144	40	685	483	840	1 290	335	1 262	2 169	1 150	1 725
9	169	125	181	53	676	459	712	1 322	249	1 346	2 184	1 159	1 737
10	77	49	70	20	280	189	300	595	105	623	1 008	560	840
11	70	56	93	32	279	195	295	580	105	645	999	539	805
12	10	10	32	17	39	30	40	85	15	160	165	94	130
13	61	46	66	20	283	205	345	575	137	580	1 014	549	821
14	20	20	26	10	108	90	150	240	67	235	460	254	376

S.P.	60	60	64	68	64	64	72	72	72	64	76	88	68
O.S.	40 320	60 480	4 032	7 560	24 192	24 192	756	1 512	12 096	80 640	1 512	1 512	30 240
LEVEL	151	153	153	155	155	155	157	157	157	159	159	161	163
DIM'N	237 289 689	378 224 640	177 161 985	175 061 250	350 830 592	407 486 464	105 489 615	182 801 850	223 092 870	715 309 056	192 347 298	108 686 270	688 400 856
	0	1	0	3	0	1	0	3	0	1	0	0	0
	0	1	0	0	0	1	0	0	0	0	0	0	1
	0 1	0 0	2 0	0 0	0 0	0 0	0 0	0 0	0 0	1 0	0 0	0 0	0 1
	0	0	0	0	1	0	0	0	1	1	0	0	0
	0	1	0	1	1	0	3	0	0	0	2	1	0
WEIGHT	2	0	2	0	1	2	0	2	3	0	2	4	0
NUMBER	39	40	41	42	43	44	45	46	47	48	49	50	51
0	126 189	196 992	87 759	83 160	168 000	196 224	45 675	80 262	95 760	322 560	77 400	38 220	281 268
1	69 453	108 192	48 839	46 104	93 920	109 472	25 880	45 247	54 535	181 664	44 440	22 380	160 642
2	37 505	58 416	26 708	25 215	51 632	60 048	14 465	25 153	30 585	100 656	25 178	12 970	90 408
3	20 025	30 672	14 454	13 365	28 080	32 512	8 010	13 803	17 175	54 576	14 310	7 630	49 698
4	19 772	31 056	14 278	13 668	27 790	32 280	7 940	13 761	16 765	54 860	13 970	7 360	50 206
5	10 077	16 320	7 417	7 443	14 560	16 960	4 280	7 455	8 905	29 376	7 565	4 085	27 546
6	10 260	16 062	7 514	7 185	14 730	17 062	4 295	7 397	9 125	29 266	7 695	4 175	27 263
7	5 129	8 200	3 820	3 781	7 560	8 776	2 259	3 908	4 764	15 352	4 083	2 267	14 642
8	2 665	3 888	2 000	1 780	4 000	4 512	1 230	1 993	2 755	7 616	2 365	1 425	7 374
9	2 489	4 062	1 891	1 926	3 770	4 406	1 151	2 022	2 411	7 894	2 103	1 190	7 700
10	1 155	1 938	910	945	1 806	2 156	560	1 036	1 155	3 990	1 044	595	3 963
11	1 131	2 010	876	1 032	1 781	2 150	565	1 060	1 145	3 972	1 029	595	4 005
12	176	480	135	342	320	480	120	325	205	928	210	130	1 010
13	1 230	1 900	948	905	1 914	2 172	604	994	1 310	3 850	1 151	695	3 850
14	565	880	458	425	928	1 008	309	464	655	1 840	593	375	1 920

1	1	.	.	.	5	4	.	16	.	22	.	.	.	.	44	40	110	80	67	75	180	104	146	325	320	242	145	528	528	184	224	495	720	433	335	750	15
	1	.	.	.	.	.	.	.	.	6	.	.	.	.	.	.	20	.	27	.	.	.	.	45	.	36	.	64	64	24	24	135	.	108	135	.	16
		1	4	5	5	6	10	25	30	15	26	19	30	10	110	79	130	245	50	267	449	250	375	535	903	425	460	865	1021	280	502	585	1905	530	320	1964	17
			1	1	1	1	3	8	10	5	9	7	11	4	39	30	49	98	20	107	193	110	167	233	402	193	210	392	458	131	230	271	894	250	155	958	18
				1	.	1	.	.	4	.	3	3	8	4	13	9	14	34	5	56	69	41	59	80	192	66	129	144	208	51	133	97	416	96	60	467	19
					1	1	.	4	.	5	.	.	.	.	15	14	35	30	20	31	75	46	61	125	144	95	70	220	222	80	102	195	344	180	135	375	20
						1	.	.	.	.	.	.	.	.	5	4	10	10	5	16	25	20	20	40	64	30	40	80	96	30	57	65	160	70	50	180	21
							1	2	3	1	4	2	3	1	12	9	15	38	6	42	78	46	75	93	168	84	90	162	196	55	104	111	414	105	65	456	22
								1	.	1	.	.	.	.	4	5	10	10	6	10	30	19	26	45	60	37	30	89	84	35	40	75	155	74	55	180	23
									1	.	1	1	2	1	3	2	3	13	1	18	27	16	27	32	74	31	48	56	82	19	51	38	182	36	22	215	24
										1	.	.	.	.	.	.	5	.	6	.	.	.	.	15	.	11	.	24	24	10	10	45	.	39	45	.	25
											1	.	.	.	.	.	.	5	.	6	9	5	15	10	24	15	15	16	30	5	21	10	80	9	5	94	26
												1	1	1	.	.	.	5	.	5	9	5	10	10	30	15	21	16	30	5	20	10	74	9	5	100	27
													1	1	.	.	.	.	.	5	.	.	.	.	15	.	21	.	15	.	20	.	29	.	.	40	28
														1	.	.	.	.	.	.	.	.	.	.	.	.	6	.	.	.	5	.	.	.	.	.	29
															1	1	2	3	1	4	10	7	9	14	24	12	14	30	32	12	18	24	68	25	18	83	30
																1	1	.	1	.	4	3	3	4	8	3	4	16	8	9	4	13	24	17	14	33	31
																	1	.	1	.	.	.	.	4	.	3	.	8	8	4	4	13	.	13	14	.	32
																		1	.	1	3	2	4	4	8	5	5	8	10	3	6	6	28	6	4	37	33
																			1	.	.	.	.	.	.	.	.	.	.	.	.	5	.	4	10	.	34
																				1	.	.	.	.	4	.	5	.	5	.	6	.	10	.	.	15	35
																					1	1	1	1	2	1	1	4	2	2	1	3	10	4	3	14	36
																						1	.	.	.	.	.	2	.	1	.	1	4	3	2	5	37
																							1	.	.	1	.	.	.	.	.	.	4	.	.	5	38
																								1	.	1	.	2	2	1	1	3	.	3	3	.	39
																									1	.	1	.	1	.	1	.	3	.	.	6	40
																										1	.	.	.	.	.	.	.	.	.	.	41
																											1	.	.	.	1	.	.	.	.	.	42
																												1	.	1	.	1	.	2	2	.	43
																													1	.	1	.	.	.	.	.	44
																														1	.	.	.	1	1	.	45
																															1	.	.	.	.	.	46
																																1	.	1	2	.	47
																																	1	.	.	2	48
																																		1	1	.	49
																																			1	.	50
																																				1	51

Table 3. *(continued)*

ALGEBRA E7, REPRESENTATIONS OF CLASS 1

NUMBER	0	1	2	3	4	5	6	7	8	9	10	11	12	13	14	15	16	17	18	19	20	21	22	23	24	25
S.P.	3	7	11	15	19	27	19	23	27	27	31	31	35	35	43	39	39	43	43	43	51	47	51	59	47	75
O.S.	56	576	1 512	4 032	1 512	56	4 032	12 096	1 512	12 096	4 032	20 160	15 120	12 096	1 512	4 032	40 320	4 032	12 096	24 192	1 512	12 096	1 512	1 512	60 480	56
LEVEL	28	50	62	76	80	82	84	94	96	102	104	110	114	116	116	118	124	126	128	128	130	130	132	134	136	136
DIM'N	56	912	6 480	27 664	51 072	24 320	86 184	362 880	320 112	861 840	885 248	2 282 280	4 522 000	3 792 096	2 273 920	3 635 840	13 069 056	9 480 240	14 910 896	21 633 248	9 405 760	17 926 272	14 220 360	11 376 288	63 431 424	2 785 552
WEIGHT	0	0	1	0	0	0	1	0	2	0	0	1	1	0	1	2	0	0	0	1	3	0	0	0	1	0
	0	0	0	0	0	0	0	1	0	0	0	0	0	1	0	0	0	0	0	1	0	0	0	0	0	0
	0 0	1 0	0 0	0 0	0 0	0 0	1 0	0 0	0 0	1 0	1 0	0 0	0 0	1 0	0 0	1 0	0 1	2 0	0 0	0 0	0 0	0 0	0 0	0 0	1 0	0 0
	0	0	0	1	0	0	0	0	0	0	0	1	0	0	0	0	0	0	1	0	0	1	0	0	0	0
	0	0	0	0	1	0	0	0	0	1	0	0	1	0	0	0	0	0	1	0	0	0	2	1	1	0
	1	0	1	0	1	3	0	1	1	0	2	0	1	0	3	0	1	1	0	1	1	2	1	3	0	5
0	1	6	27	71	111	45	171	531	432	999	934	2 145	3 735	3 024	1 655	2 701	0 529	5 733	0 694	12 708	5 009	9 675	7 155	5 130	31 041	1 060
1		1	6	21	35	15	56	196	162	399	385	910	1 645	1 344	755	1 217	4 004	2 744	4 193	6 097	2 443	4 760	3 570	2 625	15 568	560
2			1	5	10	5	16	66	57	149	150	365	695	570	335	526	1 809	1 269	1 958	2 832	1 163	2 275	1 740	1 320	7 593	295
3				1	2	1	4	20	18	51	53	139	275	231	137	220	784	562	879	1 272	539	1 042	813	633	3 596	147
4					1	1	.	5	4	16	22	44	110	80	67	75	325	242	384	528	224	495	393	335	1 621	85
5						1	.	.	.	.	6	.	20	.	27	.	45	36	64	64	24	135	108	135	261	45
6							1	5	6	15	15	50	100	90	50	92	319	235	374	556	252	445	360	285	1 649	70
7								1	1	4	5	15	35	30	20	31	125	95	154	222	102	195	160	135	718	35
8									1	.	.	5	10	10	5	16	40	30	54	96	57	65	60	50	304	15
9										1	1	4	10	10	6	10	45	37	59	84	40	75	64	55	301	15
10											1	.	5	.	6	.	15	11	24	24	10	45	39	45	106	15
11												1	2	3	1	4	14	12	19	32	18	24	21	18	120	5
12													1	.	1	.	4	3	8	8	4	13	13	14	40	5
13														1	.	1	4	5	5	10	6	6	5	4	45	1
14															1	.	.	.	.	.	.	5	4	10	.	5
15																1	.	.	.	5	6	.	.	.	15	.

NUMBER	26	27	28	29	30	31	32	33	34	35	36	37	38	39	40	41	42	43	44	45	46	47	48	49	50	51
S.P.	51	51	55	55	55	63	59	63	59	67	67	59	63	67	63	79	67	67	67	71	83	71	75	71	83	75
O.S.	24 192	40 320	20 160	20 160	60 480	576	15 120	12 096	24 192	1 512	4 032	60 480	12 096	24 192	120 960	4 032	24 192	12 096	60 480	24 192	1 512	60 480	15 120	120 960	15 120	60 480
LEVEL	138	142	144	146	146	148	148	148	150	150	152	152	154	156	158	158	160	160	162	162	164	164	166	168	168	170
DIM'N	67 395 888	86 141 440	86 184 000	100 677 808	188 972 784	32 995 248	179 262 720	101 985 312	195 102 336	93 768 192	94 057 600	339 066 000	236 888 960	370 677 384	789 703 992	133 617 120	619 736 832	462 143 232	985 944 960	656 594 400	190 466 640	1 218 986 496	1 017 592 576	2 120 058 304	846 807 808	2 176 761 600
WEIGHT	1	0	2	0	0	0	2	0	1	2	3	0	0	0	1	0	1	0	1	2	4	1	1	0	1	2
	0	1	0	0	1	0	0	1	1	0	0	0	0	0	0	0	0	2	0	1	0	0	0	1	0	0
	1 0	0 0	0 0	1 1	0 0	3 0	0 0	0 0	1 0	0 0	1 0	1 0	1 0	1 0	0 1	1 0	2 0	0 0	0 0	0 0	0 0	0 0	0 0	1 0	0 0	1 0
	0	1	1	0	0	0	0	0	0	0	0	1	0	0	0	0	0	0	1	0	0	1	0	0	0	0
	0	0	0	0	1	0	1	0	0	0	0	0	2	1	0	0	0	0	1	0	0	0	2	1	1	1
	2	0	0	0	1	0	1	3	0	3	0	1	0	2	1	4	1	1	0	1	1	2	1	0	3	0
0	30 645	36 774	34 605	38 864	12 756	11 541	64 980	35 466	68 730	30 735	30 393	114 750	75 285	110 421	237 051	35 160	174 960	130 788	270 126	175 920	45 846	313 875	247 284	517 059	188 208	503 325
1	15 645	18 998	18 060	20 524	38 451	6 225	34 685	19 160	36 870	16 765	16 513	62 265	41 335	61 432	131 390	20 020	98 145	73 233	152 292	99 274	26 250	179 130	142 499	297 619	110 243	291 916
2	7 791	9 574	9 215	10 589	19 885	3 285	18 160	10 160	19 380	8 999	8 822	33 120	22 285	33 617	71 545	11 250	54 130	40 336	84 492	55 194	14 865	100 715	81 010	168 785	63 818	167 021
3	3 753	4 708	4 610	5 340	10 021	1 695	9 290	5 236	10 004	4 699	4 652	17 220	11 759	17 972	38 258	6 170	29 319	21 852	46 076	30 263	8 334	55 584	45 247	94 285	36 238	94 304
4	1 808	2 198	2 160	2 580	4 977	850	4 640	2 771	4 882	2 503	2 287	8 810	6 129	9 708	19 909	3 510	15 528	11 472	24 606	15 978	4 454	30 587	25 202	51 520	20 944	51 849
5	432	374	360	480	1 188	180	1 080	891	900	783	416	2 160	1 609	3 024	4 860	1 350	3 942	2 808	6 516	3 888	1 080	9 315	7 892	14 120	7 650	14 193
6	1 731	2 258	2 275	2 624	4 890	850	4 650	2 590	5 096	2 379	2 453	8 700	6 040	9 311	20 064	3 270	15 570	11 661	24 658	16 442	4 683	29 976	24 755	51 820	20 078	52 628
7	790	1 029	1 050	1 240	2 340	416	2 250	1 300	2 455	1 205	1 201	4 320	3 055	4 836	10 235	1 770	8 080	6 021	12 903	8 578	2 489	16 050	13 425	27 864	11 204	28 572
8	320	464	525	565	1 045	190	1 090	570	1 240	565	672	2 000	1 455	2 281	5 145	850	4 080	3 126	6 608	4 682	1 473	8 145	6 970	14 739	5 804	15 688
9	337	458	475	575	1 068	200	1 050	606	1 170	577	586	2 075	1 490	2 389	5 105	895	4 113	3 058	6 600	4 419	1 319	8 310	7 053	14 734	5 985	15 318
10	162	169	170	220	513	80	480	366	445	327	220	990	755	1 374	2 385	605	1 947	1 428	3 275	2 034	600	4 560	3 953	7 390	3 745	7 612
11	132	196	219	256	459	92	480	260	559	261	300	942	689	1 104	2 490	420	2 037	1 530	3 285	2 297	726	4 140	3 571	7 625	3 046	8 160
12	56	72	76	96	209	35	205	142	211	133	112	434	339	600	1 129	262	937	703	1 593	1 034	326	2 169	1 928	3 764	1 796	3 988
13	50	79	90	114	185	45	205	105	252	115	137	420	305	490	1 173	185	1 001	735	1 579	1 126	367	2 007	1 745	3 860	1 503	4 209
14	16	.	.	.	44	.	40	66	.	57	.	80	75	224	180	150	146	104	320	144	40	685	628	720	840	736
15	15	29	50	40	65	15	100	35	126	50	92	150	115	180	544	70	450	360	734	646	252	906	810	1 865	684	2 274

1	1	2	2	1	3	3	3	14	1	18	28	31	42	78	17	81	51	95	51	52	183	143	246	516	105	447	330	742	501	165	996	896	1868	824	2030	16
	1	.	.	.	.	.	.	5	.	6	9	10	20	24	11	30	15	40	21	20	75	55	90	225	35	222	144	325	225	75	435	389	904	355	1001	17
		1	.	.	1	2	2	4	1	4	10	10	15	30	6	28	20	36	18	20	75	66	113	215	53	191	146	339	217	75	457	440	889	415	960	18
			1	1	.	.	.	4	.	5	10	15	15	25	6	35	15	45	20	31	65	50	85	230	35	200	156	329	267	102	430	390	884	350	1053	19
				1	.	.	.	.	.	.	.	5	.	.	.	10	.	10	5	16	.	.	.	40	.	30	30	54	96	57	65	60	155	50	304	20
					1	1	2	.	1	4	.	.	.	15	.	14	20	.	18	.	30	31	82	75	53	61	46	144	66	20	283	275	344	345	360	21
						1	1	.	1	.	.	.	.	5	.	4	5	.	4	.	10	16	32	25	22	20	20	64	26	10	108	130	160	150	160	22
							1	.	1	.	.	.	.	.	.	.	5	.	4	.	.	.	16	.	22	.	.	.	.	.	44	40	.	110	.	23
								1	.	1	3	4	6	8	3	10	5	16	6	10	26	21	35	92	15	88	64	140	106	40	186	174	411	160	478	24
									1	.	.	.	.	.	.	.	.	.	.	.	.	.	.	.	6	.	.	.	.	.	.	.	.	20	.	25
										1	.	.	.	4	.	5	5	.	6	.	10	10	25	30	15	26	19	60	30	10	110	109	155	130	175	26
											1	1	2	2	1	2	1	6	1	4	9	7	12	35	5	36	27	55	44	18	72	69	183	63	216	27
												1	.	.	.	2	.	3	1	4	.	.	.	14	.	12	12	19	32	18	24	21	69	18	120	28
													1	.	1	.	.	2	.	1	3	2	3	12	1	18	9	18	15	6	24	22	80	20	93	29
														1	.	1	1	.	1	.	3	4	8	10	5	9	7	24	11	4	39	44	68	49	76	30
															1	.	.	.	.	.	.	.	.	.	.	6	.	.	.	.	.	.	14	.	15	31
																1	.	.	1	.	.	.	.	4	.	3	3	8	8	4	13	13	24	14	40	32
																	1	.	1	.	.	.	4	.	5	.	.	.	.	.	15	14	.	35	.	33
																		1	.	1	.	.	.	4	.	5	5	5	10	6	6	5	30	4	45	34
																			1	.	.	.	.	.	.	.	.	.	.	.	5	4	.	10	.	35
																				1	.	.	.	.	.	.	.	.	5	6	.	.	.	.	15	36
																					1	1	2	3	1	4	2	8	3	1	12	14	28	15	32	37
																						1	1	.	1	.	.	4	.	.	4	10	10	10	10	38
																							1	.	1	.	.	.	.	.	4	5	.	10	.	39
																								1	.	1	1	2	2	1	3	3	10	3	14	40
																									1	.	.	.	.	.	.	.	.	5	.	41
																										1	.	.	.	.	.	.	4	.	5	42
																											1	.	1	1	.	.	4	.	4	43
																												1	.	.	1	2	3	2	4	44
																													1	1	.	.	.	.	4	45
																														1	.	.	.	.	.	46
																															1	1	.	2	.	47
																																1	.	1	.	48
																																	1	.	1	49
																																		1	.	50
																																			1	51

Table 3. *(continued)*

ALGEBRA E8, REPRESENTATIONS OF CLASS 0

NUMBER	S.P.	O.S.	LEVEL	DIM'N	WEIGHT								0	1	2	3
					row 1	row 2	row 3 (left)	row 3	row 4	row 5	row 6	row 7				
0	0	1	1	1	0	0	0	0	0	0	0	0	1			
1	2	240	59	248	0	0	0	0	0	0	0	1	8	1		
2	4	2 160	93	3 875	1	0	0	0	0	0	0	0	35	7	1	
3	6	6 720	115	30 380	0	0	0	0	0	0	1	0	140	35	7	1
4	8	240	117	27 000	0	0	0	0	0	0	0	2	120	29	6	1
5	8	17 280	137	147 250	0	0	1	0	0	0	0	0	370	111	29	6
6	10	30 240	151	779 247	1	0	0	0	0	0	0	1	1 407	455	133	34
7	12	60 480	169	2 450 240	0	0	0	0	0	1	0	0	2 960	1 056	350	105
8	14	13 440	173	4 096 000	0	0	0	0	0	0	1	1	4 480	1 624	552	174
9	18	240	175	1 763 125	0	0	0	0	0	0	0	3	1 765	645	224	74
10	14	69 120	183	6 696 000	0	1	0	0	0	0	0	0	6 000	2 296	826	275
11	16	2 160	185	4 881 384	2	0	0	0	0	0	0	0	4 104	1 584	580	198
12	16	138 240	195	26 411 008	0	0	1	0	0	0	0	1	18 688	7 504	2 864	1 028
13	18	181 440	207	76 271 625	1	0	0	0	0	0	1	0	43 065	18 081	7 279	2 790
14	20	30 240	209	70 680 000	1	0	0	0	0	0	0	2	37 680	15 959	6 504	2 535
15	20	241 920	221	146 325 270	0	0	0	0	1	0	0	0	64 470	28 441	12 103	4 938
16	22	181 440	227	344 452 500	0	0	0	0	0	1	0	1	135 540	60 948	26 537	11 142
17	22	138 240	229	301 694 976	1	0	1	0	0	0	0	0	114 816	52 040	22 840	9 658
18	24	6 720	229	203 205 000	0	0	0	0	0	0	2	0	75 720	34 328	15 112	6 434
19	26	13 440	231	281 545 875	0	0	0	0	0	0	1	2	99 555	45 480	20 216	8 720
20	32	240	233	79 143 000	0	0	0	0	0	0	0	4	25 320	11 705	5 300	2 345
21	24	483 840	241	1 094 951 000	0	1	0	0	0	0	0	1	343 160	160 929	73 396	32 437
22	26	30 240	243	820 260 000	2	0	0	0	0	0	0	1	243 600	115 255	53 116	23 765
23	26	483 840	251	2 275 896 000	0	0	1	0	0	0	1	0	607 440	292 825	137 704	62 991
24	28	138 240	253	2 172 667 860	0	0	1	0	0	0	0	2	551 700	267 939	127 107	58 755
25	28	604 800	261	4 825 673 125	1	0	0	0	0	1	0	0	1 102 885	545 629	264 019	124 642
26	30	362 880	265	8 634 368 000	1	0	0	0	0	0	1	1	1 834 880	917 616	449 568	215 346
27	34	30 240	267	3 929 713 760	1	0	0	0	0	0	0	3	778 400	393 085	194 824	94 620
28	30	483 840	271	6 899 079 264	0	0	0	1	0	0	0	0	1 357 104	688 156	341 950	166 170
29	32	17 280	273	4 076 399 250	0	0	2	0	0	0	0	0	763 890	390 286	195 573	95 918
30	32	138 240	275	8 634 368 000	1	1	0	0	0	0	0	0	1 579 520	810 080	407 584	200 792
31	36	2 160	277	2 903 770 000	3	0	0	0	0	0	0	0	494 800	256 492	130 601	65 193
32	32	967 680	279	20 288 765 952	0	0	0	0	1	0	0	1	3 545 472	1 830 192	927 584	460 810
33	34	181 440	283	21 039 669 000	0	0	0	0	0	1	1	0	3 429 720	1 788 696	916 930	461 337
34	36	181 440	285	23 592 339 045	0	0	0	0	0	1	0	2	3 678 885	1 930 131	996 260	505 242
35	34	967 680	287	45 329 752 170	1	0	1	0	0	0	0	1	7 044 570	3 702 798	1 913 630	970 920
36	38	13 440	287	17 535 336 000	0	0	0	0	0	0	2	1	2 614 080	1 380 211	717 528	366 832
37	42	13 440	289	12 692 520 960	0	0	0	0	0	0	1	3	1 774 080	944 384	495 712	256 306
38	50	240	291	2 642 777 280	0	0	0	0	0	0	0	5	332 640	179 326	95 586	50 330
39	36	1 451 520	297	85 424 220 000	0	1	0	0	0	0	1	0	11 592 720	6 217 056	3 283 210	1 705 212
40	38	181 440	299	66 393 847 000	2	0	0	0	0	0	1	0	8 619 640	4 652 312	2 473 983	1 294 634
41	38	483 840	299	83 080 364 250	0	1	0	0	0	0	0	2	10 801 770	5 826 100	3 096 477	1 619 872
42	40	30 240	301	63 513 702 720	2	0	0	0	0	0	0	2	7 903 680	4 289 629	2 295 466	1 209 737
43	38	967 680	305	124 436 480 000	0	0	1	0	0	1	0	0	15 128 960	8 244 208	4 428 704	2 342 650
44	40	967 680	309	234 550 030 000	0	0	1	0	0	0	1	1	26 775 280	14 713 832	7 977 538	4 263 270
45	44	138 240	311	110 977 024 000	0	0	1	0	0	0	0	3	11 914 240	6 598 720	3 609 088	1 947 692
46	40	1 209 600	313	220 778 105 625	1	0	0	0	1	0	0	0	24 114 825	13 342 581	7 284 620	3 920 586
47	42	483 840	319	267 413 986 840	0	1	1	0	0	0	0	0	26 862 520	15 037 259	8 312 087	4 532 864
48	42	1 814 400	319	574 197 082 368	1	0	0	0	0	1	0	1	57 768 768	32 313 184	17 851 904	9 732 748
49	44	138 240	321	223 850 628 000	2	0	1	0	0	0	0	0	21 580 080	12 147 933	6 755 148	3 707 369
50	44	181 440	321	355 647 996 000	1	0	0	0	0	0	2	0	34 358 640	19 321 176	10 736 076	5 890 067
51	46	362 880	323	508 731 738 750	1	0	0	0	0	0	1	2	47 250 270	26 699 481	14 915 944	8 232 444

1	.	7	27	56	29	77	63	336	1 008	959	1 897	4 500	3 880	2 681	3 724	1 065	13 789	10 283	27 889	26 439	57 169	100 856	45 255	78 547	45 851	96 416	31 808	223 832	227 640	252 063	482 427	184 933	131 320	26 489	869 301	665 560	833 448	627 725	1 217 944	2 244 165	1 037 904	2 076 060	2 434 067	5 232 004	2 004 541	3 189 445	4 491 081	4
	1	7	28	48	21	84	62	344	1 007	932	1 925	4 480	3 920	2 624	3 605	994	13 846	10 276	27 944	26 370	57 274	100 520	44 828	78 848	45 969	96 712	31 852	224 000	227 332	251 125	482 909	183 876	130 032	26 020	869 932	665 806	832 859	626 876	1 218 952	2 242 856	1 035 496	2 078 321	2 436 952	5 231 058	2 006 768	3 186 940	4 482 941	5
		1	6	12	6	22	17	104	338	323	705	1 717	1 508	1 027	1 442	414	5 666	4 274	11 961	11 461	25 503	45 656	20 755	36 371	21 442	45 360	15 181	106 264	109 559	122 302	234 873	90 444	64 916	13 299	435 063	335 927	420 363	319 118	622 776	1 160 387	542 296	1 083 253	1 289 469	2 768 896	1 069 636	1 699 441	2 408 180	6
			1	2	1	5	4	28	103	101	243	617	550	376	537	159	2 215	1 698	4 921	4 786	10 979	20 044	9 280	16 306	9 732	20 712	7 052	49 084	51 448	58 037	111 546	43 351	31 566	6 611	213 018	166 036	207 747	159 157	312 140	589 198	278 788	554 879	671 381	1 441 348	561 290	891 507	1 272 760	7
				1	1	.	.	6	27	34	71	216	171	139	219	74	806	630	1 916	1 962	4 487	8 605	4 190	6 941	4 194	9 004	3 133	21 993	23 625	27 297	51 300	20 685	15 550	3 405	101 547	79 952	100 849	77 949	152 537	294 138	142 260	277 407	341 383	737 440	287 886	460 911	666 360	8
					1	.	.	.	.	7	.	27	.	21	56	29	77	63	225	336	588	1 512	959	980	603	1 344	504	3 976	4 641	6 075	9 576	4 781	4 144	1 065	20 880	16 772	22 645	17 703	33 096	69 672	36 960	63 168	80 499	182 344	69 293	117 740	179 172	9
						1	1	6	28	27	77	203	190	127	183	56	818	644	1 932	1 898	4 550	8 462	3 983	7 084	4 292	9 190	3 191	21 980	23 464	26 720	51 605	20 195	14 910	3 198	101 884	80 277	100 254	77 638	153 216	292 998	140 330	278 943	343 616	736 772	289 700	459 424	660 784	10
							1	.	7	6	21	56	64	42	57	21	279	238	714	693	1 813	3 416	1 624	2 968	1 849	3 968	1 420	9 464	10 346	11 852	23 206	9 128	6 832	1 518	47 480	37 954	47 012	37 024	73 528	142 246	68 832	137 459	172 781	369 160	147 042	232 232	336 271	11
								1	6	7	21	63	56	41	63	21	280	224	714	729	1 785	3 451	1 687	2 926	1 799	3 885	1 379	9 520	10 402	12 075	23 079	9 247	7 000	1 554	47 376	37 738	47 333	37 023	73 269	142 814	69 700	136 850	171 801	369 425	146 182	232 688	338 337	12
									1	1	5	16	16	11	17	6	88	74	247	254	669	1 322	658	1 171	735	1 600	583	3 954	4 424	5 187	10 012	4 035	3 102	709	21 428	17 307	21 619	17 147	34 226	67 708	33 492	65 820	84 441	181 140	72 575	115 182	168 853	13
										1	.	6	.	5	12	6	22	17	75	104	212	528	323	390	248	544	203	1 584	1 877	2 412	4 036	1 917	1 632	414	9 206	7 477	9 896	7 849	15 184	31 918	16 832	29 943	39 171	87 128	33 995	56 572	85 887	14
											1	3	4	2	3	1	25	22	78	81	235	472	238	446	286	632	238	1 568	1 796	2 127	4 179	1 678	1 308	306	9 378	7 695	9 551	7 692	15 528	31 192	15 640	30 879	40 573	86 680	35 278	55 670	82 263	15
												1	.	1	2	1	5	4	22	28	70	170	101	145	96	208	79	594	721	912	1 622	739	622	159	3 897	3 219	4 166	3 371	6 722	14 167	7 460	13 761	18 528	40 532	16 254	26 521	40 279	16
													1	.	.	.	6	7	21	21	77	154	77	161	106	244	98	581	686	813	1 674	658	518	126	3 954	3 332	4 039	3 339	6 818	13 859	7 014	14 126	19 075	40 349	16 842	26 187	38 913	17

Table 3. *(continued)*

ALGEBRA E8, REPRESENTATIONS OF CLASS 0

NUMBER	0	1	2	3	4	5	6	7	8	9	10	11	12	13	14	15	16	17	18	19	20	21	22	23	24	25
	0	0	1	0	0	0	1	0	0	0	0	2	0	1	1	0	0	1	0	0	0	0	2	0	0	1
	0	0	0	0	0	0	0	0	0	0	1	0	0	0	0	0	0	0	0	0	0	1	0	0	0	0
	0 0	0 0	0 0	0 0	0 0	1 0	0 0	0 0	0 0	0 0	0 0	0 0	1 0	0 0	0 0	0 0	0 0	1 0	0 0	0 0	0 0	0 0	0 0	1 0	1 0	0 0
	0	0	0	0	0	0	0	0	0	0	0	0	0	0	0	1	0	0	0	0	0	0	0	0	0	0
	0	0	0	0	0	0	0	1	0	0	0	0	0	0	0	0	1	0	0	0	0	0	0	0	0	1
	0	0	0	1	0	0	0	0	1	0	0	0	0	1	0	0	0	0	2	1	0	0	0	1	0	0
WEIGHT	0	1	0	0	2	0	1	0	1	3	0	0	1	0	2	0	1	0	0	2	4	1	1	0	2	0
18																			1	1	1	.	.	6	6	20
19																				1	1	.	.	.	6	.
20																					1	.	.	.	.	.
21																						1	1	5	6	21
22																							1	.	.	6
23																								1	1	5
24																									1	.
25																										1
26																										
27																										

NUMBER	26	27	28	29	30	31	32	33	34	35	36	37	38	39	40	41	42	43	44	45	46	47	48	49	50	51
	1	1	0	0	1	3	0	0	0	1	0	0	0	0	2	0	2	0	0	0	1	0	1	2	1	1
	0	0	0	0	1	0	0	0	0	0	0	0	0	1	0	1	0	0	0	0	0	1	0	0	0	0
	0 0	0 0	0 1	2 0	0 0	0 0	0 0	0 0	0 0	1 0	0 0	0 0	0 0	0 0	0 0	0 0	0 0	1 0	1 0	1 0	0 0	1 0	0 0	1 0	0 0	0 0
	0	0	0	0	0	0	1	0	0	0	0	0	0	0	0	0	0	0	0	0	1	0	0	0	0	0
	0	0	0	0	0	0	0	1	1	0	0	0	0	0	0	0	0	1	0	0	0	0	1	0	0	0
	1	0	0	0	0	0	0	1	0	0	2	1	0	1	1	0	0	0	1	0	0	0	0	0	2	1
WEIGHT	1	3	0	0	0	0	1	0	2	1	1	3	5	0	0	2	2	0	1	3	0	0	1	0	0	2
18	54	34	45	36	64	24	206	273	351	603	302	264	74	1 557	1 296	1 697	1 413	2 846	6 150	3 312	5 997	8 344	18 350	7 354	12 214	18 846
19	27	34	.	.	.	.	71	111	216	171	184	219	74	531	432	806	630	999	2 850	1 962	2 145	3 024	8 222	2 701	5 815	10 260
20	.	7	.	.	.	.	.	.	27	.	21	56	29	.	.	77	63	.	225	336	.	.	588	.	504	1 512
21	48	27	50	34	78	32	206	255	318	620	262	218	56	1 579	1 344	1 673	1 394	2 856	6 036	3 174	6 142	8 534	18 275	7 620	12 060	18 303
22	12	6	15	10	32	17	60	80	96	224	87	72	21	607	563	624	563	1 148	2 423	1 264	2 667	3 835	7 948	3 553	5 313	8 007
23	12	7	15	11	24	10	66	86	108	218	92	78	21	601	522	646	551	1 156	2 493	1 334	2 603	3 749	7 977	3 391	5 336	8 190
24	6	7	.	.	.	.	21	35	63	56	56	63	21	196	162	280	224	399	1 099	729	910	1 344	3 430	1 217	2 444	4 206
25	2	1	4	3	8	4	18	24	30	72	26	22	6	216	199	230	207	442	961	516	1 074	1 610	3 334	1 500	2 256	3 465
26	1	1	.	.	.	.	5	10	16	16	16	17	6	66	57	88	74	149	397	254	365	570	1 364	526	984	1 657
27		1	.	.	.	.	.	.	6	.	5	12	6	.	.	22	17	.	75	104	.	.	212	.	190	528

1	1	2	1	4	5	6	22	5	4	1	71	69	76	73	160	347	186	420	664	1328	632	903	1395	28
	1	.	.	.	.	.	7	.	.	.	20	21	21	28	56	112	56	154	266	498	252	336	525	29
		1	1	.	.	.	6	.	.	.	21	28	21	27	49	106	56	161	266	497	281	343	519	30
			1	.	.	.	.	.	.	.	.	7	.	6	.	.	.	21	35	56	64	42	57	31
				1	2	3	4	3	3	1	20	18	25	22	51	131	81	139	231	510	220	369	609	32
					1	1	.	2	2	1	5	4	5	4	16	44	28	44	80	180	75	143	237	33
						1	.	1	2	1	.	.	5	4	.	22	28	.	.	70	.	67	170	34
							1	.	.	.	5	6	6	7	15	36	21	50	90	177	92	126	204	35
								1	1	1	.	.	.	.	.	6	6	.	.	20	.	27	54	36
									1	1	.	.	.	.	.	.	6	.	.	.	.	.	27	37
										1	.	.	.	.	.	.	.	.	.	.	.	.	.	38
											1	1	1	1	4	10	6	15	30	56	31	42	68	39
												1	.	1	.	.	.	5	10	16	16	11	17	40
													1	1	.	5	6	.	.	21	.	20	48	41
														1	.	.	.	.	.	6	.	5	12	42
															1	2	1	4	10	15	10	11	18	43
																1	1	.	.	5	.	6	12	44
																	1	.	.	.	.	.	6	45
																		1	3	3	4	2	3	46
																			1	.	1	.	.	47
																				1	.	1	2	48
																					1	.	.	49
																						1	1	50
																							1	51

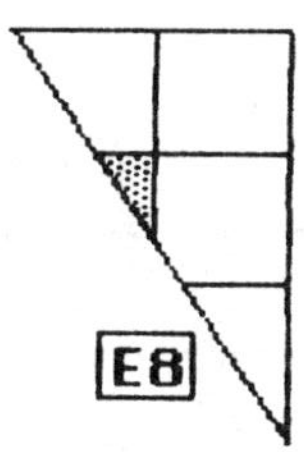

Table 3. *(continued)*

ALGEBRA E8, REPRESENTATIONS OF CLASS 0

NUMBER	52	53	54	55	56	57	58	59	60	61	62	63	64
S.P.	52	44	46	46	46	50	48	48	50	48	54	54	50
O.S.	30 240	2 419 200	138 240	967 680	1 451 520	30 240	60 480	967 680	362 880	2 903 040	6 720	181 440	967 680
LEVEL	325	329	331	333	335	335	337	337	341	343	343	343	345
DIM'N	152 883 490 500	919 045 960 000	560 213 725 500	1 198 018 560 000	1 283 242 632 840	417 532 087 000	627 099 023 250	1 349 926 375 875	1 813 461 073 920	3 233 052 753 920	492 957 660 000	1 041 872 676 000	3 191 795 712 000
WEIGHT	1 0 0 0 0 0 0 4	0 0 0 1 0 0 0 1	0 0 2 0 0 0 0 1	1 1 0 0 0 0 0 1	0 0 0 0 1 0 1 0	3 0 0 0 0 0 0 1	0 0 0 0 0 2 0 0	0 0 0 0 1 0 0 2	0 0 0 0 0 1 1 1	1 0 1 0 0 0 1 0	0 0 0 0 0 0 3 0	0 0 0 0 0 1 0 3	1 0 1 0 0 0 0 2
0	13 138 740	81 791 920	47 900 460	100 425 600	105 530 040	32 998 840	49 600 530	106 864 275	135 515 520	241 189 760	34 813 200	73 651 200	229 393 920
1	7 494 325	46 526 375	27 394 079	57 582 392	60 651 864	19 066 558	28 650 468	61 702 844	78 820 896	140 399 280	20 390 036	43 117 774	134 129 408
2	4 230 870	26 163 326	15 492 088	32 654 536	34 482 742	10 901 002	16 376 108	35 257 579	45 391 632	80 899 872	11 830 328	25 008 422	77 661 920
3	2 362 460	14 533 242	8 657 278	18 301 766	19 380 114	6 162 828	9 256 314	19 925 312	25 866 624	46 114 610	6 795 866	14 362 999	44 502 550
4	1 307 480	7 965 405	4 775 056	10 126 424	10 758 321	3 441 914	5 170 620	11 133 528	14 585 760	25 980 024	3 866 807	8 174 431	25 226 152
5	1 302 518	7 968 800	4 777 122	10 130 720	10 759 504	3 442 754	5 169 983	11 127 585	14 572 992	25 988 200	3 861 220	8 159 520	25 220 384
6	709 657	4 308 133	2 600 109	5 532 444	5 895 534	1 898 598	2 851 248	6 137 609	8 114 372	14 466 744	2 169 833	4 585 742	14 126 812
7	380 483	2 294 793	1 395 044	2 978 906	3 185 415	1 032 930	1 551 151	3 338 814	4 458 324	7 950 156	1 203 633	2 543 652	7 813 194

NUMBER	65	66	67	68	69	70	71	72	73	74	75	76	77
S.P.	56	62	72	50	52	52	52	54	56	58	54	54	56
O.S.	13 440	13 440	240	2 419 200	604 800	2 903 040	1 209 600	362 880	483 840	30 240	1 451 520	2 903 040	138 240
LEVEL	345	347	349	351	353	355	357	357	357	359	363	363	365
DIM'N	841 900 509 450	417 933 862 500	69 176 971 200	4 189 713 446 646	3 431 612 952 000	8 145 764 352 000	4 779 643 627 500	6 458 110 083 072	3 950 782 290 000	3 067 797 300 750	7 723 951 192 125	13 532 264 250 750	5 006 235 840 320
WEIGHT	0 0 0 0 0 0 2 2	0 0 0 0 0 0 1 1	0 0 0 0 0 0 0 6	0 1 0 0 0 1 0 0	2 0 0 0 0 1 0 0	0 1 0 0 0 0 1 1	0 0 1 0 1 0 0 0	2 0 0 0 0 0 1 1	0 1 0 0 0 0 0 3	2 0 0 0 0 0 0 3	1 0 0 1 0 0 0 0	0 0 1 0 0 1 0 1	1 0 2 0 0 0 0 0
0	57 329 370	26 483 940	3 944 640	284 273 766	224 294 400	523 163 520	301 048 860	399 745 152	240 453 360	180 027 630	452 225 565	793 083 150	282 730 640
1	33 717 159	15 702 659	2 365 910	167 487 957	132 768 567	310 260 304	179 032 182	238 153 680	143 486 357	107 910 522	271 362 267	475 688 205	170 401 077
2	19 652 766	9 234 988	1 409 486	97 740 776	77 862 340	182 332 816	105 502 677	140 632 304	84 890 400	64 143 842	161 424 172	282 878 246	101 831 710
3	11 346 703	5 384 755	833 609	56 464 074	45 214 668	106 129 248	61 575 982	82 272 042	49 771 261	37 793 556	95 149 386	166 706 532	60 312 186
4	6 495 944	3 119 431	491 352	32 262 048	25 976 889	61 156 704	35 562 317	47 662 776	28 920 493	22 074 786	55 524 015	97 312 473	35 372 183
5	6 481 601	3 108 196	488 384	32 273 759	25 985 680	61 147 504	35 578 200	47 654 384	28 897 596	22 056 685	55 551 160	97 311 389	35 390 076
6	3 664 451	1 777 914	284 047	18 236 719	14 768 724	34 854 516	20 334 286	27 315 708	16 611 724	12 748 078	32 098 974	56 233 686	20 557 927
7	2 045 237	1 004 133	163 129	10 183 034	8 297 185	19 638 884	11 492 717	15 482 348	9 441 858	7 287 321	18 353 353	32 154 040	11 820 527

NUMBER	78	79	80	81	82	83	84	85	86	87	88	89	90
S.P.	56	56	58	58	64	64	56	58	58	60	60	62	66
O.S.	69 120	483 840	138 240	967 680	2 160	138 240	4 838 400	1 814 400	2 903 040	1 814 400	967 680	362 880	362 880
LEVEL	365	365	367	367	369	369	371	375	377	377	379	379	381
DIM'N	3 754 721 200 320	8 715 491 428 800	4 591 371 033 000	12 737 135 385 000	976 217 435 000	4 007 202 600 000	26 125 438 976 000	29 369 472 656 250	33 699 815 424 000	33 943 999 320 000	28 854 026 240 000	26 461 348 084 080	19 994 148 864 000
WEIGHT	0 2 0 0 0 0 0 0	0 0 1 0 0 0 2 0	2 1 0 0 0 0 0 0	0 0 1 0 0 0 1 2	4 0 0 0 0 0 0 0	0 0 1 0 0 0 0 4	1 0 0 0 1 0 0 1	1 0 0 0 0 1 1 0	0 1 1 0 0 0 0 1	1 0 0 0 0 1 0 2	2 0 1 0 0 0 0 1	1 0 0 0 0 0 2 1	1 0 0 0 0 0 1 3
0	212 055 120	492 912 960	250 216 200	695 728 440	49 585 160	204 337 920	1 399 946 240	1 493 537 850	1 683 313 920	1 668 470 160	1 392 187 520	1 256 940 720	903 502 080
1	127 802 724	296 889 901	151 447 572	420 682 536	30 262 290	124 500 428	848 846 152	911 125 670	1 029 238 560	1 021 663 209	854 688 544	772 676 226	558 392 112
2	76 374 402	177 338 642	90 911 441	252 340 075	18 325 267	75 295 128	510 536 600	551 519 670	624 451 808	620 918 564	520 750 896	471 529 135	342 730 144
3	45 234 555	105 004 285	54 098 997	150 092 889	11 005 416	45 183 656	304 455 850	331 128 020	375 790 068	374 404 017	314 774 896	285 558 687	208 846 440
4	26 530 287	61 611 669	31 887 597	88 525 641	6 550 602	26 918 216	179 922 344	197 112 335	224 184 240	223 956 453	188 657 504	171 609 984	126 388 512
5	26 542 760	61 599 648	31 901 798	88 473 834	6 552 462	26 883 928	179 944 632	197 103 235	224 231 960	223 877 920	188 697 104	171 528 140	126 269 464
6	15 418 698	35 787 198	18 628 704	51 671 230	3 865 197	15 861 758	105 349 180	116 268 765	132 591 672	132 716 217	112 122 508	102 173 597	75 749 184
7	8 865 515	20 574 493	10 770 060	29 867 561	2 258 458	9 265 248	61 067 460	67 930 855	77 672 168	77 935 953	66 016 924	60 301 731	45 033 016

NUMBER	91	92	93	94	95	96	97	98	99	100	101	102	103
S.P.	74	60	62	62	62	62	64	66	64	64	68	66	68
O.S.	30 240	4 838 400	483 840	2 419 200	967 680	2 903 040	138 240	181 440	967 680	2 903 040	30 240	181 440	967 680
LEVEL	383	385	387	387	389	389	389	391	391	393	393	395	395
DIM'N	4 490 627 295 000	56 860 936 405 000	36 286 570 856 250	58 549 130 859 375	46 678 711 926 784	78 521 011 666 944	36 488 345 400 000	28 700 792 589 000	78 618 968 064 000	107 701 303 073 000	28 146 082 562 208	57 306 919 524 192	57 591 234 560 000
WEIGHT	1 0 0 0 0 0 0 5	0 0 0 1 0 0 1 0	0 0 2 0 0 0 1 0	0 0 0 1 0 0 0 2	0 0 0 0 1 1 0 0	1 1 0 0 0 0 1 0	0 0 2 0 0 0 0 2	3 0 0 0 0 0 1 0	1 1 0 0 0 0 0 2	0 0 0 0 1 0 1 1	3 0 0 0 0 0 0 2	0 0 0 0 0 2 0 1	0 0 0 0 1 0 0 3
0	186 964 680	2 607 030 280	1 608 180 570	2 596 700 175	2 034 586 624	3 421 845 504	1 564 673 760	1 189 593 720	3 315 392 640	4 468 367 960	1 129 321 008	2 300 461 632	2 276 337 920
1	116 556 902	1 609 861 968	996 964 710	1 609 280 295	1 263 637 424	2 125 375 168	973 450 890	743 080 114	2 066 556 240	2 790 115 174	707 882 445	1 441 735 722	1 428 666 288
2	72 210 362	986 868 624	613 647 907	990 309 300	779 301 264	1 310 822 464	601 446 954	461 011 656	1 279 344 864	1 730 478 946	440 792 494	897 625 112	890 920 736
3	44 444 108	600 354 146	374 893 194	604 919 805	477 070 988	802 492 168	368 922 816	283 979 948	786 357 214	1 065 738 624	272 585 994	555 029 982	551 875 368
4	27 199 704	362 266 304	227 213 490	366 681 525	289 780 400	487 440 352	224 594 091	173 608 001	479 747 912	651 597 891	167 357 397	340 774 875	339 587 360
5	27 154 638	362 314 680	227 251 449	366 646 190	289 810 504	487 515 896	224 580 884	173 631 402	479 722 936	651 502 782	167 347 824	340 715 872	339 430 664
6	16 476 120	216 816 054	136 620 047	220 429 320	174 627 656	293 758 032	135 630 805	105 328 847	290 368 456	395 217 657	101 962 698	207 590 716	207 258 976
7	9 909 962	128 611 485	81 434 462	131 383 385	104 336 728	175 517 624	81 225 867	63 376 423	174 302 420	237 785 528	61 629 293	125 469 181	125 543 240

	1	2	3	4	5	6	7	8	9	10	11	12	13	14	15	16	17	18	19	20	21	22	23	24	25	26
8	203 826	1 200 874	735 309	1 576 277	1 694 979	552 646	831 789	1 795 060	2 424 300	4 302 613	662 228	1 402 943	4 267 818	1 135 856	567 464	94 365	5 604 786	4 596 129	10 938 852	6 405 578	8 675 316	5 322 939	4 131 081	10 353 825	18 175 158	6 707 152
9	59 619	309 211	191 808	414 848	455 931	148 890	227 976	501 872	696 360	1 194 856	196 280	423 723	1 224 720	348 173	184 324	32 845	1 615 068	1 342 161	3 272 136	1 895 117	2 624 328	1 656 942	1 297 575	3 142 602	5 600 601	2 060 457
10	200 423	1 202 887	737 114	1 579 950	1 695 211	554 036	831 593	1 788 990	2 414 468	4 309 382	658 477	1 390 845	4 262 734	1 125 852	559 475	92 497	5 614 560	4 604 796	10 929 260	6 419 182	8 671 188	5 301 289	4 117 126	10 378 179	18 172 517	6 724 048
11	103 494	619 598	383 222	824 768	887 264	292 978	438 907	942 249	1 286 936	2 302 104	354 908	748 182	2 291 408	610 596	307 209	51 842	3 054 548	2 523 360	6 001 016	3 541 153	4 795 680	2 936 052	2 296 839	5 801 187	10 146 982	3 782 842
12	104 713	619 010	382 396	823 016	887 502	292 019	438 893	945 385	1 290 947	2 299 129	356 168	753 235	2 293 676	614 061	309 960	52 340	3 050 383	2 518 971	6 005 917	3 535 315	4 795 995	2 946 097	2 302 020	5 790 204	10 149 027	3 774 940
13	53 114	312 791	195 075	421 712	456 438	151 621	227 666	489 869	677 280	1 208 186	189 045	399 382	1 213 994	328 145	167 868	28 909	1 634 758	1 360 155	3 252 160	1 922 946	2 616 120	1 610 924	1 267 817	3 193 897	5 594 265	2 096 405
14	28 564	153 627	96 435	209 280	230 219	76 070	116 100	253 529	355 984	619 024	101 212	216 882	635 616	179 584	95 295	17 100	855 123	716 129	1 742 064	1 020 450	1 408 576	885 213	699 201	1 717 943	3 044 266	1 134 985
15	26 302	154 810	97 544	211 964	230 220	77 325	115 879	248 975	348 788	624 012	98 546	207 900	631 564	172 215	89 295	15 682	862 427	723 492	1 734 172	1 030 819	1 406 136	867 616	688 101	1 738 242	3 040 755	1 149 232
16	13 461	74 490	47 371	103 264	113 726	38 122	57 912	125 539	178 724	314 426	51 382	109 347	323 946	91 375	48 802	8 877	444 627	375 641	912 484	540 179	744 534	466 261	371 782	924 443	1 629 874	615 700
17	12 627	74 872	47 747	104 489	113 589	38 829	57 743	123 585	175 714	316 505	50 323	105 735	322 252	88 452	46 473	8 344	447 566	379 155	908 985	544 320	744 114	458 850	367 359	933 471	1 627 548	622 013
18	6 485	35 045	22 617	49 312	54 882	18 447	28 356	61 338	88 956	155 654	26 014	55 307	162 360	46 821	25 520	4 800	225 801	192 222	470 604	279 533	387 144	244 027	196 272	486 036	859 473	326 523
19	4 190	15 470	9 927	21 712	25 749	8 142	13 539	31 668	46 935	72 683	14 251	32 427	81 945	27 515	16 610	3 405	108 180	92 376	242 715	135 803	199 116	136 315	108 684	239 121	447 183	161 634
20	959	980	603	1 344	2 079	504	1 197	3 976	6 216	5 696	2 100	6 075	9 576	5 201	4 144	1 065	9 387	8 073	29 736	12 341	24 192	22 645	17 703	22 680	55 944	15 624
21	6 152	35 188	22 686	49 864	54 808	18 792	28 192	60 635	87 634	156 494	25 511	53 870	161 688	45 470	24 474	4 524	226 938	193 847	469 238	281 074	387 034	241 141	194 368	490 064	858 071	329 209

	27	28	29	30	31	32	33	34	35	36	37	38	39	40	41	42	43	44	45	46	47	48	49	50	51	52
8	5 030 640	11 708 831	6 147 207	17 123 394	1 303 754	5 386 180	35 010 175	39 295 285	45 008 352	45 383 608	38 457 708	35 314 056	26 623 917	5 945 898	75 545 020	48 069 324	77 640 990	61 758 962	103 864 760	48 228 201	37 778 045	103 743 299	141 933 510	36 940 404	75 252 723	75 568 321
9	1 545 852	3 672 823	1 913 352	5 501 469	416 050	1 805 916	11 093 656	12 731 055	14 570 640	15 007 517	12 587 512	11 837 574	9 184 392	2 138 941	25 234 231	16 214 562	26 398 995	20 983 760	35 220 864	16 546 923	13 005 650	35 768 600	49 340 109	12 916 533	26 431 761	26 907 944
10	5 043 104	11 697 057	6 162 053	17 074 120	1 306 748	5 353 630	35 028 378	39 288 250	45 053 382	45 309 712	38 499 500	35 242 214	26 515 278	5 905 319	75 586 449	48 106 074	77 595 865	61 786 662	103 944 610	48 214 985	37 811 116	103 719 434	141 826 951	36 940 132	75 195 378	75 399 482
11	2 837 352	6 570 312	3 488 347	9 642 465	748 862	3 055 796	19 869 584	22 479 115	25 863 936	26 058 298	22 229 512	20 381 004	15 448 656	3 483 109	43 991 172	28 151 641	45 381 945	36 247 904	60 991 616	28 352 618	22 362 669	61 151 272	83 810 419	21 950 290	44 662 162	44 865 296
12	2 831 254	6 576 029	3 480 722	9 665 250	746 921	3 070 502	19 862 269	22 481 830	25 843 476	26 091 450	22 208 262	20 411 076	15 495 108	3 499 888	43 974 679	28 135 154	45 406 845	36 236 431	60 952 164	28 358 890	22 342 824	61 161 093	83 864 732	21 945 966	44 688 392	44 947 181
13	1 572 316	3 648 343	1 945 651	5 394 504	422 810	1 734 028	11 132 166	12 717 075	14 668 000	14 842 937	12 681 040	11 679 606	8 938 770	2 045 632	25 326 409	16 297 051	26 291 715	21 046 324	35 404 456	16 513 261	13 084 917	35 711 286	49 092 798	12 917 758	26 298 901	26 509 530
14	851 352	2 006 557	1 060 439	3 012 468	233 461	994 308	6 145 568	7 111 895	8 188 960	8 407 627	7 119 152	6 667 059	5 193 704	1 218 833	14 391 371	9 307 757	15 104 230	12 072 288	20 274 176	9 530 038	7 551 036	20 658 768	28 548 213	7 525 416	15 378 820	15 641 704
15	861 902	1 996 628	1 074 105	2 970 403	236 572	966 416	6 160 460	7 106 025	8 227 164	8 342 182	7 158 888	6 604 725	5 096 832	1 182 213	14 426 580	9 339 874	15 058 910	12 095 536	20 350 892	9 514 856	7 587 184	20 637 180	28 444 171	7 529 258	15 322 664	15 478 452
16	461 776	1 080 739	579 347	1 628 275	129 341	541 684	3 357 212	3 921 575	4 541 724	4 654 068	3 975 756	3 712 754	2 908 466	689 479	8 108 527	5 280 129	8 544 640	6 865 544	11 536 648	5 428 321	4 336 968	11 803 750	16 344 901	4 340 187	8 857 319	9 007 690
17	466 515	1 076 180	586 082	1 609 860	131 049	529 832	3 363 542	3 918 825	4 558 197	4 625 257	3 995 313	3 685 815	2 866 674	673 911	8 122 808	5 293 561	8 522 745	6 874 350	11 573 198	5 420 255	4 356 485	11 795 497	16 295 265	4 344 667	8 830 380	8 933 260
18	244 902	574 731	309 234	873 711	70 059	295 396	1 804 464	2 133 520	2 478 240	2 552 828	2 183 364	2 051 694	1 625 544	392 109	4 505 346	2 953 800	4 782 940	3 854 416	6 471 824	3 059 136	2 455 095	6 669 354	9 270 669	2 474 217	5 056 410	5 160 136
19	121 236	306 432	154 146	488 064	35 402	177 420	945 495	1 146 570	1 307 736	1 423 668	1 156 629	1 159 209	963 774	246 801	2 445 633	1 608 624	2 675 320	2 119 234	3 531 348	1 713 096	1 347 531	3 738 989	5 268 489	1 391 646	2 891 790	3 043 390
20	11 844	41 783	15 372	80 193	3 750	36 960	120 120	160 140	169 344	231 364	151 256	196 056	193 536	59 619	351 435	233 289	435 400	318 864	515 424	279 567	200 242	611 072	901 971	229 152	505 581	597 632
21	246 900	572 704	312 494	865 915	70 899	290 558	1 807 356	2 132 035	2 485 084	2 540 193	2 192 864	2 039 552	1 607 096	385 211	4 511 276	2 959 036	4 773 365	3 857 362	6 488 980	3 054 543	2 464 987	6 666 576	9 247 986	2 476 997	5 042 890	5 127 600

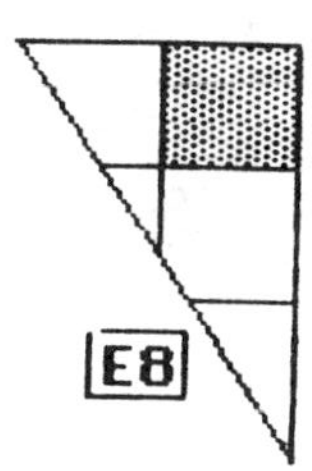

Table 3. *(continued)*

ALGEBRA E8, REPRESENTATIONS OF CLASS 0

	52	53	54	55	56	57	58	59	60	61	62	63	64
	1	0	0	1	0	3	0	0	0	1	0	0	1
	0	0	0	1	0	0	0	0	0	0	0	0	0
	0 0	0 1	2 0	0 0	0 0	0 0	0 0	0 0	0 0	1 0	0 0	0 0	1 0
	0	0	0	0	1	0	0	1	0	0	0	0	0
	0	0	0	0	0	0	2	0	1	0	0	1	0
	0	0	0	0	1	0	0	0	1	1	3	0	0
WEIGHT	4	1	1	1	0	1	0	2	1	0	0	3	2
NUMBER	52	53	54	55	56	57	58	59	60	61	62	63	64
22	2 713	16 098	10 545	23 616	25 633	9 226	13 334	28 319	41 756	76 268	12 381	25 762	78 744
23	2 812	16 110	10 543	23 292	25 747	8 939	13 413	28 765	42 316	75 848	12 515	26 365	79 118
24	1 687	6 930	4 543	9 982	11 788	3 822	6 272	14 280	21 567	34 727	6 623	14 700	38 724
25	1 203	7 145	4 752	10 656	11 709	4 214	6 175	13 134	19 688	35 954	5 918	12 363	37 650
26	658	2 980	2 004	4 432	5 207	1 746	2 812	6 229	9 634	16 097	3 008	6 507	17 803
27	323	390	248	544	879	203	510	1 584	2 572	2 528	890	2 412	4 036
28	487	3 059	2 078	4 712	5 144	1 918	2 753	5 801	8 860	16 552	2 702	5 608	17 440
29	182	1 260	890	2 016	2 156	847	1 190	2 436	3 808	7 384	1 176	2 408	7 840
30	182	1 239	855	2 064	2 142	910	1 162	2 401	3 772	7 426	1 182	2 393	7 770
31	21	154	105	384	294	238	168	315	560	1 344	202	356	1 344
32	238	1 230	848	1 904	2 204	777	1 207	2 610	4 119	7 225	1 302	2 760	7 932
33	101	470	338	736	907	303	511	1 089	1 810	3 072	600	1 247	3 430
34	101	145	96	208	351	79	209	594	1 016	1 072	363	912	1 622

	65	66	67	68	69	70	71	72	73	74	75	76	77
	0	0	0	0	2	0	0	2	0	2	1	0	1
	0	0	0	1	0	1	0	0	1	0	0	0	0
	0 0	0 0	0 0	0 0	0 0	0 0	1 0	0 0	0 0	0 0	0 1	1 0	2 0
	0	0	0	0	0	0	1	0	0	0	0	0	0
	0	0	0	1	1	0	0	0	0	0	0	1	0
	2	1	0	0	0	1	0	1	0	0	0	0	0
WEIGHT	2	4	6	0	0	1	0	1	3	3	0	1	0
NUMBER	65	66	67	68	69	70	71	72	73	74	75	76	77
22	22 083	12 021	2 289	113 587	98 631	236 236	143 600	197 928	121 849	99 743	256 149	443 072	173 650
23	22 500	12 315	2 334	113 100	97 576	237 046	142 995	197 462	123 341	100 399	254 041	444 233	172 218
24	12 642	7 560	1 554	53 354	46 102	119 314	68 684	99 170	66 493	53 788	123 592	226 898	84 492
25	10 683	5 919	1 148	55 260	48 390	116 928	71 468	98 778	61 311	50 583	129 939	225 222	88 951
26	5 695	3 397	709	25 611	22 446	57 266	33 876	48 340	31 779	26 146	62 453	112 660	43 212
27	2 107	1 632	414	4 366	3 783	13 436	5 976	11 052	9 896	7 849	11 265	26 544	7 916
28	4 906	2 748	544	26 335	23 401	56 298	34 931	48 266	29 776	24 934	65 089	111 829	45 030
29	2 163	1 218	252	12 229	11 040	26 320	16 702	23 016	14 000	11 997	31 941	54 235	22 372
30	2 142	1 218	252	12 243	11 250	26 258	16 660	23 184	13 895	11 949	32 235	53 886	22 520
31	357	217	56	2 401	2 548	5 104	3 416	5 096	2 660	2 548	7 525	11 214	5 314
32	2 449	1 455	306	11 955	10 669	26 735	16 290	22 979	14 787	12 391	30 904	54 577	21 648
33	1 142	707	159	5 311	4 766	12 234	7 490	10 600	6 937	5 874	14 493	25 925	10 330
34	824	622	159	1 957	1 719	5 848	2 800	4 892	4 166	3 371	5 439	12 210	3 903

	78	79	80	81	82	83	84	85	86	87	88	89	90
	0	0	2	0	4	0	1	1	0	1	2	1	1
	2	0	1	0	0	0	0	0	1	0	0	0	0
	0 0	1 0	0 0	1 0	0 0	1 0	0 0	0 0	1 0	0 0	1 0	0 0	0 0
	0	0	0	0	0	0	1	0	0	0	0	0	0
	0	0	0	0	0	0	0	1	0	1	0	0	0
	0	2	0	1	0	0	0	1	0	0	0	2	1
WEIGHT	0	0	0	2	0	4	1	0	1	2	1	1	3
NUMBER	78	79	80	81	82	83	84	85	86	87	88	89	90
22	130 292	297 895	166 953	451 423	38 685	152 770	957 904	1 141 605	1 340 688	1 364 158	1 195 348	1 103 452	874 044
23	129 159	298 983	164 843	455 486	38 014	155 080	956 688	1 142 130	1 337 504	1 370 430	1 189 330	1 108 574	882 244
24	63 309	156 205	81 200	247 576	18 956	89 720	493 521	603 890	697 079	748 048	622 089	613 004	508 465
25	66 738	152 807	86 160	233 889	20 276	80 541	498 192	601 622	709 668	725 397	637 328	591 282	474 042
26	32 356	78 032	41 938	123 356	9 985	44 742	252 637	312 445	364 864	386 832	328 926	319 389	265 030
27	5 938	19 957	7 756	37 289	1 899	16 832	59 240	79 820	86 136	113 108	77 668	96 803	93 696
28	33 817	76 462	44 154	117 705	10 614	41 003	254 604	311 220	370 196	377 347	335 826	309 956	250 510
29	16 856	37 268	22 232	57 680	5 487	20 244	127 512	157 760	189 840	192 276	174 048	159 096	129 528
30	16 974	37 190	22 679	57 281	5 647	20 138	127 631	157 795	189 882	191 604	175 267	158 844	128 934
31	4 108	7 958	6 005	12 159	1 687	4 368	29 960	37 905	46 928	45 780	46 096	38 927	31 696
32	16 215	38 030	21 297	60 018	5 191	21 810	126 726	158 410	187 493	196 135	171 054	163 136	135 523
33	7 688	18 380	10 156	29 373	2 511	10 970	61 770	78 905	93 488	98 926	85 894	83 397	70 494
34	2 908	9 246	3 834	16 841	955	7 460	28 462	38 820	42 808	54 022	39 064	46 756	44 460

	91	92	93	94	95	96	97	98	99	100	101	102	103
	1	0	0	0	0	1	0	3	1	0	3	0	0
	0	0	0	0	0	1	0	0	1	0	0	0	0
	0 0	0 1	2 0	0 1	0 0	0 0	2 0	0 0	0 0	0 0	0 0	0 0	0 0
	0	0	0	0	1	0	0	0	0	1	0	0	1
	0	0	0	0	1	0	0	0	0	0	0	2	0
	0	1	1	0	0	1	0	1	0	1	0	0	0
WEIGHT	5	0	0	2	0	0	2	0	2	1	2	1	3
NUMBER	91	92	93	94	95	96	97	98	99	100	101	102	103
22	212 092	2 476 141	1 637 017	2 628 645	2 139 072	3 607 840	1 695 481	1 388 194	3 718 996	5 163 009	1 399 644	2 834 406	2 878 604
23	214 874	2 474 119	1 634 850	2 635 235	2 138 402	3 597 528	1 698 613	1 380 743	3 719 744	5 176 579	1 396 113	2 841 750	2 896 460
24	130 473	1 326 801	880 852	1 450 305	1 162 378	1 940 757	937 685	749 308	2 054 829	2 897 202	774 361	1 601 740	1 673 880
25	117 065	1 338 143	891 483	1 432 278	1 169 536	1 970 976	930 759	765 972	2 047 734	2 854 799	778 116	1 578 276	1 609 082
26	68 429	708 055	474 689	774 570	627 726	1 050 364	505 864	410 880	1 113 591	1 571 483	425 402	875 617	909 974
27	28 564	182 731	122 993	223 080	168 080	272 112	145 274	106 993	318 752	471 404	121 317	266 760	308 064
28	62 740	712 508	478 890	766 875	630 284	1 063 898	502 795	418 866	1 111 282	1 552 177	427 704	864 646	881 820
29	32 844	373 080	253 281	403 585	334 376	565 712	267 492	225 974	593 880	830 032	231 896	466 332	474 880
30	32 753	372 981	253 098	402 360	333 970	567 588	266 379	227 871	594 083	828 135	233 059	465 038	472 940
31	8 358	96 166	66 555	103 600	88 368	155 008	70 036	66 248	161 728	221 718	67 368	126 798	127 504
32	35 226	371 430	251 588	406 840	333 446	559 791	268 530	222 265	594 447	838 944	230 545	471 242	487 250
33	18 848	190 608	130 457	211 110	173 804	290 600	140 697	116 581	312 096	444 218	122 350	251 909	261 814
34	13 461	92 991	63 518	111 995	86 836	140 968	74 079	56 285	163 312	241 813	63 275	138 293	156 124

77	480	341	800	878	350	489	1026	1656	3133	531	1098	3405	993	588	126	5416	4980	12065	7614	10677	6629	5718	15009	25651	10636	8000	17967	10724	28286	2705	10284	62187	78520	94518	97116	87672	81393	67635	17703	191184	130943	209505	173903	293946	139901	119021	312285	439460	123485	249006	255840	35
34	45	36	64	135	24	84	206	408	432	163	351	603	347	264	74	828	729	2448	1260	2079	1697	1413	2484	5454	1857	1338	4234	1764	7500	440	3312	13188	18480	20574	25360	18928	22455	21141	6485	46167	32112	55070	43912	70980	37071	28679	81824	122307	32250	70890	78513	36
34	.	.	.	.	.	.	71	111	.	45	216	171	184	219	74	.	.	531	.	432	806	630	.	999	.	.	934	.	2850	.	1962	2145	3735	3024	8222	2701	7470	10260	4190	8529	5733	15470	8694	12708	9927	5009	21712	35424	8142	20694	31668	37
7	.	.	.	.	.	.	.	.	.	.	27	.	21	56	29	.	.	.	.	.	77	63	.	.	.	.	.	.	225	.	336	.	.	.	588	.	504	1512	959	.	.	980	.	.	603	.	1344	2079	504	1197	3976	38
27	175	129	300	339	134	193	401	670	1282	221	453	1410	417	253	56	2333	2174	5312	3410	4764	2973	2599	6915	11797	4985	3736	8377	5061	13317	1303	4944	29620	38105	46234	47591	43332	40337	33972	9087	96245	66675	106415	88952	150216	71843	61655	161026	227803	64551	130322	134240	39
6	55	40	128	109	74	64	125	220	512	76	146	544	142	87	21	983	1014	2188	1472	2142	1194	1128	3265	5076	2355	1810	3631	2606	5714	729	2114	13830	17905	22300	22154	21780	18968	15866	4278	47346	33173	52055	44356	76768	35602	32830	81696	113678	34043	65685	66834	40
27	50	34	78	129	32	80	206	370	430	135	318	620	297	218	56	845	762	2434	1262	2094	1673	1394	2565	5406	1868	1402	4104	1885	7312	488	3174	13310	18330	20784	25065	19304	21901	20472	6152	46218	32044	54960	43856	71730	36964	29269	82144	121253	32553	70122	77400	41
6	15	10	32	39	17	30	60	120	160	45	96	224	102	72	21	350	342	952	538	888	624	563	1185	2268	863	682	1709	954	2994	273	1264	6040	8370	9840	11238	9448	9904	9096	2713	22319	15691	26175	21552	35968	17962	15249	40576	59048	16681	34609	37224	42
7	60	48	108	120	50	70	141	242	503	81	163	555	153	93	21	972	924	2227	1490	2049	1252	1128	3132	5229	2303	1722	3736	2365	5962	626	2229	13749	17950	22185	22524	21081	19257	16329	4422	47449	33333	52725	44586	75532	36099	31563	81399	115166	33232	66577	68350	43
7	15	11	24	45	10	29	66	131	162	51	108	218	107	78	21	343	315	967	545	849	646	551	1149	2310	862	638	1779	869	3098	230	1334	5998	8415	9759	11352	9200	10075	9315	2812	22382	15794	26370	21632	35469	18057	14745	40354	59722	16336	34969	37890	44
7	.	.	.	.	.	.	21	35	.	15	63	56	56	63	21	.	.	196	.	162	280	224	.	399	.	.	385	.	1099	.	729	910	1645	1344	3430	1217	3199	4206	1687	4004	2744	6930	4193	6097	4543	2443	9982	16548	3822	9842	14280	45
1	18	15	40	36	22	21	42	72	188	24	48	204	45	27	6	387	396	880	621	864	490	468	1404	2187	1046	790	1572	1132	2500	318	936	6204	8174	10392	10236	10188	8820	7476	2040	22794	16239	25323	21760	37476	17592	16176	40356	56482	17002	32992	33612	46
.	5	6	12	9	7	5	10	16	66	5	10	72	9	5	1	144	152	326	253	340	181	188	596	885	463	342	631	496	1002	143	371	2692	3580	4738	4500	4728	3889	3308	904	10667	7767	11890	10374	18016	8435	7960	19514	27057	8437	16002	16166	47
1	4	3	8	13	4	9	18	38	56	15	30	72	31	22	6	133	129	358	223	332	230	207	507	930	385	291	714	411	1223	116	516	2620	3702	4472	4920	4338	4399	4010	1203	10523	7556	12274	10360	17256	8572	7412	19432	28449	8122	16869	17898	48
.	.	.	6	.	7	.	.	.	21	.	.	21	.	.	.	49	77	106	84	154	56	77	266	294	197	160	216	280	336	98	126	1134	1505	2065	1848	2331	1624	1358	378	4746	3471	5215	4683	8799	3731	4298	9331	12208	4424	7308	7251	49
1	.	.	.	5	.	4	5	20	16	11	16	16	21	17	6	44	40	132	80	114	88	74	180	373	146	104	322	144	547	40	254	1050	1635	1860	2204	1788	2113	1992	658	4759	3481	5645	4844	7776	4004	3323	8944	13852	3739	8373	8984	50
1	.	.	.	.	.	.	5	10	.	5	16	16	16	17	6	.	.	66	.	57	88	74	.	149	.	.	150	.	397	.	254	365	695	570	1364	526	1319	1657	658	1809	1269	2980	1958	2832	2004	1163	4432	7482	1746	4552	6229	51
1	.	.	.	.	.	.	.	.	.	.	6	.	5	12	6	.	.	.	.	.	22	17	.	.	.	.	.	.	75	.	104	.	.	.	212	.	190	528	323	.	.	390	.	.	248	.	544	879	203	510	1584	52
	1	1	2	3	1	2	4	8	18	3	6	22	6	4	1	47	48	122	88	120	76	73	213	354	169	125	269	181	452	53	186	1096	1552	1986	2040	1978	1826	1644	487	4797	3529	5548	4812	8142	3969	3606	9118	13147	3940	7893	8212	53
		1	.	.	.	.	.	.	6	.	.	7	.	.	.	14	15	35	35	42	21	28	84	126	77	49	91	70	147	20	56	434	610	861	798	875	707	630	182	2107	1624	2415	2156	3710	1800	1694	4172	5817	1883	3555	3591	54

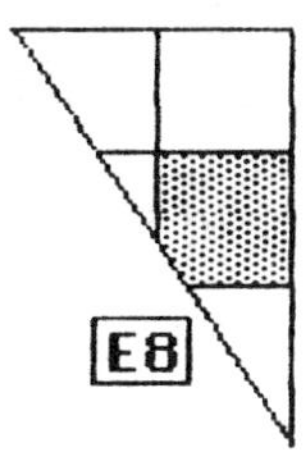

Table 3. *(continued)*

ALGEBRA E8, REPRESENTATIONS OF CLASS 0

NUMBER	52	53	54	55	56	57	58	59	60	61	62	63	64	65	66	67
	1	0	0	1	0	3	0	0	0	1	0	0	1	0	0	0
	0	0	0	1	0	0	0	0	0	0	0	0	0	0	0	0
	0 0	0 1	2 0	0 0	0 0	0 0	0 0	0 0	0 0	1 0	0 0	0 0	1 0	0 0	0 0	0 0
	0	0	0	0	1	0	0	1	0	0	0	0	0	0	0	0
	0	0	0	0	0	0	2	0	1	0	0	1	0	0	0	0
	0	0	0	0	1	0	0	0	1	1	3	0	0	2	1	0
WEIGHT	4	1	1	1	0	1	0	2	1	0	0	3	2	2	4	6
55				1	.	1	.	.	.	5	.	.	6	.	.	.
56					1	.	1	1	4	4	2	3	4	4	3	1
57						1	.	.	.	.	.	.	.	.	.	.
58							1	.	2	.	1	1	.	3	2	1
59								1	2	.	1	3	4	3	3	1
60									1	.	1	1	.	2	2	1
61										1	.	.	1	.	.	.
62											1	.	.	1	1	1
63												1	.	1	2	1
64													1	.	.	.
65														1	1	1
66															1	1
67																1
68																
69																
70																
71																
72																

NUMBER	68	69	70	71	72	73	74	75	76	77	78	79	80	81	82	83	84	85
	0	2	0	0	2	0	2	1	0	1	0	0	2	0	4	0	1	1
	1	0	1	0	0	1	0	0	0	0	2	0	1	0	0	0	0	0
	0 0	0 0	0 0	1 0	0 0	0 0	0 0	0 1	1 0	2 0	0 0	1 0	0 0	1 0	0 0	1 0	0 0	0 0
	0	0	0	1	0	0	0	0	0	0	0	0	0	0	0	0	1	0
	1	1	0	0	0	0	0	0	1	0	0	0	0	0	0	0	0	1
	0	0	1	0	1	0	0	0	0	0	0	2	0	1	0	0	0	1
WEIGHT	0	0	1	0	1	3	3	0	1	0	0	0	0	2	0	4	1	0
55	15	21	36	29	48	21	27	90	114	70	56	85	93	141	32	56	440	615
56	15	14	40	30	36	25	22	15	133	61	46	109	68	184	20	81	422	843
57	.	6	.	.	12	.	6	15	.	10	10	.	32	.	17	.	60	80
58	5	4	10	10	8	5	4	25	48	20	20	38	26	66	10	28	152	250
59	.	.	20	.	18	25	22	.	51	.	.	53	.	131	.	81	139	275
60	.	.	5	.	4	5	4	.	16	.	.	22	.	44	.	28	44	110
61	4	5	10	10	12	6	7	30	40	26	19	31	30	51	10	21	160	235
62	.	.	.	.	.	.	.	.	.	.	.	6	.	6	.	6	.	20
63	.	.	.	.	.	5	4	.	.	.	.	.	.	22	.	28	.	.
64	.	.	5	.	6	6	7	.	15	.	.	15	.	36	.	21	50	100
65	.	.	.	.	.	.	.	.	.	.	.	.	.	6	.	6	.	.
66	.	.	.	.	.	.	.	.	.	.	.	.	.	.	.	6	.	.
67	.	.	.	.	.	.	.	.	.	.	.	.	.	.	.	.	.	.
68	1	1	2	3	2	1	1	10	12	9	7	9	11	15	4	6	54	81
69		1	.	.	2	.	1	4	.	3	3	.	8	.	4	.	18	24
70			1	.	1	1	1	.	4	.	.	5	.	10	.	6	15	35
71				1	.	.	.	3	3	4	2	2	3	3	1	1	16	23
72					1	.	1	.	.	.	.	.	.	.	.	.	5	10

NUMBER	86	87	88	89	90	91	92	93	94	95	96	97	98	99	100	101	102	103
	0	1	2	1	1	1	0	0	0	0	1	0	3	1	0	3	0	0
	1	0	0	0	0	0	0	0	0	0	1	0	0	1	0	0	0	0
	1 0	0 0	1 0	0 0	0 0	0 0	0 1	2 0	0 1	0 0	0 0	2 0	0 0	0 0	0 0	0 0	0 0	0 0
	0	0	0	0	0	0	0	0	0	1	0	0	0	0	1	0	0	1
	0	1	0	0	0	0	0	0	0	1	0	0	0	0	0	0	2	0
	0	0	0	2	1	0	1	1	0	0	1	0	1	0	1	0	0	0
WEIGHT	1	2	1	1	3	5	0	0	2	0	0	2	0	2	1	2	1	3
55	846	792	926	708	624	182	2077	1554	2370	2121	3830	1731	1844	4214	5743	1970	3498	3546
56	808	833	866	868	740	230	2433	1573	2468	2108	3596	1796	1680	4876	6131	1771	3766	3020
57	120	96	224	87	72	21	304	225	330	328	832	240	563	864	869	563	558	520
58	320	330	310	326	304	101	885	678	1035	960	1552	796	704	1744	2718	767	1713	1744
59	231	510	220	506	609	238	784	562	1230	879	1272	848	539	1904	3246	777	2020	2610
60	80	180	75	210	237	101	325	242	470	384	528	338	224	736	1402	303	904	1089
61	335	307	359	282	254	77	879	682	1015	927	1625	766	778	1821	2563	852	1588	1611
62	.	20	.	54	54	34	45	36	45	64	64	36	24	64	270	24	192	206
63	.	70	.	67	170	101	.	.	145	.	.	96	.	208	351	79	209	594
64	90	177	92	176	204	77	319	235	480	374	556	341	252	800	1323	350	849	1026
65	.	20	.	27	54	34	.	.	45	.	.	36	.	64	135	24	84	206
66	.	.	.	.	27	34	.	.	.	.	.	.	.	.	.	.	.	71
67	.	.	.	.	.	7	.	.	.	.	.	.	.	.	.	.	.	.
68	128	105	138	98	88	27	353	285	408	384	678	322	334	758	1066	364	672	672
69	44	30	72	26	22	6	120	96	135	132	312	106	199	336	360	207	231	222
70	30	56	31	62	68	27	125	95	175	154	222	129	102	300	534	134	353	401
71	48	30	52	27	24	7	134	120	153	148	268	132	136	304	405	154	258	252
72	10	16	16	16	17	6	40	30	55	54	96	40	57	128	174	74	124	125

1	1	.	.	.	.	.	.	5	.	6	.	.	.	21	.	20	48	27	.	.	50	.	.	34	.	78	129	32	80	206	73
	1	.	.	.	.	.	.	.	.	.	.	.	.	6	.	5	12	6	.	.	15	.	.	10	.	32	39	17	30	60	74
		1	.	1	1	.	2	.	1	.	4	5	16	6	22	5	4	1	44	41	50	48	112	46	69	122	130	73	82	80	75
			1	.	.	1	.	2	.	1	4	10	10	15	10	17	18	7	45	37	60	59	84	48	40	108	195	50	134	141	76
				1	.	.	.	.	.	.	.	.	6	.	7	.	.	.	14	21	15	14	35	21	21	42	35	28	21	20	77
					1	.	1	.	1	.	.	.	6	.	6	.	.	.	14	15	15	14	42	21	28	42	35	27	21	20	78
						1	.	1	.	1	.	5	.	5	.	12	12	7	15	11	15	24	24	11	10	24	90	10	68	66	79
							1	.	1	.	.	.	.	.	6	.	.	.	.	.	.	.	21	.	28	21	.	27	.	.	80
								1	.	1	.	.	.	5	.	6	12	7	.	.	15	.	.	11	.	24	45	10	29	66	81
									1	.	.	.	.	.	.	.	.	.	.	.	.	.	.	.	7	.	.	6	.	.	82
										1	.	.	.	.	.	.	6	7	.	.	.	.	.	.	.	.	.	.	.	21	83
											1	2	3	3	4	3	3	1	14	12	18	19	32	15	18	40	60	22	42	42	84
												1	.	1	.	2	2	1	4	3	4	8	8	3	4	8	26	4	22	18	85
													1	.	1	.	.	.	4	5	5	5	10	6	6	12	15	7	10	10	86
														1	.	1	2	1	.	.	4	.	.	3	.	8	13	4	9	18	87
															1	.	.	.	.	.	.	.	5	.	6	6	.	7	.	.	88
																1	1	1	.	.	.	.	.	.	.	.	5	.	4	5	89
																	1	1	.	.	.	.	.	.	.	.	.	.	.	5	90
																		1	.	.	.	.	.	.	.	.	.	.	.	.	91
																			1	1	1	2	2	1	1	2	6	1	5	4	92
																				1	.	.	.	1	.	.	.	.	.	.	93
																					1	.	.	1	.	2	3	1	2	4	94
																						1	.	.	.	.	2	.	3	1	95
																							1	.	1	1	.	1	.	.	96
																								1	.	.	.	.	.	.	97
																									1	.	.	1	.	.	98
																										1	.	1	.	.	99
																											1	.	1	1	100
																												1	.	.	101
																													1	.	102
																														1	103

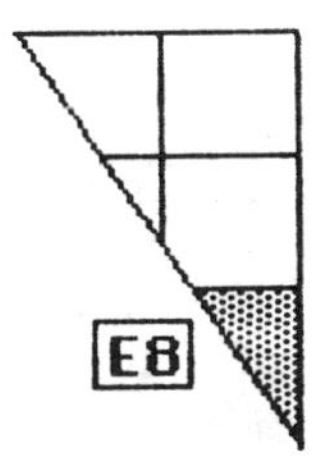

Table 3. *(continued)*

ALGEBRA F4, REPRESENTATIONS OF CLASS 0

NUMBER	0	1	2	3	4	5	6	7	8	9	10	11	12	13	14	15	16	17	18	19	20	21	22	23	24	25
S.P.	0	2	4	6	8	10	12	16	14	18	18	20	22	24	26	26	28	32	36	30	34	34	38	36	38	40
O.S.	1	24	24	96	24	144	96	24	192	24	288	144	288	96	144	192	192	24	24	576	144	288	288	288	192	144
LEVEL	1	17	23	31	33	39	43	45	47	49	53	55	59	61	61	63	65	65	67	69	71	73	75	75	77	77
DIM'N	1	26	52	273	324	1 053	1 274	1 053	4 096	2 652	8 424	10 829	19 278	19 448	17 901	34 749	29 172	16 302	12 376	106 496	76 076	107 406	119 119	160 056	205 751	160 056
WEIGHT	0 0 > 0 0	0 0 > 0 1	1 0 > 0 0	0 0 > 1 0	0 0 > 0 2	1 0 > 0 1	0 1 > 0 0	2 0 > 0 0	0 0 > 1 1	0 0 > 0 3	1 0 > 1 0	1 0 > 0 2	0 1 > 0 1	0 0 > 2 0	2 0 > 0 1	0 0 > 1 2	1 1 > 0 0	0 0 > 0 4	3 0 > 0 0	1 0 > 1 1	1 0 > 0 3	0 1 > 1 0	2 0 > 1 0	0 1 > 0 2	0 0 > 2 1	2 0 > 0 2
0	1	2	4	9	12	21	26	21	64	36	96	125	174	176	141	285	228	126	88	704	452	606	607	888	1 055	816
1		1	1	5	5	14	13	9	40	24	66	77	126	117	106	201	147	84	53	512	344	459	467	645	800	589
2			1	2	3	6	10	8	24	13	40	56	80	84	66	138	118	62	47	368	239	330	338	496	592	465
3				1	1	4	4	3	14	9	27	32	57	54	50	95	72	41	27	262	181	247	258	351	444	329
4					1	1	3	3	8	5	15	23	33	39	29	64	58	32	24	184	122	173	181	267	324	260
5						1	1	1	4	3	10	12	24	23	23	42	34	19	14	128	92	128	139	184	238	179
6							1	1	2	1	5	9	13	16	12	27	28	14	13	88	59	87	96	140	169	141
7								1	.	.	2	3	4	6	6	9	14	6	8	40	26	42	52	68	83	78
8									1	1	3	4	9	9	9	18	14	9	6	59	45	63	71	93	124	93
9										1	.	1	3	3	3	8	5	5	3	24	23	28	32	45	63	45
10											1	1	3	3	4	6	6	3	3	26	20	30	37	44	60	48
11												1	1	2	1	4	5	3	3	16	12	18	22	34	41	37
12													1	1	1	2	2	1	1	10	9	13	16	20	29	22
13														1	.	1	2	1	1	6	4	8	9	15	19	18
14															1	.	1	.	1	4	3	5	10	7	11	12
15																1	.	1	.	3	4	5	6	9	13	9

NUMBER	26	27	28	29	30	31	32	33	34	35	36	37	38	39	40	41	42	43	44	45	46	47	48	49	50	51
S.P.	42	42	50	44	50	48	46	52	52	64	50	54	54	54	56	58	58	62	66	60	72	62	68	66	66	68
O.S.	192	576	24	288	144	96	576	192	144	24	576	96	576	288	192	144	576	192	288	576	24	576	144	288	576	288
LEVEL	79	81	81	83	83	85	85	87	87	89	89	91	91	91	93	93	95	95	97	97	97	99	99	101	101	103
DIM'N	212 992	379 848	81 081	420 147	184 756	226 746	787 644	340 119	412 776	100 776	1 118 208	629 356	1 327 104	952 952	1 341 522	1 002 456	1 801 371	1 042 899	1 074 944	2 792 556	342 056	3 921 372	1 484 406	2 488 563	4 313 088	3 508 596
WEIGHT	0 0 > 1 3	1 1 > 0 1	0 0 > 0 5	1 0 > 2 0	3 0 > 0 1	0 2 > 0 0	1 0 > 1 2	2 1 > 0 0	1 0 > 0 4	4 0 > 0 0	0 1 > 1 1	0 0 > 3 0	2 0 > 1 1	0 1 > 0 3	0 0 > 2 2	2 0 > 0 3	1 1 > 1 0	0 0 > 1 4	3 0 > 1 0	1 1 > 0 2	0 0 > 0 6	1 0 > 2 1	3 0 > 0 2	0 2 > 0 1	1 0 > 1 3	0 1 > 2 0
0	1 024	1 656	345	1 803	700	906	3 180	1 263	1 560	336	4 032	2 116	4 416	3 200	4 386	3 096	5 355	3 147	2 864	8 148	944	10 836	3 918	6 339	11 328	8 748
1	776	1 304	269	1 365	567	674	2 475	936	1 193	238	3 176	1 691	3 504	2 571	3 466	2 520	4 348	2 520	2 352	6 464	745	8 811	3 106	5 234	9 216	7 098
2	584	964	198	1 074	416	552	1 904	784	954	212	2 488	1 324	2 768	1 995	2 766	1 958	3 436	2 008	1 868	5 289	610	7 076	2 592	4 200	7 472	5 860
3	438	755	156	799	338	404	1 467	570	720	149	1 940	1 050	2 178	1 593	2 168	1 590	2 774	1 600	1 531	4 153	482	5 718	2 036	3 454	6 042	4 716
4	328	545	116	623	242	331	1 116	477	575	134	1 504	813	1 704	1 220	1 724	1 221	2 167	1 272	1 204	3 382	400	4 554	1 694	2 746	4 872	3 873
5	240	427	89	456	200	237	849	343	426	94	1 160	639	1 332	969	1 331	989	1 743	1 001	988	2 628	309	3 656	1 322	2 250	3 908	3 092
6	176	301	63	354	140	194	637	289	339	85	890	489	1 034	732	1 047	747	1 348	787	771	2 133	253	2 886	1 101	1 770	3 128	2 529
7	88	164	33	195	84	114	352	177	192	58	512	284	624	422	618	452	830	471	500	1 326	158	1 796	717	1 128	1 968	1 624
8	130	232	51	252	113	134	480	198	249	56	678	382	795	580	806	603	1 072	623	624	1 633	199	2 301	842	1 441	2 499	1 998
9	72	120	32	133	61	73	264	111	146	34	384	224	456	346	484	363	638	392	380	992	132	1 422	523	905	1 584	1 269
10	64	125	26	136	66	75	262	116	138	35	386	221	472	335	470	362	654	370	398	1 000	121	1 423	537	913	1 562	1 272
11	48	80	19	103	41	61	191	97	113	32	288	165	352	248	369	263	486	294	296	803	103	1 099	441	697	1 240	1 029
12	32	63	14	70	34	40	139	63	76	20	214	126	266	193	273	212	384	222	239	597	75	865	328	566	968	797
13	22	39	9	53	20	34	99	54	60	19	158	92	198	136	211	150	284	169	178	480	62	660	271	429	756	642
14	12	33	6	36	23	21	69	39	37	14	112	67	160	100	144	124	230	119	160	352	42	513	215	347	572	488
15	18	28	9	32	16	19	72	30	45	10	114	70	141	111	159	120	216	140	138	346	51	510	192	340	600	486

1	.	1	2	1	3	5	6	7	9	8	19	3	27	12	18	48	33	30	13	82	49	114	70	111	84	166	90	113	282	34	387	174	258	444	393	16
	1	.	.	1	.	.	3	3	3	8	5	5	9	3	8	24	12	23	6	40	24	48	45	71	45	77	72	52	155	32	213	89	143	288	228	17
		1	.	.	.	2	.	.	3	.	4	.	6	6	4	9	14	6	8	16	11	40	13	24	26	54	20	52	90	10	122	78	84	136	133	18
			1	1	2	3	3	5	4	6	14	3	16	9	10	34	17	20	6	59	37	80	56	82	67	126	71	86	199	26	298	119	206	346	295	19
				1	.	.	1	2	1	4	5	3	6	3	4	16	7	12	3	28	18	36	34	47	37	61	48	42	107	19	166	63	117	212	171	20
					1	1	1	2	1	2	6	1	7	4	5	15	8	9	3	30	20	40	28	43	35	70	37	48	110	14	169	67	122	198	176	21
						1	.	.	1	.	3	.	3	4	2	6	6	3	3	12	8	26	11	17	20	40	15	37	60	6	92	48	68	104	98	22
							1	1	1	2	2	1	5	1	4	10	7	9	3	20	13	26	20	34	22	44	32	30	87	14	121	54	84	156	138	23
								1	.	1	2	1	3	1	2	6	3	4	1	14	10	18	15	23	18	34	22	23	57	9	93	35	69	114	102	24
									1	.	1	.	2	1	2	4	5	3	3	8	5	16	7	14	12	25	12	22	48	6	65	37	47	80	77	25
										1	.	1	.	.	.	3	.	4	.	5	4	6	9	13	9	14	18	10	28	9	45	16	34	72	51	26
											1	.	1	1	1	2	2	1	1	6	4	10	6	9	9	19	8	16	30	3	49	22	39	58	57	27
												1	.	.	.	.	.	1	.	.	.	.	3	3	3	.	8	.	5	5	9	3	8	24	12	28
													1	.	1	1	2	1	1	4	3	6	3	7	4	12	6	9	24	3	34	18	26	42	46	29
														1	.	.	1	.	1	.	.	4	.	.	3	5	.	10	7	.	11	12	9	12	13	30
															1	.	1	.	1	2	1	2	1	3	1	6	2	5	12	1	16	9	15	20	26	31
																1	.	1	.	2	2	3	3	5	4	8	6	6	14	3	24	9	19	34	29	32
																	1	.	1	.	.	2	.	.	1	3	.	5	6	.	7	9	6	8	11	33
																		1	.	.	.	.	1	2	1	.	4	.	5	3	6	3	4	16	10	34
																			1	.	.	.	.	.	.	.	.	2	.	.	.	3	.	.	.	35
																				1	1	1	1	2	1	4	2	3	6	1	12	4	11	15	17	36
																					1	.	.	1	.	2	1	1	2	1	6	1	5	6	9	37
																						1	.	.	1	2	.	3	3	.	5	4	5	6	7	38
																							1	1	1	.	2	.	2	1	5	1	4	10	7	39
																								1	.	.	1	.	2	1	3	1	2	6	6	40
																									1	.	.	.	1	.	2	1	2	4	3	41
																										1	.	1	1	.	2	1	3	2	4	42
																											1	.	.	1	.	.	.	3	.	43
																												1	.	.	.	1	.	.	.	44
																													1	.	1	1	1	2	3	45
																														1	.	.	.	.	.	46
																															1	.	1	1	2	47
																																1	.	.	.	48
																																	1	.	1	49
																																		1	.	50
																																			1	51

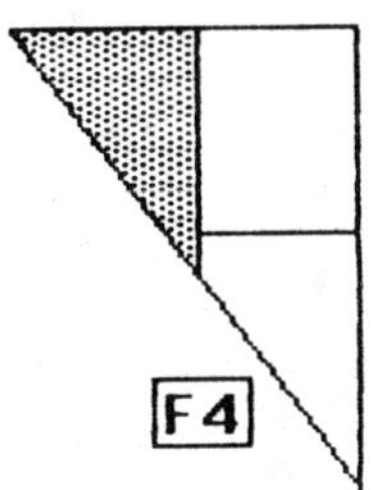

F4

Table 3. *(continued)*

ALGEBRA F4, REPRESENTATIONS OF CLASS 0

NUMBER	52	53	54	55	56	57	58	59	60	61	62	63	64	65	66	67	68	69	70	71	72	73	74	75	76	77
S.P.	70	74	72	82	70	76	74	74	76	84	78	80	100	78	86	82	86	82	98	86	84	90	90	88	90	102
O.S.	576	144	288	144	576	192	192	576	288	192	192	144	24	1 152	192	288	576	576	24	288	576	144	576	288	576	288
LEVEL	103	103	105	105	105	107	107	107	107	109	109	109	111	111	111	113	113	113	113	115	115	115	117	117	117	119
DIM'N	3 955 952	1 850 212	4 582 656	1 360 476	7 113 106	3 195 192	5 218 304	8 843 094	4 528 953	2 674 763	6 680 856	4 940 676	627 912	16 777 216	4 313 088	10 482 472	10 862 592	15 031 926	1 264 120	9 683 388	23 056 488	8 498 776	16 665 831	15 997 696	19 214 624	7 142 499
WEIGHT	2 1 > 0 1	1 0 > 0 5	2 0 > 2 0	4 0 > 0 1	0 1 > 1 2	1 2 > 0 0	0 0 > 3 1	2 0 > 1 2	0 1 > 0 4	3 1 > 0 0	0 0 > 2 3	2 0 > 0 4	5 0 > 0 0	1 1 > 1 1	0 0 > 1 5	1 0 > 3 0	3 0 > 1 1	1 1 > 0 3	0 0 > 0 7	0 2 > 1 0	1 0 > 2 2	3 0 > 0 3	2 1 > 1 0	0 2 > 0 2	1 0 > 1 4	4 0 > 1 0
0	9 392	4 420	10 704	2 820	17 002	7 032	11 840	19 758	10 209	5 411	14 424	10 452	1 128	34 432	8 640	20 416	20 352	29 262	2 272	17 676	43 800	15 064	28 695	28 624	34 304	10 995
1	7 822	3 665	8 708	2 395	13 987	5 678	9 776	16 350	8 355	4 352	11 979	8 568	885	28 688	7 192	17 156	17 064	24 658	1 919	14 958	36 610	12 832	24 398	23 952	28 904	9 435
2	6 292	2 950	7 248	1 920	11 438	4 835	8 048	13 448	6 952	3 768	9 869	7 200	799	23 840	5 976	14 256	14 280	20 398	1 586	12 482	30 759	10 642	20 424	20 332	24 284	7 934
3	5 225	2 442	5 856	1 634	9 357	3 870	6 610	11 073	5 652	3 008	8 166	5 869	624	19 764	4 962	11 932	11 928	17 133	1 345	10 524	25 583	9 050	17 316	16 928	20 390	6 801
4	4 160	1 955	4 856	1 298	7 606	3 288	5 416	9 054	4 689	2 603	6 703	4 925	564	16 336	4 120	9 862	9 936	14 089	1 116	8 730	21 413	7 460	14 417	14 328	17 078	5 691
5	3 456	1 611	3 896	1 112	6 181	2 613	4 416	7 421	3 782	2 069	5 508	3 982	442	13 480	3 396	8 218	8 280	11 795	939	7 339	17 711	6 340	12 202	11 860	14 268	4 885
6	2 729	1 276	3 219	876	4 990	2 220	3 594	6 034	3 127	1 794	4 487	3 330	402	11 088	2 802	6 752	6 874	9 635	770	6 057	14 763	5 194	10 113	10 005	11 894	4 077
7	1 784	814	2 124	600	3 218	1 492	2 344	3 985	2 043	1 244	2 952	2 232	292	7 456	1 872	4 574	4 752	6 515	526	4 166	10 056	3 602	7 071	6 928	8 180	2 933
8	2 247	1 056	2 553	743	4 035	1 736	2 920	4 911	2 511	1 399	3 688	2 676	306	9 086	2 318	5 602	5 681	8 041	657	5 064	12 153	4 396	8 508	8 231	9 922	3 474
9	1 430	698	1 640	487	2 602	1 135	1 912	3 198	1 662	934	2 464	1 783	213	6 032	1 592	3 772	3 832	5 435	468	3 447	8 265	3 017	5 846	5 652	6 876	2 433
10	1 459	673	1 664	503	2 588	1 149	1 894	3 221	1 631	952	2 416	1 774	216	6 072	1 542	3 778	3 896	5 420	446	3 468	8 236	3 036	5 917	5 660	6 799	2 482
11	1 113	531	1 354	380	2 064	970	1 528	2 574	1 353	820	1 960	1 481	197	4 920	1 280	3 061	3 176	4 367	369	2 817	6 818	2 441	4 817	4 742	5 648	2 030
12	918	431	1 058	325	1 643	743	1 220	2 072	1 057	628	1 582	1 163	147	3 992	1 032	2 518	2 608	3 620	307	2 341	5 536	2 063	4 038	3 851	4 645	1 728
13	694	328	864	244	1 297	631	974	1 647	864	546	1 263	968	135	3 224	838	2 029	2 126	2 887	247	1 893	4 554	1 646	3 282	3 223	3 817	1 413

NUMBER	78	79	80	81	82	83	84	85	86	87	88	89	90	91	92	93	94	95	96	97	98	99	100	101	102	103
S.P.	90	92	100	96	94	104	94	98	98	98	102	100	106	104	106	108	108	122	102	114	112	106	110	108	128	124
O.S.	576	576	144	96	576	144	576	576	192	576	288	576	576	192	144	96	288	144	1 152	192	192	576	576	576	24	192
LEVEL	119	119	119	121	121	121	121	123	123	123	123	125	125	125	125	127	127	127	127	127	129	129	129	129	129	131
DIM'N	28 481 544	26 476 956	7 147 140	11 955 216	38 854 452	10 044 008	34 504 704	31 702 671	28 068 768	44 355 584	18 206 370	46 743 268	28 481 544	27 625 000	20 407 140	13 530 946	33 898 527	7 822 737	97 274 034	15 611 882	26 108 082	78 962 688	66 663 324	65 609 375	4 188 834	16 016 924
WEIGHT	0 1 > 2 1	2 1 > 0 2	1 0 > 0 6	0 0 > 4 0	2 0 > 2 1	4 0 > 0 2	0 1 > 1 3	1 2 > 0 1	0 0 > 3 2	2 0 > 1 3	0 1 > 0 5	1 1 > 2 0	3 1 > 0 1	0 0 > 2 4	2 0 > 0 5	0 3 > 0 0	3 0 > 2 0	5 0 > 0 1	1 1 > 1 2	0 0 > 1 6	2 2 > 0 0	1 0 > 3 1	3 0 > 1 2	1 1 > 0 4	0 0 > 0 8	4 1 > 0 0
0	49 056	44 772	11 796	19 488	63 180	15 296	57 024	48 447	44 352	69 248	27 978	70 060	40 152	41 464	29 604	18 778	47 031	9 681	141 210	21 650	34 530	110 016	89 484	90 743	5 298	19 340
1	41 566	37 554	9 903	16 490	53 775	12 833	48 368	41 622	37 778	58 984	23 961	59 576	34 740	35 329	25 512	15 938	40 119	8 498	121 099	18 568	29 316	94 576	77 133	77 626	4 527	16 391
2	35 004	32 080	8 454	14 044	45 456	11 117	40 944	35 182	32 076	50 152	20 264	51 181	29 440	30 191	21 624	13 902	34 710	7 194	103 500	15 904	25 712	81 168	66 261	67 022	3 930	14 558
3	29 546	26 775	7 073	11 838	38 564	9 293	34 602	30 150	27 232	42 576	17 316	43 343	25 437	25 654	18 607	11 744	29 502	6 322	88 478	13 622	21 741	69 558	56 971	57 134	3 362	12 296
4	24 770	22 806	6 037	10 056	32 458	8 042	29 192	25 359	23 050	36 080	14 595	37 126	21 456	21 891	15 719	10 223	25 461	5 329	75 356	11 664	19 044	59 512	48 788	49 227	2 932	10 916
5	20 824	18 952	5 021	8 430	27 454	6 702	24 560	21 693	19 482	30 520	12 434	31 316	18 538	18 508	13 492	8 602	21 583	4 699	64 212	9 947	16 045	50 824	41 860	41 795	2 491	9 204
6	17 376	16 112	4 275	7 132	23 010	5 797	20 634	18 162	16 419	25 770	10 433	26 755	15 576	15 731	11 340	7 481	18 599	3 948	54 498	8 485	14 045	43 338	35 748	35 924	2 163	8 178
7	12 060	11 324	2 988	5 010	16 212	4 192	14 424	12 937	11 568	18 272	7 365	19 158	11 302	11 188	8 118	5 445	13 554	2 943	39 132	6 107	10 344	31 320	26 112	26 013	1 587	6 144
8	14 552	13 287	3 554	5 959	19 375	4 788	17 314	15 471	13 851	21 719	8 889	22 445	13 395	13 296	9 729	6 249	15 666	3 463	46 272	7 253	11 745	36 896	30 543	30 402	1 849	6 833
9	10 078	9 207	2 522	4 182	13 530	3 381	12 136	10 920	9 798	15 336	6 354	15 936	9 576	9 540	6 997	4 502	11 259	2 528	33 081	5 308	8 526	26 600	22 087	22 005	1 384	5 039
10	10 065	9 271	2 472	4 165	13 592	3 427	12 062	10 982	9 728	15 336	6 269	15 990	9 679	9 423	6 949	4 520	11 352	2 568	33 110	5 208	8 584	26 578	22 218	21 928	1 348	5 089
11	8 316	7 817	2 119	3 505	11 254	2 939	10 080	9 066	8 158	12 848	5 245	13 559	8 004	8 008	5 801	3 913	9 699	2 118	27 873	4 458	7 477	22 512	18 789	18 795	1 183	4 502
12	6 905	6 375	1 723	2 894	9 422	2 404	8 364	7 700	6 807	10 736	4 423	11 277	6 882	6 677	4 945	3 235	8 116	1 864	23 502	3 754	6 197	19 022	15 978	15 740	990	3 733
13	5 670	5 376	1 458	2 425	7 775	2 071	6 938	6 333	5 665	8 954	3 651	9 550	5 679	5 625	4 091	2 802	6 938	1 535	19 712	3 172	5 412	16 048	13 492	13 437	859	3 315

599 258 686 230 $\overset{1}{005}$ 487 756 $\overset{1}{339}$ 649 436 985 749 108 $\overset{2}{616}$ 652 $\overset{1}{657}$ $\overset{1}{804}$ $\overset{2}{372}$ 200 $\overset{1}{574}$ $\overset{3}{645}$ $\overset{1}{421}$ $\overset{2}{802}$ $\overset{2}{586}$ $\overset{3}{076}$ $\overset{1}{258}$ $\overset{4}{658}$ $\overset{4}{412}$ $\overset{1}{156}$ $\overset{1}{971}$ $\overset{6}{538}$ $\overset{1}{740}$ $\overset{5}{660}$ $\overset{5}{409}$ $\overset{4}{651}$ $\overset{7}{460}$ $\overset{3}{021}$ $\overset{7}{928}$ $\overset{4}{992}$ $\overset{4}{593}$ $\overset{3}{459}$ $\overset{2}{315}$ $\overset{5}{883}$ $\overset{1}{407}$ $\overset{16}{549}$ $\overset{2}{617}$ $\overset{4}{517}$ $\overset{13}{468}$ $\overset{11}{585}$ $\overset{11}{128}$ 704 $\overset{2}{805}$	14
556 282 648 204 $\overset{1}{032}$ 464 779 $\overset{1}{305}$ 693 398 $\overset{1}{041}$ 761 96 $\overset{2}{572}$ 708 $\overset{1}{652}$ $\overset{1}{703}$ $\overset{2}{391}$ 219 $\overset{1}{554}$ $\overset{3}{687}$ $\overset{1}{376}$ $\overset{2}{703}$ $\overset{2}{588}$ $\overset{3}{184}$ $\overset{1}{182}$ $\overset{4}{685}$ $\overset{4}{311}$ $\overset{1}{221}$ $\overset{1}{995}$ $\overset{6}{429}$ $\overset{1}{654}$ $\overset{5}{784}$ $\overset{5}{319}$ $\overset{4}{750}$ $\overset{7}{455}$ $\overset{3}{138}$ $\overset{7}{846}$ $\overset{4}{812}$ $\overset{4}{755}$ $\overset{3}{515}$ $\overset{2}{284}$ $\overset{5}{706}$ $\overset{1}{335}$ $\overset{16}{532}$ $\overset{2}{743}$ $\overset{4}{404}$ $\overset{13}{509}$ $\overset{11}{350}$ $\overset{11}{269}$ 743 $\overset{2}{691}$	15
441 194 552 165 790 413 602 $\overset{1}{050}$ 534 375 783 621 99 $\overset{2}{094}$ 528 $\overset{1}{326}$ $\overset{1}{444}$ $\overset{1}{876}$ 158 $\overset{1}{263}$ $\overset{2}{991}$ $\overset{1}{116}$ $\overset{2}{247}$ $\overset{2}{158}$ $\overset{2}{521}$ $\overset{1}{008}$ $\overset{3}{810}$ $\overset{3}{699}$ 981 $\overset{1}{648}$ $\overset{5}{351}$ $\overset{1}{481}$ $\overset{4}{686}$ $\overset{4}{410}$ $\overset{3}{864}$ $\overset{6}{188}$ $\overset{2}{490}$ $\overset{6}{688}$ $\overset{4}{065}$ $\overset{3}{866}$ $\overset{2}{844}$ $\overset{2}{003}$ $\overset{4}{992}$ $\overset{1}{137}$ $\overset{13}{821}$ $\overset{2}{209}$ $\overset{3}{929}$ $\overset{11}{328}$ $\overset{9}{705}$ $\overset{9}{491}$ 609 $\overset{2}{476}$	16
232 146 308 89 496 240 384 621 378 217 548 414 60 $\overset{1}{280}$ 408 834 864 $\overset{1}{229}$ 132 798 $\overset{1}{980}$ 703 $\overset{1}{400}$ $\overset{1}{428}$ $\overset{1}{796}$ 633 $\overset{2}{542}$ $\overset{2}{397}$ 750 $\overset{1}{132}$ $\overset{3}{503}$ 961 $\overset{3}{288}$ $\overset{2}{922}$ $\overset{2}{724}$ $\overset{4}{240}$ $\overset{1}{851}$ $\overset{4}{520}$ $\overset{2}{679}$ $\overset{2}{868}$ $\overset{2}{058}$ $\overset{1}{375}$ $\overset{3}{348}$ 763 $\overset{9}{558}$ $\overset{1}{723}$ $\overset{2}{694}$ $\overset{7}{944}$ $\overset{6}{640}$ $\overset{6}{854}$ 497 $\overset{1}{708}$	17
182 60 228 84 262 173 208 420 182 188 271 246 58 848 192 542 688 748 62 536 $\overset{1}{221}$ 518 $\overset{1}{056}$ 920 $\overset{1}{032}$ 536 $\overset{1}{632}$ $\overset{1}{737}$ 420 726 $\overset{2}{500}$ 786 $\overset{2}{016}$ $\overset{2}{092}$ $\overset{1}{704}$ $\overset{2}{880}$ $\overset{1}{090}$ $\overset{3}{202}$ $\overset{2}{116}$ $\overset{1}{733}$ $\overset{1}{326}$ $\overset{1}{000}$ $\overset{2}{608}$ 648 $\overset{6}{594}$ $\overset{1}{024}$ $\overset{2}{072}$ $\overset{5}{472}$ $\overset{5}{031}$ $\overset{4}{568}$ 300 $\overset{1}{416}$	18
352 163 411 136 624 299 478 825 417 267 638 479 67 $\overset{1}{664}$ 435 $\overset{1}{077}$ $\overset{1}{148}$ $\overset{1}{548}$ 137 $\overset{1}{035}$ $\overset{2}{402}$ 929 $\overset{1}{845}$ $\overset{1}{723}$ $\overset{2}{079}$ 836 $\overset{3}{138}$ $\overset{2}{942}$ 803 $\overset{1}{350}$ $\overset{4}{413}$ $\overset{1}{170}$ $\overset{3}{880}$ $\overset{3}{701}$ $\overset{3}{220}$ $\overset{5}{123}$ $\overset{2}{119}$ $\overset{5}{473}$ $\overset{3}{436}$ $\overset{3}{236}$ $\overset{2}{431}$ $\overset{1}{622}$ $\overset{4}{084}$ 983 $\overset{11}{552}$ $\overset{1}{883}$ $\overset{3}{177}$ $\overset{9}{503}$ $\overset{8}{134}$ $\overset{7}{902}$ 517 $\overset{1}{990}$	19
197 113 236 78 381 177 296 493 275 161 423 306 43 $\overset{1}{024}$ 308 678 708 $\overset{1}{007}$ 103 659 $\overset{1}{565}$ 599 $\overset{1}{174}$ $\overset{1}{128}$ $\overset{1}{426}$ 541 $\overset{2}{083}$ $\overset{1}{920}$ 577 913 $\overset{2}{919}$ 774 $\overset{2}{660}$ $\overset{2}{483}$ $\overset{2}{223}$ $\overset{3}{488}$ $\overset{1}{521}$ $\overset{3}{712}$ $\overset{2}{319}$ $\overset{2}{313}$ $\overset{1}{728}$ $\overset{1}{119}$ $\overset{2}{786}$ 678 $\overset{7}{966}$ $\overset{1}{398}$ $\overset{2}{201}$ $\overset{6}{628}$ $\overset{5}{622}$ $\overset{5}{607}$ 395 $\overset{1}{401}$	20
210 95 247 84 375 186 294 504 253 168 394 298 44 $\overset{1}{052}$ 272 693 738 990 88 679 $\overset{1}{552}$ 608 $\overset{1}{223}$ $\overset{1}{135}$ $\overset{1}{359}$ 566 $\overset{2}{086}$ $\overset{1}{960}$ 534 911 $\overset{2}{969}$ 796 $\overset{2}{600}$ $\overset{2}{532}$ $\overset{2}{185}$ $\overset{3}{480}$ $\overset{1}{438}$ $\overset{3}{764}$ $\overset{2}{385}$ $\overset{2}{215}$ $\overset{1}{676}$ $\overset{1}{137}$ $\overset{2}{849}$ 700 $\overset{7}{996}$ $\overset{1}{307}$ $\overset{2}{246}$ $\overset{6}{642}$ $\overset{5}{716}$ $\overset{5}{524}$ 366 $\overset{1}{430}$	21
141 50 162 66 206 119 164 320 138 124 219 182 35 664 154 436 528 615 52 437 966 422 851 724 842 431 $\overset{1}{338}$ $\overset{1}{345}$ 334 590 $\overset{2}{032}$ 596 $\overset{1}{660}$ $\overset{1}{744}$ $\overset{1}{412}$ $\overset{2}{354}$ 925 $\overset{2}{586}$ $\overset{1}{758}$ $\overset{1}{436}$ $\overset{1}{128}$ 795 $\overset{2}{084}$ 548 $\overset{5}{468}$ 861 $\overset{1}{636}$ $\overset{4}{560}$ $\overset{4}{149}$ $\overset{3}{762}$ 247 $\overset{1}{100}$	22
142 76 193 55 290 155 230 383 217 144 319 251 41 816 234 536 572 770 75 526 $\overset{1}{265}$ 462 943 938 $\overset{1}{130}$ 438 $\overset{1}{678}$ $\overset{1}{621}$ 473 755 $\overset{2}{374}$ 671 $\overset{2}{154}$ $\overset{2}{009}$ $\overset{1}{811}$ $\overset{2}{862}$ $\overset{1}{205}$ $\overset{3}{131}$ $\overset{1}{885}$ $\overset{1}{895}$ $\overset{1}{381}$ 975 $\overset{2}{382}$ 549 $\overset{6}{596}$ $\overset{1}{138}$ $\overset{1}{935}$ $\overset{5}{526}$ $\overset{4}{709}$ $\overset{4}{727}$ 331 $\overset{1}{256}$	23
117 60 142 47 222 110 179 297 156 100 250 183 27 644 180 433 455 629 62 432 993 388 778 735 899 366 $\overset{1}{371}$ $\overset{1}{272}$ 365 610 $\overset{1}{956}$ 525 $\overset{1}{743}$ $\overset{1}{695}$ $\overset{1}{483}$ $\overset{2}{339}$ 996 $\overset{2}{545}$ $\overset{1}{608}$ $\overset{1}{538}$ $\overset{1}{164}$ 783 $\overset{1}{944}$ 481 $\overset{5}{468}$ 934 $\overset{1}{554}$ $\overset{4}{596}$ $\overset{3}{939}$ $\overset{3}{853}$ 269 $\overset{1}{004}$	24
92 37 126 41 159 100 128 239 117 105 175 153 32 512 128 337 400 472 42 337 784 313 649 602 691 324 $\overset{1}{073}$ $\overset{1}{106}$ 291 490 $\overset{1}{613}$ 496 $\overset{1}{368}$ $\overset{1}{374}$ $\overset{1}{165}$ $\overset{1}{920}$ 761 $\overset{2}{149}$ $\overset{1}{369}$ $\overset{1}{220}$ 917 683 $\overset{1}{728}$ 419 $\overset{4}{494}$ 736 $\overset{1}{409}$ $\overset{3}{784}$ $\overset{3}{388}$ $\overset{3}{210}$ 220 959	25
57 45 70 25 132 56 105 168 111 52 169 120 15 370 140 258 261 398 51 261 630 238 469 465 633 228 875 792 282 397 $\overset{1}{233}$ 330 $\overset{1}{195}$ $\overset{1}{086}$ $\overset{1}{014}$ $\overset{1}{565}$ 735 $\overset{1}{650}$ $\overset{1}{038}$ $\overset{1}{126}$ 831 517 $\overset{1}{263}$ 323 $\overset{3}{684}$ 728 $\overset{1}{026}$ $\overset{3}{128}$ $\overset{2}{639}$ $\overset{2}{740}$ 219 674	26
74 30 88 34 122 69 100 180 84 69 137 108 20 400 98 270 308 385 34 277 611 258 527 467 549 265 877 851 223 395 $\overset{1}{311}$ 375 $\overset{1}{108}$ $\overset{1}{154}$ 955 $\overset{1}{556}$ 634 $\overset{1}{731}$ $\overset{1}{150}$ 988 772 544 $\overset{1}{383}$ 362 $\overset{3}{708}$ 604 $\overset{1}{112}$ $\overset{3}{138}$ $\overset{2}{795}$ $\overset{2}{605}$ 176 748	27
12 23 16 6 40 15 32 48 45 15 71 45 6 112 72 80 80 155 32 87 237 89 154 176 288 81 340 294 146 160 473 128 528 435 448 672 378 677 420 564 414 224 528 141 $\overset{1}{607}$ 408 447 $\overset{1}{392}$ $\overset{1}{149}$ $\overset{1}{318}$ 132 310	28
47 20 69 20 92 60 78 133 69 60 105 89 19 308 78 208 234 286 26 215 492 188 402 387 438 200 699 704 190 327 $\overset{1}{036}$ 315 898 906 781 $\overset{1}{258}$ 506 $\overset{1}{442}$ 892 823 612 472 $\overset{1}{154}$ 276 $\overset{3}{032}$ 504 965 $\overset{2}{594}$ $\overset{2}{281}$ $\overset{2}{202}$ 154 659	29
33 6 36 23 27 26 24 69 18 39 32 37 14 144 24 96 160 132 10 102 210 124 251 168 184 160 321 388 76 149 591 215 400 519 357 676 232 772 629 370 327 250 738 230 $\overset{1}{630}$ 236 580 $\overset{1}{392}$ $\overset{1}{458}$ $\overset{1}{126}$ 74 450	30
25 9 39 12 51 37 44 72 37 38 58 51 13 184 42 126 142 169 14 136 297 113 255 247 265 131 444 450 116 214 664 210 570 599 504 812 320 964 598 532 399 325 781 192 $\overset{2}{022}$ 326 674 $\overset{1}{752}$ $\overset{1}{548}$ $\overset{1}{481}$ 102 469	31
35 20 43 16 71 35 59 99 56 34 89 67 10 228 71 162 171 237 26 168 382 154 314 293 371 158 558 522 163 257 821 228 744 736 646 $\overset{1}{022}$ 450 $\overset{1}{120}$ 730 703 535 359 889 234 $\overset{2}{478}$ 451 729 $\overset{2}{126}$ $\overset{1}{850}$ $\overset{1}{818}$ 137 492	32
19 3 27 12 21 23 18 48 16 33 24 30 13 108 18 72 114 95 6 79 168 84 183 142 145 113 255 315 65 123 453 174 324 398 291 534 182 639 466 308 252 219 600 165 $\overset{1}{320}$ 193 499 $\overset{1}{140}$ $\overset{1}{155}$ 954 63 389	33
7 12 14 3 28 15 24 36 34 15 47 37 6 88 48 63 64 107 19 67 187 63 123 146 212 64 264 254 113 131 377 114 408 339 350 536 275 573 332 437 306 199 456 109 $\overset{1}{303}$ 308 404 $\overset{1}{144}$ 955 $\overset{1}{088}$ 103 285	34
4 . 6 6 . 4 . 9 . 14 . 6 8 16 . 11 40 13 . 12 24 26 54 22 20 52 39 90 10 22 122 78 48 102 48 136 27 166 182 52 60 59 228 84 330 35 184 288 420 236 15 188	35
19 9 24 9 38 21 34 53 28 20 48 35 6 136 37 99 103 138 14 107 228 93 199 183 217 102 356 329 95 168 527 148 464 489 415 652 279 748 489 444 343 247 600 161 $\overset{1}{644}$ 285 503 $\overset{1}{434}$ $\overset{1}{251}$ $\overset{1}{199}$ 88 344	36
9 4 12 4 20 11 20 27 15 10 27 18 3 76 22 61 56 77 9 66 135 54 120 111 129 60 223 198 60 109 328 89 290 312 270 408 177 484 312 290 219 165 387 104 $\overset{1}{074}$ 191 333 958 822 792 62 228	37
14 3 16 9 16 13 14 34 11 17 20 20 6 80 15 56 80 78 6 63 127 67 141 106 117 86 205 225 50 97 355 119 264 326 239 424 157 493 375 254 216 165 451 136 $\overset{1}{070}$ 164 368 931 918 765 52 278	38

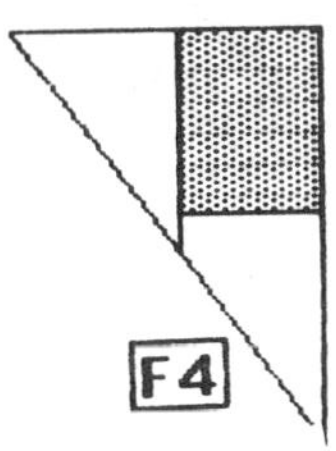

Table 3. *(continued)*

ALGEBRA F4, REPRESENTATIONS OF CLASS 0

	21>01	10>05	20>20	40>01	01>12	12>00	00>31	20>12	01>04	31>00	00>23	20>04	50>00	11>11	00>15	10>30	30>11	11>03	00>07	02>10	10>22	30>03	21>10	02>02	10>14	40>10	WEIGHT
	52	53	54	55	56	57	58	59	60	61	62	63	64	65	66	67	68	69	70	71	72	73	74	75	76	77	NUMBER
	7	9	10	3	20	9	18	26	20	9	34	22	3	68	32	51	50	87	14	55	138	54	102	107	156	53	39
	3	4	9	1	14	9	14	18	15	9	23	18	3	52	22	40	38	57	9	42	108	35	78	90	114	39	40
	5	3	6	3	8	6	8	16	7	7	14	12	3	40	12	29	36	48	6	33	76	37	71	62	80	42	41
	6	1	7	4	9	7	8	15	6	8	11	9	3	44	8	33	40	43	3	40	73	35	81	66	68	48	42
	.	4	.	.	5	.	4	6	9	.	13	9	.	14	18	13	10	28	9	15	45	16	27	34	72	15	43
	3	.	3	4	.	2	.	6	.	6	.	3	3	12	.	8	26	11	.	10	17	20	40	16	15	37	44
	2	1	5	1	6	6	6	10	6	7	9	9	3	30	8	22	26	30	3	26	59	22	52	53	58	30	45
	.	1	.	.	.	.	.	.	3	.	3	3	.	.	8	.	.	5	5	.	9	3	.	8	24	.	46
	2	1	3	1	4	3	6	6	3	3	7	4	1	22	6	10	10	24	3	21	42	18	41	37	42	23	47
	1	.	2	1	.	2	.	4	.	5	.	3	3	8	.	5	16	7	.	7	14	12	25	14	12	22	48
	1	.	1	1	2	2	2	2	1	2	3	1	1	12	2	9	10	12	1	14	21	9	25	23	20	16	49
	.	1	.	.	2	.	2	3	3	.	5	4	.	8	6	8	6	14	3	9	24	9	18	19	34	10	50
	.	.	1	.	1	2	2	1	1	2	2	1	1	8	2	7	6	6	1	9	17	4	16	18	15	9	51
	1	.	1	1	.	1	.	2	.	2	.	1	1	6	.	4	10	6	.	6	9	9	19	10	8	16	52
		1	.	.	.	.	.	.	1	.	2	1	.	.	4	.	.	5	3	.	6	3	.	4	16	.	53
			1	.	.	1	.	1	.	2	.	1	1	4	.	3	6	3	.	4	7	4	12	9	6	9	54
				1	.	.	.	.	.	1	.	.	1	.	.	.	4	.	.	.	.	3	5	.	.	10	55
					1	.	1	1	1	.	2	1	.	4	2	5	3	6	1	6	12	4	11	11	15	6	56
						1	.	.	.	1	.	.	1	2	.	1	2	1	.	3	3	1	6	6	2	5	57
							1	.	.	.	1	.	.	2	1	3	1	2	1	3	6	1	6	5	6	3	58
								1	.	.	.	1	.	2	.	2	3	3	.	3	5	4	8	5	6	6	59
									1	.	1	1	.	.	2	.	.	2	1	.	5	1	.	4	10	.	60
										1	.	.	1	.	.	.	2	.	.	.	.	1	3	.	.	5	61
											1	.	.	.	1	.	.	2	1	.	3	1	.	2	6	.	62
												:	.	.	.	.	.	1	.	.	2	1	.	2	4	.	63
													1	.	.	.	.	.	.	.	.	.	.	.	.	2	64
														1	.	1	1	1	.	2	2	1	4	3	2	3	65
															1	.	.	.	1	.	.	.	.	.	3	.	66

	01>21	21>02	10>06	00>40	20>21	40>02	01>13	12>01	00>32	20>13	01>05	11>20	31>01	00>24	20>05	03>00	30>20	50>01	11>12	00>16	22>00	10>31	30>12	11>04	00>08	41>00	WEIGHT
	78	79	80	81	82	83	84	85	86	87	88	89	90	91	92	93	94	95	96	97	98	99	100	101	102	103	NUMBER
	214	188	76	104	310	84	314	291	276	422	217	457	289	331	251	154	365	97	1058	234	314	936	789	840	75	219	39
	166	159	60	87	241	74	242	222	218	330	156	385	219	261	183	137	309	71	850	180	282	766	641	693	62	199	40
	124	125	37	61	205	63	176	194	159	268	117	301	214	184	153	104	266	78	683	128	227	608	563	525	42	170	41
	130	129	30	63	211	67	166	207	155	256	99	316	228	164	135	111	278	84	692	106	238	616	577	501	34	177	42
	69	57	45	35	97	25	132	96	115	168	111	154	96	169	120	55	123	36	422	140	111	378	307	398	51	80	43
	32	60	6	16	92	48	40	84	39	104	24	124	141	41	50	44	162	66	264	28	127	232	220	182	10	134	44
	97	104	30	51	154	54	136	146	126	204	84	253	156	145	108	94	221	55	545	98	202	494	445	434	34	153	45
	12	12	23	8	16	6	40	15	32	48	45	35	15	71	45	17	28	6	112	72	33	104	80	155	32	26	46
	73	70	20	41	122	33	105	123	101	133	69	195	130	112	89	70	165	41	434	78	141	402	352	328	26	107	47
	24	48	6	13	65	37	32	59	32	80	18	100	92	36	37	39	126	41	207	24	108	184	239	157	10	105	48
	47	40	9	24	71	22	60	80	59	88	37	124	85	63	51	48	107	34	272	42	99	254	224	201	14	75	49
	41	35	20	22	62	16	71	63	66	99	56	102	66	89	67	37	85	25	266	71	78	246	206	237	26	57	50
	35	33	9	21	51	18	46	54	48	66	28	102	55	53	35	42	86	20	210	37	87	206	170	168	14	66	51
	20	30	3	10	49	22	26	50	25	58	16	75	74	27	30	29	88	34	163	18	75	148	180	116	6	69	52
	10	7	12	6	14	3	28	15	24	36	34	27	15	47	37	11	22	6	88	48	23	84	64	107	19	18	53
	15	24	3	9	34	18	20	34	20	42	11	62	47	23	20	26	69	20	126	15	66	118	133	98	6	60	54
	.	7	.	.	11	12	.	9	.	12	.	13	33	.	6	5	36	23	27	.	26	24	69	18	.	39	55
	25	19	9	14	37	9	38	40	39	53	28	66	42	48	35	25	55	16	164	37	52	158	129	138	14	38	56
	9	12	1	5	16	9	12	21	12	20	6	38	25	13	9	19	39	12	75	8	42	72	72	58	3	38	57
	15	9	4	10	21	4	20	24	24	27	15	42	24	27	18	16	33	9	96	22	33	103	75	77	9	23	58
	11	14	3	6	24	9	16	26	17	34	11	40	35	20	20	16	43	16	98	15	39	91	99	78	6	34	59
	7	7	9	5	10	3	20	9	18	26	20	24	9	34	22	11	20	3	68	32	23	68	50	87	14	18	60
	.	6	.	.	7	9	.	6	.	8	.	11	19	.	3	5	27	12	21	.	23	18	48	16	.	33	61
	6	3	4	4	9	1	14	9	14	18	15	18	9	23	18	7	15	3	52	22	15	55	38	57	9	11	62
	3	5	3	3	6	3	8	6	8	16	7	15	7	14	12	7	14	3	40	12	18	40	36	48	6	15	63
	.	.	.	.	.	3	.	.	.	.	.	.	4	.	.	.	6	6	.	.	4	.	9	.	.	14	64
	7	6	1	4	12	4	9	16	10	15	6	25	19	11	9	11	24	9	57	8	24	57	53	43	3	20	65
	.	.	4	.	.	.	5	.	4	6	9	.	.	13	9	.	.	.	14	18	.	13	10	28	9	.	66

1	.	.	.	1	1	.	2	1	1	1	4	2	1	3	6	1	4	8	7	6	3	14	9	7	4	6	12	4	32	6	13	36	27	24	3	10	67
	1	.	.	.	.	1	2	.	.	3	.	3	.	.	5	4	.	5	.	6	.	7	14	.	3	3	16	9	16	.	13	14	34	11	.	17	68
		1	.	.	1	1	.	1	2	.	3	2	1	2	5	1	6	6	6	10	6	11	7	9	9	5	10	3	30	8	11	32	26	30	3	9	69
			1	.	.	.	.	.	.	.	.	.	1	.	.	.	.	.	.	.	3	.	.	3	3	.	.	.	.	8	.	.	.	5	5	.	70
				1	.	.	1	1	.	1	2	1	.	1	2	1	2	6	3	2	1	8	6	3	1	5	7	4	18	2	9	18	15	12	1	8	71
					1	.	.	1	1	.	2	2	1	2	3	1	4	3	5	6	3	10	3	7	4	5	9	1	22	6	11	26	18	24	3	9	72
						1	.	.	.	.	.	1	.	.	2	1	.	2	.	4	.	3	5	.	3	2	6	3	8	.	6	8	16	7	.	7	73
							1	.	.	1	.	1	.	.	2	1	.	3	.	2	.	4	6	.	1	2	7	4	9	.	7	8	15	6	.	8	74
								1	.	.	1	1	.	1	1	1	2	2	2	2	1	6	2	3	1	4	5	1	12	2	8	14	10	12	1	7	75
									1	.	.	.	1	.	.	.	2	.	2	3	3	.	.	5	4	.	.	.	8	6	.	8	6	14	3	.	76
										1	.	.	.	.	.	1	.	.	.	.	.	.	3	.	.	.	3	4	.	.	2	.	6	.	.	6	77
											1	.	.	1	1	.	1	2	2	1	1	4	2	2	1	2	3	1	8	2	4	12	6	6	1	3	78
												1	.	.	1	1	.	1	.	2	.	3	2	.	1	2	5	1	6	.	6	6	10	6	.	7	79
													1	.	.	.	.	.	.	.	1	.	.	2	1	.	.	.	.	4	.	.	.	5	3	.	80
														1	.	.	.	.	1	.	.	2	.	1	.	1	1	.	2	1	3	6	1	2	1	2	81
															1	.	.	1	.	1	.	2	2	.	1	1	3	1	4	.	3	5	6	3	.	3	82
																1	.	.	.	.	.	.	1	.	.	.	2	1	.	.	2	.	4	.	.	5	83
																	1	.	1	1	1	.	.	2	1	.	.	.	4	2	.	5	3	6	1	.	84
																		1	.	.	.	1	1	.	.	1	1	1	2	.	2	2	2	1	.	2	85
																			1	.	.	.	.	1	.	.	.	.	2	1	.	3	1	2	1	.	86
																				1	.	.	.	.	1	.	.	.	2	.	.	2	3	3	.	.	87
																					1	.	.	1	1	.	.	.	.	2	.	.	.	2	1	.	88
																						1	.	.	.	1	1	.	1	.	2	2	1	1	.	2	89
																							1	.	.	.	1	1	.	.	1	.	2	.	.	2	90
																								1	.	.	.	.	.	1	.	.	.	2	1	.	91
																									1	.	.	.	.	.	.	.	.	1	.	.	92
																										1	.	.	.	.	1	.	.	.	.	1	93
																											1	.	.	.	1	.	1	.	.	2	94
																												1	.	.	.	.	.	.	.	1	95
																													1	.	.	1	1	1	.	.	96
																														1	.	.	.	.	1	.	97
																															1	.	.	.	.	1	98
																																1	.	.	.	.	99
																																	1	.	.	.	100
																																		1	.	.	101
																																			1	.	102
																																				1	103

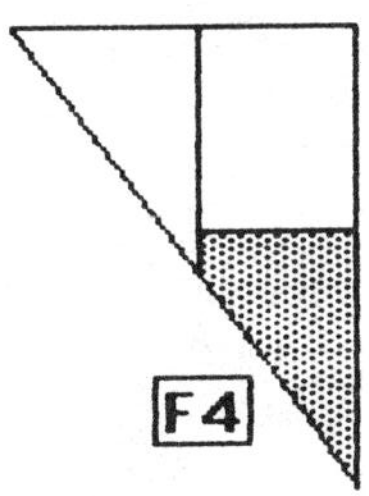

F4

Table 3. *(continued)*

ALGEBRA 62, REPRESENTATIONS OF CLASS 0

S.P.	0	2	6	8	14	18	24	26	32	38	42	50	54	56	62	72	74	78	86	96	98	98	104	114	122	128
O.S.	1	6	6	6	12	6	6	12	6	12	12	6	6	12	12	6	12	12	12	6	6	12	12	12	12	6
LEVEL	1	7	11	13	17	19	21	23	25	27	29	31	31	33	35	37	37	39	41	41	43	43	45	47	47	49
DIM'N	1	7	14	27	64	77	77	189	182	286	448	378	273	729	924	714	896	1547	1728	748	1254	2079	2926	3003	2261	2079
WEIGHT	0 < 0	1 < 0	0 < 1	2 < 0	1 < 1	3 < 0	0 < 2	2 < 1	4 < 0	1 < 2	3 < 1	5 < 0	0 < 3	2 < 2	4 < 1	6 < 0	1 < 3	3 < 2	5 < 1	0 < 4	7 < 0	2 < 3	4 < 2	6 < 1	1 < 4	8 < 0
NUMBER	0	1	2	3	4	5	6	7	8	9	10	11	12	13	14	15	16	17	18	19	20	21	22	23	24	25
0	1	1	2	3	4	5	5	9	8	10	16	12	9	21	24	18	20	35	36	16	24	39	52	51	35	33
1		1	1	2	4	4	3	8	7	10	14	11	7	19	23	16	20	32	34	13	23	37	50	48	35	31
2			1	1	2	3	3	6	5	7	12	9	7	16	19	14	16	29	30	13	20	33	44	44	30	28
3				1	2	2	2	5	5	7	10	8	5	15	18	13	16	26	28	11	19	31	43	41	30	27
4					1	1	1	3	3	5	7	6	4	11	14	10	13	21	23	9	16	26	36	35	26	23
5						1	1	2	2	3	6	5	4	9	11	9	10	19	20	9	14	23	31	32	22	21
6							1	1	1	2	4	3	3	7	8	7	8	15	16	8	11	19	26	27	19	18
7								1	1	2	3	3	2	6	8	6	8	13	15	6	11	18	25	25	19	17
8									1	1	2	2	1	4	6	5	6	10	12	5	9	14	21	21	16	15
9										1	1	1	1	3	4	3	5	8	9	4	7	12	17	17	14	12
10											1	1	1	2	3	3	3	7	8	4	6	10	14	16	11	11
11												1	.	1	2	2	2	4	6	2	5	7	11	12	9	9
12													1	1	1	1	2	4	4	3	3	7	9	10	8	7
13														1	1	1	2	3	4	2	3	6	9	9	8	7
14															1	1	1	2	3	1	3	4	7	8	6	6
15																1	.	1	2	1	2	2	4	6	3	5
16																	1	1	1	1	1	3	4	4	5	3
17																		1	1	1	1	2	3	4	3	3
18																			1	.	1	1	2	3	2	3

S.P.	126	134	150	146	152	162	158	168	182	182	186	200	194	216	206	218	222	224	242	234	254	248	258	266	266	294
O.S.	12	12	6	12	12	6	12	12	12	12	12	6	12	6	12	12	12	12	6	12	12	12	12	12	12	6
LEVEL	49	51	51	53	53	55	55	57	57	59	59	61	61	61	63	63	65	65	67	67	67	69	69	71	71	71
DIM'N	4096	5103	1729	4928	4914	3289	7293	8372	4928	7722	9177	5005	12096	3542	13090	10206	11648	15625	7371	19019	9660	19683	18304	17017	24948	6630
WEIGHT	3 < 3	5 < 2	0 < 5	7 < 1	2 < 4	9 < 0	4 < 3	6 < 2	1 < 5	8 < 1	3 < 4	10 < 0	5 < 3	0 < 6	7 < 2	2 < 5	9 < 1	4 < 4	11 < 0	6 < 3	1 < 6	8 < 2	3 < 5	10 < 1	5 < 4	0 < 7
NUMBER	26	27	28	29	30	31	32	33	34	35	36	37	38	39	40	41	42	43	44	45	46	47	48	49	50	51
0	64	75	25	68	66	43	93	104	56	90	105	55	132	38	136	102	116	151	69	179	84	177	160	145	210	54
1	60	72	22	66	63	41	91	99	56	87	100	53	128	34	133	99	112	148	67	173	84	172	154	142	205	50
2	56	66	22	60	58	38	83	93	50	81	95	49	120	34	124	93	106	138	63	165	77	163	148	134	195	50
3	52	63	19	58	56	36	81	89	50	78	90	48	116	31	121	90	102	136	61	159	77	159	142	131	190	46
4	45	55	17	51	49	32	72	79	45	70	81	43	105	28	110	82	93	124	56	146	71	146	131	121	176	43
5	42	50	17	46	45	30	65	74	40	65	77	40	98	28	102	77	88	115	53	139	65	138	126	114	167	43
6	36	43	15	40	40	26	57	66	36	58	69	36	88	26	92	70	80	105	48	127	60	127	116	105	154	40
7	33	41	13	39	38	25	56	62	36	56	65	35	85	23	90	68	77	103	47	122	60	123	111	103	150	37
8	28	35	11	34	33	22	49	55	32	50	58	32	76	21	81	61	70	94	43	111	55	113	102	95	138	34
9	24	30	10	29	29	19	43	48	29	44	52	28	68	19	73	56	63	85	39	101	51	103	94	87	127	32
10	22	27	10	26	26	18	38	45	25	41	49	26	63	19	67	52	60	78	37	96	46	97	90	82	120	32
11	16	22	7	22	21	15	32	37	22	35	40	23	54	15	59	45	52	69	33	83	42	86	78	74	107	27
12	16	19	8	18	20	13	29	34	20	31	39	20	50	16	53	43	48	64	30	79	39	80	76	68	101	28
13	14	18	6	18	19	12	28	32	20	30	36	20	48	14	52	41	46	63	29	75	39	78	72	67	98	25
14	11	15	5	16	15	11	24	28	17	27	31	18	42	12	47	36	42	56	27	68	35	71	65	62	89	23
15	8	11	4	12	11	9	17	23	12	22	25	15	34	11	38	29	36	45	23	58	28	61	56	53	76	21
16	8	10	4	10	12	7	18	20	14	19	25	13	33	10	36	30	32	46	21	55	30	57	55	50	75	20
17	7	9	4	9	10	7	15	19	11	18	23	12	30	10	33	27	31	41	20	52	26	54	52	47	70	20
18	4	7	2	8	7	6	12	15	9	16	17	11	25	7	29	22	27	35	18	44	23	48	43	43	61	16

1	.	1	1	1	2	1	4	4	3	4	7	3	9	11	8	10	16	7	20	8	21	20	20	30	13	37	20	38	40	33	53	16	19
	1	.	1	2	1	2	2	4	1	6	4	5	8	11	6	12	12	9	18	5	23	16	22	27	15	35	18	39	34	36	49	13	20
		1	1	1	2	1	3	4	2	4	6	3	9	10	8	10	14	7	19	6	21	19	19	29	13	35	20	37	37	33	51	14	21
			1	1	1	1	2	3	1	4	4	3	7	9	6	9	11	7	16	5	19	15	18	25	12	31	17	34	32	31	45	12	22
				1	.	1	1	2	1	3	2	3	4	7	3	8	8	6	12	4	15	11	16	18	11	26	12	29	26	27	37	11	23
					1	.	1	1	1	1	3	1	4	4	5	4	8	3	10	4	11	12	10	18	7	21	14	22	25	20	34	10	24
						1	.	1	.	2	1	2	2	4	2	6	4	5	8	2	11	7	12	13	9	18	9	23	18	22	28	7	25
							1	1	1	1	2	1	3	4	3	4	7	3	9	4	10	10	10	15	7	20	11	21	23	19	31	10	26
								1	.	1	1	1	2	3	2	4	4	3	7	2	9	7	9	12	7	16	9	19	17	18	26	7	27
									1	.	1	.	1	1	2	1	4	1	4	3	4	7	4	9	3	11	8	11	16	10	20	8	28
										1	.	1	1	2	1	3	2	3	4	1	7	4	8	8	6	12	6	15	12	16	19	5	29
											1	.	1	1	2	1	3	1	4	2	4	6	4	9	3	10	8	11	14	10	19	6	30
												1	.	1	.	2	1	2	2	1	4	2	6	4	5	8	3	11	8	12	13	4	31
													1	1	1	1	2	1	3	1	4	4	4	7	3	9	6	10	11	10	16	5	32
														1	.	1	1	1	2	1	3	2	4	4	3	7	3	9	8	9	12	4	33
															1	.	1	.	1	1	1	3	1	4	1	4	5	4	8	4	10	4	34
																1	.	1	1	.	2	1	3	2	3	4	2	7	4	8	8	2	35
																	1	.	1	1	1	2	1	3	1	4	3	4	7	4	9	4	36
																		1	.	.	1	.	2	1	2	2	1	4	2	6	4	1	37
																			1	.	1	1	1	2	1	3	2	4	4	4	7	2	38
																				1	.	1	.	1	.	1	2	1	4	1	4	3	39
																					1	.	1	1	1	2	1	3	2	4	4	1	40
																						1	.	1	.	1	2	1	3	1	4	2	41
																							1	.	1	1	.	2	1	3	2	1	42
																								1	.	1	1	1	2	1	3	1	43
																									1	.	.	1	.	2	1	.	44
																										1	.	1	1	1	2	1	45
																											1	.	1	.	1	1	46
																												1	.	1	1	.	47
																													1	.	1	1	48
																														1	.	.	49
																															1	.	50
																																1	51

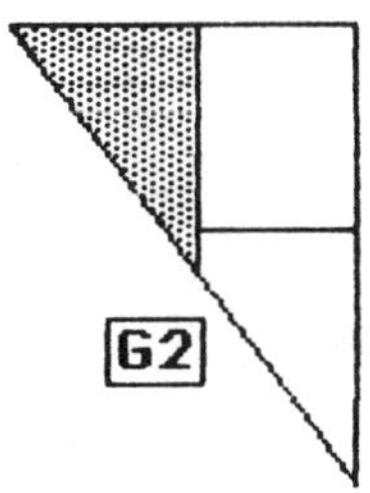

Table 3. *(continued)*

ALGEBRA 62, REPRESENTATIONS OF CLASS 0

NUMBER	52	53	54	55	56	57	58	59	60	61	62	63	64	65	66	67	68	69	70	71	72	73	74	75	76	77
S.P.	288	278	296	294	302	314	312	338	338	326	342	344	350	384	366	362	386	392	378	392	398	402	434	422	416	438
O.S.	6	12	12	12	12	12	12	12	6	12	12	12	12	6	12	12	12	6	12	12	12	12	12	12	12	12
LEVEL	73	73	73	75	75	77	77	77	79	79	79	81	81	81	83	83	83	85	85	85	87	87	87	89	89	89
DIM'N	10 556	28 672	19 278	28 652	30 107	24 192	37 961	17 472	14 756	41 769	33 495	40 579	46 656	11 571	33 592	55 614	33 858	20 196	59 136	53 599	56 133	69 160	29 667	45 696	79 002	57 344
WEIGHT	12 < 0	7 < 3	2 < 6	9 < 2	4 < 5	11 < 1	6 < 4	1 < 7	13 < 0	8 < 3	3 < 6	10 < 2	5 < 5	0 < 8	12 < 1	7 < 4	2 < 7	14 < 0	9 < 3	4 < 6	11 < 2	6 < 5	1 < 8	13 < 1	8 < 4	3 < 7
0	86	232	150	224	227	180	281	120	104	297	231	277	312	75	220	360	210	126	372	325	339	412	165	264	456	320
1	83	228	146	218	224	176	273	120	102	291	224	272	306	70	215	355	206	123	364	321	333	403	165	260	448	312
2	79	216	139	209	212	168	263	112	97	279	217	260	294	70	207	340	198	118	352	307	321	391	156	250	433	304
3	77	212	136	203	209	164	256	112	95	273	210	256	288	66	202	335	194	116	344	304	315	382	156	246	426	296
4	71	197	126	189	195	153	239	105	89	256	197	240	271	62	190	316	183	109	325	287	298	362	148	233	404	281
5	68	186	120	181	184	146	230	98	85	245	191	229	260	62	183	302	176	105	314	274	287	351	140	224	390	274
6	63	172	112	168	171	136	215	92	79	229	179	215	244	59	172	204	166	99	296	259	271	332	133	212	370	260
7	61	169	109	163	169	133	208	92	78	224	173	211	239	55	168	280	163	97	289	256	266	324	133	209	363	253
8	57	156	101	151	157	124	194	86	73	209	162	198	224	52	158	263	153	92	272	242	251	306	126	198	344	240
9	52	144	94	140	146	115	180	81	68	195	152	185	210	49	148	247	145	86	256	228	237	289	120	187	325	228
10	50	135	89	134	137	110	173	75	65	186	147	176	201	49	143	235	139	83	247	217	228	280	113	180	313	222
11	45	122	80	120	125	100	155	70	60	169	132	162	184	43	131	217	128	77	226	202	211	257	107	168	290	204
12	42	114	77	114	118	94	149	66	56	161	129	153	176	44	125	206	124	73	218	193	202	249	102	160	279	200
13	41	112	75	110	116	92	144	66	55	157	124	151	172	41	122	203	121	72	212	191	198	242	102	158	274	194
14	38	103	68	102	107	86	133	61	52	146	115	141	160	38	115	190	113	68	199	179	187	228	96	150	258	183
15	34	88	59	90	91	76	118	52	46	129	103	125	142	36	104	168	101	62	180	159	169	207	85	136	234	168
16	31	86	59	85	92	72	114	54	44	125	101	121	140	34	99	166	101	59	174	159	164	202	87	132	229	165
17	30	80	55	82	85	69	109	49	42	119	97	115	133	34	96	157	96	57	168	150	158	195	81	127	220	160
18	27	72	48	73	76	63	96	45	39	108	85	106	120	29	88	144	87	53	153	138	146	177	76	119	203	145
19	22	60	44	61	67	52	86	40	32	93	79	90	108	29	76	127	80	46	136	125	128	162	69	104	183	136
20	23	60	39	62	63	54	81	38	34	92	72	92	102	25	77	125	75	47	134	120	129	155	67	106	179	128

NUMBER	78	79	80	81	82	83	84	85	86	87	88	89	90	91	92	93	94	95	96	97	98	99	100	101	102	103
S.P.	450	434	446	486	456	458	488	482	474	494	512	494	504	542	518	518	546	546	536	554	600	578	558	566	602	584
O.S.	6	12	12	6	12	12	12	12	12	12	6	12	12	12	12	12	12	12	12	12	6	6	12	12	12	12
LEVEL	91	91	91	91	93	93	93	95	95	95	97	97	97	97	99	99	99	101	101	101	101	103	103	103	103	105
DIM'N	27 132	81 719	81 081	19 096	76 076	99 008	56 133	61 047	109 375	89 726	35 853	110 592	117 649	47 872	101 269	137 781	93 093	80 256	148 148	133 056	30 107	46 683	146 965	165 242	88 803	132 678
WEIGHT	15 < 0	10 < 3	5 < 6	0 < 9	12 < 2	7 < 5	2 < 8	14 < 1	9 < 4	4 < 7	16 < 0	11 < 3	6 < 6	1 < 9	13 < 2	8 < 5	3 < 8	15 < 1	10 < 4	5 < 7	0 < 10	17 < 0	12 < 3	7 < 6	2 < 9	14 < 2
0	150	455	441	100	410	524	285	315	565	446	177	552	577	220	487	657	429	372	686	600	131	207	661	728	375	576
1	147	449	434	95	402	518	280	310	555	442	174	544	566	220	481	648	420	366	678	592	125	204	651	721	370	568
2	142	433	420	95	390	500	271	300	540	426	168	528	552	210	466	630	411	356	658	576	125	198	635	700	360	553
3	139	427	413	90	383	494	267	295	530	422	166	520	542	210	460	621	402	350	651	568	120	195	625	693	355	546
4	132	406	393	86	364	471	254	281	506	403	158	497	518	201	440	595	385	335	624	545	115	187	600	666	341	524
5	128	391	380	86	353	454	246	272	492	388	153	482	505	192	426	578	377	326	605	530	115	182	585	646	332	510
6	121	371	361	82	336	432	235	259	469	370	146	460	483	184	407	553	361	312	580	508	111	174	561	620	319	490
7	119	366	355	78	329	427	231	255	460	367	144	453	473	184	402	545	353	307	573	501	106	172	552	614	315	483
8	113	347	337	74	313	406	220	243	438	350	138	432	452	176	384	521	338	294	549	480	102	165	529	589	302	464
9	107	329	320	71	297	386	210	231	417	334	131	412	431	169	367	498	324	281	525	460	98	158	507	565	291	445
10	104	316	309	71	288	371	203	224	405	321	127	399	420	161	355	483	317	274	508	447	98	154	494	547	283	433
11	97	295	288	64	268	348	190	210	377	303	120	374	392	154	335	454	296	258	480	422	90	146	465	519	268	410
12	93	283	278	65	259	334	185	202	366	292	115	362	382	148	323	440	291	250	464	410	91	141	453	502	262	398
13	91	279	273	61	254	330	182	199	358	289	114	356	374	148	319	433	284	246	459	404	87	139	445	497	258	393
14	87	264	258	58	241	313	172	190	340	275	109	339	356	141	305	413	271	236	438	386	83	134	426	476	247	377
15	80	239	235	55	222	284	158	175	313	250	101	312	330	128	281	381	253	220	404	358	80	125	397	440	230	352
16	77	235	232	53	215	281	157	170	306	250	98	306	323	130	276	375	249	214	399	354	77	122	389	436	229	345
17	75	225	223	53	209	269	151	165	297	239	95	296	314	123	267	363	243	209	385	343	77	119	379	421	222	336
18	70	210	206	47	194	251	140	155	275	224	90	277	291	117	252	340	225	197	363	322	70	113	356	398	209	318
19	62	187	189	46	175	228	131	139	252	206	81	252	271	108	228	313	213	180	333	300	69	103	329	368	197	293
20	63	187	182	42	174	224	125	140	247	201	82	250	262	106	229	307	204	180	330	292	64	104	325	363	191	292

21	59	42	59	66	51	82	40	32	91	75	89	105	26	74	125	78	45	132	123	126	157	69	103	179	131	61	185	185	43	171	225	128	137	246	204	80	248	264	108	226	308	207	177	329	295	65	102	323	364	194	289	21
20	54	37	55	59	48	75	36	30	84	68	84	96	24	70	116	71	43	124	114	119	146	64	98	168	122	58	175	173	40	163	212	120	131	233	192	77	236	250	102	216	292	196	170	314	280	62	98	309	347	184	278	22
18	45	30	49	48	43	65	29	27	74	59	74	83	22	64	101	62	39	112	98	108	131	55	89	151	110	54	158	155	38	150	190	108	121	214	172	71	217	229	91	199	268	181	159	288	257	59	92	288	318	170	259	23
13	39	30	39	47	34	58	30	22	64	56	63	78	20	53	93	60	33	98	95	94	121	55	78	138	104	47	143	147	35	133	179	104	108	196	166	64	198	215	90	182	251	172	144	269	245	55	84	265	303	164	239	24
15	36	23	39	38	36	51	24	23	61	46	63	68	17	54	85	51	34	94	83	93	109	48	78	130	92	47	138	133	31	131	166	94	107	186	151	64	192	199	82	178	237	158	142	258	228	51	83	257	285	152	234	25
13	36	27	38	42	33	55	26	21	61	53	60	73	20	52	87	56	32	95	88	91	116	50	75	132	100	46	137	140	35	130	170	99	105	190	157	62	192	208	84	176	242	167	141	259	236	55	82	259	291	158	233	26
12	32	22	34	36	31	47	23	20	55	44	56	64	16	48	79	49	30	86	79	85	103	46	71	121	88	43	128	127	30	121	157	90	99	175	145	59	180	190	79	167	225	152	133	244	219	49	78	243	273	147	221	27
7	22	20	23	30	20	38	20	13	41	40	40	54	16	35	63	44	22	68	68	65	89	40	54	100	80	34	103	111	29	98	134	81	80	149	128	48	150	169	70	138	196	139	112	209	196	47	66	209	241	134	189	28
11	26	16	29	28	27	38	18	18	46	35	49	52	13	43	67	40	27	75	66	75	88	39	64	105	75	39	114	109	26	109	138	78	90	156	127	54	162	168	70	152	201	135	122	221	195	44	72	222	246	132	203	29
7	22	19	22	29	20	36	20	13	40	37	40	52	14	34	62	42	22	66	67	64	85	40	54	98	76	33	102	108	26	96	132	79	79	145	126	48	148	164	70	137	192	134	110	207	192	44	65	205	238	131	187	30
9	18	11	23	19	22	29	12	15	36	27	39	40	11	36	52	31	23	62	51	63	72	30	54	87	62	34	95	90	23	94	114	65	78	134	105	47	140	144	58	132	173	117	108	191	168	40	64	196	212	115	180	31
7	20	15	21	25	19	32	17	13	37	32	38	46	12	33	57	37	21	62	60	61	78	36	52	91	69	32	97	99	24	92	123	72	76	137	117	46	141	153	65	132	181	125	106	197	180	41	63	197	225	123	180	32
7	16	11	19	18	18	27	12	12	32	26	34	38	11	31	48	30	20	56	49	56	68	29	48	81	60	30	87	86	22	86	108	63	71	125	101	43	130	138	56	122	164	113	100	180	162	39	59	184	203	111	169	33
3	11	12	11	18	10	21	14	7	23	25	23	34	10	20	40	30	13	42	47	41	59	30	35	67	56	22	70	79	20	66	96	60	55	105	96	34	107	124	55	100	145	105	81	156	150	35	49	155	186	105	142	34
6	12	7	15	13	16	19	9	11	26	18	29	29	7	27	39	23	18	46	39	49	54	24	43	68	47	27	76	71	17	75	92	52	64	107	86	39	115	116	49	110	143	95	90	160	140	32	54	164	179	97	153	35
3	10	10	11	15	10	20	11	7	22	23	22	31	10	20	37	27	13	41	42	40	56	26	34	64	53	22	67	74	20	65	90	56	54	102	89	33	104	119	50	97	139	101	80	150	143	35	48	152	177	100	139	36
5	8	4	11	8	12	13	6	9	18	12	23	20	5	22	29	16	15	36	29	39	41	18	36	53	36	23	62	55	13	63	74	41	54	88	69	34	96	95	40	94	118	78	78	136	116	27	47	141	151	81	133	37
3	9	7	10	12	10	16	9	7	20	17	21	26	7	19	33	22	13	37	36	38	48	23	33	58	44	21	63	65	16	61	82	49	52	93	80	32	98	106	46	93	128	89	76	141	130	30	46	143	164	91	133	38
1	4	7	4	9	4	11	8	3	11	16	11	20	8	10	22	20	7	24	30	23	38	20	20	42	40	13	43	54	16	42	64	44	35	71	68	22	72	90	40	67	103	80	56	110	112	29	34	111	137	81	102	39
3	7	4	9	8	9	12	6	7	16	12	19	19	5	18	27	16	12	32	28	34	39	18	31	49	35	20	56	53	13	56	70	40	48	82	67	30	88	91	39	86	112	76	71	127	112	26	43	131	145	79	123	40
1	4	6	4	9	4	10	8	3	11	14	11	19	6	10	22	19	7	23	29	23	36	20	20	41	37	13	43	52	14	41	63	42	35	69	67	22	71	86	40	67	101	76	55	109	109	26	34	109	135	79	101	41
3	4	2	7	4	8	8	3	6	12	8	15	13	4	16	19	11	11	26	19	29	30	12	27	39	27	18	46	41	11	49	55	31	43	69	52	27	76	75	30	75	94	63	64	108	93	23	39	116	121	66	110	42
1	4	4	4	7	4	9	6	3	10	11	11	16	5	10	20	15	7	22	25	22	32	17	20	38	32	13	41	46	12	40	58	37	34	65	60	22	68	79	36	65	94	69	54	104	100	24	33	105	126	72	98	43
2	2	1	4	2	6	4	2	5	8	4	11	8	2	12	13	7	9	18	13	23	20	9	22	29	18	15	36	30	7	39	42	23	36	53	40	23	62	57	24	63	75	48	54	89	74	17	34	96	99	53	94	44
1	3	2	4	4	4	7	3	3	9	8	10	12	4	10	16	11	7	20	18	21	27	12	19	33	26	13	37	38	11	38	49	30	33	59	49	21	63	69	29	61	84	60	52	94	87	22	32	99	111	63	93	45
.	1	3	1	4	1	4	5	1	4	8	4	10	4	4	11	12	3	11	18	11	21	14	10	23	25	7	24	34	10	23	40	30	20	43	47	13	44	59	30	42	68	56	35	73	79	20	22	73	97	60	68	46
1	2	1	3	2	4	4	2	3	7	4	9	8	2	9	12	7	7	16	13	19	19	9	18	27	18	12	32	29	7	34	40	23	31	49	39	20	56	55	24	56	71	47	48	83	72	17	30	88	95	52	86	47
.	1	2	1	3	1	4	3	1	4	7	4	9	4	4	10	10	3	11	15	11	20	11	10	22	23	7	23	31	10	23	37	27	20	42	42	13	43	56	26	41	65	53	35	70	74	20	22	72	91	56	67	48

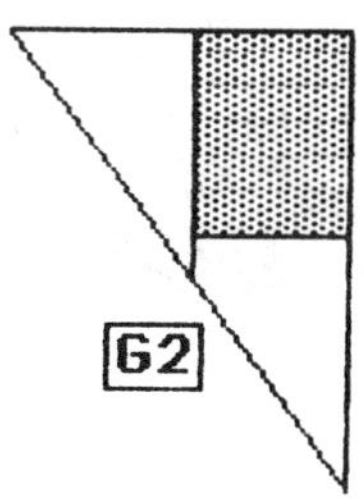

62

Table 3. *(continued)*

ALGEBRA 62, REPRESENTATIONS OF CLASS 0

12<0	7<3	2<6	9<2	4<5	11<1	6<4	1<7	13<0	8<3	3<6	10<2	5<5	0<8	12<1	7<4	2<7	14<0	9<3	4<6	11<2	6<5	1<8	13<1	8<4	3<7	15<0	10<3	5<6	0<9	12<2	7<5	2<8	14<1	9<4	4<7	16<0	11<3	6<6	1<9	13<2	8<5	3<8	15<1	10<4	5<7	0<10	17<0	12<3	7<6	2<9	14<2	WEIGHT
52	53	54	55	56	57	58	59	60	61	62	63	64	65	66	67	68	69	70	71	72	73	74	75	76	77	78	79	80	81	82	83	84	85	86	87	88	89	90	91	92	93	94	95	96	97	98	99	100	101	102	103	NUMBER
1	1	.	2	1	3	2	1	3	4	2	7	4	1	8	8	4	6	12	8	15	13	6	16	19	12	11	26	20	5	29	30	16	27	39	29	18	46	42	18	49	56	36	43	69	56	13	27	76	77	41	75	49
.	1	1	1	2	1	3	2	1	4	4	4	7	2	4	9	7	3	10	12	11	16	9	10	20	17	7	22	26	7	22	33	22	20	38	36	13	41	48	23	40	59	44	34	66	65	16	22	68	83	49	65	50
.	.	1	.	1	.	1	2	.	1	4	1	4	3	1	4	7	1	4	9	4	11	8	4	11	16	3	11	20	8	11	22	20	10	24	30	7	24	38	20	23	42	40	20	44	54	16	13	45	64	44	42	51
1	.	.	1	.	2	1	.	2	2	1	4	2	1	6	4	2	5	8	4	11	8	3	12	13	8	9	18	13	4	23	20	11	22	29	19	15	36	31	12	39	42	27	36	53	42	11	23	62	58	31	63	52
	1	.	1	1	1	2	1	1	3	2	4	4	1	4	7	4	3	9	8	10	12	6	10	16	12	7	20	19	5	21	27	16	19	33	28	13	37	39	18	38	50	35	33	59	53	13	21	63	71	40	61	53
		1	.	1	.	1	2	.	1	3	1	4	2	1	4	6	1	4	9	4	10	8	4	11	14	3	11	19	6	11	22	19	10	23	29	7	24	36	20	23	41	37	20	44	52	14	13	44	63	42	42	54
			1	.	1	1	.	1	2	1	3	2	1	4	4	2	3	7	4	9	8	3	9	12	8	7	16	13	4	19	19	11	18	27	19	12	32	30	12	34	40	27	31	49	41	11	20	56	56	31	56	55
				1	.	1	1	.	1	2	1	3	1	1	4	4	1	4	7	4	9	6	4	10	11	3	11	16	5	11	20	15	10	22	25	7	23	32	17	23	38	32	20	42	46	12	13	43	58	37	41	56
					1	.	.	1	1	.	2	1	.	3	2	1	3	4	2	7	4	2	8	8	4	6	12	8	2	15	13	7	16	19	13	11	26	20	9	29	30	18	27	39	30	7	18	46	43	23	49	57
						1	.	.	1	1	1	2	1	1	3	2	1	4	4	4	7	3	4	9	8	3	10	12	4	11	16	11	10	20	18	7	22	27	12	22	33	26	20	38	38	11	13	41	49	30	40	58
							1	.	.	1	.	1	1	.	1	3	.	1	4	1	4	5	1	4	8	1	4	10	4	4	11	12	4	11	18	3	11	21	14	11	23	25	10	24	34	10	7	24	40	30	23	59
								1	.	.	1	.	.	2	1	.	2	2	1	4	2	1	6	4	2	5	8	4	1	11	8	4	12	13	8	9	18	13	6	23	20	12	22	29	20	5	15	36	31	16	39	60
									1	.	1	1	.	1	2	1	1	3	2	4	4	2	4	7	4	3	9	8	2	10	12	7	10	16	13	7	20	19	9	21	27	18	19	33	29	7	13	37	40	23	38	61
										1	.	1	1	.	1	2	.	1	3	1	4	3	1	4	7	1	4	9	4	4	10	10	4	11	15	3	11	20	11	11	22	23	10	23	31	10	7	24	37	27	23	62
											1	.	.	1	1	.	1	2	1	3	2	1	4	4	2	3	7	4	1	9	8	4	9	12	8	7	16	13	6	19	19	12	18	27	20	5	12	32	30	16	34	63
												1	.	.	1	1	.	1	2	1	3	2	1	4	4	1	4	7	2	4	9	7	4	10	12	3	11	16	9	11	20	17	10	22	26	7	7	23	33	22	23	64
													1	.	.	1	.	.	1	.	1	2	.	1	4	.	1	4	3	1	4	7	1	4	9	1	4	11	8	4	11	16	4	11	20	8	3	11	22	20	11	65
														1	.	.	1	1	.	2	1	.	3	2	1	3	4	2	1	7	4	2	8	8	4	6	12	8	3	15	13	8	16	19	13	4	11	26	20	11	29	66
															1	.	.	1	1	1	2	1	1	3	2	1	4	4	1	4	7	4	4	9	8	3	10	12	6	11	16	12	10	20	19	5	7	22	27	16	22	67
																1	.	.	1	.	1	2	.	1	3	.	1	4	2	1	4	6	1	4	9	1	4	10	8	4	11	14	4	11	19	6	3	11	22	19	11	68
																	1	.	.	1	.	.	2	1	.	2	2	1	.	4	2	1	6	4	2	5	8	4	2	11	8	4	12	13	8	2	9	18	13	7	23	69
																		1	.	1	1	.	1	2	1	1	3	2	1	4	4	2	4	7	4	3	9	8	3	10	12	8	10	16	13	4	7	20	19	11	21	70
																			1	.	1	1	.	1	2	.	1	3	1	1	4	4	1	4	7	1	4	9	6	4	10	11	4	11	16	5	3	11	20	15	11	71
																				1	.	.	1	1	.	1	2	1	.	3	2	1	4	4	2	3	7	4	2	9	8	4	9	12	8	2	7	16	13	7	19	72
																					1	.	.	1	1	.	1	2	1	1	3	2	1	4	4	1	4	7	3	4	9	8	4	10	12	4	3	11	16	11	11	73

1	.	.	1	.	.	1	1	.	1	3	.	1	4	.	1	4	5	1	4	8	1	4	10	4	1	4	11	12	4	74
	1	.	.	1	1	.	.	2	1	.	3	2	1	3	4	2	1	7	4	2	8	8	4	1	6	12	8	4	15	75
		1	.	.	1	1	.	1	2	1	1	3	2	1	4	4	2	4	7	4	4	9	8	2	3	10	12	7	11	76
			1	.	.	1	1	.	1	2	.	1	3	.	1	4	3	1	4	7	1	4	9	4	1	4	10	10	4	77
				1	.	.	.	1	.	.	2	1	.	2	2	1	.	4	2	1	6	4	2	1	5	8	4	2	11	78
					1	.	.	1	1	.	1	2	1	1	3	2	1	4	4	2	4	7	4	1	3	9	8	4	10	79
						1	.	.	1	1	.	1	2	.	1	3	2	1	4	4	1	4	7	2	1	4	9	7	4	80
							1	.	.	1	.	.	1	.	.	1	2	.	1	4	.	1	4	3	.	1	4	7	1	81
								1	.	.	1	1	.	1	2	1	.	3	2	1	4	4	2	1	3	7	4	2	9	82
									1	.	.	1	1	.	1	2	1	1	3	2	1	4	4	1	1	4	7	4	4	83
										1	.	.	1	.	.	1	2	.	1	3	.	1	4	2	.	1	4	6	1	84
											1	.	.	1	1	.	.	2	1	.	3	2	1	.	3	4	2	1	7	85
												1	.	.	1	1	.	1	2	1	1	3	2	1	1	4	4	2	4	86
													1	.	.	1	1	.	1	2	.	1	3	1	.	1	4	4	1	87
														1	.	.	.	1	.	.	2	1	.	.	2	2	1	.	4	88
															1	.	.	1	1	.	1	2	1	.	1	3	2	1	4	89
																1	.	.	1	1	.	1	2	1	.	1	3	2	1	90
																	1	.	.	1	.	.	1	1	.	.	1	3	.	91
																		1	.	.	1	1	.	.	1	2	1	.	3	92
																			1	.	.	1	1	.	.	1	2	1	1	93
																				1	.	.	1	1	.	.	1	2	.	94
																					1	.	.	.	1	1	.	.	2	95
																						1	.	.	.	1	1	.	1	96
																							1	.	.	.	1	1	.	97
																								1	.	.	.	1	.	98
																									1	.	.	.	1	99
																										1	.	.	1	100
																											1	.	.	101
																												1	.	102
																													1	103

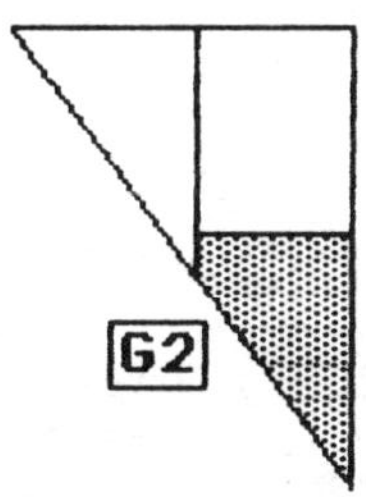

References

1. V. K. Agrawala, J. G. Belinfante, Weight diagrams for Lie group representations: A computer implementation of Freudenthal's algorithm in ALGOL and FORTRAN, BIT *9* (1969), 301-304.
2. M. I. Krusemeyer, Determining multiplicities of dominant weights in irreducible Lie algebra representations using a computer, BIT *11* (1971), 310-316.
3. R. E. Beck, B. Kolman, A computer implementation of Freudenthal's multiplicity formula, Indag. Math. *34* (1972), 350-352; Computers in Lie algebras: I. Calculation of inner multiplicities, SIAM J. Appl. Math. *25* (1973), 300-312.
4. W. G. McKay, J. Patera, D. Sankoff, The computation of branching rules for representations of semisimple Lie algebras, in *Computers in Non-Associative Rings and Algebras* (R. Beck and B. Kolman, eds.), Academic Press, New York, 1977, pp. 235-277.
5. R. V. Moody, J. Patera, Fast recursion formula for weight multiplicities, Bull. Amer. Math. Soc. *7* (1982), 237-242.
6. R. V. Moody, J. Patera, R. T. Sharp, Character generators for elements of finite order in simple Lie groups A_1, A_2, A_3, B_2, and G_2, J. Math. Phys. *24* (1983), 2387-2397.
7. R. V. Moody, J. Patera, Characters of elements of finite order in Lie groups, SIAM J. Algebraic Discrete Methods *5* (1984), 359-383.
8. R. V. Moody, J. Patera, General charge conjugation operators in simple Lie groups, J. Math. Phys. *25* (1984), 2838-2847.
9. R. V. Moody, J. Patera, Computation of character decompositions of class functions on semisimple Lie groups, to appear.
10. J. E. Humphreys, *Introduction to Lie Algebras and Representation Theory,* Springer, New York, 1980.
11. H. Freudenthal, H. deVries, *Linear Lie Groups,* Academic Press, New York, 1969.
12. N. Bourbaki, *Groupes et algèbres de Lie,* Chapitres 4-6, Hermann, Paris, 1968.
13. F. W. Lemire, J. Patera, Congruence number, a generalization of SU(3) triality, J. Math. Phys. 21 (1980), 2026-2027.
14. H. S. M. Coxeter, W. O. J. Moser, *Generators and Relations for Discrete Groups,* Springer, New York, 1972.
15. J. Dieudonné, *La géometrie des groupes classiques,* Springer, New York, 1963.